NATURAL HISTORY
UNIVERSAL LIBRARY

西方博物学大系

主编：江晓原

A NIEWE HERBALL

本草新说

[法] 伦贝特·多东恩斯 著
[英] 亨利·赖特 译

华东师范大学出版社

图书在版编目(CIP)数据

本草新说 = A Niewe Herball：英文 / (法) 伦贝特·多东恩斯著；(英) 亨利·赖特译. — 上海：华东师范大学出版社, 2018
 (寰宇文献)
 ISBN 978-7-5675-7723-7

Ⅰ.①本… Ⅱ.①伦…②亨… Ⅲ.①药用植物学–英文 Ⅳ.①Q959

中国版本图书馆CIP数据核字(2018)第094120号

本草新说
A Niewe Herball
(法) 伦贝特·多东恩斯著　　(英) 亨利·赖特译

特约策划　黄曙辉　徐　辰
责任编辑　庞　坚
特约编辑　许　倩
装帧设计　刘怡霖

出版发行　华东师范大学出版社
社　　址　上海市中山北路3663号　邮编 200062
网　　址　www.ecnupress.com.cn
电　　话　021-60821666　行政传真　021-62572105
客服电话　021-62865537
门市(邮购)电话　021-62869887
地　　址　上海市中山北路3663号华东师范大学校内先锋路口
网　　店　http://hdsdcbs.tmall.com/

印 刷 者　虎彩印艺股份有限公司
开　　本　16开
印　　张　52.5
版　　次　2018年6月第1版
印　　次　2018年6月第1次
书　　号　ISBN 978-7-5675-7723-7
定　　价　898.00元(精装全一册)

出 版 人　王　焰

(如发现本版图书有印订质量问题，请寄回本社客服中心调换或电话021-62865537联系)

《西方博物学大系》总序

江晓原

《西方博物学大系》收录博物学著作超过一百种，时间跨度为15世纪至1919年，作者分布于16个国家，写作语种有英语、法语、拉丁语、德语、弗莱芒语等，涉及对象包括植物、昆虫、软体动物、两栖动物、爬行动物、哺乳动物、鸟类和人类等，西方博物学史上的经典著作大备于此编。

中西方"博物"传统及观念之异同

今天中文里的"博物学"一词，学者们认为对应的英语词汇是Natural History，考其本义，在中国传统文化中并无现成对应词汇。在中国传统文化中原有"博物"一词，与"自然史"当然并不精确相同，甚至还有着相当大的区别，但是在"搜集自然界的物品"这种最原始的意义上，两者确实也大有相通之处，故以"博物学"对译 Natural History 一词，大体仍属可取，而且已被广泛接受。

已故科学史前辈刘祖慰教授尝言：古代中国人处理知识，如开中药铺，有数十上百小抽屉，将百药分门别类放入其中，即心安矣。刘教授言此，其辞若有憾焉——认为中国人不致力于寻求世界"所以然之理"，故不如西方之分析传统优越。然而古代中国人这种处理知识的风格，正与西方的博物学相通。

与此相对，西方的分析传统致力于探求各种现象和物体之间的相互关系，试图以此解释宇宙运行的原因。自古希腊开始，西方哲人即孜孜不倦建构各种几何模型，欲用以说明宇宙如何运行，其中最典型的代表，即为托勒密（Ptolemy）的宇宙体系。

比较两者，差别即在于：古代中国人主要关心外部世界"如何"运行，而以希腊为源头的西方知识传统（西方并非没有别的知识传统，只是未能光大而已）更关心世界"为何"如此运行。在线性发展无限进步的科学主义观念体系中，我们习惯于认为"为何"是在解决了"如何"之后的更高境界，故西方的分析传统比中国的传统更高明。

然而考之古代实际情形，如此简单的优劣结论未必能够成立。例如以天文学言之，古代东西方世界天文学的终极问题是共同的：给定任意地点和时刻，计算出太阳、月亮和五大行星（七政）的位置。古代中国人虽不致力于建立几何模型去解释七政"为何"如此运行，但他们用抽象的周期叠加（古代巴比伦也使用类似方法），同样能在足够高的精度上计算并预报任意给定地点和时刻的七政位置。而通过持续观察天象变化以统计、收集各种天象周期，同样可视之为富有博物学色彩的活动。

还有一点需要注意：虽然我们已经接受了用"博物学"来对译 Natural History，但中国的博物传统，确实和西方的博物学有一个重大差别——即中国的博物传统是可以容纳怪力乱神的，而西方的博物学基本上没有怪力乱神的位置。

古代中国人的博物传统不限于"多识于鸟兽草木之名"。体现此种传统的典型著作，首推晋代张华《博物志》一书。书名"博物"，其义尽显。此书从内容到分类，无不充分体现它作为中国博物传统的代表资格。

《博物志》中内容，大致可分为五类：一、山川地理知识；二、奇禽异兽描述；三、古代神话材料；四、历史人物传说；五、神仙方伎故事。这五大类，完全符合中国文化中的博物传统，深合中国古代博物传统之旨。第一类，其中涉及宇宙学说，甚至还有"地动"思想，故为科学史家所重视。第二类，其中甚至出现了中国古代长期流传的"守宫砂"传说的早期文献：相传守宫砂点在处女胳膊上，永不褪色，只有性交之后才会自动消失。第三类，古代神话传说，其中甚至包括可猜想为现代"连体人"的记载。第四类，各种著名历史人物，比如三位著名刺客的传说，此三名刺客及所刺对象，历史上皆实有其人。第五类，包括各种古代方术传说，比如中国古代房中养生学说，房中术史上的传说人物之一"青牛道士封君达"等等。前两类与西方的博物学较为接近，但每一类都会带怪力乱神色彩。

"所有的科学不是物理学就是集邮"

在许多人心目中,画画花草图案,做做昆虫标本,拍拍植物照片,这类博物学活动,和精密的数理科学,比如天文学、物理学等等,那是无法同日而语的。博物学显得那么的初级、简单,甚至幼稚。这种观念,实际上是将"数理程度"作为唯一的标尺,用来衡量一切知识。但凡能够使用数学工具来描述的,或能够进行物理实验的,那就是"硬"科学。使用的数学工具越高深越复杂,似乎就越"硬";物理实验设备越庞大,花费的金钱越多,似乎就越"高端"、越"先进"……

这样的观念,当然带着浓厚的"物理学沙文主义"色彩,在很多情况下是不正确的。而实际上,即使我们暂且同意上述"物理学沙文主义"的观念,博物学的"科学地位"也仍然可以保住。作为一个学天体物理专业出身,因而经常徜徉在"物理学沙文主义"幻影之下的人,我很乐意指出这样一个事实:现代天文学家们的研究工作中,仍然有绘制星图,编制星表,以及为此进行的巡天观测等等活动,这些活动和博物学家"寻花问柳",绘制植物或昆虫图谱,本质上是完全一致的。

这里我们不妨重温物理学家卢瑟福(Ernest Rutherford)的金句:"所有的科学不是物理学就是集邮(All science is either physics or stamp collecting)。"卢瑟福的这个金句堪称"物理学沙文主义"的极致,连天文学也没被他放在眼里。不过,按照中国传统的"博物"理念,集邮毫无疑问应该是博物学的一部分——尽管古代并没有邮票。卢瑟福的金句也可以从另一个角度来解读:既然在卢瑟福眼里天文学和博物学都只是"集邮",那岂不就可以将博物学和天文学相提并论了?

如果我们摆脱了科学主义的语境,则西方模式的优越性将进一步被消解。例如,按照霍金(Stephen Hawking)在《大设计》(The Grand Design)中的意见,他所认同的是一种"依赖模型的实在论(model-dependent realism)",即"不存在与图像或理论无关的实在性概念(There is no picture- or theory-independent concept of reality)"。在这样的认识中,我们以前所坚信的外部世界的客观性,已经不复存在。既然几何模型只不过是对外部世界图像的人为建构,则古代中国人干脆放弃这种建构直奔应用(毕竟在实际应用中我们只需要知道七政"如何"运行），又有何不可？

传说中的"神农尝百草"故事，也可以在类似意义下得到新的解读："尝百草"当然是富有博物学色彩的活动，神农通过这一活动，得知哪些草能够治病，哪些不能，然而在这个传说中，神农显然没有致力于解释"为何"某些草能够治病而另一些则不能，更不会去建立"模型"以说明之。

"帝国科学"的原罪

今日学者有倡言"博物学复兴"者，用意可有多种，诸如缓解压力、亲近自然、保护环境、绿色生活、可持续发展、科学主义解毒剂等等，皆属美善。编印《西方博物学大系》也是意欲为"博物学复兴"添一助力。

然而，对于这些博物学著作，有一点似乎从未见学者指出过，而鄙意以为，当我们披阅把玩欣赏这些著作时，意识到这一点是必须的。

这百余种著作的时间跨度为 15 世纪至 1919 年，注意这个时间跨度，正是西方列强"帝国科学"大行其道的时代。遥想当年，帝国的科学家们乘上帝国的军舰——达尔文在皇家海军"小猎犬号"上就是这样的场景之一，前往那些已经成为帝国的殖民地或还未成为殖民地的"未开化"的遥远地方，通常都是踌躇满志、充满优越感的。

作为一个典型的例子，英国学者法拉在（Patricia Fara）《性、植物学与帝国：林奈与班克斯》（*Sex, Botany and Empire, The Story of Carl Linnaeus and Joseph Banks*）一书中讲述了英国植物学家班克斯（Joseph Banks）的故事。1768 年 8 月 15 日，班克斯告别未婚妻，登上了澳大利亚军舰"奋进号"。此次"奋进号"的远航是受英国海军部和皇家学会资助，目的是前往南太平洋的塔希提岛（Tahiti，法属海外自治领，另一个常见的译名是"大溪地"）观测一次比较罕见的金星凌日。舰长库克（James Cook）是西方殖民史上最著名的舰长之一，多次远航探险，开拓海外殖民地。他还被认为是澳大利亚和夏威夷群岛的"发现"者，如今以他命名的群岛、海峡、山峰等不胜枚举。

当"奋进号"停靠塔希提岛时，班克斯一下就被当地美丽的土著女性迷昏了，他在她们的温柔乡里纵情狂欢，连库克舰长都看不下去了，"道德愤怒情绪偷偷溜进了他的日志当中，他发现自己根本不可能不去批评所见到的滥交行为"，而班克斯纵欲到了"连嫖妓都毫无激情"的地步——这是别人讽刺班克斯的说法，因为对于那时常年航行于茫茫大海上的男性来说，上岸嫖妓通常是一项能够唤起"激情"的活动。

而在"帝国科学"的宏大叙事中，科学家的私德是无关紧要的，人们关注的是科学家做出的科学发现。所以，尽管一面是班克斯在塔希提岛纵欲滥交，一面是他留在故乡的未婚妻正泪眼婆娑地"为远去的心上人绣织背心"，这样典型的"渣男"行径要是放在今天，非被互联网上的口水淹死不可，但是"班克斯很快从他们的分离之苦中走了出来，在外近三年，他活得倒十分滋润"。

法拉不无讽刺地指出了"帝国科学"的实质："班克斯接管了当地的女性和植物，而库克则保护了大英帝国在太平洋上的殖民地。"甚至对班克斯的植物学本身也调侃了一番："即使是植物学方面的科学术语也充满了性指涉。……这个体系主要依靠花朵之中雌雄生殖器官的数量来进行分类。"据说"要保护年轻妇女不受植物学教育的浸染，他们严令禁止各种各样的植物采集探险活动。"这简直就是将植物学看成一种"涉黄"的淫秽色情活动了。

在意识形态强烈影响着我们学术话语的时代，上面的故事通常是这样被描述的：库克舰长的"奋进号"军舰对殖民地和尚未成为殖民地的那些地方的所谓"访问"，其实是殖民者耀武扬威的侵略，搭载着达尔文的"小猎犬号"军舰也是同样行径；班克斯和当地女性的纵欲狂欢，当然是殖民者对土著妇女令人发指的蹂躏；即使是他采集当地植物标本的"科学考察"，也可以视为殖民者"窃取当地经济情报"的罪恶行为。

后来改革开放，上面那种意识形态话语被抛弃了，但似乎又走向了另一个极端，完全忘记或有意回避殖民者和帝国主义这个层面，只歌颂这些军舰上的科学家的伟大发现和成就，例如达尔文随着"小猎犬号"的航行，早已成为一曲祥和优美的科学颂歌。

其实达尔文也未能免俗，他在远航中也乐意与土著女性打打交道，当然他没有像班克斯那样滥情纵欲。在达尔文为"小猎犬号"远航写的《环球游记》中，我们读到："回程途中我们遇到一群黑人姑娘在聚会，……我们笑着看了很久，还给了她们一些钱，这着实令她们欣喜一番，拿着钱尖声大笑起来，很远还能听到那愉悦的笑声。"

有趣的是，在班克斯在塔希提岛纵欲六十多年后，达尔文随着"小猎犬号"也来到了塔希提岛，岛上的土著女性同样引起了达尔文的注意，在《环球游记》中他写道："我对这里妇女的外貌感到有些失望，然而她们却很爱美，把一朵白花或者红花戴在脑后的髮髻上……"接着他以居高临下的笔调描述了当地女性的几种发饰。

用今天的眼光来看，这些在别的民族土地上采集植物动物标本、测量地质水文数据等等的"科学考察"行为，有没有合法性问题？有没有侵犯主权的问题？这些行为得到当地人的同意了吗？当地人知道这些行为的性质和意义吗？他们有知情权吗？……这些问题，在今天的国际交往中，确实都是存在的。

也许有人会为这些帝国科学家辩解说：那时当地土著尚在未开化或半开化状态中，他们哪有"国家主权"的意识啊？他们也没有制止帝国科学家的考察活动啊？但是，这样的辩解是无法成立的。

姑不论当地土著当时究竟有没有试图制止帝国科学家的"科学考察"行为，现在早已不得而知，只要殖民者没有记录下来，我们通常就无法知道。况且殖民者有军舰有枪炮，土著就是想制止也无能为力。正如法拉所描述的："在几个塔希提人被杀之后，一套行之有效的易货贸易体制建立了起来。"

即使土著因为无知而没有制止帝国科学家的"科学考察"行为，这事也很像一个成年人闯进别人的家，难道因为那家只有不懂事的小孩子，闯入者就可以随便打探那家的隐私、拿走那家的东西、甚至将那家的房屋土地据为己有吗？事实上，很多情况下殖民者就是这样干的。所以，所谓的"帝国科学"，其实是有着原罪的。

如果沿用上述比喻，现在的局面是，家家户户都不会只有不懂事的孩子了，所以任何外来者要想进行"科学探索"，他也得和这家主人达成共识，得到这家主人的允许才能够进行。即使这种共识的达成依赖于利益的交换，至少也不能单方面强加于人。

博物学在今日中国

博物学在今日中国之复兴，北京大学刘华杰教授提倡之功殊不可没。自刘教授大力提倡之后，各界人士纷纷跟进，仿佛昔日蔡锷在云南起兵反袁之"滇黔首义，薄海同钦，一檄遥传，景从恐后"光景，这当然是和博物学本身特点密切相关的。

无论在西方还是在中国，无论在过去还是在当下，为何博物学在它繁荣时尚的阶段，就会应者云集？深究起来，恐怕和博物学本身的特点有关。博物学没有复杂的理论结构，它的专业训练也相对容易，至少没有天文学、物理学那样的数理"门槛"，所以和一些数理学科相比，博物学可以有更多的自学成才者。这次编印的《西方博物学大系》，卷帙浩繁，蔚为大观，同样说明了这一点。

最后，还有一点明显的差别必须在此处强调指出：用刘华杰教授喜欢的术语来说，《西方博物学大系》所收入的百余种著作，绝大部分属于"一阶"性质的工作，即直接对博物学作出了贡献的著作。事实上，这也是它们被收入《西方博物学大系》的主要理由之一。而在中国国内目前已经相当热的博物学时尚潮流中，绝大部分已经出版的书籍，不是属于"二阶"性质（比如介绍西方的博物学成就），就是文学性的吟风咏月野草闲花。

要寻找中国当代学者在博物学方面的"一阶"著作，如果有之，以笔者之孤陋寡闻，唯有刘华杰教授的《檀岛花事——夏威夷植物日记》三卷，可以当之。这是刘教授在夏威夷群岛实地考察当地植物的成果，不仅属于直接对博物学作出贡献之作，而且至少在形式上将昔日"帝国科学"的逻辑反其道而用之，岂不快哉！

2018年6月5日
于上海交通大学
科学史与科学文化研究院

《本草新说》 出版说明

伦贝特·多东恩斯
（1517-1585）

伦贝特·多东恩斯（Rembert Dodoens）生于法兰德斯的梅赫伦（现属比利时），是当地的名医兼植物学家。他十三岁就进鲁汶大学修习医学、地理。1538年在老家开业行医，后转居巴塞尔。母校鲁汶大学曾邀请他出任教授，西班牙国王腓力二世则请他出任御医，均被他婉拒。自1575年起，多东恩斯在维也纳担任神圣罗马皇帝鲁道夫二世的御医，1582年被聘为莱顿大学医学教授，三年后于任期内去世。

多东恩斯一生著作颇丰，其中最有名的当属1554年在安特卫普刊行的《本草新说》（*Des Cruydeboeck*）。受学界前辈莱昂哈特·福克斯的影响，他在书中将植物分为六类群，配以七百余幅插画，涉及植物中有很多药草。除荷兰语及拉丁文版本外，也被卡洛斯·库尔修斯译为法语版《植物的历史》（*Histoire des Plantes*），随后也涌现出其他语种译本。

当时，这本书的译本种类之多，仅次于《圣经》，在问世后的二百年间一直是被广为参考的医药文献。

亨利·赖特
（1529-1607）

本书最有名的译本——也是最完善的版本，当属英国植物学家兼文物学家亨利·赖特（Henry Lyte）的英译本。亨利·赖特1529年生于英格兰西部的萨默塞特郡，1546年进入牛津大学，算得上是与多东恩斯同时代。这位兴趣广泛的学者起先主修逻辑学与哲学，继而云游四方，开拓眼界。归家后帮父亲将家庭产业打理得井井有条，还做过萨默塞特的郡守——从玛丽一世时代一直做到伊丽莎白一世登基。

植物学是亨利·赖特的主要兴趣之一，库尔修斯于1557年将《本草新说》译为法语本后引起他的注意，遂着手将其翻译为英文，添加了副标题"亦即，植物的历史"。《本草新说》英译本的初版为对开本，于1578年在安特卫普刊行，题献给英国女王伊丽莎白一世。这个版本不只是单纯翻译文本，更在原本基础上加入了很多增补文字与图片，洋洋779页，配以870幅版画。之后四十年间，又出过两个版本，但都删去了所有版画。

1607年10月16日,赖特在自己出生的赖特卡里大宅去世。身后,其后代保管了一册《本草新说》作为纪念,它如今已是英格兰国家名胜古迹信托收藏的文物。

Allusio ad Insignia Gentilitia Henrici Leiti, Armigeri, Somersetensis, Angli.

*Tortilis hic lituus, niueusq́; Olor, arguit in te
Leite animum niueum, pectus & intrepidum.*

LÆTITIA ET SPE IMMORTALITATIS.

Lyke as the Swanne doth chaunt his tunes in signe of ioyfull mynde,
So Lyte by learning shewes him selfe to Prince and Countrie kynde.

To the most High, Noble, and

Renovvmed Princesse, our most dread redoubted
Soueraigne Lady Elizabeth, by the grace of God, Queene of
Englande, Fraunce, and Irelande, defendour of the fayth, &c.
Your graces most humble, loyall, and faythfull sub-
iect Henry Lyte, vvisheth long life, perfect health,
florishing raigne, and prosperous successe to
Gods good pleasure, in all your
most Royall affaires.

TWO thinges haue mooued me (most noble Princesse) hauing newly translated into English this Herball or Historie of Plantes (not long sithence, set foorth in y⁶ Almaigne
or Douche tongue, by that paynefull and learned Physition D. Rembert Dodoens, and sithence that, agayne by
the trauayle of sundry skylfull Herbarians into diuers
other languages translated) to offer the same vnto your
Maiesties protection. The one was that most cleare, amiable and chearefull countenaunce towardes all learning and vertue: whiche
on euery syde most brightly from your Royall person appearing, hath so enfla-
med and encouraged, not onely me, to the loue and admiration thereof: but al
suche others also, your Graces loyall subiectes, whiche are not to to dull of vnderstanding: that we thinke no trauayle to great, whereby we are in hope
both to profite our Countrie, & to please so noble & louing a Princesse: whose
whole power and endeuour we see therto bent, that vertue & knowledge (the
two most beautifull ornamentes of a wel gouerned kingdome) may florish and
beare sway: Uice and ignorance (the foes of all goodnesse) may vanish & giue
place. The other was, that earnest zeale, and feruent desire that I haue, and a
long time haue had, to shewe my selfe (by yeelding some fruite of painefull dili-
gēce) a thankeful subiect to so vertuous a Soueraigne, & a fruitful member of
so good a cōmon Weale. The first of these, hartened or emboldened me against
those perswasions of mine owne vnworthynes: which (vndoubtedly) had put
me to vtter silence, had I not bene susteined both with the comfortable remem-
braunce of your highnesse clemencie, and withall consydered, that no gift may
lightly be more acceptably presented to y⁶ head, then that, which wholly tēdeth
to the preseruation of the rest of the body. Of whiche sort, when I consydered
this Historie of Plantes to be, I feared the lesse to present it vnto your Maiestie. Knowing that by your Princely clemēcie y⁶ same being receiued, & by your
high wisdome & aucthoritie alowed, shal take such place in your body politike,
as in the natural, those do that by the head (whiche by reasons rule gouerneth
the whole) are knowen to be approoued and condignely allowed. The seconde
pricketh me continually forwarde with this or the lyke perswasion: That as a
thankeful hart towardes a natural mother cannot be better testified, then by
loue shewed and practised towardes her deare children: nor a more acceptable
fruitfulnesse be required of any one branche, then that which may redounde to
the ornament of the whole stocke: so I in no wise should be more able to shewe
my thankful minde towardes your highnesse (the most louing and tender mother of this cōmon Weale) then in publishing this historie to y⁶ benefite of your
most louing subiectes, as being the best token of loue and diligence that I am

The Epistle to the Queene.

at this time able to shew vnto either. And (doubtlesse if my skill in the translation were answearable to the worthynesse eyther of the Historie it selfe, or of the Authours therof, I doubt not, but I should be thought to haue honoured your Maiestie with an acceptable present. As touching the worthinesse of the Historie it self, truely that thing may not iustly be thought vnmeete to be offered vnto a Prince, the knowledge wherof, beside that it is by daily experience knowen to be both profitable to al, and pleasant to many, is aboue al other faculties (the diuine knowledge wherby the soule liueth) only excepted, with so high commendations in the holy Scriptures extolled, that not onely the professours therof are accounted worthy of admiration & honor: but euē Salomon that royall and wise kyng, for that he had the knowledge of the natures of Plantes, & was able to dispute therof, from the highest to the lowest, from the Cedar in *Libanon* to the Hysope that springeth out of the wall, is therefore in the sacred Bybel highly dignified & renowmed. I wyll say nothing of Mithridates, Lysimachus, Gentius, Artemisia, and such noble & mighty Princes: whose delight and lyking towardes this knowledge of the nature of Plantes was such, that as by their diligent inquisitiō they wittily found out the vse of many of them, so, hauing found the same, they disdayned not to denominate and imparte therto their owne names, which euen to this day many of them do styll retayne. But argumentes to this purpose, are before your most excellent Maiestie needelesse to be alleaged: aswel because your highnes is dayly cōuersant in the most cleare light of al both diuine & humayne knowledge, whereby you farre more easily see the whole compasse, then men of meane estate are able to conceiue a parte: as also for that the professours of this facultie be with your highnesse had in such price and estimation, that they are not onely by your Maiestie and your most noble Progenitours, with sundry Priuileges & liberties endowed, with many & great stipendes and pentions in your Graces Vniuersities and Schooles fostered and mainteyned: but also as they shalbe founde to haue laudably profited therein, so are they aduaunced & called to the charge of your person, & of the persons of your Nobles. And arte being by honor nourished, encreaseth dayly, & putteth al men out of doubt, that they which so embrace the professours thereof, do both well lyke and thinke of the facultie, and sufficiently vnderstande both the vse and the excellencie thereof. As touching the Authour of this worke which I haue trāslated, how paineful a man he is, how skilful, and how luckely he hath atchieued this his businesse, as it shal best appeare by diligent reading ouer his workes: so also may it easily be knowen by the testimonies & iudgementes of the most learned Physitions of this age. Of whom, some are by their owne workes alredy extāt, notable & renowmed, & others (by the great trauaile that they haue bestowed in translating him out of his tongue wherin he wrote into diuers other languages) are made euidēt and famous: but none before this into English. Which hath made me desyrous (folowing their example) to make my Countreymen partakers of such knowledge, as other learned and wise men in other Countries haue thought meete to be made knowen in the natiue tongues of their commō Weales. Touching my selfe this onely I haue to promise, that in this translation I haue vsed my most skil and diligence to please and pleasure al such as delight in this so honest and profitable a knowledge. Most humbly crauyng a fauourable acception hereof at your Maiesties handes, and pardon, if any poynt I haue giuen iust occasion of blame and deserued reprehension.

1.Reg.4.

From my poore house at Lytescarie within your Maiesties Countie of Somerset, the first day of Ianuarie, M.D.Lxxviij.

Your Maiesties most humble and faithfull subiect, Henry Lyte.

⁋ To the friendly and indifferent Reader.

IF thou be ignoraunt (gentle Reader) and desyrous to knowe, either how profitable this Historie of Plantes is, or how worthy to be studied, either how harde & how highly in times past esteemed, what be the causes of the hardnesse therof, how they may be remedied, and why the Authours hereof (after so many learned both auncient and late writers) tooke vpō him the setting forth of the same: or why in his Annotations & last edition he hath reuoked certayne thinges which in the first escaped him: for thy instruction & resolution in these matters, I referre thee to the same Authours two Prefaces, wherin he learnedly, & as briefly as the nature of the matters will permitte, discourseth therof sufficiently. But if thou wouldest know of me, why I haue takē vpon me the translation & publication of ye same in this our natiue tonge, as I might without any great labour yeelde thee many iust and reasonable causes of my so doing, if I thought it greatly expedient or necessarie so to do: so I thinke it sufficient for any, whom reason may satisfie, by way of answeare to alleage this action & sententious position: Bonum, quo communius, eo melius & præstantius: a good thing the more common it is, the better it is. Seing then ye my translation shall make this good & profitable historie (which hitherto hath lien hid from many of my Countriemen, vnder the vayle of an vnknowen language) familiar and knowen vnto them: and if it be good (as no good man wil denie) to enlarge a good thing, and to make many partakers thereof: then can there not lacke iust cause to be alleaged of this my doyng: neither thinke I, that any will mislike or repine thereat, except such, as either enuie the Weale of others whom they accompt simpler then them selues, and therfore recken vnworthy to be in their owne language made partakers therof: or els are so studious of their owne priuate gaine, that they feare, least by this meanes some parte therof may be lessened: whyles others vnderstanding the nature and vertues of Plantes and herbes, shalbe the lesse beholding to their scrupulous skill. But the good and vertuous Phisition, whose purpose is rather the health of many, then the wealth of him selfe, will not (I hope) mislike this my enterprise, whiche to this purpose specially tendeth, that euen the meanest of my Countriemen (whose skill is not so profounde that they can fetche this knowledge out of strange tongues, nor their habilitie so wealthy, as to entertaine a learned Phisition) may yet in time of their necessitie, haue some helpes in their owne, or their neighbours fieldes and gardens at home. If perchaunce any list to picke a quarrell to my translatiō, as not being either proper or not ful, if I may obteine of him, to beare with me til he him selfe shall haue set foorth a better, or til the next impression, and the meane while (consydering that it is easier to reprehend a mans doings, then to amend it) vse me as a whet stone to further him selfe, I wil not muche striue: for I seeke not after vayne glorie, but rather how to benefite and profite my Countrie.

Fare well.

(∴)

VV. B.

Ermani fateor Dodoneo plurima debent,
    Nec debent Angli(Lite)minora tibi.
Ille suis etenim plantarum examina scripsit,
    Tuque tuis transfers, quæ dedit ille suis.
Quodque opus ijs solis priuatum scripserat, illud
    Tu commune Anglis omnibus esse facis.
Crede mihi plantas quia transplantaueris istas,
    Belgica quas primûm solaque terra dedit,
Inque Britannorum lętas adduxeris oras;
    Lite tuæ laudis fama perennis erit.
Dono te nobis Dodoneum(Lite)dedisse,
    Donum est, quo nullum gratius esse potest.
Nam terræ insignes fœtus, plantasq; potentes,
    Pœoniasq; herbas, hac ratione seris.
Inde etiam lites medicorum(Lite)resoluis,
    Aegrotisque offers Phœbus vt alter opem.
Quid superest? (medici)Lito, hunc præstare fauorem,
    Si quando affectus, sit grauiore modo,
Confluit, & Litum gratis curate, nec illum
    Læthoimmaturo vos sinitotè mori.

EIVSDEM.

Gratum opus est, dignumque tuo sub nomine ferri,
(ELIZABETHA potens) cuius moderamine solo,
Pax iucunda Anglis, atque Arbor pacis Oliua
Sic viget, vt passim per apricum incedere possit.
Gens Britonum, & tutò fragrantes carpere flores.
Quid igitur Litus plantas tibi ferret & herbas
Omnigenas, donoque daret, cui porrigat herbam,
Rex quicunque tenet spatiosum sceptra per orbem.

Thomas Nevvtonus, Cestreshyrus.

PErpetuum tibi ver liber hic philomuse ministrat,
Ac paradisiaci germania læta soli.
Herbarũ huic thesaurus inest, florumq́, suppellex,
Alcinoi hic hortos Hesperidumq́, vides.
Nec flos hic desit, nec floris grata venustas,
Nec vires, nec odor, nec medicina valens.
Hoc viuunt, viuentq́, libro Podalyrius, Alcon,
  Hippocrates, Pæon, Musa, Galenus, Arabs,
Phillyrides Chiron, Epidaurius, aq́, Melampus,
  Gentius, Euphorbus, Iosina, Lysimachus,
Telephus, ac Mithridates, Artemisia, Achilles,
  Alcibides, Hieron, Attalus, atque Iuba,
Pamphilus, Atrides, Nicander, Bassus, Iollas,
  Crateias, Glaucon, & Cato, Pythagoras,
Rasis & ipse Dioscorides, Auicenna, Machaon,
  Serapio, Celsus, Menecratesq́, tumens,
Aetius, Aegineta, Ruellius ac Theophrastus,
  Tragus, Auerrhoys, Plinius, Agricola,
Macer, Oribasius, Mesue, & Brunfelsius Ottho,
  Manardus, Zerbus, Fuchsius atque Sethi,
Ginus, Humelbergus, Matthæolus ac Columella,
  Fernelius, Pineus, Pena, Eliota, Lobel,
Copho, Taranta, Leonicerus, Iberq́, Lacuna,
  Mago, Varignanus, Varro, Ioannicius,
Soranus, Constantinus, Merula, Aurelianus,
  Guido, Godaldinus, Curtius, Encelius,
Moschio, Philotheus, Cleopatra, Bonaciolusq́,
  Arnaudus, Rocheus, Ferrius, Albucasis,
Hildegardis, Trotula & Albicusq́, Torinus,
  Pandulphus, Suardus, Manlius & Diocles.
Thurinus, Dimocles, Guilandinus, Philaretus,
  Bucius, Eudoxus, Garbus, Aphrodiseus,
Montius, Aubertus, Fallopius atq́, Biesus,
  Belfortis, Bayrus, Montuus, Akakia,
Lemnius & Cordus, Rondletius atque Dryander,
  Cardanus, Vidius, Iunius, Hermoleos.
Hinc Collimitius, Fracastorius, Gemusæus,
  Clusius ac Stephanus, Scaliger atque Kyber,
Saracenus, Mizaldus, Sauonarola, Erastus,
  Cum Bacchannello Cellanoua atque Rota,
Rhegius, Erotes, Montagnana atque Aquilanus,
  Manfredus, Baccus, Wolphius, Arculeus,
Ioubertus, Trincauelius, Pictorius, Euax,
  Gesnerus, Brunswich, Langius atque Cocles
Turnerus, Caius, Bullenus, Linacrus, Askham,
  Guintherius, Vasseus, Kraut, Lonicerus item,
Brissotus, Polybus, Clementinus, Mari ab alto,
  Landulphus, Phairus, Quiricus, Hollerius,

Cubba, Dasmascenus, Gatinaria, Crato, Rulandus,
  Hallus, Culmannus, Ruff, Paracelsius Hoheim,
Augerius, Landus, Galeottus, Oroscius, Oddi,
  Struppus, Heresbachius, Gratalorusq́; pius,
Atq; Fauentinus, Merenda, Wierus, Amatus,
  Cum Quercetano, Placotomoq́; graui,
Syluius, Honterus, Cornarius ac Morisotus,
  Cumq́; Argenterio Frerus & Hatcherides,
Fumanellus, Trallianus, Bellonius, Isack,
  Musinus, Riccus, Villanouanus item,
Pantinus, Gaynerus, Clinolus ac Bruyrinus,
  Kiffus, Mantinus, Plancius, Emericus,
Compluresq̀; alij: quos nec numerare necesse est,
  Nec scio, si coner, qua ratione queam.
Id sed Apollineo Rembertus acumine præstat,
  Quem suus ornat honos, gloria, fama, decus.
Herculeo exantlans molimina tanta labore,
  Quæ non sunt vllo deperitura die.
Vtile alexicacon qui promit Pharmacopolis,
  Vtile Chirurgis, vtile Philiatris:
Vtile opus docto, indocto, iuueniq́; seniq́;,
  Diuitibus simul ac vtile pauperibus.
Nec tu Leite tuo certe es faudandus honore,
  Qui tantas Anglis sponte recludis opes.
Macte animi: sic fama polum tua scandet ad altum,
  Sic te, sic patriam nobilitare stude.

Thomas Newton.

In commendation of this worke, and the Translatour.

If all Dame Enuyes hatefull broode hereat should hap to prye,
  Or Momus in his cankred spight, should scowle with scorning eye.
Yet Mawgre thē this worthy worke the Authors name shal rayse,
  And paynefull toyle so wel employd: shal reape renowmed prayse.
Not onely he whose learned skyll and watchfull payne first pende it,
  And did with honor greete (in Douche) to Countrie his commende it:
But also he whose tender loue to this his natiue soyle,
  For vs his friendes hath first to take almost as great a toyle,
A trauell meete for Gentlemen and wightes of worthy fame:
  Whereby great Princes heretofore haue got immortall name.
As Gentius, Lysimachus, and also Mythridates,
  With Iuba, Euax, Attalus, and Dioscorides.
And many noble wightes besydes, and great renowmed Kinges,
  Haue so bewrayde their skyll in this (besydes all other thinges)
By registring their names in Herbes, as though therby they ment,
  To testifye to all degrees their toyle and trauell spent
In suche a noble facultie, was not a slauishe thing:
  But fyt for worthy Gentlemen, and for a noble King.
For if by Herbes both health be had and sicknesse put to flight:
  If health be that, without the which there can be no delight!
Who dare enuie these worthy men, that haue employde their payne,
  To helpe the sore, to heale the sicke, to rayse the weake agayne!
No fye of that, but Dodoneus aye shall haue his dewe,
  Whose learned skyll hath offered first, this worthy worke to vewe.
And Lyte whose toyle hath not bene light, te dye it in this grayne,
  Deserues no light regarde of vs : but thankes and thankes agayne.
And sure I am, all Englishe hartes that lyke of Physickes lore,
  Wyll also lyke this Gentleman : and thanke hym muche therefore.

FINIS.   VV. Clovves.

To the Reader, in commendation
of this vvorke.

WHere vertue shines, and deepely seemes to rest,
Where ayde appeares, to helpe the health of man,
Where perfect proofe assignes vs what is best,
Where counsell craues, eache willing minde to skan,
Where learning lyes to helpe vs nowe and than:
There best is deemde for man to spende his dayes,
Though it be reapt with toyell ten thousande wayes.

Then blame not him, whose carefull hande first pende,
This worthy worke, whiche nowe is brought to light,
But it embrace, and double thankes him lende,
Whose dayly toyle deserues the same by right:
For vertue shines herein to eache mans sight.
Whose ayde for health, with proofe and counsell graue,
Whose learned liues, ought sure due prayse to haue.

Well: Rembert Dodoens, wrote this first in Douche,
Whiche since in Frenche was turnde by others toyle,
And nowe by Lyte: whose trauayle hath bene suche,
For ease of all, within this natiue soyle.
Where (loe) to Momus mates, he giues the foyle,
And here presentes in the Englishe tongue,
To comfort all that are both olde and young.

The worke it selfe, of sundrie trees intreate,
Besides of Herbes, Flowers, Weedes, and Plantes that growes,
Setts downe their vertues sure which are so great,
That we may say therein great learning flowes.
The Author hath (so farre foorth as he knowes
By skilfull Iudgement) vnto eache disease,
Set downe a cure, the sicke and sore to ease.

Great was his toyle, whiche first this worke dyd frame.
And so was his, whiche ventred to translate it,
For when he had full finisht all the same,
He minded not to adde, nor to abate it.
But what he founde, he ment whole to relate it.
Till Rembert he, did sende additions store,
For to augment Lytes trauell past before.

Whiche last supply so come to Lyte his hande,
He fitly furnisht, euery peece in place.
The worke agayne he wrote I vnderstande,
For feare if ought therein shoulde breede disgrace.
And did as much as one coulde in this case.
For English names, to euery herbe and plant,
He added hath, whereby is nothing skant.

This

This rare deuice, eache one may well esteeme,
Which bringeth ayde, and comfort vnto man.
The learned wyll accept the same I deeme.
Wherfore I craue yf ought espie thou can,
(As none can be so watchfull nowe and than
But faultes may scape, for want of Argus eyes.)
To mende the same, and nought herein despise.

With willing minde, good Reader here I craue,
Accept this worke, thus written for thy sake,
And honor him that seekes mans health to saue.
Yeelding him thankes, whiche it dyd vndertake:
And vnto Lyte due thankes thou hast to make.
His paynefull pen deserues thy good report.
Whose toyle was great, to ende it in this sort.

        T. N. Petit ardua virtus.

Iohannis Hardingi in laudem tam Auctoris
quam interpretis Duodecasticon.

Edant Turneri pingues simûlatq; Lobeli,
 Horti ac egregij gloria summa Tragi.
Vnicus hic reliquis longe est Præstantior hortis,
 Quem pia iam Liti cura laborq; dedit.
Illorum tenues abijt decor omnis in auras,
 Huius at æterno gloria viua manet.
Aspice quam virides insultent vndiq; plantæ,
 Quales viderunt sæcula nulla priûs.
Tantum igitur Lito debes gens Anglica docto,
 Quantum Remberto Teutonis ora suo.
Nec plus Remberto letantur Meclinienses,
 Quam te Lite tui Candide Murotriges.

   Fato prudentia maior.

REMBERTI DODONÆI MECH-
LINIENSIS MEDICI, IN SECVNDAM COM-
MENTARIORVM SVORVM, DE STIRPIVM
*Historia, editionem, ad studiosos Medicinæ
Candidatos, Præfatio.*

STIRPIVM ac vniuersæ materiæ Medicæ cognitionem potentissimis Regibus, antiquissimis Heroibus, Præstantissimis Medicis ac Philosophis olim in pretio habitam, vtilem ac necessariam Medicę arti iudicatam, sommo studio, nec minori diligentia excultam, adeo manifestum est, vt multis assertionibus opus non sit. Præsertim non paucis herbis Regum ac Heroum, qui has vel primi inuenerunt, vel in frequenti medendi vsu habuerunt, nomina retinentibus, vt Mithridatium, Eupatorium, Gentiana, Lysimachia, Achillea, Centaurium Alcibiadium, Telephium, Arthemisia, aliæque plures: & veteres ipsos, atque inter eos Hippocratem, Medicorum omnium longè principem, Galenum, nõnullosq; alios, longinquas peregrinationes, cognoscendæ materiæ Medicæ causa suscepisse, & propria & aliorum scripta testentur.

Eandem verò scientiam, à posteriorum nostroque tempore vicinorum seculorum medicis ac philosophis, planè neglectam & contemptam fuisse, res ipsa quoque euidenter docet. Solæ enim illæ Medicinæ partes, ab illorum seculorum Medicis coli visæ sunt, quæ ex rationum physicarum fontibus deductæ, hinc dubitandi disputandiq; vberem materiam præberent: aliæ verò, vsu atque experientia constantes, vt steriles ac ieiunæ spretæ. Cuiusmodi ipsa βοτανικὴ est, Physicarum rationum subsidium vel nullum vel exiguũ admittens. Quamobrem eius omni notitia, mulieribus, herbarijs analphabetis, vel indoctis pharmacopœis relicta, indignum professione sua ac magnificis titulis parum decorum infœlices illi Medici existimabant, cognoscendæ alicuius materiæ medicæ herbæ aut stirpis causa, vel minimum operæ laborisque sumere: extra vrbes ad montes, conualles, prata, suburbanaque loca excurrere.

Tantam studiorum dissimilitudinem, fatalis ille seculorum ordo peperit, qui & maximas vrbes, potentissima regna, latissimè patentia imperia, & hanc Medicinæ partem, nonnullasq; alias pessundedit, ac propemodum extinxit. Vnde factum, vt quæ olim facillimè, ac nullo propemodum negocio materiæ Medicæ ac Stirpium cognito percipi poterat, difficilis ac obscura reddita sit.

Tradebant eam Dioscorides, Galenus, eiusq; ætatis Medici, veluti per manus à maioribus acceperant, seruatis eousq; nominibus, quæ magna ex parte incorrupta ad illorum tempora venerant.

Nobis ea fœlicitas denegata, multis modis veris ac genuinis appellationibus, corruptis, péruersis ferè abolitis, barbaris in earum locum suppositis, multo tempore intermissa Stirpium notitia, solis veterum descriptionibus relictis, ex quibus hęc disciplina & requirenda & restituenda est. Id quàm difficile sit, licèt alio loco scripserimus, tamen hìc repetere visum fuit non alienum.

Non leuis autem difficultatis huius scientię, aut vna aliqua causa est, sed maxime exæque præcipuæ duę; innumera videlicet multitudo stirpiũ, immensaq; varietas: & eorũ qui de harum historia, aut materia Medica scripta reliquerunt, breuitas, incuria, negligentia, subinde varia atq; dissimilis apud diuersos descriptio: & vtinã non quorundã errores obscuritatem nõ exiguam in eã intulissent.

Stirpium siquidem herbarumq; infinita sunt genera, variè per orbem terrarum sparsa immensa multitudo, vt non vno loco paucisq; regionibus requirere eas liceat, sed ad eas omnes cognoscēdas omnium ferè regnorũ ac prouinciarũ peragratione, longi temporis peregrinatione opus videatur. Sunt nonnullæ quibusdam vel insulis dicatæ vel regionibus propriæ, quæ in alia quæuis loca transferri nequeunt, vel tellure cœloq; mutatis mutatur, vt Theophrastus libro quarto ait. Alię pluribus quidem terris communes, non omnes tamen passim aut crebrò obuię, sed certis tractubus peculiares, vel montibus scilicet, promontorijs, præruptis rupibus, saxosis aut niualibus locis, collibus, densis syluis, vmbrosis lucis, arborum caudicibus, aruis, campestribus, apricis, læto pinguiq; solo, macro & sterili, humidis, vliginosis, riguis, paludibus, stagnantibus aquis, fontibus, fluminibus, fluuiorum ripis, maris littoribus, scopulis, vel ipso deniq; mari addictæ. Harum autem istæ eandem ferè vbiq; formam retinent, vel exiguam mutationem assumunt: illæ in diuersis regionibus pro cœli soliq; varietate, aliam formam & magnitudinem induunt. Quod segetum, fabarũ, nucum, aliorumq́ue apud Indos nascentium (si Herodoto fides) exemplis manifestum est, quorum longè maior magnitudo, quàm in Ægypto nascentium. Segetum enim culmi instar harundinum crassescunt: fabæ triplo maiores Ægyptijs, sesamum miliumq; eximiæ magnitudinis: nuces tantæ molis vt miraculi loco in templis suspendantur. Strabo quoque in extremo Mauritanorum quodã tractu, iuxta

PRÆFATIO AD

creditum Nili exortum vitem tantæ crafsitudinis nafci tradit, quam vix duo homines complecti queant: omnem herbam cubitalem: Staphylinorum, Hippomarathri, Solymi caules duodenum cubitorum, crafsitudine quatuor palmorum reperiri. Et tātus quidem Stirpium numerus immensa latifsimè fparfa multitudo, varia ac multiplex natura.

Auctorum verò ipforum quanta fuerit negligentia, vel incuria in multarum præfertim vulgo notarum, formis differentiisq; defcribendis, cùm multarum extent nomina, quarum formæ non funt expreffæ, aut leuiter tantùm defcriptę, nemo ferè eft qui ignoret, Ruellio idipfum fcriptis fuis teftante.

Eofdem verò non femper conuenire, atq; interdum inter fefe difsidere, dum fub vno eodemq; nomine aliùs aliam herbam vel fruticem defignat, vel eandem alia nomenclatura exprimit, Diofcoridis cum Theophrafti aliorumq; fcriptis diligens collatio oftendit, fuppeditabit & huius varietatis Hiftoria noftra non pauca exempla.

Defcriptiones verò quorūdam erroribus effe confperfas, quod ad Plinianas attinet manifeftifsimum eft, Leoniceni enim libri de Plinij erratis pafsim proftant, & omnes ferè noftri feculi, qui de materia Medica aut Stirpibus fcripfère, in redarguendis ac notandis Plinij lapfubus plurimi funt.

Verùm de Diofcoride id nemo forfitan expectauerit aut fufpicatus fuerit, Galeni teftimonio atq; fcriptis commendato. Reperiuntur tamen in eius commentarijs nō exigui errores. Alias enim difsimilium Stirpium, eiufdem apud diuerfos auctores nominis, delineationes in vnam hiftoriam contrahit: alias eandem non ijfdem nominibus nuncupatam, veluti membratim diuülfam diuerfis locis defcribit, vt ijs commentarijs quos in Stirpium hiftorias, quæ apud Diofcoridem extant meditamur, oftenfuros (fi Deus vitam, valetudinem, ociumq; concefferit) nos fperamus.

Nec tamen hi errores impediunt, quo minus Diofcorides alijs omnibus longè præftet, cum omnes vel imperfectiorem multò hiftoria, vel pluribus, maioribus erroribus, ac fabulis, prę̧ftigijsque plena fcripta reliquerint. Theophraftus reliquos omnes in fuo fcribendi genere fuperans, formas ex profeffo non defcripfit, fed Stirpium multiplicem differentiam aliaq; philofopho homine digna profequi ftuduit. De Plinij fcriptis quid iudicandum, iam fcripfimus. Nicandri, Θηριακά καὶ ἀλεξιφάρμακα folum reperiuntur. Apuleius de paucis tātummodo egit. Galenus, Paulus Aëtius, figuras à Diofcoride expreffas omiferunt. Alij veteres Græci & Latini, Philofophi, Medici, Poëtæ, Hiftoriographi, Architecti, Agriculturæ fcriptores, Hippiatri fiue Veterinarij, quorum lectione ad quarundam plantarum notitiam peruenimus, non nifi obiter quarundam meminerunt. Bithyni Iolæ, Heraclidis Tarentini, Crateuæ herbarij, Andreæ medici, Iulij Bafsi, Nicerati, Petronij Nigri, Diodoti, Pamphyli, Mantęę, Apollonij antecefforum Diofcoridis & Galeni fcripta, neque ad pofteritatem peruenerunt, neque perfectum aliquid tradiderunt. Pleriq; horum anilibus fabulis aut prę̧ftigiaturis Ægyptijs, coniurationibúfue pleni funt, alij de vna aliqua materia, aut paucis tantùm, vt Galenus fcribit, egerunt: vniuerfam verò materiam complecti non ftuduerunt.

Quibus de eaufis illorum omnium fcriptis pofthabitis, vni Diofcoridi fummam laudem auctoritatemque Galenus tribuit, quam illi quoque deberi nemo negare poteft, abfque eius fiquidem fcriptis, Stirpium materięque Medicæ cognitio reftitui nulla ratione poteft.

Non enim idcirco veluti parum vtilis abdicandus, quòd in plerifque locis lapfus fit, cùm nec Plinium multo grauius & pueriliter fępè hallucinatum minimè negligi oporteat ad Stirpium notitiam plurimum conferentem. Neque enim eam ob caufam de erratis eius cœpimus admonere, fed vt huius fcientię & ftudij difficultas ab omnibus intelligatur maiorique cum attentione, in ftirpium cognitionę ftudiofi incumbant: diligentius omnes notas expendant: leuibus coniecturis contenti facile iudicium non promant, minus admirentur fi poft complures in hoc ftudio verfatos, multa in notitiam nondū perducta adhuc lateant: plures quotidie exoriantur in reftituenda hac Medicinæ parte laborantes, aut quod hi qui in lucem fubinde nonnulla dedère, fententiam alicubi mutent. Nam hæc omnia, haud dubio, difficultati huius fcientiæ magis, quàm negligentię, incuriæ, aut temeritati huius ætatis fcriptorum accepta referre æquum eft.

Si enim Diofcorides exercitatifsimus vir, qui multum ftudij, laboris, in Stirpium, materięque Medicę cognitionem impendit, eo feculo quo nomina magna ex parte incorrupta vulgo retinebantur, ipfarum notitia à maioribus accepta veluti per manus tradebatur, vel immenfo numero, multipliciq; earum varietate, vel auctorum imperfectis, varijs, fabulofis defcriptionibus detentus, errorem vitare non potuit: qua ratione nunc quifquam, antiquis nomenclaturis ac appellationibus vix receptis, aut plurimum deprauatis, Stirpium cognitione longo tempore intermiffa atque contempta, in tanta veterum (vti diximus) negligentia, incuria, varietate, atque erroribus fubinde implicata obfcuritate, vel facile vel abfque magno labore, diligentifsima inquifitione earum notitiam confequi fe poffe fperabit?

Quum igitur tanta huius fcientię vel magnitudo vel difficultas fit, vt non nifi diligentifsimo omnium ftirpium maturoq; examine plurimorum veterum auctorum lectione exactifsima, id eft,

multo

multo labore, diutinis peregrinationibus, continuo studio comprehendi queat: ac fieri vix possit, vt his omnibus, vnius hominis aut paucorum vita diligentiaq; par sit. Citra omnem admirationem esse debet, post multos recentiorum in hac materia diligenter versatos, alios indies exoriri, qui hâc augere studiant, & nostros quoque de Stirpium historia libros prodire.

Nemine siquidem hanc scientiam ad perfectionem perducente, sed omnibus plurima prætermittentibus, occasio posteris relinquitur, priorum inuentis ac obseruatis plurima adijciendi atque stirpium cognitionem locupletandi. Quod recentiorum non paucis præstantibus, priuato studio, peregrinatione, aliáue occasione comperta in commune proferentibus & mei officij fore iudicaui, vt eam quam existimabam me huic scientiæ posse accessionem facere, aut emendationem adhibere, in publicum mitterem, atque veræ Medicinæ Stirpiumq; studiosis communicarem.

Non frustra autem vel inutiliter hunc laborem vel recentiores vel nos suscepimus. Pertinet enim hęc scientia ad præcipuas & principes duas Medicinę partes διαιτητικὴν καὶ φαρμακευτικὴν. Illa victus ratione: hæc medicamentis sanitati hominis consulit: vtraque herbis, frugibus, stirpibus, earumq; seminibus, fructibus, radicibus, succis, veluti necessarijs & materia & instrumentis vtitur. Hæc enim vt plurimum, ad artis opera Medici vel impermixta, vel alijs aut inter se commixta adhibent. Si enim nemo illum bonum fabrum aut artificem dixerit, qui malleum, incudem aut ferrum, aliáve artis suæ instrumenta vel materiam nō nouerit: Medicum quis habebit doctum, qui Betam à Blito distinguere nesciat, in crassa supinaque omnium Stirpium Medicæque materiæ ignorantia versetur.

At multum, fortasse dicet aliquis, inter medicum & reliquos artifices interest, neque enim manum medici operibus apponunt, sed veluti architecti tantum præcipiunt, omnem Stirpium & Medicæ materiæ notitiam, præparationem, variam mistionem pharmacopœis relinquunt.

Fatemur à multis annis medicos præparandorum miscendorumque pharmacorum morem omisisse, ac à veterum consuetudine recessisse, quos abunde constat nulla pharmacopœorum opera vsos, medicamenta etiam proprijs & discipulorum manibus miscuisse, neque vt pharmacopœi rursus fiant requirimus, & quærendis, terendis, tundendis, præparandis, miscendis, medicamentis occupentur, sed Stirpium & materię Medicæ, quarum potissimum frequentior ac quotidianus vsus, notitiam exigimus: non secus ac in Architecto omnis materiæ ædificiorum cognitio requiritur. Qui enim eam ignorauerit, bonus Architectus esse non potest. Sic etiam neque doctus aut perfectus medicus, qui artis suæ materiam non cognorit. Quod si fabri ex non conuenienti ligno vel trabes vel ædium contignationes struant, Latomi luto pro cæmento parietes compingant aut reliqui artifices alijs modis imposturas moliantur, admittet aut dissimulabit hæc bonus Architectus? Non existimo quenquam fore, qui non putet harum rerum curam ad eum pertinere. Cur igitur conniuebit Medicus, si pharmacopœus spuria legitimis genuinis adulterata recentibus exoleta, calida frigidis, frigida calidis, alexi pharmacis deleteria substituat, ac deficiente vno, alias hoc, modo illud, absque vllo iudicio aut delectu supponat? Conniuere autem cum huiusmodi imposturis aut grauissimis erratis oportebit, si huius scientiæ rudis imperitus fuerit, cum nulla ratione dolum deprehendere poterit.

Omnes igitur Medicinæ studiosos in hac Stirpium materiæque Medicæ notitia sese exercere conuenit, atque operam & diligentiam summam adhibere, vt harum cognitionem reliquis Medicæ artis partibus adiungant: veterum antiquissimorum, probatissimorumque Medicorum huius scientiæ studiosissimorum, vestigijs hac in parte insistat, ac Galeni clarissimi & maximę auctoritatis medici, præceptis & consilio obsequundent, qui omnes Medicos & iuuenes artis candidatos ad Stirpium & Medicamentorum materiæ exactam notitiam admonet, atque incitat, Medicus (inquit libro de antidotis primo) omnium Stirpium, si fieri potest, peritiam habeat, consulo: sin minus, plurium saltem quibus frequenter vtimur. Item tertio de Medicamentis secundum genera. Hinc puto bonę indolis iuuenes incitatum iri, vt medicamentorum materiam cognoscant, ipsi met inspicientes, non semel aut bis, sed frequenter, quoniam sensibilium rerum cognitio sedula inspectione perficitur. Et ibidem paulò infra: Vos ergo admoneo amici, vt in hoc quoque me sequamini si artis opera pulchrè obiri velitis. Nouistis enim quomodo ex omni natione, præstantissima quotannis medicamenta mihi adferantur, eò quòd perditi illi omnigerarum rerum coëmptores (Greci ῥωποπῶλας vocant) varijs modis ea contaminant. Præstiterat fortassis non hos solum, sed multò magis etiam mercatores, qui illa aduehunt, incusare: atque his multò magis ipsos herbarios: item nihil minus eos, qui radicum liquores, succos, fructus, flores & germina ex montibus in vrbes conferunt. Hi siquidem omnium primi in eis dolum exercent. Quisquis igitur auxiliorum vndique copiam habere volet, omnis materiæ Stirpium, animalium & metallorum, tum aliorum terrestrium corporum, quæ ad Medicinæ vsum ducimus, expertus edo, vt ex eis & exacta & notha cognoscat. Deinde in commentario meo, quem de simplicium medicamentorum facultate prodidi, sese exerceat. Nisi enim hoc modo instructus ad præsentis operis præsidia veniat, verbotenus quidem medendi methodum

* ij

thodum sciet, opus verò nullum ipsa dignum perficiet. Hactenus Galeni verba. Ex quibus manifestum est ac indubitatum relinquitur, hanc scientiam medico & vtilem, & necessariam, vt qui absq; huius peritia nihil possit medendo certi assequi, aut eximium quicquam ex arte præstare, herbariorum, myropolarum, pharmacopœorumq; dolis, imposturis, ac subinde crassa vel pertinaci ignorantia delusus. Quod omnes medicos diligentissimè cauere cum salutis ac valetudinis suæ fidei concreditorum, tum propriæ existimationis causa, maximè decet.

Neque scientiæ huius difficultas, quæ ipsius penè immensam magnitudinem ostendit, quenquam ab eius studio absterrere debet, sed potius ad auxiliares ei manus conferendas omnes studiosos excitare, accendere, inflammare: ne tam necessaria humanæ vitæ scientia diutius vel neglecta, vel contempta iaceat, sed plurimorum communi labore atq; diligentia crescens, ad perfectionem veniat, ab interitu vindicetur, medicinæq; reliquis partibus adiungatur, ac veluti postliminio restituatur. Quò enim difficultatis ac magnitudinis scientia aliqua amplius habet, hòc magis bona ingenia in ea occupari libentius solent. Ignauorum existimatur in paruis ac facilibus versari: industriorù verò ac diligentiù in grauibus, magnis, ac difficilibus. Paruæ ac faciles res nulli opinionem aut auctoritatem pariunt. Difficiles & magnę, honores & gloriâ conferunt. Magnæ enim rei, quantumcunque quis possederit, participem fieri, non minima est gloria, vt Columella ait.

Sed vt ad hos de Stirpium commentarios veniamus. Contraxeramus in hos, quum primù ederemus, quicquid herbarum plantarumq; in cognitionem nostram venerat. Secunda hac editione seuera animaduersione adhibita, omnia recognouimus, pleraque mutauimus, nonnulla transtulimus, totum opus non exigua accessione locupletauimus & auximus, multarum Stirpium nemini quod sciam adhuc depictarum imagines adiecimus. Vtrobique formas omnium, qua potuimus diligentia tradidimus: nomina Græca, Latina, officinis recepta, Germanica, Gallica, & nobis Brabatis ac vicinis Fládris aut Hollandis Frisiisve vernacula, singularum historijs adscripsimus. Temperamenta deinde ac vires ex probatissimorum Medicorum scriptis subiunximus, haud præteritis ijs quæ recentiorum experientia repperit, maximè earum quas in veterum cognitionem non venisse neoterici putant. Atq; hæc omnia breuissimè complectentes non elementorù ordine stirpes digessimus, sed vel forma, vel viribus, vel alia ratione congeneres ac similes coniungere studuimus.

His autem describendis non tantum nostro studio vel, si quę est, industria profecimus, sed antecessorum quoq; scriptis plurimum adiuti sumus, Leoniceni videlicet, Hermolai, Manardi, Ruellij, Cordi vtriusq;, Hieronymi Tragi, aliorumq;. Leonharti Fuchsij imagines in priores nostros magna ex parte omnes recepimus (vt in ipsa imaginù nostrarù prima editione adiecta causa scripsimus) non sic tamen vt sententiam eius in omnibus sequeremur, sed adhibito iudicio & animaduersione, verisimiliorem amplecteremur.

In recognoscendis verò & noua accessione augmentandis, licet nostro labore plurimù creuerint, non tamen Petri Bellonij obseruationibus profecisse nos inficiari possumus: aut ex Petri Andreæ Matthioli commentarijs quædam mutuatos. Quibus tamen multum pepercimus, propterea quod eæ, quas reliquimus, in conspectum nostrum non venissent. Annisi enim sumus ad hoc, vt vix alias describeremus, quàm oculis nostris aliquando subiectas & conspectui exhibitas. Itaq; paucissimas ex eius commentarijs accepimus, idq; ferè non absq; ea mentione, quę videlicet propter naturę affinitatem, vel nominis similitudinem cum alijs à nobis descriptis, negligi vix poterat: vel in opinione apud nostros sic versabantur, vt aliena pro veris supponerentur. Quod si autem quæ aliæ sunt nobis cum Matthiolo communes, eas nostra cura depictas fuisse, figuræ magna ex parte aliæ, atque vernaculi commentarij prius quàm Matthioli ad nos venirent editi, facile testabuntur. Nec defuit nobis locupletandis nostris Doctiss. Andreas Lacuna, qui Corrudam & Palmam in opus nostrum intulit. Profuit etiam industria Caroli Clusij cognitioni vniuersæ materiæ Medicæ, tum ipsius artis studiosiss. qui & raras quasdam stirpes nobis suppeditauit, & conuertendis commentarijs hisce in Gallicum Idioma, benignam ac diligentem suam operam exhibuit.

Reliquum est studiosi iuuenes, vt nostris hisce commentarijs, cùm in Stirpium herbarumq; cognitione facilius assequenda adiuti, tum huius scientię & Medicinæ partis non minimè, vtilitate necessitatéque prouocati, excitatis, accensis, inflammatis animis, omni studio in hoc diligentissimè incumbatis, vt non solùm quæ à nobis descripta sunt, ac per icones expressæ stirpes, in notitiam vestram veniant, verùm etiam earum quæ apud veteres supersunt, nondum satis notarum, ac vniuersæ materiæ Medicæ peritiam assequamini, vel saltem auctarium aliquod hactenus repertis ac traditis adijciatis, quo multorum communi studio maius ac maius incrementum hæc scientia accipiens, ad ἀκμὴν tandem ac perfectionem perueniat. Valete,

Mechliniæ, Quinto Id. Iulias.

EPISTOLA AD LECTOREM.
REMBERTI DODONÆI DE
RECOGNITIONE SVORVM COMMEN-
TARIORVM AD LECTORES EPISTOLA CVM
imaginum eius parte altera olim edita.

FVTVRVM omnino auguror candide Lector, vt simul ac nasutiores & morosiores aliqui, nostros de re Herbaria commentarios aut imagines viderint, studium statim nostrum sint suggillaturi: quod post tam multos doctos viros, in hoc studij genere summa cum diligetia versatos, melius me aliquid inuenire, & eorum inuentis superaddere, posse sperauerim. Vbi vero in annotationes inciderint, quas hoc loco adiecimus, & in ijs quædam retractata, nonnulla in dubium reuocata à nobis offenderint, multo magis temeritatem nostram sint damnaturi: vt qui mox ab editione, aliam sententiam in nonnullis sequar, vel non satis perspecta atque comperta in publicum dare voluerim. His responsum cupio, huic studio hoc vnicè proprium esse multorum operam atque laborem desiderare: nempe in quo non exigua sit difficultas, nec minor varietas, quæ summam etiam diligentissimorum industriam fatigent. Infinita enim sunt stirpium quę vel sine nominibus, vel cum barbaris & peregrinis nobis sese offerunt genera, quibus vetera & antiqua reddere nomina instituti nostri præcipua & maxima pars est. Quarum etsi veteres Herbariæ rei & stirpium historiæ scriptores, differentias ac notas omnes, summa diligentia descriptas nobis reliquissent, impossibile tamen foret, de omnibus facile aut citra summum laborem & indefessum studium, veritatem assequi, cum non vno loco, sed per vniuersum orbem spersas requirere & cognoscere oporteret. Non vno enim loco aut eadem in regione omnes pluresue, sed aliæ alijs vel regionibus vel locis addictæ sunt. Dictamnum Cretæ proprium est. Rha supra Bosphorum regiones & paucæ aliæ suppeditant. Thus Sabæorum gignit prouincia. Balsamum sola Palestina producit. Et vt plures tales regionibus quibusdam solis proprias prætereamus, ex his que in pluribus terris inueniuntur, nonnullæ nisi in coualibus proueniunt: sunt aliæ montibus familiares & propriæ. Amant hæ aprica loca: illa vmbrosa aut densas sylvas. Inter saxa, lapidosis locis aut in præruptis rupibus reperiuntur quædam: aliæ in arborum caudicibus nascuntur. Lætum pinguęque solum desiderant nonnullæ: in sterili agro magis proficiunt aliæ. Delectant quasdam arua: alias vineta: illas horti: istas prata: nonnullæ in vliginosis & riguis oriuntur. Sunt quas temerè alio loco quam in maris littoribus requiras. Ad quas omnes cognoscendas & perquirendas cum preter diligentem veterum lectionem, diuturna & longi temporis per infinita loca, per omnes ferè orbis partes peregrinatione opus sit, multi labores sudoresque perferendi, infinita pericula subeunda, fieri non potest, vt vna hominis vita his omnibus satis sit, vt interim omittamus quam multa superueniunt incommoda, quæ peregrinandi occasionem aut adimunt aut multum impediunt, veluti bella incogniti diuersarum gentium, ritus, mores & lingua, horridę, incultæ, squalidæ regiones, & ad hæc maximè rerū vel publicarū, vel priuatarum & domesticarum curatio, aut ferendis maximis sumptibus impar fortuna. Nunc autem cum his omnibus grauissimis impedimentis, accedat etiam, quod veteres multarum stirpium veluti vulgo cognitarum formas non expresserint: aliarum tam leuiter descripserint: vel attigerint, vt non videantur tradidisse: iam & in nonnullarum descriptionibus non vulgares sed maximi auctores varient, veluti in Asphodelo & alijs quibusdam, quis non summam in hac disciplina difficultatem esse affirmet, quæ multorum quantumuis industriorum & studiosorum indefessos labores & maximam diligentiam requirat? nec solum requirat verum etiam superet? Cum igitur tam infinita stirpium sunt genera, singulorumque multiplices differentiæ, tam diuersa & natura & situ dissidentia in quibus gignuntur loca, quę adire omnia non solum difficilè verum etiam impossibile fuerit, & ad hæc mutilę, imperfectæ ac confusæ veterum descriptiones, vt propter hæc grauissima impedimenta, de absoluta stirpium cognitione desperandum videatur. Nulla certè presentior via, commodior ratio, aut expeditius consilium, quo hoc studium, hæc scientia in lucem reuocari & crescere possit, quam vt multorum laboribus & lucubrationibus adiuuetur. Plurimorum enim poterit industria quod paucorum nequit præstare opera. Dum enim hic quædam in lucem adfert, alij quædam adijciunt, nonnulla corrigunt, alia supplent, non exiguum herbarum studium & simplicis medicinæ cognitio incrementem capit Hanc rationem videntur mihi insequuti Leonicenus, Manardus, Ruellius, Cordus vterque, Musa, Tragus, Fuchsius, & quotquot in hac disciplina non omnino infęliciter hac ętate versantur, Neque enim quisquam istorum, aut stirpium historiam absoluit, aut perfectam sibi eius cognitionem vendicat: cum multas etiam de industria prętereant. Sed quod quisque sibi suo labore, sua industria per ocium peperit peregrinatione inuenit, aut alia ratione cognouit, hoc in commune proferre, & veritatis amatoribus communicare studet, & quo quisque

EPISTOLA AD LECTOREM.

posterior hoc maiorem huic scientiæ accessionem facit, dum antecessorum opera ac laboribus etiã adiuuatur. Mouit certè nos & hęc ratio, cum enim multas stirpes in omnibus recentioribus desiderari animaduertissem, in quibusdam deceptos eos obseruassem, & non paucas me supplere & aliorum inuentis adijcere, nec non in quibus erratum videbatur, veritatem aperire posse sperassem, volui vt post multorum doctorum virorum qui in Stirpium historia versati sunt commentarios, mei quoque labores, mei conatus, in publicum prodirent, non quod laudem ac gloriam mihi hinc aliquam postulem, sed vt nostris inuentis & studijs aliquo etiam modo, stirpium herbarumq́ue cognitio & simplicis medicinę studium promoueatur. Desinant igitur morosi censores, frustra aut temere hunc laborem à nobis susceptum criminari, quando nostra industria & opera huic scientiæ non mediocris plantarum, fruticum ac arborum numerus accesserit, antea à nemine quod sciam, recentiorum traditarum, præter omnes quę ab errore vindicatæ sunt quarum non exiguus quoque numerus est. Quod verò paucula quædam, post editos commentarios à nobis retractentur, facit summa huius scientiæ & maxima, vt diximus, difficultas, quæ nos ita tenet Cimmerijs quasi tenebris immersos, vt vix etiam summo studio, & frequenti ipsarum plantarum collatione, veritatem queamus inuenire. Si enim in ijs scientijs aut artibus (vt Socrates alicubi inquit) errores etiam subinde committantur, quæ vel ex naturalium rationum fontibus deducuntur, vel certis præceptionibus, regulis aut methodo constant, quo non tandem modo, in Herbaria disciplina, nullis regulis, nulla methodo firmata, ex naturalibus rationibus minimum, imo pene nullum subsidium admittente, sæpius & nolentibus, & non sentientibus nobis errores irrepent? Docet id scripta Leoniceni, Hermolai, Manardi, Cordi, Ruellij, Musæ, Tragi, Fuchsij & aliorum recentiorum, quorum iudicia vel à seipsis sæpius reuocata, vel ab alijs retractata & correcta sunt. Non reputo me his diuiniorem, & ego homo sum, decipi & errare possum, præsertim in re tam multis de causis difficili ac obscura, vt si alicubi ferendus aut dissimulandus est error, hic dissimulari & tolerari debeat. Quamobrem non tam pudet nos horum errorum, quàm pœnitet. Quando igitur mihi hoc commune cum alijs est, vt in quibusdam minus veritatem assequutus fuerim, non habeo aliud præsentius remedium, quàm vt mihi ipsi medear meosque errores ipse è medio sustollam. Et præstat sanè me mei ipsius correctorem esse, quouis alio, quamuis & aliorum animaduersiones non nisi æquissimo animo accepturus sum. Cùm enim publicæ vtilitatis causa hunc laborem susceperim, & eiusdem intersit sicubi à me erratum, idipsum corrigi ac notari, nulla in re magis mihi gratificari poterunt veritatis studiosi, quàm si nostra omnia ad examen ducant, cumq́; veterum descriptionibus diligenter conferant, ac vbi me veritatem minus assequutum deprehenderint, amicè & synceriter admoneant. Atqui sic morosis & seueris istis censoribus responsum esto, quos optauerim ab ista calumniandi tentigine, & doctorum huius seculi scriptorum, suggillandi & reprehendi studio, ad meliorem frugem, & bonarum artium ac scientiarum studia conuersos in hoc totos esse, & omnibus ingenij viribus certare, vt, vel mediocris eruditionis viros, doctrina æquent, si superare se posse diffidunt. Cæterum quod ad annotationes istas attinet, sequuti in his sumus, secundam commentariorum nostrorum editionem, quam ob causam quædam retractauimus, de nonnullis videlicet aliter iudicantes, de alijs vel nostram vel aliorum huius ætatis doctorum virorum coniecturam indicantes: vnum aut alterum Dioscoridis locum aut vitiosum aut confusum ostendimus: figuras complures adiecimus, omnes videlicet quæ secundæ ac posteriori æditioni accesserunt. Sunt autem ex his non paucæ quidem nouæ, id est, antea aut prius non depictę: nonnullæ infeliciter prius expressæ, nunc ædificiosius & elegantius formatæ paucissimę ex Doctiss. Pet. And. Matthioli commentarijs translatæ, quas nempe cognatio vel similitudo cùm alijs à nobis descriptis, non sinebat prætermitti, vt etiam in commentariorum nostrorum præfatione scripsimus, Nam à reliquis, quas forte alius in suos commentarios traduxisset propterea abstinuimus, quòd in conspectum nostrum non venissent. Illud enim nobis in primis curę fuit, vt quam paucissimas describeremus, quas non aliquando oculis coram cernere contigit, & maxima nouarum figurarum pars ad viuarum plantarum imitationem depingeretur, vt ipsæ stirpium descriptiones, & imagines alię nec aliorum similes facile testabuntur.

Vale, atque presentibus fruere, dum succisiuis horis ocium nacti, alia, his locupletiora, meditamur.

(∴)

De

APPENDIX.

De his qui Latine vſus Herbarum ſcripſerunt,
& quando ad Romanos notitia earum peruenerit.
Item de Herbarum inuentione, & antiqua medicina,& quare hodie minus exerceantur earum remedia, ex Plinij lib.25.cap.2.

Inus hoc quam par erat, noſtri celebrauere, omnium vtilitatum & virtutum rapa‑ ciſsimi. Primúlq; & diu ſolus idem ille M. Cato, omniũ bonarum artium magiſter, paucis duntaxat attigit. Boum etiam medicamina non omiſſa. Poſt eum vnus illu‑ ſtrium tentauit C. Valgius, eruditione ſpectatus, imperfecto volumine ad diuum Auguſtum, inchoata etiam præfatione religioſa, vt omnibus malis humanis illius potiſsimum principis ſemper medicetur maieſtas. Ante condiderat ſolus apud nos, quod equidem inueni, Pompeius, Lenæus, Magni Pompei libertus, quo primum tempore hanc ſcientiam ad noſtros peruenisse animaduerto. Nam quam Mithridates, maximus ſua ætate regum, quem debellauit Pompeius, omnium ante ſe genitorum diligentiſsimus vitę fuiſſe argumentis præ‑ terquam fama intelligitur. Vni ei excogitatum, quotidie venenum bibere, præſumptis remedijs, vt conſuetudine ipſa innoxium fieret. Primo inuenta genera antidoti, ex quibus vnum etiam nomen eius retinet. Illius inuentum autumant, ſanguinem anatum Ponticarum miſcere antidotis, quoniam veneno viuerent. Ad illum Aſclepiadis medendi arte clari, volumina compoſita extant, cum ſollicitatus ex vrbe Roma, præcepta pro ſe mitteret. Illum ſolum mortalium Mithridaten. 22. linguis lo‑ cutum certum eſt: nec de ſubiectis getibus vllum hominem per interpretem appellatum ab eo an‑ nis 56. quibus regnauit. Is ergo in reliqua ingenij magnitudine medicine peculiariter curioſus, ab hominibus ſubiectis, qui fuere pars magna terrarum, ſingula enquirens, ſcrinium commentationũ harum & exemplaria, effectúſq; in arcanis ſuis reliquit. Pompeius autem omni regia præda potitus, transferre ea ſermone noſtro libertum ſuum Lenæum, grammatice artis doctiſsimum, iuſsit: vitæq; ita profuit non minus quam reipublicæ victoria illa. Præter hos Græci auctores medicinę prodi‑ dere, quos ſuis locis diximus. Ex his Euax rex Arabum, quid de ſimplicium effectibus ad Neronẽm ſcripſit: Crateias, Dionyſius, Metrodorus orõne blandiſsima, ſed qua nihil penè aliud quam rei difficultas intelligatur. Pinxere namq; effigies herbarum, atque ſcripſere effectus. Verum & pictu‑ ra fallax eſt ex coloribus tam numeroſis, præſertim in æmulatione naturæ, multumq́ue degenerat tranſcribentium ſors varia. Præterea parum eſt ſingulas earum ætates pingi, cum quadripartitis varietatibus anni faciem mutent. Quare cæteri ſermone eas tradidere. Aliqui effigie quidem indi‑ cata, & nudis quidem plærumq́ue nominibus defuncti: quoniam ſatis videbatur, poteſtares vim‑ que demonſtrare quærere volentibus. Nec eſt difficile cognitu. Nobis certe, excecptis admo‑ dum paucis, contigit reliquas contemplari ſcientia Antonij Caſtoris, cui ſumma auctoritatis erat in ea arte noſtro æuo, viſendo hortulo eius in quo plurimas alebat: centeſimum ætatis annum ex‑ cedens, nullum corporis malum expertus, ac ne ætate quidem memoria, aut vigore concuſsis. Nec aliud mirata magis antiquitas reperietur. Inuenta iampridem ratio eſt prænuncians ho‑ ras, non modo dies ac noctes, ſolis lunæque defectum. Durat tamen tradita perſuaſio in magna parte vulgi, veneficijs & herbis id cogi, in eo namq́ue fœminarum ſcientiam præualere. Certe quid non repleuere fabulis Colchis Medea, aliéque, imprimiſq́ue Italica Circe, dijs etiam adſcriptæ? Vn‑ de arbitror natum, vt Æſchylus è vetuſtiſsimis in poëtica re, refertam Italiam herbarum potentia proderet. Multique Circæios agros, vbi habitauit illa, in magno argumento etiamnum durante in Marſis, à filio eius orta genere, quos eſſe domitores ſerpentium conſtat. Homerus quidem primus doctrinarum & antiquitatis parens, multus alias in admiratione Circes, gloriam herbarum Ægypto tribuit, tum etiam cum rigaretur Ægyptus illa, non autem eſſet, poſtea fluminis limo inuecta. Herbas certe Ægyptias à regis vxore traditas ſuæ Helenæ plurimas narrat, ac nobile illud Nepen‑ thes, obliuionem triſtitiæ veniamq́ue afferens, & ab Helena vtique omnibus mortalibus propinan‑ dum. Primus autem omnium quos memoria nouit, Orpheus de his herbis curioſius aliqua prodi‑ dit. Poſt eum muſæus & Hæſiodus Polion herbam in quantum mirati ſunt, diximus. Orpheus & Heſiodus ſuffitiones commendauere. Homerus & alias nominatim herbas celebrat, quas ſuis locis dicemus. Ab eo Pythagoras clarus ſapientia, primus volumen de earum effectu compo‑ ſuit: Apollini, Æſculapioq́ue, & in totum dijs immortalibus inuentione & origine aſsignata: com‑ poſuit & Democritus, ambo peragratis Perſidis, Arabiæ, Æthiopiæ, Ægyptique magis. Adeoque ad hæc attonita antiquitas fuit, vt affirmaret etiam incredibilia dictu. Xanthus hi‑ ſtoriarum auctor, in prima earum tradidit, occiſum draconis catulum reuocatum ad vitam à pa‑ rente herba, quam Balin nominat: eademq́ue Tillonem, quem draco occiderat, reſtitutum ſaluti.

Mithridates.

Euax.

Antonius Caſtor.

Homerus.

APPENDIX.

Et Iuba in Arabia herba reuocatum ad vitam hominem tradit. Dixit Democritus, credidit Theophrastus esse herbam, cuius contactu illatae ab alite, quam retulimus, exiliret cuneus à pastoribus arbori adactus. Quae etiam si fide carent, admirationem tamen implent: coguntq́; confiteri, multum esse quod vero supersit. Inde & plerosque video existimare, nihil non herbarum vi effici posse, sed plurimarum vires esse incognitas. Quorum numero fuit Herophilus clarus in medicina: à quo ferunt dictum, quasdam etiam fortasis calcatas prodesse. Obseruatum certe est, inflammari vulnera ac morbos superuentu eorum, qui pedibus iter confecerint. Haec erat inter antiqua medicina, quae tota migrabat in Graeciae linguas. Sed quare nunc non plures nascuntur caussae? Nisi quod eas agrestes, litterarumq́ue ignari experiuntur, vtpote qui soli inter illas viuant. Praeterea securitas quaerendi, obuia medicorum turba. Multis etiam iuuentis nomina desunt, sicut illi quam retulimus in frugum cura, scimusq́ue defossam in angulis segetis praestare, ne qua auis intret. Turpissima caussa raritatis, quod etiam qui sciunt, demonstrare nolunt tanquam ipsis periturum sit quod tradiderint alijs. Accedit ratio inuentionis anceps. Quippe etiam in repertis, alias inuenit casus, alias (vt vere dixerim) Deus. Insanabilis ad hosce annos fuit rabidi canis morsus, pauorem aquae, potusq́; omnis afferens odium. Nuper cuiusdam militantis in praetorio mater vidit in quiete, vt radicem syluestris rosae, quam cynorhodon vocant, eblanditam sibi aspectu pridie in fruteto, mitteret filio bibendam in lacte (in Lusitania res gerebatur, Hispaniae proxima parte) casuq́ue accidit, vt milite à morsu canis incipiente aquas expauescere, superueniret epistola orantis vt pareret religioni: seruatusq́ue est ex insperato: & postea quisquis auxilium simile tentauit. Alias apud auctores cynorhodi vna medicina erat, spongiolae, quae in medijs spinis eius nascitur, cinere cum melle alopecias capitis expleri. In eadem prouincia cognoui in agro hospitis nuper ibi repertum dracunculum appellatum, caule pollicari crassitudine, versiculoribus viperarum maculis, quem ferebant contra omnium morsus esse remedium. Alius est quem nos in priori volumine eiusdem nominis diximus, sed huic alia figura, aliudq́ue miraculum exeuntis è terra ad primas serpentium vernationes bipedali fere altitudine, rursusq́ue cum ijsdem in terram se condentis: nec omnino occultato eo apparet serpens, vel hoc per se satis officioso naturae munere, si tantum praemoneret, tempusq́ue formidinis demonstraret.

Alia herbarum laus, ex eodem Plinio libro vicesimoseptimo, cap. 1.

CRESCIT APVD me certe tractatu ipso admiratio antiquitatis: quantoq́ue maior copia herbarum dicenda restat, tanto magis adorare priscorū in inueniendo curam, in tradendo benignitatem subit. Nec dubie superata hoc modo posset videri etiam rerum naturae ipsius munificentia, si humani operis esset inuentio. Nunc vero deorum fuisse eam apparet, aut certe diuinam, etiam cum homo inuenerit: eandemq́ue omnium parentem genuisse haec & ostendisse, nullo vitae miraculo maiore, si verū fateri volumus. Scynthicam herbā à Maeotidis paludibus, & euphorbiam è monte Atlante, vltraq́; Herculis columnas, & ipso rerum naturae defectu, alia parte Britannicam ex oceani insulis extra terras positis: iteq́; aethiopidem ab exusto sideribus axe alias praeterea aliunde vltro citroq́ue humanae saluti in toto orbe terrarum portari, immensa Romanae pacis maiestate, non homines modo diuersis inter se terris gentibusq́ue, verum etiam montes, & excedentia in nubibus iuga, pastusq́ue pecorum & herbae quoque inuicem ostentant. AEternum quaeso deorum sit munus istud. Adeo Romanos, velut alteram lucem, dedisse rebus humanis videatur.

De laude Agriculturae ex Marco Catone initio operis sui.

EST INTERDVM praestare populo, mercaturis rem quaerere, ni tam periculosum fiet, & item foenerari, si tam honestum fiet: maiores enim nostri sic habuerunt, & ita in legibus posuerunt, furem duplici condemnari, foeneratorem, quadrupli. Quanto peiorem ciuem existimarūt foeneratorem, quam furem, hinc licet existimari. Et virum bonum cum laudabant ita laudabant, bonum agricolam, bonumq́ue colonum amplissime laudari existimabatur, qui ita laudabatur. Mercatorem autem strennuum, studiosumq́ue rei querendae existimo, verum, vt supra dixi periculosum, & calamitosum. At ex agricolis, & viri fortissimi, & milites strennuissimi gignuntur, maximeq́; pius quaestus, stabilissimusq́ue consequitur, minimeq́ue inuidiosos, minimeq́ue male cogitantes sunt, qui in eo studio occupati sunt.

Quod antiquis maximum studium Agricultura fuerit, & de cultura hortorum singularis diligentia, ex Plinio libro 18. cap. 1.

SEQVITVR natura frugum hortorumq́ue ac florum, quaeq́ue alia praeter arbores aut frutices benigna tellure proueniunt, vel per se tantum herbarum immensa contemplatione, si quis aestimet varietatem, numerum, flores odores, coloresq́ue, & succos ac vires earum, quas salutis aut voluptatis hominum gratia gignit: qua in parte primum omnium patrocinari terrae, & adesse cunctorum parenti iuuat, quanquam inter initia operis defense. Quoniam tamen ipsa materia intus accendit ad reputationem eiusdem parientis & noxia, nostris eam criminibus vrgemus, culpamq́ue nostram

APPENDIX.

nostram illi imputamus. Genuit venena, sed quis inuenit illa præter hominem? Cauere ac refugere alitibus ferisq́ue satis est. Atq; cùm in arbores exacuant limentq́ue cornua elephanti, & duro saxo rhinocerotes, & vtroque apri dentium sicas, sciantq́ue ad nocendum se præparare animalia, quod tamen eorum tela sua excepto homine venenis tingit? Nos & sagittas vngimus, & ferro ipsi nocentius aliquid damus. Nos & flumina inficimus & rerum naturæ elementa. Ipsum quoque quo viuitur aërem in perniciem vertimus. Neq; est vt putemus ignorari ea ab animalibus, quæ quidem quæ præpararent contra serpentium dimicationem, quæ post prælium ad medendum excogitarent, indicauimus. Nec ab vllo præter hominem veneno pugnatur alieno. Fateamur ergo culpam, ne ijs quidem quæ nascuntur contenti: etenim quando plura earum genera humana manu fiunt. Quid? non & homines quidem ad venena nascuntur? Atra hominū ceu serpentium lingua vibrat, tabesq́; animi contrectata adurit culpantium omnia, ac dirarum alitum modo, tenebris quoque & ipsarum noctium quieti inuidentium gemitu (quæ sola vox eorum est) vt in auspicatarum animantiū vice obuij quoq; ventent agere, aut prodesse vitæ. Nec vllum aliud abominati spiritus præmium nouere, quàm odisse omnia. Verùm & in hoc eadem naturæ maiestas tanto plures bonos genuit ac frugi, quáto fertilior in ijs quæ iuuat aluntq́;: quorū æstimatione & gaudio nos quoq; relictis æstuatione suæ istis hominum turbis, pergamus excolere vitam: eoque constantius, quo operæ nobis maior quàm famæ gratia expetitur. Quippe sermo circa rura est, agrestesq́ue vsus, sed quibus vita honosq́; apud priscos maximus fuerit.

De ijs qui in agri & hortorum cura Romæ illustres fuerunt,
ex Plinij lib.18, cap.3.

QVAE NAM ergo tantæ vbertatis caussa erat? Ipsorum tunc manibus Imperatorum colebantur agri (vt fas est credere) gaudente terra vomere laureato, & triumphali aratore, siue illi eadem cura semina tractabant, qua bella, eademq́ue diligentia arua disponebant, qua castra, siue honestis manibus omnia lætius proueniunt, quoniam & curiosius fiunt. Serentem inuenerunt dati honores, Serranum, vnde cognomen. Aranti quatuor sua iugera in Vaticano, quæ prata Quintia appellantur, Cincinnato viator attulit dictaturam, & quidem (vt tradit Norbanus) nudo plenoq́ue pulueris etiamnum ore. Cui viator, Vela corpus, inquit, vt proferam senatus populiq́ue Romani mandata. Tales tum etiam viatores erant, quibus idipsum nomen inditum est, subinde ex agris senatum ducesq́ue accersentibus. At nunc eadem illa vincti pedes, damnatæ manus, inscripti vultus exercent: non tamen surda tellure, quæ parens appellatur, coliq́ue dicitur & ipsa, honore hinc assumpto, vt nunc inuita ea, & indignè ferente credatur id fieri. Sed nos miramur ergastulorum nō eadem emolumenta esse quæ fuerunt Imperatorum. Igitur de cultura agri præcipere principale fuit & apud exteros. Siquidem & reges fecere Hieron, Philometor, Attalus, Archelaus, & duces Xenophon, & Pœnus etiam Mago: cui quidem tantum honorem senatus noster attribuit Carthagine capta, vt cum regulis Africæ bibliothecas donaret, vnius eius duodetriginta, volumina censeret in Latinam linguam transferenda, cum M. Cato præcepta condidisset, peritisq́; linguæ Punicæ dandum negocium: in quo præcessit omnis vir tum clarissimæ familiæ D. Syllanus, sapientiæ compositissimæ. Quos sequeremur prętexuimus in hoc volumine non ingratè nominando M. Varronem, qui octogesimum primum vitæ annum agens, de ea re prodendum putauit.

Laus Agricolarum, & quæ obseruanda in agro parando,
ex Plinij lib.18, cap.5.

FORTISSIMI viri & milites strennuissimi ex agricolis gignuntur, minimeq́; malè cogitantes. Prædium ne cupide emas. In re rustica operi ne parcas, in agro emendo minimè. Quod malè emptum est, semper pœnitet. Agrum paraturos, ante omnia intueri oportet, aquarum vim, & vicinum. Singula magnas interpretationes habent, nec dubias. Cato in conterminis hoc amplius æstimari iubet, quo pacto niteant. In bona est, inquit, regione bene nitere. Attilius Regulus ille Punico bello bis consul, aiebat, neque fœcundissimis locis insalubrem agrum parandum, neque effœtis saluberrimum. Salubritas loci non semper incolarum colore detegitur, quoniam assueti in pestilentibus durant. Præterea sunt quædam partibus anni salubria: nihil autem salutare est, nisi quod toto anno salubre. Malus est ager, cum quo dominus luctatur. Cato inter prima spectari iubet, nū solum sua virtute valeat qua dictum est positione. Vt operariorum copia prope sit, oppidumq́; validum: Vt nauigiorum euectus vel itinerum: vt bene ædificatus & cultus. In quo fallv plerosq́ue video. Segniciem enim prioris domini pro emptore esse arbitratur. Nihil est damnosius deserto agro. Itaq; Cato, de bono domino melius emi, nec temerè contemnendam alienam disciplinam: agroq́ue vt homini, quamuis quęstuosus sit, si tamen & sumptuosus, non multum superesse. Ille in agro quęstuosissimam iudicat vitem: non frustra, quoniam ante omnia de impensæ ratione cauit. Poxime hortos irriguos: nec id falso, si & sub oppido sint. Et prata, quæ antiqui prata dixêre. Idemque Cato interrogatus, quis esset certissimus quæstus? respondit. Si bene pascat, quis proximus?

si me-

APPENDIX.

si mediocriter pascat. Summa omnium in hoc spectando fuit, vt fructus is maximè probaretur, qui quam minimo impendio constaturus esset. Hoc ex locorum occasione aliter alibi decernitur. Eodemq; pertinet, quod agricolam vendacé oportere esse dixit. Fundum in adolescentia conserendum sine cunctatione, ædificandū non nisi consito agro. Tunc quoq; cunctater, optimumq; est (vt vulgo dixere) aliena insania frui, sed ita, vt villarū tutela non sit oneri. Eum tamē qui bene habitet, sepius ventitare in agrum: frontemq; domini plus prodesse quam occipitium, non mentiuntur.

De hortorum cura ex Plinio lib. 19. cap. 4.

AB HIS superest reuerti ad hortorum curam, & suapte natura memorandam. Et quoniam antiquitas nihil prius mirata est, quam & Hesperidum hortos, ac regum Adonis & Alcinoi, itemq; Pensiles, siue illos Semiramis, siue Assyriæ rex Cyrus fecerit, de quorum opere alio volumine dicemus. Romani quidem reges ipsi coluere. Quippe etiam superbus Tarquinius nūcium illum sęuum atq; sanguinarium remisit ex horto. In duodecim tabulis legum nostrarum nusquam nominatur villa, semper in significatione ea hortus: in horti verò hæredium. Quam rem comitata est & religio quædam: hortosq; & fores tantum contra inuidentium fascinationes dicari videmus. In remedio saturnica signa, quanquam hortos tutelæ Veneris assignāte Plauto. Iam quidam hortorum nomine in ipsa vrbe delitias, agros, villasq; possident. Primus hoc instituit. Athenis Epicurus, hortorum magister. Vsq; ad eum, moris non fuerat in oppidis haberi rura. Romę quidem per se hortus ager pauperis erat. Ex horto plebi macellum, quanto innocentiore victu. Mergi enim credo in profunda satius est, & ostrearum genera naufragio exquiri: aues vltra Phasidē amnem, peti & fabuloso quidē terrore tutas, imo sic preciosiores, Alias in Numidia atq; Æthiopia in sepulchris aucupari, aut pugnare cum feris, mandi ab eo cupientem quod mādat alius. Ad hercle, quam vilia hæc, quam parata voluptati satietatiq;, nisi eadem quæ vbiq; indignatio occurreret. Ferendum sane fuerit exquisita nasci poma, alia sapore, alia magnitudine, alia mōstro, pauperibus interdicta, inueterari vina saccisq; castrari: nec cuiquam adeo longam esse vitam, vt non ante se genita potet. E frugibus quoq; quoddā alimentum excogitasse luxuriam, ac medullam tantum earum superq; pristinarum operibus & cęlaturis viuere, alios pane procerum, alios vulgi, tot generibus vsq; ad infimam plebem descendente annona. Etiámne in herbis discrimen inuentum est? Opesq; differentiam fecere in orbe, etiam vno asse venali. Et in his aliqua quoque sibi nasci tribus negant, caule in tantum saginato, vt pauperis mensa non capiat. Syluestres fecerat natura corrudas, vt quisq; demeteret passim. Ecce altiles spectantur asparagi. Et Rauenna ternis libris rependit. Heu prodigia ventris, Mirum esset non licere carduis pecori vesci, non licet plebi, Aquæ quoque separantur. Et ipsa naturæ elementa vi pecuniæ discreta sunt. Hi niues, illi glaciem potant poenasq; montium in voluptatem gulæ vertunt. Seruatur algor æstibus, excogitaturq;, vt alienis mensibus nix algeat. Decoquunt alias quas mox & illas hyemant. Nihil itaq; homini sic quomodo rerum naturæ placet. Etiámne herba aliqua diuitijs tantum nascitur? Nemo sacros, Auentinosq; montes, & iratæ plebis secessus circūspexerit? Mors enim certe ęquabit quos pecunia superauerit. Itaque hercle, nullum macelli vectigal maius fuit ROMAE clamore plebis incusantis apud omnes Principes, donec remissum est portorium mercis huius cōpertumque, non aliter quæstuosius censum haberi aut tutius, ac minore fortunæ iure cum credatur pensio ea pauperum. Is in solo sponsor est, & sub die redditus, superficiesque coelo quocunque gaudens. Hortorum CATO pręd icat caules. Hinc primum agricolæ existimabantur prisci, & sic statim faciebant iudicium, nequam esse in domo matremfamilias (etenim hæc cura fœminæ dicebatur) vbi indiliges esset hortus. Quippe carnario aut macello viuendum est, nec caulus vt nunc maximè probabant, damnantes, pulmentaria quæ egerent alio pulmentario, Id erat oleo parcere. Nam carnis desyderia etiā erant exprobratione. Horti maximè placebant, quia non egerent igni, parcerentque ligno. Expedita res & parata semper, vnde & acetaria appellabantur, facilia concoqui, nec oneratura sensum cibo, & quę minime accederent ad desyderium panis. Pars eorū ad condimenta pertinens, fatetur domi versutam fieri solitam, atque non Indicum piper quæsitum, quęque trans maria petimus. Iam quoque in fenestris suis plebis vrbana in imagine hortorum quotidiana oculis rura præbebant, antequam præfigi prospectus omnibus coëgit multitudinis innumeratæ sęua latrocinatio. Quare obrem sit aliquis & his honos, néue auctoritatem rebus vilitas adimat, cum pręsertim etiam cognomina procerum inde nata videamus, Lactucinosque in Valeria familia non puduisse appellari: & contingat aliqua gratia, operi curæque nostrę, Virgilio quoque confesso, quam sit difficile verborum honorem tam paruis perhibere. Hortos villæ iungendos non est dubium, riguósque maximè habendos, si contingat profluo amne. Si minus, è puteo pertica, organísue pneumaticis, vel tollenonum haustu rigandos. Solum proscindendum à fauonio, in autumnum præparandum est post quatuordecim dies, iterandumque ante brumam. Octo iugerum operis palari iustum est. Fimum tres pedes alte cum terra misceri, areis distingui, easq; resupinis puluinorum toris ambiri singulis tramitum sulcis, qua detur accessus homini, scatebrisque decursus.

APPENDIX.

In his hortis nascentium alia bulbo commendantur, alia capite, alia caule, alia folio, alia vtroq;, alia semine, alia cortice, alia cute, aut cartilagine, alia carne, alia tunicis carnosis. Aliorum fructus in terra est, aliorum & extra, aliorum non nisi extra. Quędam iacent crescuntq;, vt cucurbitę & cucumis. Eadem & pendent, quanquā grauiora multo etiam ijs quę in arboribus gignuntur. Sed cucumis cartilagine, Cortex huic vni maturitate transit in lignum. Terra conduntur raphani, napiq;, & rapa, atq; alio modo inulæ, siser, pastinacæ. Quædā vocabimus ferulacea, vt anethū, maluas. Namq; tradunt auctores in Arabia maluas septimo mēse arborescere, baculorum vsum prębere extemplo. Sed & arbor est malua in Mauritania Lixi oppidi æstuario, vbi Hesperidum horti fuisse produntur 200. passum ab oceano, iuxta delubrum Herculis, antiquis Gaditano vt ferunt. Ipsa altitudinis pedes 20. crassitudinis, quam circumplecti nemo possit. In simili genere habebitur & cānabis. Nec nō & carnosa aliqua appellabimus, vt spongias in humore pratorum enascentes. Fungorum enim callum, in ligni arborumq; natura diximus, & alio genere tuberum pauló ante.

¶ *Ratio rigandorum hortorum, & quæ translatè meliora fiant. Item de succis hortensiorum & saporibus, ex Plinio lib. 19. cap. 12.*

HIs HORAE rigandi matutina atq; vespera, ne inferuescat aqua sole. Ocimo tantum & meridiana etiam. Satum celerrimè erumpere putant inter initia feruenti aqua aspersum. Omnia autem translata meliora gradioraq; fiunt, maximè porri, napiq;. In translatione & medicina est, desinuntq; sentire iniurias, vt gethyum, porrum, raphani, apium, lactucæ, rapa, cucumis. Omnia autem syluestria ferè sunt & folijs minora & caulibus, succo acriora, sicut cunila, origanū, ruta. Solummodo ex omnibus lapathum syluestre melius. Hoc in satiuum rumex vocatur, nasciturq; fortissimum. Traditur semel satum durare, nec vnquam vitiari, terra maximè iuxta aquam. Vsus eius cum ptisana tantum in cibis leuiorem gratioremq; saporem præstat. Syluestre ad multa medicamenta vtile est. Adeoq; nihil omisit cura, vt carmine quoq; comprehensum repererim, in fabis caprini fimi singulis cauatis, si porri, erucę, lactucę, apij, intubi, nasturtij semina inclusa serantur, mirè prouenire. Quæ sunt syluestria, eadem in satiuis sicciora intelliguntur & acutiora. Namq; & succorum saporumq; dicenda differentia est, vel maior in his quàm pomis. Sunt autem acres cunilæ, origani, nasturtij, sinapis. Amari, absynthij, centaureæ. Aquatiles, cucumeris, cucurbitæ, lactucæ. Acuti tantum cunilæ Acuti & odorati, apij, anethi, fœniculi. Salsus tantùm è saporibus non nascitur, alioquin extra insidit pulueris modo, & circulis tantum aquæ vt intelligatur vana, ceu plærumq; vitę persuasio. Panax Piperis saporem reddit, magis etiam siliquastrum, ob id piperitidis nomine accepto Libanotis odore thuris, murrha myrrhę. De panace abundè dictum est. Libanotis locis putridis & macris ac roscidis seritur semine. Radicem habet olusatri, nihil à thure differentem. Vsus eius post annum stomacho saluberrimus. Quidam eam nomine alio rosmarinum appellant. Et smyrnium olus seritur ijsdem locis, myrrhamq; radice resipit Eadem & siliquastro satio. Reliqua à cęteris odore & sapore differunt, vt anethum. Tantaq; est diuersitas atq; vis, vt non solùm aliud alio mutetur, sed etiam in totum auferatur. Apio eximi in coctis obsonijs aceto, in eodem cellario in saccis odorem vino grauem. Et hactenus hortensia dicta sint, ciborum gratia duntaxat. Maximum quidem opus in ijsdem naturæ restat, quoniam prouentus tantum adhuc, summasq; quasdam tractauimus. Vera autem cuiusq; natura non nisi medico effectu prænosci potest, opus ingens occultumq; diuinitatis, & quo nullum reperiri possit maius. Ne singulis id rebus contexeremus iusta fecit ratio, cùm ad alios medendi desyderia pertinerent, longius vtriusq; dilationibus futuris si miscuissemus. Nunc suis quæque partibus constabunt, poteruntq; à volentibus iungi.

¶ *De morbis hortorum, & remedijs circa formicas, & culices, ex Plinio lib. 19. ca. 10.*

MORBOS HORTENTIA quoque sentiunt, sicut reliqua terræ sata. Namq; & Ocimum se necat, degeneratque ritè in Serpillum, & sisymbrium in calamintam. Et ex semine brasicę veteris, rapę fiunt. Atque inuicem enecatur cyminum ab imo dorso, nisi repurgetur. Est autem vnicaule, radice bulbo simili, non nisi in solo gracili nascens. Alias priuatim cymini morbus, scabies. Et ocimum sub canis ortu pallescit. Omnia vero accessu mulieris menstrualis flauescunt. Bestiolarum quoq; genera innascuntur, Napis culices, raphano erucę, & vermiculi. Item lactucis & oleri. Vtrisq; hoc amplius limaces & cochleæ. Porro verò priuatim animalia quæ facillimè stercore iniecto capiuntur condentia in id se. Ferro quoque non expedire tangi rutam, cunilam, mentam, ocimum, auctor est Sabinus Tiro in libro Cepuricon, quem Mecœnati dicauit. Item contra formicas non minimum hortorum exitium, si non sint rigui, remedium monstrauit, limo marino, aut cinere obturatis earum foraminibus. Sed efficacissimè heliotropio herba necantur. Quidam & aquam diluto latyce crudo inimica eis putant. Naporum medicinæ sunt, siliquæ vna seri, sicut olerum cicer, arcet enim eucas. Quæ si omisso iam natę sint, remediū absynthij succus decocti inspersus & sedi, quam aizoun vocant, genus hoc herbæ diximus Semen olerum, si succo eius madefactum seratur, *Sabinus Tiro.*

olera

APPENDIX.

olera nulli animalium obnoxia futura tradunt. In totū verò nec erucas, si palo imponãtur in hortis ossa capitis ex equino genere foeminæ duntaxat. Aduersus erucas & cancrum fluuiatilem in medio horto s. ipsūm auxiliari narrant. Sunt qui sanguineis virgis tangant ea, quæ nolūt his obnoxia esse. Infestant & culices hortos, riguos præcipuè, si sunt arbuiculæ aliquæ. Hi galbano accéso fugantur.

¶ De inuestigandis qualitatibus Herbarum ex colore, odore, & succis ex Plinij lib. 21. cap. 7.

TRoianis temporibus ei iam erat honos. Et hos certè flores Homerus treis laudat, loton, crocō, hyacinthum. Omnium autem odoramentorum, atq; adeo herbarum differentia est in colore, & odore, & succo. Odorato sapor raro vlli non amarus, econtrario dulcia rarò odorata: itaq; & vina odoratiora multis, syluestria magis omnia satiuis. Quorundam odor suauiore longinquò est, propius admotus hebetatur, vt violæ. Rosa recens à longinquo olet, sicca propius. Omnis autem verno tempore acrior, & matutinis. Quicquid ad meridianas horas diei vergit, hebetatur. Nouella quoq; vetustis minus odorata. Acerrimus tamen odor omnium æstate media. Rosa & crocum odoratiora, cum serenis diebus leguntur: & omnia in calidis magis quàm in frigidis. In Ægypto tamen minimè odorati flores, quia nebulosus & roscidus aër est à Nilo flumine. Quorundam suauitati grauitas inest. Quædā cùm virent non olent, propter humorem nimium: vt buceros, quod est foenum græcum. Acutus odor non omnium sine succo est, vt violæ, rosæ, croco. Quæ verò ex acutis succo carent, eorum omnium odor grauis, vt in lilio vtriusque generis. Abrotonum & amaracus acres habent odores. Quorundam flos tantum iucundus, reliquæ partes ignauæ, vt violæ ac rosæ, Hortēsium odoratissima quæ sicca, vt ruta, menta, apium & quæ in siccis nascuntur. Quædam vetustate odoratiora, vt cotonea eademq; decerpta, quam in suis radicibus. Quædam non nisi defracta, aut ex attritu olent. Alia non nisi detracto cortice. Quædam verò non nisi vsta, sicut thura, myrrhęq;. Flores triti omnes amariores quàm intacti. Aliqua arida diutius odorem continent, vt melilotos. Quędam locum ipsum odoratiorem faciunt, vt iris, quin & arborem totam cuiuscumque radices attingunt. Hesperis noctu magis olet, inde nomine inuento. Animalium nullum odoratum, nisi de pantheris quicquam dictum est, si credimus.

Buceros,
fœnum
græcum.

Quibus temporibus maximè legendæ sunt herbæ, ex Dioscoridis præfatione.

VERVM IN PRIMIS curam impendere oportet, vt suis temporibus singula & demetantur, & recondantur. Intempestiuè enim decerpta, conditáue, aut nullo, aut euanido munere funguntur. Serena etenim cœli constitutione demetenda sunt. Magni siquidem refert inter colligendum, si vel squallores, vel imbres infesti sint: quemadmodum si loca in quibus prodeunt cliuosa, & ventis exposita sint, & perflata, frigidaq;, & aquis carentia: in his enim locis vires eorum longè validiores intelliguntur. Contra, quæ in campestribus, riguis & opacis, ceterisq; locis à vento silentibus es nascuntur, plærunque degenerant, & minus viribus valent: multóque magis, si non suis horis pers opportunè colligantur, aut si per imbecillitatem intabuerint. Neque ignorandum, quod sæpe præcoci, aut serotina loci natura, aut anni clementia, maturius, aut celerius adolescūt. Nonnulla propria vi hyeme florent & folia pariunt, quędam bis anno florifera. Quare cui in animo est, horum peritiā assequi, necesse est ijs prima germinatione solo emergentibus, adultis, & senescentibus adesse. Nam qui pullulanti herbæ duntaxat astiterit, adultam cognoscere non potest: neque qui adultam tantum inspexerit, nuper erumpentem noscet. Quo fit, vt propter mutatam foliorum faciem, caulium proceritates, florum, seminisque magnitudinem, nonnulli qui olim has ætatum varietates non perspexerunt, magno in errore versentur. Quæ caussa etiam nonnullis scriptoribus imposuit, qui herbas quasdam, verbi gratia, Gramen, Quinquefolium, & Tussilaginem, emittere florem, fructum, & caulem negant. Ergo qui sæpius ad visendas herbas, & earum loca se contulerint, earum cognitionem maximè consequetur. Scire etiamnum conuenit, sola ex herbaceis medicaminibus Veratri genera, nigrum inquam & candidum, multis edurare annis: reliqua à trimatu inutilia. Quæ verò ramis scaeat, sicut Stichas, Trixago, Polium, Abrotonū, Seriphium, Absynthium, Hyssopum, & alia id genus semine prægnantia, demetantur. Flores quoque antea quàm sponte sua desidant, Fructus autem vt maturi excutiantur necesse est, & semina vbi siccari cœperint, priusquā destuant, Herbarum succus, & foliorum elici debet, germinantibus adhuc cauliculis. Lac & lachrymæ excipiantur, inciso per adolescentiam caule. Radices, & liquamenta, corticésue, vt recondantur, eximere conuenit, cùm herbę suis folijs exuuntur. Siccantur etiam expurgatæ inibi, locis asperginem non redolentibus: sed quæ luto, aut puluere sunt obsitæ, aqua elui debent. Flores, & omnia quæ iucundum odorem esflant, arculis tiliaceis nullo situ obductis reponantur. Nonnunquam charta, aut folijs semina, vt perennent, aptè inuoluuntur. Liquidis medicaminibus densior materia, argentea, vitrea, aut cornea conuenit. Fictilis, etiamsi modò rara non sit, accommodatur, & lignea, præsertim è buxo: sed ænea vasa liquidis oculorum medicamentis, quæ aceto, pice liquida, aut cedria componuntur. Adipem autem & medullas stagneis vasis recondi conuenit.

The first parte of the Historie of Plantes,

Conteyning the kyndes and differences, with the proper figures, and liuely descriptions of sundry sortes of Herbes and Plantes, their naturall places, times, and seasons: Their names in sundry languages, and also their temperature, Complexions, and vertuous operations.

Compiled by the learned D. Remberte Dodoens, now Physition to the Emperour his Maiestie.

Of Sothrenwood. Chap.i.

❧ The Kyndes.

There be two sortes of Sothrenwood (as Dioscorides sayth) the one called female Sothrenwood, or the great Sothrenwood, the other is the male kinde, or small Sothrenwood, and are both meetely common in this Contrie.

Abrotonum fœmina.
Great Sothrenwood.

Abrotonum mas.
Small Sothrenwood.

❧ The Description.

1. The great Sothrenwood, doth oftētimes surmount the heigth or stature of a tall man, (especially being well guyded, & stayed in the growing vp) so that it seemeth as a littell tree: his twigges & branches be hard, about the which there groweth many small grayish leaues, much cut and iagged, the whiche do perish and vade in winter, like the leaues of diuers other trees, and do renew and spring againe in Aprill. The flowres be like vnto small buttons, yellow as golde, growing alongst the braunches like Wormwood floures.

The small Sothrenwood doth neuer grow very highe: his braunches or twigges are small, weke and slender, for the most parte so springing vp from the roote: The leaues be greener, longer, tenderer and more iagged and cut, than the leaues of the great Sothrenwood, the whiche do vade and fall of at winter, & renew and spring againe in May out of the same old branches, & also fró the new springs. It doth seldom flower in this countrey: it is of a stronger sauour then the great Sothrenwood. The roote is tender, creeping alongst ỹ groũd, about ỹ which there cõmeth forth diuers outgrowings & new springs.

Besides the two forenamed, there is founde a thirde kinde, the whiche is much like the smal Southrenwood in his growing & branches, but his leaues are like the great Southernwood, sauing that they be somwhat tenderer and not so white. This kinde is of a very pleasant sauour, not muche vnlike the smell of garden Cypres. Wherof shalbe written in his conuenient place.

❧ *The Place.*

The two first kinds grow not in this countrey, but only in gardens wheras they are planted: neither the thirde kinde, whiche is more seldome founde, and lesse knowen then the other. ❧ *The Tyme.*

They flower in August, and their seede may be gathered in September.

❧ *The Names.*

Southrenwood is called in Greeke ἀβρότονον: in Latine & in Shoppes Abrotonum: in Italion and Spanish *Abrotano*, yet some of them call it *Hyerua Lombriguera*: in high Douch Stabwurtz, Gertwurtz, Garthagen, Shoswurtz, Kuttelkraut, Affrusch: in base Almaigne Aueroone: in French *Auronne*.

The great Southrenwood, is called in Greeke ἀβρότονον θῆλυ: in Latine Abrotonum foemina, that is to say, female Southrenwood: in french *Auronne femelle*: in high douch Stabwurtz weiblin: in base Almaigne Aueroone wijfke.

The small Southrenwood is called in Greeke ἀβρότονον ἄρρεν: in Latine Abrotonum mas: in French *Aronne masle*: in high Douch Stabwurtz menelin: in base Almaigne Aueroone manneken, and clein Aueroone.

The thirde kinde seemeth to be that whiche Dioscorides calleth in Greeke ἀβρότονον σικελιωτικόν: in Latine Abrotonum Siculum, which is a kinde of female Southrenwood: the highe Almaignes do call it Wolrieckende Stabwurtz, that is to say, sweete smelling, or sauering Southrenwood.

❧ *The Nature.*

All the Sothrenwoods, are hoate & dry in ỹ third degree, & of subtill parts.

❧ *The Vertues.*

The seede of Sothrenwood either greene or drie made into pouder, or boyled in water or wine, & dronken, is very good and greatly helpeth suche as are troubled with shortnesse of winde, and fetching of breath, by meanes of any obstruction or stopping about the Breast, and is good against the hardnesse, bursting & shrinking of Sinewes. It is good against the Sciatica, the difficultie and stopping of vrine, and for women that cannot easily haue their termes, or natural floures: for by his subtill nature it hath power to expell, waste, cõsume and digest all colde moysture, and tough slime and fleume stopping the splene, kidneys, bladder, and Matrix.

Sothrenwood dronken in wine is good against such venome as is hurtfull vnto man, and destroyeth wormes.

The perfume thereof driueth away all venemouse beastes: and so doth the hearbe in all places whereas it is layde or strowen.

The asshes of Sothrenwood, mingled with ỹ oyles of Palma Christi, rapes, or old oyle Oliue, restoreth the heare fallen from the head, if the head be rubbed there-

therewithall, twise a day in the Sunne, or against a fyre.

If the saide asshes be mingled with any the aforesaide oyles, & the chinne be rubbed therewithall, it causeth the bearde to come forth speedely.

Sothrenwood pounde with a rosted Quince, & laide to the eyes in manner of a plaister, is very good & profitable against all the inflammation of the eyes.

The same pounde with Barley meale, and boyled togither, dothe dissolue & waste all colde humours or swellings, being applied or layde thereupon.

Sothrenwood stieped or soked in oyle, is profitable to rubbe or annoint the body, against the benomming of members taken with colde, and the brusing or shyuering coldes that come by fittes, like as in Agues.

Plinie writeth þ if it be layde vnder the bedde, pillow or bolster, it prouoketh carnall copulation, & resisteth all enchantments, which may let or hinder such businesse, & the inticements to the same. ✲ *The Daunger.*

Sothrenwood is a very hurtfull enimie to the stomacke: wherefore Galen the chiefest of Physitions, neuer gaue the same to be receiued into the body.

Of Wormwood. Chap. ij.

✲ *The Kindes.*

THere be three sortes of Wormwood (as Dioscorides saithe). The first is our cōmon Wormwood. The second is sea Wormwood: the thirde kinde is that, which is called *Santonicum*. And bysides these, there is founde an other kinde, which is called in this countrey Romaine wormwood.

Absynthium Latifolium.	Seriphium.
Common Wormwood.	Sea Wormwood.

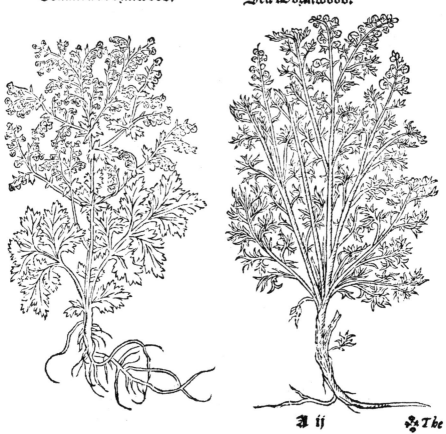

The first Booke of

✤ *The Description.*

1. The cōmon Wormwood hath leaues of a grayishe ashe colour, very much cut & iagged, & very bitter: The stalke is of a wooddy substance, of two cubites high or more, full of braunches: & alongst the braunches groweth litle yellow buttons, wherin when they are ripe & ready to fall, is found small seede like to the seede of garden Tansie, but farre smaller. The roote is likewise of a wooddy substance, and full of small threedes, or hearie rootes.

There is also founde in the gardens of some Herboristes of this countrey, an other sorte of this kinde of Wormwood, the whiche is named of some men Abſynthiũ Ponticum, much like to our cōmon wormwood, sauing the leaues are much more iagged and finelier cut, and not so bitter (at the least way) as that whiche is set and sowen in this countrey.

2. The second kinde, whiche is the Sea Wormwood is also of a whitishe or gray colour, and hath many whitish leaues much like to cōmon Wormwood, but much smaller, tenderer and whiter, & finelier cut, it hath many floures like to small buttons, & the seede ioyning to the braunches, like as in the common Wormwood. It groweth to the heigth of a foote and halfe or more, it is of a strong smell, salt, & of a straunge & bitter taste, being gathered in his naturall & proper place: but being remoued into gardens, or into groundes which are naturally holpē wt sweete waters, it doth marueloussly alter both in sauour & nature, as diuers other herbes, but especially such as grow in salt groundes, & are remoued frō their naturall soyle, to some other place of a cōtrary kinde.

3. The thirde kinde of Wormwood called Santonicum, is almost like to Sea Wormwood, in his small tender and iagged leaues, but the colour of this is whiter, and the smell thereof is not so ranke.

4. Wormwood Romayne is like the Wormwood aforesaide, sauing that it is lower and smaller, the leaues be also smaller and finer, and not so white as the cōmon Wormewood, but chaunging more towardes greene, yet they turne somwhat grayish and ashe coloured. It putteth forth yellow buttons, the whiche afterwarde do bring forth both floures & seede. The roote is full of hearie threedes, trayling here and there, and putting foorth on euery side much encrease of new springs.

5. The fifth kinde of Wormwood is like vnto Sea Wormwood in his smal and tender leaues, also it is like in the stalke of floures: but it is of a sadde or deeper colour, and it hath neyther bitter taste nor sauour.

6. The sixth kinde of Wormwood, his leaues be long and narrow, and of a whitish colour muche like the leaues of Lauender, and somwhat like it also in sauour. The stalkes also be of wooddishe substance, in the toppes whereof there groweth both floures and seede, like as in the reste of the Wormwoodes, but smaller.

✤ *The Place.*

1. The common Wormwood groweth naturally in stony places and rough mountaynes, & in dry, rude

Abſynthiũ Põticum Galeni.
Wormwood gentle, or Romayne.

the Historie of Plantes.   5

2   Sea wormwood groweth in salte ground, and in places adioyning to the Sea. It groweth plentifully in Zeland and Flaunders, alongst the sea coast, and in some places of Brabant, as about Barowgh.

3   The third kind groweth in some places of Zwiserland, vnder the hilles or at the foote of mountaines, as Conrade Gesner, that famous Clerke writeth.

4   Wormwood Romayne groweth plentifully in Hungarie, & places neare about Constantinople, & in some places of Almaigne, also vpon mountaynes, & about sandy wayes. Yet it groweth not in this countrie, except it be planted.

5.6   The other twayne are not common in Base Almaigne: sauing onely in the gardens of certaine diligent Herboristes. ❦ *The Tyme.*

All the sortes of Wormwood, are in flowre in July or August, or somwhat later: And shortly after, the seede is ready to be gathered.

❦ *The Names.*

Wormwood is called in Greeke ἀψίνθιον, & Βαβύπικρον, or Βαρύπικρον, bycause of his bitternesse: in Latine Absynthiū, whiche name it hath retayned in shops euen vntill this present time. Apuleius calleth it Absynthium rusticum: in Italian *Assenzo*: in Spanishe *Axensios, y Assensios, y Alosna*. The high Douchmen do cal it Wermut, & Werommout, or Acker Werommout, that is to say, Field wormwood: the base Almaignes do call it Alsene: in French it is called *Aluyne*.

The first kind (which is our common Wormwood) may be rightly named Absynthium Latifolium, that is to say, great or broade leaued wormwood.

That wormwood that is most like vnto the aforesaid, is called of the Herboristes, Absynthium Ponticum, and Absynthium Græciæ, & is a kinde of the first sort of wormwood: & so is the wormwood of Cappadocia, & the wormwood of mount Taurus, and likewise that wormwood that groweth alongst by the old walles at Roome. Whereof the good religious fathers, that wrote the Commentarie vpon Antidotarium Mesue, haue writen. For all these sortes of wormwood are of the first kinde, and may well be called Absynthia Latifolia, as a difference from the other Wormwoods, whose leaues be a great deale finer and smaller: for there is no great diuersitie betwixt these wormwoods, sauing in respect of the places where as they growe.

2   The second kinde of wormwood is called in Greeke ἀψίνθιον Θαλάσσιον καὶ σέριφον: in Latine Seriphium, and Absynthium Marinum: vnknowen of the Apothecaries. In English Sea wormwood.

3   The third is called of some Herba alba: & without question, is the true Absynthium Santonicum. Miratur fortasse, hæc legens, Santonicum a Santonibus vt Dioscorides scribit cognominatū, apud Heluetios requiri, at hunc magis admirari æquum est, Santonum prouinciam à quoquam in ea Galliæ parte, quæ alpibus vicina est, reponi. Santones enim Aquitanię populi sunt, ad oram maritimam Oceani, infra Garumnam fluuium versus Septentriones siti, longissimè ab Alpibus, procul etiam à Pyrenęis. Quam ob causam mendosum hunc Dioscoridis locum esse oportet: aut Dioscoridem, vel eum ex quo hęc transcripsit, Geographiæ fuisse rudem & imperitum.

4   The fourth kinde of wormwood is called of Galen, in the .xj. booke of his Methode in Greeke ἀψίνθιον ποντικόν: in Latine Absynthiū Ponticum: of the Apothecaries of Brabant Absynthium Romanum: in Frenche, *Aluyne Romaine* or *Pontique*: in base Almaigne, Roomsche Alsene, bycause this is a straunge herbe & not cōmon in that countrey. For they do cōmonly call al such straunge herbes as be vnknowen of the cōmon people, Romish or Romayne herbes, although the same be brought frō Norweigh, which is a coūtrey far distant frō Roome.

The

The first Booke of

5 The fifth is called in Latine Absynthium fatuum, & Absynthium insipidum.

6 The sixth is called Absynthiũ angustifolium, & it is thought of some to be a kinde of Lauender, bicause his leaues hath smal leaues like Lauender: it may be called in English Lauender-wormewood, or narrow leaued wormewood.

❧ *The Nature or Temperament.*

1 Our common Wormwood is hoate in the first degree, and dry in the thirde, bitter, sharp and astringent: wherefore it clenseth, purgeth, comforteth, maketh warme and dryeth.

2 Sea wormwood is hoate in the second degree, & dry in the thirde, & of subtile parts, & of the same nature is Santoni wormwood, or French wormwood.

3.4 Wormwood Romayne is in temperature not muche vnlike the common wormwood, neuerthelesse it is more astringent.

❧ *The Vertues.*

1 The common Wormwood is a profitable & excellent medicine against the payne of the stomacke, that is oppressed or charged with hoate Chollericke humors: for it expelleth them partly by the stoole, & partly by vrine, besides that it comforteth the stomacke. Yet notwithstanding it will not serue, to purge ye stomacke that is charged with fleume and colde humors, neither can it mundifie and cleanse the breast and lunges that are stopped and charged with the saide humors, as Galen sayeth. A

Likewise it doth both by seige & vrine purge Cholerike humors, compact & gathered together in the vaynes and liuer: wherefore the infusion or decoctiõ thereof, taken day by day, cureth the Iaundise or Yealowsought. B

If it be taken fasting in the morning, it preserueth frõ drõkennes that day. C

It is good against the windinesse and blastings of the belly, against the paynes and appetite to vomit, and the boyling vp or wamblings of the stomacke: if it be drunken with Annis seede or Seseli. D

The same drunken with vineger is good for such as are sicke, with eating venemous Champions or Tode stooles. E

The same taken with wine, resisteth all venom, but chiefly Hemloke, and the bitings and stingings of spiders and other venemouse beasts. F

Wormwood mingled with hony, is good to be layde to the dimnesse of the sight, and to the eyes that are bloudshotten, or haue blacke spottes. And with the same boyled in Bastarde, or any other sweete wine, they vse to rubbe and strake painefull bleered eyes. G

The same pounde or mengled with figges, salte peter, and Iuray meale, & layde to the belly, sides or flankes, helpeth the dropsie, & such as are splenitike. H

The same layde in chestes, presses & wardrobes, keepeth clothe & garments from mothes and vermine. And with the oyle of Wormwood, a man may annoynt & rubbe any place to driue away fleas, flies, knattes, and wormes. I

Inke made with the infusion or decoctiõ of Wormwood, keepeth writings from being eaten with Mice and Rattes. K

Some do vse to make Wormwood wine, very excellent for all the diseases aforesayde. L

2 Sea Wormwood boyled by it selfe or with Rice, or with any other foode or meate, and eaten with Hony, sleeth both long and flatte wormes, and all other kindes whatsoeuer, loosing the belly very gentilly. It is of like operatiõ being layde too, outwardly vpon the belly or nauell, and for this purpose it is of more strength and vertue, then all the other kindes of Wormwood: but it is more hurtefull to the stomacke. M

3 The seede of Sea Wormwood also, is very excellent against all sortes of wormes N

wormes engendred within the bodie.

Dioscorides writeth, that such Beeues, Sheepe and Cattell, as feede vpon ye Sea Wormwood do waxe very fatte.

4   Wormwood Romayne is singular against all inflammation, and heate of ye the stomacke & liuer, passing for this purpose, all other kindes of Wormwood as Galen writeth.

Of Buglosse, or common langue de Beufe.   Chap.iij.

❧ *The Kindes.*

He common Buglosse, or langue de beuf (as it is now called) is of diuerse kindes, whereof the first is the greatest, and it is familiar and common in gardens: The three others are small: The fifth is the wilde Buglosse, or Sheepes tongue.

Lycopsis.	Anthusæ genus.
Garden Buglosse.	Alkanet.

❧ *The Description.*

1   The first kinde called of vs great Buglosse of the garden, hath lõg, rough swartegreene, hearie & sharpe leaues, almost like to the leaues of Lettice, but longer & sharper at the ende. The stem is rough and pricking, of two or three foote high, whereupon groweth many proper littell floures, eche one parted into fiue small leaues, like to littell wheeles, of a fayre purple colour at the first, but afterwardes azure. When they are fallen, ye may see in the rough huskes, three or foure long gray seedes, full of riffes and wrinckles. The

I iiij   roote

roote is long and single and blackish in the outside.

2.3 The lesser Buglosses in their rough and hearie leaues and stalkes, and also
4 in their rootes are like to the aforesaide: sauing they be lesse: for their stalkes be shorter, their leaues smaller and narrower: their littell floures are in proportion like to the others, sauing they be smaller, and one is of a cleere blew or skie colour, an other is of a browne violet, or a blew like to a Cyanus, the third is yellow, and in proportion long and hollow. The seede also is like the other sauing it is smaller and blacker. The rootes of the Buglosses and especially of the firste kinde of the lesser Buglosses, are of a diepe redde colour, and are vsed to die, and colour things withall.

5 The wilde kinde of Buglosse is like to the small Buglosses, & specially like to the second kinde, sauing the leaues be rougher, smaller, and narrower. The floures also be like the aforesaide, sauing they be a great deale smaller & blew. The seede is small and browne. The roote long and slender.

❧ *The Place.*

Lycopsis Syluestris.
Wilde Buglosse.

1 The great garden Buglosse, groweth in some places of his owne accord, as in the countrie of Lorraine, aboute Nancie in fertile and chāpion places, amongst the corne. It groweth not wilde in this countrey, but onely in gardens.

2.3 The smaller Buglosses grow in Italie,
4 Spayne and Fraunce, and in diuers other countreys or regions: and that which beareth blewe violet floures groweth also in some places of Germanie: but they be not very cōmon in Flaunders, neither are they to be seene or founde but in certaine mens gardens.

5 The wilde groweth in moste places of this countrie, in barren soyle, and grauelly grounde.

❧ *The Tyme.*

They floure in June, July, and August, and forthwith they deliuer their seede.

❧ *The Names.*

The three herbes are called (in shoppes) all by one name, that is to say, Buglossa or Lingua bouis: in French, *Buglosse or Langue de buef*: in high Douch, Ochsenzung: in base Almaigne, Buglosse and Ossentonghe: in English, Buglosse and Oxetongue: Albeit it is not the true Buglosse, for that is our common Borage, wherof we shall write in his proper place. Wherebnto agreeth Leonicenus, Manardus, and diuers other learned men of our time.

1 The first kinde is called in Greeke λύκοψις: in Latine Lycopsis: in þ shoppes of this countrey Buglossa, & Buglossa domestica maior, that is to say, the great garden Buglosse, & of some it is called Buglossus Longifolia. Peraduenture it is that kinde of Anchusæ, which Paulus Aegineuscalleth in Greeke χαιροσπέ- λιθον, Chœrospelethon.

2.3 The smal Buglosses are called in Greeke ἄγχουσαι, in Latine Anchusæ. The
4   first

the Historie of Plantes.

firſt is called in Greeke Ἄγχουσα ὀνοκλεια, Anchuſa onoclea: in French Orchanette: in Engliſh Alkanet, or Orchanet. The other is called in Greeke ἄγχουσα ἀλκιβιάδιον καὶ ὀνοχειλὲς: in Latine Anchuſa Alcibiadium, & Onocheles. This ſhould be the ſecond kinde of Anchuſa or Orchanette: in Engliſh Alkanet.

5 The fifth kinde is wilde, and may be called Lycopſis Sylueſtris, the Apothecaries call it Bugloſſa Sylueſtris. The French men cal it Bugloſſe or Langue de buef Sauuage. The baſe Almaignes, Wilde Oſſentonghe, & ſome call it Scaepſtonghe, that is to ſay, Sheepes tongue, and it may be Pſeudanchuſa Plinij.

❈ The Nature.

1 The great garden Bugloſſe, but ſpecially his roote, is of temperature ſomwhat colde and drie, but in degree not farre of from the meane temperature.

2.3 The others are of the like complexion, but ſomewhat hoater.

❈ The Vertues.

1 The roote of great Bugloſſe, pounde, and mengled with oyle and waxe, is A good to be layde too againſt ſcalding or burning with fyre, againſt woundes and old ſores. With fine wheate meale it cureth the diſeaſe called the wilde fyre, and of ſome ſaint Anthonies fyre. And layde too with vinegre it healeth fretting ſores, foule ſcuruines and hoate itchings.

2.3 4 The ſmall Bugloſſes haue greate vertue againſt all the venim of ſauage B and wilde beaſtes, and ſpecially againſt the poyſon of Serpents and Vipers, howſoeuer it be taken, whether in meate or drinke, or whether it be caried about you.

5 The roote of the wilde Bugloſſe dronken with Hiſope and Creſſes, doth C kill and driue out all flat wormes engendred in the bodie of man.

The Phyſitions of our tyme do affirme, that theſe herbes (but eſpecially D the greateſt) do comforte and ſwage the heauineſſe of the harte, driuing away all penſiueneſſe, eſpecially the garden Bugloſſe, and that the floures, ſteped in wine, or made into a Conſerue, cauſeth ſuch to reioyce and be gladde, as were before heauie and ſadde, full of anger, and melancholique heauineſſe.

Of Echium or Vipers Bugloſſe. Chap. iiij.

❈ The Deſcription.

1 Echium hath long rough and hearie leaues, much like to the leaues of Bugloſſe, but ſmaller than the leaues of the firſt Bugloſſe. The ſtalke is rough, full of littell braunches, charged on euery ſide with diuerſe ſmall narrow leaues, ſharp pointed, and of a browne greene colour, ſcattered or ſpredde like littell feathers, and very ſmall towardes the height or toppe of the ſtalke: betwixt whiche leaues are the floures of a ſadde blew or purple colour at the firſt, but whan they do open, they ſhew a fayre Azure colour, long and hollow, with foure or fiue littell ſmall blewe threedes: nothing anſwering the floures of the other Bugloſſes, but onely in the colour. After that the floure is fallen, the ſeede is blacke and ſmall, like to the head of an Adder or Viper. The roote is long and ſtraight, and redde without.

2 Of this ſorte there is an other kinde, whoſe leaues, ſtalkes, rootes, and floures, are very like vnto the foreſaide: but his floures are of a light redde or purple colour.

❈ The Place.

1 It delighteth in fruitefull places, and fertile ſoyle, as aboute Bruſſels, and Louayne, and diuers other places of Brabant.

2 But that which beareth purple or light redde floures, groweth in Fraunce eſpecially about Montepelier.

❈ The

10 The first Booke of

Echion siue Alcibiacum.

❧ *The Tyme.*

It floureth almoste all the Somer long, & oftentimes or at sundry seasons it bringeth forth seede as the other Buglosses.

❧ *The Names.*

It is called in Greeke ἔχιον καὶ ἀλκιβιάδιον: in Latine Echium Alcibiacum. Apuleius calleth it ἐυριόριζον ἐχίδνιον in Greeke: Viperina and Serpentaria in Latine: in Spanishe Yerua della biuora: in French l'Herbe aux Vipers, and l'Herbe aux Serpens: in base Almaigne Slanghencruyt: it is called in English wilde Buglosse the lesser: it may be also called Vipers herbe, or Vipers Buglosse.

❧ *The occasion of the name, Alcibiacum.*

This herbe was called Alcibiacum, & Alcibiadion of one Alcibiades the first finder out of the vertues of this herbe, a present remedie against the bitings of Serpets. For as the auncient Nicander writeth, Alcibiades (being asleepe) was hurt with a Serpent: wherefore whan he awoke and saw this hearbe, he tooke of it into his mouth and chewed it, swalowing downe the iuyce thereof: after that he layed the herbe being so chewed vpon the sore, and was healed. Others name it Echion, Echidnion, Viperina, &c. Whiche is asmuch to say as Vipers herbe, which names haue bene giuen to this plante, bycause it is very good against the bitings of Serpents and Vipers, and bycause also his seede is like the head of an Adder or Viper.

❧ *The Nature.*

It is of the same nature that Buglosse is of: but that it is somwhat hoater and more subtile. ❧ *The Vertues.*

The roote boyled in wine and dronke, doth not onely helpe such as are hurt by Serpents, but also, after that a man hath taken it in manner aforesaide, it will preserue hym from being so hurte. The like vertue hath the leaues & seede.

It swageth the payne of the raynes or loynes.

Also being dronken with wine or otherwise, it causeth plenty of milke in womens breastes.

❀ Of Dogges tunge. Chap. v.

❧ *The Description.*

The common Houndes tongue, hath a harde, rough, browne stalke, of two or three foote high: the leaues be long much like the leaues of the great garden Buglosse, but narower, smaller, and not rough, but hauing a certaine fine horenesse vpon thē like veluet. At the toppe of the braunches it beareth many floures, of a darke purple colour. The seede is flat and rough, three or foure together like to a trueloue, or foure leaued grasse, the whiche

the Historie of Plantes.

Cynoglossos altera Plinij.

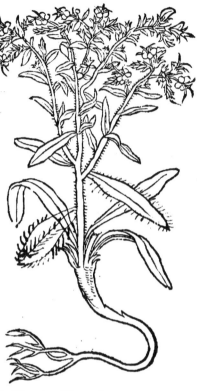

whiche do cleaue faste vnto garments, whan they are ripe, like vnto Aegrimonie and other rough seedes. The roote is long & thicke, & blacke withoutside.

※ *The Place.*

It groweth almoste euery where in waste and vntilled places, but specially in sandie countreys, about pathes and high wayes.

※ *The Tyme.*

It floureth in June, and his seede is ripe in July.

※ *The Names.*

It is called in Greeke κυνόγλωσσον καὶ κυνόγλωσσος: in Latine and in Shoppes Cynoglossum, Cynoglossa, and Lingua canis: whereof also the Italians call it *Lingua de Cane*: the Spaniardes call it *Lengua de perro*: in English Dogs tunge or Houndes tongue: in French *Langue de chien*: in high Douch Hundtzung: in base Almaigne Hondtstonghe. This is that second kinde of Cynoglossa, whereof Plinie wrote in the eight chapter of the.xxv.Booke: it should seeme also to be a kinde of Isatis syluestris, whiche a man shall finde described in some examples of Dioscorides, in the Chapter Isatis. And of Aëtius in his.x.booke and.viij.Chapter Limonium.

※ *The Nature.*

Houndes tongue, but specially his roote, is colde and dry, yea colder than the great garding Buglosse.

※ *The Vertues.*

The roote of Houndes tongue is very good to heale woundes: and it is A with good successe layde to the disease called the wilde fyre, whan it is pounde with Barley meale.

The water or wine wherin it hath bene boyled, cureth old sores, woundes B and hoate inflammations, and it is excellent against the Ulcers & grieuances of the mouth.

For the same purpose, they make an oyntment, as followeth. Firste they C boyle the iuyce therof with hony of Roses, than whan it is well boyled, they mingle Turpentine with it, sturring it harde, vntill all be well incorporate togither, than they apply it to woundes.

The roote rosted in hoate imbers, and layde to the fundament, healeth the D inwarde Hemerrhoydes.

Of Borage.   Chap.vi.

※ *The Description.*

Orage hath rough prickely leaues, broade & large, of a swart greene colour, at the first comming vp bending, or rather spreading themselues abroade flatte vpon the ground, in proportion like to an Oxe tongue. The stalke is rough and rude, of the heigth of a foote & half,

parting

parting it selfe at the toppe into diuers small braunches bearing fayre & pleasant floures in fashion like Starres, of colour blew or Azure, and sometimes white. The seede is blacke, and there is founde twoo or three togither in euery huske, like as in the common Buglosse, but it is smaller and blacker then Buglosse seede.

2 There is also an other kinde of Borage which indureth the winter like to the cōmon Buglosse, and is like to the aforesaide Borage in proportion, sent, sauour, and vertues, but his floures be very small and like to the common Buglosse floures, but smaller.

❧ *The Place.*

It groweth in all gardens, and in sandie champion countreys.

❧ *The Tyme.*

It beginneth to floure in June, and continueth flouring all the Somer.

❧ *The Names.*

The auncient fathers called it in Greeke βούγλωσσον: in Latine Lingua bubula, Libanium, or Lingua bouis, that is to say, Langue de beuf ou vache: in English Oxe tongue: Plinie calleth it ἐυφρόσυνον, bycause it maketh men gladde and merie: the Apothecaries name it Borago: and accordingly it is called in Italion Borragine, in Spanish Borraia, & Borraienes, in English Borage: in Frēch Bourroche, or Bourrache, in Highdouche Burretsch: in base Almaigne, Bernagie or Bornagie.

Buglossum verum.

❧ *The Nature.*

It is hoate and moyste.

❧ *The Vertues.*

Ye may finde this written of Borage, that if the leaues or floures of Borage be put in wine, and that wine dronken, it wil cause men to be gladde and mery, and driueth away all heauy sadnesse, and dull Melancholie. [A]

Borage boyled with honied water, is very good against the roughnesse or hoarsenesse of the throte. [B]

Dioscorides writeth that he hath heard say, that if one pound Borage, that hath but onely three braunches, togither with his roote and seede, and afterward a man giue the same to drinke, to him that hath a Tertian ague, cureth the same. Also that of foure braunches prepared after the same manner is good to be giuen to drinke against the feuer Quartayne. [C]

Of Anthyllis.   Chap.bij.

❧ *The Kindes.*

Anthyllis (as saith Dioscorides) is of two sortes. Whereof one may be called great Anthyllis, and the other small Anthyllis.

Anthyllis

the Historie of Plantes. 13

Anthyllis prior.
Great Anthyllis.

Anthyllis altera, Kali species.
Small Anthyllis.

❧ *The Description.*

1. The first Anthyllis in his stalke & leaues, is not much vnlike vnto Lentill, sauing that it is whiter, softer, and smaller. The stalke is of a foote high, white and softe, with leaues spred broade white and softe also, but smaller & thicker then Lentill leaues: the floures clustering togither at the toppe of the stalke, of a yellow or pale colour. The seede is in small huskes. The roote is small and of wooddy substance.

2. The second is not much vnlike Chamæpythis. It hath fiue or sixe small branches or more, creeping or trayling alongst the ground, thicke set, with little small narrow leaues, betwixt whiche & the stalkes there riseth small purple floures, with seede according. The roote is small, and of the length of a fingar. The whole herbe is full of sape, & salt like Tragus, whereof we shal speake hereafter, and of this herbe they make Arsen, whiche is vsed for the making of glasses.

❧ *The Place.*

It groweth in salt sandy grounds, as in Zeland alongst the coast, where there is store of it. ❧ *The Tyme.*

It floureth in Iune, and the seede is rype in Iuly.

❧ *The Names.*

1. The first kinde is called of Dioscorides in Greeke ἀνθυλλίς. And we haue named it Anthyllis prior, as a difference from the second Anthyllis. Plinie calleth it in Latine Anthyllon, Anthyllion, and Anthycellon: vnknowen of the Apothecaries. Some Arboristes do call it Glaudiola, the which worde is deriued from Glaux, and some iudge it to be Glaux, albeit it is not the right Glaux.

2. The second is named in Greeke Ἀνθυλλὶς ἑτέρα: in Latine Anthyllis altera, as a diffe-

a difference from the first Anthyllis: some of our time do call it Borda.

❧ *The Nature.*

It is dry, and serueth properly, to heale and close vp woundes.

❧ *The Vertues.*

If one drinke halfe an ounce of the first Anthyllis: it shall preuayle much against the hoate pisse, the Strangury or difficultie to make water, and against the payne of the Reynes. A

The same mingled with milke and oyle of Roses, is good for the Matrix or Mother being charged and oppressed with colde humors, to be applied or layde outwardly to the belly. B

Also it cureth woundes by it self, being layde vpō them, or being mixte with salues, oyntments, or oyles. C

The other Anthyllis taken with Oximell (that is honied Vineger) is good for them that haue the falling sickenesse. D

Of the Clote Burre. Chap. viij.

❧ *The Kindes.*

Here be two sortes of Clote Burres in this countrey: the one is the great Burre, & the other ye lesser Burre, the whiche Dioscorides described aparte. Neuerthelesse we haue reduced both into one chapter, bycause of the likelihood that is betwixt them both in name & fashion.

Arcium siue Personata.
Great Clote Burre.

Xanthium.
Louse Burre, or the lesser Clote.

 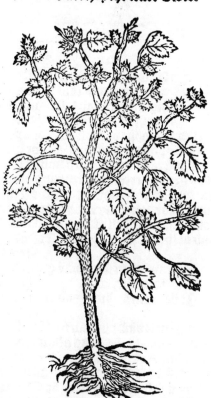

the Historie of Plantes.

❧ The Description.

1. The great Clote hath leaues very large and long, greater than Gourde leaues, of a swarte greene colour, but of a grayish colour on the side next ye ground. The stalke is round & hollow, of colour somwhat white & redde, with diuers side bowghes & braunches set ful of small leaues: vpon the braunches there groweth small bullets or rounde balles, garnished full of little crookes or hookes, wherewithal they take holde or cleaue fast, and hang vpon garments: at last the sayde bullets or knoppes do open and put forth a fayre purple, thromde, or veluet floure. The roote is single, long, blacke without, white within, and in taste bitter.

2. The lesser Clote Burre hath grayish leaues like vnto Orache, iagged or snipte round about the edges. The stalke is a foote and half long, full of blacke spottes, diuiding it selfe into many branches or winges. Betwixt the leaues and the sayde branches, there groweth three or foure small Burres in a cluster, somewhat long, like to a small Oliue, or Cornell berry, prickly, and cleauing fast vnto garments. In the middell of those small Burres, there groweth forth as it were a little Crownet, somewhat aboue the Burres, vpon whiche groweth small floures, the which do perish after their opening, and do fall with their Crowne: than commeth the little Burres with long seede: the which afterward do neuer open, nor floure otherwise than is aforesayde. The roote is redde, and full of small threedes or hearie strings.

❧ The Place.

The Clote Burres delight to grow by the way side, about the borders of fieldes, in vntilled places, and dry Diches.

❧ The Tyme.

Theyr season is in July, and August.

❧ The Names.

1. The great Burre called in Greeke ἀρκτιον καὶ προσώπιον: in Latine Personatia, Personata and Arcium: of Apuleius Dardana: in Shoppes Bardana maior, and Lappa maior: in Italiã *Lappola maggiore*: in Spanish *Lampazos, yerua dos pegamazos, pagamacera mayor*: in English the great Burre or great Clote Burre: in French *Bardane la grande, & Lappe grande: grand Glouteron* or *Gleteron*: in high Douch Grosz kletten: in base Almaigne Groote Clissen.

2. The lesser is called in Greeke ξάνθιον καὶ φάσγανον: in Latine Xanthium: in Shoppes Lappa minor, and Lappa inuersa: in Italian *Lappola minore*: in Spanish *Pagamacera menor*, that is to say, the small Burre, & the Burre turned in & out: in French *Le petit Glouteron* in high Douch Bettlerttz leusz, and Spitz kletten, that is to say, Rams lyce, or Beggers lysse, and the poynted or sharpe Burre: in base Almaigne cleyn Clissen: in English Diche Burre, and lowse Burre.

❧ The Nature.

The Clote Burres haue power to dry vp, consume, or dissolue: but the lesser is the hoater.

❧ The Vertues.

The iuyce of the great Burre dronken with Hony prouoketh bryne, and swageth the payne of the bladder.   A

The same dronken with olde wine, healeth the bitings and stingings of venemous beasts.   B

The leaues pound with a littell salte, is with great profite layd vnto the bitings and stingings of Serpents, madde Dogges, & other benemous beasts.   C

The seede made into pouder & taken with the best wine that may be gotten by the space of fortie dayes, is very profitable for such as haue the Sciatica.   D

A dramme

A dramme (which is the eigth parte of an vnce) of the roote, pound with the kernelles of Pine apple, and dronken, is a soueraigne medicine for such, as spit bloud and corrupt matter.

It is good for such as haue ache or payne in their ioyntes, by reason that the sayde ioyntes or bones haue bene before out of ioynt, broken or hurte.

The greene leaues pounde with the white of Egges, cureth burnings and olde sores, being layde thereto.

The iuyce of the lesser Burre dronken with wine, is much vsed against the bitings of venemous beasts, and also against the grauell and the stone.

The fruite pounde & layde vnto colde swellings (called in Greeke Oedema) consumeth the same, and scattereth or wasteth all colde humors: and is specially good against the kings euell, called Strumas and Strofulas.

Of Mugworte. Chap.ix.

❧ *The Description*. Artemisia communis.

Mugworte hath broade leaues, all iagged & torne like the leaues of Wormwood, but something smaller, & specially those whiche grow about y stalke, they are of a browne greene colour aboue, and white hoare or gray vnderneath. The stalke is long and straight & full of branches. The floures are smal round buttons, growing alongst the branches, like Wormwood, smelling whan they begin to waxe ripe somewhat after Mariorum. The roote is of a wooddy substance & hath small hearie strings. Of this herbe there be twoo kindes moe, differing onely in colour.

1. The one hath redde branches & floures, and is called redde Mugworte.
2. The other hath greenish branches, changing towardes white, and is called white Mugworte, in all things els like one to another. ❧ *The Place*.

Mugworte groweth in the borders of fieldes, & about highwaies, and the bankes of brookes or quiet standing waters.

❧ *The Tyme*.

It floureth in July & August, and sometimes later. ❧ *The Names*.

This herbe is called in shops Artemisia, & of some Mater herbarum: in Spanish *Artemya*: in English Mugworte: in French *Armoyse, l'herbe S. Ian*: in high Douch Beyfusz, Bucken, & S. Johans gurtel: in base Almaigne Byuoet, & S. Jans cruyt, the which is this kind of Mugwort, whiche is called in Greeke ἀρτεμισία λεπτόφυλλος: in Latine Artemisia tenuifolia, the which is the fourth kinde in Dioscorides, and the third kinde in Apuleius.

❧ *The cause of the Name*.

Mugworte as Plinie saith, had this name of Artemisia Queene of Halicarnassus and wife of Mausolus king of Carie, who chose this herbe & gaue it her name, for before that it was called παρθενίς. Parthenis, that is to say, Virginal:

the Historie of Plantes.

some say that Artemisia was so called of the Goddesse Diana who was also called Artemis, & for bycause this herbe is singular for womens diseases, who are all under the gouernment of Diana, as the Heathen do imagine and dreame.

❧ *The Nature.*

Mugworte is somewhat astringent, and not to hoate.

❧ *The Vertues.*

Mugworte pound with oyle of sweete Almondes, and layd to the stomake as a playster, cureth all the payne and griefe of the same. A

Also if one do annoynt his ioyntes, with the iuyce thereof mengled with oyle of Roses, it cureth the ache, shaking, and drawing togither of Sinewes. B

If it be hanged or cast into barrels or hoggesheads of Bier, it will preserue the same from sowring. C

Whosoeuer shal carrie this herbe about him (as Plinie saieth) no venemous beast, or any like thing shall hurte him, and if he trauell vpon the way, he shall not be weary. D

Of Tansie.   Chap.x.

❧ *The Kindes.*

There be two sortes of Tansie. The one great and yellow, the other small and white.

Tanacetum maius.   Tanacetum minus.
Great Tansie.   White Tansie.

The first Booke of

❧ The Description.

1. The great or common Tansie hath a blackishe stalke, three or foure foote high, diuided at the top into many single braunches, at the end wherof are round tuftes, bearing yellow floures like small round buttons, or like the middle of the floure of Cammomill, but greater and of stronger sauour. The leaues be long & made of many small leaues, set directly one against an other, and spread abroade like wings, the whiche be also iagged and snipte like small feathers, especially round aboute the edges: the roote is slender casting it selfe here and there.

2. The small Tansie hath broade leaues, much iagged and cut, well like the leaues of Feuerfew, but smaller and more cut and iagged. The stalke is small, of the length of a foote or more, vpon the which groweth small tuftes, bearing little white floures, much like to the floures and tuftes of the white Mylfoyll or common Yarrow. The roote is harde, and sometimes parted into two or three: all the herbe is much like in smell and sauour to the other Tansie, sauing that it is not so strong.

❧ The Place.

1. The first groweth about high wayes, hedges, and the borders of fieldes, and is very common in this countrie.

2. The second groweth in some places of Italie: in this countrey ye shall not finde it but in the gardens of certayne Herborstes.

❧ The Tyme.

They do bothe floure in July and August.

❧ The Names.

The first is now called in shoppes Tanacetum, and Athanasia: in Englishe Tansie: in French *Athanasie*, in high Douch Reinfarn: in base Almaigne Reynuaer, and Wormcruyt. Some learned men iudge it for to be the third kinde of Artemisia, called in Greeke ἀρτεμισία μονόκλωνος: in Latine Artemisia vnicaulis, of Apuleius Artemisia Fragantes, or Tagetes.

The second without doubt is also a kinde of Tansie, the whiche some learned (and especially the famous Matthiolus of Siena,) do thinke it to be right Milfoyle, called in Greeke ἀχίλλεος. But if this herbe shoulde be the right Achillea, the common Tansie should be also without doubte a kinde of Achillea, for they are very much like one an other, not onely in smell and taste, but also in vertues and operation, as we haue written in our Annotations.

❧ The Nature.

Tansie is hoate in the second degree, and dry in the third, as it doth well appeere by his strong smell, and bitter taste.

The small Tansie is of the like operation, or facultie.

❧ The Vertues.

A. The seede of Tansie is a singular and proued medicine against wormes: for in what sorte soeuer it be taken, it killeth and driueth forth wormes.

B. The same pounde and afterwarde mengled with oyle, is very good against the payne and swelling of Sinewes.

C. If before the comming of fittes of the Ague, the body be annoynted with the iuyce of Tansie mengled with the oyle of Roses, it will cause the Ague to be gone.

D. The same dronken with wine, is good against the payne of the bladder, and whan one cannot pisse but by droppes.

E. The roote condited or preserued with hony and taken of them that be sicke, doth ease & helpe very much, such as are troubled with the goute in their feete.

Of Feuerfew. Chap.xi.

❋ *The Description.*

Euerfew hath many tēder leaues much torne & iagged of a grayishe or white greene colour, in colour and fashion, like to the first & nethermost leaues of Coriander: the stalkes be two or three foote long, vpon whiche groweth many smal floures yellow in the middest, and compassed aboute as it were with a little pale of small white leaues, like to the order of Cammomil floures, of a strong smell and bitter taste: whan the floures be past, the knoppes be ful of seede, like to the knops of Camomill. The roote is of wooddy substāce with diuers hearie threedes or strings hanging by.

Parthenium.

❋ *The Place.*

It groweth well in dry places, by olde walles, and such like rough places.

❋ *The Tyme.*

Feuerfew floureth in July & August, and almost all the Sommer.

❋ *The Names.*

It is called in Greeke παρθένιον, of Galen, and Paule ἀμάρακον: in Latine Parthenium and Amaracus: in shoppes, and of Serapio Chap. 253. Matricaria, of some Amarella or Marella: in English, Feuerfew, & of some Whitewurte, also S. Peters wurt: in French *Espargoutte*, or *Matricaire*: in high Douch Mutterkraut, and Meidt blumen: in base Almaigne Mater & Moedercruyt.

❋ *The Nature.*

It is hoate in the third degree, and dry in the second degree.

❋ *The Vertues.*

A   Feuerfew dryed and made into pouder, and two drammes of it taken with hony, or other thing, purgeth by siege Melancholy and fleume: wherefore it is very good for such as haue the giddinesse & turning in the head or swimming, for them that are pursy or troubled with the shortnes of winde, and for Melancholique people, and such as be sadde and pensiue and without speach.

B   The herbe without his floures, boyled in water is good to be dronken of such as haue the stoone.

C   The same is good against the Suffocation of the Matrix (that is, the stopping and hardnesse of the Mother) to be boyled in wine, and applied to the nauell, the harte, or the side.

D   The broth also, or decoction of Feuerfew, is very good for women to bathe and sitte in against the hardnesse of the Mother, and the Matrix that is ouercharged or swollen.

E   The greene leaues with the floures of Feuerfew stamped, is good to be layde to the disease called the wilde fyre or Saint Anthonies fyre, and other cholerike inflammations.

Of Fole foote, or Horse houe. Chap. xij.

❧ The Description.
Bechion, Tussilago.

Fole foote hath greate broade leaues, growing out into many corners, or indeted angles, with many vaynes, like to a Horse foote, sixe or seuen leaues springing out of one roote, of a white, hoare, or grayish colour next to the ground, and greene aboue. The stem or stalke is white, and as it were cottoned with fine heare of a span long, at the end wherof are fayre yellow floures and full, which do suddenly fade, and chaüge into downe, or cotton, which is carried away with the winde, like to y̆ head of Dandelion. The roote is white and long creping here and there.

❧ The Place.
Fole foote groweth well in watery places and moyst fieldes.

❧ The Tyme.
It putteth forth his wolly stalke without leaues, at the beginning of March & April. At the toppe of the stalke is the yellow floure: After the floures the leaues spring out from the roote: then vanisheth away the stalke and the floures, so that one shall seldome finde the leaues and floures altogether at one time.

❧ The Names.
It is called in Greeke βήχιον καὶ χαμαιλεύκη: in Latine Tussilago: in shoppes Farfara, and Vngula Caballina: in Italian *Vnghia di cauallo*: in Spanishe *Vña de asno*: in English Fole foote, Horse houe, Coltes foote, and Bull foote: in French *Pas de Cheual*, of some *Pas d'asne*: in high Douch Roßhub, or Brandtlattich: in base Almaine, Hoefbladeren, Peerdts clauw, Brant lattowe, and Saint Catrinus cruyt.

❧ The Nature.
The greene and fresh leaues are moyst, but whan they are dry they become sharpe or sower, and therefore are of a drying nature.

❧ The Vertues.
The greene leaues of Fole foote pounde with Hony, do cure and heale the hoate inflammation called Saint Anthonies fyre, and all other kindes of inflammation.

The parfume of the dryed leaues layde vpon quicke coles, taken into the mouth through the pipe of a funnell, or tunnell, helpeth suche as are troubled with the shortnesse of winde, and fetche their breath thicke or often, & do breake without daunger the impostems of the breast.

The roote is of the same vertue, if it be layde vpon the coles, and the fume therof receiued into the mouth.

Of Butter Burre.   Chap.xiij.

❊ *The Description.*

Petasites.

Vtter Burre hath great round leaues, at the firste lyke the leaues of Folefoote, the which do afterwardes waxe so great, that with one leafe, one may couer a smal rounde table, as with a carpet. Of a greene colour vpon the outside, and of a gray whitishe colour nexte the grounde. It putteth forth a hollow stalke of a span long, set full of small incarnate floures at the toppe, as it were clustering thicke togither: the which togither with the stalke do perish and vanish away. The roote is thicke, white within & hollow, of a strong smell and bitter taste.

❊ *The Place.*

It groweth well in freshe and moyste places, bysides small riuers and brookes.

❊ *The Tyme.*

The floures do appeare at the beginning of Marche, and do vanish away in Aprill: then the leaues come forth, and remayne all the Somer.

❊ *The Names.*

It is called in Greeke πιτασιτης: in Latine Petasites, vnknowen in shoppes: yet some call it Bardana maior: in Englishe, Butter Burre: in French *Herbe aux tigneux*: in high Douch Pestilentz wurtz: in base Alnaigne Dockebladeren, and Pestilentie wortel.

❊ *The Nature.*

Butter Burre is dry in the thirde degree.

❊ *The Vertues.*

Butter Burre dried, and made into powder and than dronken in wine, is a A souueraigne medicine against the Plague, and Pestilent feuers, bycause it prouoketh sweate, and for that cause it driueth from the harte all venim, and euill heate. It killeth wormes, and is of great force against the Suffocation, and strangling of the Mother to be taken in the same sorte.

It cureth all naughty Vlcers, or olde filthie, fretting, sores, or consuming B Pockes, and inflammations, if the powder be strewed thereon.

The same cureth the Farcyn, in Horses, howsoeuer it be ministred, whether C it be giuen inwardly to receyue, or applied outwardly.

Of Britannica or Bistorte.   Chap.xiiij.

❊ *The Kyndes.*

There is two sortes of Bistorte, as Leonard Fuchs, and Hierome Bock, (men of great knowledge and learning) haue lately written: the one called the Great Bistorte, the other the Small Bistorte.

Bistorta

The first Booke of

Biſtorta maior.
Great Biſtorte.

Biſtorta minor.
Small Biſtorte.

❧ *The Deſcription.*

1. The great Biſtorte hath long leaues, like Patience, but ſmaller, and not ſo ſmothe or playne, but wrinkled or drawen into rimples, of a ſwart greene colour vpon one ſide, and of a blewiſhe greene on the ſide next the ground. The ſtalke is long, ſmothe and tender, hauing a ſpiked knap at the ende, ſet full of ſmall incarnate floures cluſtering togither. The ſeede is angled and browne. The roote is great and long, wounden and turned backe, or crokedly turning togither like a Snayle, blacke and hearie without, and ſomewhat redde within, in taſte like an Oke kernell.

2. The ſmall Biſtorte is like the other in leaues, knap, floures, ſeede & ſtalke, but ſmaller, his leaues alſo are ſmother and playner. The roote is ſhorter and more roundly turned togither without any ſmall threeds, or hearines, browne without, and of a darke redde colour within, in taſte like the firſt.

❧ *The Place.*

They grow well in moyſt & watery places, as in medowes, and darke ſhadowy wooddes.

❧ *The Tyme.*

They floure in May and Iune.

❧ *The Names.*

The learned do call the herbes Biſtortæ and Serpentariæ: in French *Biſtorte*: in high Douch Naterwurtz: in Brabant Hertſtonghen. This ſhould ſeeme to be Dracunculus Latinorū, wherof Plinie wrote in the. 6. chap. of the. 24. Booke.

The firſt is called of ſome in Latine Colubrina, & of Leonard Fouchs, Naterwurtz weiblin, that is to ſay, female Adderwurte or Snakeweede: in
French

the Historie of Plantes.

French *Grande Bistorte:* and *Serpentair femelle:* in base Almaigne Hertstonghe.

2 The second is the small Bistorte: & is called in some places of England Oysterloyte: of the same Leonard Fouchs Naterwurtz menlin, that is to say, male Adderwurte or Snakeweede.

❧ *The Nature.*

Bistorte doth coole and dry in the third degree.

❧ *The Vertues.*

The roote of Bistorte boyled in water or wine, and dronken, stoppeth the laske, and is good against the bloudy flixe.

It stoppeth the ouermuch flowing of womens termes or floures, and all other issue of bloud.

Also if it be taken as is aforesayd, or if it be made into pouder and dronken with redde wine, it taketh away the desire to vomite or parbrake.

The decoctiō of the leaues is very good against all sores, & inflāmatiō of the mouth & throote, & it fasteneth loose teeth, if it be oftē vsed, or holdē in ẏ mouth.

Of Fumeterre.   Chap.xv.

❧ *The Kindes.*

THere is two kindes of Fumeterre, (as Plinie writeth in the.xiiij.chap. of the.xxv.booke of his naturall History.) Wherof the first is the common Fumetory the which was knowen & vsed in Medicine, of Galen, Paule, & other the Greeke Physitions. The second is an other herbe, onely knowen of Plinie: the whiche both are knowen in this countrey.

Capnos fumaria.   Capnos {Plinij. Phragmites.
Fumeterre.   Hedge Fumeterre.

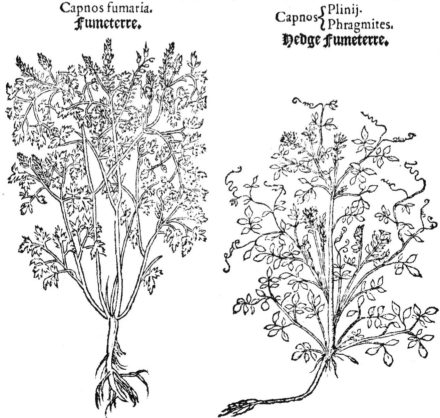

✤ The Description.

1. The common Fumeterre hath a square stalke, beset with small leaues, very tender, weake, and finely iagged, & somewhat gray like asshie colour, like to the leaues of Coriander but much smaller: the floure is small and purple, growing togither like a littell cluster, and changeth into littell small knops or beries, wherein is very small seede. The roote is but simple with a very few small heares or strings about the same.

2. Small Fumeterre, hath also many slender branches, vpon whiche groweth small iagged leaues, in colour, taste, and in fashion also, somewhat like the Fumeterre aforesayde. It hath also certaine small threedes or clasping tendrels, by the whiche it taketh holdfast in all places by Hedges, and other herbes. The floures are small and clustering togither, of a white colour mixed with a littell blew: after the floures there commeth forth small huskes or coddes, in which is conteyned the seede. The roote is single and of the length of a fingar.

✤ The Place.

Fumeterre groweth best amongst wheate & Barley, also it groweth in gardens amongst pot herbes, in Uineyardes, and such other open places.

Small Fumeterre groweth vnder hedges, in the borders of fieldes, and about olde walles.

✤ The Tyme.

They do bothe floure in May and June.

✤ The Names.

1. The first of these herbes is called in Greeke καπνός, κάπνου, καὶ καπνίτης: in Latine Fumaria and Capnium: in Shoppes Fumus terræ: in Spanish Palomilla, y palomina, y yerua malariña: in English Fumeterre: in French Fumeterre: in high Douch Erdtrauch, Taubencropff, Katzenkorbel: in base Almaigne, Grysecom, Duyuekeruel, and Eerdtroock.

2. The second is called of Plinie Capnos, & Pes Gallinaceus: Therfore Capnos Plinij, and this is that whiche is called Hermolaus, of Aëtius, καπνος χελιδόνιος, in Latine Capnum Chelidoniũ, not knowen in shoppes, some following Plinie do call it in Latine Pes gallinaceus: in French, Pied de geline: in base Almaigne cleyn Eerdtroock: in English Hedge Fumeterre, and Hennes foote.

✤ The Nature.

Fumeterre is hoate and dry, almost in the second degree, and so is Hennes foote, as one may know by the sharpnes, and bitter taste.

✤ The Vertues.

The iuyce of Fumeterre dropped into the eyes, doth sharpen and quicken the sight, the same mengled with gumme, and layd to the eye liddes, will cause that the heare that hath bene ones pulled of, shall not grow againe. A

The decoction of Fumeterre dronken, driueth forth by vrine & siege all hoate Cholerique, burnte, & pernicious humors. Bysides this it is very good against the foule scurffe, and rebellious olde sores, and the great Pockes. B

The iuyce of Fumeterre dronken worketh the like effect, & for this purpose is of greater power, than the Decoction of Fumeterre. C

1. Henfoote or hedge Fumeterre (as Plinie sayth) is of the same nature & vertue as the other Fumeterre: and is a singular medicine against the weakenesse of the sight, especially for such as seeme to see small strawes, if the iuyce thereof be dropped into the eyes. D

Of Germander. Chap.xvi.

✤ The Description.

Germander is a shorte herbe, of a spanne or foote long, bringing foorth from his roote many tender stemmes or branches. The leaues are smal & tender, indented

the Historie of Plantes.

Chamædrys. Germander.

indēted & cut about, much like the leaues of certayne Okes, but farre smaller. The floures are small of a broune blew colour compassing round the toppe of the stalke. The seede is small, blacke, and rounde. The roote is small and slender, creping vnder the earth, here and there.

❧ The Place.

Germander groweth luckely in stony hilles & mountaynes, & such like places, also it groweth in wooddes, it is to be found growing in certayne wooddes of Brabant, and it is planted in gardens.

❧ The Tyme.

Germander floureth in June & July.

❧ The Names.

The first is called in Greeke χαμαίδρυς: in Latine Chamædrys, Trixago, & of som Quercula minor, & Serratula: in Shoppes Chamædryos: of the Italians *Querinola, Chamedrio, Chamandrina*: in Spanish *Chamedreos yerua*: in French *Germandreé*, or *Chesnette*: in English Germander, & English Treacle: in high Almaigne Gamanderlein and Kleyn Bathengel: in base Almaigne, Gamanderlijn.

❧ The Nature.

It is hoate & dry in the third degree.

❧ The Vertues.

Germander with his floures boyled in water and dronken, deliuereth the body from all obstructions & stoppings, and cutteth of tough and clammy humors: & therfore being receiued as is before sayde, it is specially good for them that haue the cough & shortnesse of breath, the Strangury or stopping of vrine, and for such as begin to haue the Dropsie. A

It bringeth downe womens naturall sicknesse. B

If it be dronken with vineger, it is good against the hardnesse and stopping of the Milte or Splene. C

The iuyce of the leaues mengled with oyle, and straked vpon the eyes, drieth away the white Cloude, called the Hawe or Pearle in the eye, and all maner dimnes of the same. D

Of Paules Betony. Chap.xvij.

❧ The Kindes.

THere is two kindes of Vetonicæ, or Betonicę Pauli. The one is p right Veronica the which is called Veronica mas: The other is a small herbe very like the right Veronica, and is called Veronica fœmina.

❧ The Description.

THe male Veronica is a small herbe, & crepeth by the ground, with smal reddish, & hearie braunches or stalkes. The leafe is something long, and somwhat greene, a little hearie, & dented or snipte roūd about the edges like a sawe. The floures are aboue about p top of the branches, smal, & of a light blew mengled w purple: the seede is in smal flat pouches. The roote is smal & hearie. C

Betonica.

{ Betonica Pauli.　　　　　Veronica fœmina.
{ Veronica mas.

Paules Betony. Herbe Fluellyn, or Speedewell. Groundhele. Laudata Nobilium.

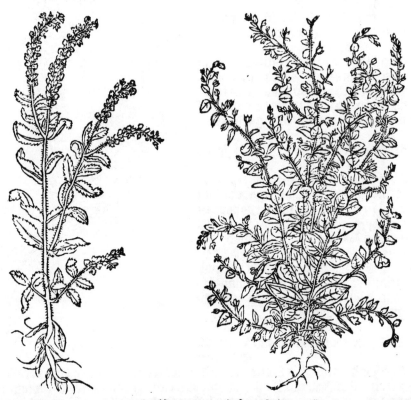

2　The female Veronica doth also creepe and spread vpon the grounde, it hath slender stemmes and somwhat large leaues, a littell hearie and pleasantly soft. The floures be yellow, with small croked tayles, like the floures of Larkes claw, or Larkes spurre. The seede is in small rounde huskes, like the seede of Pympernell.

❧ *The Place.*

1　The male Veronica groweth in rough sandy places, aboute the borders of fieldes and wooddes.
2　The female groweth in low moyst places.

❧ *The Tyme.*

They floure in June and July.

❧ *The Names.*

1　The first Veronica is called of Paulus Aegineta Lib. vij. in Greeke βετωνική, that is to say, in Latine Betonica: and therefore Doctor William Turner and I do call it Betonica Pauli: The common Herboristes do call it in Latine Veronica: in high Douch Erenbreiſz mennlin, and Grundheyl: in base Almaigne Eerenprijs manneken.

2　The second is called Veronica fœmina of the Latinistes: in Frenche *Veronique femelle*: in high Douch Erenbreiſz weiblin: in base Almaigne Eerenprijs wijfken.

❧ *The Nature.*

Veronica or Paules Betony, is dry and somewhat hoate.

❧ *The Vertues.*

1 Veronica (as Paule witnesseth) is specially good for the stoppings, & paynes of the kidneys.

The Decoction of Veronica dronken, doth soder and heale all fresh, and old woundes, and clenseth the bloud from all euill corruptions, and from all rotten and aduste humors: and for that cause it is good to be dronken for the kidneys, and against scuruinesse and foule spredding Tetters, and consuming or fretting sores, the small Pockes and Meselles.

The water of Veronica distilled with wine, and so often new drawen vntill it ware of a reddish colour, is much vsed against an old Cough, the drynesse, and harmes of the lunges: for men say that it will heale all vlcers, inflammations and harmes of the Pulme or Lunges.

2 The female Veronica is of the like operation, but much weaker, and not so good as the Male.

Of Ground Pyne, or Iua Moscata. Chap. xviij.

❧ *The Kindes.*

There be three sortes of the herbe called in Latine Chamæpitys, (as Dioscorides sayth) the one like the other in smell and fashion.

Chamæpitys prima.	Chamæpitys altera.
The first Grounde Pyne.	The second Ground Pyne.

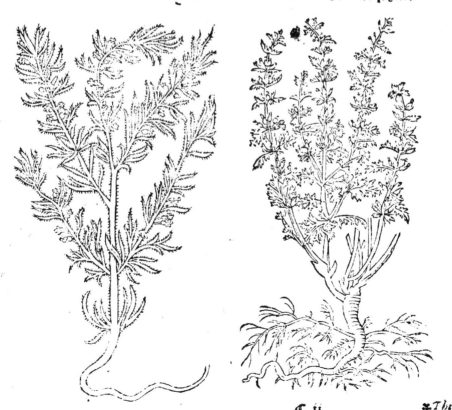

The first Booke of

❧ The Description.

1. The first kinde of these herbes, is a small herbe and tender, creping vpon the ground: it hath small braunches, & something croked: the leaues be small, narrow & hearie, of the sauour of the Pyne, or Fyrre tree: The floures be small, pale, yellow, or white, the roote is sleight or single, & of wooddy substãce.

2. The second hath also small braunches, browne, hearie, and tender, croking in, after the fashion of an ancker, out of which braunches groweth small hearie leaues, much clouen and cut crosse wise: The little floures be of a purplishe colour, and grow about the stalkes in tuffes like garlãds or crownets. The seede is blacke and rounde, and the whole plante sauoureth like to the other.

3. The thirde is the least of all, and hath small, white, rough leaues, the floures be yellow: and in smell like to the others.

Chamępitys tertia.
The third Ground Pyne.

❧ The Place.
These herbes loue to growe in stony groundes and mountaynes: in this countrey it is sowen and set in gardens.

❧ The Tyme.
They floure in July and August.

❧ The Names.
These three herbes be all called by one Greeke name χαμαιπίτυς: in Latine Aiuga, Abiga, and Ibiga: in shoppes Iua, and Iua Artetica, or Iua moscata: in Spanishe *Pinillo*, in English also Chamæpitys, Groũd Pyne, Herbe Iue, Forget me not, & field Cypres: in Frenche *Iue musquée*: in highe Douch Veit Cypres, & of some Hoelangher hoe liener.

❧ The Nature.
They are hoate in the second degree, and dry in the thirde.

❧ The Vertues.

The leaues of Chamæpitys droken in wine by the space of seuen dayes, healeth the Iaundes, & dronken with Meade or Melicrat by the space of fortie dayes, it healeth the Sciatica, that is to say, the payne of the hippe or hocklebone.

It is also good against the stoppings of the liuer, the difficultie of vrine, and B causeth women to haue their termes or naturall sicknesse.

Chamępitys greene pound, and mengled with Honie, and layde vpon great C woundes, and virulent, and corrupt vlcers, cureth the same.

Also the same being greene pound, and layde to womens breasts or pappes, D dissolueth the hardnesse of the same.

And being ordered as is beforesaide, and layde to the bytings or stingings E of Serpents, Vipers, and such other venemouse beasts, is of great vertue and much profitable against the same.

The Decoctiõ of Chamępitys dronken, dissolueth clottie & congeled bloud. F And the same boyled in vineger and dronken, deliuereth the dead childe.

If the body be rubbed or annoynted with the iuyce thereof, it causeth much G sweating.

The

The like vertue haue the two other kindes, but it is weaker and not of so great efficacy.

Of lauender Cotton, or Garden Cypres. Chap.xix.

The Kyndes.

There be sundry sortes of garden Cypres, growing in the gardens of this countrey.

Chamæcyparissus.

The Description.

1. The first and the most common Cypres, is a small tree or shrubbe of wooddy substance, with vpright braunches, bringing forth small, narrow, long and roūd, ragged or purled leaues, at the top of the braunches or stems groweth fayre Orenge-colour floures, like the floures of Tansey, but greater. The roote is of wooddy substance, with many strings or threddes hanging at it.

2. The other Cypres is much like to the first in stalkes, leaues, floures, & fashion, sauing that the braunches that bare the leaues are smaller, & set or couered with long small leaues, the floures be paler & smaller, and the whole herbe is not of so strong a sauour, but smelleth more gentilly, and pleasantly.

3. The third kind his leaues be smaller, & shorter, almost like the leaues of heath.

4. The fourth kinde his leaues be more single, and like the leaues of the Cypresse tree, but they are white.

5. The fifth hath softe wollie leaues, as it were layde with a certayne downe or fine Cotton: with stalkes creeping alōgst the ground. The floures of these three kindes, are not vnlike the floures of the first kinde.

The Place.

They grow not in this coūtrey, but in the gardens where as they are plated.

The Tyme.

They do both floure in July, and August.

The Names.

1. Plinie calleth this herbe in Greeke χαμαικυπάρισ⊙ : and in Latine Chamæcyparissus: some of the later writers do call it Santolina, and Camphorata: vnknowen in shoppes: some call it in English Lauender Cotton, and som Garden Cypres: in French Cypres de iardyn: in Douch Cypres.

2. The others without doubte are of the kindes of Cypres, and not Cedre, as some call it. The seede of this herbe is called in shoppes, Semen cōtra lumbricos, Semen Santonici, & Semen sanctum.

The Nature.

It is hoate and very dry.

The Vertues.

Plinie writeth that Chamæcyparissus droken in wine is good against Serpents, and Scorpions, and other kinde of poyson.

The first Booke of
Of Celandine, Figwoꝛte, and Marſhe Marigolde. Chap.xx.

❧ The Kindes.

There be two kindes of the herbe called in Greeke Chelidonium, wherof the one is the great Celandyne, the other is ſmall Celandyne, in Latine Strophularia minor.

Chelidonium maius.	Chelidonium minus.
Great Celandyne.	Small Celandyne.

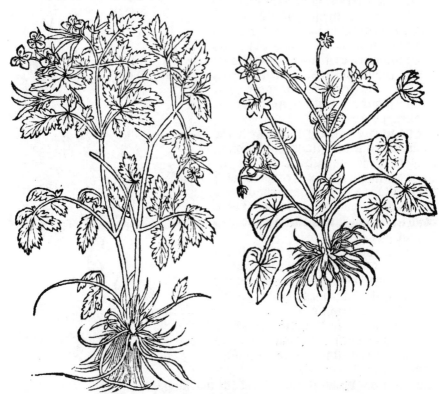

❧ The Deſcription.

1. Great Celandyne hath a tender ſtalke, round, hearie, and full of bꝛaunches, euery bꝛaunche hauing diuers ioyntes and knottes. The leaues much like vnto Colombyne, but tenderer & deeper iagged oꝛ cut, of a grayiſh colour by one ſide, and greene vpon the other ſide ſomewhat dꝛawing towards blew. The floure is at the toppe of the bꝛaunches fayꝛe and yellow like the wall Gyllofer, & turneth into long coddes oꝛ huſkes, in them is the ſeede, whiche is ſmall and pale. All the herbe is of a ſtrong ſmell: and the iupce (whereof the floures, the leaues, the ſtalke, and the roote is full, and commeth foꝛth whan they be either bꝛuſed oꝛ broken) is yellow as Saffron, ſharpe and bitter, but that of the roote ſpecially, the which is yellow as golde. The roote hath many ſmal ſtrings oꝛ thꝛeddy laces hanging thereby.

2. The ſmall Celandyne is a low herbe growing by the ground, hauing a little ſmall bꝛowniſh ſtem, the leaues be ſmall and ſomewhat round, like Iuie leaues, but much ſmaller, tenderer, ſofter, and ſmother. The flower is yellow

like

the Historie of Plantes. 31

like to a golde cup, or Crowfoote floure. The roote is full of small threddes, or hearie laces, with diuers knottes in them like to wheate or barley cornes.

3 There is an other herbe muche like to small Celandyne in leaues & floures, the which we may call Marsh Mary-golde, or Braue Celandyne, the leaues be of a swarte greene colour, somwhat round, and shining, like to a Popler leafe, but larger & a little cut, or purlde about the edges. The stalke is round, and diuided into many braunches, vpon which are the pleasant yellow floures, like to yellow Crowfoote or golde Cup, but larger and fayrer to behold. The floures being gone or fallen, yee shal see three or foure small huskes or cods, like to the huskes of Colombyne, wherein is côteyned smal yellow seedes. The roote is great and thicke, with many threddy strings.

Caltha Palustris.
Marshe Marigolde.
Dotterbloemen Belgarum.

※ *The Place.*

1 The great Celandyne groweth in dry places, about old rotten walles, and by the way sides, and vnder Hedges & quicksets.

2.3 The small Celandyne, and the Braue Bassinet, or Marsh Marigold, do grow in moyst medowes, vpon the bankes and borders of ditches.

※ *The Tyme.*

1 The great Celandyne beginneth to floure in Aprill, and lasteth flouring all the Sommer.

2 The small bringeth forth his floure bytimes, about the returne of Swallowes, in the ende of February. It remayneth flouring all Marche, euen vntill Aprill, and after it doth so vanish away, that a man shall seldome see it in May.

3 The Braue Bassinet, floureth in May and Aprill.

※ *The Names.*

1 The great Celandyne is called in Greeke χελιδόνιον: in Latine Chelidonium maius, and Hirundinaria maior: in shoppes Chelidonia: & of some as Athenæus writeth, Anemone: in Spanishe *Cheliduñea, yerua d'anduriña y yerua de las golundriñas*: in English Celandyne, Swallowurte, and of some Tetterwurte: in French *Cheledoine*, or *Esclaire*: in high Douch Grosz Schelwurtz, grosz Schwalbenkraut, and Schelkraut: in base Almaigne Gouwortel, & Groote Gouwe.

2 The lesser is called in Greeke χελιδόνιον μικρόν: in Latine Chelidoniũ minus, and Hirundinaria minor: in shoppes Scrofularia minor, and Ficaria: in Italian *Fauoscello*: in Spanish *Scrofularia menor*. in English Pyleworte, or Figworte: in frenche *Scrofulaire*, or *Petite Esclaire*: in high Douch klein Schelwurtz, klein Schwalbenwurtz, feigwartzen, or Blaternkraut, Pfaffenhodlin, & Meyenkraut: in base Almaigne, Cleyn Gouwe, and cleyn Speencruyt.

Caltha Palustris so named of certaine late writers, of some Tussilago altera, and Farfugium, whereunto notwithstanding it is but a littell like, may well be Englished Marshe Marigolde: in French *Bassinet de prez*, or *Bassinet de marés*: in high Douch Moszblumen, Dotterblumen, Geelweiszblumen, and Martenblumen:

C iiij

The first Booke of

blumen: in base Almaigne, groote Booterbloemen, and Dotterbloemen.

❧ The occasion of the Names.

1. The great Celandyne is named in Greeke χελιδόνιον, Chelidonium, that is to say, Swallow-herbe, bycause (as Plinie writeth) it was first found out by Swallowes, and hath healed the eyes, and restored sight to their yong ones, that haue had harme in their eyes, or haue bene blinde.

2. The small Celandyne was so called, bycause that it beginneth to spring & to floure, at the comming of the Swallowes, and withereth at their returne.

❧ The Nature.

The two Celandynes are hoate and dry in the thirde degree: and the small Celandyne is the hoatest.

The Braue Bassinet, or Marshe Marigolde, is also of a hoate nature, but not exceeding.

❧ The Vertues.

The iuyce of Celandyne mingled with Hony, & boyled in a vessell of copper or brasse, cleareth the sight, and dropped into the eyes, taketh away the spots, scarres or blemisshes, bloudshotten, and webbe of the eye. A

If with the same iuyce and wine, one washe fretting, and consuming sores, it will consolidate and heale them. B

The roote boyled with Anise seede in white wine, openeth the stoppings of the Liuer, and healeth the Iaundice. C

The same roote chewed in the mouth, taketh away the tooth-ache. D

The smal Celandyne pound, & layde vnto rough & corrupt nayles, causeth ye same to fall away, & fayrer or better to grow in their places: And if it be pound in vryne or wine, especially the roote, and after applied and layde to the Hemorrhoides, it doth dissolue and heale them: so doth the iuyce, if it be mingled with wine or vrine, and the Hemorrhoides be wasshed therewithall. E

The decoction of this herbe in wine gargarised, doth purge the head from naughtie fleume & euill humors, and causeth the same to be easily spitte out. F

The iuyce of the roote mingled with honie, and snifte or drawen vp into the nose, purgeth the brayne from superfluous moystures, and openeth the stoppings of the nose. G

3. The Marshe Marigolde, is not vsed in Physicke. H

Of Peruincle. Chap. xxi.

Clematis Daphnoides.

❧ The Description.

Peruincle hath many small & slender long branches with ioyntes, wherby it spreadeth abroade vppon the ground, creeping & trayling hither and thither. The leaues be greater thā the leaues of Boxe, muche like to Bay leaues in colour & fasshion, sauing that they be far smaller. The floure most cōmonly is blew, & sometimes white, & tawnie, but very seldome: it is parted into fiue leaues, somewhat like the floure of great Buglosse, but larger & pleasanter to beholde, yet without sauour. The roote is hearie and yellow.

❧ The Place.

Peruincle groweth wel, in shadowy, moyst

moyst places, as in the borders of wooddes, and alongst by hedges.

※ *The Tyme.*

It floureth most commonly in March and Aprill, but it remayneth greene all the yeare.

※ *The Names.*

It is called in Greeke κληματίς δαφνοειδὲς: in Latine Clematis Daphnoides: Plinie in a certaine place nameth it Clematis Aegyptia: & in an other place Chamædaphne: in shoppes Peruinca, and Vinca peruinca: in Italian Prouenqua, in Spanish Peruinqua: in English Peruincle: in French Peruenche, and du Lisseron. in high Douch Ingruen, & Syngruen: in base Almaigne Vincoorde, Ingroen, and Maechden palm.

※ *The Nature.*

Peruincle is dry and astringent.

※ *The Vertues.*

The decoction of this herbe sodde in wine, and dronken, stoppeth the laske, and the bloudy flixe: it stayeth the immoderate course of the floures, spitting of bloud, and all other fluxe of bloud. A

The same mengled with milke, and oyle of Roses, & put into the Matrix, in a pessarie or Mother suppositorye, taketh away the paynes of the same. B

The same chewed healeth the tooth-ache, & al stinging of venemouse beasts, if it be applied thereto. C

The same bruised and put into the nose, stoppeth nose bleeding. D

Of Bastarde Saffron. Chap.xxij.

※ *The Description.*

Cnicus. Carthamus.

Wilde Saffron hath a rounde stalke of three Cubites long or more, decked with lóg, narrow, dented & sharp pricking leaues: at the toppe of the braunches, are small round prickley heades or knoppes, the whiche at their opening, do bring forth a pleasant Orenge colour floure, of a good sauour, & colour like to the threds of right Saffron: whan the floure is withered and past, there is found within the prickly heades or knoppes, a white long cornered seede, wrapped in a certayne hearie downe, or chaffe.

※ *The Place.*

They vse to plante it in gardens.

※ *The Tyme.*

It floureth in July and August.

※ *The Names.*

It is called in Greeke κνίκος: in Latine Cnicus: of the Apothecaries, and of Mesue, & of Serapio, Cartamus: of some Crocus Hortēsis, & Crocus Saracenicus: in Italian Saffrano Sarracinesco: in Spanish Alaçor, Açafran del huerto, y semente de Papagaios: in English Bastard Saffron: in Frēch Saffran sauuage, or Bastard: in high Douch

Douch Wilden garten Saffron: in base Almaigne Wilden Saffraen.

❀ The Nature.

The seede of Bastarde Saffron (as Mesue writeth) is hoate in the first degree, and dry in the second.

❀ The Vertues.

The iuyce of the seede of Saffron bruised and pound, and dronken with Honied water, or the brothe of a Chicken or pullet, prouoketh the stoole, and purgeth by siege slymie fleumes, and sharpe humors: Moreouer it is good against the Colike, that is to say, the payne, and stopping of the bowels or guttes, and also against the payne in fetching of breath, the cough, & stopping of the breast, and it is singuler against the Dropsie.

Also the iuyce of the same seede put into milke, causeth the same milke to congeale and crudde, and maketh it of great force, to lose and open the belly.

The floures dronke with Honied water, openeth the Liuer, and are very good against the Iaundise. Also the same floures are very good to be vsed in meates to giue them a yellow colour.

❀ The Daunger.

The seede of Bastard Saffron is very hurtfull to the stomacke, causing a desire to vomite, and is of harde and slowe operation, remayning long in the stomake and entrailles.

❀ The Amendement.

Ye must put to the same seede, somethings comfortable to the stomake, as Anise seede, Galangall, or Mastike, or some other good thing to hasten his operation, as Gynger, Sal gemme, common salte, &c. And if it be vsed after this manner, it shall not hurte the stomacke at all, and his operation shall be more speedy.

Of Conyza, or Flebane. Chap. xriij.

❀ The Kindes.

There are two sortes of Conyza, as Dioscorides & Theophrastus writeth: The one called the great or male Conyza: the other the small or female Conyza: Ouer and bysides these, there is a thirde kinde, the which is called the middle or meane Conyza.

❀ The Description.

1. The greate Conyza hath leaues somewhat large, almost like Cowslippe leaues, sauing that they are browner and softer. The stalke is round, couered with a safte Cotton or fine Downe, of a foote and halfe long or more, towardes the toppe spreading abroade into many small branches, vpon which groweth long buddes whiche turne into yellow floures, the whiche also do afterward chaunge into Downie heads, fleeing away with the wind. The roote is somewhat thicke.

2. The small Conyza groweth not aboue the heigth of a spanne, or foote, and differeth not from the first, sauing that it is a great deale lesse. The floures be of a darke yellow, almost like the floures of Tansie, or like to the middell of the floures of Cammomill: they are both of a strong sauour, but the sauour of the greater is more then the small.

3. The third and middell kinde of Conyza, hath a round white wollish stalke, of a foote and a halfe long, the leaues be long & cottony, or wolly. The floures at the top of the stalke, like to Cammomill, but greater, & not onely of a broune yellow colour in the middell, but also round about.

Conyza

the Historie of Plantes. 35

Conyza maior.
Great Conyza.

Conyza media.
Middell Conyza.

❧ *The Place.*

The great Conyza, for the most parte groweth in dry places. The two others grow in valleys, that are moyst and grassie, and by water sides.

❧ *The Tyme.*

They floure in the end of July and August.

❧ *The Names.*

1 These herbes are called in Greeke κονύζαι: Plinie in some place calleth them Cunilagines: Theodor Gaza calleth them Policariæ, and Pulicariæ: vnknowen in shops: one kinde of it is called in English Flebane: some call it in high Douch Durwurtz, and Donnerwurtz: in Spanish *Attadegua*.

2 Theophrast calleth the great, Conyza the male: and the smaller Conyza the female.

❧ *The Nature.*

The great and the small Conyzæ, are hoate and dry in the third degree. The third is of the like substaunce, but not so hoate.

❧ *The Vertues.*

The leaues and floures of Conyza boyled in wine and dronken, haue great A power to prouoke the floures, and to expell the dead childe.

They haue also great power against the hoate pisse, and Strangury, against B the Jaundise, and the gnawing or gryping paynes of the belly.

The same taken with vineger is good for the Epilepsie, or falling sicknesse. C

The Decoction of Conyza is very profitable to women against the diseases D
and

The first Booke of

and payne of the Mother, if they sitte ouer it in a close vessell or stewe.

The leaues bruised and layde vpon the bitings, or stingings of venemouse beasts, are very good: also they are good to be layde vpō woundes & œdemes, that is, harde lumpes or colde swellings.

The same mingled with oyle, is good to annoynt the body, to take away all colde shakings and bruisings.

The same layde strowed or burned in any place, driueth away al venemouse beasts, and killeth gnattes and flees.

Of Sterrewurte or Sharewurte.   Chap.xxiiij.

❧ *The Description.*   After Atticus.

1 STerrewurte hath a browne, hearie, and woodish stalke, the leaues be lōg, thick, hearie, and of a browne, or swartgreene colour. At the toppe of the branches groweth three or foure shining floures, after the fashion of Camomill, yellow in the middle and set rounde about with small purple leaues, in order and fashion like a Sterre, whiche at length do turne into downe, or Cotton, & the pluine is carried away with the winde. The roote is bearded with hearie strings.

2 There is an other kinde of this herbe whose floures are not onely yellow in the middle, but the small leaues also growing about the edges in order like the Camomill floure, are also of yellow colour, but otherwise like to the first.

❧ *The Place.*

Sterrewurte groweth vpon small hillockes, barrowes, or knappes, in Mountaynes and high places, and sometimes in wooddes, and in certaine medowes, lying about the riuer of Rheyne.

❧ *The Tyme.*

It doth most cōmonly floure in August.

❧ *The Names.*

This herbe is called in Greeke Ἀστὴρ ἀττικός, καὶ βούβωνιον : in Latine After Atticus, & Inguinalis : of Vergill Flos Amelli : of some Stellaria : in Italian *Alibio* : vnknowen in Shoppes : in English Sharewurte or Sterrewurte : in French *Aspergoutte menue, or Estoille* : in high Douch Megerkraut, Scartenkraut, and Sternkraut : in base Almaigne Sterrecruyt.

❧ *The Nature.*

It doth refresh and coole, and is almost of temperature like the Rose.

❧ *The Vertues.*

A It is very good against the ouer much heate and burning of the stomake, being layde to outwardly, vpon the same: And being greene stamped, and layd to the botches or impostumes, about the share or priuie members preuayleth much against the same.

B It helpeth and swageth the rednesse and inflammation of the eyes, and fundament

dament oʒ siege, and the falling downe of the Arſe gutte.

The blew of the floure, dʒonken in water is good to be giuen to yong chil-
dʒen, againſt the Squinancie, and the falling ſickneſſe.

Some men ſay that this herbe putteth away all tumoʒs & ſwellings of the
ſiege, share, and fundament, yea whan it is but onely carried about a man.

Of Pennywurte. Chap.xxv.

❧ The Kyndes.

WE ſhall deſcribe in this Chapter, three ſortes of Penniewurte, oʒ Cotyledon: wherof two kindes were well knowen of the Auncients, as they be alſo in many countries, at this day: The thirde, bycauſe of a certayne ſimilitude oʒ likeneſſe that it hath with Pennywurte of the wall, we do call water Pennywurte.

Cotyledon vera.
Wall Pennywurte.

Cotyledon altera Matthioli.
Thicke Pennywurte.

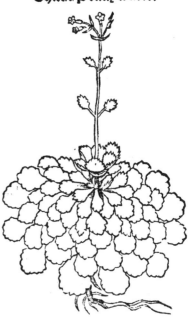

Cotyledon aquatica.
Water Pennywurte.

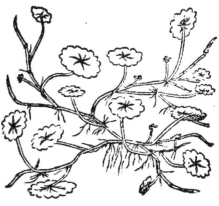

❧ The Deſcription.

1 The leafe of the firſt kind of Pennywurte, is rounde and thicke, much lyke to Iuie leaues, but rounder, & ſomewhat bluntly indēted about, with ſome hollownes oʒ concauitie aboue, & a ſhoʒte ſtem vnderneath in the middell of the leafe. The ſtalke is ſmall and hollow, aboute a ſpanne long, with diuers littell long floures, of a whitiſhe oʒ incarnate colour. The roote is white, and rounde, like an Olyue.

2 The ſecond kinde hath broade thicke and ſomewhat rounde leaues, ſpread abroade,

abroade, round about the stalke like to Syngreene or Houslike, from the middell whereof, springeth vp the tender stalke, bearing small floures.

3 Water Pennywurte hath littell smothe leaues, rounde and hollow aboue, but not very much, euen as it were a small shollow plate, the stem is vnderneth in the middest of the leafe, somewhat drawing towardes the proportion of Wall Pennywurte, but it is smaller, smother and of a swarter colour, and and somewhat deeper natched or dented, but yet bluntly also. The floures be very small and white, and grow beneth, or also vnder the leaues. The rootes be smal and hearie, creeping and putting forth vpon euery side many smal yong leaues.

❧ The Place.

1 Pennywurte, as Plinie saith, groweth in stonie places neare the Sea: but it groweth not in many coūtreys, except it be planted or set in gardens. It groweth pietifully in some parts of England, in Sommersetshyre, & about Welles.

2 Mountayne or Syngreene Pennywurte, is a rare plante, it groweth in some places of the Alpes and other mountaynes beyond the Sea.

3 Pennywurte of the water groweth plentifully in this countrey, in low medowes, and moyst valeys, whereas water standeth in the winter.

❧ The Tyme.

Wall Pennywurte, floureth in May & June, but Pennywurte of the water floureth in July.

❧ The Names.

1 This herbe is called in Greeke κοτυληδών: in Latine Cotyledon, and Vmbilicus veneris, and Acetabulum. And of Plinie Herba Coxendicum. Iacobus de Manlijs in Luminari maiori, calleth it Scatum Cœli & Scatum cellus: in Italian Ombilico di venere, Cupertoiule. in Spanish Scudetes, Coucillos, Capadella, Ombligo de venus: in English great Pennywurte, and wall Pennywurte: in French Nombril de venus: in base Almaigne Nauelcrupt.

2 The second is called in Greeke κυμβάλιον καὶ κοτυληδών ἑτέρα: in Latine Cymbalium, Acetabulum alterum, & Vmbilicus veneris alter: in base Almaigne Dat ander, or dat tweede Nauelcrupt: in English, the second Pennywurte: and Mountayne Pennywurte.

3 Pennywurte of the water, is called in the shops of this countrey, Vmbilicus Veneris & Scatū cœli, although it is not the right kinde, as is beforesayd: ye base Almaignes do call it Penntinckcrupt: in English Sheepe killing Pennygrasse.

❧ The Nature.

The wall Pennywurte, which is the right kinde, is cold & moyst: the Pennywurte of the water, is not without heate as may be perceiued by the taste.

❧ The Vertues.

1 The iuyce of Pennywurte of the wall, is a singular remedy against all inflāmation, and hoate tumors, S. Anthonies fire, & kybed heeles to be annoynted therewithall: and being applied to the stomacke it refressheth the same. A

The leaues and roote eaten, do breake the stone, prouoke vrine, & are good against the Dropsie. B

2 The second kinde is of vertue like to the great Syngreene, or Houselike. C

The vertue of the water Pennywurte, or Peny grasse is not yet knowen: albeit the ignorant Apothecaries do dayly vse it in steede of ye right Cotyledon, wherein they do naught, and commit manifest errour, for the right Cotyledon is the great Pennywurte, called of some Pennywurte of the wall, bycause it groweth euer in old walles & stonie places. But this groweth in low grouds and Marishes, and is a hurtefull herbe vnto Sheepe. D

the Historie of Plantes.

Of Orpyne. Chap. xrvi.
❧ *The Description.*

ORpyne hath a roũd grosse brittell stem, set full of thicke leaues, grosse & full of sappe & somwhat dented about the edges. At the top of ye stalke groweth many fayre purple floures, of fashion like the floures of S. Johns wurte, called in Greeke Hypericum. The roote is white and very knobby, or knottie.

There is a kinde of this herbe whose floures are white: and also a thirde kinde whose floures are yellow, the residue is agreeable to the first. ❧ *The Place.*

Orpyne proueth wel in moyst shadowy places. The people of the countrey delight much to set it in pots & shelles on Midsomer Eue, or vpõ timber slattes or trenchers dawbed with Clay, & so to set, or hang it vp in their houses, where as it remayneth greene a long season and groweth, if it be somtimes ouer sprinckled with water.

❧ *The Tyme.*

It floureth most commonly in August.

❧ *The Names.*

They do now call this herbe Crassula maior, some call it Fabaria, & Faba crassa: in English Orpyne, & Liblong, or Liuelõg: in French Orpin, & Chicotrin. In high Douch Wundkraut, knabenkraut, Fotzlwang, and Fotzwein: in base Almaigne Wondencrupt, and Smeerwortele.

❧ *The Degree or Nature.*

Orpyne cooleth in the thirde degree.

❧ *The Vertues.*

Orpyne in operation & vertue is like to Houselike or Syngreene.

Of Eyebright. Chap. xxvij.
❧ *The Description.*

1 EYebright is a proper small low herbe, not aboue a span long, ful of branches, couered wt little blackish leaues, dẽted or snipt roũd about like a saw: the floures be small and white, sprincled & poudered within, with yellow and purple speckes. The roote is littell, small and hearie.

2 There is yet an other herbe, whiche some do call Eyebright (although it be not the right Eyebright): it groweth to the heygth of a foote or more: The stalkes

Crassula maior.

Eufrasia.

stalkes be round, parted into many collaterall or side braunches, vpon whiche are littell small leaues, long and narrow, most commonly bending or hanging downwards. The floures be redde: The roote is smal as the other Eyebright roote. This I thought necessary to declare, to the intent that men may learne to know the diuersitie betwixt them both, & that they shoulde not take the one for the other: for this last kinde hath not the vertue of the true Eyebright.

❧ The Place.

Eyebright groweth in dry medowes, greene & grassie wayes, and pastures standing against the Sunne. ❧ The Tyme.

Eyebright beginneth to floure in August, and floureth still vntill September, and in forwarde yeares, it is found to floure in July. It must be gathered and dryed whiles it is in floure. ❧ The Names.

Some call this herbe in Latine Euphrasia: ὀφθαλμική, Ophthalmica & Ocularis: some ἐυφροσύνη, Euphrosyne: in English Eyebright: in Frēch Euphrase: in high Douch Augentrost: in base Almaigne Ooghentroost, that is to say, in Latine Oculorum solamen. ❧ The Degree or Nature.

It is hoate and dry, almost in the second degree.
❧ The Vertues.

1 Eyebright pound and layde vpon the eyes, or the iuyce thereof with wine dropped into the eyes, taketh away the darknesse of the same, & cleareth ye sight. **A**

So doth a powder made of three partes of Eyebright dried, and one parte of Macis, if a sponefull of it be taken euery morning by it selfe, or with sugar, or wine. And taken after the same sorte, it comforteth the memory very much. **B**

Eyebright boyled in wine and dronken is good against the Iaundice. **C**

2 That other Eyebright is vnprofitable, and therfore not vsed in Physicke. **D**

❧ Of Filipendula or Dropworte.   Chap. xxviij.
❧ The Description.

Filipēdula.

Filipēdula hath lōg leaues, spread abrode like feathers, made of many smal & little leaues, al dēted, snipte, & iagged roūd about, growing by a lōg string or smal stem, not much vnlike the leaues of wild Tāsey, or Burnet, but lōger, his stalke is round, about the height of two or three foote, at the top whereof are many faire white floures, euery one parted in sixe small leaues, like a little Sterre. The seede is smal, & groweth togither like a button. The rootes be small & blacke, whereon is hāging certaine small knops or blacke pellets, as in the rootes of the female Pionye, sauing ye they be a great deale smaller. ❧ The Place.

Filipēdula groweth in Almayne, Fraūce & England vpon stony moūtaines & rough places. It is also plāted in diuers gardens.
❧ The Tyme.

It floureth in May, June, and July.
❧ The Names.

Som cal this herb in latin Saxifraga rubea: in shops Filipēdula, or Philipēdula: in Italiā & Spanish Filipendola: in french Filipende, or Filipendule:

Filipendule: in high Almaigne Rotsteinbrech, & wilde Garben: in base Almaigne Roode steenbreeck: in English Filipendula, Dropwurte, & Redde Saxifrage.

The Nature or temperament.

Dropwurte is hoate and dry, but not full out in the thirde degree.

The Vertues.

The roote boyled in wine and dronken is good against the Droppisse, or Strangury, and against all the paynes of the bladder, it causeth one to make water, and breaketh the stone.

The same (as Mathew Syluaticus, & Symon Genuensis do write) is very profitable against the diseases springing of colde, windinesse, and blastings of the stomacke, to be made in powder, and taken in wine with Fenell seede.

If the pouder of the roote of Filipendula or Dropwurte, be often vsed to be taken or eaten with meate, it will preserue a man from the falling sicknesse.

Of Medewurte, or Goates bearde. Chap. xxix.

The Description.

Barba Capri siue Vlmaria.

MEdesweete or Medewurte which is called in Latine Vlmaria, and Barba Capri, hath great long brode leaues like Egrimonie, sauing they be larger and longer, rough, boysteous and harde, crompled, and wrinckled, like to the leaues of Byrche or Elme trees. The stalke is hollow, square, & reddish, sometimes as long as a man, and beareth at the toppe a great many of small floures, clustering & growing togither like the blowing of Filipendula, of colour white and sauour pleasant, the whiche do chaunge or turne into small seedes, whiche be as they were wrenched or writhen about, and grow three or foure togither, like to a little warte. The roote is long & blacke without, and browne-red or incarnate within, of a strong sauour & astringent taste, like Ake-kernels.

The Place.

It groweth in medowes, and mosty groundes, also in shadowie wooddes.

The Tyme.

This herbe floureth most commonly in July and August.

The Names.

This herbe is called in Latine Barba Capri, Vlmaria, and Regina prati: in English Medewurte, and Medesweete, and of some after the Latine name Goates bearde: in French *Barbe de Cheure*: in Douche Reynette, and grooten Gheytenbaert.

The Nature.

Medewurte doubtlesse dryeth much, and is astringent, wherefore it restrayneth, and bindeth manifestly.

The Vertues.

The rootes of Medesweete boyled, or made into pouder, and dronken, stoppeth

42　　　　　　　　　The first Booke of

peth the laſke, and all iſſue of bloud.

The floures boyled in white wine and dronken, cureth the feuer Quartayne.

Of Thalietron or Baſtard Rewbarbe.　　Chap. xxx.

✤ The Kyndes.

Of the falſe & Baſtard Rewbarbes, there are at ye leaſt foure or fiue kindes, and of them ſome be great, and one is ſmall.

Thalietrum magnum.	Thalietrum paruum.
The great Baſtard Rewbarbe.	The ſmall Baſtard Rewbarbe.

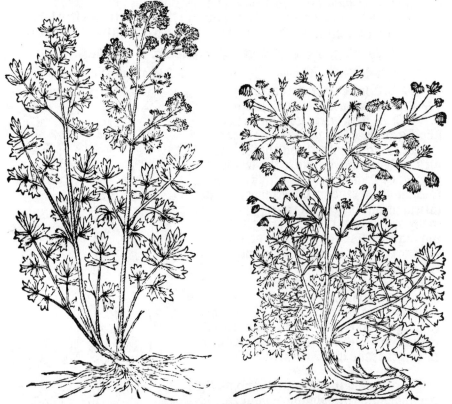

✤ The Description.

1. The firſt great Thalietron or Baſtard Rewbarbe hath large leaues parted, or diuided into diuers others, ſomwhat nickt, or dented about the edges: the ſtalkes are ſtraked and creſted, of a redde purpliſh colour: in the toppes of theſe ſtalkes groweth many ſmall and hearie white floures: after them commeth ſmall narrow huſkes like coddes, foure or fiue growing togither: the roote is yellow, long, round, and knotty, and it groweth farre abroade in many places. The colour of the vpper parte of the leafe, is a browne greene or deepe greene, and ſome are more darker and blacker than ſome, but vnder they are of a lighter colour.

2. The ſecond kinde of great Thalietron or Baſtard Rewbarbe his leaues be of a blewiſh greene colour, his floures be yellow, and his ſtalkes longer, & the ſauour more grieuous: but otherwiſe it is like to the afoꝛſayde.

3. The thirde is very well like to the firſt, ſauing that his ſmall floures are of a light

the Hiſtorie of Plantes.

a light blew colour.

4 The ſmall Thalietron is like vnto the abouesayde, but in all reſpects leſſe, his ſtalkes be of a ſpanne long, his leaues be thinne & tender, & the rootes are ſmall & ſlender, the little ſloures grow togither in ſmall bundels or tuftes, of a light yellow colour almoſt white: and it is alſo of a very grieuous ſauour.

❊ The Place.

1 The firſt kinde oftentimes groweth in moyſt medowes, & it is alſo founde in gardens.

2.3 But that whiche hath the yellow, and violet colour floures, are brought to vs as ſtraungers, as that kinde alſo is with the blackiſh greene leaues.

4 The ſmal kinde is found in Zealand, & other coaſtes bordering vpõ the ſea.

❊ The Tyme.

They floure moſt commonly in July, and Auguſt.

❊ The Names.

In certayne Apothecaries ſhoppes they call this kinde of herbe Pigamum, and do erroniouſly vſe it for Rue, which is called in Greeke Peganon: The common ſorte call it Rhabarbarum, and therefore it is called falſe or Baſtard Rewbarbe: but many learned men call it in Greeke θαλίκτρον, in Latine Thalietrum, and do vſe it for the ſame.

4 But the ſmal Thalietrum, is not Hypecoon, as we haue thought it earthis.

❊ The Nature.

Baſtard Rewbarbe is of complexion hoate and dry.

❊ The Vertues.

The leaues of Baſtard Rewbarbe, taken in meate or otherwiſe loſeth the belly.

The rootes alſo ſhould ſeeme to be of the ſame nature and vertue: and for this conſideration partly they were called Rewbarbe, & partely alſo they were ſo called, bycauſe their rootes are yellow like Rewbarbe.

Scrophularia maior. A
B

Of water Betony, or Brounewurte. Chap. xxxj.

❊ The Description.

1 Brounewurte hath a ſquare, browne, hollow ſtalke, large leaues, natched or dented rounde about, very like vnto Nettell leaues, but ſmother or playner, and nothing ſtinging or burning at all. The floures grow about the toppe of the ſtalkes, and are ſmall and tawney, hollow like a helmet, or a ſnayle ſhell. The ſeede is ſmall rounde, poynted like to ſome prety pellots or buttons. The roote is white and knobby, like the roote of Orpyn or Lyblong, wherof we haue ſpoken Chap. 26.

2 There is an other kinde of this herbe, like to the firſt, in ſtalkes, leaues, floures, and huſkes, or ſeede veſſelles, but it differeth in the roote: for his roote is not knobby or ſwollen like to the other, but full of threddiſh ſtrings: otherwiſe there is no difference betwixt this kinde and the other, which they call Scrophularia maior: for ỹ ſtalke is alſo ſquare, and the leaues like to Nettell leaues, and are cut, & dented round about in like manner: the floures are like to open helmets alſo, &c. ſo that oftentimes, thoſe

D iiij   that

that take not hede to the differéce in the rootes, do gather the one for the other.

3 There is yet a thirde kinde which is nothing like to the others, sauing only in the floures and seede, wherein it is very like to the other Scrophularies: wherefore wee haue thought good to make mention of it in this place: his stalke is right, or straight and rounde. The leaues are like to Roquet leaues, but smaller and browner. The floures are like to them aforesayde, sauing they be smaller and of a blewe colour, straked with small strakes of white. The roote is threddy, like the roote of the second kinde of Scrophularia, and is euerlasting, putting forth yearely new springs, as also doth the rootes of the other two Scrophularies.

❀ *The Place.*

The two firste kindes do grow very plentifully in this countrey, in the borders of fieldes, and vnder hedges, and about lakes and ditches.

The thirde is not found here, but onely planted in gardens.

❀ *The Tyme.*

They floure in June and July.

❀ *The Names.*

1 The first is called in Shoppes, and of the Herboristes, Scrophularia maior, & of some Castrangula, Ficaria, Millemorbia, Ferraria: in English Broune wurte, and Water Betony: in high Almaigne Braunwurtz, Sauwurtz, and grosz Feigwartzen kraut: in base Almaigne groot Speencruyt & Helmcruyt. Some thinke it to be the herbe that is called in Greeke γαλίοψις καὶ γαλιόδολοψ: in Latine Galeopsis and Vrtica labeo.

2 The second hath no certayne name in Latine, nor of the Apothecaries: but in base Almaigne it is called Beeckscuym, and S. Anthuenis cruyt: this should be κλύμινορ: Betonica Aquatica Septentrionalium: in English Water Betony.

3 The thirde is vnknowen and without name, notwithstāding it may be taken for a kinde of Galeopsis, bycause his floure is like to an open Helmet.

❀ *The Nature.*

Scrophularia is hoate and dry in the third degree, and of subtill partes.

❀ *The Vertues.*

1 The leaues, stalke, seede, roote, & iuyce of the right Galeopsis, or Brounewurte, doth waste and dissolue al kindes of tumors, swellings, and hardnesse, if it be pound with vinege, rand layde thereupon two or three times a day. A

The leaues stampte and layde to old, rotten, corrupt, spreading and fretting Vlcers or Pockes, doth heale them, it doth also heale Cankers, if it be pound with Salte and layde thereto. B

If a man washe his face with the iuyce of this herbe, it taketh away the rednesse of the same. C

The roote eaten drieth vp and healeth the Hemorrhoides: the like vertue it hath to be pound and layde too outwardly. The seede of Brounewurte dronken killeth wormes. D

2 The second kinde (whiche is the right water Betony) is also very good against all corrupt vlcers and consuming sores, being layde too, as the first. E

3 The third is not onely vnknowen in name, but also in vertues. F

Of Herbe Roberte, Pynke needle, and Storkes bill, with other of the same kinde. Chap.xxxij.

❀ *The Kindes.*

THere is found in this contrey diuers sortes of herbes, whose seedes be long & sharpe like to a Hearons beake or byl, the which for the self same cause, are

all

the Historie of Plantes.

all comprehended vnder the name and kindes of Hearons bill. The twoo first are described by Dioscorides, and other of the auncient wryters: The fiue other are set foorth by the later wryters, and learned men of our time.

❦ *The Description.*

1 The first kinde of Geranion or Storckes bill, his leaues are cut and iagged in many peeces, like to Crowfoote, his stalkes be slender, and parted into sundry braunches, vpon which groweth smal floures somwhat like roses, or the floures of Mallowes, of a light murrey or redde colour: after them commeth little round heades, with smal long billes, like Nedels, or like the beakes of Cranes and Hearons, wherein the seede is contayned: The roote is thicke, round, shorte, and knobby, with certayne small strings hanging by it.

Geranium alterum.
Doue foote.

Geranium tertium.
Storckes bill, or Acus Moschata.

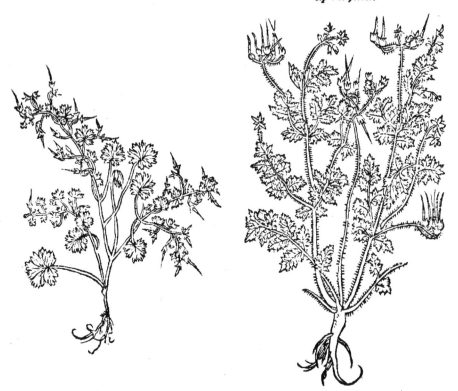

2 The seconde whiche they call Doue foote, hath also smal, tender, hearie, and browne stalkes: the leaues are like to the small Mallow, cut rounde about. The floures be small, of a cleare purple colour, and do likewise turne into little knappes, or heads, with billes, but yet not so great & long as the first Geraniũ.

3 The thirde kinde also hath tender stalkes, rounde, and somewhat hearie, small leaues, cut as it were in little iagges or peeces, and before the growing vp of the stalkes, the leaues lie spreading vpõ the ground: the floures are smal, of a pleasant light redde: after these floures followeth certayne small narrow peakes or beakes as in the others: The roote is white, of the length of a finger like to Rampions.

Sideritis

Sideritis tertia, aut
Geranium Robertianum.
Herbe Roberte.

Geranium gruinale.
The fourth Cranes bill.

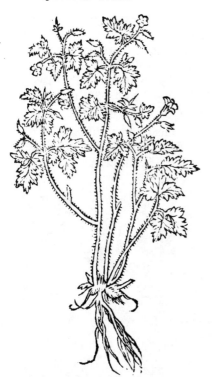 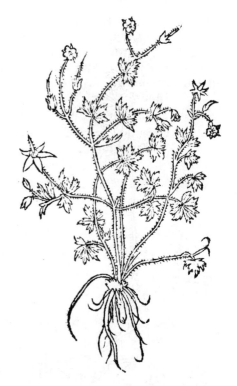

4. The fourth hath hearie stalkes like the other, but all redde, with diuers ioyntes and knots, the leaues are much cut and iagged, like to Cheruill, or Coriander leaues, but redder & of a more lothsome smell. The floures be redde, and bringeth forth small bullets like littell heades, with sharpe billes. The roote is somewhat greene of colour.

5. The fifth is like to the aforesayde, in his hearie stalkes, redde floures, and sharpe billes, sauing that his leaues are much more, and deeper cut, and his floures be somewhat greater.

6. The sixth is like the fourth, in small, weake, tender, heary stalkes, in leaues deepely cut, in floures, and braunches, sauing that the stalkes of the fifth kinde do grow longer and higher, the leaues be greater, and the floures larger lyke vnto littell Roses. The roote is long and most comonly all redde and sanguine within.

7. The seuenth hath also long reddish, hearie stalkes, and great leaues, lyke Crowfoote, but larger, his floures are blew, after whiche there commeth forth small beckes or billes, as in the other kyndes. The roote is thicke & long with many small strings.

✤ *The Place.*

1. 2. These herbes do grow of themselues, in barren sandy groundes, by high
3. 4. way sides, and borders of fieldes. Herbe Roberte likewise groweth about olde walles, and olde tyled, or stone healed houses.

5. 6. The twoo last kindes are not found in this countrey, sauing in gardens where as they be planted.

Geranium

the Historie of Plantes. 47

Geranium hæmatites.
Sanguin Geranium, or Blood Roote.

Geranium, batrachiodes.
Gratia Dei, or **Baſſinet** Geranium, or **Crowfoote** Geraniũ.

 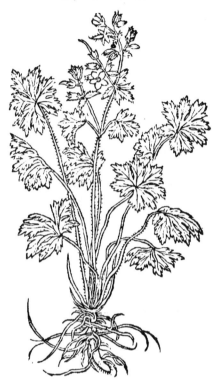

❧ *The Tyme.*

They floure moſt commonly in May and June, and ſometimes alſo in Aprill, eſpecially the firſt kinde. ❧ *The Names.*

All theſe herbes are called by one Greeke name γεράνιον, that is to ſay, in Latin Geranium, Gruina, or Gruinalis: in Italian *Roſtro di Grua*: in Spaniſh *Pico de Cigueña, Aguyas pampillos.*

1. The firſt kinde is called Geraniũ tuberoſum, Acus paſtoris, & Acus Moſchata: and Geranium ſupinum: in Engliſh **Storkes byll, Pinkeneedell,** and of ſome **Moſchata:** in high Almaigne **Storkenſnabel:** in French *Bet de grue:* in baſe Almaigne **Oyeuaerttſbeck, or Cranenbeck.**

2. The ſecond is called Geranium alterum, Geranium Columbinum, and Pes Columbæ: in Engliſh **Doue foote:** in French *Pied de Pigeon:* in high Douch **Daubenfuſz:** in baſe Almaigne **Duyuenuoet.**

3. The third is called in ſhoppes Roſtrum Ciconiæ, and Geranium ſupinum: in Engliſh **Hearons bill, or Storkes bill:** in high Douch **Storkenſchnabel:** in baſe Almaigne **Oyeuaerſbeck, or Cranenbeck.**

4. The fourth kinde of theſe herbs, is a kinde of Siderits of the Auncients, & is called of Dioſcorides Siderits tertia, and Siderits Heraclea: now they call it Ruberta, Herba Roberti, & Robertiana, & Geranium Robertianũ: in Engliſh **Herbe Robert:** in French *Herbe Robert.* in high Douch **Rubrechtzkraut, Schartenkraut,** and of ſome klein **Scholwurtz,** in baſe Almaigne **Robrechts cruyt.**

The

48 The first Booke of

5 The fifth is called Gruinalis, & Geranium gruinale: in English Cranes bill: in high Douch kranichhals: in base Almaigne Craenhals.

6 The sixth is called in high Douch Blutwurtz: in base Almaigne Bloet wortele, that is to say, the Sanguine roote, or Bloud roote: and Geranium Hæmatodes, for the same cause.

7 The seuenth is called Gratia Dei: in English also Gratia Dei: Bassinet Geranium, and Crocfoote Geranium: in high Douche Gottes gnad, that is to say, the Grace of God: in base Almaigne Godts ghenade, and blauw Booterbloemen, and Geranium batrachiodes. ❧ *The Nature.*

The most part of these herbes, are of a drying temperature, some also are clensing, & haue power to ioyne togither or soulder, but it is not much vsed to that purpose. ❧ *The Vertues.*

1 The roote of the first taken in wine, driueth away and healeth al blastings, and windinesse of the Matrix or Mother, it prouoketh vryne, and is very good for them that haue the stone.

2 The second (as ye Auncients say) is not good in Medicyne. Notwithstãding at this time, it is much vsed against al woundes, & vlcers, being layd thervnto.

3 Herbe Roberte doth stanche the bloud of greene woundes, to be bruised and layde thereto, as Dioscorides saith.

The same herbe (as hath bene proued sithence Dioscorides tune) is singuler against the sores & vlcers of the Pappes, & the priuie mẽbers, especially of men, if it be pounde & layde thervnto, or if the iuyce therof be dropped or poured in.

The decoction of Herbe Roberte cureth the corrupt vlcers, and rotten sores of the mouth, and amendeth the stinking of the same.

The rest are not vsed in medicine.

Of Sea Trifoly and Mylkewurte.   Chap.xxxiij.

❧ *The Kindes.*

There be two kyndes of Mylkewurte, differing both in name and figure: whereof one is called Glaux, and the other Polygala.

Glaux.   Polygala.   Milkewurte.

Milkewurte, or Sea Tryfoly.

*The

the Historie of Plantes.

❧ The Description.

1. The first Milkewurte hath many smal stemmes, cōming forth of one roote, the sayde stalkes be weake and tender, & of halfe a foote high, vpon which groweth small long leaues, like the smallest leaues of Lentilles, and are whiter vnderneth the leafe than aboue. The floures amongst the leaues, are like to Gillofloures, but smaller, of colour purple and incarnate. The roote is smal, full of hearie threedes, and creeping alongst the grounde.

2. The second kinde of Milkewurte called in Latine Poligala, is a small herbe, with slender pliant stemmes of wooddy substance, as long as a mans hande creeping by the ground, the leaues be small and narrow, like the leaues of Lentill or small Hysope. The floures grow somwhat thicke about the stemmes, not much differing from the floures of Fumitory, in figure, and quātitie, sometimes tawney, sometimes blewe, and sometimes white as snow, without smell or sauour, after whiche floures, there commeth small coddes, or purses, like to them of Bursa Pastoris, but smaller, and couered by euery side with small leaues, like littell winges. The roote is slender and of wooddy substance.

❧ The Place.

1. The first Milkewurte groweth in lowe salte marshes, and watery places nigh the Sea thoroughout all Zealand.

2. The second groweth in Dry Heathes, and commons, by the high way sides.

❧ The Tyme.

1. Glaux floureth in June and July.

2. Polygala floureth in May about the Rogation, or Gang wecke, the which the Almaignes call Cruysedaghen, & therefore they call them Cruysbloemkens, as Tragus that countreyman wryteth.

❧ The Names.

1. The first is called in Greeke γλαῦξ, καὶ γάλαξ, ἢ γλᾶξ: in Latine Glaux, and Glax, that is to say, in English Milkewurte: in Frēch Herbe au laict: in Douch Milchkraut, and Melckcrupt. Turner calleth it Sea Trysoly.

2. The second is called πολύγαλον, Poligala, that is to say, the herbe hauing plenty of milke, by which name it is not knowē, for the Almaynes call it Cruysbloeme.

❧ The Nature.

Both these herbes are hoate and moyst, as Galen sayth.

❧ The Vertues.

A. The firste taken with meats, drinke, or potage, ingendreth plenty of milke: therefore it is good to be vsed of Nurses that lacke milke.

B. The same vertue hath Polygala, taken with his leaues and floures.

Of Pellitory of the Wall. Chap. xxiiij.

❧ The Description.

Pellitory or Paritory hath rounde tender, thorough shining, & browne redde stalkes: the leaues be rough & somwhat broade, like

Helxine, Parietaria.

☾ Mercury

Mercury but nothing snipte or dented about. The floures be small ioyning to the stemme, amongst the leaues. The seede is blacke and very small, couered with a littell rough huske or coate, whiche hangeth faste vpon garments. The roote is somewhat redde.

❧ *The Place.*

It delighteth to growe about hedges, and olde walles, and by way sides.

❧ *The Tyme.*

It floureth most commonly in July.

❧ *The Names.*

This herbe is called in Greeke ἐλξίνη καὶ περδίκιον: in Latine Muralium Perdicium, and Vrceolaris, and of some Parietaria, Muralis, & Perdicalis: in Shoppes Paritaria: in Italian Lauirreola: in Spanish Yerua del muro, Alfahaquilla del muro, Alfahaquilla de culebra: in English Parietary, Pellitory of the Wall: in high Douch, Tag vnd nacht, S. Peters kraut, Glaszkraut, Maurkraut: in base Almaigne Parietarie, and Glascruyt.

❧ *The Nature.*

Parietarie is somewhat colde & moyst, drawing nere to a meane temperature.

❧ *The Vertues.*

Parietorie is singuler against cholerike inflammations, the disease called Ignis sacer, S. Anthonies fyre, spreading and running sores, burnings, and all hoate vlcers, being stamped and layde thereupon. [A]

An oyntment made with the iuyce of this herbe and Ceruse, is very good against all hoate vlcers, spreading and consuming sores, hoate burning, scuruy, and spreading scabbes, and such like impediments. [B]

The same iuyce mengled with Deare sewet, is good to annoynt the feete against that kinde of goute, which they call Podagra. [C]

The same iuyce mengled with oyle of Roses, and dropped into the eares, swageth the paynes of the same. [D]

The decoction or brothe of Parietorie dronken, helpeth suche as are vexed with an olde Cough, the grauell and stone, and is good against the difficultie and stopping of vrine, and that not onely taken inwardly, but also layde to outwardly vpon the region of the Bladder, in maner of a fomentation or a warme bathe. [E]

Of Chickeweede. Chap. xxxv.

❧ *The Kindes.*

ALthough Dioscorides and Plinie, haue written but of one kinde of Alsine, or Chickeweede, neuertheleste a man may finde in most places of this countrey, diuers sortes of herbes comprehended vnder the name of Alsine or Chickeweede, ouer and bysides that whiche is found in salt groundes: whereof the first, and right Alsine is that whiche Dioscorides and the Auncients haue described.

❧ *The Description.*

The great Chickeweede hath sundry vpright, rounde, and knobby stalkes. The leaues growe at euery ioynt or knotte of the stalke, alwayes twoo togither, one directly standing agaynst an other, meetely large, sometimes almost of the breadth of twoo fingers, not much vnlike Parietory leaues, but longer and lesse hearie: about the toppe of the braunches, amongst the leaues groweth small stemmes, with littell knoppes, the whiche chaunge into small white floures diepely cutt and snipte, after the floures yee shall perceyue huskes or Coddes somewhat long and rounde, wherein lieth the seede. The whole herbe dothe not differ much from Parietory, for his stemmes also be

thorough

through ſhyning, and ſomewhat redde about the ioyntes, and the leaues be almoſt of the ſame quantitie: ſo that Dioſcorides ſayth, that this herbe ſhould be Parietory, but that it is ſmaller and baſer or lower, and that the leaues be longer, and not ſo hearie.

Alſine maior.	Alſines ſecundum genus.
Great Chickeweede.	**The ſecond Chickeweede.**

2. The ſecond is like to the great Chickeweede, ſauing that it is ſmaller and groweth not vpright, but lieth and ſpreadeth vpon the ground. The leaues are much ſmaller, growing twoo and twoo togither at euery ioynt. The floures, huſkes and ſeede is like the great Chickeweede. The roote hath many ſmall hearie threddes.

3. The thirde and ſmalleſt Chickeweede, is not much vnlike the ſecond, but a great deale ſmaller in all reſpectes, in ſo much that his ſtemmes be like vnto ſmall threddes, and his leaues no bigger then Tyme, otherwayes it is lyke to the ſecond.

4. The fourth kinde (called of the baſe Almaignes Hoenderbeet) that is to ſay Henbit, hath many rounde & hearie ſtemmes. The leaues be ſomewhat round, hearie, & a little ſnipt or iagged about the edges, otherwiſe not much vnlike the leaues of great Chickeweede. The floures be blew or purple, & do bring forth ſmall cloſe knappes or huſkes, in which is incloſed the ſeede.

5. The fifth kinde is like to the aforeſayd, in his hearie ſtemmes, his leaues be longer & narrower, and iagged rounde about, the floures of a cleare blew, the ſeede is in broade huſkes, as the ſeede of Veronica or Paules Betony.

E ij   Alſines

The first Booke of

Alsines tertium genus.
The third Chickeweede.

Alsines quartum genum.
The fourth Chickeweede.

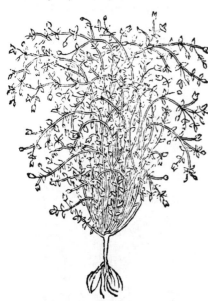

Alsines quintum genus.
The fifth Chickeweede.

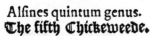

There is yet a sixte kynde of Chickeweede, which groweth onely in Salte ground, like to the others in leaues and knotty stemmes, but chiefly like to the second kinde, sauing that his stemmes are thicker & shorter, and the knots or ioyntes stande nearer one to an other. The leaues are thicker, & the huskes be not long but flat, rounde, and somewhat square or cornered, like a great hasting or gardē pease, euery huske hauing three or foure browne seedes, almost of the quantitie of a vetche.

✻ *The Place.*

The great Chickeweede groweth in moyst shadowy places, in hedges & busshes, amongst other herbes: in such like places ye shall finde the rest, but the sixth groweth not, except onely in salte groundes, by the sea side.

✻ *The Tyme.*

These herbes do most cōmonly floure about Midsomer. ✻ *The Names.*

The great Chickeweede is called in Greeke ἀλσίνη, in Latine Alsine, and of some late writers Hippia maior: in Italian *Panarina*, and *Centone*: vnknowen to the Apothecaries.

The second & third, are called of the Apothecaries Morsus gallinæ, & Hippia minor:

the Historie of Plantes.   53

minor: in English Middle Chickeweede: in high Douch Vogelkraut, and Hunerbitz: in base Almaigne Vogelcruyt, Hoenderbeet, and cleynen Muer.

4 The fourth also is called of some Morsus gallinæ: in high Douch Hunerbitz: in base Almaigne Hoenderbeet: it may also be called in French *Morgeline Bastarde.*

5 The fifth is called of the high Almaignes Huners crb, of the base Almaines Hoender crue, that is to say, the Hennes right, or Hennes inheritance: it is also called in French *Moron Bastard, Moron violet,* and *Oeil de Chat.*

6 The sixth, whiche groweth in salte groundes, wee may call Alsine marina, that is to say, Sea Chickeweede.

※ *The Nature.*

Chickeweede is colde and moyst, in substance much lyke Parietorie, as Galen writeth.

※ *The Vertues.*

1 The great Chickeweede pounde, and layde to the eyes, or the iuyce thereof straked vpon the eyes, is good against inflammations, and the hoate vlcers of the eyes.

The same vsed in manner aforesayde, and layde to the place, is good against B all hoate vlcers, that be harde to cure, but especially those aboute the pryuie partes.

The iuyce thereof dropped into the eares, is good agaynst the payne and C griefe of the same.

2.3 The small Chickeweede, and specially the second kinde, boyled in water and D salte, is a soueraigne remedie against the scuruy heate and itche of the handes, if they be often wasshed or bathed in the same.

Sea Chickeweede, serueth to no knowen vse.   E

Of Mouse eare.   Chap.xxxvi.

※ *The Description.*

1 Mouse eare, (as Dioscorides saith) hath many small and slender stemmes, somewhat redde bylow, about the whiche groweth leaues, alwayes two togither standing one directly against an other, they are small, blackishe, and somewhat long, and sharpe poynted, almost like to the eare of a Mouse or Ratte: betwixte the leaues there groweth forth small braunches, whereupon are blew floures, like the floures of female Pimpernell. The roote is as thicke as a fingar.

2 There is yet an other herbe, whiche some holde for Mouse eare: This is a low herbe most commonly spreading vpon the ground, enuironned & set about with a fine and softe heare, the reste is very like the second Chickeweede, for it hath many hearie stemmes, comming forth of one roote, of a reddishe or tawnie colour bylow. The leaues be long rough, & hearie, much like to a Mouse eare, the small floures be white. The huskes somewhat long, like Chickeweede huskes. The roote is very threddy.

3 Bysides these two there is yet a kinde of Mouse eare, whiche spreadeth or creepeth not vpon the ground, but standeth vpright, growing amongst other herbes, lyke to the others in stemme and leaues, but it is greater and of colour white, couered ouer with a clammy Downe, or Cotton, in handling as though it were bedewed or moystened with Honie, and cleaueth to the fingers. The floures come forth of small knoppes or buttons, as in the second kinde. The Coddes, wherein is the seede, are almost like to the seede vessels of wylde Rose Campion.

E iij   Auricula

The first Booke of

Auricula muris Matthioli.
Mouse eare.

Auricula muris, quibusdam.

❀ *The Place.*

Mouse eare (as Matthiolus writeth) groweth in medowes, and is common in Italy.

The two other kindes grow in this countrey vnder hedges, about the borders of fieldes, and by the way side, as Chickeweede doth.

❀ *The Tyme.*

They floure in June and July.

❀ *The Names.*

Mouse eare is called in Greeke μυὸς ὦτα: in Latine Auricula muris, that is to say, Mouse eare: in Douch Meuszorlin: in base Almaigne Muysooren.

The two others are counted of some for Mouse eare, yet they should seeme rather to be of the kindes of Alsine or Chickeweede.

❀ *The Nature.*

Mouse eare, drieth without any heate.

❀ *The Vertues.*

Mouse eare pounde helpeth much against the fistulas, and vlcers, in the corners of the eyes, to be layde thereto.

A man may finde amongst the wrytings of the Egyptians, that if a body be rubbed in the morning early, before he hath spoken, at the first entrance of the moneth of August, with this herbe, that all the next yeare he shal not be greued with bleared or sore eyes.

Of Pimpernell.   Chap.xxxvij.

❀ *The Kindes.*

There be two sortes of Pimpernell: the one hath redde floures, and is called Male Pimpernell, the other hath blew floures, and is called Female Pimpernell

pernell, but otherwise there is no kinde of difference betwixt them.

Anagallis mas.   Anagallis fœmina.
The male Pimpernell.   The female Pimpernell.

❦ *The Description.*

Both Pimpernelles haue small, tender, square stalkes, with diuers ioyntes, and it spreadeth or creepeth vpon the ground. The leaues be small, like the leaues of Middle Chickeweede, but rounder, and greene aboue, but vnderneth of a grayishe colour, and poudred full of small blacke speckes. The floures of the male kinde be redde, but the floures of the female kinde are of a fayre Azure colour. The seede is contayned in small round littell bollikens, or knoppes, whiche spring vp after the floure.

❦ *The Place.*

It groweth plentifully in tylled fieldes, and also in gardens amongst pot herbes, and euery where by way sides.

❦ *The Tyme.*

It floureth all the Sommer, but most in August.

❦ *The Names.*

This herbe is called in Greeke and in Latine ἀναγαλλίς, and of some (as Plinie sayth) Corchorus: in Spanish Muruges: in English Pimpernell: in French Moron: and that whiche beareth the purple floures is called also in Greeke κοράλλιον, Corallium as Paulus Aegineta in his seuenth Booke writeth: in high Douch Gauchheyl: in base Almaigne Guychelheyl.

❦ *The Nature or temperament.*

Pimpernell is hoate and dry, without any acrimonie, or byting sharpnesse.

❦ *The Vertues.*

Pimpernell boyled in wine and dronken, is singuler against the bytings of venemouse beasts, and against the obstructions, and stopping of the liuer, and the payne and griefe of the kydneys.

The iuyce of Pimpernell snifte into the Nosethrilles, draweth downe from the head phlegmatique and naughtie humors, and openeth the conductes of the Nose: also it healeth the tooth ache, whan it is put into the Nose on the contrary side of the griefe.

Pimpernell layde vpon corrupt and festered vlcers, or fretting sores, dothe clense and heale the same. Also it draweth forth thornes & splinters or shiuers, if it be bruised and layde vpon the place.

It is also very good against the inflammation, or heate of the eye.

The iuyce of the same mingled with Hony, and straked, or often put into the eyes, taketh away the dimnesse of the sight.

It is written of these herbes, that the Pimpernell with the blew floures, doth settell & stay the falling downe of the siege or great gutte: And the other with the redde floure draweth it forth of his place.

Of Francke or Spurry.   Chap.xxxviij.

❧ *The Description.*

Spergula.

SPurry hath round stalkes, with three or foure knots or ioyntes, about the whiche groweth a sorte of very narrow small leaues, compassing the ioynts in fashion of a Starre: at the top of the stalkes it bringeth foorth many small white floures, after them there cometh small pellets or bullets like Line seede, wherein is contayned blacke seede. The roote is slender, and of a finger length.

❧ *The Place.*

Spurry groweth most commonly, in fieldes, whereas they vse to sowe it.

❧ *The Tyme.*

It floureth for the most parte in May & Iune.   ❧ *The Names.*

This herbe is called in Englishe Francke, bicause of the propertie it hath to fat cattell. It is also called in English Spurrie, & so it is in Frenche & Douch: whereof sprang the Latine name Spergula, vnknowen of the Apothecaries, & the oldest writers also, wherfore it hath none other name that is knowen vnto vs.

❧ *The Vertues.*

Spurry is good fourage or fodder for Oxen & Kyen, for it causeth kyen to yeelde store of milke, and therefore it is called of some Polygala, and other properties it hath not, that are as yet knowen.

Of Agrimonie.   Chap.xxxix.

❧ *The Description.*

THe leaues of Agrimonie, are long, & hearie, greene aboue, & somwhat grayish vnder, parted into diuers other small leaues, snipte round about ỹ edges, almost like the leaues of Hemp. The stalke is of two foote & a halfe lõg, or therabouts, rough & hearie, vpon whiche groweth many small yellow floures, one aboue

the Historie of Plantes.

Eupatorium.

aboue an other vpwardes towardes the toppe, after the floures cōmeth the seede somewhat long and rough like to small Burres, hāging downewards, the which being ripe, do hang fast vpon garments, whan one doth but scarsly touche it. The roote is meetely great, long, and blacke.

❁ *The Place.*

Agrimonie groweth in places not tylled, in rough stony moūtaynes, in hedges and Copses, and by way sides.

❁ *The Tyme.*

Agrimonie floureth in Iune, and Iuly. The seede is ripe in August. The Agrimonie that is to be occupied in medicine, must be gathered, and dryed in May.

❁ *The Names.*

Agrimonie is called in Greeke ἐυπατόριον καὶ ἡπατώριον: in Latine Eupatorium, and Hepatorium: in shoppes Agrimonia: of some Ferraria minor, Concordia, and Marmorella: in Spanish *Agramonia*. in English Agrimonie: in French *Eupatoire*, or *Aigremoine*: in high Douch Odermenich, Bruchwurtz: in base Almaigne Agrimonie, & of some Leuercrupt, that is to say, Liuerwurte.

❁ *The Nature.*

Agrimonie is of fine and subtill partes, without any manifest heate, it hath power to cut in sunder, with some astriction.

❁ *The Vertues.*

The Decoction or brothe of Agrimonie dronken, doth clense and open the stoppings of the liuer, and doth strengthen the same, & is specially good against the weakenesse of the same.

Agrimonie boyled in wine and dronken, helpeth against the bytings of venemous beasts: the same boyled in water stoppeth the pissing of bloud.

The seede therof dronken in wine, is singuler against the blouddy flixe and daungerouse laske.

The leaues of Agrimonie pounde with Swines grease, and layde too hoate, doth cure and heale olde woundes, that are harde to close or drawe to a Scarre.

Of Bastarde Agrimonie.   Chap.xl.

❁ *The Kyndes.*

There be sundry kindes of herbes called in Latine Hepatica or Iecoraria, that is to say, Lyuerwurtes, whiche are commended, and founde good agaynst the diseases of the Lyuer, whereof wee shall describe three kindes in this Chapter vnknowen to the old wryters. The two first kindes are Bastarde Agrimonie. The third is Three leaued Agrimonie, or Noble Lyuerwurte.

The first Booke of

Pseudohepatorium mas.
Baſtard Agrimonie the male.

Pseudohepatorium fœmina.
Baſtard Agrimonie female.

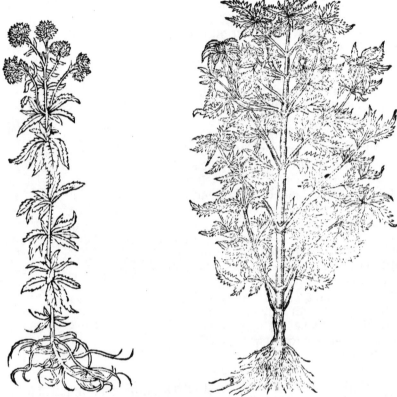

❧ *The Deſcription.*

1. THe male Baſtarde Agrimonie, hath a long round ſtalke, full of white pith within, at the whiche groweth long blackiſh leaues, ſomewhat rough and hearie, ſnipte and cut round about, almoſt like the leaues of Hempe, and bitter. At the toppe of the ſtalkes groweth many ſmall floures, of incarnate colour, cluſtering or growing thicke togither in tuftes, the whiche being withered and chaunged into ſeede, it fleeth away with the winde. The roote is full of threddy ſtrings.

2. The female Baſtard Agrimonie, hath alſo a round purple ſtalke, about three foote long and full of braunches. The leaues be long and dented or ſnipt round about, like the leaues of Hempe or of the other Agrimonie, ſauing that they be a littell larger. At the toppe of the branches, and round about the ſtalke, groweth three or foure ſmall leaues growing harde one by an other after the faſhion of a ſtarre, amongſt whiche is a knap or button that bringeth forth a yellow floure intermengled with blacke, within whiche being withered, is conteyned the ſeede whiche is long, flatte, and rough, and hangeth vpon garments whan it is rype.

3. The leaues of Hepatica are broade, and diuided into three partes, not much vnlike the leaues of Cockow bread, ſower Tryfoly, or Alleluya, but larger. Amongſt the leaues groweth fayre azured or blew floures, euery one growing vpon a ſingle ſtemme, the whiche do change into ſmall bullets or bolyns, wherin the ſeede is conteyned. The roote is blacke and full of ſmall hearie ſtrings.

The

the Historie of Plantes. 59

Hepatica siue Hepaticum Trifolium.
Noble Lyuerwurt, or threeleaued
Lyuerwurte.

❧ *The Place.*

The Bastard Agrimonies do grow in moyst places, by diches, and standing pooles. Hepatica groweth not of his owne kinde in this countrey, but it is planted in gardens.

❧ *The Tyme.*

The Bastard Agrimonies do floure in July and August, but the Noble or great Lyuerwurt floureth in Marche.

❧ *The Names.*

1. The male Bastarde Agrimonie, is called in Shoppes Eupatorium, and is wrongfully taken of them for the right Agrimonie, the which is described in the former chapter. The learned mē in these dayes do call it Pseudohepatorium, and Eupatorium aquaticum, or Adulterinum: Of Baptistus Sardo, Terzola: in highe Douch Kunigundkraut, Wasserdost, & Hirssenclee: in base Almaigne Coninghinne cruyt, Hertsclaueren, and Boelkens cruyt manneken.

2. We haue named the second Pseudohepatorium foemina: in base Almaigne Boelkens cruyt wijfken: it is thought to be that Agrimonie whereof Auicen writeth Chap. cclxiiij. and therfore some haue called it Eupatorium Auicennæ.

3. The third, which is called at this day in Latine Hepatica, and of some Herba Trinitatis: may be called in English Hepatica, Noble Agrimonie, or Three leafe Lyuerwurte: in French *Hepatique*: in high Douch Leberkraut, Edel leuer cruyt. We know of none other name except it be βάλαρις, Balaris, whereof Hesychius writeth.

❧ *The Nature.*

The two Bastard Agrimonies are hoate and dry, as their bitternesse doth manifestly declare. Hepatica doth coole, dry, and strengthen.

❧ *The Vertues.*

1. The male Bastarde Agrimonie boyled in wine or water, is singuler good against the old stoppings of the Lyuer, and Melte or Splene. Also it cureth old feuer tertians, being dronken.

The Decoction thereof dronken, healeth all hurtes, & woundes, for whiche purpose it is very excellent, and to heale all manner woundes both outwarde and inwarde.

2. The female Bastard Agrimonie is of the same operation, and is vsed more than the other in wounde drenches.

3. The Hepatica, or Noble Lyuerwurte, is a soueraigne medicine, against the heate and inflammation of the Lyuer, and all hoate Feuers or agues.

Of Tornesole.   Chap. xli.

❧ *The Kyndes.*

THere be two kindes of Heliotropium or Tornesol: The one called the great Tornesol: and the other the small Tornesol.

Helio-

Heliotropium magnum. Heliotropium paruum.
Great Tornesol. **Small Tornesol.**

❧ *The Description.*

1. The great Tornesol, hath straight round stalkes, couered with a white hearie cotton, especially about the toppe. The leaues are whitish, softe, and hearie like veluet, and fashioned like Basill leaues. The floures be white, at the toppe of the stalke growing thicke togither in rewes by one side of the stem, the which at the vpper end, do bend & turne againe like a Scorpios tayle, or the tayle of a Lobster, or riuer Creuis. The roote is small and harde.

2. The small Tornesol carrieth only but one stem, of the length of a foote or somewhat more, the which diuideth it self into many branches. The leaues be whitishe, almost like to the first, but somewhat drawing towardes the leaues of the small Clote Burre. The floures be yellow and small, growing thicke togither, and perish or vanish away without the bringing foorth of any fruyte like the floure of Palma Christi. The seede is grayishe, inclosed in triangled huskes or Coddes, like the huskes of Tithymall or Spurge, hanging downe vnderneth the leaues, by a single stem: they come forth without floure, for the floure is vnprofitable as is before sayde.

❧ *The Place.*

1. The great Tornesol (as Ruellius saith) groweth in France, in frutefull tylled groundes: but in this countrey it is onely found in gardens.

2. The small Tornesol, groweth in lowe, sandie, and waterie places, and is found very plenteously in diuers places of Languedock.

❧ *The*

The Tyme.

1.2 The Tornesolles, do floure about Mydsomer, and in July.

The Names.

1 The great Tornesol is called in Greeke ἡλιοτρόπιον μέγα, καὶ σκορπίουρον: in Latine Heliotropium magnum: of the new, or late wryters Verrucaria maior, and Herba cancri, Solaris herba, Scorpionis herba, and therefore the base Almaignes do call it Creeftcruyt, and great Creeftcruyt.

2 The small Tornesoll is called ἡλιοτρόπιον μικρόν, Heliotropium paruum: of Aëtius Heliotropium tricoccum, of some it is called Verrucaria: in Spanishe *Tornasol:* in French *Tournesol:* in base Almaigne Cleyn Creeftcruyt, and cleyne Sonnewendt.

The Nature.

The Tornesols, are hoate and dry in the thirde degree.

The Vertues.

1 A handfull of the great Tornesoll boyled & dronke, expelleth by opening the belly gentilly, hoate Chollericke humors, and tough, clammy, or slimie flegme.

The same boyled in wine and droke is good against the stingings of Scorpions, it is also good to be layde too outwardly vpon the wounde.

They say, that if one drinke foure graynes of ye seede of this herbe, an houre before the comming of the fitte of the feuer Quartayne, that it cureth the same: And three graynes so taken cureth the feuer Tertian.

The seede of this herbe pounde, & layde vpon Wartes, and such like excrescence, or superfluous out growings, causeth them to fall away.

The leaues of the same pounde, and layde too, cureth the Goute, with bruisings, burstings, and dislocation of members.

2 The small Tornesoll and his seede boyled, with Hysope, Cressis, and Sall Nitri, and dronke, casteth foorth wormes both round and flat.

The same bruised with salte, and layde vpon Wartes, driueth them away.

With the seede of the smal Tornesoll (being yet greene) they die and stayne old linnen cloutes and ragges into a purple colour (as witnesseth Plinie in his xxj. booke, Chap. vii.) wherewithall in this coūtrey men vse to colour gellies, wynes, fine Confections, and Comfittes.

Of Scorpioides, or Scorpions grasse.
Chap. xlij.

The Description.

1 Scorpioides is a small, base, or lowe herbe, not aboue the length of ones hande, the stemmes are small, vpon whiche groweth fiue or sixe narrow leaues (and somewhat long after the fashion of a Hares eare, which is the cause that some Douche men call it Hasen oore). The floures be small and yellow, after whiche commeth the seede, whiche is rough & prickley, three or foure cleauing togither, distinguished by ioyntes, and turning rounde, or bending like a Scorpions tayle.

2 Matthiolus describeth an other Scorpioides, with sleder stalkes, and round leaues sometimes three togither. At the toppe of the stalkes groweth two or three little small long hornes togither, the whiche also do shewe as they were separated by certayne ioyntes.

Bysides these two kindes of Scorpioides, there is yet twoo other small herbes whiche some do also name Scorpion grasse, or Scorpion worte, although they be not the right Scorpion grasse. The one of them is called Male Scorpion, the other female Scorpion.

Scorpioides.

The first Booke of

Scorpioides.
Scorpion Grasse.

Scorpioides Matthioli.
Matthiolus Scorpion grasse.

3 ❧ The male Bastard Scorpioides groweth about the length of a mans hand, or to the length of a foote, his stalkes are crested, and crokedly turning, aboue at the top, whereas the knoppes, buddes, and floures do stande, euen like to a Scorpions tayle, the leaues be long, narrow, and small. The floures be fayre and pleasant, being of fiue littell leaues set one by an other, of Azure colour, with a little yellow in the middell.

4 ❧ The female Bastarde Scorpioides is very much like to the male, sauing that his stalkes and leaues be rough and hearie, & his floures smaller. The toppes of the stalkes be likewise croked, euen as the toppes of the male.

✠ *The Place.*

1.2 Scorpioides groweth not of his owne kinde in this countrey, but is sowen in the gardens of certaine Herboristes.

3 The male Bastarde Scorpioides, groweth in medowes, alongst by running streames and watercourses: and the nearer it groweth to the water, the greater it is, and the higher, so that the leaues do sometimes grow to the quantitie of willow leaues.

4 The female Bastarde Scorpioides, groweth in the borders of fieldes and gardens.

✠ *The Tyme.*

1.2 Scorpioides floureth in June and July.

3.4 The Bastarde kindes, do begin to floure in May, and continue flouring the most parte of all the Sommer.

the Historie of Plantes.

Pseudoscorpioides mas.
Bastard Scorpioides the male.

Pseudoscorpioides fœmina.
Bastard Scorpioides the female.

※ *The Names.*

1. The first is called in Greeke σκορπιοειδἑς: in Latine also Scorpioides: in English also Scorpioides, Scorpion wurte, or Scorpion grasse: in French *Herbe aux Scorpions*: in base Almaigne Scorpioencruyt, and of some, Hasen oore, that is to say Auricula leporis.

2. The other is iudged of Matthiolus, for a kinde of Scorpioides, wherfore it may be called Matthiolus Scorpioides, or Trefoyl Scorpioides.

3.4 The Bastard Scorpioides haue none other knowen name, but some count them to be Scorpion herbes, as hath bene before sayde.

※ *The Nature and Vertues.*

Scorpioides or Scorpion grasse, is very good to be layde vpō the stingings of Scorpions, as Dioscorides saith.

Of S. Johns worte.   Chap.xliij.

※ *The Description.*

1. Saynt Johns worte hath a purple, or browne redde stalke full of branches. The leaues be long and narrow, or small, not much vnlike the leaues of garden Rue, the whiche if a man do holde betwixt the light and him, they will shewe as though they were pricked thorough with the poyntes of needels. The floures at the toppe of the branches are fayre and yellow, parted into fiue small leaues, the whiche being bruised, do yeelde a redde iuyce or liquor: after the floures commeth forth small huskes, somewhat long and sharpe poynted, like Barley cornes: in which is conteyned the seede, whiche is small and black, and senting like Rosin. The roote is wooddish, long and yellow.

2. There is also an herbe much like to S. Johns worte aforesayde, but it is very small and lowe, not growing about the length of a spanne, whose stalkes be very tender, and the leaues small and narrow, yea smaller then Rue, in all partes else like to the aforesayde.

f ij   ※ *The*

The first Booke of

Hypericum.

❧ *The Place.*

1. S. Johns worte groweth by way sides, & about the edges or borders of fieldes.

2. The other smal herbe groweth in fieldes, amongst the stubble, & harde by the wayes.

❧ *The Tyme.*

Saint Johns worte floureth most commonly in July and August.

❧ *The Names.*

S. Johns worte is called in Greeke ὑπερικόν: in Latine & in Shoppes Hypericum, and of some Perforata, and Fuga Dæmonum: in Spanish *Coraiouzillo*, and *Milfurado, yerua de San Iuan*: in English as is beforesayde, S. Johns worte, or S. Johns grasse: in high Douch S. Johans kraut, & of some Hart-haw: in base Almaigne S. Jans cruyt.

❧ *The Nature.*

S. Johns worte is hoate and dry in the thirde degree.

❧ *The Vertues.*

S. Johns worte with his floures and seede, boyled and dronken prouoketh the vrine, and causeth to make water, & is right good against the stone in the bladder: it bringeth downe womens floures, and stoppeth the laske.

The same boyled in wine and dronken, driueth away feuer Tertians, and Quartaynes.

The seede dronken by the space of fortie dayes togither, cureth the payne in the hanches whiche they call the Sciatica.

The leaues pound are good to be layde as a playster vpon burnings: The same dryed and made into pouder, and strowen vpon woundes, and naughtie, olde, rotten and festered vlcers, cureth the same.

Of S. Peters worte, or Square S. Johns Grasse.   Chap. xliiij.

❧ *The Description.*

1. This kinde of S. Johns worte, in his leaues and stemmes differeth not much frō Hypericum, sauing that it is greater. The stalke is long without branches or springs, the leaues are like the other S. Johns Grasse, but longer, browner, and greener, for the most parte vnderneth, it is ouerlayde and couered with fine softe heare, sweete in taste, and do not shew thorow holed or pricked as the other. The floures are like to Hypericum, but paler and with longer leaues. The buddes before the opening of the floures, are spotted with small blacke speckes. The seede is in huskes like the seede of Hypericum: and smelleth likewise, almost like Rosin.

2. There is yet an other kinde of this herbe, the which the base Almaignes do call Conraet, very like to the aforesaide, sauing p̄ his leaues be greater, whiter, & not so hearie or softe, but better like S. Johns worte, although they appeere not thorow prickt or holed. The floures are like to p̄ aforesaid, & ar also specked in the

the Historie of Plantes. 65

in the knappes and buddes, with small blacke spottes. The roote is woodishe like the other.

Ruta syluestris. &c.
Great S. Johns worte.

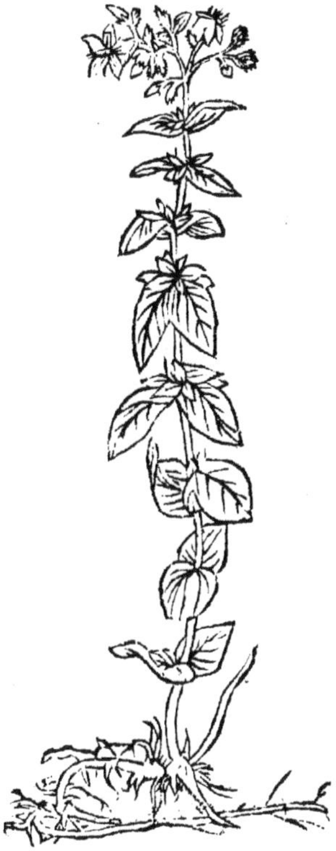

Ascyrum.
S. Peters worte.

❧ *The Place.*
These herbes grow in rough vntilled places, in hedges, and Copses.

❧ *The Tyme.*
They floure in July, and August.

❧ *The Names.*
1 The first is called in Greeke πήγανον ἄγριον, that is, wild Rue: yet this is none of the kindes of the grieuous sauored or stinking Rue: it is also called of some ἀνδρόσαιμον, Androsæmum.

2 The second is called in Greeke ἄσκυρον, and in Latine Ascyrum: both are vnknowen in Shoppes: in English Square S. Johns grasse, great S. Johns worte: & most cómonly S. Peters worte: in high Douch Harthew, & Waldt Hoff: The secōd is called of some Kunratz: in base Almaigne Herthoy, & Coenraet.

❧ *The Nature.*
They are hoate and dry, and lyke to Hypericum.

❧ *The Vertues.*
The seede of S. Peters worte, or square S. Johns grasse, dronken the weight of two Drammes with Honied water, and vsed a long space, cureth the Sciatica, that is the payne in the hanches.

The same pounde is good to be layde vpon burnings.

The wine wherein the leaues therof haue ben boyled, hath power to cósolidate, & close vp woundes, if they be oftentimes washed with the sayde wine.

F iij Of

The first Booke of
Of Tutsan or Parke leaues. Chap.xlv.
Androsæmon.

❧ The Description.

ANdrosemon is like to Saint Johns worte, & S. Peters grasse. It hath many rounde stalkes comming out of one roote, the whiche do bring forth leaues, muche larger than the leaues of S. Johns wurte, in y beginning greene, but after that the seede is ripe, they waxe redde, & than being brused betwixt ones fingers, they yeeld a redde sappe or iuyce. At the toppe of the stalkes groweth small knoppes or round buttons, the whiche in their opening do bring forth floures like to S. Johns grasse, but greater: whan they are fallen or perisshed, there appeareth littell small pellots or round balles, very red at the beginning, but afterward of a browne, and very darke redde colour whan they be ripe, like to the colour of clotted or congeled drie bloud, in whiche knops or beaties is conteyned the seede, which is small and browne, the roote is harde and of wooddie substance, yearely sending forth new springs.

❧ The Place.

This herbe groweth not in this countrey, except in gardens where as it is sowen and plated. The Authors of Stirp. Aduers. noua, do affirme that Androsemon groweth by Bristow in England in S. Vincentes Rockes and woody Cleues beyond the water. But if Androsemon be Tutsan or Parke leaues, it groweth plentifully in woodes and parkes, in the west partes of England.

❧ The Tyme.
It floureth in July, and the seede is ripe in August.

❧ The Names.
It is called in Greeke ἀνδρόσαιμον, in Latine Androsæmum: vnknowen to the Apothecaries. It hath none other common name that I know: yet some do also call it Androsæmum fruticans.

Tutsan so called in French and in English, is thought of some late writers to be Clymenon, and is called of the Clymenon Italorum, siue Siciliana: of our Apothecaries Agnus castus. ❧ The Nature.

It is hoate and dry like S. Johns grasse, or S. Peters wurte.

❧ The Vertues.
Androsemon his vertues are lyke to S. Peters wurte, & S. Johns grasse, as Galen saith. Tutsan is much vsed in Baulmes, Drenches, and other remedies for woundes.

Of Woad or Pastel. Chap.xlvi.
❧ The Kindes.

Here be two sortes of Woad: the one is of the garden, and cometh of seede, and is vsed to colour and die cloth into blew: The other is wilde Woad, and commeth vp of his owne kinde.

Isatis

the Historie of Plantes.

Iſatis ſatiua.
Garden Woad.

Iſatis ſylueſtris.
Wilde Woad.

❧ *The Deſcription.*

1 Garden Woad hath long, broade, ſwartegreene leaues, ſpread vpon the ground, almoſt lyke the leaues of Plantayne, but thicker, and blacker: the ſtalke riſeth vp, from the middeſt of the leaues of two cubites long, ſet full of ſmaller and ſharper leaues, the whiche at the toppe diuideth and parteth it ſelf into many ſmall branches, vpon the whiche groweth many littell floures, very ſmall and yellow, and after them long broade huſkes, like littell tunges, greene at the firſt and afterward blackiſhe, in whiche the ſeede is conteyned. The roote is white, ſingle and ſtraight, and without any great ſtore of threeds or ſtrings.

2 The wilde is very like to the garden Woad, in leaues, ſtalke, and making, ſauing that the ſtalke is tenderer, ſmaller, and browner, and the huſkes more narrow, otherwiſe there is no difference betwixt them.

❧ *The Place.*

1.2 Garden Woad is ſowen in diuers places of Flanders, & Almaigne, in fertill & good grounds. The wilde groweth of his owne kinde in vntilled places.

❧ *The Tyme.*

Both do floure in May and June.

❧ *The Names.*

This herbe is called in Greeke ἰσάτις: in Latine alſo Iſatis, and Glaſtum, of the late writers Guadum, and of ſome Luteum herba: in Engliſh Woad, or Paſtel: in French *Gueſde* or *Paſtel*: in Spaniſh alſo *Paſtel*: in Italian *Guado*: in high

f iiij   Douch

The first Booke of

Douche, Weidt and Waydt: in base Almaigne, Weedt.

❧ *The Nature.*

1 Garden Woad is dry without any sharpnesse.
2 The wilde dryeth more, and is more sharpe and byting.

❧ *The Vertues.*

1 Garden or sowen Woad, bruised is good to be layde vpon the woundes of A mightie strong people, whiche are vsed to dayly labour and exercise, and vpon places to stop the running out of bloud, & vpon fretting vlcers & rotten sores.

It scattereth and dissolueth all colde empostumes being layde therevpon. B

2 The wilde Woad resisteth moyst and flowing vlcers, and consuming rotten C sores, being layde therevpon: but against the other grieffes, for which the garden Woad serueth, it is of lesse strength, and serueth to smal purpose, bycause of his exceeding sharpnesse.

The Decoction of wilde Woad dronken, is very good for such as haue any D stopping, or hardnesse in the Milte or Splene.

Of Dyers weede. Chap.clvij.

❧ *The Description.* Herba lutea.

THe leaues of this herbe are long, narrow & blackish, not much vnlike the leaues of Woad, but they are smaller, narrower, & shorter, from the middest whereof commeth vp the stalke to the length of three foote, couered bylow with small narrow leaues, and aboue with little pale yellow floures, thicke set, and clustering one aboue an other, the whiche do turne into small buttons, cut crossewise, wherein the seede is conteyned, whiche is small and blacke. The roote is long and single.

❧ *The Place.*

They sowe it in sundry places of Brabant, as about Louayne, and Brussels: it groweth also of it selfe in places vntilled, and by way sides.

❧ *The Tyme.*

It floureth in May, and soone after the seede is ripe.

❧ *The Names.*

This herbe is called in Latine (as Ruellius wryteth) Herba lutea, and of some Flos tinctorius: in base Almaigne Wouwe, & of some Orant, but not without error: for Orant is an other herbe nothing like vnto this, as shall be more playnely declared in the processe of this booke.

❧ *The Nature.*

It is hoate and dry.

❧ *The Vertues.*

Seing that Herba lutea is not recevued, for any vse of A Physicke, and is vnknowen of the Auncients, we be able to wryte nothing els of this herbe, sauing that it is vsed of Dyers, to colour and dye their clothes into greene, and yellow.

Of S. James worte. Chap.clviij.

❧ *The Kindes.*

Saynt James worte, or as some call it, Saynt James floure is of twoo sortes.

Iacobea.

the Historie of Plantes.

Iacobea.	Iacobea marina.
S. James woꝛte.	S. James woꝛte of the Sea.

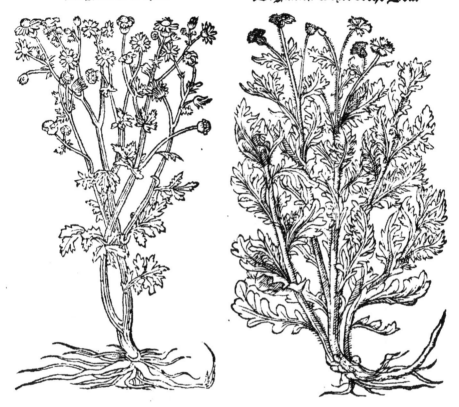

❧ *The Description.*

1. The firſt kinde of S. James woꝛte, hath long, bꝛowne, red, creſted, oꝛ ſtraked ſtalkes, two oꝛ three foote long. The leaues be great & bꝛowne, much clouen and cut, not much vnlike the leaues of Woꝛmewood, but longer, larger, thicker, and nothing white. The floures be yellow, growing at the top of the ſtalkes, like to Cammomill, in the middeſt whereof is the ſeede, gray, and woolly, oꝛ downy, and fleeth away with the winde. The roote is white and ful of ſtrings.

2. The ſecond kinde called S. James woꝛte of the Sea, is much like to the firſt, but the ſtalke is nothing redde. The leaues be ſmaller, whiter, and moꝛe deeper, and ſmaller iagged. The floures be like to the firſt kinde, but moꝛe pale oꝛ bleaker. The roote is long, thꝛeddy, and creeping, and bꝛingeth foꝛth round about him, new ſpꝛings.

❧ *The Place.*

1. S. James woꝛte, groweth almoſt euery where, alongſt by wayes and wateriſh places, and ſometimes alſo in the boꝛders of fieldes.

2. Sea S. James woꝛte groweth in trenches and diches, and like places, adioyning to the Sea. ❧ *The Tyme.*

They floure in July and Auguſt.

❧ *The Names.*

The firſt is now called in Latine Iacobea, Herba S. Iacobi, and Sancti Iacobi
flos:

flos: in Englishe Saynt James worte: in French *Herbe ou fleur S. Iaques:* in high Douch S. Jacobs bluom: in base Almaigne S. Jacobs cruyt, and S. Jacobs blomen.

2 The second without doubte is a kinde of S. James worte.

❧ *The Nature.*

They are both hoate and dry in the third degree.

❧ *The Vertues.*

S. James herbe hath a speciall vertue to heale woundes, wherfore it is very good for all old woundes, fistulas, and naughtie vlcers. A

Some affirme, that the iuyce of this herbe gargeld, or gargarised, healeth B all inflammations, or swellings, and empostems of the throote.

Of Flaxe, or Lyn. Chap. xlix.

❧ *The Description.*   Linum.

FLax hath a tender stalke, couered with sharpe narrow leaues, parted at the toppe into small shorte branches, the whiche bringeth foorth fayre blewe floures, in steede wherof being now fallen there commeth vp round knappes, or buttons, in whiche is contayned a blackishe seede, large, fatte, and shining.

❧ *The Place.*

Flaxe is sowen in this countrey, in fatte, and fine ground, especially in lowe moyst fieldes.

❧ *The Tyme.*

It floureth in May and June.

❧ *The Names.*

Flaxe is called in Greeke λίνον: in Latine Linum, and in Shoppes it is wel knowen by the same name. And here ye may perceyue the cause why the base Almaignes do vse the worde Lyn, to all things made of Flaxe, or Lyne, as Lijnendoeck and Lijnen laken, that is to say, Lyneclothe, or clothe made of Lyne: in Englishe Flaxe, or Lyne: in Frenche *Lin*: in highe Douche Flaschsz: in base Almaigne Ulas.

❧ *The Nature.*

The seede, whiche is muche vsed in medicine, is hoate in the firste degree, and temperate of moysture and drynesse.

❧ *The Vertues.*

The seede of Lyn boyled in water and layde too in manner of a pultis, or playster, appeaseth all payne. It softeneth all colde tumors, or swellings, the empostems of the eares, and neck, and of other partes of the body.

Lynseede pound with figges, doth rypen and breake all Empostems, layde B there vpon: and draweth foorth Thornes, and all other things that sticke fast in the body, if it be mingled with the roote of the wilde Cocomber.

The same mingled with Cresses and Hony, and layde vnto rough, rug- C ged, and euill fauoured nayles, aswell of the handes, as of the feete, causeth them that be corrupt to fall of, and cureth the partie. The same rawe, pounde and layde to the face, clenseth and taketh away all spottes of the face.

The wine wherein Lynseede hath bene boyled, preserueth the vlcers & old D
sores, y shalbe washed in the same, frō corruption, festering or inward rāckling.

The water wherein Lynseede hath bene boyled, doth quicken and cleare the E
sight, if it be often dropped or stilled into the eyes.

The same taken in glisters swageth the gryping paynes of the belly, and of F
the Matrix or Mother, and cureth the woundes of the bowelles, and Matrix,
if there be any.

Lynseede mengled with Hony, & taken as an Electuary, or Lochoch, clean- G
seth the breast, and appeaseth the Cough, and eaten with Raysons, is good for
such as are fallen into Consumtions, and feuer Hetiques.

❧ The Daunger.

The seede of Lin, taken into the body, is very euill for the stomacke: it hin-
dereth the digestion of meates, and engendreth much windinesse.

Of Hempe.   Chap.I.

❧ The Kyndes.

There are two kindes of Hempe, the one is frutefull and beareth seede: The
other beareth but floures onely.

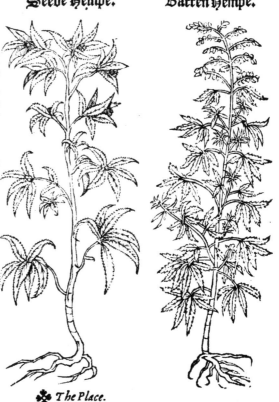

Cannabis semen faciens.
Seede Hempe.

Cānabis absq; semine.
Barren Hempe.

❧ The Description.

1. The first kind of Hempe, hath a rounde hollow stalk foure or fiue foote long, full of branches, & like to a little tree: at the top of the branches groweth little smal round bags, or huskes, wherin is cōteyned the seede which is round. The leaues be great, rough, & blackishe, parted into seuen, nine, ten, and sometymes into moe partes, long, narrow, and snipt or dented round about with notches, like the teeth of a Saw. The whole leafe with all his partes is like to a hand spread abroade.

2. The secōd is also in leaues like to the first, and it hath a thicke stalke, out of whiche by the sides groweth foorth sundry branches: but it bea-reth neither seede nor frute, sauing small white floures, the whiche like duste or pou-der is caried away with the winde.

❧ The Place.

1. 2   These two sortes of Hempe are sowen in fieldes, and (whiche is a thing to
be marueled at) they do both spring of one kinde of seede. A mā shall sometimes
finde the male Hempe growing in the borders of fieldes, and by the wayes.

❧ The Tyme.

The

The seede of the male Hempe is ripe at the end of August, and in September. The female Hempe is ripe in July.

✿ The Names.

Hempe is called in Greeke κάνναβις, ἀσέριον, καὶ χοινοστρόφον: in Latine, and in Shoppes Cannabis: in Italian Cannape. in Spanish Cañamo, Canauo: in English Hempe, Neckeweede, & Gallowgrasse: in French Chanure, Chenneuis, or Cheneue: And here ye may perceyue the cause why the Normans and others do call the Clothe made of Hempe, Chenneuis, or Canuas, for it soundeth so after the Greeke, Latine, and French: the high Douchmen call Hempe Zamerhauff: in base Almaigne Kempe.

✿ The Nature.

Hempe seede is hoate and dry in the thirde degree.

✿ The Vertues.

Hempe seede doth appeace, and driue the windinesse out of the bodie, and if a man take a littell to much of it, it drieth vp Nature, & the seede of generation, and the Milke in wemens brestes.

The seede stamped and taken in white wine, is highly commended at this day, against the Jaundice, and stopping of the Lyuer.

The iuyce of the leaues of greene Hempe put into ones eare swageth the payne of the same, and bringeth forth all kinde of vermine of the same.

The roote of Hempe boyled in water doth help and cure the Sinewes, and partes that be drawen togither and shronken, also it helpeth against the Goute, if it be layde thereupon.

✿ The Daunger.

Hempe seede is harde of digestion, and contrary to the stomacke, causing payne and griefe, and dulnesse in the head, and engendreth grosse and naughtie humors in all the body.

Of Lysimachion, Willow herbe or Lous strife.
Chap. LI.

✿ The Kindes.

There are now diuers kindes of Herbes comprehended vnder the name of Lychimachia, but especially foure, vnder whiche all the Lysimachies shalbe comprysed. The first is the right Lysimachion. The second is the red Willow herbe with coddes. The third is the second kinde of redde Willow herbe without Coddes. The fourth is a kinde of blew Lysimachion.

✿ The Description.

1. The first Lysimachus, or the yellow Lysimachus, hath a rounde stalke, very littell crested or straked, of a Cubite or two long. The leaues be long & narrow like willow, or wythie leaues, nothing at all cut or snipt about the edges, but three or foure leaues standing one against an other round about the stalke, at the ioyntes. The floures be yellow and without smell, and grow at the toppe of the branches, in steede wherof whan they are fallen away, there groweth rounde seede, like Coriander seede. The roote is long and slender, creping here and there, and putteth forth diuers yong springs, whiche at their first comming vp are redde.

2. The second Lysimachus in leaues and stalkes is like to the other, sauing that his leaues be not so broade, and are snipt about the edges, much like vnto Willow leaues. The floures in colour and making, are somewhat like the floures of the common wilde Mallow or Hock, that is to say, it hath foure little broade round leaues standing togither, and lying one ouer an others edges, vnder whiche

the Historie of Plantes.

whiche there groweth long huskes or Coddes, like to the huskes of stocke Gillofers, whiche huskes do appeare before the opening of the floure: the whiche huskes or seede vessels, do open of themselues, and cleaue abroade into three or foure partes, or quarters, whan the seede is rype, the whiche bycause it is of a woolly or cottony substance, is carried away with the winde. The roote is but small and threddy.

Lysimachion verum.
Yellow Lysimachion or Louse strysse.

Lysimachium purpureum primum.
The first purple red, willow herbe, or Lysimachium, also the Sonne before the Father.

There is an other smal kinde of this sorte, like to the other in stalke, leaues, floures and huskes, sauing that it is in all partes smaller, and the stalke is so weake, that it can very seldome grow straight. The floures be of carnation colour, like to Gillofloures, but somewhat smaller.

Yet there is a thirde kinde of redde Lysimachus, very like to the first redde kinde. The floures do grow also at the top or end of the huskes, but they be paler, and in making not so well like the other, but rather like to Gillofers parted into foure small leaues, whiche are set crossewise.

The second kinde of redde Lysimachus is like to the aforesayde, in stalkes & leaues: sauing that his floures do grow like crownes or garlandes rounde about the stalke like to Penny royall: of colour redde, & without Huskes, for the seede doth grow in the smal Corones, frō whence the floures fell of. The stalke is square and browne. The roote is very browne and thicke, of a wooddishe substance, and putteth forth yearely new springs.

Lysima-

Lysimachion purpureum alterum.	Lysimachium cæruleum.
Partizan Lysimachion, or Spiked Lysimachion.	**Blew Lysimachion.**

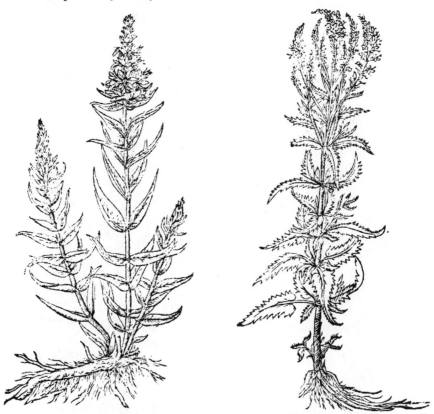

4 ⁂ The blew Lysimachus also in stalke and leaues is like the others: his blew or Azured floures are growing at the end of the stalkes, spike fasshion, or eared like Spike or Lauander beginning to bloow bylow, & so flouring vpwarde, after whiche there cometh small round Coddes or purses, wherein the seede whiche is very small is conteyned. The roote is threddy.

※ *The Place.*

The yellow & redde do grow in waterish & moyst places, in low medowes, and about the brinckes and borders of water brookes and diches. The blew is not found in this countrey but in the gardens of such as loue Herbes.

※ *The Tyme.*

They do all floure most commonly in June, and July, and their seede is ripe in August.　　　　※ *The Names.*

Lysimachia is called in Greeke λυσιμάχιον καὶ λύτρον: in Latine Lysimachium, Lysimachia, and of some Salicaria, vnknowen in Shoppes: in English Lysimachia, Willow herbe, and Louse strife: in Spanish *Lisimacho yerua*.

1　　The first which we may call Golden or yellow Lysimachus, Willow herbe, and Louse strife, is called in Fraunce *Cornelle, Souscy d'eauë, Pellebosse*, or *Chassebosse*: in high Almaigne Geelwelderich: in base Almaigne Geelwederick.

2　　The second is called of some, in Latine Filius ante Patrem, that is to say, the sonne before the father, bycause ẏ his long huskes in which the seede is côteined

do come

do come forth and waxe great, before that the floure openeth: in Englishe, the first red Lysimachus, or wythie herbe, or Louse stryfe: in French *Lysimachie rouge*: in high Douch Braun or Rod weiderich: in base Almaigne Root wederick.

3 The third is called in Brabant Particke. It may be called in English Partizan or sharpe Lysimachus, or poynted willow Herbe with the purple floure.

4 The fourth hath none other name, but *Lysimachium cæruleum* in Latine: in English Blew or Azured Lysimachus.

✻ The cause of the Name.

This herbe tooke his name of the valiant & noble Lysimachus, the friend & Cosin of Alexander the great, king of Macedonia, who first found out the propertie of this herbe, and taught it to his posteritie or successours.

✻ The Nature.

The yellow Lysimachus or golden Louse stryfe, is colde, dry, and astringent.

The temperament of the redde and blew Lysimachia, is not yet knowen.

✻ The Vertues.

The iuyce of the leaues of the yellow Lysimachus stoppeth all fluxe of A bloud, and the Dysenteria or bloudy fluxe, being eyther taken inwardly, or otherwise applied outwardly.

The same stayeth the inordinate course of wemens floures, being put with B a pessarie of wool or cotton into the Matrix, or secrete place of women.

The herbe brused & put into the nose, stoppeth the bleeding of the same, and C it doth ioyne togither and close vp all woundes, and stoppeth the bloud, being layde thereuppon.

The perfume of this herbe dryed, driueth away all Serpents, & venemous D beasts, and killeth flies, and knattes.

✻ The Choyse.

Whan ye will vse Lysimachus, for any griefe aforesayde, ye shall take none other but of that kinde with the yellow floure, which is the right Lysimachus: for although the others haue now the selfe same name, yet haue they not the same vertue and operation.

Of Mercury.   Chap.lij.

✻ The Kindes.

THere be two sortes of Mercury: the garden, and wilde Mercury: the which againe are diuided into two other kindes, a Male, and Female, differing onely but in seede.

✻ The Description.

1 The male garden Mercury, or the French Mercury, hath tender stalkes, ful of ioyntes & branches, vpon the which groweth blackish leaues, somwhat long almost like the leaues of Parietory, growing out from the ioynts, frõ whence also, betwixt the leaues and the stem there cõmeth forth two little hearie bullets, ioyned togither vpon one stem, eche one conteyning in it selfe a small round seede. The roote is tender and full of hearie strings.

2 The female is like to the male, in stalkes, leaues, and growing, and differeth but onely in the floures and seede, for a great quantitie more of floures and seede, do grow thicke togither like to a small cluster of grapes, at the first bearing a white floure, and afterwarde the seede, the whiche for the most parte, is lost before it be ripe.

3. 4 The wilde Mercury is somewhat like to the garden Mercury, sauing yͭ his stalke is tenderer & smaller, and not aboue a span long, without any branches,

G ij   the

The first Booke of

the leaues be greater and standing farder a sunder one frō an other. The seede of the male, is like to the seede of the male garden Mercury, and the seede of the female, is like the seede of the female garden Mercury. The roote is with hearie strings, like the roote of the garden Mercuries.

Mercurialis mas. Phyllon Theophrasti.
French Mercury.

Mercurialis fœmina.
Mercury female.

5 ¶ There is yet an other herbe founde called Noli me tangere, the whiche also is reduced and brought vnder the kindes of Mercury. It hath tender rounde knobbed stalkes, with many hollow wings, and large leaues, like to the Mercury in stalke and leaues, but much higher and greater, the floures hang by small stemmes, they are yellow, broade, and hollow before, but narrow behind, and croking like a tayle, like the floures of Larkes spurre, after the whiche there commeth foorth small long round huskes, the whiche do open of them selues, and the seede being ripe, it spurteth and skippeth away, as soone as it is touched.

6 ¶ One may well describe and place, next the Mercuries (but especially them of the garden) the herbe whiche is called Phyllon, bycause that some do thinke that Phyllon and Mercury are but one herbe, but by this treatice they may know that they be diuers herbes. Now therfore there be two sortes of Phyllon (as Crateuas writeth) the Male and the Female. It hath three or foure stalkes, or more, the leaues be somewhat long and broade, something like the leaues of the Olyue tree, but somewhat larger and shorter. All the herbe his stalkes and leaues, is couered with a fine softe white wool or Cotton. The seede of the female Phyllon, groweth in fashion like to the seede of the female
Mercurie:

the Historie of Plantes. 77

Mercurie: and the seede of the male groweth like to the male Mercurie.

Cynocrambe.　　　　　　Phyllon Thelygonon.
Wilde Mercury.　　　　　Childzen Mercury.

❧ *The Place.*

The garden Mercurie groweth in vineyardes, and gardens of pot herbes. The wilde groweth in hedges and Copses. The fifth kinde groweth in deepe moyst vallies, and if they be ones planted, they come vp againe yearely afterward, of their owne accorde, or of their owne sowing.

Phyllon is founde, growing thorough out all Languedock, and Prouince.

❧ *The Tyme.*

They floure in June, and continue flouring all the Somer.

❧ *The Names.*

1.2　Garden Mercurie is called in Greeke λινόζωσις, and of some παρθένιον, καὶ ἑρμῦ βοτάνιον: of Theophrastus φύλλον: in Latine and in Shoppes Mercurialis: in Italian *Mercurella*: in Spanish *Mercuriales*: in English Mercury, and of some in French *Mercury*: in high Douch Zamen Bingelkraut, kuwurtz, and Mercurius kraut: in base Almaigne Tam Bingelcruyt, & Mercuriael. And that that hath the round seedes is called Mas the Male. And the other is called the female. Some do also take it, for wilde Mercury.

3.4　The wilde Mercury is called in Greeke κυνοκράμβη, κυνία, καὶ λινόζωσις ἄρρικ: in Latine Canina brasica, and Mercurialis syluestris: in English wilde Mercury, and Dogges Call: in French *Mercurialle sauuage, Chou de Chien*: in high Douch

G iij　　　　　　　　　　　　　　　　　.wilde

wilde Bingelkraut, and Hundszkol: in Brabant wildt Bingelcruyt, and wilden Mercuriael.

5 The Noli me tangere, was vnknowen of the Auncients, wherefore it hath none other name in Greke or Latine. They cal it in high Douch Springkraut: in Brabant Springcruyt, and Cruydeken en ruert my niet: and for that cause men in these dayes do call it, Noli me tangere: that is to say, touche me not.

6 Phyllon is called in Greeke φύλλον καὶ ἐλαιόφυλλον: in Latine Phyllum. The male is called ἀρρενογόνον, whiche may be Englished Barons Mercury or Phyllon, or Boyes Mercury or Phyllō. And the female is called in Greeke Θηλυγόνον: and this kinde may be called in English Gyrles Phyllon or Mercury, Daughters Phyllon, or Mayden Mercury. And we can giue it none other name as yet. This is Dioscorides Phyllon, but not Theophrastus Phyllon. For Theophrastes Phyllon, is nothing els but Dioscorides Mercury. And for to be knowen from the Mercuries, this Phyllon is also called Elæophyllon, Oliue Phyllon.

❧ *The Nature.*

The Mercuries, are hoate and dry in the first degree, as Auerroys saith.

❧ *The Vertues.*

A Mercury boyled in water and dronken, loseth the belly, purgeth, & driueth forth colde phlegmes, and hoate and cholerique humors: & also the water that is gathered togither in the bodies of such as haue the Dropsie.

B For these purposes, it may be vsed in meates, and potages, and they shall worke the same effect, but not so strongly.

C The same pound with Butter, or any other greace, and layde to the fundament, prouoketh the stoole or siege.

D The Barons Mercury, or male Phyllon dronken, causeth to engender male children, and the Maydē Mercurie, or gyrles Phyllon dronken, causeth to engender Gyrles, or Daughters.

Of Mony worte/ or Herbe two pence. Chap. liij.

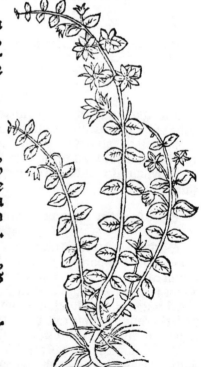

Nummularia.

❧ *The Description.*

Monyworte hath small slender stalkes, creeping by the grounde, vpon euery side whereof groweth small rounde leaues, and somewhat large, almost like to a pēny. The floures be yellow almost lyke to gold cuppes. The roote is smal and tender.

❧ *The Place.*

This herbe groweth in moyst medows, about ditches, & watercourses, & in Copses that stand lowe.

❧ *The Tyme.*

It beginneth to floure in May, and continueth flouring all the Somer.

❧ *The Names.*

This herbe is now called in Latine Nūmularia, Centummorbia, & of some Serpentaria, and also Lunaria grassula: in English Herbe two pence, two penny grasse, and

and Monyworte: in French *Herbe à cent maladies* in high Douch Pfenningkraut, Egelkraut, & clein Naterkraut: in base Almaigne Peninckruyt, & Eghelcruyt.

❦ *The Nature.*
Two penny grasse is dry in the thirde degree.

❦ *The Vertues.*
The later wryters do say, that if this herbe be boyled in wyne and dronken with Hony, that it healeth and cureth the woundes and hurtes of the Lunges, & that it is good against the Cough, but specially against the dangerous Cough in yong children, to be taken as is afore sayde.

Wilde Flare, or Tode Flare. Chap. liiij.
Linaria.   Osyris.

❦ *The Description.*
1. Stanworte, wilde flare, or Tode flar, hath small, slender, blackish stalkes, out of which groweth many leaues togither long, and narrow, much like to the leaues of Lyn. The floures be yellow, large, & close before, like to a frogges mouth, and narrow behinde, & croked like to Larkes spurre, or Larkes clawe. The seede is large and blacke, conteyned in small rounde huskes, the whiche commeth forth, after the falling of, of the floure.

2. There is an other kinde of this herbe, the which is not common, and it beareth fayre blewe floures, in all other things lyke to the other, sauing that his stalkes, floures, and leaues are smaller, and tenderer, but yet it groweth vp to a higher stature.

3. To these kindes of wilde Flare or Linarie, it were not amisse to ioyne that

80   The first Booke of

herbe, which is called in Italy Beluedere. This plante hath diuers small shutes or scourges bearing small narrow leaues almost like to the leaues of Flaxe: the floures be small and of a grasse colour, and do grow at the toppe of the stalkes.

❧ *The Place.*

1.2 They grow wilde in vntilled places, about hedges, & the borders of fieldes.
3 Osyris groweth in many places of Italy and Lombardie.

❧ *The Tyme.*

They floure most commonly in July, and August.

❧ *The Names.*

1.2 This herb is called in Shoppes Linaria, and of some Pseudolinum, and Vrinalis: in English Tode flaxe, and wilde flaxe: in French *Linaire* or *Lin sauuage*: in high Douch Lynkraut, Flaschkraut, Harnkraut, vnser frawen Flasch, wild Flasch, krotten flasch: in base Almaigne Wildt vlas.

3 The third kinde is called in Greeke ὄσυρις: in Latine Osyris: but in this our age it is called in Greeke ἄζυρις: and as we haue sayde it is called in Italy Beluedere: in English Stanneworte.   ❧ *The Nature.*

Stanworte is hoate and dry in the thirde degree.

❧ *The Vertues.*

The Decoction of Osyris, or Tode flaxe dronken, openeth the old, cold stoppings of the Liuer & Milte, & is singuler good, for such as haue the Iaunders, without Feuers, especially whan the Iaunders is of long continuance.   A

The same doth also prouoke vrine, and is a singuler medicine for suche as can not pisse, but droppe after droppe, and against the stoppings of the kidneys, and Bladder.   B

Of Shepherds purse.
Chap. lv.

❧ *The Description.*

Pastoria bursa.

BVrsa Pastoris hath round, tough, and pliable braunches, of a foote long: with long leaues, depely cut or iagged, like the leaues of Seneuy, but much smaller. The floures are white, & grow alongst by the stalkes, in place whereof whan they are gone there riseth small flatte Coddes, or triangled pouches, wherein the seede is conteyned, whiche is small, and blacke. The roote is long, white, and single.

❧ *The Place.*

Sheepeherds pouche groweth in streates and wayes, & in rough, stonie, and vntilled places.

❧ *The Tyme.*

It floureth most commonly in June and July.

❧ *The Names.*

This herbe hath neither Greeke nor Latine name giuen to him of the Ancient writers, but the later writers, haue called it in Latine Pastoria bursa, Pera & Bursa

Burſa paſtoris: in Engliſh Shepherds purſe, Scrippe, or Pouche: and of ſome Caſſeweede: in French Labouret, or Bourſe de bergers: in high Douch Deſchelkraut, and Hirten ſechel: in baſe Almaigne Teſkens or Borſekens crupt.

※ *The Nature.*

It is hoate and dry in the thirde degree.

※ *The Vertues.*

The Decoction of Shepherdes purſe dronken, ſtoppeth the laſke, the bloudy flixe, the ſpitting and piſſing of bloud, womens termes, and all other fluxe of bloud, howſoeuer it be taken: for whiche it is ſo excellent, that ſome write of it, ſaying, that it will ſtanche bloud if it be but only holden in the hande, or carried about the body.

Of Cinquefoyle, or Fiue fingar graſſe. Chap. lvi.

※ *The Kyndes.*

There are foure ſortes of Pentaphyllon, or Cinquefoyle: two kindes therof beareth yellow floures, wherof the one is great, the other ſmal. The third kinde beareth white floures, and the fourth kinde redde floures, all are like one an other in leaues and faſhion.

Pentaphyllon luteum maius.	Pentaphyllon luteum minus.
Yellow Cinquefoyle the greater.	Yellow Cinquefoyle the leſſer.

※ *The Deſcription.*

1. The great yellow Cinquefoyle, hath rounde tender ſtalkes, creeping by the ground, and running abroade, like the ſtalkes or branches of wilde Tanſie, and

and taking holde in diuers places of the ground, vpon whiche slender branches groweth long leaues snipt or dented round about ye edges, alwaies fiue growing togither vpon a stem, or at the ende of a stem. The floures be yellow, and parted into fiue leaues. The which whan they are vanished do turne into small round, and harde bearies, like Strawberies, in which groweth the seede. The roote is blackish, long and slender.

2 The small yellow Cinquefoyle is much like the other, in his leaues, & creeping vpon the ground, also in his stalkes, floures, and seede, sauing that it is a greate deale smaller, and doth not lightly take holdfast & cleaue to the ground, as the other doth. The leaues are smaller then the others, and of a whitish colour vnderneth, next to the grounde.

Pentaphyllon album.
White Cinquefoyle.

Pentaphyllon Rubrum.
Redde Cinquefoyle.

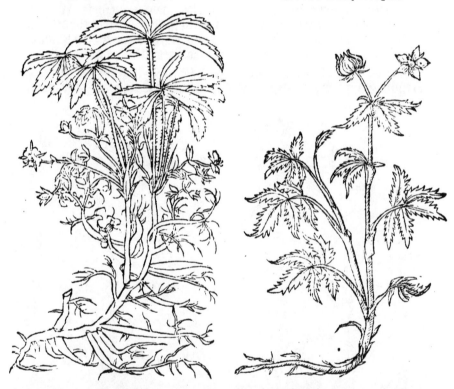

3 The white Cinquefoyle, is like the great yellow Cinquefoyle, in his small and slender branches creeping by the grounde, and in his leaues diuided into fiue partes, but that his stalkes or branches be rough. The leaues be long, and not snipt or dented rounde about, but before onely. The floures be white, and the roote is not single, but hath diuers other small rootes hanging by.

4 The redde Cinquefoyle also, is somewhat like to the others, especially like the great yellow kinde. The leaues be also parted in fiue leaues, and nicte or snipte round about, the whiche are whitish vnderneth, and of a swarte greene colour aboue. The stalke is of a spanne or foote long, of colour browne, or reddishe, with certayne ioyntes or knots, but not hearie. The floures grow at the toppe of the stalkes, most comonly two togither, of a browne redde colour, after the

the whiche there commeth vp small round beries, of a swarte redde colour like Strawberies, but harder: within whiche the seede is conteyned. The roote is tender, and spreading about here and there.

❧ *The Place.*

Pentaphyllon or Cinquefoyle groweth low and in shadowie places, sometimes also by water sides, especially the redde kinde, whiche is onely founde in diches, or aboute diches of standing water.

❧ *The Tyme.*

Cinquefoyle floureth in May, but chiefly in June.

❧ *The Names.*

Cinquefoyle is called in Greeke πωτάφυλλον: in Latine & in Shoppes Pentaphyllum, and Quinquefolium: in Italian Cinquefolio: in Spanish Cinco en rama: in English Cinquefoyle, or Sinkefoyle, of some Fyueleaued grasse, or Fiuefingred grasse: in French *Quintefueille*: in high Douch Funfffingerkraut, & Funfftblat: in base Almaigne, Uyfvingercruyt.

❧ *The Nature.*

Cinquefoyle is dry in the thirde degree.

❧ *The Vertues.*

The roote of Synkefoyle boyled in water vntill the thirde parte be consumed, doth appease the aking, and raging payne of the teeth. Also if one hold in his mouth the decoction of the same, and the mouth be well wasshed therewithall, it cureth the sores and vlcers of the same. A

The same decoction of the roote of Sinkefoyle dronken, cureth the bloudy flire, and all other flure of the bellie, and stancheth all excessiue bleeding, and is good against the goute Sciatica. B

The roote boyled in vineger, doth mollifie and appease fretting and consuming sores, and dissolueth wennes and colde swellings, it cureth euilfauored nayles, and the inflammation and swelling about the siege, and all naughtie scuruinesse, if it be applied thereto. C

The iuyce of the roote being yet yong and tender, is good to be dronken against the diseases of the Liuer, the Lunges, and al poyson. D

The leaues dronken in honied water, or wine wherein some Pepper hath bene mengled, cureth Tertian, and Quartaine feuers: And dronken after the same maner, by the space of thirtie dayes, it helpeth the falling sicknesse. E

The leaues pound & layde too healeth filme burstings, or the falling doune of the bowelles or other mater into the Coddes, and mengled with salte and Hony, they close vp woundes, Fistulas and spreading vlcers. F

The iuyce of the leaues dronken doth cure the Jaunders, and comforte the Lyuer. G

❧ Of Tormentill, or Setfoyle.   Chap.lvij.

❧ *The Description.*

Tormentill is much like vnto Sinckefoyle: it hath slender stalkes, rounde, and tender, fiue or sixe springing vp out of one roote, and creeping by the ground. The leaues be small, fiue, or most commonly seuen growing vpon a stem, much like the leaues of Sinckefoyle, and euery leafe is likewise snipte and dented rounde about the edges. The floures be yellow, much like the floures of wilde Tansie, and Sinckefoyle. The roote is browne, redde and thicke.

❧ *The Place.*

Tormentill groweth in low, darke & shadowy woodes, & in greene wayes.

❧ *The*

The first Booke of

❧ *The Tyme.*

It floureth oftētimes, al the somer long.

❧ *The Names.*

This herbe is now called in Shoppes and in Latine Tormentilla, and of some in Greeke ἑπτάφυλλον: in Latine Septifolium: in English Setfoyle and Tormentill: in French *Tormentille*, & *Souchet de bois*: in high Douch Tormentill, Brickwurtz, and Rot Heylwurtz: in base Almaigne Tormētill. The markes and notes of this herbe do approche very neare to the description of Chrysogonum.

❧ *The Nature.*

It dryeth in the third degree.

❧ *The Vertues.*

Tormentilla.

A The leaues of Tormentill with their roote boyled in wine, or the iuyce thereof dronken prouoketh sweate, and by that meanes it driueth out all venim from the harte: moreouer they are very good to be eaten or dronken against all poyson, and against the plague or pestilence. The same vertue hath the dryed rootes, to be made in pouder and dronken in wine.

B Also the roote of Tormentill made into pouder, & dronken in wine whan one hath no feuer: or with the water of a Smythes forge, or water wherein Iron, or hoate & burning steele hath bene often quenched, whan one hath a feuer, cureth the blouddy flixe, & al other fluxes or laskes of the belly. It stoppeth the spitting of bloud, the pissing of blond, and the superfluouse running of womens floures, & all other kindes of fluxe, or issue of bloud.

C The Decoction of the leaues & roote of Tormentill, or the iuyce of the same dronken is good for all woundes, both inwardly, and outwardly: it doth also open and heale the stoppings and hurtes of the Lunges, and the Lyuer, and is good against the Iaunders.

D The roote of the same made into pouder, and tempered or knoden with the white of an Egge, and eaten, stayeth the desire to vomitte, and is good against the disease called Choler or Melancholy.

E The same boyled in water, and afterward the mouth being washed therewithall, cureth the noughtie vlcers, and sores of the same.

Of Strawberries.   Chap.lviij.

❧ *The Description.*

The Strawberrie with his small and slender hearie branches, creepeth alongst the ground, and taketh roote and holdefast, in diuers places of the ground like Sinckefoyle, the leaues also are somewhat like Sinckfoyle, for they be likewise cut and snipte round about, neuerthelesse it bringeth forth but onely three leaues growing togither vpon each hearie stem or footestalke. The floures be white, & yellow in the middel, somewhat after the fasshiō of Cinquefoyle, the whiche being past it beareth a pleasant round fruite, greene

at the

the Historie of Plantes.

at the firste, but redde whan it is rype, sometimes also ye shall finde them very white whan they be ripe, in taste and sauour very pleasant. ❧ *The Place.*

Strawberies growe in shaddowy wooddes, & deepe trenches, and bankes, by high way sides: They be also muche planted in gardens.

❧ *The Tyme.*

The Strawbery floureth in Aprill, and the frute is ripe in June.

❧ *The Names.*

The Strawbery is called in Latine Fragaria, Fragula: in English Strawbery & Strawbery plante: in French *Fraisier*: in high Douch Erdtbeerē kraut: in base Almaigne Erdtbesiencruyt. The frute is called in Latine Fraga: in French *Des fraises*: in high Douch Erdtbeer: in base Almaigne Erdtbesien.

❧ *The Nature.*

The Strawbery plante or herbe, with the greene and vnripe Strawberies, are colde & dry. The ripe Strawberies are colde and moyst.

❧ *The Vertues.*

A The Decoction of the Strawbery plante dronken, stoppeth the laske, & the superfluouse course of womens floures.

B The same decoction, holden & kept in the mouth comforteth the gummes, & cureth the naughtie vlcers and sores of the mouth, & auoydeth ye stinking of ye same.

C The iuyce of the leaues cureth the rednesse of the face.

D Strawberies quench thirst, & the continual vse of them is very good, for them that feele great heate in their stomacke.

Of Siluer weede, or wilde Tansie. Chap.lix.

❧ *The Description.*

THe wilde Tansie, is much like to the Strawbery plante, and Cinquefoyle, in his small and slender branches, and in his creeping alongst and hanging fast to the grounde, his stalkes be also small and tender. The leaues be long, deepely cut euen harde to the stem and snipt round about, much like to the leaues of Agrimonie, of a whitish shining colour next the ground, & of a faint greene aboue. The floures be yellow, much like

Fragaria.

Argentina.

like the floures of Cinquefoyle. The roote hath hearie strings.

❦ The Place.

Wilde Tansie groweth in moyst, vntilled, and grassie places, & about diches, but especially in clay groundes, that are left from tillage.

❦ The Tyme.

It floureth most commonly in June and July.

❦ The Names.

This herbe is now called in Latine Potentilla, and Argentina, and of some Agrimonia syluestris, or Tanacetum syluestre: in English wilde Tansie, Siluer weede, and of some wilde Agrimonie: in French Tanasie sauuage, or Bec d'oye, and Argentine: in high Douch Grensigh, Grensing, or Genserich, and according to the same in Latine Anserina: in base Almaigne Ganserick, and Argentine.

❦ The Nature.

Wilde Tansie is dry in the thirde degree.

❦ The Vertues.

Wilde Tansie boyled in wine or water and dronken stoppeth the laske, the bloudy flixe, and all other fluxe of bloud, and preuayleth muche against the superfluous course of womens floures, but specially agaynst the white floud, or issue of floures.

The same boyled in water and salte and dronken, dissolueth all clotted and congeled bloud, and is good for suche as are squatte and brused with falling from aboue.

The Decoction of wilde Tansie, cureth the vlcers, and sores of the mouth, the hoate humors that are fallen downe into the eyes, and the strypes that perishe the sight, if they be washed therewithall.

Wilde Tansie hath many other good vertues, as against the stone, inward woundes and corrupt or fretting vlcers of the gummes, and priuie or secrete partes, it strengtheneth the bowelles, and closeth vp greene woundes, it fasteneth loose teeth, and swageth tooth-ache. The distilled water of this plante is good against the feeckles, spottes, and pimpels of the face, and to take away Sunne burning.

Of common Mouse eare. Chap.lx.

❦ The Kindes.

Of the herbe called in Latine Pilosella, there is found at this time two kindes: The one called the great Pilosella, the other small Pilosella, the whiche some men do also call Mouse eare, howbeit they are not the right Mouse eare.

❦ The Description.

1. The leaues of great Pilosella are spreade vpon the grounde, white hoare, and hearie much lyke a great Mouse eare. The stalke is also hearie, aboute a spanne long, and beareth double yellow floures, the whiche do change into a certayne hearie downe or Cotton seede, and is caried away with the winde. The roote is of the length of a fingar, and hath many hearie strings.

2. The small Pilosella is like to the other, sauing that it is much lesse. The leaues be small and little, and white hoare nexte to the ground, and hearie also. The floures grow at the toppe of the stalkes, many togither, and are of colour sometimes yellow, sometimes redde, and sometimes browne, and sometimes speckled. The roote is small and threddy.

Pilosella

the Historie of Plantes.

Pilosella maior.
Great Pilosella.

Pilosella minor.
Small Pilosella.

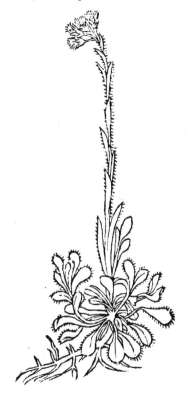

❧ *The Place.*

1 The great Pilosella groweth vpō small grauely or dry sandy mountaynes, and vpon dry bankes about the borders of fieldes.

2 The small Pilosella groweth in dry Heathes and Commons, and such like waste and vntilled places. ❧ *The Tyme.*

They floure in May and June.

❧ *The Names.*

1 The great is now called in Latine Pilosella maior: in English also Great Pilosella, & of some Mouse eare: in high Douch Nagelkraut, and of some also Meutzor: in base Almaigne groote Piloselle, and Naghelcruyt.

The small is called Pilosella minor, and of the high Almaignes Meutz-orlin and Hasenpfattin, it may be called in English small Pilosella, or Heath-mouse eare. ❧ *The Nature.*

The Pilosellas be hoate and dry.

❧ *The Vertues.*

The Decoction of the leaues and roote of this herbe dronken, doth cure, and A heale all woundes both inward and outward, and also Hernies, Ruptures, or burstings.

The leaues of Pilosella dryed, and afterward made into pouder, & strowen B or cast into woundes, is able to cure and heale the same.

The iuyce of the great Pilosella, dropped into the eares, cureth the payne C of the same, and clenseth them from all filth and corruption.

H ij   The

The same Pilosella eaten or taken in meates, doth clense & clarifie the sight and cureth the rednesse of the eyes.

Of Golden floure Stechados, and Cotton weede. Chap. lxj.

❧ *The Kindes.*

Here be two principall kindes of herbes, whiche beare white, softe, and woolly, or Cottony leaues, whiche some men call Pilosellæ, or Filagines. The one hauing fayre golde yellow and sweete smelling floures. The other hath pale yellow floures without smell, & is of three sortes, as it shall appeare hereafter.

❧ *The Description.*

Ageratum Aurelia.
Golden Stæchas or Moth weede.

Filago.
Cotton weede.

1 The first of these herbes which the Almaignes do call Rheinblumen hath slender stalkes, round and cottonie, harde, & whitish, or of a hoare gray colour, of a spanne long, with small, narrow, & very softe rottonie leaues, in quantitie & making like þ leaues of Hyssop: at the toppe of the stalkes groweth small tuftes, or as it were nosegayes of ten or xij. floures or more, whiche are round in fashion like to small buttós, the which do not lightly perishe or vade, but may be kept a long time in their estate, & colour, neither are they of an vnpleasant sauour, but are somewhat bitter. The roote is small, shorte, and blacke.

The other kinde which is called Filago, or Cotton weede, is of three sortes, as is before sayde.

2 The first of them is like to the aforesayde, but it is greater and higher, sometimes growing to the heigth of two foote long, or more. The stalkes be small, rounde and grayishe, couered with a certayne fine wooll or Cotton, three or foure growyng vp from one roote, straight, and most commonly without any branches. The leaues be long, narrow, whitish, softe, and woolly, like the leaues of Golden or yellow Stæchas, sauing that they be longer, and broader, and somewhat of a greener colour. The floures be rounde, and after the fashion of buttons, growing at the toppe of the stalkes, a great many togither, but nothing so yellow, as the floures of Golden Stæchas, neither so long lasting, but are carried away with the winde whan they be ripe, like diuers other floures.

The

the Historie of Plantes.

3 The other kinde of Cotton weede, in stalkes and leaues is much lyke to the aforesayde, the floures also be like to the aforesayde: howbeit they grow not in tuffetes at the toppe of the stalke, but betwixt the leaues alongst by the stalke, and this is the greatest difference, betwixt this and the other.

4 The fourth kinde of these herbes, is like to the two other, last recited Cottonweedes, in stalkes and whyte cottony leaues, but it is altogither tenderer, smaller, and lower, seldome growing to the length of ones hande. The floures grow at the toppe of the stalkes, in small round buttons, of colour and fasshion like to the other Cotton weedes.

❀ *The Place.*

The first kinde groweth in sandy playnes, and dry Heathes, and is plentifully founde in sundry places by the riuer of Reene. In this countrey they sow it in gardens. The other three kindes groweth in this countrey in sandy groundes, about dry Diches, and in certayne moyst places, and in wooddes.

❀ *The Tyme.*

They floure most commonly, in June, and July. Cotton weede floureth often, and againe in August.

❀ *The Names.*

1 The firste kinde of these herbes is called of Theophrastus in his nienth Booke, and xxj. Chapter, in Greeke ἐλειόχρυσον: in Latine of Theodore Gaza Aurelia: And of Dioscorides ἀγήρατον Ageratum: in Shoppes Sticas citrina, and Sticados citrinum: Of some Tinearea, and Amaranthus Luteus: in English Golde floure, Mothewort, or Golden Stechados, and of Turner Golden Floureamor: in high Douch Rheinblumen, Mottenblumen, & Mottenkraut: in base Almaigne Rheynbloemen and Rhijnbloemen.

There is yet an other herbe descrybed by Dioscorides, called ἐλίχρυσον Elichrysum and Amaranthus, the whiche is nothing like to Sticas citrina. For ἐλειόχρυσον Eliochryson of Theophrastus, and ἐλίχρυσον Elychryson of Dioscorides, are two seuerall herbes. And therefore they are greatly deceyued that thinke Sticas citrina to be the Elichryson of Dioscorides.

2 The other three are all called (at this tyme) by one name, in Latine Filago: in Spanish *Yerua Golandrina*: in high Douch Rhurkraut: in base Almaigne Rhuercruyt, or Root melizoen cruyt, that is to say, Bloudy Flixeworte.

The first of these three is called of Plinie, Herba impia, bycause that his last floures, do surmount and grow higher than the first.

Some would haue these three herbes, to be that whiche the Gretians call γναφάλιον: and the Latinistes Centunculum, Centuncularis, and Tomentitia, but yet their iudgement is not right, as it shall appeare in the Chapter nexte following.

❀ *The Nature.*

These herbes be of a drying nature. The Golde floure or golden Stæchas, is hoate also, as it may be perceyued by his bitternesse.

❀ *The Vertues.*

Golden Stæchas boyled in wine and dronken, killeth wormes, and bringeth them forth, and is good agaynst the bytings and stingings of venemouse beastes.

The same boyled in lye, clenseth the heare from Lyce, and Nittes. The same layde in Warderoobes and Presses, keepeth apparell, and garmentes from Mothes.

The first Booke of

Of Gnaphalion, or Small Cotton. Chap.lxij.

❧ *The Description.* Gnaphalium.

Naphalion is a base or low herbe, with many slender softe branches, and small leaues, couered all ouer w^t a certaine white cotton or fine wooll and very thicke: so that ye would say it were all wooll or Cotton. The floures be yellow, & growe like buttons, at the top of the stalkes, as ye may perceyue by the figure.

❧ *The Place.*

This herbe groweth no where, but by the sea coast, there is plenty of it in Languedock, and Prouince.

❧ *The Tyme.*

It floureth in June and July.

❧ *The Names.*

This herbe is called in Greeke γναφάλιον: in Latine Gnaphalium, Centunculus, Centuncularis, Tucularis, Albinum, & of some Gelaso, Anaphalis, Anaxiton, Hires, and Tomentitia. Also Bombax humilis: in English of Turner, Cudweed, Chafeweed, Cartaphilago. It may be called also Pety Cotton, or small Bobase: in Frēch *Petit Coton, & l'herbe borreuse,* or *Cotoniere.* Pena in his Stirp. Aduers. noua, calleth it Chamæzylon.

❧ *The Nature.*

Gnaphalion is dry and astringent.

❧ *The Vertues.*

The leaues of Gnaphalion, boyled in thicke red wine, are good against the blouddy flire, as Dioscorides, and Galen doth witnesse.

Of Plantayne or Waybrede. Chap.lxiij.

❧ *The Kyndes.*

Here is found in this coūtrey, of three sortes of Plantayne great plenty, bysides whiche there is yet an other founde, which groweth in salte grounde, all whiche in figure, do partely resemble one an other.

❧ *The Description.*

1 The great Plantayne hath great large leaues, almost like to a Beete leaffe, with seuen ribbes behinde, on the backeside, which do al assemble & meete togither, at the ende of the leafe next the roote. The stemmes be round, of the heigth of a foote or more, sometimes of a reddish colour and comming vp from the roote in the middell amongst the leaues, the which stalkes or stēmes, from the middle vpwarde towardes the toppe, are couered rounde about with small knoppes or heads (whiche first of all do turne into small floures, and afterward into smal huskes conteyning a blackish seede) like to a Spike eare, or a little

the Historie of Plantes.

little Torche. The roote is shorte, white, and of the thickenesse of a fingar, hauing many white hearie threedes. Of this kinde, there is founde an other, the Spikes, eares, or torches wherof, are very dubble, so as in every Spyky eare, in steede of the little knappes or heades, it bringeth foorth a number of other small torches, wherof eche one is lyke to the spike or torch of great Plantayne.

Plantago maior.
Great Plantayne.

Plantago media.
Middle Plantayne.

2. The seconde kinde of Plantayne, is like to the first, sauing that his leaues be narrower, smaller, and somewhat hearie. The stalkes be round, and somewhat cottony or hearie, and bringeth foorth at the toppe, spiked knoppes, or torches, a great deale shorter than the first Plantayne, the blowings of it are purple in white. The roote is white and longer than the firste.

3. The thirde kinde of Plantayne is smaller than the seconde. The leaues be long and narrow, with ribbes like the leaues of the other Plātayne, of a darke greene colour, with small points, or purles, set, here and there alongst the edges of the leaues. The stalkes be crested or straked, and beare at the toppe fayre spiked knappes with white floures or blossoms, like the spykie knoppes of the middle Plantayne. The roote is shorte and very full of threedy strings.

4. The Sea Plantayne is narrower, it hath long leaues very narrow, and thicke: the stalkes be of a spanne long, full of small graynes or knops, from the middle euen vp to the top, much like to the torche of the great Plantayne. The roote is also threddy.

❧ *The Place.*

The three first Plantaynes grow almost every where in this countrey, in pastures and leases, about wayes, and moyst places.

Plantago

The first Booke of

Plantago minor.
Small Plantayne, Ribworte.

Plantago marina.
Sea Plantayne.

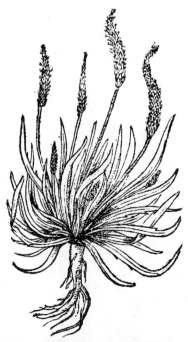

The Sea Plantayne groweth in salt groundes, vpon the bankes and borders of salte water streames, as in Zealand, & Barowgh in great plenty, by the water Zoom. ❧ *The Tyme.*

The Plantaynes do floure most commonly in this countrey, in the monethes of Iune and Iuly. The seede is ripe in August.

❋ *The Names.*

Plantayne is called in Greeke ἀρνόγλωσσος, that is Lingua Agnina, Lambes tungue: ἄρνειον, προβάτειον, πολύνευρον, καὶ ἑπτάπλευρον: in Latine and in Shops Plantago: in Italian *Plantagine*: in Spanish *Tamchagen, Lengua de oueja*: in English Plantayne: in high Douch Wegrich or Schaffzungen; in base Almaigne Wechbree.

1 The first kinde is now called in Latine Plantago maior, and Plantago rubra: in English Great Plantayne: in French *Grand Plantaine*: in high Douch Roter Wegrich: in base Almaigne Roode wechbree.

2 The second kind is called in Latine Plantago media: in French *Plantain moien*: in English Middle Plantayne: in high Douch Mittel and breyter Wegrich: in base Almaigne Breet wechbree.

3 The thirde is called of some in Greeke πωτάνευρον, that is to say, in Latine Quinqueneruia: otherwise it is now called in Latine Lanceolata, and Lanceola: in English Ribbeworte: in French *Petit Plantain, Lanceole*, and *Lanceolette*: in high Douch Spitzer wegrich: in base Almaigne Cleyn wechbree, & Hontsribbe.

4 We call the fourth Plantago marina: in English Sea Plantayne: in French *Plantain de mer*: in base Almaigne Zee wechbree.

❧ *The*

✿ *The Nature.*

Plantayne is colde and dry in the second degree.

✿ *The Vertues.*

The leaues of Plantayne eaten with meates, or otherwise are very good against the falling downe of Reumes & Catarres, they comfort the Stomacke, and are good for such as haue the Phthisike (which is a disease in the lunges, with a consumption of all the body.) And against the Cough. A

The Decoction of leaues of Plantayne dronken, stoppeth the bloudy flire, and other fluxes of the belly, also it stoppeth the spitting of bloud, the pissing of bloud, and the superfluous flowing of womens termes, and all other issue of bloud. B

The iuyce of Plantayne dronken, stoppeth and appeaseth the great desire to vomitte, and stancheth all fluxe of bloud, aswell as the leaues and seede. C

The roote of Plantayne by himself, or with his seede boyled in sweete wine and dronken, openeth the Conduytes, or passages of the Lyuer and kidneys being stopped, and is good against the Jaunders, and the ulceration of the kidneys, and bladder. D

Some haue writen, that three rootes of Plantayne, taken with wine and water, doth cure the Feuer tertian: and foure rootes so taken do cure the Feuer quartayne. E

The vse of Plantayne is good against all euill, corrupt, and running sores and vlcers, and against woundes both old and new, all hoate emposteins, and inflammations, against Cankers, Fistulas, & the foule euill or French Pockes, and all scuruinesse. It is good against the byting of Madde Dogs, to bruse the leaues of Plantayne and lay therevpon, or to poure of the iuyce of Plantayne into the woundes, or if it be mixed with emplaysters, and oyntments, that be made for such purpose. F

The leaues of Plantayne do asswage, and mitigate the paine of the Goute, and are excellent to be layde vpon swollen members, that are full of heate and payne or anguish. G

The iuyce of Plantayne dropped or stilled into the eares, is very good against the payne in the same. And to be dropped into the eyes against the inflammation, and payne of the eyes. H

The same iuyce or the Decoction of the leaues or rootes of Plantayne, do cure and heale the naughtie Ulcers of the mouth, the tooth ache, and the bleeding of the gummes or Jawes, whan the mouth is oftentimes washed with the same. I

The leaues of Plantayne pounde or stamped with salte, and layde to the emposteins, wennes or harde swellings about the eares and throte, cureth the same. The roote also, is good to be carried or hanged about the necke, for the same purpose, as some men wryte. K

Of Buckhorne Plantayne, or Coronop Plantayne.   Chap.lxiiij.

✿ *The Kyndes.*

3 There is founde in this countrey at this present, two kindes of herbes, both comprehended vnder the name of Crowfoote.

✿ *The Description.*

The first Crowfoote or Hartshorne, hath long narrow and hearie leaues, & bringeth forth vpon each side of the leafe three or foure shorte startes or branches, almost like to the branches of a Hartes horne. It lieth spread vpon

The firſt Booke of

vpon the ground like a ſtarre. Frō the middle of thoſe leaues, groweth vp ſmall round hearie ſtemmes, bearing long ſpiked knappes, oꝛ toꝛches, like the middle Plantayne. The roote is long and thꝛeddy.

Pſeudocoronopus.
Buckhoꝛne Plantayne.

Coronopus Ruellij.

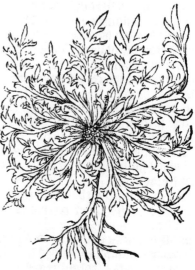

Coronopi ſpecies peregrina.

3 **The ſecond Crowfoote**, hath thꝛee oꝛ foure ſtemmes oꝛ bꝛanches, creeping vpon the ground, & alwayes lying flat vpon the earth, but neuer mounting oꝛ riſing higher, & are ſet full of long, narrow, & iagged leaues, much like to the leaues of the other Crowfoote Plantayne, but ſmaller, and nothing hearie. The floures be ſmall & white, & growing betwixt the leaues & the ſtalke, well faſtened to the ſtēme, whan they are decayed, there cōmeth foꝛth ſmall flat purſes, bꝛoade & rough, in whiche the ſeede is conteyned. The roote is white of the length of ones fingar, in taſt lyke to garden Creſſis.

One may alſo place amongſt yͤ kindes of Coronopus, a certayne herbe, whiche we ſhall now offer vnto you (the which is a ſtranger, & but little knowen in this countrey)

countrey) seing that it is very well like to Harte Horne. The leaues be long & narrow, branched with shorte startes, altogither like to the leaues of Hartes horne, sauing that sometimes they be bigger. They lie also flat, & spread round vpon the ground, and are somewhat rough, & hearie, like the leaues of Hartes horne: so that it is harde to know one from an other, whan they are both without stalkes and floures. But whan this herbe beginneth to haue stalkes and floures, than the difference is easily marked: for this herbe bringeth forth two or three rounde stalkes, parted into sundry branches, at the toppe whereof are placed knoppes and buttons, like to Cyanus or Corne floure, sauing that the scales of the knappes or heades, be not so closely couched, and layde one vpon an other, & the sayde scales seeme cleare and thorough shining, especially whan the floure is fallen of and withered. The floures come forth of the sayde knops or heads, in colour and making like the floures of Cychorie, but smaller. The roote is long and slender.

The Place.

1. The first kinde groweth in Brabant & Flaunders, in vntilled, sandy places.
2. The second also groweth about wayes, and dry sandy pathes, and vpon bankes and rampiers, especially in certayne places about Antwarpe, where as it groweth so plentifully, that almost one shall see none other herbe.
3. The third which is a strange herbe, groweth not of his owne kinde in this countrey, but it is planted in gardens. It groweth plentifully in Languedock, in stony and dry places.

The Tyme.

1. 2. The two first kindes, do floure in May, and June.
3. The third floureth, in July, and August.

The Names.

1. The first is called in Latine Cornu ceruinum, or Herba Stellæ, and Stellaria: in English of Turner, and Cooper Herbe Iue, and Crowfoote Plantayne: of Pena, Buckhorne. We may also call it Hartes horne Plantayne, Buckehorne Plantayne, or Coronop Plantayne: it is called in French Corne de Cerf, or Dent au chien: it is vnknowen in Shoppes: The Brabanders, do call it Hertshoren, and Crayenuoet cruyt: Some late writers call it in Greeke κορωνόπους: in Latine Pes cornicis: in high Douch Kraenfuz, supposing it to be the same Coronopus, whiche Dioscorides hath described in the.123. Chap. of his second booke, although in deede they be not like at all: And therefore we haue called it ψευδος κορωνόπους, Pseudocoronopus, that is to say, Bastard Crowfoote.

2. The second kinde draweth neare to the description of Coronopus made by Dioscorides, albeit that notwithstanding, the learned Ruellius, Doctor in his time at Paris, could not be made beleeue, that this was the right Coronopus. Wherefore for the same Ruellius sake, who made a liuely description of this herbe, we do now call it Coronopus Ruellij: in base Almaigne Crayenuoet, or Rauenuoet. They call it at Paris Verrucaria: in some places of England they call it Swynescressis. We may also call it Ruellius Coronopus.

3. This strange herbe hath no name as yet, knowen vnto vs, sauing that the Herboristes of Languedock take it to be a kind of Scabious, or for the Cornefloure called Cyanus, in English blew Bottell. A man may doubt whether this be not Dioscorides Coronopus, bycause that Hartes horne should seeme to be a kinde of Plantayne. But bycause the Description of Coronopus is very short, we are not able to assure you. This may also be a kinde of Condrilla.

The Nature.

1. Hartes Horne is colde and dry in temperature much like Plantayne.

The first Booke of

2 The Swines Cresses, or Ruellius Corónopus (as it is euident by the taste) is hoate and dry, like to garden or towne Cressis, but not althing so hoate.

❧ *The Vertues.*

Hartes horne is in vertue like to Plantayne, whereof it is a kinde, and may be vsed in all things whereto Plantayne serueth. Also it hath bene proued singuler against the pissing of bloud, the grauell & the stone, to be taken in meates or otherwise.

If Swines Cressis, or Ruellius Coronop, be the true Coronopus, than the roote thereof rosted in the imbres or hoate Ashes, and eaten in meates is very good against the laske proceeding from the coldnesse of the stomacke, whiche is the cause of slimie humors in the Guttes: for whiche purpose the very sent, and taste of the roote here described, declareth the same to be very good, bycause it is hoate and somewhat astringent.

Of Bloud strange, or Mouse tayle. Chap. lxv.

❧ *The Description.* Myosouron.

Mouse tayle is a small low herbe, with small leaues and very narrow, emongst whiche springeth vp from the roote small stemmes, garnisshed with very small whitish floures, and afterward with little lōg torches, much like to a Mouse or Rattes tayle, & like the seede or torches of Plantayne, before it blooweth, in whiche is contepned very small and browne seede.

❧ *The Place.*

Mouse tayle groweth in good pastures, and certayne medowes, and sometimes also by high way sides.

❧ *The Tyme.*

It floureth in Aprill, and the torches and seede is ripe in May, & shortly after the whole herbe perissheth, so that in June, ye shall not finde the dry or withered plante.

❧ *The Names.*

It is called in English Mouse tayle, & Bloud strange: in French *Queue de souris*: and accordingly in Greeke μυὸς ὀυρὰ, ἢ μυόσουρ⊙: in Latine Cauda murina, and Cauda muris: in high Douch Tausent korn: in base Almaigne Muyse steertkens. This is not Holosteum, neither Denticula Canis Ruellij, as some do iudge.

❧ *The Nature.*

The leaues of this herbe do coole, and differ not muche from the nature of Plantayne.

❧ *The Vertues.*

The operation and vertues of this herbe, are not yet knowen, howbeit, as farreforth as men may iudge by the taste and sente thereof, it is much like in facultie to Plantayne.

Of Water Plantayne. Chap. lxvi.

❧ *The Description.*

Water Plantayne is a fayre herbe, with large greene leaues, not muche vnlike the leaues of Plantayne, with a stalke full of branches, & small white floures, diuided into three partes, and after them it bringeth forth tryangled huskes or buttons, the roote is of threddy strings.

❧ *The Place.*

This

the Historie of Plantes.

Plantago Aquatica.

This herbe groweth about the borders and brinkes of diches and pondes, & somtimes also in riuers and brookes.

❧ *The Tyme.*

It floureth from June till August.

❧ *The Names.*

This herbe is now called in Latin Plātago aquatica: in English water Plátayne: in French *Plantain d'eau*: in high Douche wasser Wegrich, and Frochloefelkraut: in base Almaigne water Wechbree.

❧ *The Nature.*

Some men write of this herbe, that it is of temperament colde and dry.

❧ *The Vertues.*

A   Some lay store of the leaues of water Plantayne, vpon the shanks or shinnes of such as haue the Dropsie, supposing that ye water in the belly shall by that meanes be drawen downe to the shinnes or shanks.

B   The learned men of our time do write, that it hath the same vertues, & faculties as the other Plantayne, wherof we haue alreadie written in the lxij. Chapter.

Of Knotgrasse. Chap.lxvij.

❧ *The Kyndes.*

There be two kindes of this herbe as Dioscorides writeth, the Male, and the Female: the Male is called in Englishe Swynes grasse, and Knot grasse, but the Female is called small Shaue-grasse.

❧ *The Description.*

1   Knot grasse hath many round, weake & slender branches, full of knots and ioyntes, and creeping alongst the grounde, it hath long narrow leaues, not much vnlike the leaues of Rew, sauing that they be loger. The floures be small, growing alongst the branches betwixt the leaues and the ioyntes, of colour sometimes white, sometimes purple or incarnate, after them commeth a triangled seede, like to sorrell seede. The roote is round and reddish with many strings.

2   The second kinde whiche they call female knot grasse, hath three or foure vpright, round, and euen stemmes, without branches, full of ioyntes, and much like to the stalkes and ioyntes of Hippuris, or Horse tayle, but not so rough, and about the ioyntes groweth many small, and narrow little leaues, like to a Starre, and not much vnlike the leaues of Rosemary. The roote is white and runneth alongst the grounde, putting forth many new shutes or springs.

3   Amongst the kindes of knot grasse, we may well recken that herbe, whiche doth so wrap & enterlace it self, & is so ful of ioynts, that the base Almaignes cal it Knawel, that is to say, knot weede, it groweth to the heigth of a mans hand, & bringeth forth many tender braches full of knotty ioynts, entagled & snarled, or wrapped one in an other. The leaues be smal & narrow, well like to Juniper

I   leaues

The first Booke of

Polygonum mas.
The male knot graſſe, or Swines graſſe.

Polygonum fœmina.
Female knot graſſe, or ſmall Shauegraſſe.

Polygonum tertium.
The third knot graſſe.

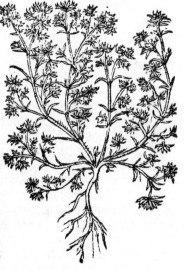

leaues, ſauing they be ſmaller and not prickly: amongſt which groweth little floures after the faſſhion of ſtarres, in colour like to the ſtemmes and leaues, which are grayiſh. The roote is hearie, and as long as ones fingar.

❧ *The Place.*

1   The Male knot graſſe groweth, in fieldes about wayes and pathes, and in ſtreates.

2   The Female groweth in moyſt places, about þ brinkes & borders of runing waters.

3   The third groweth about chāpion fields, & places not well huſbanded, eſpecially in a moyſt yeare.   ❧ *The Tyme.*

The Male knot graſſe, & the third kind do floure, from after June vntill the end of Somer. The female is found moſt commonly in July & Auguſt.   ❧ *The Names.*

Knot graſſe is called in Greeke πολύγονον : in Latine Sanguinaria, Sanguinalis, and Proſerpinata.

1   The firſt kinde is called in Greeke πολύγον ἄῤῥεν, καὶ πλίγονον, ἢ πολύκαρπον: in Latine Seminalis: in Shoppes Centumnodia, & Corrigiola: of ſome Sanguinaria, Sanguinalis,

the Historie of Plantes.

Sanguinalis, Proserpinaca: in Italian *Corrigiola*: in Spanish *Corriola, y cien nudos y rua* in English male knot grasse: in French *Renouee, & Corrigiole*. in high Douch Weggras, and Wegtritt: in base Almaigne Wechgras, Uerkens gras, and Duysent knoop manneken.

2 The second is called in Greeke πολύγονον θῆλυ: in Latine Sanguinalis fœmina: in base Almaigne Duysentknoop wijfke: in English of Turner Medow shauegrasse, and small Shauegrasse.

3 The thirde kinde is called in base Almaigne Knawel, the whiche without doubte is a kinde of knot grasse, albeit Dioscorides hath described but twoo kindes: Neither do we take it to be Polygonon of Dioscorides, but for one of the foure kindes of Polygonon, whereof Plinie hath writen in the xxvij. booke of his History.

※ *The Nature.*

All these herbes are colde in the second degree, and dry in the thirde, astringent, and making thicke.

※ *The Vertues.*

1 The iuyce of knotgrasse dronke, is good against the spitting of bloud, the pissing of bloud, and all other fluxe or issue of bloud, and is good against vomiting and laskes. A

The same dronken in wine, helpeth against the biting of venemouse beasts. B

It is also good against tertian feuers, to be dronken, an houre before the fit. C

The leaues of knotgrasse boyled in wine or water and dronken, stayeth all D maner of laskes and flures of the belly. The bloudy flire, and womens floures, the spitting of bloud, and all flure of bloud, aswell as the iuyce.

The iuyce of knotgrasse, put with a Pessarie into the naturall places of wo- E men, stoppeth the floures, and the inordinate course of the same: and put into the Nose, it stancheth the bleeding of the same: poured into the eares, it taketh away the payne of the same, and dryeth vp the corrupt matter and filth of the same.

The same boyled in wine and Honie, cureth the vlcers, and inflammations F of the priuie or secrete partes.

The greene leaues being layde too, preuayle much against the great heate & G burning of the stomacke, hoate swellings & empostems, the consuming & burning of S. Anthonies fire, and all greene or fresshe woundes.

Dioscorides also saith, that knotgrasse prouoketh vrine, & is good for such H as pisse drop after droppe: the whiche is founde true, whan the vrine is hoate and sharpe.

2 The female knotgrasse hath the same vertue, as the male knotgrasse (as I Dioscorides saith) but not so strong.

3 And the third kinde also, his vertues be much like to the Male knotgrasse.

Of Horse tayle, or Shauegrasse. Chap. lxviij.

※ *The Kindes.*

There be twoo sortes of Horse tayle, or Shauegrasse, as Dioscorides and Plinie writeth.

※ *The Description.*

When the great Shauegrasse or Horse tayle beginneth to spring, it bringeth foorth rounde naked, and hollow stemmes, rough and full of ioyntes: yea their roughnesse is such, that Turners, Cutelers, & other Artificers, do vse them to polish, & make playne, & smoth their workes, as the heftes of knyues & Daggers &c. At the top of those Asparagus, shutes or ſtemmes, groweth ſmal, round, and blacke knoppes oꝛ tuffets. Afterwarde the ſtemmes do waxe bꝛowne and reddiſhe, and bꝛinge foorth rounde about euery knot oꝛ ioynte, diuers little, ſmal, ſlender, and knottie ruſſhes. It mounteth ſo high, that with his hanging ruſſhes, oꝛ ſmall bꝛanches, it is not much vnlyke to a Hoꝛſe tayle. The roote is white and hath ioyntes oꝛ knottes lyke the ſtalke oꝛ ſtemme.

Maioris Equiſeti aſparagus.	Equiſetum minus.	Equiſeti minoris flores.
The.j.ſpꝛings oꝛ ſhutes of Hoꝛſetayle, oꝛ ſhauegraſſe.	Smal ſhauegraſſe oꝛ Hoꝛſetayle.	The floures of ſmal Shauegraſſe oꝛ Hoꝛſe tayle.

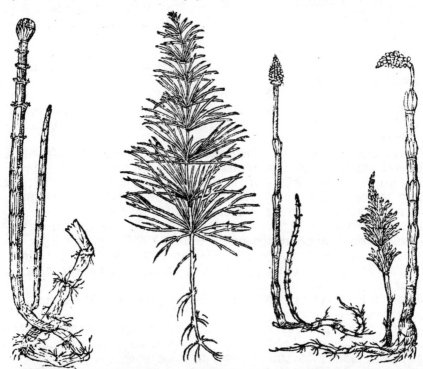

2 The ſmall Shauegraſſe oꝛ Hoꝛſe tayle, is not much vnlike to the great. It bꝛingeth foꝛth whan it beginneth to ſpꝛing, bare and naked ſtemmes, that be alſo round, hollow and knotty, at the toppe oꝛ ende of thoſe ſtemmes it hath as it were, a ſpiked eare oꝛ knop, of ſmall white floures, whiche periſh incontinently: Than ſpꝛingeth vp from the roote other ſhutes oꝛ bꝛanches, full of knottes oꝛ ioyntes, about the whiche alſo there groweth round knotty ruſſhes, like as in the great Hoꝛſe tayle oꝛ Shauegraſſe, but not ſo great noꝛ ſo rough, but moꝛe ſofte and gentell, ſo that they are nothing woꝛth to poliſhe withall. The roote is ſmall, blacke, and ſlender.

❧ *The Place.*

The great Shauegraſſe, groweth in diches, and pondes, and very moyſt places. The ſmall Hoꝛſetayle oꝛ Shauegraſſe groweth in low ſhadowy places, and alſo in dꝛy ſandie fieldes. ❧ *The Tyme.*

The naked ſtemes of the great Hoꝛſetayle, do ſpꝛing vp in May. The ſhutes and bloſſoms of the ſmall Hoꝛſetayle do ſpꝛing in Apꝛill, ꝛ ſhoꝛtly after cōmeth vp the vp the stemmes, set full of small rushes.

❧ *The Names.*

These herbes are called in Greeke ἱππουρις, & of some ἔφυδρον καὶ ἀνάβασις: in Latine Equisetū, Equiseta, Equiselis, Equinalis, and Salix Equina: in Shoppes Cauda equina: in Italia *Asprella*, *Codo di canallo*, *prela*: in Spanish *Cola de mula*, *Rabo de mula*: in English Horse tayle, and Shauegrasse: in highe Douch Schaffthew: in base Almaigne Peertsteert.

The greater kinde is called Equisetum maius, & of some Asprella: in English great Shauegrasse, and Horse tayle: in high Douch grosz Schaffthew, Roszschwātz, Pferdtschwantz, Roszwadel, Kannenkraut: in base Almaigne groot Peertsteert and Kannencruyt.

The small is called in Greeke ἱππουρις ἕτερα κὶ ἐκύτιορ: in Latine Equisetum minus, aut alterum, and Equitium. And of some as Anthonius Musa writeth, Sceuola: in English smal Shauegrasse, and of some Tadpipes: in high Douch kleyn Schaffthew, Katzenwedel, Ratzenschwantz, Katzensaghel: in base Almaigne cleyn Peertsteert, and Cattensteert.

❧ *The Nature.*

These two Shauegrasses or Horse tayles, are colde in the first degree, and dry in the second, astringent, and drying without sharpnesse.

Equisetum maius.
The great Shauegrasse, or Horse tayle.

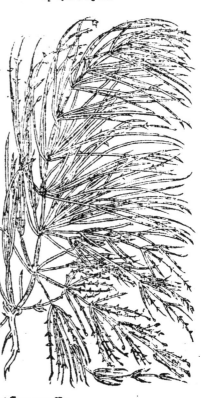

❧ *The Vertues.*

The decoction of Horse tayle, in wine or water dronken, stoppeth all fluxe of A bloud, & al other extraordinary fluxes, especially the inordinate issue of floures, it doth also cure the bloudy flixe and dangerous laske, and all other kinde of laskes. And for all the aforesayde ententes it is a soueraigne remedie (as Galen writeth). The iuyce of this herbe dronken alone or with wine, is of the same operation and effect.

Horse tayle or Shauegrasse, being taken in manner aforesayde, is most cō- B uenient and profitable, for all vlcers, sores, and hurtes of the kidneys, the bladder and bowels, and against all burstings.

Horse tayle with his roote boyled, is good against the Cough, the difficultie C and payne of fetching breath, and against inwarde burstings as Dioscorides and Plinie writeth.

The iuyce thereof put into the Nose, stancheth the bleeding of the same, and D with a Pessarie or Mother Subpositorp conueyed into the naturall places of women, stoppeth the floures.

The same pounde and strowed vpon freshe and greene woundes, ioyneth E them togither and healeth them, also it preserueth them from inflammation. And so dothe the powder of the same herbe dryed, and strowed vpon new, and greene woundes.

The first Booke of
Of white Roote, or Solomons seale. Chap.lxix.

❧ The Kindes.

White roote or Salomons seale is of two sortes. The one called the great or broade Seale of Salomon: The other is the small and narrow Salomons seale.

Polygonatum latifolium. Polygonatum angustifolium.
Broade leaued white roote. Narrow leaued white roote.

 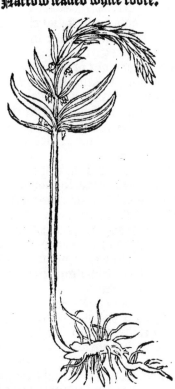

❧ The Description.

1. The great Salomons seale, hath long rounde stalkes: the leaues be long and greene, larger, longer, & softer then bay leaues, betwixt the whiche leaues and the stalke, vpon short stemmes, hang pleasant white greene floures, long and hollow, three or foure togither, so that euery stalke doth commonly bring forth, moe floures than leaues. The floures perished, they turne into rounde bearies, the which be greene at the first, and afterward blacke, like Iuy beries or whortes. The roote is long of the quantitie of ones fingar, full of knobbes or ioyntes, and of colour white, with many hearie stringes, in taste at the first sweete, but afterward somewhat sharpe and bitter.

2. The smal Salomons seale, doth not much varie from the other, sauing that his leaues be narrower, & do not grow alone, or seuerally one by one, but foure or fiue grow out of one knot or ioynte, rounde about the stalke, almost starre fashion. The floures are greener, and the frupte is blacker than the other. The roote is smaller and slenderer, in all poyntes els like to the aforesayde.

❧ The Place.

the Historie of Plantes.

1. The great Salomons seale, groweth in this country in dry wooddes, standing vpon mountaynes.
2. The second also groweth in mountaynes and wooddes, especially in Almaigne. A man shall not lightly finde it in this countrey, except in the gardens of such as haue pleasure in herbes.

❧ *The Tyme.*

They do both floure in May and June.

❧ *The Names.*

Salomons seale is called in Greeke πολυγόνατον: in Latin Polygonatum: in Shoppes Sigillum Salomonis: in Italian *Frassinella*: in Spanish *Fraxinella*: in English also Scala cœli: White roote, or white wurte: in high Douch Weißwurtz: in French *Signet de Salomon*: in base Almaigne Salomons seghel: in the Tuscane tunge Fraisinella. ❧ *The Nature.*

Salomons seale is of Nature hoate and dry, abstersiue, or clensing, & somewhat astringent. ❧ *The Vertues.*

The roote of Salomons seale pound, doth close vp, and heale the woundes A whereupon it is layde.

The same being freshe and new gathered, to be pounde and layde vpon, or if B one be annoynted with the iuyce thereof, it taketh away all spottes, freckles, & blacke and blew markes that happen by beating, falling, or brusing, whether it be in the face, or in any other parte of the body.

This herbe, neither yet his roote, is good to be taken into the body, as Galen writeth. C

Of Flea Worte, or Fleabane. Chap.lxx.

❧ *The Description.*

Psyllion.

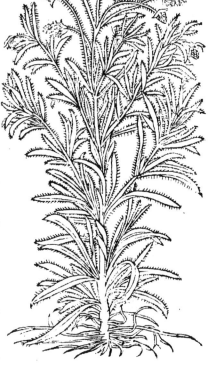

THe leaues of Fleebane, be long, narrow, and hearie, amongst whiche springe vp rounde and tender branches, set ful of leaues like them aforesayde, but smaller, & garnished at the top, with little, long, round, spikie knappes like eares, with greenish floures or blossoms, which do afterward change into a browne and shyning seede, in proportion colour and quantitie like vnto fleas.

❧ *The Place.*

This herbe groweth in fieldes, and deserte places, as Dioscorides saith. In this countrey men sow it in gardens, and wher as it hath bene once sowē, it groweth continually afterwarde of his owne sowing, or sheding of seede.

❧ *The Tyme.*

It floureth in July and August: and sometimes also the seede is ripe.

❧ *The Names.*

This herbe is called in Greeke ψύλλιον: in Latine Psyllium, and Herba Pulicaris: in Shoppes Psyllium: in Italian *Psillio*:

I iiij in

in Spanish Zargatona, in English Fleawurte, and Fleabane: in French Herbe aux poulces: in high Douch Flohekraut: in base Almaigne Vloycrupt.

❧ The Nature.

The seede of Psyllium or Fleaworte, (whiche is chiefly vsed in medicine) is colde in the second degree, and temperate in moysture and drynesse: As Galen and Serapio writeth.

❧ The Vertues.

The seede of Fleabane boyled in water, or stiped & dronken, purgeth downe- **A** wardes Aduste and Cholerique humors: bysides this it swageth payne, and slaketh the inflammation and heate of the entrayles, or bowels, and is good agaynst hoate Feuers or burning Agues, and all inwarde heates, and against great drouth, and thirst.

The same seede somewhat brused but not broke, parched at the fire, is good **B** against the bloudy flixe, and vehement laske, especially whan they proceede of taking strong and violent medicines.

The seede therof mengled with oyle of Roses & vineger, or water, is good to **C** be straked or applied vnto hoate griefes of the ioynts, & apostems & swellings behind the eares, and other hoate swellings: also it is good against head ache.

The same layde too with vineger is good against the going out of the Na- **D** uell, and the bursting of yong children.

The water wherin the seede hath bene soked or stiped, is good to be layd to **E** the burning heate called S. Antonies fire, and to all hoate swellings. It is also good to be dropped into running eares, and against the wormes in the same.

Some holde, that if this herbe whiles it is yet greene, be strowed in the **F** house, that Fleas will not come nor ingender where as it is layed.

❧ The Daunger.

Too much of Fleabane seede taken inwardly, is very hurtfull to mans nature: it engendreth coldnesse and stiffenesse through out the body, with pensiue heauinesse of the harte, so that such as haue dronken thereof, do sometimes fall into great distresse.

❧ The Remedie.

Whan one hath taken too much of the seede of Fleabane, so that he feeleth some noyance or harme, aboue all things it shalbe good for him, to prouoke vomite, with medicines conuenient, to cast vp if it be possible that whiche hath bene before take. Afterward giue him to drinke of the best & most sauoury old wine that may be gotten, by it selfe, or boyled with Wormewood, or wine mengled with hony and a little lie, or the Decoction of Dyll as Serapio writeth: And bysides this ye may giue him all things that is good against the dangers that happen of eating greene Coliander.

Of certayne Herbes, that fleete or swimme vpon the water. Chap.lxxj.

❧ The Kyndes.

There be diuers sortes of herbes that growe in & aboue water, whereof the greatest parte shalbe described, in other places, & other Chapters: so that in this present Chapter, wee shall intreate but onely of foure or fiue sortes of them that grow vpon the water.

❧ The Description.

The first and most notable of these kindes of floting herbes, the whiche is called water spyke, or most comonly Podeweede, hath long roud & knotty branches. The leaues grow vpō smal short stems, & are large great & flat, layde

the Historie of Plantes. 105

layde and carried vpon the water, somewhat like to great Plantayne, but a great deale smaller. The floures grow at the toppe of the branches, aboue the water vpon long purple spykie knoppes like to the eares or spikes of Bistorte, the which being perished, there commeth vp round knoppes, wherein the seede is inclosed, whiche is harde.

Potamogeiton.
Ponde weede.

Viola Palustris.
Water violet, or Gyllofer.

2   The second kinde, hath long small stemmes: The leaues be long and iagged very small, spred abroade vnderneth the water, alwayes fiue or sixe standing directly one against an other, as ye leaues of Madder, or Woodrow, euery leafe like to Tansie or Yerrow leaues, but smaller, and more iagged than the leaues of Tansie, and greater and broader then the leaues of Yerrow or Milfoyle, but not so finely cut as Milfoyle. It bringeth forth his floures, vpon stalkes or stemmes, growing aboue the water, alwayes three or foure floures set one against an other, parted into fiue leaues like to a little wheele, or like stocke Gillofers, or like the floures of common Buglosse, of colour white, and yellow in the middell. The rootes be nothing else, but like to long small blacke threedes, and at that ende whereby they are fastened to the ground, they are white and shyning like Cristall.

3   The third herbe swimming vpon the water is called Morsus Ranæ, or Frog bitte, and it hath round leaues layde flatte and spread vpon the water, like the leaues of Asarabacca or Folefoote, but smaller, & tied vpon shorte stemmes comming out from the roote. The floures grow amongst the leaues, and are white, and a little yellow in the middell, parted into three leaues, much like in figure
to the

The first Booke of

to the floures of water Plantayne, & the floures of water Milfoyle or Crabs clawe. The roote is thicke and shorte with many long threedes or strings, like the roote of water Milfoyle.

4 There is also carried vpon the water, certayne little small greene rounde leaues, not much larger then the seede of the pulse called Lentilles, hauing vnder them for rootes, very small white threddy strings, & are called water Lentils, Duckes meate and Grayues.

5 Amongst the fleeting herbes, there is also a certayne herbe whiche some call water Lyuerworte, at the rootes whereof hang very many hearie strings like rootes, the which doth oftentimes change his vppermost leaues according to the places where as it groweth. That whiche groweth within the water, carrieth, vpon slender stalkes, his leaues very small cut, much like the leaues of the common Cammomill, but before they be vnder the water, and growing aboue about the toppe of the stalkes, it beareth small rounde leaues, somewhat dented, or vneuenly cut about. That kinde whiche groweth out of the water in the borders of diches, hath none other but the small iagged leaues. That whiche groweth adioyning to the water, & is sometimes drenched or ouerwhelmed with water, hath also at the top of the stalkes, small rounde leaues, but much more dented than the round leaues of that whiche groweth alwayes in the water. The floures of these herbes are white, and of a good sent or smell, with a certayne yellow in the middel, like the floures of Crowfoote, golde Cuppes, or Strawbery floures: whan they are gone, there commeth rounde, rough, and prickley knoppes, like the seede of Crowfoote, or Golde knappes.

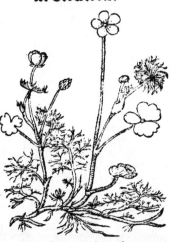

Polyanthemum palustre.
White Crowfoote, or water Crowfoote.

❧ *The Place.*
These herbes grow in standing waters, and diches.

❧ *The Tyme.*
Water Spike, and Frogge bitte, do floure most commonly in June. The others in May.

❧ *The Names.*

1 The first is called in Greeke ποταμογείτων καὶ σταχνίτης: in Latine Fontalis, & Fontinalis, & of some Spicata, vnknowen in Shoppes: in English Water spike, and Podeweede: in French *Espi d'eaue*, and *Bete Aquatique*. in high Douch Zamkraut: in base Almaigne Fonteyncrupt.

2 The second is counted of some of the wryters in these dayes, for a kinde of the herbe called in Greeke μυριοφυλλον: in Latine Millefolium. Some call it in French *Gyroflee d'eaue*: in Latine Viola palustris: in base Almaigne Water Filieren: in English Water Gillofer.

3 The thirde is called Morsus ranæ, that is to say, Frogge bitte, & it hath none other Greeke nor Latin name that I know: it is called in base Almaigne Uorschen Beet, & Cleyn plompen, that is, Paruum Nymphæa, or small Water lyllie.

4 The water Lentyll is called in Greeke φακῶ, καὶ φακῶ ὁ ἐπὶ τῶ τελμάτων: in Latine Lens palustris, or Lacustris: in Shoppes Lenticula aquæ: in English Water Len-

the Historie of Plantes. 107

ter Lentils, Duckes meate, and Graynes: in high Douch Meerlinsen: in base Almaigne water Linsen, and of some kynde geuen.

5 The fifth whiche is like to Golde cuppe in his floure and seede, seemeth in sight to be a kinde of Ranunculus or Crowfoote, called in Greeke Polyanthemon: Therefore it may be well called in Latine Polyanthenium palustre, or Aquaticum: in English white Crowfoote, & water Crowfoote: in base Almaigne Witte or water Boterbloemen. The Apothecaries of this time do call it Hepatica, and Hepatica aquatica, or Palustris: And do very erroniously vse it for Hepatica.

✱ *The Nature.*

Pondeweede doth coole, and so doth Frogge bitte, and water Lentill or Graynes.

✱ *The Vertues.*

1 Pondeweede or water Spyke is good to be layde to rotten and consuming or fretting sores, and to sores that runne in the legges, if it be layde to with hony and vineger, as Plinie saith.

2 The Decoction thereof boyled in wine is good to be dronken against the bloudy flire and all other laskes, and hath the vertue like knotgrasse, as Galen wryteth.

3 Water Lentils or Graynes mengled with fine wheaten floure, and layde too, preuayleth much against hoate swellings, as Phlegmons, Erisipeles, and the paynes of the ioyntes.

4 The same doth also helpe the falling downe of the siege or Arsegut in yong children. It is also good against the bursting of young children.

The three other kindes are not vsed in Medicine.

Of Alysson. Chap. lxxij.

✱ *The Description.*

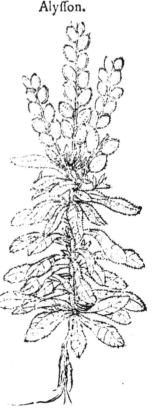

Alysson.

THe stem of this herbe is right & straight, parting it self at the top into three or foure smal branches. The leaues be first round, and after long, whitish and rough, or somewhat woolly in handling. It bringeth foorth at the top of the branches little yellow floures, & afterward, small, rough, whitish, and flat huskes, and almost round, fashioned lyke Bucklers, wherein is conteyned a flat seede, almost like to the seede of Castell or Stocke Gillofers, but greater.

✱ *The Place.*

Alysson, as Dioscorides wryteth, groweth vpõ rough mountaynes, & is not founde in this countrey but in the gardens of some Herboristes.

✱ *The Tyme.*

It floureth in this countrey in June, and the seede is ripe in July.

✱ *The Names.*

This herbe is called in Greeke ἄλυσσον: in Latine also Alyssum, & this is Dioscorides Alysson: for Alyssa of Galen and Plinie are vnlike to this, & of some late wryters Lunaria maior. This is the right Alysson of Dioscorides: for the Alysson of Galen and Plinie, is not like vnto this.

✱ *The Nature.*

Alysson is of a drying nature as Galen wryteth.

✱ *The*

The Vertues.

Alysson dronken, or holden to the Nose to smell at, driueth away yexing, or the Hicket.

The same taken with other meates, cureth the rage or madnesse, caused by the byting of a madde Dogge.

The same hanged in the house, or at the gate, or entry, keepeth both man and beast from enchantments, and witching.

Of Scabius. Chap. lxxiij.

The Kyndes.

There are found in this countrey three kindes of Scabius, like one to an other: alwell in the floures, as in the leaues.

| Scabiosa Communis. | Scabiosæ tertium genus. |
| Scabius. | Sheepes Scabius. |

The Description.

1. The first kinde which is the most common & the greatest, at his first cōming vp, his leaues be long and small, of a grayishe hore colour, and hearie, spread abroade vpon the ground, amongst the which springeth vp round, and hearie shootes or stēmes, bearing leaues very iagged, of a hoare grayishe colour, & hearie also, in fashion somewhat like to the leaues of the great Valerian, whiche we call Setwall. At the toppe of the stalkes groweth blewish floures in thicke tuffets, fashioned like to a littell flat rounde Hatte. The roote is white, long and single.

Of this sorte there is found an other kinde, in all poyntes like to the aforesayde

the Historie of Plantes.   109

sayde, sauing that at euery head or knap, there groweth in the steede of floures, many other small knoppes, or littell tuffets of floures, hanging downe by long stemmes, after the same maner, as one may also sometimes see, in some kindes of Daysies, and Marigolds.

2 The second kinde of Scabious is the smallest or least amongst the kindes of Scabious, no higher than ones hande, much like vnto the great Scabious, both in his leaues and floures, sauing that it is smaller, and the leaues be more deeper cut and iagged.

3 The third kinde is as it were a meane betwixte the other twayne, smaller than the greatest, and bigger than the smallest, in floures much like the other twayne. The leaues be long, hearie and grayish, snipt, and cut rounde aboute, but nothing so much or so deepely gaysht, as the two others. The roote is long and slender like the roote of the first and greatest Scabious.

4 There is also an herbe like vnto Scabious, growing to the heigth of a foote & half or two foote long, with long narrow leaues, like to the leaues of the greater Scabious, or Diuels bitte, the which be somwhat snipt, and bluntly cut about the edges. The stalkes or stemmes be round, vpon the toppes whereof groweth small round knappes or bollines, couered with scales, like to the knops of blew Bottell, or Corneflloure, but much greater, out of the middest wherof groweth purple hearie floures, like to the middell parte of Cyanus or Blew bottell. The roote is thicke, shorte, & croked, with many threedy strings.

Iacea nigra.
Materfilon or Knapweede.

❧ The Place.

The great Scabiouse and Iacea nigra, do grow in medowes and pastures. The smaller Scabious groweth in medowes and watery groundes that stande lowe. Sheepes Scabiouse groweth in the fieldes, and by the way sides.

❧ The Tyme.

They do all floure in June and July.

❧ The Names.

These herbes were not described of the Auncient writers (as far as I can learne) and therfore they haue no Greeke nor Latine name to vs knowen.

1 The first is now called in Shoppes Scabiosa: and of some χόρα: in English Scabious: in French Scabieuse: in Douch Apostemkraut, Pestemkraut, and Grindtkraut: in base Almaigne Scabiose.

2 The second is now called Scabiosa minor, that is to say, small Scabious.

3 The third is called in English Sheepes Scabious: in French Scabieuse de brebis: in base Almaigne Schaeps Scabiose.

4 The fourth is now called in Shoppes Iacea nigra, and Materfilon: and it hath none other name knowen vnto vs.

❧ The Nature.

All the Scabiouses are hoate & dry, digesters & diuiders of grosse humors.

K   ❧ The

The first Booke of

❧ *The Vertues.*

Scabious boyled by it self, or with his roote, in wine or water and dronken, doth clense the breast, and the lunges, and is good against an old Cough, & the impostems of the breast, and all other inward partes, as in the clensing, riping, sodering, & healing of the same. The same effect hath the Conserue made with the floures of Scabiouse and sugar to be vsed dayly.

Scabious is also good against all itch & scuruinesse, to be pound and layde to the same, or to be mixte with oyles and oyntments fit for the same.

The lye wherin Scabious hath ben boyled or stiped, doth clense the heare fro all bran or white scurffe, (whiche is small duste or scales, which falleth from the head) whan the head and heare is wasshed therewithall.

The Decoction of Iacea nigra gargeled, or whan the mouth is often wasshed therewithall it doth waste & consume the impostems of the mouth and throte, that are yet fresh and new, and doth ripe and breake them that be olde.

The small Scabious and the sheepes Scabious, are not vsed in medicine.

Of Deuels bitte.   Chap. lxxiiij.

❧ *The Description.*   Morsus Diaboli.

THe stalkes of Deuels bitte, are round, and of two or three foote log bearing broade leaues very little or nothing at al snipt about the edges. The floures be of a darke purple colour, & sometimes white, growing round & thicke togither, like the croppe or floure of Hoppes, after the falling away whereof, the seede is carried away with the winde. The roote is blacke & harde, short & thicke, with many threddy strings by the sides, the whiche in the middell, or as it were about the hart of the same, seemeth as it were bitten of.

❧ *The Place.*

Deuels bit groweth in dry medowes and woodes, and about way sides.

❧ *The Tyme.*

This herbe floureth most comonly in August, the which being in floure is easie to be knowen, otherwise it is somewhat harde to be knowen, bycause it doth resemble Scabious, or Iacea nigra.

❧ *The Names.*

It is called in Shops Morsus diaboli: in English Deuels bit: in French Mors de diable: in high Douch Teuffels abbitz: in base Almaigne Duyuels beet. Of some late writers Succila in Latine. And it hath none other names whereby it is yet knowen.

❧ *The Nature.*

Deuels bitte is hoate and dry like vnto Scabious.

❧ *The Vertues.*

The decoctio of Deuels bit, with his roote, boyled in wine & dronken, is good against al the diseases, that Scabious serueth for, & also against the Pestilence.

The

The same decoction diſſolueth clotted bloud in the body, by meanes of any **B** bruſe or fall.

Diuels bitte freſh and greene gathered, with his roote and floures pounde **C** or ſtamped, and layde to Carboncles, Peſtilential ſores and Botches, doth ripe and heale the ſame.

The decoction of the roote boyled in wine and dronken, is good againſt the **D** payne of the Matrix or Mother, and againſt all poyſon.

Of Scordium, or water Germander. Chap.lxxb.

❧ *The Description.* Scordium.

His herbe hath ſquare hearie or cottony ſtalkes, creeping by the ground, and ſet vpon euery ſide with ſofte, crimpled, and round, whitiſh leaues, nickt, & ſnipt roūd about the edges like a ſaw, betwixt which and the ſtalke groweth littell purple floures, like to the floures of dead Nettell, but ſmaller. The roote hath threedy ſtrings creeping in the ground.

❧ *The Place.*

This herbe groweth in moyſt medowes, neare about diches, & is found in ſome partes of the countrey of Brabant.

❧ *The Tyme.*

Scordion floureth moſt commonly in June & July, & thā is the beſt gathering of it.

❧ *The Names.*

This herbe is called in Greeke σκόρδιον: in Latine Scordiū, & Trixago paluſtris, of ſome Mithridatium: in high Douch waſſer Batenig, and of ſome Lachen Knoblauch: in baſe Almaigne Water loock: in Engliſh alſo Scordion, & water Germander.

❧ *The Nature.*

Scordion is hoate & dry in the thirde degree.

❧ *The Vertues.*

Scordion broken with wine, openeth the ſtoppings of the Liuer, the Milte, **A** the kidneys, the Bladder, and the Matrix: it prouoketh vrine, and is good againſt the ſtoppings of vrine, and ſtrangury, whan a man cannot piſſe but drop after drop: it moueth and prouoketh womens floures.

The ſame taken in manner aforeſayde, is good againſt the biting of Ser- **B** pents, and al other venemous beaſts, and for them that haue taken any poyſon, and for them alſo whiche are burſten, or hurte inwardly.

Dry Scordion made into pouder, & taken in the quantitie of two drames, **C** with honied water, cureth and ſtoppeth the bloudy flixe, and is good for the paynes of the ſtomacke.

The ſame made into pouder, and mengled with Hony, and eaten, clenſeth **D** the breaſt from all fleume, and is good againſt an old Cough.

Freſhe and greene Scordion pounde, and layde vppon greate greene **E** woundes, cureth the ſame. The ſame dryed and tempered or mixte with Hony, or made into pouder and cast into olde woundes, and corrupt, and rotten vlcers, cureth the same, and doth eate, and waste the prowde, and superfluouse flesshe.

This herbe boyled in water or Vineger, and layde vpon the payne of the ioyntes easeth the griefe, causing it the sooner to departe.

Of Teucrion, or wilde Germander. Chap. lxxvi.

❧ *The Description.*

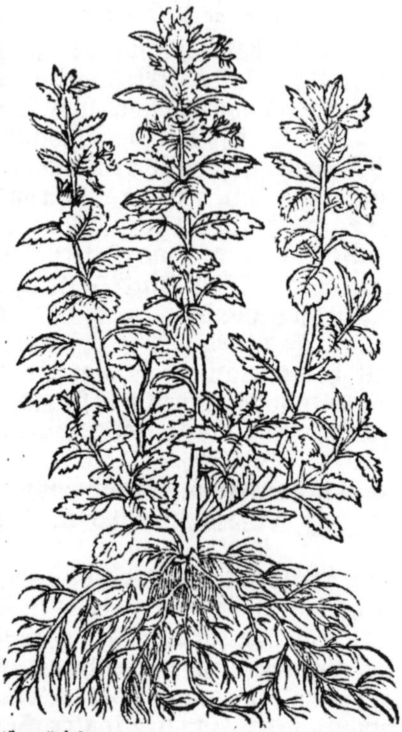

Teucrion hath browne stemmes, bringing forth rounde, & wrinkled leaues, snipt and cut round about the edges, much like to the leaues of Germander afore described in the xvi. Chapter. The little small floures, are of a sadde purple, or browne redde colour, like to the floures of Germander. The roote is whyte and of hearie or threddy strings.

❧ *The Place.*

This herbe, as Dioscorides saith, is found in Cilicia: in this countrey it is not to be found, but sowen or planted in the gardens of certayne Herboristes.

❧ *The Tyme.*

That which groweth in this coūtrey is seene in floure in June, and July.

❧ *The Names.*

This herbe is called in Greeke τώς κριον, καὶ τεύκρις: in Latine Teucrium: vnknowen in Shoppes: in English wilde Germander: in high Douch it is called of some Grosz batengel: that is to say, great Germander.

❧ *The Nature.*

Teucrion as Paulus Aegineta saith, is hoate in the second degree, and dry in the thirde.

❧ *The Vertues*

Teucrion boyled in wyne and dronken, openeth the stoppings of the Milte or Spleene, and cureth the swelling and hardnes of the same, for whiche purpose it is very good, and hath a singuler propertie. The herbe pounde with figges and Uineger worketh the same effect, being layde vpon the place of the Spleene in maner of a playster. A

Teucrion onely mengled with vineger, is good to be layde to the bytings and stingings of venemous beasts. B

Of Houselyke and Sengreene. Chap. lxxvii.

❧ *The Kindes.*

Sengreene, as Dioscorides wryteth, is of three sortes. The one is great: the other small: and the thirde is that whiche is called Stone Croppe, and Stone hore.

Semper-

the Historie of Plantes. 113

Semperuiuum maius.
Houselike, or Sengreene.

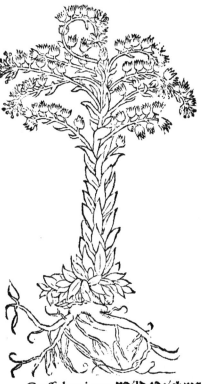

Semperuiuum minus.
Prickmadam.

Crassula minor. Wild Prickmadam. Great Stonecrop.

Illecebra.
Stone crop, or Stone Hore.

❧ *The Description.*

THe great Sengreene hath great, fat, and thicke leaues, as large as a mans thombe, and sharpe at the end fasshioned like
K iij   a toung,

a tounge, emongst whiche leaues, there groweth vp a stalke of the length of a foote or more, beset and decked roūd about with leaues like to the first, parting it self afterward about the toppe, into diuers other branches, alongst the which groweth a great many of browne, or reddish floures.

2 Prickmadame hath small narrow thicke and sharpe poynted leaues. The stalkes be great and tender of a spanne long, beset round about with the round and sharpe poynted leaues aforesayde, the whiche do bring forth at the top, smal yellow, and starre like-floures. The roote is small and creepeth by the ground.

3 Amongst the kindes of Sengreene also, at this time there is conteyned, the herbe (called Crassula minor) whiche is great stone Crop, called of some wilde Prickmadam, or wormegrasse, the which hath tender stalkes, and leaues somwhat long, all rounde, and reddishe, like vnto small wormes, euery worme lyke to a wheate corne. The floures be white, and like the floures of Prickmadam but smaller.

4 Small Stonecrop is somewhat like to wilde Prickmadam or Vermicularis, & the ignorant Apothecaries do gather it in steede of Vermicularis or Crassula minor, not without great errour, and to the perill and daunger of the sicke and diseased people, in so vsing it in steede of Crassula minor. It hath tender stalkes, couered or set full of very small, short and thicke leaues, growing neare togither. The floures at the toppe of the stemmes are yellow, and like to the floures of Prickemadame, but greater.

5 There may be also placed amógst the kindes of Sengreene, a certayne smal herbe very like to the aforesayd in making and growth, sauing that his leaues are somewhat larger & thicker, the whole herbe is eger or sharpe, with white floures. ❧ *The Place.*

1 The greater Sengreene or Houselike, groweth in many places vpon olde walles and houses, where as it hath bene planted.

2 The small Sengreene, whiche we call Prickmadam, groweth not in this countrey but onely in gardens, where as it is planted.

3.4 The great and small Stonecroppe, groweth in stonie and sandy countries, and vpon olde walles.

5 The fifth kinde also groweth vpō old walles: but not here in this countrey.
❧ *The Tyme.*

Houselike or great Sengreene, floureth in July and August. The other kindes floure in May and June.
❧ *The Names.*

Sengreene is called in Greeke ἀείζωον: in Latine Sedum, and Semperuiuum, of Apuleius, Vitalis.

1 The first is called in Greeke ἀείζωον μέγα: in Latine Sedum, & Semperuiuum magnum, of Apuleius ςέργηθρον καὶ ζωοφθαλμον: in Shops Barba Iouis: in Italian *Semper viua*. in Spanish *Terua pruntera*: in English Houselike and Sengreene: in French Ioubarbe, and *grande Ioubarbe*: in high Douch Hauswurtz, and grosz Donderbart: in bas Almaigne Donderbaert.

2 The second is called in Greeke ἀείζωον μικρον: in Latine Semperuiuum, or Sedum minus, of some τριφαλὲς, of Apuleius Erithales: in English Prickmadam: in French *Triquemadame*: in high Douch klein Doderbart: in base Almaigne cleyn Donderbaert.

3 The third kinde is called in Shoppes Crassula minor, and Vermicularis: in Italian *Herba grauelosa*, *Vermicolare*: in Spanish *Vuas de perro*, *vermicular*: in English wilde Prickmadam, great Stone Croppe, or Wormegrasse: in base Almaigne Bladeloose and Papecullekens.

4 The fourth is called in Greeke ἀνδράχνη ἀγρία καὶ τηλέφιον: in Latine Illecebra: in English Stone Crop, and Stone Hore, & of some it is called Wall Pepper: in French *Pain d'oyseau:* in high Douch Maurpfeffer, & Katzentreublin: in base Almaigne Muerpeper.

5 The fifth is called of the later writers, Capraria, and we know none other name to call it by.

※ *The Nature.*

The great and small Sengreene, and the fifth kinde (called Capraria) are colde and dry in the third degree. The great and small Stone Crop, are hoate and dry almost in the fourth degree.

※ *The Vertues.*

1 The Decoction of the great Sengreene, or the iuyce thereof droken is good A against the bloudy flire, and all other flires of the belly, and against the byting of Phalanges, whiche is a kinde of fielde Spyders.

The iuyce thereof mengled with parched Barlie meale, and oyle of Roses, B is good to be layde to the paynes, or aking of the head.

The same iuyce dropped into eyes is good against the inflammation of the C same: and so is the herbe brused, and layde outwardly therevnto.

The iuyce of Sengreene, conueyed into the Matrix with a Pessary of cot- D ton or wooll, stoppeth the running of the floures.

Sengreene brused alone, or mengled with parched barlie meale, is good to E be layde to S. Anthonies fire, and to hoate burning & fretting vlcers or sores, and vpon scaldings and burnings, and all inflammations: It is also good to be layde to the goute comming of hoate humors.

2 The small Sengreene or thrifte Stone crop, hath the like vertue. F

3.4 The iuyce of small Stone crop or wall Pepper taken with vineger, causeth G vomite and to cast out by vomiting, grosse and slymie flegmes, and hoate Cholerique humors: Also it is good against Feuers, and all poyson taken within the body: but yet it may not be ministred, except vnto strong and lustie people.

This Stone crop mingled with Swynes grease, dissolueth and driueth a- H way wennes, and harde swellings being layde therevnto.

The herbe alone layde vpon the bare skinne causeth the same to waxe red, I and to rise full of wheles and blisters, and pearceth the whole flesh.

5 It hath bene tried by experience, that Capraria, brused with (pourcelets) cal- K led in Greeke ὀνιοκόι, and oyle of Roses, cureth the blinde Hemorrhoides that are not open or pearced, if it be applied thereto.

Of the kindes of Kali, or Saltworte. Chap.lxxviij.

※ *The Description.*

1 The herbe named of the Arabians Kali, or Alkali hath many grosse stalkes, of halfe a foote or nine inches long: out of them groweth small leaues, somewhat long & thicke, not much vnlike the leaues of Prickmadam, sauing they be longer, and sharpe poynted, with a harde prickley toppe or poynt, so that for this consideration the whole plant is very rough and sharpe, and his leaues be so dangerous and hurtfull by reason of their sharp prickles, that they cannot be very easily touched. Amongst the leaues groweth small yellow floures, and after them followeth small seede. The roote is somewhat long, weake and slender. This herbe is salte and full of iuyce or sap like Anthyllis altera, which is before described in the seuenth Chapter.

2 There is an other herbe in nature much like vnto this, the whiche is called Salicornia, the same hath stalkes without leaues, and diuideth it selfe agayne into

The firſt Booke of

into ſundꝛy and diuers other bꝛanches with many knottes and ioyntes, eaſie to be pluckte of, oꝛ bꝛoken away: euery of the ſayde ioyntes are of the quantitie of a wheate Coꝛne. This plante is alſo ſalte in taſte and full of iuyce like Kali.

Kali.
Saltewoꝛte.

Salicornia.
Sea grape, oꝛ knotted Kali.

 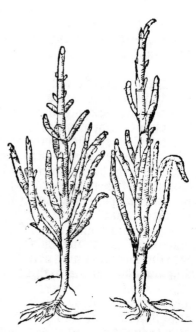

Of theſe two plantes are made Alumen Catinum, and Sal Alcali, whiche is much vſed in the making of glaſſes, and foꝛ diuers other purpoſes.

❀ *The Place.*

Theſe herbes grow in ſaltiſh groundes, by the Sea ſide oꝛ Coaſte, in Zealand, and England.

❀ *The Tyme.*

Theſe herbes are found in their naturall places, in Sommer.

❀ *The Names.*

1 The firſt is called in Italian Soda: in Spaniſh Barilla, and Soda Barilla: and it is the right Kali, oꝛ Alkali of the Arabians: ſome call it in Engliſh Salte woꝛte, we may alſo call it Kali, oꝛ Pꝛickled Kali.

2 The ſecond is now called Salicornia, ⁊ it is a certaine kinde of Kali. Some call it in Engliſh Sea grape, and knotted oꝛ ioynted Kali.

The Arſen oꝛ aſſhes, whiche are made of burnt Kali, is called in Latine of the Alcumiſtes and Glaſſemakers Alumen Catinum, but the Salte whiche is made of the ſame Arſen, is called Sal Alcali: And that which fleeteth oꝛ ſwimmeth vpon the ſtuffe whereof Glaſſes are made, is now called in Shoppes Axungia vitri: in Engliſh the fatte oꝛ floure of Glaſſe: in French Suin de voirre: in Douch Smout van ghelaſen: in Italian Fior de Criſtallo: that is to ſay, in Latin Flos Cryſtalli: in Engliſh the Creame oꝛ floure of Cryſtall.

❀ *The Nature.*

Theſe herbes be ſalte, and therefoꝛe dꝛie.

the Historie of Plantes.

Of Sophia, or Flixeweede. Chap.lxxix.

Thalietrum.

❧ The Description.

Sophia or Flixweede, his leaues be much iagged, like to ye leaues of Coliander, or Wormewood Romayne. The stalkes be rounde and harde like to the stalkes of Rue, and bringeth forth at the toppe, small pale or bleake yellow floures, and after them little long and tender Coddes or huskes, in which is conteyned a small reddish seede. The roote is of a wooddishe substance, long and straight.

❧ The Place.

Sophia groweth alongst by wayes, in vntilled places, and specially where as there hath bene in times past any buyldings. And where as it hath bene ones sowen, it cometh vp yearely of his owne accorde.

❧ The Tyme.

This herbe beginneth to floure in Iune, and continueth so flouring vntill September, & within this space the seede may be gathered.

❧ The Names.

This herbe is now called Sophia: in English Sophia, & Flixewort: in French *Argentine*: in high Douch Welsomen: in base Almaigne Fieecruyt and Rootmelizoen cruyt.

❧ The Nature.

Sophia dryeth without any sharpnes, or manifest heate.

Cochlearia.

❧ The Vertues.

A The seede of Flixeweede or Sophia dronken w wine or water of the Smithes forge, stoppeth the bloudy flixe, the laske, and all other issue of bloud.

B Sophia brused, or pounde, and layde vpon old vlcers, and sores, closeth & healeth them vp, and that bycause it dryeth without acrimonie or sharpnesse.

Of Spoonewortte. Cha.lxxx.

❧ The Description.

Spoonewortte, at the first his leaues be broade and thicke, & somwhat hollow aboue like to a little Spoone, and somwhat crested about the edges, almost like the leaues of Romayne sorrel, sauing that they be not so softe and tender, nor so white, but harde and of a browne greene colour.

colour. The stemmes also be somewhat crested, of the length of ones hande, or a foote long. The littell floures be white, and growe at the toppe of the stalkes alongst the braches: whan they are gone, there followeth the smal seede which is reddish, and inclosed in little huskes. The roote is threedy.

❧ *The Place.*

Spoonewort groweth in many places of Holland, and Friseland, and the countries adioyning about diches and in medowes. In Brabant they sowe it in gardens.

❧ *The Tyme.*

Spoonewort floureth in Aprill, May, and afterwardes.

❧ *The Names.*

This herbe is called in Holand, and Flaunders Lepelcrupt: in French Herbe aux cuiliers: in English Spoonewort, and accordingly it is called in Latine Cochlearia: in high Douche Leffelkraut.

❧ *The Nature.*

Spoonewort is hoate & dry, & of a sharpe & biting tast, almost like kresses.

❧ *The Vertues.*

Spoonewort boyled in water is a singuler medicine, against the corrupt & rotten vlcers, and stench of the mouth, if it be often washed therewithall. This is also a singuler remedie against the disease of the mouth called of Hipocrates Volnulus hæmatites, of Plinie Stomacace, and of Marcellus Oscedo, and of the Hollanders and Friselanders Scuerbuyck, against whiche euill it hath bene lately proued to be very good, and is in great estimation and muche vsed of the Hollanders and Friseans.

It is in vertue like Telephium, wherfore if it be layde with vineger vpō the body, it taketh away the white and blacke spottes, and Lentils or freckles.

Also the herbe alone pounde, and onely layde vpon such spottes and markes by the space of sixe houres, taketh them cleane away, but yet those spottes must be playstered afterwardes with Barly meale.

Of Mullepne, or Hygtaper. Chap.lxxxi.

❧ *The Kyndes.*

THere be foure sortes of Mulleyne, as Dioscorides writeth: wherof ẏ two first are white Mulleyne, and of them one is Male, and the other female: The third is blacke Mulleyne: The fourth is wilde Mulleyne.

❧ *The Description.*

1 THe white male Mullepn (or rather Wollepn) hath great, broade, long, white, softe, & wolly leaues, from the lowest parte vpward, euen to the middell of the stem or somewhat higher: but the higher, the smaller are the leaues. From the leaues vpwarde, euen to the top of the stalke, it is thicke set round about with pleasant yellow floures, each floure parted into fiue smal leaues, the whole top with his pleasant yellow floures sheweth like to a waxe Candell or taper cunningly wrought. The roote is long and single, of a woddy substance, and as thicke as ones thombe.

2 The other white Mullepne called the female Mullepn, hath white leaues fryled with a soft wooll or Cotton, the stalkes and roote are like to the aforesayde, sauing that the floures be white, and parted into sixe littell leaues.

3 The third Mullepn, which is also of the female kind, is like to ẏ abouesayd in stalkes, leaues, & floures, sauing that his leaues be larger, & his floures are of a pale yellow colour, with small redde threedes in the middell, fashioned almost like to a littell Rose. The roote is long and thicke like the others.

The

Verbascum album mas.
White male Mulleyne.

Verbascum album foemina albo flore.
White female Mulleyne, with the white floure.

4 ¶ The Blacke Mulleyn, hath great, blacke, rough leaues, of a strong sauour, and not softe or gentill in handeling. The floures be yellow, in fashion like the others, but a great deale smaller, the stalke and roote is like to the others.

5 ¶ The wilde Mulleyn, is very much like Sage, aswel in stalkes as in leaues. It hath many square twigges and branches of wooddy substance, alwayes two growing togither out of a ioynt, standing directly one against an other.

¶ The leaues be soft and whitishe, like to the leaues of Sage, but much greater and softer. The floures grow at the toppe of the branches, and are of yellow colour.

❧ *The Place.*

¶ The Mulleynes grow about the borders of fieldes, by the high way sides, and vpon bankes.

5 ¶ The wilde Mulleyn, is not common in this countrey, but we haue seene it in the pleasant garden of James Champaigne, the deere friende and louer of Plantes. ❧ *The Tyme.*

¶ The Mulleyns do floure most commonly in July, August, and September, and the wilde kinde floureth againe more later.

❧ *The Names.*

¶ Mulleyn is called in Greeke φλόμος : in Latine Verbascum, of Apuleius Lychnitis, and Pycnitis, and of some Candela regis, Candelaria, and Lunaria: in **Shoppes** Tapsus barbatus: in **Italian** *Tassobarbasso*: in **English also** Tapsus barbatus.

The first Booke of

Verbascū albū fœmina luteo flore.
White female Mulleyne, with yellow floures.

Verbascum nigrū.
Wilde Mulleyne.

Verbascum syluestre **Wild Mulleyne.**

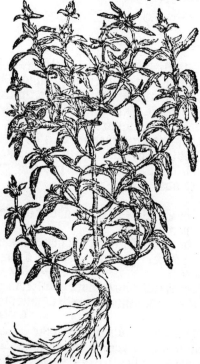

batus, Mulleyne, or rather Wulleyn, Hig-taper, Torches, and Longworte: in high Douch Wullkraut, Kertzenkraut, Brēkraut, Himelkraut, Unholdenkertz, and Kunningskertz: in base Almaigne Vollecrupt, Wollebladeren, and Tortsecrupt.

✻ *The Nature.*

The Mulleyns be dry, without any manifest heate.

✻ *The Vertues.*

A   The roote of white Mulleyne boyled in redde wine, and dronken, stoppeth and healeth the dangerous laske, and bloudy flixe.

B   The same boyled in water & dronken, is good for them that are broken, & hurte inwardely, and against an old Cough of long continuance.

C   The decoction of the roote swageth tooth ache, & is good against the inflammations, and vlcers of the Aulmondes, or kernels of the throte, to be kept warme in the

the Historie of Plantes.

in the mouth, and the mouth to be wasshed and clensed, by often gargeling of the same.

We do read, that if dryed figges be wrapt in the leaues of the white female Mulleyn, it shall preserue them a long time from corruption.

The leaues of Mulleyne are also good against the Hemorrhoides, whan they be wiped and clensed therewith, and it is good to wasshe the mouth with the decoction of the same.

The blacke Mulleyn with his pleasant yellow floures, boyled in water or wine, and dronken, is good against the diseases of the brest, and the lunges, and against all spitting of corrupt and rotten matter. The leaues of the same boyled with Rue do appease the payne of the side.

The leaues of blacke Mulleyn boyled in water, are good to be layde vpon colde swellings (called Oedema) and vpon the vlcers and inflammations of the eyes. The same leaues pounde with hony and wine, do cure naughtie and mortified vlcers: and with vineger, it cureth the inflammation of woundes.

The golden floures of Mulleyn stiped in lye, causeth the heare to waxe yellow, being wasshed therewithall.

The seede of Mulleyne is good to drinke (as saith Plinie) against the bursting and falling out of ioynte of members, for it taketh away the swelling and swageth the payne.

The wilde Mulleyne stamped, is good to be layde vpon burnings and scaldings made with fire or water and otherwise.

Apuleius saith, that Mercury gaue Mulleyn to Vlysses, whã he came neare to the inchanteresse Circe, to the ende that by the vertue of Mulleyn he might be preserued against all the enchantments or witchings of Circe.

Blattaria.

Of Blattaria / or Mothe Mulleyn. Chap. lxxxij.

❦ The Description.

The leaues of this herbe are greene, smooth, long, iagged or snipt round about, and spread abroade vpon the ground, somewhat like to the leaues of Veruayne, from the middest of those leaues doo spring vp two or three stems, bearing fayre yellow floures, (and sometimes also it beareth purple floures,) so lyke to the floures of Mulleyn in smel, fasshion and quantitie, that oftentimes (as witnesseth Plinie) this herbe hath bene gathered for wilde Mulleyne. After the floures, there arise small knoppes or bullets, in whiche the seede is conteyned, smaller than the seede of Mulleyn. The roote is shorte and of wooddy substance.

❦ The Place.

This herbe groweth by way sides, in Vineyardes, and certayne fieldes, also about Riuers, and is seldome founde in this countrey.

122 The first Booke of

✤ *The Tyme.*

It floureth in June, and July.

✲ *The Names.*

Plinie calleth it in Latine Blattaria, & some call it Verbascum Leptophyllon: it may be called in English Purple, or Mothe Mulleyn: it is called in French *Herbe aux mites, Herbe vermineuse*, and *Blattaire* : in high Douch Schabenkraut, & Goldtknopflin, and of some in base Almaigne Mottencruyt.

✲ *The Nature.*

As it may be well perceyued by the bitter sauour, the herbe is hoate & dry, almost in the third degree.

✤ *The Vertues.*

As concerning the vertues of this herbe, we finde none other thing wryten of it, sauing that the Mothes, and Battes do incontinently come to this herbe, wheresoeuer it be strowen or layde.

Of Petie Mulleyn, or the kindes of Primeroses. Chap. lxxxiij.

✲ *The Kyndes.*

Petie Mulleyn (whiche we call Cowslippe and Primrose) is of two sortes great and small. The great is also of two sortes, the one hath yellow sweete smelling floures, the other hath pale floures. The smaller sorte which we call Primerose, is of diuers kindes, as yellow and greene, single and dubble.

Verbasculum odoratum. Verbasculum album.
Cowslippe. **Oxelippe.**

✲ *The*

the Historie of Plantes.

❊ The Description.

1. The firste kinde of petie Mulleyn, hath white leaues, crumpled and wrinckled, somwhat like to the leaues of Bittayne, but whiter and greater, and not so snipt or indented about the edges, amongst the whiche there ariseth bare and naked stemmes, of the length of a mans hande, bearing at the toppe a bunch, or as it were a bundell, of nine or ten yellow floures, of a good sauour and hanging lopping downewardes: after whiche floures past, ye shall finde in the huskes wherein they stoode, littell long bulleyns wherein the seede is conteyned. The roote is white and of threedy strings.

2. The Orelip, or the small kinde of white Mulleyn, is very like to the Cowslippe aforesayde, sauing that his leaues be greater and larger, and his floures be of a pale or faynt yellow colour, almost white and without sauour.

3. The Prymerose, whiche is the very least & smallest Mulleyn, hath small whitishe, or yellowish greene leaues in all partes like to the leaues of Orelippe, amongst the whiche there riseth vp littel fine hearie stemmes, eche stemme bearing but one, onely floure like to the floures of Orelippe both in smell, colour, & proportion. The roote is also small and threedy like the roote of Orelippe. Of this kinde some be very fayre and dubbell.

4. There is yet an other sorte whiche is very like the laste recited kinde in all partes, sauing that it bringeth forth greenish floures, of colour like to the leaues of the Prymerose herbe or plante.

Verbasculum minus.
Prymerose.

❊ The Place.

Cowslippes, Orelippes, and Prymeroses, grow in lowe moyst wooddes, standing in the pendant or hanging of hilles and mountaynes, and in certayne medowes. The white is common in this coūtrey, and so are al the rest, especially the greene & dubble kindes whiche are planted in gardens.

❊ The Tyme.
These herbes do floure in Aprill, and somtimes also in March, & February.

❊ The Names.

The petie Mulleyns are called in Greeke φλομίϛις: in Latin Verbascula: in Shoppes Primulæ veris, and Herbæ paralysis, and of some Artheticæ: in English Cowslippes, Primeroses, & Orelips: and dubble Cowslips, Primeroses, and Orelips: in high Douch Schlusselblumen: in Brabant Sluetelbloemen.

1. The first kind is now called in Latine Herba S. Petri: in English Cowslips: in French of some Coquu, prime vere, & Brayes de Coquu: in high Douch Hunelschlussel, S. Peters kraut, geel Schlusselblumen, & wolrieckende Schusselblumen: in base Almaigne S. Peeters cruyt, and welrieckende Sluetelbloemen.

2. The second kinde is called in Shoppes Primula veris, & Herba Paralysis: in English Orelips: in high Douch wilde Schlusselblumen, & weis Himelschullel: in base Almaigne Witte Sluetelbloemen, and of some witte Betonie.

3. The thirde kinde is called in Latine Verbasculum minus: in Shops Primula veris minor: in English Primerose, and wood Primerose: in base Almaygne cleyn witte Betonie, or enkel Sluetelbloemen, and cleyn Sluetelbloemen.

❊ The Nature.

The small or petie Mulleyns, are dry in the third degree, without any manifest heate.

❊ The Vertues.

The petie Mulleyns, that is to say, the Cowslips, Primeroses, & Orelips, are now vsed dayly amongst other pot herbes, but in Physicke there is no great accompt made of them. They are good for the head & synewes, and haue other good vertues, as Pena and Mattiolus write.

Of Aethiopis.   Chap. lxxxiij.

❊ The Description.

Aethiopis.

Aethiopis hath great brode woolly leaues, like to the leaues of Mulleyn, but rougher & better cottoned or fryſed, and not ſo rounde by the edges, but more torne with deeper cuttes in, aboute the borders, and roundly spread abroade vpon the ground, amongſt the whiche there ſpringeth vp a square rough & heary ſtalke, diuiding it ſelf abrode into ſundry branches, alongſt þ which rounde aboute certayne ioynts, it bringeth forth many white floures almost like to the floures of dead Nettell, but a great deale bigger. The roote is long and thicke lyke the roote of Mulleyn.

❊ The Place.

This herbe groweth not in this countrey, but in the gardens of certayne Herboriſtes.

❊ The Tyme.

Aethiopis floureth in May.

❊ The Names.

This herbe is called in Greeke ἀιθιοπις, & in Latine also Aethiopis, and other name than Aethiopis we know not.

❊ The Nature.

Aethiopis is meanely hoate and dry.

❊ The Vertues.

Aethiopis is good for those that haue the Pleuresie: and for ſuch as haue their breaſts charged with corrupt and rotten matter: and for ſuch as are grieued with the aſperitie and roughneſſe in the throote: & also against the Sciatica, if one drinke the decoction of the roote thereof.

For the sayde diseases of the breast, & lunges, it is good to licke oftentimes of a confection made with the roote of this herbe and hony.

Of Sage of Jerusalem. Chap. lxxxv.

❧ *The Description.* Pulmonaria.

Sage of Jerusalem hath rough, hearie, & large, browne greene leaues, sprinckled with diuers white spots like drops of milke. Amōgst the sayd leaues springeth vp certaine stalkes of a span lōg, bearing at the top many fine floures growing togither in a bunch like Cowslip floures, of colour at the first, redde or purple, and somtimes blew: after the floures it bringeth foorth small buttons, wherein is the seede. The roote is blacke, long and thicke, with many threedy strings.

❧ *The Place.*

This herbe groweth in moyst shadowie places, & is planted almost euery where in gardens.

❧ *The Tyme.*

It floureth betimes, in March and Aprill, and shortly after the seede is ripe.

❧ *The Names.*

This herbe is called of the Apothecaries, and Herboristes of this countrey Pulmonaria & Pulmonalis, in Latine Pulmonis herba, that is to say Lungewurt, or the herbe for the lunges: and of some it is called in Latine Symphitum Sylnestre, whiche may be Englished wilde Comfrey: the Picards call it *Herbe de cueur:* we call it in English Sage of Jerusalem, & Cowslip of Jerusalem: in French *Herbe aux poulmons:* in base Almaigne Onser vrouwen melck crupt, and Onser vrouwen spin, that is to say, Our Ladies Milkeworte, bycause the leaues be full of white spottes, as though they were sprinckled with milke. There is yet an other Lungeworte, whereof we shall write in the third Booke.

❧ *The Nature and Vertues.*

This herbe hath no particular vse in Physicke, but it is much vsed in meates and Salades with egges, as is also Cowslippes and Prymeroses, wherunto in temperature it is much like.

Of Veruayne. Chap. lxxxvi.

❧ *The Kyndes.*

There be two kindes of Veruayne: the one called in Latine Verbena recta, that is to say, Vpright or straight Veruayne: The other is called Verbena supina, that is to say, Low and base Veruayne, the whiche againe is diuided into two sortes, the male and female.

❧ *The Description.*

1 The straight or vpright Veruayne, hath vpright and straight stemmes, of the heigth of a foote and more, full of braunches: with small blewishe floures growing vpon the same: The leaues be greene, dented about, and in some places deepely cut or torne lyke an Oken leafe. The roote is short and hath many threedy strings.

Verbe-

The firſt Booke of

Verbeneca recta. **Upright Veruayne.** Hiera Botane mas. **Flat Veruayne.**

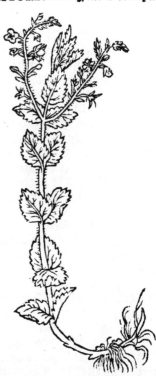

Hiera Botane fœmina.
The female flat or low Veruayne.

The flatte or creeping Veruayne, hath tender, hearie and square ſtalkes or branches of the length of a foote, or a foote & a halfe creeping by the grounde, with roundiſh leaues, dent or ſnipt round about, like to Oken leaues, or the leaues of Germander deſcribed in the xvj. Chapter of this booke, but far ſmaller then Oken leaues, & greater than the leaues of Germander: the floures be fayre and blew growing alongſt the branches at the top. After which there commeth ſmall flat coddes or purſſes like the ſeede of Paules Betony whiche we cal Speede well. The roote is thredy.

The ſecond kinde of flatte or creeping Veruayne, whiche is alſo the female low Veruayne, is very like to the aforeſayd, ſo that (as Plinie in the xix. Chap. of his xxv. booke writeth). Some haue made no difference betwixt the Male and Female, and to ſay the truth there is but ſmall differēce betwixt theſe two herbes: for the female is very well like to the male, aſwell in ſtēs,

as

as in the leaues, floures, and rootes, sauing that the stemmes of the female, are rounder: his leaues be somwhat smaller, and hath more store of branches comming vp from the roote. The floures also grow thicker or nearer togither than the floures of the male flat Veruayne.

❧ *The Place.*

The first kinde of Veruayne groweth in rude places, about hedges, walles, wayes, streates and diches. The second kinde groweth in gardens, and lowe shadowy places, and of this sorte the male is more common than the female.

❧ *The Tyme.*

The Veruaynes floure most commonly in July.

✤ *The Names.*

1. The first kinde of Veruayne is called in Greeke περιστεριών, & of some περιστεριών ὀρθίος, in Latine Verbeneca Columbina, Columbaris, Herba sanguinalis, Crista gallinacea, Exupera, and of some Feria, or Ferraria, Trixago, Verbena recta, and Columbina recta: in Shoppes Verbena: in Italian *Vermina tola, vrgibaon y Macho*. in English Veruayne, or Varueyn: in French *Veruaine*: in high Douch Eisernkraut, Eisernhart, & Eisernrich: in base Almaigne Verbene, Yserckruyt, and Yserhert.

2. The second kind is called in Greeke ἱερὰ βοτάνη: & at this time περιστεριών ὕπτιος, of Pythagoras Erysisceptrum, and of some others Demetria: in Latine Sacra herba, Verbenaca supina, and Cincinalis, of Apuleius Licinia, Lustrago, Columbina supina, and Militaris: in Shoppes (very erroneously) it is called Chamedryos, or Chamedrys: in English Base or flat Veruayne: in high Douch Erdtweirauch, and of some following the errour of the Apothecaries Gamanderle, and Blawmenderle : in base Almaigne it may be called Neere oft cruypende Verbene, that is to say, in French *Veruaine basse, ou se trainant par terre*.

❧ *The Nature.*

These two kindes of Veruayne, are of a drying power.

❧ *The Vertues.*

The leaues of vpright Veruayne, or the roote alone, or both together boyled in water are very good for the sores and vlcers of the mouth and iawes, if the mouth be washed with the same Decoction. [A]

The Decoction of the herbe or of his roote, swageth tooth ache, & fasteneth loose teeth, to be often gargled withall or kept a good space within the mouth. The same dronken continually by the space of fiue dayes, cureth the grypings of the belly. [B]

Veruayne mengled with oyle of Roses and vineger, or boyled in oyle & layd to the head after the manner of a playster, cureth the head ache. The same vertue hath a garlande or Corone of Veruayne against head ache, to be worne vpon the head, as Archigenes saith. [C]

The leaues of Veruayne pound with swynes grease or oyle of Roses, doth mitigate & appeace the paynes of the Mother or Matrix to be applied thereto. [D]

The same pound with vineger are good to be layde to S. Anthonies fyre, and naughtie scuruie and rotten sores: and stamped or pounde with Hony, it healeth greene woundes, and closeth vp olde. [E]

The flat and base Veruayne is good against all benim and poyson, against the bytings and stinging of Serpents, and other venemous beasts, to be dronken in wine, or layde vpon the greefe. [F]

The leaues thereof dronken in olde wine, the weight of a dram and halfe, to asmuch Frankēcens, by the space of fortie dayes, fasting, cureth the Jaundes. [G]

It is good to washe the mouth with the Decoction of the leaues and roote thereof [H]

The first Booke of

thereof boyled in wine, against the fretting & festering sores of the mouth and iawes, or the almondes or kernels vnder the throte.

The greene leaues pound & layd too, taketh away the swelling & the paine of hoate imposteins and tumors, and clenseth corrupt and rotten vlcers.

Some write that the water wherin this Veruayne hath bene stiped, being cast or sprinckled about the hall or place whereas any feast or banket is kepte, maketh all the company both lustie and merie.

And that a branche of three knottes or ioyntes of this herbe is good to be dronken against a feuer tertian, and a branche of foure ioyntes is good against a feuer quartayne.

Of Nettell.   Chap. lxxxvij.

❧ *The Kindes.*

THere be two kindes of Nettels. The one is the burning and stinging Nettell. The other is the dead Nettell whiche doth not burne, nor sting at all. And each of these kindes is of diuers sortes. For of the hoate and stinging Nettell there be three kindes, that is to say, the Greeke or Romayne Nettels, and the great, the small, & the burning Nettels: whereas againe they are diuided into two kindes, to wit, the Male and the Female, so that the Romayne Nettell is the Male, and the other twayne are the Female. The dead Nettell shalbe described in the next Chapter.

❧ *The Description.*

1. THE Romayne Nettell hath round, rough, hollow, and hearie stalkes. The leaues be long, rough, burning or stinging, & deepely natched, or dented aboute, betwixt the leaues & þ stalke: it bringeth foorth small rounde and rough buttós, or pellettes, full of browne, flatte, & shining seede, like vnto lyne-seede, but rounder & smaller.

2. The second kind whiche is our common great Nettell, is like the aforesayd in heigth and in his rough and stinging stémes. The leaues be also rough and stinging, and déted rounde aboute, but

Vrtica syluestris.
The wilde Nettell, or Romayne Nettell.

Vrtica maior.
The great cómon Nettell.

not

the Historie of Plantes.

not so deepely as the others, most commonly of a swarte greene colour, & sometimes reddish. The seede groweth by long smal threedes, hanging downeward, & is somewhat like the seede of Hirse or Millet, sauing it is smaller. The roote is long, small and yellow, spreading it self here, and there vnder the ground.

3 The small Nettell is like to the Nettels aforesayd, but it is much smaller, not exceeding in length a foote, or a foote and a halfe. The stalkes be round and rough, and the leaues be like to the other, sauing they be smaller and greener: The seede is bigger and the roote is shorter.

Vrtica minor.
The small Nettell.

❧ *The Place.*

The Romayne Nettels are found in some woodes of this countrey, as the wood of Soignie, but not very commonly: it is also sowen in the gardens of Herboristes. The other kindes grow in all places, as by hedges, quicke settes and walles.

❧ *The Tyme.*

Nettell seede is ripe in August.

❧ *The Names.*

The Nettell is called in Greeke ἀκαλύφη, κỉ κνίδη: in Latine & Shoppes Vrtica: in Italian Ortica: in Spanish Ortiga: in French Ortie.

1 The first kinde is now called Vrtica Romana, and Vrtica mas: in English, Greeke or Romayne Nettell, or the male Nettel: in French Ortie Griesche ou Romaine: in high Douch Welsch nessel: in base Almaigne Roomsche Netelen.

2 The second kinde is called Vrtica cõmunis, Vrtica foemina, and Vrtica maior: in English Great common nettel: in French Ortie: in high Douch Heyternesse: in base Almaigne groote Netelen.

3 The smallest kinde is called of Plinie Cania, and now Vrtica minor: in English the small Nettell, and the small burning Nettell: in French Petite Ortie, and Ortie brulante: in high Douch Brennessel, & Habernessel: in base Almaigne heete Netelen.

❧ *The Nature.*

The burning or stinging Nettels, are hoate and dry & of thinne substance.

❧ *The Vertues.*

The seede of Romayne Nettell tempered or mēgled with Honie, and often A times licked, clenseth the breast from tough and slimie fleumes, & other corrupt and rotten humors. Also it is good for the shortnesse of breath, the troublesome and behement cough that children be often vexed withall, the inflammation of the lunges, and the old Pleuresie or long sought.

The same dronken with sweete wine, doth stirre vp bodely pleasure, and is B good against the blasting and windinesse of the stomacke.

The seede of Romayne Nettell, dronken with Meede, the waight of a scru‑ C ple, at night after supper, causeth one to vomit or cast vp very easily.

The leaues thereof boyled with Muscles and dronken, do soften the belly D and prouoke vrine.

The decoctiõ of the leaues of al ỹ kinds of Nettels, dronken with Myrrhe E prouoketh the Menstruall floures. And so doth nettle seede dronken wt sweete wine.

The

The iuyce of the leaues gargarised, helpeth much against the falling downe of the Uuula and the inflammation of the same.

The leaues of Nettels pound with salt, are good to be layde to the bitings of madde Dogges, virulent and malignant vlcers, as Cankers, and suche like corrupt and stinking vlcers or sores, and vpon all harde swellings, impostumes and botches behinde the eares.

The same mengled with oyle and waxe, and layde to the hardnesse of the Melte or Spleene, cureth the same.

The same pound and layde to the Nose and forehead, stoppeth the bleeding of the nose, and put into the nose, causeth the same to bleede.

Nettell leaues pounde with Myrrhe, and reduced to the order of a Pessarie (whiche is a mother suppositorie) and put into the Matrix, prouoketh the floures.

Of Archangell, or Dead Nettel. Chap. lxxxviij.

✤ *The Kyndes.*

There be two kindes of Dead Nettel. The one which, sauoreth or smelleth but little, the other whiche hath a strong & stinking sauour, otherwise there is but small difference betwixt the one & the other: and the first kinde of these herbes is of three sortes, the one with white floures, the second with yellow floures, and the third with reddissh floures. Also the second kinde is of two sortes, and differeth but onely in the colour of the floure.

✤ *The Description.*

1. The first kinde of Dead nettels, is not much vnlike the stinging or burning Nettels, his leaues be long and dented round about like to the other nettel leaues, sauing they be whiter, and they styng not. The stalke is square, rond about the which groweth, white, yellow, or red floures, betwixt the leaues and the stemme, fashioned like to a hoode, or open helmet. The roote hath three dy strings.

2. The second kinde, which is the stincking Dead nettell, is like to the other, & like the common nettell, sauing that his leaues be smaller, & somewhat rounder. All the herbe is of a very euill, & strong stincking sauour. The floures of one kinde are pale, and the floures of the other kinde are of a browne redde colour, smaller than the floures of the first Dead nettell.

✤ *The Place.*

Dead nettell groweth euery where about hedges, quicke settes and wayes, and also in gardens.

✤ *The Tyme.*

The Dead nettell floureth the most part of all the Somer, from May forwarde.

✤ *The Names.*

Plinie calleth the Dead nettell in Latine Lamium.

Lamium.
Dead Nettell or Archangel.

the Historie of Plantes.

Lamium, and Anonium, or Aononium, at this present it is called Vrtica iners, or Vrtica mortua: in Italian Ortica morte, and Ortica fœtida: in Spanish Ortiga muerta: in English Dead nettell, Blinde nettell, and Archangel: in French Ortie morte: in high Douch Todtnessel & Taubnessel: in base Almaigne Dooue, and Doode Netelen. ✤ *The Nature.*

The Dead Nettell is of temperament, like to the other Nettels.

✤ *The Vertues.*

Dead Nettell pounde or bruised with salte, doth dissolue and cure harde wennes, botches, and impostems, being layde thereupon: and in vertue is very like the other nettels.

Of Motherworte. Chap.lxxix.

✤ *The Description.* Cardiaca.

Motherworte hath square browne stalkes, the leaues be of swarte greene colour, large and deeply gapssht or cut, almost like to Nettell, or Horehound leaues, but a great deale larger, blacker, and more deeply cut, somewhat approching towards the proportió of ye Oke leaues. The floures grow like garlandes or Cronets rounde about the stalke, like the floures of Horehound, of purple colour, not much differing fró the floures of Dead Nettell, sauing they be smaller: after the floures commeth the seede, which is smal & browne, conteyned in littell prickley huskes. The roote is small, & diuided into many small threedy partes. ✤ *The Place.*

It delighteth to grow in rough, vntilled, & vneuen places, about old walles & wayes. ✤ *The Tyme.*

Motherworte floureth in June, July, and August, within whiche time, the seede is also ripe. ✤ *The Names.*

This herbe is nowe called in Latine of suche as haue pleasure in herbes Cardiaca: in English Motherwort: in Frèch *Agripaulme:* in high Douch Hertzgspan, and Hertzgsper: in base Almaigne Hertzgespan.

This is a kinde, of the three herbes, whiche are called in Greeke σιδηρίτιδες: in Latine Sideritides, & of some Heracleæ. And it is the first kinde of the sayde herbes. Therefore it may be well called in Latine Sideritis prima. Whereof we shall write againe in our second Booke in the Chapter of Horehounde.

The herbe which Matthiolus setteth forth for the Sideritis prima, is a kind of Horehounde, and is called in this countrey Marrubium palustre, that is to say, Marrish or water Horehound. ✤ *The Nature.*

Motherworte is of a temperate heate, and yet not without bitternesse: and therfore it is also abstersiue or clensing.

✤ *The*

132 The first Booke of

❧ *The Vertues.*

Motherworte bruſed and layde vpõ woundes, keepeth them both from inflammation and apoſtumatiõ or ſwelling: it ſtoppeth the bloud, and doth cloſe, cure, and heale the ſame.

Of Bugle/and Prunell. Chap.xc.

❧ *The Kindes.*

There be two kindes of Prunell. The firſt is called Bugle. And the ſecond retepneth ſtill the name of Prunell.

Bugula. **Bugle.**   Prunella. **Prunell.**

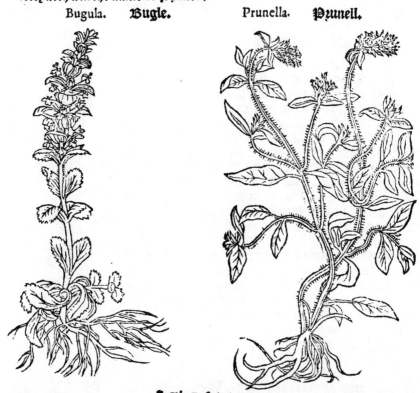

❧ *The Deſcription.*

1 Bugle ſpreadeth & creepeth alongſt the ground, like to Monyworte, or Herbe twopence: it hath ſomewhat long leaues, and broade afore, or at the top, ſofte, wrinckled and blackiſh: his ſtalkes be ſmal & tender, creeping alongſt the ground, & taking holdfaſt in certaine places here and there: and from them againe ſpring other ſquare & ſtraight ſtemmes of a ſpan long, bringing forth bright floures, amongſt certayne littell leaues, compaſſing the ſtemme about, of colour moſte commonly blew, and in ſome plantes white as ſnow. The rootes are threedy and tender.

2 Prunell hath ſquare hearie ſtalkes of a ſpanne long or more. The leaues be ſomewhat long, hearie, and ſharpe poynted. The floures grow at the top of the ſtalkes thicke ſet togither, like to an eare or ſpikie knap, of a browne colour and mirt with blewe, and ſometimes alſo very white. The roote is ſmall and very threedy.

❧ *The Place.*

They grow both in certayne Medowes, paſtures, & wooddes. Alſo Bugle
is much

❧ *The Tyme.*

Bugle floureth in Aprill. And Prunell oftentimes all the somer vntill July.

❊ *The Names.*

1. The first kinde of these herbes is now called Consolida, & Solidago, & for a difference from other herbes, whiche be also called by the same name, it is called Consolida media: in English **Middell Consounde**, or **Middle Comfery**, and **Bugle**: in French *Consoulde moyenne*, and *Bugle*: in high Douch **Gunzel**, and **gulde Gunzel**: in the Shoppes of this countrey, they call it *Bugula*, and in base Almaigne, **Senegroen**.

2. The second kinde is also called Consolida media, but most commonly *Prunella*, or *Brunella*: in English **Prunell**, **Carpenters herbe**, **Selfe heale**, & **Hooke heale**: in French *Prunelle*, and some do also call it *Herbe au Charpentier*, some call it *Oingtereule*: in high Douch **Brunellen**, and **Gottheyl**: in Brabant, **Bruynelle**.

❊ *The Nature.*

These two herbes be dry: moreouer Bugle is hoate, and Prunell temperate betwixt heate and colde, or very littell colde.

❊ *The Vertues.*

1. The decoction of Bugle dronken, dissolueth clotted & congeled bloud within the body, it doth heale and make sounde all woundes of the body, bothe inwarde and outwarde.

The same openeth the stoppings of the Liuer and Gaule, and is good to be dronken against the Jaundise, and Feuers that be of long continuance.

The same decoction of Bugle, cureth the rotten vlcers, & sores of the mouth and gummes, whan they be washed therewithall.

Bugle greene & fresh gathered, is good to be layde vpon woundes, galles, or scratches: for it cureth them, & maketh them whole & sounde. And so doth the pouder of the same herbe dryed, to be cast and strowen vpon the wounde.

The iuyce of Bugles cureth the sores & vlcers of the secrete or priuie partes, being often dropped in, and so doth the herbe brused and layde vpon.

2. The decoction of Prunell made with wine or water doth ioyne together and make whole and sounde all woundes both inwarde and outward as Bugle doth.

It is good to wassh the mouth often with the decoction of Prunell, against the vlcers of the mouth, and it is also a soueraigne remedie against that disease whiche the Brabanders do name (den Bruynen) that is, whan the tongue is inflamed and wareth blacke and is much swollen, so that the generall remedies haue gone before.

Prunell brused with oyle of Roses and vineger, and layde to the foreparte of the head, swageth and cureth the aking of the same.

Of Auens, or Sanamunda. Chap. xci.

❧ *The Description.*

The leaues of **Sanamunda**, **Auens**, or **Herbe Bennet**, are rough, blackishe, and much clouen or deepely cut, somewhat like to the leaues of Agrimonie. The stalke is round and hearie of the length of a foote and half, diuiding it self at the top into other branches, which bringeth forth yellow floures, like to the floures of Crowfoote, Goldcup, or Goldknap, & afterward littell round rough heads or knoppes, set full of seede, the which being ripe will cleaue or hang fast vnto garments. The roote is short and reddish within, with yellow threedy strings, and smelleth somewhat like Cloues, especially if it be gathered in Marche.

The first Booke of

❧ The Place.
This herbe groweth wilde in woods, and by hedges and quicklettes, it is also planted in gardēs, but that which groweth wilde is the greater, and his floures be yellower than the other.

❧ The Tyme.
It floureth in May and June.

❧ The Names.
This herbe is now called in Latine Garyophyllata, bycause his roote smelleth like Cloues, and of some Sanamunda, Benedicta, and Nardus rustica: in English Auens, herbe Bennet, and of some Sanamunda: in French *Benoitte*: in high Douch Benedictenwurtz : in Brabant Gariophyllate.

❧ The Nature.
Herbe Bennet or Auens, is hoate & dry in the second degree.

❧ The Vertues.
A The decoction of Auens made with water, or with wine and water togither and dronken, resolueth congeled and clotted bloud, & cureth all inwarde woundes and hurts. And the same decoction cureth outwarde woundes if they be washed therewithall.

B The decoction made of the roote of herbe Bennet in wine, & dronken, comforteth the stomacke & causeth good digestion: it openeth the stoppings of the lyuer, and clenseth the breast, and purgeth it from grosse and Phlegmatique humors.

C The roote dryed and taken with wine is good against poyson, & against the payne of the guttes or bowelles, whiche we call the Colique.

Of Pyrola. Chap.xcij.

❧ The Description.
Pyrola hath nine or tenne greene, tender leaues, not muche vnlike the leaues of Bete, sauing they be a great deale smaller, amongst the whiche commeth vp a stalke set with pleasant little white floures, muche like to the sweete smelling floures of lillie Conuall or May lillies. The roote is small & tender, creeping here and there.

❧ The Place.
Pyrola groweth in shadowy places, and moyst wooddes.

❧ The Tyme.
Pyrola is to be found in winter and somer, but it floureth in June and July.

Garyophyllata.

Pyrola.

❉ The Names.

Pyrola is called in Shops Pyrola: in high Douch Wintergrun, Holtzmangolt, Waldmangoldt: in base Almaigne Wintergruen: in English also Pyrola, and Wintergreene: in French Bete de prez, and Pyrole.

❉ The Nature.

Pyrola is dry in the third degree, and colde in the second.

❉ The Vertues.

The leaues of Pyrola, alone by themselues, or with other healing herbes, is good to heale woundes, and boyled in wine and dronken, they heale both inward and outward woundes, fistulas, and malignant vlcers. **A**

Greene Pyrole is also good to be layde vpon woundes, vlcers, & burnings: and so is the pouder thereof to be strowed vpon, and it is good to be mixt with oyntments and playsters, seruing for the purposes aforesayde. **B**

Of Serpents tonge, or Adders tonge. Chap.xciij.

❉ The Description. Ophioglosson.

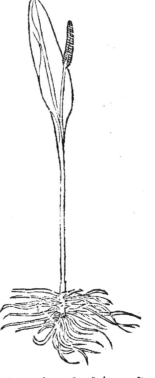

Adders tonge is an herbe of a maruelous strage nature, it bringeth forth but one leaf of the legth of ones finger, in which groweth a littell stemme, bearing a littell long, narrow, tonge, like to a Serpent, or (as my Author saith) like to the tonge of a Serpent.

❉ The Place.

Adders tonge is founde in this countrey, in certayne moyst and frutefull medowes.

❉ The Tyme.

This leafe is founde with his littell tonge, in Aprill and May: the whole herbe vanisheth away in June.

❉ The Names.

Plinie (as some learned men iudge) calleth this herb Lingua, Linguace, and Lingulace: it is now called in Greeke ὀφιόγλωσσον: in Latine Lingua serpentis, & in some countries Lancea Christi: and in other places Lucciola: in English, Adders tonge, & Serpents tonge: in French Langue de serpent: in highe Douch Naterzunglin: in Brabāt, Ons Heeren speercruyt, and Natertonghesken.

❉ The Nature.

Adders tongue is dry in the third degree, and of Nature very like Pyrola.

❉ The Vertues.

Adders tonge is also good & very singuler to heale woundes, both inward and outwarde, it is also good against burstings or Ruptures, to be prepared, & taken in like sorte as Pyrola. **A**

The Decoction of the same made with water and dronken, is good against hoate feuers, the inflammations of the liuer, and against all inwarde and outwarde heates. **B**

The same incorporated or mengled with Swynes grease, is good against burning and spreading sores or the disease called the wilde fire, also against burnings, and all hoate tumors and impostems. **C**

Of Lunaria. Chap.xciiij.

❦ *The Description.*

THe small Lunarie also, bringeth forth but one leafe, iagged & cut on both sides into fiue or sixe deepe cuttes or natches, not much vnlike the leaues of the right Scolopendria, but it is longer, larger, and greener. Vpō the sayde leafe groweth a stem of a span long, bearing at the top many smal seedes clustering together like grapes. The roote is of threedy strings. ❦ *The Place.*

This herbe groweth vpon high dry and grasie moūtaines or hilles, by dales & heaths.

❦ *The Tyme.*

The small Lunarie is founde in May and June, but afterward it vanisheth away.

❦ *The Names.*

This herbe is now called in Latine Lunaria, & Lunaria minor, of some in Greeke σκλινιτις: in English Lunarie, or Moonewort: in Frēch *Petite Lunaire:* in high Douch Monkraut, and klein Monkraut: in base Almaigne Maencruyt, & cleyn Maencruyt. The people of Sauoy, do call it Tore, or Taure.

❦ *The Nature.*

It is colde & dry of temperature, very like to Pyrola, and Adders tonge.

❦ *The Vertues.*

A This herb is also very good & singuler to heale woūdes, of vertue & facultie like to Pyrola, & Serpents tonge, very conuenient for all such griefes as they do serue vnto: the Alchimistes also do make great accōpt of this herbe about their Science.

Lunaria minor.

Of Thorow waxe, or Thorow leafe. Chap.xcv.

❦ *The Description.*

THorowleafe hath a round slēder stalke ful of branches, p̄ branches passing, or going thorow the leaues, as if they had bene drawē thorough the leaues, whiche be rounde, bare, & tender, at the top of the branches growe the floures, as it were crownes amōgst small & little leaues, of a pale or faint yelow colour, the which do afterwards chāge into a broune seede. The roote is single, white & somwhat threddy. ❦ *The Place.*

This herbe groweth in many places of Germany and England, in the Corne fieldes amongst the wheate & rye. They do also plant it in gardens. ❦ *The*

Perfoliatum.

the Historie of Plantes.

� *The Tyme.*

It floureth in July and August.

✱ *The Names.*

This herbe is now called in Latine Perfoliatum, and Perfoliata: in English Thorowware, and Thorowleafe: in French *Persefueille*: in high Douch Durchwachsz: in base Almaigne Duerwas. It is very doubtful, whether this be Cacalia of Dioscorides. ✱ *The Nature.*

Thorowware is of a dry complexion.

✱ *The Vertues.*

The decoctiõ of Thorowware boyled in water or wine, healeth woundes: and so doth the greene leaues brused and layde thereupon.

Thorowware mengled with waxe, or with some oyle or oyntment, fitte to cure woundes, healeth burstings or Harmes of yong Children, being layde thereupon.

The same herbe whan it is yet greene, brused and pounde with meale and wine, and layde vpon the Nauels of yong Children, keepeth vp the bowels, drawing them into their naturall place, and setieth them that fall too much downe, and slaketh the same whan they are blasted vp and swollen. And so doth the seede also made into pouder, and layde too after the like manner.

Of Burnet, or Pimpinell.   Chap.xcvi.

✱ *The Kyndes.*

Pimpinell is of two sortes, the great and wilde: and the small garden Pimpinell.

Pimpinella maior. Pimpinella minor.
Wilde Burnet. Sideritis altera.   Garden Burnet.

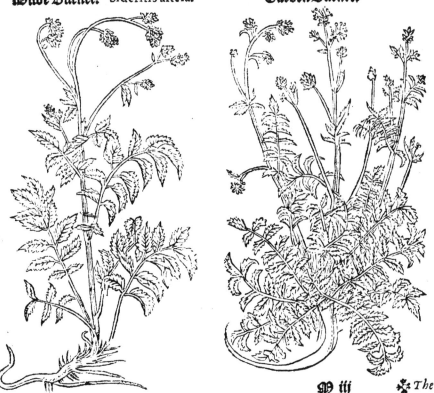

✱ *The*

✣ The Description.

1. The great wilde Pimpinell or Burnet, hath long round stemmes, two or three foote high, vpon the whiche groweth leaues, somewhat long, dented round about, and tied by long stemmes, tenne or twelue leaues growing by a stemme, standing displayed directly one against an other like vnto winges. At the top of the stalkes are round knops or heads, compact togither as it were of small purles or buttons, the which at their opening bring forth small floures of a browne redde colour: after them commeth a triangled seede. The roote is long and thicke.

2. The small or garden Pimpinell, is very much like vnto the wilde, but it is in all points smaller, and of sauour and smell more amiable, or pleasant. It hath softe and tender stalkes of a foote high or somewhat more, set with a softe and fine heare or Cotton. The leaues be like vnto the other, sauing they be a great deale smaller, greene aboue, and blewishe vnderneath. The floures be not so browne, but of an incarnate or liuely redde, with small yellow threedes, hanging forth of the middest of them. The roote is like to the other, but a great deale smaller.

✣ The Place.

The wild or great Pimpinell, groweth in dry medowes, & there is stoore of it found growing about Uiluorde. The small Pimpinell is commonly planted in the gardens of this countrey.

✣ The Tyme.

They do both floure in June, and sometimes sooner, and oftentimes vntill August.

✣ The Names.

Pimpinell is now called in Latine Pimpinella, Bipennula, Pampinula, and of some Sanguisorba, & Solbastrella: in Spanish *Frexinna*: in English Burnet, and Pimpinell: in high Douch Kolbleskraut, Hergotsbartlin, Blutkraut, and Megelkraut: in base Almaigne Pimpinelle. This herbe seemeth to be very well like to Sideritis altera of Dioscorides.

✣ The Nature.

Pimpinell is dry in the third degree, and colde in the second, & astringent.

✣ The Vertues.

The decoction of Pimpinell dronken cureth the bloudy flixe, the spitting of bloud, the pissing of bloud, and the naturall issue of women, and all other fluxe of bloud. The herbe and the seede made into pouder, and dronke with wine or water, wherein Iron hath bene often quenched, doth the like, and so doth the herbe alone being but onely holden in a mans hande, as some haue writen. [A]

The greene leaues brused and layde vpon woundes, keepe them from inflammation and apostumation. Moreouer they are good to be layde vpon phlegmons, whiche are hoate tumors, swellings, and vlcers. [B]

Pimpinell also is very good to heale woundes, and is receyued in drinkes that be made for woundes, to put away inflammation, and to stanche bleeding to much. [C]

The leaues of Pimpinell stiped in wine and dronken, doth comfort & reioyce the hart, and are good against the trembling and shaking of the same. [D]

Of Sanicle, or Sanikell. Chap. xcvij.

✣ The Description.

Sanicle hath browne, greene, plaine, shining, and roundish leaues, parted into fiue partes with deepe cuttes, like vnto vine leaues, (or rather like Maple leaues) amongst whiche there springe vp two stemmes, of the heigth of a foote,

the Historie of Plantes. 139

foote, bearing many small round buttons at the toppe, full of littell white floures, whiche do turne into smal rough burres, which is the seede. The roote hath threedy strings, and is blacke without, & white within.

❧ *The Place.*

Sanicle is founde in moyst woodes, and stony bankes, in hilly or mountayne countries Northerly.

❧ *The Tyme.*

Sanicle floureth in May and June.

❧ *The Names.*

This herbe is now called in Latine Sanicula, & of some Diapensia: in English Sanicle: in French Sanicle: in high Douch Sanicle. This is none of the kindes of Sinckfoyle or Pentaphillon, as some would haue it.

❧ *The Nature.*

Sanicle is dry in the thirde degree, & astringent. ❧ *The Vertues.*

A The iuyce of Sanicle dronken, doth make whole & sound all inward, and outwarde woundes and hurtes, so that (as Ruellius wryteth) it is a common saying in Fraunce, *Celuy qui Sanicle à, De Mire affaire il n'a.* That is to say, who so hath Sanicle needeth no Surgean.

B Sanicle boyled in water or wine, and dronken, stoppeth the spitting of bloud, & the bloudy flixe, and cureth the vlceratiõs and hurtes of the kidneys.

C The same taken in like manner, or the iuyce thereof dronken, cureth burstings, especially whan the herbe is also layd vppon the greefe, eyther brused or boyled.

D The leaues thereof, & the roote boyled in water & hony and dronken, healeth the perished lunges, and al malignant vlcers, & rotten sores of the mouth, gummes and throote, if the mouth be washed or gargled therewithall.

Of Ladies mantell, or great Sanicle. Chap. xcviij.

❧ *The Description.*

THis herb hath large roũd leaues, with fiue or sixe corners, finely dented round about, the whiche at their first cōming vp out of the ground, are folden togither or as it were playted.

M iiij   Amongst

Sanicula.   Alchimilla.

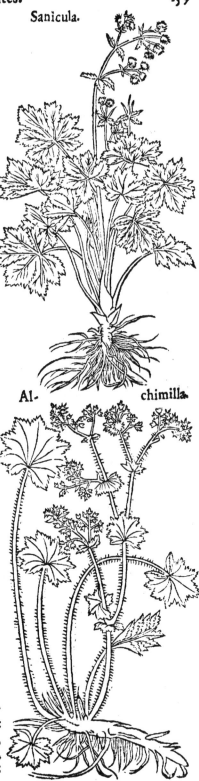

The first Booke of

Amongſt them groweth ſmall round ſtemmes halfe a foote long, ſet here and there with little leaues, and bringeth foorth at the top ſmall floures, cluſtering thicke togither, of a yellowiſh greene colour, with a ſmal yellow ſeede, no greater then Purſelane or Poppie ſeede, incloſed in ſmall greene huſkes. The roote is thicke, as long as ones fingar, browne without, and hath threedy ſtrings.

❧ The Place.

Great Sanicle or Ladies Mantell, groweth in ſome places of this coūtrey, as in certayne medowes, in the hanging of Hilles, whereas the ſoyle is of potters clay, fat and redde.

❧ The Tyme.

This herbe floureth in May, and June.

❧ The Names.

The latter wryters do call this herbe in Greeke ∂ροσέρα, Δρόσιον ∤ειάλτορ: in Latin Achimilla, Alchimilla Stellaria, Pláta leonis, Pes leonis, & of ſome in Greke λεοντοπόδιον, howbeit this is not the right Leontopodium whereof Dioſcorides writeth: in Engliſh Ladies mantell, great Sanicle, and Padelion: in French Pied de Lion: in high Douch Synnan, Lewentapen, Lewenfuſz, Unſer frauwē Mantell, & groſz Sanickel: in baſe Almaigne folowing the high Almaignes Synnaw, Onſer vrouwen mantel, and groote Sanikel.

❧ The Nature.

It is dry like Sanicle, but colder.

❧ The Vertues.

Ladies mantell is much like to Sanicle in facultie, and ſerueth for all diſeaſes wherevnto Sanicle is good. Moreouer it taketh away the payne & heate of all woundes inflamed, vlcers, and Phlegmons being applied thereto.

The ſame pound & layde vpon the Pappes or Dugges of wiues or maydens, maketh them harde and firme.

Of Sarraſins Conſounde.
Chap. xcix.

❧ The Deſcription.

SArraſines Conſounde, hath a round browne, redde, holow ſtalke, three or foure cubites high as Pena writeth, all alongſt the whiche from the loweſt parte euen vp to the harde toppe, there growe long narrow leaues like to Wythie, or Peach leaues: dented round aboute with ſmall denticles. At the toppe of the ſtalkes growe bleake or pale yellow floures, the whiche being ripe, are carried away with the winde. The roote is very threedy.

❧ The Place.

Sarraſines Conſounde groweth in ſhadowy woodes, and eſpecially there whereas it is ſomewhat moyſt.

❧ The Tyme.

This herbe is found with his floures moſt commonly in Auguſt.

Solidago Sarracenica.

❧ The Names.

This herbe is now called in Latine Solidago Sarracenica, & Consolida Sarracenica, of some Herba fortis: in English Sarrasines Consounde, or Sarrasines Comfery: in French Consoulde Sarrasine: in high Douch Heidnisch wundkraut: in base Almaigne Heydensch wondtcruyt.

❧ The Nature.

Sarrasines Consounde is almost dry in the third degree, and not without heate, in taste bitter and astringent.

❧ The Vertues.

Sarrasines Consounde healeth all sortes of woundes and vlcers, both inwarde and outward, to be ministred in the same manner as the other Consolidatiue or healing herbes are, whether it be giuē in drinke, or applied outwardly with oyntments, oyles, or emplaisters.

The same boyled in water and dronken, doth restraine and stay the wasting of lyuer, and taketh away the oppillation and stopping of the same, & of the bladder and gaule, and is good agaynst the iaundise, & feuers of long continuance, and for such as are falling into a dropsie.

The decoction of the same is good to be gargled against the vlcers, and stinking of the mouth, and against the vlceration of the gummes, and throte.

❦ Of Golden rodde. Chap. r.

❧ The Description.

Virga aurea.

Golden rodde at the firste hath long broade leaues, spredde abroade vpon the ground, amongst the which springeth vp a reddish or browne stalke of the length of a foote and half, with leaues like to the first, but smaller, it spreadeth it selfe at the toppe into diuers small branches, charged or loden, with small yellow floures, the whiche also whan they are ripe, are carried away with the winde, like to the floures of Sarrasines Consounde. The roote is browne and hath threedy strings.

❧ The Place.

This herbe groweth in wooddes, vppon mountaynes, and in frutefull soyle.

❧ The Tyme.

It floureth most commonly in August.

❧ The Names.

This herbe is now called in Latine Virga aurea, that is to say, Golden rodde: in French Verge d'or: in base Almaigne Golden roede: and we know not as yet whether it hath any other name.

❧ The Nature.

The taste of this herbe is very like to Sarrasines Consounde, and therefore it is of like nature.

❧ The Vertue and Operation.

Golden rod is also an herbe apt to heale woundes, and hath the same vertues

tues whiche Sarrasines Consounde hath, and may be vsed in all diseases for the whiche the sayde Consounde is good.

The same boyled in wine and dronken, is very good agaynst the stone B namely in the reynes. For it breaketh the same, and maketh it to descend with the water or vrine: and so doth also the water of this herbe distilled with wine, and dronken by some space of time, as wryteth Arnoldus de Villa Noua.

Of water Sengreene and knights yerrow, or Woundworte. Chap. cj.

❧ The Kindes.

Vnder the title of Stratiotes, that is to say, knights woundworte, or water yearrow, Dioscorides describeth twoo herbes, well knowen in this countrie. The one called Crabbes clawe: The other water Milfoyle or Yearrow.

Sedum aquatile.
Water Sengreene.

Stratiotes potamios.
Knights woundeworte.

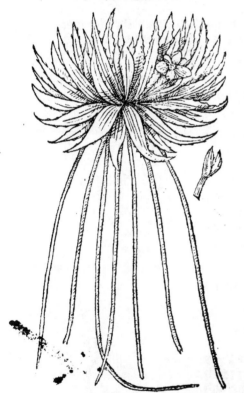

❧ The Description.

1. The first whiche is called knights Woundworte, or water Sengreene, is a water herbe whiche fleeteth vpon the water, not muche vnlike the great Sengreene, before that he bringeth foorth his stalke, but that it is greater. The leaues be narrow of halfe a foote long, hauing vpon each side sharpe teeth and prickley poyntes or indented corners, like to Bitter Aloes, or Sea

the Historie of Plantes. 143

or Sea aygreene, but muche smaller, narrower and shorter. The floures are vpon short stemmes, and grow foorth by the sides of the leaues, and are white and diuided into three, with a certayne throm or hearie yellow in the middell, and growe out of a clouen huske like to a Crabbes Clawe. It hath none other roote sauing a very shorte stemme, broade and thicke aboue, and very small and tender vnderneath, from whence springe vp the leaues: by the sayde shorte stemme vnderneath the leaues, growe long threedes (like to very fine and small lutestrings) here and there stretching themselues euen to the bottom of the water, by the whiche it taketh holde and draweth sustenance from the grounde. Certayne deceytfull and naughtie Rogues that would be taken for cunning Physitions, with their Treacles, Scammonie, and Playsters, do gather of the fine strings and hearie rootes aforesayde, and put them into Phiols or Glasses full of water, and set them openly in their shoppe windowes or standings, to be seene of the people, whereby they make the people to beleeue, that that they be wormes, whiche they haue caused men to auoyde with theyr pouders, Sugar and Oyntments.

2 The second kinde called knights Milfoyle (bycause of the great multitude and number of leaues) hath long, small, and narrow leaues, deepely cut in vpon bothe sides, like to the winge fethers of some small byrdes. For as the feathers of byrdes haue as it were a stemme, or a certayne ribbe in the middell, from whence there grow out vpon eache side long narrow barres, plumes or fine heares: euen so in like manner, these small leaues, haue also a ribbe or sinew in the middell, from whence there growe out vpon bothe sides small and narrow leaues, euery leafe like to the hearie barres or plumes of such smal fethers. Amongst the sayde leaues groweth vp a stalke or stemme of a span long bearing leaues like to the aforesayde, and at the top a fayre tufte, bushe, or nosegay of many small yellow floures like the common Yearrow or Milfoyle. The roote is tender and threedy.

❧ The Place.

The first kinde groweth in this countrey in pondes and pooles, & is found in diuers diches that are neare to the Riuers of Eschauld and Dele, in the countrey of Brabant.

The other groweth in very good and ranke medowes, but a man shall finde it very seldome.

❧ The Tyme.

Water Sengreene floureth in May. The other in August.

❧ The Names.

1 The first is called in Greeke ϛρατιώτης, καὶ ϛρατιώτης ποτάμιΘ-: in Latine Militaris: and it may be well called Sedum aquatile: in English knights worte, knights woundeworte, or knightes water woundworte, knights Pondeworte, and of some knights water Sengreene: in base Almaigne Crabbenclaw, and after the Greeke, Ruyters cruyt, or water Ruyters cruyt.

2 The second is called in Greeke ϛρατιώτης χιλιόφυλΘ-: in Latine Militaris millefolia: in English knights Milfoyle: souldiers Yerrow, and yellow knighten Yerrow: in French *Herbe militaire à millefueilles*, and *Millefueille iaulne*: in base Almaigne Geel Geruwe.

❧ The Nature.

Knights woundeworte of the water is colde and dry: The other with the thousand leaues, called knighten Mylfoile, is dry and somewhat astringent.

❧ The Vertues.

1 The first knights worte boyled in water and dronken, stoppeth the pissing

of bloud, and cureth the woundes and vlcers of the kidners, and the vse of it is good against all inwarde woundes.

The leaues therof pounde, and layde vpon greene woundes, keepeth them from inflammation and apostumation or swelling.

The same layde too with vineger, cureth the wilde fyre, or S. Anthonies fire, with other hoate tumors, as Phlegmons, &c.

2 The yellow knights worte, or Souldiers Milfoyle, is singuler good against all kinde of olde and new vlcers: it cureth Fistulas, it stancheth bloud, it soudereth, bringeth togither, and cureth woundes, whether it be pounde or brused and so layde vpon, or mixt with oyles, oyntments, and emplaysters that are made for such purposes.

Of Yarrow, or common Milfoyle.
Chap. cij.
The Description.

Achillea.

Milfoyle hath round hollow stalkes of a foote and halfe long: the leaues be long and very fine, and deepely iagged vpon both sides, euen harde vnto the middell ribbe or sinew, very wel like to the smallest leaues of Coriander or Southrenwood. The floures grow in fayre rounde tuffets or busshes at the toppe of the stalke, and are most commonly all white, sometimes also in this countrey of a purplish colour, and as Dioscorides writeth, sometines all yellow, the whiche as yet hath not bene seene in this countrie. The roote is blacke and threedy.

The Place.
Milfoyle groweth plentifully in this countrey, about paths, high wayes, and the borders of fieldes.

The Tyme.
It floureth from June to September.

The Names.
This herbe is called in Greeke ἀχίλλεως: in Latine Achillea, and Achillea sideritis, of Apuleius Myriophyllon, Myriomorphos, Chiliophyllon, Stratioticon, Heracleon, Chrysitis, Supercilium Veneris, Acron syluaticum, Militaris, and of some Diodela: in Shoppes at this present Millefolium: in Italian Millefoglio: in Spanishe Yerua Milloyas: in English also Milfoyle, Yerrow, and Nose bleede: in French Millefueille: in high Douch, Garben, Schaffgrasz, Schaffrip, and Tausenblaet: in base Almaigne, Geruwe.

Some count Achillea, to be that kinde of Tansie, whiche we before in the tenth Chapter of this present booke haue named the small white Tansie, as it is there declared.

the Historie of Plantes. 145

❧ *The occasion of the Name.*

This herbe had his name Achillea, of the noble and valiant knight Achilles, whose valiant actes & noble Historie were described by Homer. The sayde Achilles vsed this herbe very much, and it was firste taught him by the Centaure Chiron. With this herbe Achilles cured the woundes and sores of Telephus the sonne of Hercules.

❧ *The Nature.*

Milfoyle is very dry and astringent.

❧ *The Vertues.*

The Decoction of Milfoyle dronken doth cure and stoppe the bloudy flixe, A and all other laskes.

The same dronken stoppeth all fluxes, but especially the redde flure in wo- B men that floweth to abundantly. It worketh the same effect being applied to the secrete partes, or if one sitte or bathe in the decoction thereof.

The same brused and layde vpon woundes stoppeth the bloud, and keepeth C the same from inflammation and swelling, and cureth the same.

Of Comfrey. Chap. ciij.

❧ *The Description.*

Comfrey hath rough hearie stalkes, and long rough leaues, much like the leaues of commō Buglosse, but much greater and blacker. The floures be rounde and hollow like little belles, most commonly white, and sometimes reddish. The roote is blacke without and white within, very clammy or slimie to touche.

Symphytum magnum.

❧ *The Place.*

Comfrey groweth alongst by diches, and in moyst places.

❧ *The Tyme.*

It floureth in June and July.

❧ *The Names.*

This herbe is called in Greeke σύμφυτον καὶ σύμφυτον μέγα: in Latine Symphytũ magnum, & Solidago: in Shoppes Consolida maior: in Italian Consolida maggiore: in Spanish Suelda mayor, Consuelda mayor: in English Comfrey, and Comferie: in Frẽch Consyre: in high Douch Walwurtz, grosz Beinwel: in base Almaigne Waelwortel.

❧ *The Nature.*

Comfrey is hoate and dry in the second degree.

❧ *The Vertues.*

The rootes of Comfrey pound and dronken, are good for them that spitte A bloud, and healeth all inwarde woundes, and burstings.

The same also beyng brused and layde to in manner of a playster, do heale all greene and fresshe woundes: and are so glutinatiue, that if it be sodde with chopte or minsed meate, it wil reioyne and bring it all togither againe into one masse or lumpe.

The rootes of Comferie boyled and dronken, do clense the breast from flegmes, and cureth the grieffes or hurtes of the Lunges. They haue the lyke vertue, being mengled with sugar, syropes, or Honny, to be often taken into the mouth or licked.

The same with the leaues of Grounswell, are good to be layde vpon all hoate tumors or inflammations, especially to the inflammations of the fundament or siege.

The same also are good to be pounde, & layde vpon burstings or ruptures.

⁋ The ende of the first parte of Dodoneus Herball.

The seconde parte of the Historie

of Plantes, intreating of the differences, proportions, names, properties, and vertues, of pleasant and sweete smelling floures, herbes and seedes, and suche like. Written by that famous D. Rembertus Dodoneus now Physition to the Emperour.

Of Marche Violets. Chap. i.

❧ *The Kyndes.*

There be two sortes of Violets: the garden and the wilde Violet. The Garden violets are of a fayre darke or shining deepe blewe colour, and a very pleasant and amiable smell. The wilde Violets are without sauour, and of a fainte blewe or pale colour.

❧ *The Description.*

Viola Nigra.
The blacke, or purple Violet.

1 The sweete Garden or Marche Violet, creepeth alongst ye ground like the Strawberie plante, fastening it selfe and taking roote in diuers places: his leaues be rounde and blackish like to Iuye leaues, sauing they be smaller, rounder, and tenderer: emongst the whiche leaues there springeth vp fayre & pleasant floures of a darke blew colour, eache floure growing alone by him selfe, vpon a little small and tender stemme. The floures are diuided into fiue small leaues, wherof the middle of the floures, with the tippes or poynted endes of the leaues are speckled or spotted with a certayne reddish yellow. After the floures there appeareth round bullets, or huskes full of seede, the whiche being ripe do open and diuide themselues into three partes, the roote is tender & of threddish strings.

Of this sorte, there is another kinde planted in gardens, whose floures are very double, and full of leaues.

There is also a thirde kinde, bearing floures as white as snow.

And also a fourth kinde (but not very common) whose floures be of a darke Crymsen, or old reddish purple colour, in all other poyntes like to the first, as in his leaues, seede, and growing.

2 The wilde is like to the garden Violet, but that his leaues are far smaller, his floures are somwhat greater, but much paler, yea sometimes almost white, and without sauour.

❧ *The Place.*

The sweete garden Violet, groweth vnder hedges, and about the borders of fieldes and pastures, in good ground and fertyle soyle, and it is also set and planted in gardens. The wilde kinde whiche is without smell, groweth in the borders of dry, leane, and barren fieldes.

❧ *The*

❧ The Tyme.

The garden violet floureth in Marche and Apꝛill. The wilde also doth floure in Apꝛill, and afterwardes.

❧ The Names.

The sweete Violet is called in Greeke ἴον πορφυρέον: in Latine Viola nigra, Viola purpurea: of Virgil Vaccinium: in Shoppes Viola: in English Violets, the garden Violet, the sweete Violet, and the Marche violet: in Italian *Viola porporea*, and *Viola mammola*: in Spanish *Violetas*: in Frenche *Violette de Mars, ou de quaresme*: in high Douch Blauw veiel, oꝛ Mertzen violen: in base Almaigne Violetten: the Violet plante oꝛ herbe is called in Shoppes Violaria, and Mater violarum.

❧ The cause of the Greeke name.

The sweete Violet (as the Emperour Constantine wꝛyteth) was called in Greeke Ion, after the name of that sweete guirle oꝛ pleasant damosell Io, which Jupiter, after that he had gotte her with childe, turned her into a trim Heaffer oꝛ gallant Cowe, bycause that his wife Juno (beyng bothe an angry and Jelous Goddesse) should not suspect that he loued Ion. In the honour of which his Io, as also foꝛ her moꝛe delicate and holsome feeding, the earth at the commaundement of Jupiter bꝛought foorth Violettes, the whiche after the name of his welbeloued Io, he called in Greeke Ion: and therefoꝛe they are also called in Latine, as some do wꝛyte, Violæ, quasi vitulæ & Vaccinia. Nicander wꝛyteth, that the name of Ion was giuen vnto Violettes, bycause of the Nymphes of Ionia, who firste of all pꝛesented Jupiter with these kindes of floures.

❧ The Nature or Temperament.

Violets are colde in the first degree, and moyst in the second.

❧ The Vertues.

The Decoction of Violets is good against hoate feuers, and the inflammation of the Liuer, and all other inwarde partes, dꝛiuing foꝛth by siege the hoate and cholerique humoꝛs. The like pꝛopertie hath the iuyce, syꝛupe, oꝛ conserue of the same. **A**

The Syꝛupe of Violets is good against the inflammation of the lunges and bꝛeast, and against the Pleurisie, and cough, and also against feuers oꝛ Agues, but especially in yong childꝛen. **B**

The same Syꝛupe cureth all inflammations and roughnesse of the thꝛote if it be much kept oꝛ often holden in the mouth. The sugar of violets, and also the conserue, and iuyce, bꝛingeth the same to passe. **C**

That yellow whiche is in the middest of the floures, boyled in water, is good to be gargled in the thꝛote agaynst the squinancie oꝛ swelling in the thꝛote: it is also good to be dꝛonken agaynst the falling sickenesse in yong childꝛen. **D**

Violets pounde and layde to the head alone, oꝛ mengled with ople, remoueth the extreame heate, swageth head-ache, pꝛouoketh sleepe, and moysteneth the bꝛayne: it is good therefoꝛe against the dꝛynesse of the head, against melancholy, and dulnesse oꝛ heauinesse of Spirite. **E**

Violets bꝛused oꝛ stamped with barlie meale, are good to be layde vpon phlegmons, that is to say, hoate impostumes oꝛ carbuncles, and they heale the inflammation and paine of the eyes, also the hoate vlcers, and the inflammation that comꝛeth with the falling downe of the fundament. **F**

The seede of Violettes, dꝛonken with wine oꝛ water, is good agaynst the stingings of Scoꝛpions. **G**

The herbe or plante is very good against hoate feuers, and the inflammations of the liuer, and looseth the belly.

The wilde Violets are almost of the same vertue, but they be a great deale weaker, and therefore they are not vsed in Medicine.

Of Pances, or Hartes ease. Chap.ij.

❦ *The Description.* Viola tricolor.

Pances hath triangled stemmes, with many ioynts: his leaues are blackish, and dented, or toothed rounde about like a sawe, betwixte the whiche leaues there growe vp from the stalke, small naked or bare stemes: bringing foorth fayre & pleasant floures, parted into fiue littell leaues, like to a Violet, each floure being of three diuerse colours, whereof the highest leaues for the most parte are of a violet, and purple colour, the others are blewishe or yellow, with blacke and yellow streekes alongst the same, and the middell hearie: afterwarde there appeare small Bollyns or knoppy huskes, wherin the yellow seede is inclosed.

❦ *The Place.*

These floures do grow in gardens, & there is many of them found growing amongst the stubble in corne fieldes.

❦ *The Tyme.*

They begin to floure incontinent after the Violets, and remayne flouring al the sommer long.

❦ *The Names.*

This floure is called in Greeke φλόξ καὶ φλόγιον: in Latine Viola flammea, Flamma, & at this time Viola tricolor, Herba Trinitatis, Iacea, and Herba Clauellata: in English Pances, Loue in idlenes, and Hartes ease: in Frēch Penseé, and Penseé menue: in high Douch Freyscham, Freyschamkraut, and Dreyfeltigkeytblumen: in base Almaigne Dryebuldicheyt bloemen: and Penseen.

❦ *The Temperament.*

Panses are dry and temperate in colde and heate.

❦ *The Vertues.*

These floures boyled and dronken, do cure and stay the beginnings of the falling euill or the disease of young children that foome and cast vp froth, wherfore it is called in high Douch, Freyscham.

The same floures boyled with their herbe or plante, and giuen to be droken, doth clense the lunges and breast, and are very good for feuers, and inward inflammations or heates.

Planta hæc maximè probatur ad glutinanda vulnera, tā exterius illita, quàm interius sumpta: adhæc ad enterocælas. In quem vsum puluerem eius, mensura dimidij cochlearis, ex vino austéro, fœlici succeffu propinant.

The second Booke of
Of the Wall floure. Chap. iij.

✻ *The Description.*   Viola lutea.

The yellow Gillofer or Wallfloure, is a littell shrubbe or bushe, that is greene both winter & somer, whose stalkes are harde & of a woody substance, and full of branches: the leaues growing thereon are somwhat thicke set, long, narrow, and greene: at the top of the stalkes or branches, growe the floures, whiche be very yellow, and fayre, of a pleasant smell, euery floure diuided into foure small leaues, the whiche perished there commeth vp long Coddes or huskes, wherein is conteyned seede whiche is large, flatte, and yellow.

✻ *The Place.*

The yellow Gillofer or Wall floure, groweth vpon olde walles, & stonehilled houses, & is comonly planted in gardens.

✻ *The Tyme.*

The yellow Gillofer doth chiefly floure in March, Aprill, and May.

✻ *The Names.*

The yellow Gillofer is a kinde of violets called in Greeke λευκόϊα, the which are also called in Latine Leucoia lutea, and of Serapio and the Apothecaries Keyri: & of Plinie (who hath seuered them from Leucoion, that is to say, from the stocke Gillofer, or rather the white violet) Violæ luteæ: in Italian *Viola giala:* in Spanish *Violetas amarillas.* in English Yellow Gillofers, Wall floures, and Hartes ease: in French *Violes iaunes, Giroflée iaulne:* in high Douch Geel veiel: in Brabant geel Vilieren, steen Vilieren.

✻ *The Nature.*

Wall floures are hoate and dry, and of subtill partes.

✻ *The Vertues.*

Wall floures dryed and boyled in water prouoketh vrine, and causeth wemen to haue their termes, it cureth the Scirrhos, or harde impostems of the Mother, whan the same is stewed or bathed therewith.   A

The same floures with oyle and waxe, brought into a playster do heale the choppes or riftes of the siege and fundament, or falling downe of the Arse-gut, and closeth vp olde vlcers.   B

The Wall floure mengled with Hony, cureth the naughtie vlcers, and swellings of the mouth.   C

The quantitie of two drames of the seede of Wallfloures dronken in wine, bringeth downe womens floures, deliuereth the Secondyne, and the dead childe. It doth all the same very well, being conueyed into the Matrix or Mother in a Pessarie.   D

The iuyce of this Gillofer, dropped into the Eyes, doth wast and scatter all spottes and dimnesse of the same.   E

The

The roote stamped with vineger, cureth the hardnesse of the Splene or Melte, being applied thereto.

Of Stocke Gillofers, or Garnesee Violets. Chap.iiij.

❧ *The Kyndes.*

There are found two kindes of these Gilloflowres. The one is great and called the Castell, or stocke Gillofer, the whiche may be kept both winter and somer. The other is not so bigge, and is called the small stocke Gillofer, the whiche must be yearely sowen againe, and bringeth forth his floure and seede the same yeare.

❧ *The Description.*

Leucoion.

These two kindes of Violets or Gillofers, are not muche vnlike Walfloures sauing that their leaues be whiter and softer.

1 The great Castell, or stocke Gillofer his stalkes be harde and straight, of the heigth of two or three foote, with long narrow and softe leaues like Molyn, far greater, longer & larger than the leaues of Walfloures, or yellow Gillofers. The floures be of a fragrant or pleasant smel, in fashion and smell like to Hartes ease or Walfloures, but much larger, of colour sometimes white, sometimes ashe colour, sometimes Carnation, Stamell, or Scarlet colour, sometimes redde, and sometimes Violet, after whiche floures commeth long huskes or Coddes, wherin is flat or large seede.

2 The small Castell or stocke Gillofer, is like to the great in his stalkes, & whitish, wollie softe leaues, also in the sweete smel and fragrant sauour of his floures, in the diuersitie of colours, in his coddes and seede, sauing that it is smaller in all respectes, not exceeding the length of a mans foote, of small continuance, and perisshing euery yeare.

❧ *The Place.*

These kindes of Gillofers, are sowen in the gardens of this coūtrey: of this sorte there is found an other kinde in places neare the sea coast, as in Zealand not farre from the shore, but the same is smaller and lower than that whiche groweth in gardens.

❧ *The Tyme.*

The great Castell gillofer floureth in Marche and Apzill, a yeare after the sowing. The smaller floureth in July and August, the same yeare that it is first sowen.

❧ *The Names.*

These Violets, especially the greater kind are called in Greeke λευκόϊα: in Latine Violæ albæ, and is so called bycause his leaues be white, but not the leaues

152　　　The second Booke of

of the floures, for they be of diuers colours as is before sayde, they be called in Italian *Viola biancha*, in Spanish *Violetas blancas*. Some of the late writers do call them Violæ matronales, that is to say, Dames violets: but this name doth rather belong to an other sorte of Violets, whereof we shal intreate in the next Chapter folowing. But if we ought to call these Violets by the aforesayde name, the name will best agree with the small Castell Gilofer. The greater sorte is called in English Garnesie Violets, white Gilofer, Stocke Gilofer, & Castell Gilofer the smaller kinde, may be so called also. The greater sorte is called in base Almaigne Stock Vilieren, and the smaller sorte is also called of them Heeten Vilieren.

※ *The Temperament.*

These Violets are hoate and dry, & of nature somwhat like to Walfloures.

※ *The Vertues.*

The floures of stocke Gillofers, boyled in water & dronken, is good against A the difficultie of breathing, and the cough.

These Violets do likewise prouoke the floures, and vryne, and do cause to B sweate, if one do sitte ouer a bathe or stewe full of the decoction thereof.

To conclude, they are of nature very like to the yellow, or Walgilofer: The C whiche yet notwithstanding is in all respectes better & fitter in Medicine than the stocke Gilofers.

Of Dames violets, or Gilofloures. Chap.v.

※ *The Description.*

Violæ Matronales.

1　Dames Gillofers hath greate large leaues of a browne greene colour, somwhat snipt or dented rounde aboute the edges: Amongst the whiche springeth vp a stemme beset with the like leaues full of branches, whiche beareth sweete and pleasant floures at the toppe, in proportion like to the Gillofers aforesayde, most commonly of a white colour, sometimes carnation, and somtimes reddish, afterwardes come vp long rounde coddes or huskes, in whiche the seede is conteyned.

2　Of this kinde of Damaske Violets or Gillofloures, are they also which at now called *Vetarias*: wherof there be ij. sortes.

The first hath fiue leaues or moe, like hempe growing vpon one litle stem, the stalkes be smal and short, not much aboue the heigth of nine inches: vpon the grow smal floures of a violet colour in proportion like to Garnesee violets or Dames Gillofloures: after them comme huskes & seede like to them. The rootes be somewhat thicke, & vneuen, and as they were couered with certayne scales.

The other his leaues grow alongst the little stalkes, & are spread abroade like to the leaues of the Ashe, or Walnut trees, sauing they be smaller. The
floures

the Historie of Plantes.

floures be almost white, & the huskes or cods are like to the huskes of garnesey violets: the rootes be rough & vneuen, much like to the rootes of the first kind.

❊ *The Place.*

The violets or Gillofers are very common almost in all gardens.

❊ *The Tyme.*

They floure in May, and oftentimes else, whiles Somer lasteth.

❊ *The Names.*

These floures be now called in Latine Violæ Matronales: in English Damaske violets, Dames violets or Gillofers, and Rogues gillofers: in French Violettes de Dames: in high Douch Winter violen, wherefore some do also call them in Latine Hyberna viola, or Viola hyemalis: in base Almaigne Mastbloemen, and after the Latine name they call it Joncfrouwen bilieren, whiche may be Englished Dames violets.

The other kinde is knowen by the name of Dentarie: and is not otherwise knowen to vs. ❊ *The Temperament and Vertues.*

These floures are not vsed in medicine, therefore their temperature, and naturall operation, is yet vnknowen.

Of Bolbanac, or strange Violets. Chap. vi.

❊ *The Description.*

Viola Latifolia.

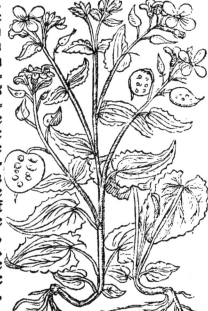

1 Bolbanac hath hard round stés, full of branches, his leaues be great & large, dented or tothed of a browne greene colour, and snipt or dented about the edges, not much vnlike the leaues of the Fylberte or Hasseltree. The floures be like to the floures of Damaske violets, of a pale purple colour, the whiche being vanisshed, there commeth vp white huskes, which be flat rounde, and very large, of the quantitie of a groote, or Testerne, wherein is conteyned a browne seede, after the fasshion of the Moone, the whiche may be seene thorough the thinne huskes or skinne of the Codde. The roote is white, & a litle thicke, and somwhat knottie or knobbie, which after y̅ it hath borne sede perisheth.

2 Yet there is founde a certayne kinde whose roote dieth not yearely, and that kinde both in his leaues & broad huskes, is smaller than the abouesayde.

❊ *The Place.*

This herbe is founde sowen in certayne gardens of this countrey.

❊ *The Tyme.*

Bolbonac floureth in Aprill and May, the next yeare after the sowing.

❊ *The Names.*

Forasmuch as these floures are somewhat like vnto violets, therefore they are now placed amongst the kindes of Violets, and are called in Latine Violæ Latifoliæ, of some Violæ peregrinæ. For vnder the name of ἴον in Greeke: and Viola in Latine, are commonly comprehended all sortes of floures whiche

whiche be any thing like vnto Violets. The Herboristes and certayne Apothecaries do call this herbe by a certayne barbarous and strange name Bolbonac: the Brabanders or base Almaignes, do call it Penninckbloemen, that is to say, Penny floure, or mony floure, and they call it also Paeschbloemen. The Auncients did account it for a kinde of Thlaspie, especially for that kinde descrybed by Crateuas, whiche some do call Sinapi Persicum, and of Dioscorides Thlaspie Crateuæ. ❊ *The Temperament and Vertues.*

The seede of this herbe is sharpe and biteth the tongue, and of a drying qualitie, and therefore is in vertue like the other Thlaspi.

Of Gillofers.　　　Chap.vij.

❊ *The Kyndes.*

Vnder the name of Gillofers (at this time) diuerse sortes of floures are contayned. Whereof they call the first the Cloue gillofer whiche in deede is of diuerse sortes & variable colours: the other is the small or single Gillofer & his kinde. The third is that, which we cal in English sweete Williams, & Colminiers: whereunto we may well ioyne the wilde Gillofer or Cockow floure, which is not much vnlike the smaller sort of garden Gillofers.

Vetonica altilis.　　　　　　　Vetonica altilis minor.
Carnations, and the double-　　The single Gillofers, Soppes in
cloaue Gillofers.　　　　　　　wine, and Pinkes, &c.

❊ *The Description.*

1. The Cloue gillofer hath long small blades, almost like Leeke blades. The stalke is round, and of a foote and halfe long, full of ioyntes and knops, & it beareth

the Historie of Plantes.

beareth two leaues at euery ioynt or knot. The floures grow at the top of the stalkes or stemmes, out of long round, smooth huskes and dented or toothed aboue like the spice called cloaues, or like to a littell crownet, out of the whiche the small feathered leaues do grow rounde about, spread in compasse, whereof some be of colour white, some carnation, or of a liuely fleshe colour, some be of a cleare or bright redde, some of a darke or deepe redde, and some speckled, and do all smell almost like Cloues. When the floures be past, there groweth in the sayde round cuppes or huskes, other long poynted huskes like barlie cornes, in which the small blacke seede is inclosed.

Armerius flos primus.
Sweete Williams.

Armerius flos tertius.

Vetonica syluestris.
Wilde Williams, or Cockow Gillofers.

2   The Pynkes, and small feathered Gillofers, are like to the double or cloaue Gillofers in leaues, stalkes, & floures, sauing they be single and a great deale smaller. The leaues be long & narrow, almost like grasse, the smal stemmes are slender and knottie, vpon whiche growe the sweete smelling floures, like to the Gillofers aforesayde, sauing eache floure is single, with fiue or sixe small leaues, deepe and finely snipt, or frenged like to small feathers, of white, redde, and carnation colour, after whiche floures there groweth also in the rounde huskes, other sharpe huskes, or as it were long pellottes, in the which the seede is conteyned.

3   The first sweete William or Colmenier (which is now called in Latine Armerius flos) is also somwhat like to the cloaue Gillofers, their leaues be narrow, their stalkes ioyntie, & their floures small, like to littell Gillofers, grow-
ing

ing three or foure togither at the toppe of the stalkes, & somtimes nine or tenne togither, like to a nosegay or small bundell of floures, of colour sometimes red, and sometimes spotted with white, and somtimes (but very seldom) all white.

There is an other kind of Armeriorum, whose leaues be broade, almost like the leaues of floure Constantinople. The stalkes of this kinde, with the nuber of small floures growing togither, which are of colour redde and white, & speckled or sprinckled with small spots, are very like vnto the aforesayde Armerijs.

There is also a certaine thirde kinde Armeriorum, with thinne whitishe or faynte greene leaues, and slender smooth knottie stalkes, whiche in handling seemeth to be somwhat fatte or clammy, in the toppe of the sayde stalkes grow small floures clustering or growing rounde togither, of a fayre washed purple redde colour, after them commeth narrow seede vessels, or small huskes like as in the other Gillofers, wherein the seede is conteyned.

The wilde Gillofers are somewhat like to Armeria or Colmeniers: they haue also small knottie stalkes, & narrow leaues, but yet they be larger, shorter, & a great deale whiter greene, than the leaues of the gillofers or Pinkes. The floures be most comonly redde, & somtimes also white, & deepely cut or iagged, almost like to white Pinkes or Soppes in wine, but without sauour. The floures gone, the seede groweth in long huskes like to Pynkes, or feathered Gillofers.

※ *The Place.*

The Cloaue gillofers, and the smaller, or single Gillofers, with the sweete Williams, and Colmeniers, are set and planted in the gardens of this countrie. The wilde Williams or Cockow gillofers, do grow of them selues in all medowes, and moyst grassie places.

The other kinde of Armerius groweth in Germanie, in certaine rough hillie places that stande open against the Sunne.

In Flaunders also there is sometimes found a certaine wilde floure, like to the Gillofers and Armerijs, sauing it is very small.

※ *The Tyme.*

All these sortes of floures, do most commonly floure all the somer time, from after May vntill September.

※ *The Names.*

The two first sortes are now called flores Garyophyllis, & of some in Greeke βετ'ονικου: in Latine Vetonicę: some iudge them to be Cantabricam, whereof Plinie writeth in the .xxb. Booke.

1. Whereof the first is also called Ocellum, Ocellum Damascenum, Ocellum Barbaricum, & of some it is called Vetonicam altilem, & Vetonicā Coronariam: in English garden Gillofers, Cloaue gillofers, and the greatest & brauest sorte of them are called Coronations, or Cornations: in Italian *Garofoli*: in high Douch Grasblumen, Negelblumen, and Neglin: in base Almaigne Ginoffelen: in French *Gyrofflees*, and *Oeilletz*, or *Oilletz*.

2. The second sorte, is also of the kinde of Vetonicarum, or gillofers, and may well be called Vetonica altilis, or Vetonica Coronaria minor: in English single Gillofers, wherof be diuers sortes great & small, & as diuers in colours as the first kindes, & are called in Englishe by diuers names, as Pynkes, Soppes in wine, feathered Gillofers, & small Honesties: they are called in high Douche Mutwille, & of some Hochmut: & accordingly they be called in Latine Superba, that is to say, Gallant, prowde, & gloriouse: in base Almaigne Pluymkens, and cleyn Ginoffelen, some call them also in French *des Armoiries*, or *des Barberies*.

3. That sorte which are called in English sweete Williams, are counted also to be of the kindes of the garden or Cloue gillofers (called in Latine Vetonica or

Canta-

Cantabrica, but now they be called in Latine Flores Armerij, yet some esteeme them to be a certayne kinde of Herbe tunice: the Germaynes call them Donderneglin, Feldtneglin, Heidenblumen, and Blutstroppsle: in base Almaigne Keykens: of the Frenchmen *des Armoires*. There is a kinde of this herbe which is common in the countrey gardens, and they call it Colmeniers.

4 ❧ The fourth is a kinde of wilde Vetonica, and therefore it is called Vetonica sylueſtris: in English wilde Williams, Marshe gillofers, or Cockow gillofers: in high Douche Gauchblum: in Brabant Crayebloemkens, and Coeckcoeckbloemkens: it may be called also in Latine Armoraria sylueſtris vel pratensis: or Flos Cuculi: and in French *des Barbaries sauuages*.

※ *The Nature.*

For the most parte all these kindes of floures, with their leaues and rootes, are temperate in heate and drynesse.

※ *The Vertues.*

A ❧ The Conserue of the floures of the first kinde, made with Sugar, comforteth the harte, & the vse thereof is good against hoate feuers & the Pestilence.

Of floure Constantinople. Chap. viij.

※ *The Description.*   Flos Constantinopolitanus.

THe floure Constantinople hath two, three, or foure, long holow and vpright stemmes, full of knees, or ioyntes, (with a certaine roughnesse). At euery ioynt groweth two leaues, which be somwhat long and large, and of a browne greene colour, the floures grow at the toppe of the stalkes, many cluſtering togither after the manner of Tol-me-neers, or sweete Williams, but somewhat larger, of the colour of Red-lead, or lyke to the colour of the Orenge pill that is throughly ripe. The floures be very pleasant and delectable to looke on, but they are without any pleasant sente or sauour. The leaues and ſtalkes be somewhat rough. The roote is whyte, and diuided into diuers other long and slender rootes, in taste somewhat sharpe.

※ *The Place.*

The Herboriſtes and suche as haue pleaſure in the ſtrāge varietie of floures, do plant these in theyr gardens.

※ *The Tyme.*

These floures do florishe from Midsomer, vntill it be almost winter.

※ *The Names.*

This pleaſant floure is called of the Herboriſtes Flos Constantinopolitanus, that is to say, floure Constantinople.

※ *The Nature.*

The roote of this herb is hoate & dry, as it doth manifeſtly appeare by ye taſt.

The second Booke of
Of Rose Campion. Chap.ix.

❧ The Description.

Lychnis satiua.

Rose Campion his stalkes be round, woolly, and knotty, hauing at euery knot or ioynt, a couple of long softe woollie leaues like ÿ leaues of Molin or higtaper, but much smaller, & narrower. The floures growe at the top of ÿ stalkes, out of long crested huskes, whereof some be of an excellent shining, or Orient redde, & some be white. The single floures are parted into fiue or sixe leaues, with little sharpe poynts in the middell of the floures, whereunto the smaller endes of the little leaues of the sayde floures are ioyned. When the floures are perished, there groweth within the playted or crested huskes, other coddes or huskes, whiche be somewhat long and round, wherein the seede whiche is blacke is conteyned. The roote is long and small.

❧ The Place.

These floures are planted in the gardens of this countrie.

❧ The Tyme.

They floure in June, July, and August.

❧ The Names.

These kinde of floures are called in Greke λυχνὶς στεφανωματική: in Latine Lychnis coronaria, and Lychnis satiua, of some Athanatos, and Acydonium, of Plinie Iouis flos: in English Rose Campion: in French Oeillets, & Oeillets Dieu: in high Douch Margenroszlin, & Marien rosen, and accordingly they are now called in Latine Rosa mariana: in base Almaigne they are most commonly called Christus ooghen.

❧ The Nature.

The floures are hoate and dry.

❧ The Vertues.

The seede with the floure, or either of them alone dronken, are good against the stinging of Scorpions.

Of wilde Campion. Chap.x.
❧ The Kindes.

There be two sortes of these floures, that is to say, a white and a redde, whereof the white kinde is the greater and of a larger grothe. The redde is smaller and lesse.

❧ The Description.

The wilde white Campion, hath a rough white stemme: The leaues be white & cottony, much like to the leaues of Campions, sauing that the stalkes be slenderer, and the leaues narrower and not so white. The floures growe out of a rough huske, greater then the huske of the garden Rose Campion, and the proportion of the floure is muche like to the same, but

but more indented aboute the edges, and without any sharpe poynted peake in the middell: the floures being vanished, there commeth after them rounde bollettes or pellets in whiche the seede is conteyned. The roote is ordinarily of the length of a foote and halfe, and as thicke as a finger.

Lychnis sylue stris alba.
The white wilde Campion.

Lychnis sylue stris purpurea.
The purple wilde Campion.

2 ❧ The redde wilde Campions, are in all things like to the white, sauing that they grow not so high, and their roote is not so long, but is for the moste parte shorter and hearie. The floures be redde, and in proportion like to the other.

✤ *The Place.*

These floures grow in vntilled groūdes, in the borders of fieldes, & alongst the wayes: some also vse to set them in gardens, and it commeth to passe, that by often setting they waxe very double.

✤ *The Tyme.*

They floure most commonly from May vntill the ende of Somer.

✤ *The Names.*

The wilde Campions, are called in Greeke λυχνὶς ἀγρία: in Latine Lychnis syluestris, of some Tragonatum, Hieracopodium, or Lampada: in the Shoppes of this countrie Saponaria, howbeit this is not the right Saponaria: in English wilde Campion, or wilde rose Campion: and of some Crowesope: in high Douch Lydweyck, wilde Margenrößlin, and in some places widerstoß: in Brabant Jennettekens.

✤ *The Nature.*

These floures with their plante, are in temperament like to garden rose Campions.

✤ *The Vertues.*

The seede and floures, with the whole herbe, of the wilde Campions, are very

Of Cockle, or fielde Nigella.   Chap.xi.

Anthemon.

❧ The Description.

Ockle or fielde Nigelweede, hath ſtraight ſlender hearie ſtemmes, the leaues be alſo long, narrow, hearie, & grayiſh. The floures be of a browne purple colour, changing towardes red, diuided into fiue ſmall leaues, not much varying from the proportion of the wilde Campions, after the which there groweth rounde boileyns or cups, wherein is cotepned plenty of ſeede (of a broune or ruſſet colour.)

❧ The Place.

Theſe floures grow in the fieldes, amongſt the Wheate, Rye, and Barley.

❧ The Tyme.

It floureth in May, June, and July.

❧ The Names.

This floure is now called amongſt the learned men Githago, or Nigellaſtrum, or Pſeudolanthium, of ſome flos Micancalus, as Ruellius writeth: in Engliſh field Nigella or Cockle: in high Douch Raden, Groſzraden, and Kornroſz: in Brabant Corenrooſen, and Negelbloemen: in French *Nielle*.

❧ The Temperament and Vertues.

The vertues, & temperament of this herbe, are not yet knowē, bicauſe it is not in vſe, ſauing of certayne fonde people, whiche do vſe it in the ſteede of Vuray or Darnell, or for the right Nigella, to the great daunger and perill of the ſicke people.

Of Blew Bottell, or Corneflouze.   Chap.xij.

❧ The Deſcription.

1   Cyanus hath a creſted ſtalke, vpon the whiche growe narrowe, ſharpe poynted & grayiſhe leaues, whiche haue certayne natches or cuts about the edges, & ſharpe corners like teeth. About the toppe of the ſtalkes, it beareth ſmall round buttons whiche be rough & ſcalie, out of the whiche grow pleaſant floures, of fiue or ſixe ſmall iagged leaues, moſt commonly blew (eſpecially the wilde kinde.) Sometimes alſo thoſe that grow in gardens, do beare grayiſh, purple, crimſen, and white floures: the whiche being vaniſſhed, there groweth within the ſcalye huſkes & heades, certayne long ſeede, whiche is incloſed in a hearie downe or Cotton.

2   There is alſo in certayne gardens, an other kynde of Cyanus, whoſe floures be lyke to the aforeſayde, it hath greate broade leaues, larger than the leaues of the garden Roſe Campion, the whiche bee alſo ſofte and woolly, lyke the leaues of Mullen. The floures of this hearbe are lyke to the

the Historie of Plantes.

Cyanus.
Corne floure.

Cyanus maior.
Great Corne floure.

the other Cyanus floures both in his Scaly knopped buttons, as also in his iagged, or frenged leaues, & seede: but a great deale larger, and of colour blew, in the middle turning somwhat towardes redde, or purple. The roote is of long continuance, and sendeth forth new stemmes and springs yearely.

❧ The Place.

Cyanus or Blew bottell groweth in the fieldes amongst the wheate, but specially amongst Rie. Those which haue the white and purple floures, and the great Cyanus, are sowen and planted in gardens.

❧ The Tyme.

These floures do flowrish, from May vntill August.

❧ The Names.

1   This floure is called of Plinie in Latine Flos Cyanus, of some later wryters Baptisecula, or Blaptisecula: in Italian *Fior Campesi*: in English of Turner Blewbottell, and Blewblaw, it may also be called Hurte Sicle, and Cornefloure: in French *Aubifoines, Bleuets, Perceles,* and *Blaueoles*: in high Douch Kornblumen: in Brabant Corenbloemen, and Roghbloemen.

2   The second kinde is called Cyanus maior, and is counted of the learned for a kinde of Verbascum, and therefore they call it Thryallis and Lychnitis: in high Douch it is called Waldt kornblumen: and in Brabant groote Corenbloemen: we may also call it in English great Cornefloure, and wilde Cornefloure.

❧ The Temperament.

Cyanus or Blewblaw, is colde and dry.

O iij

The second Booke of

❊ *The Vertues.*

¶ This Corneſloure bruſed or pound, is profitably layde vnto the rednesse, the inflammation and running of the eyes, or to any kinde of Phlegmon or hoate tumor about the eyes.

¶ The diſtilled water of Cyanus, cureth the rednesse and payne of the eyes, whan it is either dropped into the eyes, or elſe that the eyes be waſſhed therewithall.

Of Marygolds. Chap. liij.

❊ *The Description.* Calendula.

THe Marygolde hath three or foure ſtalkes of a foote and a half long, ſet with leaues ſomewhat long & large, and of a white greene colour: at the toppe of the ſtalkes growe pleaſant bright & ſhining yellow floures, ſomewhat ſtrong in ſauour, the whiche do cloſe, at the ſetting downe of the Sunne, and do ſpread and open againe at the Sunne riſing. Each floure hath in the middeſt thereof a yellow or browne crowne (like to a ſhauen Crowne) about the circuyt or compaſſe wherof, there are ſet many littell ſmall yellow leaues. Whan the floures are vaniſſhed, there groweth in the places, from whence they fell, certayne round knops like vnto great buttons, cōpact of many crooked ſeedes growing togither into a knop like a button, each ſeede alone is croked like to a halfe Circle, or the new Moone. The roote is white and threddy.

❊ *The Place.*

¶ Theſe floures do grow in euery garden where as they are ſowē, and they do yearly ſpring vp a new of the fallen ſeede.

❊ *The Tyme.*

¶ They floure almoſt euery moneth in the yeare, but eſpecially from May vntill winter.

❊ *The Names.*

¶ They be now called in Latine Calendula, and of ſome Caltha, and Calthula: in Engliſh Marygoldes, and Ruddes: in Italian *Fior rancio:* in French *du Soucy,* and *Sousie:* in high Douch Ringelblumen: in baſe Almaigne Goutbloemen. (Pena calleth it in Latine Caltha poetarum, and Chryſanthemon.)

❊ *The Nature.*

¶ The Marygolde in complexion is hoate and dry.

❊ *The Vertues.*

¶ The floures by them ſelues, or togither with their plante, boyled in wine & dronken, prouoketh the Menſtruall fluxe.

¶ The ſame with their herbe dryed, and ſtrowed vpon quicke coles, draweth forth the ſecondyne or afterbirth, with the dead childe, the fume thereof being receyued at the conuenient place.

¶ The diſtilled water of Marygoldes, put into the eyes, cureth the redneſſe, and

the Historie of Plantes.

and inflammation of the same.

The conserue that is made of the floures of Marygoldes, taken in the morning fasting, cureth the trembling and shaking of the harte, it is also good to be vsed against the Plague, and corruption of the ayre.

Of Horse floure, or Cowe wheate. Chap.xliij.

❧ *The Description.*   Melampyrum. Triticum vaccinum.

HOrse floure hath a straight stemme of a foote long, w̄ three or foure braunches by the sides, couered with long narrow leaues: at the toppe of the braunches growe fayre spiked eares, full of floures and small leaues, deepely cut and iagged, in proportion not much vnlike to a Foxetayle. This eare beginneth to floure below, & so it goeth flouring by little and littell vpward. Before the opening of the floures the small leaues & buddes of the floures, are all of a fayre blewish purple colour: and immediatly after the opening of the floures, they are of a yellow colour mixed with purple, and after the falling away of the floures, those small purple leaues do also loose their colour and waxe greene, and in steede of the floures, there commeth broade huskes, wherein commonly are inclosed two seedes, not much vnlike vnto wheate cornes but a great deale smaller and browner. The roote is slender and of woody substance.

❧ *The Place.*

This plant groweth amongst wheate and Spelt, in good frutefull groundes.

❧ *The Tyme.*

Melampyrum floureth in June, and somtimes in July.

❧ *The Names.*

They call this herbe now, in Latine Triticum vaccinum, or Triticum bouinum, that is to say, Cow wheate, or Oxe wheate: in French *Bled noir*: that is to say, Blacke wheate, or Corne: in high Douch Kuweyssen, and of some Braun fleischblumen: in Brabant Peertsbloemen: that is to say, Horse floure: And it should seeme to be that vnprofitable herbe whereof Theophrastus writeth in his viij. booke Chap. b. And Galen Primo de alimentorum facultatibus, Cap. vltimo, called in Greke μελάμπυρον: in Latine Melampyrum, which as they do write is but a weede, or vnprofitable plante growing amongst wheate, and so called bycause of the seede, whiche is blacke and proportioned like wheate. Yet this is not the Melampyrum of Dioscorides, the whiche also is called Myagrion.

❧ *The Nature.*

Horse floure, or Blacke wheate, especially the grayne or seede, is hoate, and rayseth vp fumes.

❧ *The Vertues.*

The seede of this herbe taken in meate or drinke troubleth the braynes, causing headache and dronkennesse, yet not so much as Duray or Darnell. Vaccis pabulo grata & inuocua.

Of Larckes ſpurre. Chap.xv.

The Kyndes.

There be two ſortes of Conſolida regalis: wherof one kinde groweth in gardens, and the other is wilde.

Delphinium.
Garden Larkes ſpurre.

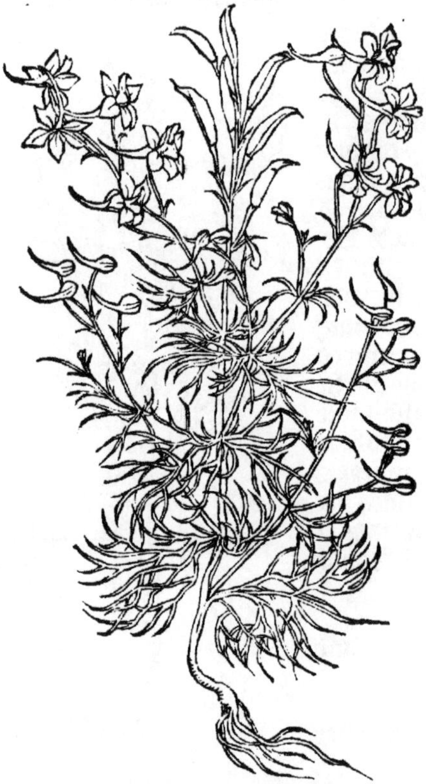

Bucinum.
Wilde Larkes ſpurre.

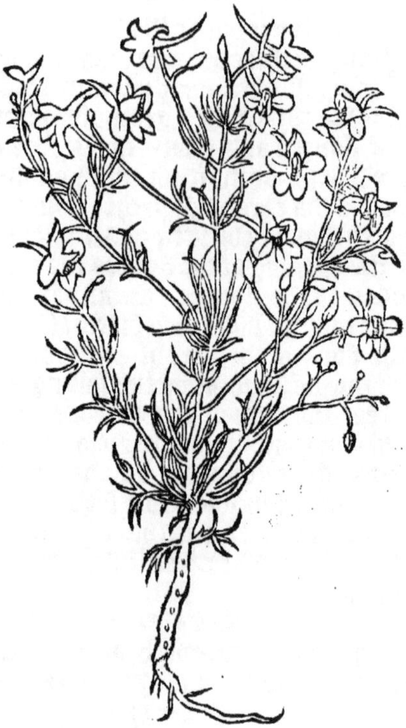

The Deſcription.

1 The garden Larkes ſpurre hath a round ſtraight ſtem full of branches, ſet with tender leaues, all iagged and cut very ſmall, or frenged much like to the leaues of the ſmaller Southrenwood. The floures grow alongſt the ſtalkes at the toppe of the branches, and are compacte of fiue littell leaues growing togither, ſomewhat like to the March violet, ſauing that one of the leaues of this floure is long and hollow, hauing behind it a croked ſpurre or tayle, turning like the floure of wilde Lyn, or Toode flaxe. Theſe floures are of colour, ſometimes purple blewe, ſometimes white, and ſometimes Carnation: after the falling of, of theſe floures, there commeth vp long coddes, wherein is conteyned browne ſeede.

2 The wilde Larkes ſpurre is lyke the other, but a great deale ſmaller in his ſtalkes and leaues, and in length ſhorter. Theſe floures are like to the aboueſayde, but they be much ſmaller, and grow not ſo well togither, of a fayre purple blew colour like vnto Violets, and after them alſo commeth vp coddes, wherein the ſeede is contayned.

The

the Historie of Plantes. 165

❧ *The Place.*

The garden Larkes spurre, is sowen in this coūtrie in the gardens of Herboristes. The wilde groweth amongst corne, in fertill countries.

❧ *The Tyme.*

The garden Larkes spurre floureth all the Somer long. The wilde floureth in June, and July. ❧ *The Names.*

The garden Larkes spurre is called in Greeke Δελφίνιον, and in Latine Delphinium, of some late wryters Flos regius, or Flos equestris. Also Calcatrippa: in Italian *Sperone de Caualliere*.

The wilde is called in Greeke Δελφίνιον ἕτερον, Delphinum alterum, & Βουκίνους, Bucinum, it is now called in Latine Cōsolida regia aut regalis: in English Kings Consounde, wilde Larkes spurre, or Larckes Claw: in French *Consoulde royale*, and *Pied d'alouette*: in high Douch Rittersporn, and according to the same in base Almaigne Riddersporen, that is to say, knightes spurre.

❧ *The Nature.*

Larkes Claw in complexion is temperately warme.

❧ *The Vertues.*

The seede of the garden Larckes spurre dronken is very good agaynst the stinging of Scorpions, & in deede his vertue is so great against their poyson, that the only herbe throwen before the Scorpions, doth cause them to be without force or power to do hurte, so that they may not moue or sturre, vntill this herbe be taken from them.

The seede of wilde Larkes spurre, is of vertue like to the garden Larke spurre, but not so strong.

Of Columbyne.   Chap. xvi.

❧ *The Description.*

Aquilegia.

Columbyne hath great broade leaues, with ij. or iij. deepe cuts or gasshes in the leaues, like to the leaues of the great Celondyne, but whiter (& in some kindes of a darke sage colour) but of no strong sent or sauour, neither yelding forth any such yellow iuyce, sappe, or liquor, whan it is broke or bruised, as the Celondyne doth. The stalkes be round, & playne or smoth, of ij. or iij. foote long, vpon which growe the floures, cōpact of two kindes of little leaues, wherof one sorte, are small & narrow, & the others growing with them ar hollow, tō a long croked tayle like larkes Claw (& bending somwhat towards the proportiō of the necke of a Culuer). The floures are somtimes single, & somtimes dubble, & of colour somtimes blew, somtimes white, sometimes skie colour, somtimes red, somtimes speckled, & intermēgled with blew & white. After the vanisshing of the sayde floures, there commeth foorth iiij. or fiue sharpe huskes or cods, growing ioyntly togither, wherein is cōteyned a blacke (shining) seede. ❧ *The*

❧ The Place.

They sowe, and plante them here in gardens, and they do also grow in high woodes, and rockes, but not in this countrie.

❧ The Tyme.

They floure most commonly in May and June.

❧ The Names.

This floure is now called in Latine Aquilegia, or Aquileia, and of the later writers Columbina, vnknowen of the Auncients, howbeit some late wryters make a question, whether it be Ponthos Theophrasti, siue Desiderium, after the interpretation of Gaza: it is called in English Columbine of the shape & proportion of the leaues of the floures whiche do seeme to expresse the figure of a Doue, or Culuer: in French *Ancoly*, in high Douch Agley, and Ageley: in base Almaigne Akeley.

❧ The Nature.

Columbine is temperate in heate, and moysture.

❧ The Vertues.

This floure as Ruellius writeth, is not vsed in Medicine: howbeit some of the new wryters do affirme it to be good against the Jaundice, and sounding, and it openeth the wayes of the Liuer, and the people vse it against the inflammation, and sores of the iawes and windepipe. These floures mengled with wheaten meale, make a good playster against scratches and gaules.

Of Goates bearde, or Iosephs floure.   Chap.xvij.

Barba hirci.   Scurzonera.

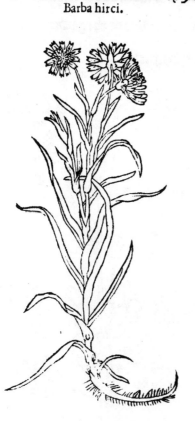
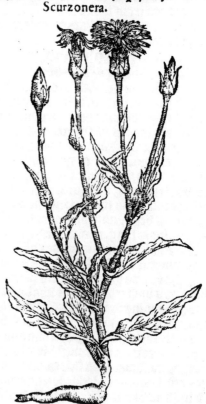

❧ The Description.

1 Oates Bearde hath a round straight knottie stem, couered with long narrow leaues almost like to Garlycke leaues. At the toppe of the stemmes, it beareth fayre double floures, and full: of colour sometimes blewishe purple, with golden threedes in the middell, and sometimes yellow, the whiche in the morning at Sunne rising do open and spreade abroade, and do turne & bende towards the Sunne, and do close agayne and go togither at noone: After the vanishing of whiche floures, out of the knoppes or heades, from whence the floures are fallen, there groweth a certayne long seede with a hearie tufte at the toppe. And whan this seede is ripe, his knoppie head openeth, and is changed or turned into a rounde hearie baule, lyke to the heads of Dantedelyon, which fleeth away with the winde. The roote is long, and as thicke as a finger, in taste sweete. The whole herbe with his stemmes, leaues, floures and roote, is full of white sappe, or iuyce like milke, the whiche commeth forth, whan the plante is broken or bruised.

2 The Spaniardes Scurzonera seemeth also to be a kinde of Tragoponon, or Buckes bearde, it hath long, broade leaues, and somewhat thicke, and vneuen aboute the borders or edges, a slender stemme parted into diuers branches, whereon groweth floures very like to the floures of Tragopogon, and of a yellow colour: the roote is long and thicke and white within, and couered with a thinne blacke barke or rinde.

❧ The Place.

Goates Bearde groweth in certayne medowes, & it is also planted in gardens for the beawtie of his floures.

1.2 Scurzonera groweth in Spayne vpon diuers shadowie mountaynes, and in moyst watery places: it is also often founde in Thoringia, a countrie of Germanie.

❧ The Tyme.

1.2 The floures of bothe these kindes of Plantes, come foorth in May and June.

❧ The Names.

1 This herbe is now called in Latine Barba hirci, and is taken for the herbe whiche the Auncients called in Greeke τραγοπώνον ἢ κόμη: in Latine Barbula hirci, and Coma: in English Goates bearde, Josephs floure, Starre of Hierusalem, and Go to bedde at Noone: in French Barbe de bouq, and Barbe de Prestre: in high Douch Bocksbart, Gauchbroot: in base Almaigne Bockbaert, and Josephs Bloemen.

2 The Spaniardes call the other Scurzoneram, whiche is ἔχιον, Echium in Greeke, and in Latine Viperinam.

❧ The Nature.

Goates bearde is temperate in heate and moysture.

❧ The Vertues.

A 1 The roote of Goates bearde, boyled in water & dronken, swageth paynes, and cureth the prickings, and empostems of the side.

B The sayde roote also, is very good to be vsed in meates and Salades, to be taken as the rootes of Rampions.

C 2 Scurzonera is thought to be maruelous good, against the bytings of Vipers and Snakes and other venemous beastes.

The second Booke of
Of floure Gentill, or purple veluet floure. Cap.xviij.

❧ *The Description.*

Amaranthus purpureus.

1 Floure Gentill hath rounde stalkes full of branches: the leaues be long and large, somewhat like the leaues of Pettie Morrell, or Nightshade, but much greater, amongst ye which groweth, alongst the branches, fayre long spiked eares, or floures of Crimsen purple colour, softe and gentill in handling, lyke Crymson veluet, the whiche doth not lightly fade or decay, but may be preserued and kept a long time in theyr colour and beautie, especially if they be dryed in an ouen that is halfe hoate. The seede groweth in the spikie tuftes, or eares, and is small, and all blacke.

2 There is an other kinde altogither like to the aforesayd, in stalkes, leaues, seede, and roote, sauing that his eares, or spikie tuftes are not fashioned like the others, but are larger and not so thicke set, and do bende, & bowe downe againe at the toppe lyke feathers, so that it maketh a gallant shew: and doth also keepe his Crymson colour like to the other.

3 There is yet a third kynde like to the others, but it groweth to the heigth of nine or ten foote. All his leaues are much larger, his stalkes are thicker and harder, and straked or crested, with ribbes standing foorth: his spikie tuftes, or earie floures are greater, longer, and fuller, but not of so fayre a colour, or pleasant hew, for it fadeth incontinent, and turneth into a greenish herbelike colour, as soone as it is gathered.

❧ *The Place.*

These kindes of herbes grow not in this countrey, except they be sowen or planted in gardens. The wemen of Italie make great accoumpt of the second kinde, bycause of his pleasant beautie, so that ye shall not lightly come into any garden there, that hath not this herbe in it.

❧ *The Tyme.*

They bring foorth their floures, or spikie tuftes in August, and the seede is rype in September.

❧ *The Names.*

These pleasant tufts, or floures, are called of Plinie libro. 21. Cap. 8. Amaranthus, and of some late writers Flos Amoris, and Amaranthus purpureus: in Italian *Fior vellino*: in English floure Gentill, Floramor, & Purple veluet floure: in French *Passeuelours*: in high Douch Samatblumen, Floramor, and Dausent schoon: in base Almaigne Flouweelbloemen.

❧ *The Nature.*

These floures are of complexion colde and dry.

❧ *The Vertues.*

Floure Gentill, or Floramor boyled in wine and droken, stoppeth the laske, and the blouddie flixe.

the Historie of Plantes.

Of Dapsies. Chap.xix.

❧ *The Kyndes.*

There are two kindes of Dapsies, the great and the small. The small also is of two kindes, whereof the one groweth in gardens, & the other groweth wilde.

❧ *The Description.*

Bellis maior.
The great wild Daysie,
or Maudelynwurte.

Bellis minor hortensis.
The small garden Daysie.

1 The greate wilde Dasie, hath grene leaues somwhat lõg, & dented rounde about: the stem is round, and set with like leaues, & groweth somtimes to y heigth of ij. foot lõg, at the top whereof it beareth fayre floures in the middell, and set rounde aboute with a little border of small white leaues, in maner of a pale, not much vnlike the floures of the cõmon Camomill, but much greater, and without sauour. Whã they perishe, the littell smal white leaues fall downe, & the yellow in y middell, which is the seede, swelleth vp.

2 The small garden Dasie hath his leaues somewhat like to the abouesayde, but they are smaller and not so much dented. It sendeth foorth his floures from the roote, vpon shorte small stemmes, somewhat like the floures of the great Daysie, sauing that the small leaues, whiche in the great Daysie do compasse the yellow in the middle, are so thicke sette, or so double that a man shall perceiue very littell of the yellow in the middell, or none at all. And these floures are sometimes white, & somtimes very redde, & sometimes speckled or partie coloured of white and redde. There growe also sometimes aboute the compasse of the sayde littell floures, many more as it were small floures growing vpon small stemmes, out of the knops or cuppes of the sayde floures. The roote is white and thredby.

3 The small wilde Daysie, is like to the small garden Daysie in his leaues. His littell floures do also spring vp from the roote, vpon short stemmes: they be also yellow in the middell, and set aboute with little white leaues, after the order of the great Daysie, but they are a great deale smaller, and without sauour, as all the other sortes of Daysies be. The roote is like to the roote of the small garden Daysie.

✿ The Place.

The great Daysie, and the small wilde Daysie, do grow in medowes, and moyste pastures. The fayre double garden Daysie is planted and set in gardens.

✿ The Tyme.

The great Daysie floureth most commonly in May. The small garden Daysie floureth from May all the Sommer long. The small wilde Daysie floureth very timely in March, and sometimes sooner, and continueth flouring vntill Aprill and somewhat later.

✽ The Names.

These floures are called of Plinie in Latine Bellis and Bellius, and now they are called in Latine Consolida minor, and Herba Margarita, of some Primula veris, (especially the small wilde Daysie) in English Daysies: in French Marguerites or Pasquettes in high Douch Maszlieben, Massuselen, and in some places Seitloszlin: in Brabant Madelieuen, and Kersouwen.

Bellis minor syluestris.
The small wilde Daysie.

✿ The Temperament.

These floures and herbes, are of nature colde and moyst.

✿ The Vertues and effects.

The decoction of the small Daysies, with their leaues or boyled alone in water, is good to be dronken against Agues, the inflammation of the Liuer and all other inwarde partes.

The herbe taken in meates or potages, doth loose the belly gentilly.

Mawdelenwurte, or the herbie parte of the wilde Daysie is good against all burning vlcers and impostems, and against the inflammation and running of the eyes, being applied thereto.

The same layde vnto woundes, keepeth the same from inflammation, and impostumation.

Of Canterbury Belles, or Haskewurte. Chap. xx.

¶ The Kyndes.

THere be diuers herbes whiche haue floures like Belles, whereof this Throtewurte or Haskewurte is a kinde, of whiche we shall speake in this Chapter, and it is also of three sortes, that is to say, the great and small, and the creeping kinde.

✿ The Description.

THe great Belfloure hath square, rough, ¶ hearie stalkes, vpon whiche growe sharpe poynted leaues, dented rounde aboute like to Nettell leaues, the floures grow alongst the stalkes lyke Belles, and like the floures of Rampions, but farre greater, and rough hearie within, of colour sometimes white, sometimes blew, and sometimes Carnation or fleishe colour. It beginneth to floure at the toppe of the stalke and so goeth florishing downewarde. The floures past, the seede whiche is small and graye, commeth vp in long knoppie huskes, like the Rampion seede. The roote is white ¶ much wrythen and interlaced.

The

the Historie of Plantes.

Trachelium maius.   Trachelium minus.   Auicularia.
Great Haſkewurte o₂ Belfloure.   The leſſe Haſkewurte,
                                            o₂ Belfloure.

2   The ſmall Belfloure in ſtalkes is like to the great, ſauing that it groweth not ſo high, the leaues be ſomewhat long, ſmaller & whiter, and not ſo deepely dented as the leaues of the greater Belfloure, but very well like vnto Sage leaues. The ſmall Belles are violet, and purple, growing at the toppe of the ſtalke, and cluſtering thicker then the floures of the great Belfloure. The roote is ſlender and very threedy.

3   The third in his leaues & ſtalkes is lyke to the firſt, but his leaues be ſmaller and not ſo deepely cut. The floures hang downewardes, and grow almoſt harde by the ſtalke, of a light violet colour, in proportion and making like to the others. The rootes moſt commonly are ſlender and crokedly creeping alongſt the ground, putting foorth new ſprings & plantes in diuers places, fró whiche groweth ſmall long and thicke rootes, not muche vnlike Rampions, whereof both this and the former ſortes are a certayne kinde.

4   There may be very well ioyned vnto theſe Belfloures, the pleaſant floures whiche are called at Paris Auicularia, ſeing that they be ſomewhat lyke to the floures of Haſkewurte o₂ Belfloure. The plante that beareth theſe floures groweth to the heigth of a hande breadth o₂ twayne, the ſtalkes are ſmall and tender, and ſet full of ſmall leaues. The floures growe at the toppe of the ſtalkes of a fayre purple colour, almoſt faſſhioned lyke a Bell o₂ Cymball, with a ſmall white clapper in the middle. They open after Sunne ryſing and cloſe agayne towardes Sunne ſette: and whan they be cloſe, they haue fyue creſtes o₂ playtes like the Belfloures, o₂ Couentrie Marians,

P ij

The second Booke of

or wilde Rapes, or lyke to Rampions, and such other floures before theyr opening.

❧ *The Place.*

1.2 Both these Belfloures, grow of their owne kinde in certayne dry meades and pastures, and they be also planted in gardens.

3 The thirde is founde in diuers Champion places, and sweete pastures of Zealand. And it is also planted in gardens, where as it prospereth ouermuch: for it doth so spread abroade and multiplie, that it hurteth other herbes, and cannot easily be weeded or ouercome.

4 Auicularia groweth in good ground, in fields amongst wheate, or where as wheate hath growen.

❧ *The Tyme.*

They floure most commonly in July.

❧ *The Names.*

1.2 The Belflower is called in Greeke τραχήλιον: and in Latine Trachelium, Ceruicaria, and Vuularia, according to the Douch name: in English they be called Belfloures, and of some Canterbury Belles. The Plante may be very wel called Haskewurte, or Throtewurte: in French *Gantelée*: in high Douch Halszkraut: in base Almaigne Halscrupt: And they are like the kindes of Rampions, as the Couentrie Marians violet or wilde Rape is, whereof shall be written here vnder.

3 The thirde kinde is vnknowen in the Shoppes of this countrie. The Herboristes of Fraunce do call it Auicularia: the Brabanders call it Vrouwen spiegel. And I know none other name, except it be ye herbe that is called in Greeke ονόβρυχις: & in Latine Onobrichis, that is to say the braying, or sounding againe of the Asse, whervnto it hath some small proportion or similitude.

❧ *The Nature.*

Belfloure is of a complexion colde and dry, like to Rampion, wherfore it may be vsed in meate as the Rampions. ❧ *The Vertues.*

3 The Belfloure boyled in water, is soueraigne to cure the payne and inflammation of the necke, and inside of the throte, and it is good against all vlcerations of the mouth, if one do gargle or wash his mouth therewithall.

Of Autumne Belfloures, or Calathian Violets. Chap. xrj.

❧ *The Description.*

Amongst all ye kindes of Belfloures, there is none more beutiful in colour then this: it hath small straight knottie stemmes, & at euery knot or ioint, it hath two leaues set directly one against an other, whiche be long & narrow: by each side whereof, as also at the top of ye stalke, groweth forth pleasant floures, whiche be long & hollow, alwayes bending outwardes, like to a small long bell, with two or three small white threedes in the middle. They are of a blew colour, so cleare and excellent, that they seeme to passe, the azured skies. When they are paste, there cometh vp in the middle of the floure a round long huske, full of long small seede.

❧ *The*

❧ The Place.

These pleasant floures grow in moyst medowes, & low vntilled groundes, standing in frutefull soyles.

❧ The Tyme.

They are in floure about the end of August and September.

❧ The Names.

Plinie calleth these floures in Latine Campanulæ Autumnales, & Viola Autumnalis: we may also cal them in English Autumne Belfloures, Calathian violets, or Autumne violets: in high Douch they are called Lungen blume: for the which cause Cordus calleth them Pneumonanthe: and truly it seemeth to be a certayne kinde of Gentian: in base Almaigne it is called blauw Leliekens, and Duysent schoon.

❧ The Temperament and vertues.

The temperament, nature and propertie of these pleasant little floures are very like vnto Gentian, as the bitter taste declareth.

Of Marians violet/or Couentrie Belles. Chap.lxij.

❧ The Description.   Viola Mariana.

This braue & pleasant floure, hath his first leaues whiche grow next the ground, long, broade and somwhat hearie, not much vnlike the leaues of wild rose Campions, from the middest whereof springeth vp the second yeare after the sowing or planting one stalke or moe, full of branches, set with suche like leaues, but somwhat smaller: there grow vpon the sayde branches, many fayre and pleasant hollow floures, most commonly of a cleare purple colour, and sometimes white, in proportion very well like to the common Belfloure, but much larger and rounder, and not so deepely cut about the brimmes or edges, the whiche also before their opening are folden togither as it were with fiue crested playtes or edges. Whan they are past there cometh vp small rounde buttons or huskes, with fiue rough endes, or tayles, whiche be hollow, short, plyed, or turned backe, in all things else like to the knops or huskes of Rampion, or the common Belfloure. The seede is in the middle of the sayde knoppis huskes, & it is small & broune, coloured like a Chestnutte. The roote is white and thicke, and putteth forth by the sides diuers other rootes.

❧ The Place.

These pleasant floures grow about Couentrie in England, and are founde sowen in the gardens of Herboristes, and are not yet very common.

❧ The Tyme.

They floure from July vntill September, and afterwarde, and notwithstanding,

standing, though they seeme alwayes to floure, yet they do also beare seede, so that oftentimes as soone as this herbe beginneth to floure, one may alwayes finde vpon the same buddes, floures, and ripe seede.

❧ *The Names.*

Men do now call these pleasant floures in Latine Violæ Marianæ: that is to say in English, Marianes violets, we may also cal them Couentrie Rapes: in base Almaigne Marietes: of the old writers in Greeke γογγύλη ἀγρία. In Latine Rapum sylueſtre. Of this kind also are y̅ Belfloures, described afore in the xx. Chapter of this Booke.

❧ *The Nature.*

These floures, and their roote specially are colde and dry.

❧ *The Vertues.*

Their vertue is all one, with the other Belfloures, and may be vsed in like sorte.

They vse about Couëtrie in England where as great store of these plantes do grow, to eate their rootes in Salads, as Pena writeth in his booke intituled Stirpium aduersaria noua. Fol. 138.

Of Blew belles. Chap. xxiij.

❧ *The Description.*

Campanula cærulea satiua.

1 These floures whan their plant beginneth first to spring vp out of the ground, haue small rounde leaues like to Marche violets, amongst the whiche springeth vp a long high hollow stalke, set with long narrow swartgreene leaues, amongst the whiche also at the top of the stalke grow fayre Belles or hollow floures, greater than the floures of Rampion, of colour blew turning towardes purple most commonly, but sometimes also they be white. Whan they are fallen away, the seede is founde in small bullets, or huskes like Rampion seede. The roote is small and threedie. The whole plante is full of white sappe or iuyce like milke, the whiche commeth foorth whan the herbe is broken or brused, and tasteth like Rampions.

2 There is also a wild kinde of these floures, the which is like to the aforesaid, in growing, leaues, stalkes, floures, and seede. Neuerthelesse it is a great deale and in all respects smaller, and it peeldeth a white iuyce also like the first.

3 There is also a certayne thirde kinde of this Blew belfloure muche greater than the first: his stalkes be long and high: his leaues be somewhat large: and it hath very many floures growing alongst the stalkes, as it were littell small Belles of a fayre blew colour: and after them certayne hollow little huskes or Celles: his roote at the first is long and slender, but whan the plante wareth olde, the roote is full of knots and knobbes, and diuided into sundry branches: and finally this herbe is full of white sape like to the first.

❧ *The Place.*

They

the Historie of Plantes. 175

They plante the first kinde in gardens.
And the smal wild kinde groweth in the borders of fields, & vnder hedges.

❧ *The Tyme.*

They floure in June & July. And the wilde doth also floure vntill August.

❧ *The Names.*

These floures be now called Fayre in sight: in French *Belle videre*: in Douch Blauw clockkens, that is to say in Latine Campanula cærulea. All these three plantes are very like that herbe whiche is called of Theophrastus in Greeke ἰασιόνη, and in Latine of Plinie Iasione.

❧ *The Nature and Vertues.*

These floures be not vsed in medicine, wherefore the temperature and vertues thereof are vnknowen.

Of Fore gloue. Chap. xxiiij.

❧ *The Description.*

Digitalis.

FOre gloue hath long broade swartgreene leaues, somwhat dented about the edges, & somwhat like the leaues of wilde Mulleyne, amongst the whiche springeth vp a straight rounde stem of twoo Cubites long or there aboute, by one side whereof, from the middle to the very toppe, there growe fayre long round hollow floures, fashioned like finger stalles, of colour sometimes carnation, and speckled, in the inside with white spots, and sometimes all white, & sometimes yellow. Whan they are fallen of, there appeareth rounde sharpepoynted huskes in which is conteyned the seede, of a bitter taste. The roote is blacke & full of threedy strings.

❧ *The Place.*

It groweth in stony places & mountaynes, in darke shadowie valleys or coombes, where as there hath bene myning for Iron and Smithes cole. It is also planted in certayne gardens.

❧ *The Tyme.*

Fore gloue floureth chiefly in July and August.

❧ *The Names.*

This herbe is now called in Latine Digitalis, Campanula syluestris, and Nola syluestris: in English Fore gloue: in French *Gantz nostre Dame*, and *Digitale*: in high Douch Fingerhut, Fingerkraut, Waldt glocklin, & Waldt schell: in base Almaigne Vingerhoetcruyt. This as some do write, is that kinde of Verbascū, whiche the Greekes call λυχνίτις καὶ θρυαλλίς, of the Latinistes Lychnitis, and Thryallis, whereunto it is much like.

❧ *The Nature.*

Fore gloue is hoate and dry.

The Vertues.

Foxe gloue boyled in water or wine and dronken, doth cut and consume the thicke toughnesse of grosse and slimie humors. Also it openeth the stoppings of the lyuer, & Spleene or Mylte, and of other inwarde partes.

The same taken in the like maner, or else boyled with honied water, doth scoure and clense the breast, and ripeth, and bringeth forth tough and clammy flegme.

Of Turkie, or Aphrican Gilofers. Chap. xxv.

The Kyndes.

There be two sortes of these floures found in this countrie: one great & the other small, the great (Othanna) groweth to the height of a man, and floureth very late. The small groweth low, and floureth betimes.

The Description.

Flos Aphricanus.

1 The great Aphrican floure hath a long broune red, crested & knottie stalke, ful of branches, & groweth viij. or ix. foote high, hauing at euery knot or ioynt, two braches, set with great long leaues, coposed of many small lōg narrow leaues, nickt & tothed roūd about, & spred abrode as it were winges, & set one ouer against an other, altogither like Athanalia or garden Tansie. The floures grow at ye ende of the branches, out of long round huskes, of a browne Drège colour aboue, and of a faynt or pale yelow vnderneath. After the falling of the floures, the seede whiche is inclosed in the aforesayde round huskes, is long, narrow and blacke.

2 The small Aphrican floure is like vnto the abouesaide, in his stalkes, leaues, floures, & seede, sauing it is in al respects smaller, & groweth not very much higher than a foote. They are both in their leaues and floures of a naughtie strong & vnpleasant sauour, especially whā they be either rubbed or bruised betwixt ones fingers.

The Place.

These floures grow in Aphrica, & from thence they where brought into this countrey, after that the mightie and Noble Emperour Charles the fifth, wan the Towne and Countrie of Thunes, they are planted here in gardens.

The Tyme.

2 The small African Gilofer, beginneth to floure in Aprill or in May, and from thence forth all the Sommer.

1 The great Othonna beginneth not to floure before August.

The Names.

This floure may be called in Latin Flos Aphricanus, for it was first brought out of Aphrica into the countreys of Germany and Brabant. We do call this floure Turkie Gillofers, and French Marygoldes, Aphrican floures, or
Aphrican

the Historie of Plantes.　177

Aphrican Gillofers: the Frenchmen do call these floures *Oillets de Turque*, and *Oillet d'Inde*: and from thence it commeth to passe that the Latinists do cal it Flos Indianus: in high Douch Indianisch Negelin: in base Almaigne Thuenis bloemen: of Valerius Cordus Tanaceum petunianū. Some learned men thinke that this herbe hath bene called of the Auncient wryters ὀθόννα, Othonna, and that it should be the Othonna, wherof Dioscorides hath writen, which groweth in Arabia about Egypt, whose leaues be holy, asthough they had bene eaten with Locustes, Paulmers or Snayles, which thing almost may be perceyued in the leaues of this Indian Gillofer, if a man looke vpō them against the light. But in my iudgement it is better like to be that herbe, whiche Galen in his fourth booke of Symples calleth Lycopersium, or Lycopersion.

※ *The vile Nature and euill qualitie of this Herbe.*

The Indian Gillofer is very dangerous, hurtfull, and venemous, both to man & beast, as I haue tried by experience, namely vpon a yong Catt, whereunto I haue giuen of these floures to eate, very finely pound with greene or fresh Cheese: whereupon she blasted immediatly, and shortly after died. And I was moued to make this experience, by the occasion of a yong childe who had gathered of these floures & put them into his mouth, so that straight waies his mouth & lippes did swell exceedingly & within a day or two after, they became very sore and scabbed, as also it doth often happen to them, that put into their mouthes the pipes, or hollow stalkes of Hemlocke. Wherfore it is manifest that this herbe with his floure is very euil and venemous, and of complexion much like vnto Hemlocke, the whiche also may be partely perceyued by his foule and lothsome sauour, whiche is very strong and stinking, not muche differing from the rancke and noysom smell of Hemlocke.

Of May Lillie / or Lillie Conuall. Also of Monophillon. Chap.xxbj.

※ *The Description.*

Lillie Conuall hath two greene smooth leaues, like to the leaues of ye common white Lillie but smaller and tenderer, betwixt whiche there springeth vp a naked stalke of a span long, or thereabout, at the which stalke there hangeth seuen or eight, or moe, proper small floures, as white as Snowe, and of a pleasant strong sauour, smelling almost like the Lillie. Whan the floures be past, theyr commeth in their steede certayne redde bearies, like to the frute or bearies of garden Asparagus. The roote is threedishe, creeping here and there.

It should seeme that Monophillon were a kinde of Lylie Conuall, it hath a leafe not much vnlike the greatest leaues of Iuie, with many ribbes or sinewes alongst the same, like to a Plantayne leafe: the whiche one leafe, or single leafe, doth alwayes spring vp out of the grounde alone, sauing whan the herbe is in floure and seede: for than it bareth two leaues vpon a rounde tender stalke like to the other, but smaller & standing one aboue an other, aboue the sayde leaues groweth the small white floures like to Lylie Conuall, but not of so strong a sauour, after whiche there riseth small bearies or rounde frute, whiche is white at the firste and afterward redde. The roote is very slender and creepeth in the grounde.

※ *The Place.*

Lyllie Conuall and Monophillon, groweth in shadowie wooddes.

※ *The Tyme.*

They do both floure in May.

※ *The*

The second Booke of

Lilium Conuallium.
Lillie Conuall.

Vnifolium.
Monophillon.

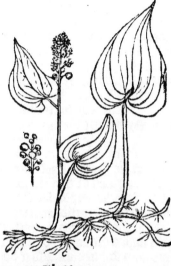

❋ *The Names.*

Lyllie Conuall, is now called in Latine Lilium conuallium, that is to say, the Lyllie of the vallie: in English Lyllie conuall, May blossoms, May lyllies, & Lyryconfancy: in Frēch *Grand Muguet*: in high Douch Meyenblumlin: in base Almaigne Meybloemkens.

2 **Monophillon** is now called in Latine Vnifolium: it may be also called in English, one Leafe, one Blade, or Singleleafe: in high Douch Einblat: and in base Almaigne Eenbladt, and it should seeme to be a kinde of Lillie conuall, seing that it is so well like vnto it in floures and seede.

❋ *The Nature.*

They be in complexion, hoate and dry, like the Lillies.

❋ *The Vertues.*

They write that the water of the floures of Lyllie conuall, distilled with A good strong wine, and dronken in the quantitie of a sponefull, restoreth speach to them that are fallen into the Apoplexie, & that it is good for them that haue the Paulsie, and the Goute, and it comforteth the Harte.

The same water as they say, doth strengthen the Memorie, and restoreth it B agayne to his naturall vigor, whan thorough sicknesse it is diminished.

Besides this they say also that it is good to be dropped in, against the inflā- C mation, and watering of the eyes.

2 The roote of Monophillon is counted of some late writers, for a soueraigne D and speciall remedie against the Pestilence and al poyson, whan the weighte of halfe a Dragme of the pouder of the sayde roote is giuen in vineger, or good wine, or in both mixte togither, according to the nature or complexion of the sicke, so that vpon the recepte thereof, they go to bedde and sweate well.

Monophillon is good to be layde with his roote, vnto greene woundes, to E preserue them from inflammation and Apostumation.

Of

the Historie of Plantes.　　　179

Of Calfes snowte, or Snap Dragon.　　Chap.xxvij.

❧ *The Kindes.*

THere are in this countrie two sortes of this herbe, the one great and the other small. The great hath broade leaues, and it is the true Antirrhinum of Dioscorides. The smaller kinde hath long narrow leaues.

Antirrhinon.　　　　　　　　　　　Orontium.
The great snap Dragon, or Calues snowte.　　Small Calues snowte.

❧ *The Description.*

1. THe great Antirrhinon hath straight round stemmes, & full of branches, the leaues be of a darke greene, somewhat long and broade, not muche vnlike the leaues of Anagallis or Pimpernell, alwayes two leaues growing one against an other, like the leaues of Anagallis. There groweth at the top of the stalke alongst the braches certayne floures one aboue an other, somwhat long and broade before, after the fashion of a frogs mouth, not muche vnlike the floures of Tode flaxe, but muche larger, and without tayles, of a faint yellowissh colour. After them comme long round huskes, the foremost part whereof are somwhat like to a Calfes snowte or Moosell, wherin the seede is conteyned.

There is also an other kinde of great Antirrhinum, whose leaues be long & narrow, almost like to the leaues of Tode flaxe, whiche beareth sometimes a redde floure, sometimes a faynt redde, and sometimes a white floure: else in all things like to the aboue saide.

2. The small Antirrhinum his stalkes be small and tender, not very full of branches,

braunches, his leaues be long and narrow, betwixte whiche and the stalkes, growe the small red floures, like to the aforesayde floures, but a great deale smaller. Whan they are past, there riseth vp small rounde heades or knappes, with little hooles in them, like to a dead scull, within whiche is conteyned smal seede.

❧ *The Place.*

1.2 The first and great Antirrhinum, groweth not in this countrey, but in the gardens of certayne Herboristes where as it is sowen. The second groweth in some fieldes of this countrie, by high wayes, and vnder hedges.

❧ *The Tyme.*

The great Antirrhinum floureth in August and July. The small Antirrhinum beareth floures in July.

❧ *The Names.*

1 The first kinde is called in Greeke ἀντίῤῥινον καὶ ἀντίῤῥιζον: in Latine Antirrhinum, and Syluestris Anagallis: in English Calfes snowte, and Snapdragon: in French *Grand Antirrhinum*, and *Moron violet*: in Douch Orant, and of some Calfs nuese.

2 The second kinde is called of some in Greeke ὀρόντιον: in Latine Orontium: in English small Snapdragon, or Calfs snowte: in French *Petit Antirrhinum*: in Douch cleyne Orant, of this kinde Galen hath made mention in lib 9. de Medicamentis secundùm loca, amongst the Medicines whiche Archigenes made for them that haue the Jaunders. And it seemeth to be ῇ Phyteuma of Dioscorides, called in Greeke φύτευμα.

❧ *The Nature.*

1 The great Antirrhinum is hoate, and of like nature and complexion vnto After Atticus, called in English Sharewurte, as Galen wryteth.

2 The small is hoate and dry, and of suttell partes.

❧ *The Vertues.*

Some haue writen, that who so carrieth about him the great Antirrhinum, A cannot take harme or be hurte with any venim or poyson whatsoeuer.

The small Antirrhinum doth scatter away, and consume the yellow colour B of the bodie, whiche remayneth after one hath had the Jaundice, if one be well washed with the decoction thereof.

Of water Lyllie.   Chap. xxviij.

❧ *The Kyndes.*

There be two kindes of water Lyllies, that is to say, the yellow, & the white, not onely differing in floure but also in roote.

❧ *The Description.*

1 The white water Lillie, hath great broade roundishe leaues, sometimes fleeting or swimming aboue the water, and somtimes vnder, the which all do spring vp from the roote, vpon long rounde smooth stalkes. The floures do also growe vpon suche like stemmes comming from the roote, and they haue in the middle many yellow threedes, or thromnes, compassed round about with xxvj. or xxviij. white leaues set in very good order, each leaf almost as large as ones finger, or like in proportió to the leaues of Houselike or Sengreene. Whan the floures be past, there come in their steede rounde knoppes or bolliens, wherin the seede lieth, which is large and swarte. The roote is blacke and rough, sometimes of the bignesse of ones arme with many threedy strings.

2 The yellow water Lyllie his leaues be very muche like to the white, his floures be yellow and smaller then the floures of the white, the whiche being fallen, there commeth in their place round long knoppes or bolliens, narrow at the

the toppe, like to a small glasse or phiall. The roote is white and of a spongie substance, of the greatnesse of ones arme, ful of knobbes and knottes, with certayne great stringes hanging by it.

Nymphæa alba.
White water Lillie.

Nymphæa lutea.
Yellow water Lillie.

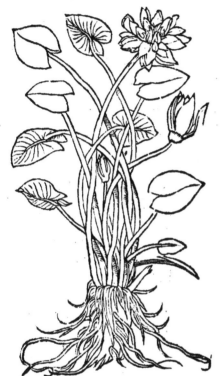 

❧ *The Place.*

These floures do grow in Riuers and Pooles, and other standing waters.

❧ *The Tyme.*

Water Lillie floureth in June, and sometimes sooner.

❧ *The Names.*

1. The first kinde of these floures, is called in Greeke νυμφαία: in Latine Nymphæa, of some Clauus Veneris, and Papauer palustre: of the Apothecaries Nenuphar: in English White water Lillie, Water Rose, and white Nenuphar: in Italian Nenuphar biancho: in Spanish Adarguas del Rio, Escudettes del Rio, Figues del Rio blanquos: in French Nenuphar blanc, or Blanc d'eaue: in high Douch Seeblumen, wasser Gilgen, Wassermahen, Horwurtz, Horstang: in Brabant Plompen, and witte Plompen.

2. The second kinde is called in Latine Nymphæa lutea, and Nenuphar citrinum: in English Yellow Nenuphar, or Water Lillie: in Italian Nenuphar giallo: in Spanish Figues del Rio amarillos, Golfan Amarillo: in French Nenuphar iaulne, or Iaulne d'eaue. The floure thereof, as Dioscorides wryteth, is called in Greeke Βλέφαρα, Blephara.

❧ *The Nature.*

Both sortes of Nenuphar, and specially the roote are in temperature colde and dry without any acrimonie or sharpnesse.

❊ The Vertues.

The roote or seede of the white water Lillie, boyled in wine and dronke, is good for them that haue the laske, the blouddie flire and Tenasme, whiche is a desire to go often to the stoole and may do nothing. ₳

The same roote boyled in white wine, cureth the diseases of the Milte and Bladder. ℬ

The roote & seede of the white water Lillie are very good agaynst Venus, or fleshly desires, if one drinke the Decoction thereof, or vse the pouder of the saide seede and roote in meates: for it dryeth vp the seede of generation, and so causeth to liue in chastitie. The same propertie is in the roote as Plinie writeth, if it be brused and applied outwardly to the secrete partes. ℂ

The Conserue of the floures thereof, is also very good for all ye aforesayd diseases, moreouer it is good against hoate burning feuers, & the head ache, & it causeth sweete and quiet sleepe, and putteth away all venereous dreames. ⅅ

The roote thereof brused or stamped, is good to be layde to the payne and inflammation of the stomacke, and the bladder. ℰ

The same roote pounde with water, taketh away all the spottes of the skin whan it is rubbed therewithall, and being mengled with Tarre, it cureth the naughtie scurffe of the head. ℱ

The roote of water Lillie being yet greene, pound & layde vpon woundes, doth stanche the bloud, as Theophrastus writeth. ℊ

The roote of yellow water Lillie, boyled in thicke redde wine and dronken, stoppeth the inordinate course of the floures, especially the white fluxe. ℋ

Of Chamomill. Chap.xxix.

❊ The Kyndes.

Camomill, as Dioscorides and other of the Auncients haue written, is of three sortes. The one hath white floures. The other hath yellow floures. And the third whiche is the greatest of the three, hath floures betwixt redde & purple. Yet at this time there be diuers other sortes found, and first there be two sortes of Chamomill which are very sweete and of stróg sauour, called Romaine Camomill. The one hath white floures, the other yellow, and bysides these there be others, whiche do (for the moste parte) growe in deserte places, and therefore we haue named them Camomill of the Forest, or wildernesse.

❊ The Description.

1 The first kinde of Camomill hath diuers long rounde stalkes, creeping alongst the grounde, and taking roote in diuers places, very seldome growing higher than ones hande. It hath diuers small tender leaues very small cut, or finely iagged.

2 The second kinde is much like vnto the first, sauing his leaues be smaller, his floures be nothing else but certayne yellow buttons, like the middle of the floures of the other Camomill, without any small leaues growing about it, as ye may perceyue by the figure, but otherwise it is like to the first Camomill.

Of the number of these two kindes, there is yet an other, wheih hath small yellow leaues growing rounde aboute the small yellow knoppes or buttons, and are altogither like to the first, in leaues, sauour, and fashion, sauing his floures be altogither yellow.

These two kindes of Camomil (that is to say) the white & the yellow, haue a very pleasant sauour, like the smell of a Cytron, whereof they firste tooke

their

the Historie of Plantes. 183

their name in Greeke Chamæmelum.

Chamæmelum leucanthemum.
White Romaine Camomill.

Chamæmelum chrysantemum.
Yellow Romaine Camomill.

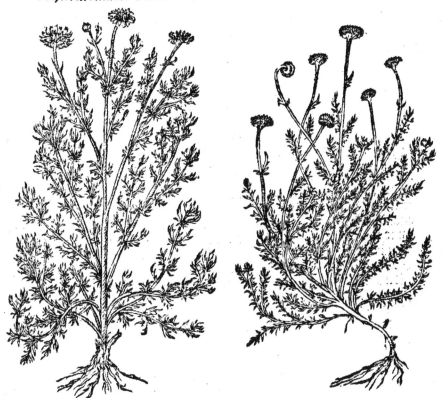

3 ❦ The third kinde of Camomill which beareth red purple floures, & groweth higher than the two others, is not yet knowen vnto vs, except it be that floure whiche some call flos Adonis, and other Anemone.

❦ *The Place.*

None of the sweete Romaine Camomils groweth in this countrie, of their owne kindes, but are planted in the gardens of some diligent Herboristes, and are come hither as strangers. ❦ *The Tyme.*

These Camomils do floure in June, & July, & sometimes also sooner. They last all the winter, and may very well abide the colde.

❦ *The Names.*

The Camomill is also called in Greeke ἀνθεμίς καὶ χαμαίμηλον: in Latine Chamæmelum, and as Apuleius writeth Bene olens, at this day Camomilla: in English Camomill: in French *Camomille*: in Douch Camille.

1 The first kind of sweete Camomill with the white floure is called in Greeke λευκάνθεμον: in Latine Chamæmelum album: in base Almaigne Roomsche Camille: in English white Camomill: in French *Camomille blanche.*

2 The second kinde of sweete smelling Camomill with the yellow floure is called in Greeke χρυσάνθεμον: in Latine Chrysanthemum, and Chamæmelum luteum: in English Yelow Camomill: in French *Camomille iaulne*: in Douch geele Roomsche Camille.

3   The third kinde is called in Greeke ἠράνθεμον: in Latine Eranthemum, and Chamæmelum purpureum. It may be called in English Purple Camomill: in French Camomille à fleur purpurée: in Douch Roode Camille.

※ *The Temperament.*

The Camomill, especially the white, is hoate and dry in the first degree, and hath power to dissolue, & make subtill. But the Romaine Camomils are hoater, and more drying.

※ *The Vertues.*

The Decoction of the floures, herbe, and roote of Camomill, being dronken causeth women to haue their termes, driueth foorth of the belly the dead frute, prouoketh vrine, & breaketh the stoone. It is of the like vertue, if one do bathe in a bath of the same Decoction.   A

The floures and herbe of Camomill boyled in wine and dronken, driueth forth windinesse and cureth the cholicke, that is to say, the paine in the bowels and bellie.   B

Camomill taken in the same sorte doth purge & beautifie those that haue an euill colour remayning after the Jaunders, and cureth them that haue any greefe or impediment of the liuer.   C

Camomill pound with his floures, and taken in the quantitie of a Dragme with wine, is very good against the biting of Serpents, and all other venemous beastes.   D

The Decoction of Camomill made in water and applied outwardely vpon the region of the bladder, taketh away the payne of the same, prouoketh vrine, and driueth forth grauell.   E

Camomill chewed in the mouth, cureth the vlcers & sores of the same. Of like vertue is the decoction to wash the mouth withall.   F

Camomill also closeth vp al woundes, and old vlcers, especially those which happen about the corners of the eyes, whan it is brused and layde vpō, or if one wash such woundes and sores with the decoction thereof.   G

Camomill mēgled with oyle & taken in glister, is singuler against all feuers whiche happen by meanes of the obstruction or stopping of the skinne.   H

The oyle of Camomill doth asswage and mitigate all payne and ache, it cureth weried & brused partes, it looseth and softeneth all that which is hard and stretched out or swollen: it doth mollifie and make soft all that whiche is hard, and openeth all that is stopped.   I

Of wilde or common Camomill.   Chap. xxx.

¶ *The Kyndes.*

There are foure kindes of wilde Camomill. The first kinde is the common Camomill: the second is the Cotula fœtida: the thirde is the greate wilde Camomill called Cotula non fœtida: the fourth is the wilde Camomill with the yellow floures called in Latine Cotula Lutea.

※ *The Description.*

1   The common Camomill hath slender, tough & hard stemmes: the leaues be tender, and very small cut and iagged. The floures growe at the toppe of the branches, and are yellow in the middell, and set rounde aboute with many small white leaues, altogither lyke the floures of garden Camomill with the white floures, and also of a meetely pleasant sauour, but nothing so strong nor pleasant in smell as the garden Camomill.

2   Stinking Camomill or Cotula fœtida, hath a thicke greene stemme, and full of iuyce, whiche breaketh quickly whan it is troden vpon. The leaues be greater

the Historie of Plantes.

Chamæmelum, Leucanthemum
commune & syluestre.
The common wilde Camomill.

Cotula fœtida.
Mathers or stinking Camomill.

be greater and greener than the leaues of the common Camomill. The floures be much like vnto the aforesayde. The whole herbe is of a very strong vnpleasant stincking sauour, and of a sufficient bitter taste.

3 Vnsauery Camomill, or Cotula non fœtida, hath small tender pliant stems, many growing vp from one roote, the leaues be long, greater and whiter than the leaues of the common Camomill. The floures are like to the two kindes aforesayde, but they are a great deale greater and without any manifest smell. The roote is great and very threddie, the which dieth not lightly at winter but springeth vp yearely a newe.

4 Golden Cotula is like to Cotula non fœtida in his stalkes, leaues, & floures sauing that his leaues be greater and whiter, drawing towards Asshie colour, and his floures be not onely yellow in the middle, but also they are set round about with smal yellow leaues, in fashion like the other Camomilles, & without smell like to Cotula nõ fœtida. Also it doth not lightly die or decay, but springeth vp yearely out of the olde rootes.

❧ *The Place.*

The iij. first kinds do grow most comonly in this coūtrie in euery corne field.
The golden Cotula groweth in suche like places in France and Germanie, but not in this countrie, except in the gardens of Herboristes.

❧ *The Tyme.*

All these kindes of Camomill do floure in June, & from thence forth all the Somer long.

Cotula

The second Booke of

Cotula non fœtida.
Unſauerie Maydeweede.

Cotula lutea.
Golden Cotula.

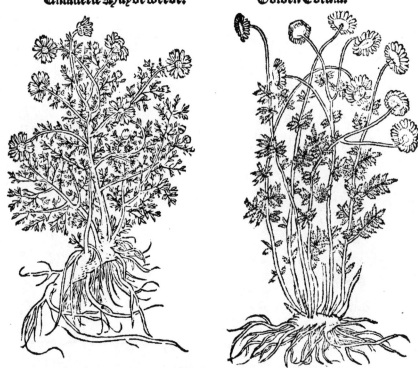

❧ *The Names.*

1. The firſt kinde of wilde Camomill is now called Chamæmelum album: in Shoppes Chamomilla, whereas it is aptly vſed for Leucanthemum: in Engliſh common Camomill: in Italian *Camamilla*: in Spaniſh *Macella, Manzamilla*. in French *Camomille vulgaire*: in high Douch Chamill. Albeit this is not the right Camomill. Wherefore we call it Chamæmelum ſylueſtre, that is to ſay, wilde Camomill.

2. The ſecond kinde is now called in Latine and in Shoppes Cotula fœtida, of ſome Cauta and Camomilla fœtida and in Greeke κυνανθεμις, Cynanthemis, and κυνοβοτανή, Cynobotane. that is to ſay, Dogges Camomill: in Italian *Druſaculo*: in Spaniſh *Maguarca*. in high Douch Krottendill, Hunſzdill, Hundiſzblum, and wilde Chamill: in Brabant Paddebloemen, and ſtinckende or wilde Camille: in Engliſh Mathers, Mayweede, Dogges Camomill, Stincking Camomill, and Dogge Fenell: and peraduenture it is Parthemium mucrophyllon of Hippocrates.

3. The thirde kinde is called Cotula non fœtida, Camomilla fatua, and Camomilla inodora, of ſome in Greeke βούφθαλμον, Buphthalmum. that is to ſay in Latine, Oculus bouis: in high Douch Rudill, and Rundſaug & Kueaug: in French *Oeil de beuf*· howbeit this is not the right Buphthalmum, as one may ſee in the Chapters following, and therefore it may better be called Cotula non fœtida, or Cotula alba, than to call it by a name not beloging vnto it. I haue Engliſhed it Unſauerie Camomill, fooliſh Mathes, and white Cotula without ſauour.

4. The fourth kinde may wel be called Cotula lutea, ſeeing it is ſo well like vnto the

the Cotules abouesaide: in English Golden Cotula: in high Douch Streich-blumen, and Steinblumen, and according to the same it is called in base Almaigne Strijck bloemen. Some whiche thinke that this is the second kinde of Camomill, do call it Chrysanthemum, that is to say, yellow Camomill: in French Camomille iaulne: in high Douch geel Camille, but they are deceyued, and their opinion is not like to be true, bycause this herbe hath no speciall smell. Moreouer the fasshion of the leaues is nothing like to the leaues of garden Camomill, neither yet like the common Camomill.

❧ The Temperament.

1 The common Camomill is of complexion hoate and dry, and not so feruent as the Romaine Camomill, but more pleasant and gentill.

2 Cotula fœtida is hoate and dry, as his smell and sauour declareth.

3.4 The other two kindes are of complexion somwhat like, but not so strong.

❧ The Vertues.

1 As the common Camomill is very like in complexion to the right Camomill, so is it like in his faculties and operation, sauing that it is not althing so strong in operation.

This Camomill hath bene proued to be very good against the Cholicke and the Stoone, and also it prouoketh vrine, to be vsed in like manner as the Romaine or right garden Camomill, and it is more conuenient, and agreable vnto mans nature than the Romaine Camomill.

And surely this Camomill also is right excellent in all kindes of mollifiyng and softning playsters, that serue to swage payne & to dissolue tumors & swellings: for it easeth and swageth all paynes, and dissolueth & scattereth tumors, causing the same to vanish away: & therefore it is very good to be vsed in such clysters as are made against the Colique and the stone.

The oyle of this Camomill is singuler against all kinde of ache and payne, and against brusings, shrinkings, hardnesses, and stoppings, like the oyle of the garden Camomill. Moreouer it is better, and more conuenient to be put into Clisters, whiche are made against the Feuer, than that oyle that is made of the floures of garden Camomill.

2 Cotula fœtida is good for such women, whose Matrix is loosed, and falling downe from one side to an other, if one do washe their feete with a decoction thereof made in water.

It is also good against the Suffocations of the Matrix, if you giue it to be eaten or smelt too, and it is of like vertue to Castorium, as the learned writers of our time haue found out by experience.

3 The operation and vertues of the two others are not yet knowen, but accordingly as one may iudge, they are in facultie not muche vnlike the Camomils, sauing that they be altogither feebler.

4 Some do write, that golden Cotula boyled in wine and dronken, is good against the Jaunders, and restoreth the good & liuely colour, whiche is a signe that it is of like vertue vnto Camomill, for Camomil worketh the same, as we haue declared in the former Chapter.

Of Passe floure, or Redde Mathes. Chap.xxxi.

❧ The Description.

This herbe hath thicke greene stalkes, and leaues very small cutte and iagged, much like bothe in stalkes and leaues, and also in smell and sauour, vnto Cotula fœtida. The floure is of a fayre purple red colour, of fasshion and making like vnto the golde cup, or the floure of Crowfoote: whan they are past, there come vp rouñd rough knops, like ye knops of Crowfoote, but somwhat longer, wherein is the seede (like to Spinache seede).

❧ *The Place.*

These fayre & pleasant floures grow in some places in the cómon corne fieldes as in Prouence and Languedoc, and in some places of England, in some countries they grow not but in gardens.

❧ *The Tyme.*

This herbe beginneth to floure in May, and remayneth flouring all the Somer.

❧ *The Names.*

Heranthemum forte.

The stalkes & leaues with the whole herbe that beareth these floures, which is like vnto some of the Camomilles in sauour, smell, and proportion, are sufficient enough to proue this herbe to be a kinde of Camomill, and especially the thirde kinde called Heranthemum: the floures onely, whiche are not very like vnto Camomill floures, causeth me to doubt. For if the floures were like fasshioned vnto Camomill, I would without doubte mayntayne this herbe to be the thirde kinde of Camomill, which is the true Heranthemum, whiche Dioscorides describeth to be greater than the two other kindes, & to haue a purple floure, vnto whiche description this herbe draweth neare, sauing only in the fasshion of his floure. For the whole plant is greater and higher then Camomill, but otherwise very like it, and the floures be of a fayre purple red colour. But whatsoeuer this herbe is, it is better like to be the thirde kinde of Camomill, than Consolida regia, whiche we call Larckes spurre, is, or After Atticus whiche we call Sharewurte, which haue bene both described of some writers for this kinde of Camomill, although they were nothing like Camomill, neyther in their leaues, floures, nor smell, and they beare not redde floures but blew, whiche is against the description of Heranthemum, whose floures (as it is aboue sayde) Dioscorides writeth to be of a red purple colour. Wherfore this herbe may better be called Heranthemum, then either Larckes spur, or Shareworte: it may be called in English purple Camomill, Redde Mathes, and Passe floure: it is also called in French *Passefleur*: the Brabanders call it Bruynettekens.

Some would haue it to be flos Adonis, but their opinion seemeth not to be very likely, bycause that Flos Adonis should seeme to be none other, then a kind of Anemone.

❧ *The Temperament.*

The taste and smell of this herbe doth manifestly declare it to be of complexion hoate and drie like the Camomill, but chiefly like to Cotula fœtida.

❧ *The Vertues.*

The vertues and operation of this herbe are yet vnknowen vnto vs: but if this herbe be Heranthemum, it is singuler against the stoone, as we haue alredie written in the xxix. Chapter of this Booke.

Of

Of Buphthalmos, or Oxe eye. Chap.xxxij.

❧ The Description.

Buphthalmum.

Buphthalmos is a braue plante, with pleasant floures & stems, of a span or a halfe cubite long: it hath three or foure stalkes, set with tender leaues very small cut and iagged, not muche vnlike vnto Fenell leaues, but a great deale smaller, and very well like to the leaues of the small Sothrenwood, sauing they be greener. The floure is of a fayre bright yellow colour, and large, with many small thrommes or yellow threedes in the middle, almost like to the floures of Marigoldes sauing they be much larger, & haue not so many small leaues set round about the golden knops or yellow heades. The floure perished, there commeth in steede thereof a rounde knop almost like the sedie knop of Passe floure, the roote is blacke & very threedie.

❧ The Place.

This herbe as witnesseth Dioscorides groweth in ye fieldes without the towne: in this countrie the Herboristes do plant it in their gardens.

❧ The Tyme.

It beareth his floures in Marche and Aprill.

❧ The Names.

This herbe bycause of his floures, whiche be of the quantitie and fashion of an Oxe eye, is called in Greeke βούφθαλμ{Θ} ἢ βούφθαλμον: in Latine Buphthalmum, & Oculus bouis: in high Douch Rindsaug, Kuaug: in base Almaigne Rundsooge, and Coeooghe: some call it also Cachla, Cauta, or Caltha. This is the right Oxe eye described by Dioscorides.

In certayne places the Apothecaries do sell, and vse the rootes of this plant in steede of the roote of blacke Hellebor, and from hence it cometh that certaine studious Herboristes haue called this plant Helleborum nigrum, and do count it for a very naughtie and vehement plante, howbeit that of it selfe it hath not in it any speciall malice or force, neither will it prouoke the stoole as some haue proued by experience. Therfore some haue called it Helleborine tenuifolia: some others call it Helleborastrum, or Consiligo, whereunto it is nothing like.

❧ The Temperament.

Buphthalmos or Oxe eye is hoate and dry, of a more sharper and cutting nature than Camomill.

❧ The Vertues.

The floures of Buphthalmos pounde, and mergled with oyle and waxe, & layde to colde and harde swellings, dissolueth and wasteth the same.

Some do affirme, (as witnesseth Dioscorides and Serapio) that Buphthalmos or Oxe eye cureth the Jaunders, & causeth the body to be of good colour, if one drinke it boyled in wine, after his comming out of a bath.

The second Booke of
Of Goldenfloure, or the wild Marygolde. Chap. xxxiii.

❧ *The Description.*   Chrysanthemon.

1. This herbe hath rounde smooth stems diuided into many branches. The leaues be long and deepely iagged round about, as if they were rent or torne. The floures grow at the top of the branches in fasshion like the floures of Camomill, but they be a great deale larger, and not only yellow like fine gold in the middle, but also round about, and of a pleasant smell. The roote is white and threddie.

2. There is yet an other kinde of this herbe in all things like to the same, as in his stalkes, colour, floures, sauour, and fasshion, but his leaues be a great deale more deepely cut and iagged, euen harde to the middle ribbe or sinew. The which I thought good to note, to the ende that by this one may know and vnderstand, how one kind of herbe may often change his shape and proportiō, according to the nature of the soyle or place where it groweth, as first of all we may learne by this herbe, the which in some places hath not his leaues so much clouen and iagged, and therefore it approcheth not so neare to the description of Dioscorides his Chrysanthemum: as it doth whan it groweth in some other places, whereas it beareth leaues, very much clouen and iagged, and than it is agreable in all respects to the true description of Chrysanthemum.

❧ *The Place.*

This herbe groweth amongst the Corne, and in householde gardens amongst other herbes, and by the high way sides.

❧ *The Tyme.*

It beginneth to floure in June, and from thence forth almost vntill winter.

❧ *The Names.*

This herbe is called in Greeke Χρυσάνθεμον, and in Latine Chrysanthemum, that is to say, Goldenfloure, & Caltha, and of some Buphthalmum: in Italian *Chrispula herba*: in Spanish *Mequeres amarillo*: in French *Camomille Saffranée*: in high Douch S. Johans blum, & Gensblum: in base Almaigne Uokelaer, geel Gansebloemen, Hontstroosen: vnknowen in shoppes as many other good herbes be.

❧ *The Nature.*

This herbe is hoate and dry, not much differing from Camomill.

❧ *The Vertues.*

Chrysanthemum boyled in wine, cureth the Jaunders, & restoreth good colour, whan one doth drinke it, after that he hath bene often & long in the bath.   A

The seede of the same dronken in wine by it selfe, or pound with his floures   B
doth also cure the Jaunders, as the later writers haue proued.

The

¶ The floures of this herbe pound with oyle and waxe, and applied in maner of a playster, dissolueth colde swellings whiche chaunce to be on the head.

¶ The leaues and tenderest braunches of Chrysanthemum, may be well vsed in potage and Salades, as other herbes of like nature: for in time past our elders haue so vsed it.

Of the Indian Sunne / or Golden floure of Perrowe. Chap.xxxiiij.

❦ *The Description.*   Chrysanthemum Peruuianum.

The Indian Sunne, or the golden floure of Perrowe is a plante, of suche stature and talnesse, that in one Somer it groweth to the length of thirtene or fouretenne foote, and in some places to the heigth of foure & twentie, or fiue and twenty foote, his stalkes be right straight and thicke, and his leaues are very many, especially they that grow vpmost, for the vnder leaues do quickly fall and vanissh: especially those great broade leaues whiche before the springing vp of the stalke, are in quantitie almost as large as the leaues of the Clote Burre. In the very top of the sayde high stalke there groweth a very large & most excellent floure most likest to Camomill, or Chrysanthemum, but much larger, & in quantitie almost like to a prettie broade Hatte, so that oftentimes whan the circuit, or vttermost Compasse of the sayde floure is measured, it is founde to be of the breadth of halfe a foote. The middle of the floure in whiche the seede groweth, is like to a fine cloath wrought as it were with needle worke: the small leaues whiche grow in compasse aboute, are of a bright shining yellow colour, and euery one of them are in quantitie like the leaues of the Lyllie floures, or rather greater, and are almost fiftie in number or moe. The seede is flat and long, and somwhat browne or swarte, in quantitie like to the Gourde seede. The rootes are like to the rootes of Reedes or Canes.

❦ *The Place.*

This plante groweth in the Weste India, the whiche is called America, and in the Countrey of Perrowe: and being sowen in Spayne, it groweth to the length of foure and twentie foote, and it beareth floures lyke to the aboue sayde: in base Almaigne it groweth not aboue xij. or xiij. foote high, and it doth scarsly bring foorth his floure, and if it chaunce sometimes to beare his floures, yet than they be smaller and very little, and they come foorth agaynst winter, so that they can come to no perfection.

❦ *The Names.*

This

The second Booke of

This floure is called Sol Indianus, and Chrysanthemum Peruuianum: in base Almaigne Sonne van Indien: we may also call it the Indian Sunne, or the Golden floure of Perrowe.

❧ The Nature and Vertues.

Of the vertue of this herbe and floure, we are able to say nothing, bycause the same hath not bene yet found out, or proued of any man.

Of Floure Deluce, or Iris. Chap. xxxv.

❧ The Kyndes.

There be many kindes of Iris, or floure Deluce: whereof some are great & tal, and some are little and small. The greater sortes are knowen one from an other by their colours, and so be also the smaller sortes. There is also a certayne kinde with narrower blades, in sauour somewhat lothsome or grieuous, almost of the sauour of Spatulæ fœtidæ, or Gladyn, bysides the Dwarffe Ireos, the stincking Iris, and the yellow Iris.

❧ The Description.   Iris.

1. The greater Iris, or floure Deluce his leaues be lög & large, not much vnlike to the blade of a two edged swoorde, emongst the which there springeth vp playne and smooth little stalkes of two foote long or more, bearing floures made of six leaues ioyned togither, wherof the three that stande vpright, are bent inward one towards an other: and most commonly in the leaues that hang downewardes, there are certaine rough or hearie weltes lyke vnto a mans browes, growing or rising from the nether parte of the leafe vpwarde, almost of a yellow colour. The rootes be thicke, long and knobby, with many strings, as it were hearie threedes hanging at them.

One kinde of these beareth floures betwixte purple and blewe, with a certayne changeablenes, especially in the nethermost leaues.

The other kinde his leaues that hang downewardes, are of a fayre violet colour, but those that grow vpright, and bende inwardes, are of a fainte blew.

The third floure is altogither or wholly of a fainte blewe.

The fourth kinde his floures be all white.

The fifth kinde his leaues be of a very fayre deepe violet colour, and his smell is moste delectable, and the hearie or rough weltes of this kynde are white.

2. The smaller floure Deluces, or Ireos, are in all things like to the greater, sauing that their stemmes be very shorte, and their flagges or blades, are also shorter and smaller than the others. Their floures are like to the greater, most commonly of a yellow colour, and sometimes of a fainte colour, and sometimes betwixte purple and skie colour: and the same is in some kyndes of them

sadder,

the Historie of Plantes.

sadder, and in some lighter.

3 The narrow leaued Ireos, his flagges be long and narrowe, but yet they be shorter then the leaues or blades of the greater Iris, and of a blewishe greene colour, of sauour somewhat greeuous, but nothing so horrible or lothsome as Spatula fœtida. The stemmes growe to the height of halfe a foote, at the toppes whereof growe cleare blewe or skie coloured flowers, lyke to the other flower Deluces, sauing that their litle leaues are smaller and narrower, and the vpper leaues do not bende inwarde, one towarde another. After the sayde flowers folowe certayne triangled great coddes or huskes, separating them selues into three partes when they are rype: in them is playne sede which is very thicke & flat or thrust togither. The rootes also grow crokedly lyke the others, but they be smaller, harde, and knottie, in the outsyde of a Chesnut colour, and white within, or somewhat yellowe.

❧ *The Place.*

1 The flower Deluces or Irices do growe in diuers Countries, most commonly in lowe groundes about the bankes of riuers and waters.

The three first kindes are meetely common in Englande, Brabant, and Flaunders.

The fourth also is sometimes founde in gardens.

But the brauest of them, with the flowers twixt purple & violet, commeth to vs from Spayne and Portingale.

2 The smaller flower Deluces, are but strangers with vs, neyther doo they growe of them selues amongst vs.

3 The narrow leaued Ireos groweth in certayne playnes of Germanie, and in lowe moyst places, also it is founde in open feeldes.

❧ *The Tyme.*

The Irides or flower Deluces do most commonly flower about May: and the smaller somwhat before the others, and the narrow leaued flower Deluce last of all. But in Portingal and Spayne they flower at the later ende of Autumne, a little before winter. ❧ *The Names.*

This herbe is called in Greeke ἶρις. and as Atheneus, and Theophrastus write ἶρις: in Latine, Iris, Consecratix, Radix Naronica.

That kinde whose flower is of purple and blewe is called of some Iris Germanica: in Shops, Iris, of others Lilialis, and Spatula: in English also Iris: and of some blew flower Deluce: and garden flagges: in Italian, *Giglio azuro, Giglio celeste*: in Spanishe, *Lirio Cardeno*: in high Douche, Blauw Gilgen, Blauw Schwertel, Himmel Schwertel: in base Almaigne, Blauw Lisch: in french, *Flambe*.

That kinde with the white flower, is called of the most part Iridem florentinam: in Shoppes, Ireos, (especially the dryed rootes) by the which name it is knowen of the Clothworkers and Drapers: for with these rootes they vse to trimme their clothes to make them sweete and pleasant: in English, White flower Deluce, and of some Iris florentine: and the rootes be comonly called Ireos: in Italian, *Giglio bianche*: in french, *Flambe blanche*: in high Douch, Violwurtz, weisz Violwurtz: in neather Douchland, Wit Lisch: and the rootes of this white flower Deluce, are iudged for the best Ireos, especially when we shall haue neede to vse of the dryed rootes.

That kinde which beareth the faire purple flower, is now called in Latine, Lusitanica Iris, and Iris serotina, that is to say, Portingale Iris, and late Iris: in Douche, Spade Lisch, and Lisch van Portegall.

3 Finally, that kinde with the narrow leaues, is called in Latine, Iris angustifolia,

folia, or Iris tenuifolia, and Iris Cærulea: in English, Narrowe bladed Ireos: in high Douche, Blo Schwertel, that is, blewe Lillie.

❋ The Nature.

The Ireos rootes being yet greene and newe gathered, are hoate and dry in the thirde degree, & they burne in the mouth or throte when they are tasted: but when they be dry they are euer or alwaies hoate but in the second degree: neuerthelesse they be euer drie in the thirde degree.

❋ The Vertues.

The greene and new gathered rootes of Iris, and specially the iuyce therof, doo purge downewarde mightily, and bring foorth yellowe choler, and almost al waterish humours, and are therfore good against the dropsie: but they may not be taken but in smal quantitie, and yet they ought to be well mingled with thinges that coole: for otherwise they wil inflame the very bowels. A

But the same roote dried, prouoketh not the belly, but it prouoketh vrine, and breaketh the stone. B

The rootes of Iris bring foorth the flowers, whether the same be receiued into the body, or conueyed in with Pessaries, or els mingled in bathes and stewes made for the purpose. C

The same rootes doo clense the breast and the lunges, and ripe tough fleme and slimie humours, and they loose the same and make them thinne, & they are good against the shortnesse of breath, and an old cough to be mixed with sugar or honie, and often taken into the mouth or licked on. D

The same rootes dronken with vineger or water, are good against the bitinges and stinginges of Scorpions and other venemous beastes. E

This roote is very good for them that are troubled with the paine and stopping of the milt or splene, & for them that haue any member shronken, or sprong out of ioynt, or displaced, or taken with the Crampe, stiffe or benummed. F

The same roote or the powder thereof put into the nose, causeth Sternutation or niesing, and draweth foorth tough, colde, and slymie humours. G

The same roote mingled with hony, doth mundifie and clense corrupt and filthy vlcers, and draweth foorth shiuers, and splinters of wood, and broken bones, out of the fleshe, it doth also regenerate and increase newe fleshe, it is very good against the vlcers and blisters of the fingers and toes, that rise about the naples aswell in the handes as in the feete, & with conuenient oyles and oyntmentes it helpeth the impostumes, and chappes or riftes of the fondement. H

The rootes of Iris, and the rootes of white Hellebor, with twise asmuche honie is good to annoynte the face, against the lentiles, freckles, pimples, and all other spottes and blemishes of the face, for they clense the same. I

The same mingled with oyle of Roses is good against the headache, when it is annoynted therwith. K

❋ The choyse.

The best and most conuenient in medicine, are the Ireos rootes whiche growe in Sclauonia: the next is the Iris of Macedonia, and the thirde best is that whiche groweth in Africa, as Dioscorides and Plinie write, but the African Ireos is muche discommended of Galen. At this day the white Ireos is taken for the best, especially the Ireos of Florence, whiche is called in shops Ireos, and Ireos Florentina of the base Almaignes.

the Historie of Plantes. 195

Of small floure Deluce, or dwarffe Ireos. Chap.xxxvi.

❧ The Description.
Chamæ-iris.

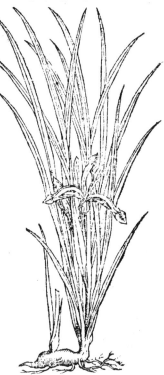

That kinde of flagge, whiche we do nowe call the small floure Deluce, hath narrow long blades, almoste like the leaues of the right Gladin, but of a browner greene, & somewhat thicker. The stalkes are shorter than the leaues, but onely of a span long, the which do beare two or three small floures vpon shorte stes, standing all togither at the very top of the sayd stalkes, and not one aboue an other as other flagges. These floures are almost like the floures of the other flagges, sauing that they be smaller, & the three first leaues that hange downeward, haue not such hearie strakes or lines as are to be perceyued in the other floure Deluces. Their colour for the most parte is a cleare blewe, straked in certayne places with small lines & points, of white & yellow, alongst the sides of the leaues that hang downewardes. They be of a pleasant sauour, sweeter and stronger than any of the other floure Deluces. The roote is harde, browne without, and white within.

❈ The Place.
This kinde of a flagge is founde in this countrie in the gardens of Herborists.

❈ The Tyme.
It floureth here in May and June.

❈ The Names.
This floure Deluce may well be called in Greeke χαμαιιρις, Chamæiris: that is to say, Dwarffe Ireos, or the smallest floure Deluce, bycause it is the least of all the flagges. The Herborists do now call it Iris Illyrica. And so doth also Hermolaus Barbarus in Corollario. But Antonius Musa in Examine Simpliciũ, doth very well declare, that this is not Iris Illyrica.

The Temperament and Vertues.
This flagge also is hoate and dry, leauing (whan it is chewed) a certayne heate vpon the tongue, as the rootes of all the other flagges do.

Of wild Ireos, stincking Gladin, or Spourgewort. Chap.xxxvij.

❈ The Description.

The stincking flagge or Gladyn hath long narrow bladed leaues like to the leaues of Ireos or the floure Deluce, but a great deale smaller and of a darke greene colour, of a lothsome smell or stincke, almost like vnto the stincking worme called in Latine Cimex. The stalke is rounde, vpon which groweth floures like to the floure Delice, but smaller and of a gray, or ashye colour: whan they are gone, there appeare great huskes or coddes, wherein is round red seedes, eche grayne or bearie of the quantitie of a little rounde pease.

Xyris.

The roote is long and very threedy.

✤ *The Place.*

This herbe is a stranger in Brabant, for it is seldome found in that countrey out of the gardens of Herboristes. It is very common in England, especially neare to the sea side, growing in stonie places by hedges and the borders of woodes.

✤ *The Tyme.*

It floureth in August, and the seede is ripe in September.

✤ *The Names.*

This herbe is called in Greeke ξύρις: in Latine Xyris, and Iris syluestris: in Shoppes Sphatula foetida: in Spanishe *Lirio Spadanal:* in English Stinking gladyn, Spourgeworte, & wilde Ireos: in Frenche *Glaieul puante:* in high Douche Welsch Schwertel, Wandtleutzkraut: in base Almaigne Wädtlupscruyt, wilde Lisch, and stinckende Lisch.

✤ *The Nature.*

It is hoate & dry in the third degree, of power to cut and make subtill.

✤ *The Vertues.*

The seede of the stinking Gladyn, taken in weight of half a dram prouoketh vrine mightyly, & taken with vineger it doth wast and cure the hardnesse and stopping of the Melte or Spleene.

The roote of stinking Gladyn pounde with a little Verdegris, a little of the B roote of the great Centory, & a little Hony, draweth forth al kindes of thornes, splinters, and broken bones, and is very good for the woundes, and bruses of the head, to draw foorth the broken bones.

The same mengled with vineger doth consume and waste cold tumors and C swellings being layde therevpon.

This herbe dryeth away and killeth the stinking wormes or Mothes cal- D led Cimici, if the place whereas they haunt or ingender, be rubbed with the iuyce thereof.

Corne flagge, or Gladioll. Chap.xxxviij.

✤ *The Description.*

This Gladyn or Corne flag hath long narrow blades, like to the blades of Ireos, & the rest of the flagges, but a great deale smaller & narower, amongst the which there springeth vp a round stalke of a cubite lõg, at the toppe whereof there hangeth in order fayre purple floures, one aboue another, after whiche there commeth roundish huskes, diuided in three partes, almost like to the huskes of Hyacinthe or Iacinthe, in whiche the seede is conteyned. The roote is like vnto two round bullettes set one vpon an other.

✤ *The Place.*

This Gladyn is not found in this coũtrey, but in the gardẽs of Herborists.

✤ *The Tyme.*

This Gladyn floureth in this countrie in May and June.

the Historie of Plantes. Gladiolus.

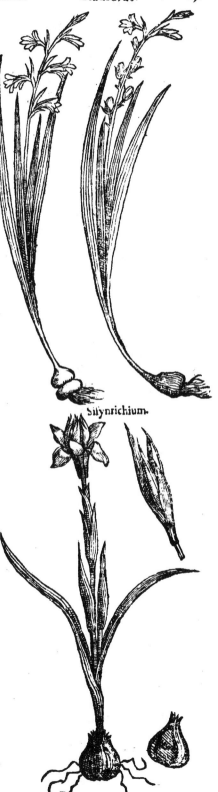
Sisyrichium.

¶ *The Names.*

This herbe is called in Greeke ξιφιον, of some μαχαιρωνιον και φασγανον: in Latine, Gladiolus, of Apuleius Gladiolus segetalis, and Lingua ceruina: vnknowen in shops: in Italian, *Monacuccie*: in Spanish, *Gladiolo di entres los panes*, of some Victorialis: in Douche, Aller man harnisch: we may cal it in English, Corne Gladin: Corne flag, and right Gladin.

The Nature.

The roote of Corne Gladin, especially the vppermost, doth drye & make subtil, and hath a litle drawing qualitie, as Galen writeth. *The Vertues.*

A The vpper roote of this Gladin pound with Frankensence and wine, draweth foorth thornes, and thinges that sticke fast in the fleshe.

B The same roote mingled with Iuray meale and honyed water (called Hydromell) doth waste and make subtil harde lumpes or swellinges.

C They say also that the vpper roote dronken in wine, prouoketh Venus, or bodily pleasure and the lower roote causeth barrennesse.

Of Sisynrichion. Chap. xxxix.

The Description.

Sisynrichion hath two or three long, narrow, litle leaues, from which growe vp rounde stems, about halfe a foote long, on the toppes of them, growe very faire little flowers of a light blew or skie colour, so growing by course one after the other, the one of them is euer open and spread, and that standeth alwayes at the top, in fashiō almost lyke the flowers of Ireos, but smaller, and somewhat differing in proportion. After the sayde litle flowers there appeare small, long, rounde knops or huskes, wherein the seede groweth. The roote doth almost make two round heades, lyke Onyons or Bulbos, most commonly placed one vppon another, which are inclosed as it were in certaine litle houses. *The Place.*

This plante groweth in Portingale and Spaine: & is very seldome found in Flaunders, sauing in the gardēs of some diligent Herboristes. R iij *The

❋ *The Names.*

The Grecians call this plante σκυριχιον: it is called also in Latine of Plinie, Sifynrichium: in Shoppes, and Portingal, Nozelhals.

❋ *The Nature and Vertues.*

Sifynrichium is of a temperate complexion, and good to be eaten: The Auncientes dyd accompt it amongst the number of rootes that may be eaten, and the Spaniardes and Portingales at this day, do vse it for foode or meate.

Of Ireos Bulbosa. Chap.xl.

¶ *The Kyndes.*

There are founde three kindes of Iris Bulbosa.

❋ *The Description.*

Bulbosa Iris.

THE first kinde of Bulbus Ireos, his blades be lōg, narrowe, and straked, or crested, wel like the leaues of ẏ pellowe Asphodil: his stalke is almost of a cubite long, in the toppe whereof growe beautiful flowers, in fashion like the flowers of Ireos, of a braue and excellēt colour, betwixt purple and skie colour: after them commeth long and thicke coddes or huskes in whiche the seede groweth. The roote is after the manner of Bulbus, that is round lyke a Saffron head or Onyon, ẏ which when it is in flower, diuideth it selfe in twayne, or two Bulbus rootes.

2 The other in leaues is like to the first, but his flowers are partie coloured, for the leaues of the litle flowers that hang or turne downewardes are somewhat white, & the leaues that grow vpward, are of a cleare or light blewe colour, also the litle leaues of the sayd smal flowers are lesse then the others, and the coddes be longer and thinner.

3 The thirde is like to the other, but it beareth a flower altogither of a pleasant yellow colour.

❋ *The Place.*

1 The first kinde is founde in Englande.
2.3 The other twayne growe in Spayne and Portingale.

❋ *The Tyme.*

The flowers of these strange plantes, doo shewe them selues commonly in June, in base Almaigne where as they are scantly knowen or hardly founde, sauing in the gardens of some diligent Herborists.

❋ *The Names.*

This flower is called now in Latine, Bulbosa Iris, bycause it hath a Bulbus roote, and a flower lyke Ireos. But it seemeth to be Apuleius Bulbus, called in Greke ιερβολβ⊙, and Hieribulbus: they call this plante in Spayne, especially that with the yellowe flower *Reilla Buen*: and we may call it Bulbus Ireos in English.

the Hiſtorie of Plantes.

✣ *The Nature and Vertues.*

The nature of this kinde of Bulbus or flower, with his vertues are not yet knowen, bycauſe there is no experience made of it as yet.

Of the yellowe wilde Ireos, or Flower Deluce. Chap.xli.

✣ *The Deſcription.* Pſeudoiris Lutea.

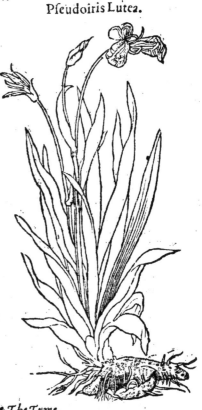

THe wild yellow Iris or flower Deluce, hath long narrowe flagges or blades, almost lyke to the right Iris or garden flagge, but a great deale longer and narrower very lyke to the blade of a long and narrowe double edged ſworde. The ſtalkes be rounde, ſmooth, and holow, at the toppe whereof groweth the yellowe flower with the three leaues hanging downewardes, like to ye gardē flower Deluce, & three mounting vpwardes, but they are ſmaller thē the leaues yt hāg downewardes. Whē they are paſt there come vp thicke triangled coddes or huſkes, in which is conteyned large yellow ſeede. The roote is thicke & ſpreadeth here and there, and ſometimes it hath other ſmall rootes hanging by it, and many thready ſtrings, of a fleſhly colour within, and of a rough aſtringent or binding taſte.

⸿ *The Place.*

This wilde yellowe Iris groweth in moyſt places, and low medowes, and in the borders and brinkes of Riuers, pondes, and lakes: very common in England, Flaunders, & other Countries. ✣ *The Tyme.*

This flower Deluce or wilde Iris flowreth in May and Iune.

⸿ *The Names.*

The wilde yellow Iris is nowe called in Latine, Pſeudoiris Lutea: and of ſome Sylueſtris Iris lutea, it hath bene called in Shoppes, Acoron, and hath ben taken in medicine for the ſame, not without great errour, loſſe, and danger of the ſicke, as it is of diuers learned men now very wel noted: and for that cauſe it is alſo called Pſeudoacorus, that is to ſay, falſe or baſtarde Acorus: in Douch, Geel Schwertel, geel wald Schwertel, & Drakenwurtz: in baſe Almaigne, Geel wilt Liſch, and Bore boonen: in French, *Glayeul baſtarde*, & *Flambe baſtarde*: in Engliſhe, the yellow wilde Iris, the yellow flower Deluce, wilde flagges, water flagges, and Lauers, or Leuers.

⸿ *The Nature.*

The yellowe baſtarde Iris his roote is colde and drie in the thirde degree, & of aſtringent or binding facultie, lyke to the rootes of Tormentill & Biſtorte.

✣ *The Vertues.*

The roote of yellowe flower Deluce, or baſtarde Iris boyled in water and dronken, ſtoppeth the bloody flixe, and other fluxes of the belly: and ſtoppeth

R iiij blood

blood from whence so euer it floweth, & womens flowers in what sort soeuer it be taken, yea if it be ministred but outwardly onely eyther in playsters or in bathes.

Of the white Lillie. Chap.clij.

❧ *The Description.*   Lilium Candidum, &c.

He white Lillie his leaues be long and broade, and somewhat thicke or fat, amongst ẏ which springeth vp a straight stemme or stalke of three foote long or more, set and garnished with leaues from the roote to the toppe, which by litle and litle as they grow vp toward the top, do waxe smaller, & smaller. In the top of the sayd garnished stemme growe the pleasant, beautiful, white, and sweete smelling Lillies, diuided into sixe small, long, and narrowe leaues, whiche haue in the outsyde of euery leafe, a certayne strake or ribbe, but within they are altogither of an excellent shynyng & pure white colour, bending somewhat backwardes at the top, in the middle amongst these leaues, ther hang vpō sixe very smal steins, sixe smal yellow pointes or litle markes, as it were tongues, in the middle amongst these also, there groweth another long bright and triangled stemme, thicker then the rest, and lyke to the Clapper of a Bell. The roote is lyke to a great Onyon, or rather a garlike head compacte and made of diuers cloues or kernelles.

❧ *The Place.*

The white Lillies be very common not onely in this Countrie, but in all places els where in gardens.

❧ *The Tyme.*

This kinde of Lillies doth flower at the beginning of June or there about.

❧ *The Names.*

The white Lillie is called of the Grecians κρίνον καὶ λείριον, of some καλλίριον, and κρινάνθεμον: the plante is called κρινωνία. It is called in Latine, Lilium, and Rosa Iunonis: in Shoppes, Lilium album: in Italian, *Giglio*, and *Giglio biancho*: in Spanish, *Azucena* in Douche, Weiſʒ Gilgen, or weiſʒ Lilgen: in Frenche, *Lys blanc*.

❧ *The cause of the Name.*

Constantine writeth this of the Lillie, that when Jupiter had begotten Hercules vpon Alcumena, and being desyrous to make him immortall, he caryed him to sucke Juno his wife, whiles she was sleeping, and when he perceiued the childe to haue suckt his fyll, he drewe him from her breast, by meanes whereof there fell great store of mylke from the breastes of Juno, the greatest parte whereof was spilt in heauen and fell vppon the Skies, whereof the signe and marke remayneth at this day, that is to say, that white and milkie way that goeth through heauen, from the North to the South (called in

Latine

Latine Via lactea): The rest fell vpon the earth, whereof sprang these Lillies, in the floures whereof, there remayneth the very whitenesse of the sayde milke: and hereof it came to passe, that this floure was called in Latine Iunonis rosa, that is to say, Iunos rose.

❧ The Nature.

The floures of the white Lillie are hoate, and partely of a subtile substance. The roote is dry in the first degree, and hoate in the second.

❧ The Vertues.

The roote of the white Lillie sodde in honied water and dronken, dryueth A forth by the siege all corruption of bloud, as Plinie sayeth.

The same rosted, or pounde and well mengled with oyle of Roses, doth sof- B ten the hardnesse of the Matrix, & prouoketh the monethly termes, being layde therevpon.

The same pounde with Hony, ioyneth togither sinewes that are cut, consu- C meth or scoureth away the vlcers of the head called Achores, and cureth all maner of naughtie scuruinesse, aswell of the head as of the face, and is good to be layde to all dislocations or places out of ioynt.

The roote of the white Lillie mengled with vineger or the leaues of Hen- D bane, or Barley meale, cureth the tumors and impostems of the genitors.

The same boyled in vineger, causeth the Cornes which be in the feete to fall E of, if it be kepte vpon the sayde Cornes as a playster by the space of three dayes without remouing.

The same mengled with oyle or grease, bringeth the heare agayne vpon pla- F ces that haue bene either burned or scalded.

The same roote rosted in the embers, or well pounde with oyle of Roses, is G good against the foule breaking out called the wild fire. It cureth all burnings, and closeth vp vlcers. The same vertue haue the leaues. Moreouer they are good to be layde vpon the bytings of Serpents.

The iuyce of the leaues boyled with vineger and hony in a brasen pipken or H skillet, is very good to heale & mundifie both olde vlcers and greene woundes.

With the floures of Lillies there is made a good Oyle, to supple, mollifie & I digest, excellent to soften the synewes, and to cure the hardnesse of the Matrix or Mother.

The seede of Lillies is good to be dronken against the biting of Serpents. K

Of the Orenge colour, and redde purple Lillies. Chap. cliij.

❧ The Kyndes.

There be three kindes of redde or purple Lillies, wherof the first is the small and common redde Lillie, the second is great, and the thirde is of a meane sise or quantitie.

❧ The Description.

1 The small purple Lillie, his stalkes be almost of the length of halfe a foote, set full of narrow darke greene leaues: the floures in fashion are like the floures of the white Lillie, sauing they are without sauour, and of a fyrie redde colour, sprinckled or poudered with blacke speckes: the rootes be also round, and with cloues or kernels like to the rootes of the white Lillies.

2 The greater red Lillie groweth to the heigth of the white Lillie, and there groweth oftentimes vpon one stalke twenty, fiue & twentie or thirtie floures, or moe, of a shyning yellowish redde colour, & speckled with very small blacke spottes,

spottes, or little poynted markes as the other. The roote is also like the other, sauing it is somwhat smaller.   Liliũ purpureũ.

3   The third redde Lillie is in grouth higher than the first, yet not so high & tal as the seconde. This kinde of Lillie beareth at the toppe of the stalke, and also amongst his leaues as it were certayne pypes or clysters, whiche if they be set in the ground, will grow, and after three or foure yeares they will beare floures.

❧ The Place.

These kindes of Lillies are planted in some gardens, especially in Flaunders and Germany, but in some countries they grow wilde in rough and harde places.

❧ The Tyme.

They floure in May and June.

❧ The Names.

The red purple Lillie is called in Greeke κρίνον πορφυροῦν: in Latine Lilium rubrum, Lilium rufum : and of Ouide it is called Hyacinthus. Pausanias calleth one of these kindes Comosandalon: the Italians Giglio saluatico, & some call the greatest kinde Martagon: it is called in Douche Root golt Gilgen.

❧ The cause of the Name.

Of the redde Lillie Ouide wryteth this, that it came of the bloud of the Boy Hyacinthus, the whiche Apollo (by misfortune slue) in playing with him, so as the grasse and herbes were bedewed and sprinckled with the bloud of him. Whereupon it came to passe immediately by the commaundement of Apollo, that the earth brought forth a floure altogither like to a Lillie, sauing it was redde, as Ouid wryting in the tenth Booke of his Metamorphoseos, saith.

  Ecce cruor, qui fusus humo signauerat herbas,
  Desinit esse cruor, Tyrioq; nitentior ostro
  Flos oritur, formamq; capit quam Lilia: si non,
  Purpu eus color his, argenteus esset in illis.

And for a perpetuall memorie of the Boy Hyacinthus, Apollo named these floures Hyacinthes.

❧ The Nature and Vertues.

The nature and vertues of the redde Lillies are yet vnknowen, bycause they are not vsed in medicine.

Of the wilde Lillie.   Chap. xliiij.

❧ The Description.

The wilde Lillie hath a straight rounde stemme set full of long leaues, at the toppe whereof there grow fayre pleasant floures, in proportion much like to the Lillie, diuided into sixe small, thicke, and fleshie leaues, bending or turning backwardes almost like a ring, of an olde purple or dimme incarnate colour, poudered or dashte with small spottes, and without any speciall smell. The roote is like to the common garden Lillie, sauing it is smaller and yellow as golde.

❧ The Place.

This herbe groweth in some places of Almaigne, as in the woodes, & medowes whose situation or standing is vpon Mountaynes: but in this countrey

the Historie of Plantes.

trie they plante them in gardens.

※ *The Time.*

The wilde Lillie floureth in Maye and June.

❧ *The Names.*

This floure is called of some in Greeke ἡμεροκαλλίς: in Latine, Lilium syluestre: and in some places Affodillus, amongst the Apothecaries, and is vsed for the right Asphodelus (but very erroniously: in Englishe, Wilde Lillie: in Frenche, *Lis sauuage*: the Italians call it *Martagon*: and the Spaniardes, *Amarillis* in high Douche, Goldwurtz, and Heydnischblumen: in base Almaigne, Lelikens van Caluarien, Heydens bloeme, and Wilde Lelien: some take it for ἡμεροκαλλίς, Hemerocallis, howebeit the flower is not yellow.

※ *The Nature and Vertues..*

The wilde Lillie also is not vsed in medicine, & therfore his nature & vertues are as yet hidden, & vnknowen.

Of Dogges tooth. Chap.xlv.

※ *The Description.*

This lowe base herbe, hath for the most parte but two leaues, speckled with great redde spottes, betwixt whiche there springeth vp a little tender stalke or stemme with one flower at the toppe hanging downeward; whiche hath certayne small leaues growing togither lyke an arche or vaute, and like the wilde Lillie, of colour white or pale purple, like to a Carnation or flesh colour: out of the middest of this flower, there hange also sire smal thrommes, or short threds, with little titles or pointed notes, like as in the Lillies. After the flower there foloweth a roud knop or litle head, in which the sede is côteyned. The roote is long & slender lyke to a Chebol, with certayne hearie threddes, or stringes hanging at it.

※ *The Place.*

It groweth in diuers places of Italy, but chiefely on the hilles & mountaynes of Bononia and Mutinens, and the Countrie theraboutes: it groweth not in Brabant sauing in ye gardens of certayne diligent Herboristes.

❧ *The Names.*

This herbe is nowe called Denticulus canis, and Dens caninus, of some it is also called Pseudohermodactylus, of others Satyrio Erythroniū, wherwithall notwithstanding it hath no similitude: but it seemeth to be Ephemeron nō lethale, of Dioscorides, whiche is also called κρίνον ἄγριον, that is, Lilium syluestre: and it may wel be called Lilium syluestre: bycause that the flower when as it hangeth downeward towardes ye ground, is much like to the Lillies, & especially the wilde Lillies, sauing it is euer smaller. ❧ *The*

Lilium syluestre. Martagon Italorū Amaryllis Hispanorum.

Denticulus canis. Ephemeron nō lethale.

❦ The Nature and Vertues.

Of the nature and vertues of this herbe we can affirme nothing, but if it be Ephemeron as it seemeth to be, then it is good for the teeth, as Dioscorides saith, for as he writeth, the water wherein the roote is boyled is wholesome and specially good for the teeth.

The leaues of this herbe boyled in wine and layde to, do scatter and driue away all small tumours and wheales, and pushes of the body.

Of Lillie non Bulbus. Chap. xlvi.

❧ The Kindes.

There be two sortes of this Lillie, whereof one hath a yellowe flower, the other a darke Crimsin or purple flower.

❦ The Description.

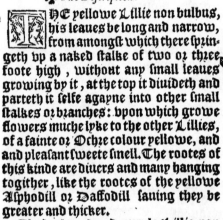

Lilium non Bulbosum.
Lillie non bulbus.

1. THE yellowe Lillie non bulbus, his leaues be long and narrow, from amongst which there springeth vp a naked stalke of two or three foote high, without any small leaues growing by it, at the top it diuideth and parteth it selfe agayne into other small stalkes or branches: vpon which growe flowers muche lyke to the other Lillies, of a fainte or Ochre colour yellowe, and and pleasant sweete smell. The rootes of this kinde are diuers and many hanging togither, like the rootes of the yellowe Asphodill or Daffodill sauing they be greater and thicker.

2. The darke red and purple Lillie non bulbus, in stalke & rootes is like to the other, but his flowers be of a darke or dim red purple colour, somewhat larger then the flowers of the yellow kinde the leaues be also larger and rougher. The flowers of both kindes do last but a very small time, not aboue a day at y furthest, especially the purple whiche fadeth very lightly, & withereth often times before Sonne set. ❦ The Place.

These Lillies are strange in this Coūtrie & Flaunders, and are not founde sauing in gardens, wheras they grow easily, and prosper wel. ❦ The Time.

They flower, with the other Lillies, and somwhat after, and somtimes they flower againe in Autumne when the whether is milde and pleasant.

❦ The Names.

The Latinistes do call this kinde of Lillies, Lilium nonbulbosum. And it seemeth to be that kinde of Lillie which the Grecians call ἡμεροκαλλίς, Hemerocallis: for as Atheneus writeth, it is called Hemerocallis only, bycause it lasteth but a day. Moreouer Hemerocallis is called 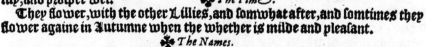 in Latine, Lilium syluestre, and Lilium marinum, whiche names are most agreeable vnto these kindes of Lillies.

the Historie of Plantes. 205

✲ *The Nature and Vertues.*

These kindes of Lillies are neither vsed in meate nor medicine, and therefore their nature and vertues are yet vnknowen.

Of the Lillie of Alexandria. Chap.xlvij.

✲ *The Description.*

Ornithogalum maius.

The leaues of this kind of Lillie are long and narrow, amongst whiche riseth vp a litle smooth, tender stalke, at the top whereof there growe diuers faire and pleasant flowers, of a shining white colour, and proportioned like to a little Lillie, in the middle whereof, ouer and aboue certayne smal threddy stalkes or thrommes, there commeth foorth one somewhat greater then the rest, lyke to an aglet, or triagled huske, in which after the falling of, of the flowers the seede groweth. The roote is rounde after the manner of Bulbus and somewhat great, & white of colour: diuiding it selfe easily into diuers other rootes.

❦ *The Place.*

This is also a stranger with vs. And it seemeth that it was first brought from Alexandria into Italie and these regiōs or Countries. **✲** *The Names.*

This Lillie is called Lilium Alexandrinum: but of Dioscorides in Greeke ὀρνιθόγαλον: in Latine, Ornithogalum: and bycause there is yet another Ornithogalum, described in the fifth parte of this worke, this is therefore called Ornithogalum maius. **✲** *The Nature and Vertues.*

Dioscorides writeth of Ornithogal, that the bulbus, or round roote thereof may be eaten and vsed for meate either rawe or sodden.

Of the Hyacinthes. Chap.xlviij.

✲ *The Kindes.*

There be two sortes of Hyacinthes, yet ouer and aboue diuers others whiche are also counted Hyacinthes, whereof we will write in the next Chapter. **✲** *The Description.*

The first Hyacinthes which are common in the lower Germanie, haue long narrowe leaues: amongst which spring vp smooth stalkes, which being loden litle flowers from the middle euen vp to the very top, are with the waight and burden of the same, made crooked, or forced to fal, bende, or stoupe. The litle flowers are long and holowe, and afterwarde somewhat spread abrode like vnto Lillie Conuall, not so strong in smell, but yet pleasant and sweete, of colour most commonly blew lyke azure, and sometimes purple, and sometimes as white as snowe, gray, or ashe coloured: when these flowers are fallen, there folow triangled huskes or coddes, wherein the small rounde seede is conteyned. S The

Hyacinthus vulgaris &c.   Hyacinthus Orientalis &c.

2 ❧ The Oriental Hyacinthes are much like to the aforesayde, but his leaues stalkes and rootes are greater: and the flowers be also larger, & of an excellenter blewe colour.

❧ *The Place.*

The common Hyacinthes do grow about the borders of fallowed feeldes and pastures in sandy or grauely ground, and are founde in many places, especially about Wincaunton, Storton, & Mier, in y West partes of Englande, &c.

❧ *The Tyme.*

1 The common Hyacinthe flowreth about the ende of Maye, and in June, or somewhat rather.

2 The Oriental Hyacinthes do flower before the common sort, sometimes in Marche.   ❧ *The Names.*

These are called in Greeke ὑακίνθοι: in Latine, Hyacinthi, & as some thinke, Vaccinia: in Englishe also Hyacinthe or Crowtoes: but these be not those Hyacinthes wherein the notes or mourning markes are printed: for they are in the red purple Lillies, as before is sayde.

❧ *The Nature.*

The roote of Hyacinthe is drie in the first degree, and colde in the seconde: but the seede is drie in the thirde degree, yet temperate twixt heate and colde.

❧ *The Vertues.*

The roote of Hyacinthe boyled in wine and dronken, stoppeth the belly, prouoketh vrine, and helpeth much agaynst the venemous bitings of the feeld Spidder.

The seede is of the same vertue, and is mightier in stopping of the laske: it helpeth them that haue the bloody flixe, and if it be dronken in wine, it is very good agaynst the falling sicknesse.

the Historie of Plantes. 207
Of other sortes of Hyacinthes. Chap.xlix.
¶ The Kyndes.

BEsydes the two sortes of Hyacinthes (whiche in deede are the right Hyacinthes) described in the former Chapter: there be also diuers flowers, which are also taken for Hyacinthes and are now reckned amongst them.

Hyacinthus neotericorum primus.   Hyacinthus neotericorum tertius.

❧ The Description.

1. THE first of these kindes of Hyacinthes, hath long, narrowe, greene leaues: amongst which are slender stalkes, löger then a hand breadth, bearing many trimme flowers, growing togither, about the top of the stalke in a cluster or bundel lyke to a nosegay or litle bunche of grapes, especially before the opening or spreading abrode of the flowers. The roote is rounde after the order of Bulbus or Onyons, and doth quickly encrease acd multiplie diuers others. The flowers are not muche vnlike to Lillie conual, most commonly of an azure or skie colour, wherof some are more shining & cleare, & some are of a deeper colour: sometimes they be also white, and sometimes ye shall see of them changing towardes a carnation or flesh colour: whereof the white are of a very sweete and pleasant sauour.

2. The second is somewhat lyke to the aforesayd: but his leaues be larger and thicker, and they lye strowen or spread vpon the grounde. The flowers be also greater, and doo stande further apart or asunder one from another, of colour somwhat white. The round or Bulbus roote also for his quantitie is greater.

3. The third his leaues also are longer and broder than the abouesayd, much like vnto Leeke blades: the stalke of a foote long, carying many small holowe

S ij   flowers,

flowers, growing so thicke about the top: that they shewe like a brush or holywater sprinckle, at the first of a faire violet colour, but when they beginne to wither, of a decayed or olde worne color, & sometimes but very seldome white. Finally the round and bulbus roote of this kind of Hyacinthe is greater, and of colour somewhat red or purple without.

Hyacinthus Autumnalis.

4   The fourth whiche is called Hyacinthus Autumnalis, is the least of these Hyacinthes, yea it is lesse then the first: it hath litle, narrowe, small, and tender leaues: and small slender stemmes of halfe a span long, at the whiche growe very smal flowers, of a cleare azure or skie colour, and fashioned, when they are open, like litle starres, with certayne fine, small, and short threddes growing in the middest of them. The seede is inclosed in a smal triangled huske. The roote is smal, yet of the fashion of an Onyon or Bulbus.

5   The last of al which is described of Fuchsius amongst the Hyacinthes, hath sometimes two, and sometimes three small leaues, amongst whiche there springeth vp a a little stemme, bearing fiue or sixe, or mo flowers at the very toppe, euery one of them growing vpon a small stalke by it selfe: eche flower hath sixe smal leaues, fashioned lyke a starre when they are spread abrode and open: of a skie colour and sometimes white. After these folow rounde knoppes wherein the seede is conteyned. The rootes are small and Bulbus fashioned, like the rest, and lyke vnto litle Onyons, but lesse.

✿ *The Place.*

1   The first kind of these base Hyacinthes do grow in the woods of Artoys that are next to the lowe Countrie of Germanie, in moyst, wet, and lowe groundes: and they be also often set and planted in gardens: whereof the blew sort is meetely common, but the white are geason, and rare to be founde.

2.3.   The seconde and thirde do also grow in suche lyke places of Italy and Germanie.

4   The fourth sort doth growe in Fraunce, especially neare about Paris.

5   The fifth is meetely common in Germanie, it delighteth most in good fatte groundes, but especially in pastures and vntoyled places.

✿ *The Time.*

1   The flowers of the first kinde, do shewe bytimes, as in Marche or before, if the weather be milde, and surely one kinde of these flowers, especially that with the perfect azure or deepe colour putteth foorth his leaues before winter, and the rest assoone as winter is gone.

2.3.   The seconde and thirde do flower afterwarde.

4   The fourth flowreth last of all at the ende of sommer, and beginning of Autumne.

5   The last flowreth bytimes, as in Marche or Februarie.

✿ *The Names.*

1   The first bastarde Hyacinthe is of that sort of Bulbus, whiche of the Auncientes was vsed in meates, and called in Latine by the surname of Bulbi esculenti.

the Historie of Plantes. 209

Hyacinthus Fuchsij bifolius.   Hyacinthus Fuchsij trifolius.

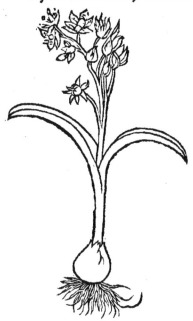 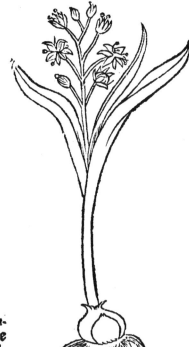

lenti. In these dayes some call them Hyacinthi Botryodes, or Hyacinthi racemosi: of the Italians (the white kinde especially) is called, Dipcadi, or Tipcadi.

2 The seconde is also in the number Bulborum esculentorum.

3 And so is the thirde also, whiche seemeth to be Bulbina, in Greeke Βολβίνη: in high Douche it is called Breunling, & of some Honds knoblauch: in English, Dogges Leekes, and bushe or tufte Hyacinthe.

4 The fourth kinde of bastarde Hyacinthe, is nowe called in Latine, Hyacinthus Autumnalis: in English, Autumne Hyacinthe.

5 The fifth Hyacinthe described of Fuchsius, is called in Douche, Mertzelblumen, and Hoornungblum: in Englishe, Our Ladyes flower.

❧ The Nature and Vertues.

These bastarde Hyacinthes are not vsed in medicine, and therfore of their nature and vertues is nothing written. They are planted in gardens onely for their flowers.

Of Narcissus. Chap.I.

❧ The Kindes.

FIrst of all there are two very faire and beautifull kindes of Narcissus, one with a Crimsin or red purple circle in the middle of the flower, the other hauing a yellow circle, or as it were a Crownet or cup in the middle of the flower.

❧ The Description.

THe first Narcissus hath small narrowe leaues lyke Leeke blades: with a crested bare naked stalke without leaues, of a foote or niene inches long,

Narcissus medio purpureus.
Narcissus with the purple edged circle in the middle.

Narcissus medio luteus primus.
Narcissus with the yellowe garlande or crownet in the middle.

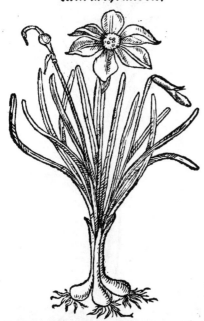
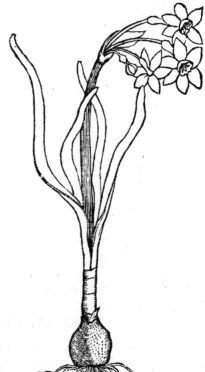

long, with a flower at the top, growing out of a certayne fylme, as it were a skinne, most commonly growing single or alone, and sometimes two togither, meetely large and sweete, made and fashioned of sire litle white leaues growing togither (almost lyke a Primerose) in the middle wherof is a certayne rounde wrinckled litle hoope, or cuppe, bordered or edged about the brinkes, with a certayne rounde edge, within which are certayne small threddes, or stemmes, with yellowish tipppes hanging vpon them: after the flower there appeare angled huskes, wherin groweth the sede whiche is blacke. The roote is rounde and bulbus, lyke an Onyon.

2   The other Narcissus with the yellow cup or circle in the middle, his blades be somewhat longer and broader and not althing so greene as the first: his stalkes be longer and thicker, and vppon euery of them three or foure flowers lyke vnto the first, sauing they be all yellowe in the middle.

There is also a kinde of Narcissus, that is also yellow in the middle, and it beareth a great many mo flowers, smaller then they before described.

And also another sorte, whiche beareth double flowers.

3   Moreouer there be other sortes of Narcissus found, whose garland or circle in the middle of the flowers is white, but these be very rare and daintie.

✳ *The Place.*

The two first kindes grow plentifully in diuers places of Fraunce, as Burgundie, and Languedoc, in medowes: but in this Countrie they growe not at al sauing in gardens, whereas they are sowen or planted.

The

❊ The Tyme.

All the Narcissus for the most part do flower in Aprill, sauing one of the first kindes is somwhat rather, and there is another whiche flowreth not vntill the beginning of May.

❊ The Names.

These pleasant flowers are called in Greeke νάρκισσος: and in Latine, Narcissus, of some as witnesseth Dioscorides λείριον, βολβὸς ἐμετικός, Bulbus vomitorius, and Anydros: vnknowen in shoppes: in Englishe, Narcissus, white Daffodill, & Primerose pierelesse: in high Douch, of some, Narcissen Roszlin: in base Almaigne Narcissen, and Spaensche Jennettekens.

❊ The cause of the Name.

These flowers tooke their name of the noble youth Narcissus, who being often required and much desyred of many braue Ladies, bycause of his passing beautie he regarded them not: wherfore being desyrous to be deliuered frō their importunate sutes and requestes, he went a hunting, and being thirstie came to a fountaine, in which when he would haue dronken sawe his owne fauour and passing beautie, the whiche before that time he had neuer seene, and thinking it had bene one of the amorus Ladyes that loued him, he was so rapt with the loue of him self, that he desyred to kisse and embrace him self, and when he cold not take hold of his owne shadow or figure, he dyed at last by extreme force of loue. In whose honour and perpetuall remembrance, the earth (as the Poetes fayne) brought foorth this delectable, and sweete smelling flower.

❊ The Nature.

Narcissus, but especially his roote, is hoate and drie in the seconde degree.

❊ The Vertues.

The roote of Narcissus boyled, rosted, or otherwayes taken in meate or drinke, causeth one to haue a desire to vomite.

The same pounde with a little honie, is good to be layde vnto burninges, it cureth the sinewes that be hurt, and is good against dislocations, and places out of ioynt, and easeth all olde greefe and payne of the ioyntes.

The roote of Narcissus taketh away all lentiles, and spottes of the face, being mingled with Nettel seede and vineger: it mundifieth corrupt and rotten vlcers, and ripeth and breaketh harde impostumes, if it be tempered with the flower or meale of Vetches and honie: and it draweth foorth thornes and splinters, if it be mixt with the meale of Juray and hony.

The seconde Booke of

Of rushe Narcissus. Chap.li.

❧ *The Description.* Narcissus iuncifolius.

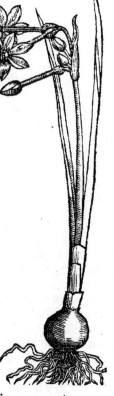

Iunquillias (as the Spaniardes call it) is also of the kindes of Narcissus, but their leaues be narrowe, thicke, rounde, tough, and plyant, smooth and playne, almost lyke rushes, they be also long and of a swarte greene colour. The stalkes grow vp to the length of a foote, at the top, whereof growe foure or fiue, or mo flowers, lyke the flowers of Narcissus, sauing they be smaller and of a yellow colour. It hath also a round Bulbus roote lyke to the rest of the Narcissis, but couered with a thinne blacke skinne or velme.

¶ *The Place.*

It groweth in sundrie places of Spayne, and from thence it was brought hither.

❧ *The Tyme.*

It flowreth in Aprill with the rest of the Narcissis.

❧ *The Names.*

It is called in Spanish, *Iunquillias:* and in Latine, Iuncifolius : bycause of the similitude it hath with rushes: we may also call it rush Narcissus: it is called of Dioscorides in Greke, βολβὸς ἐμετικός, that is in Latine, Bulbus vomitorius.

✱ *The Nature and Vertues.*

This roote eaten prouoketh vomit, as the roote of Narcissus dooth, wherevnto in nature it is very lyke: and therefore, as Dioscorides writeth, it cureth the diseases of the bladder.

Of Tulpia, or Tulipa, Lilionarcissus sanguineus poene.
Chap.lij.

¶ *The Kyndes.*

There be two sortes of Tulpia, a great and a small.

❧ *The Description.*

1. THe great Tulpia, or rather Tulipa, hath two or three leaues, which are long, thicke, and broade, and somewhat redde at their first springing vp, but after when they waxe elder they are of a whitishe greene colour, with them riseth vp a stalke, whereby the sayde leaues are somewhat aduaunced. It hath at the top a faire large & pleasant flower, of colour very diuers and variable, sometimes yellowe, sometimes white, or of a bright purple, sometimes of a light red, and sometimes of a very deepe red: and purfled about the edges or brimmes with yellowe, white, or red, but yellow in the middle and bottome of the flower, and oftentimes blacke or speckled with blacke spottes, or mixt with white and red: most commonly without smell or sauour. The Bulbus roote is lyke the roote of Narcissus.

2. The lesse Tulpia is smaller, and hath narrower leaues, and a shorter stem, the flower also is smaller, and more openly disclosed, or spread abroade. The Bulbus roote is also smaller, and may be diuided and parted in twayne or more

the Hiſtorie of Plantes.

moꝛe: when the ſtemme groweth vp, that which ſpꝛingeth in the neather part of the ſtalke is lyke to the ſtem of the great Tulpia, growing next the roote.

Tulpia maior.
Great Tulpia.

Tulpia minor.
Smal Tulpia.

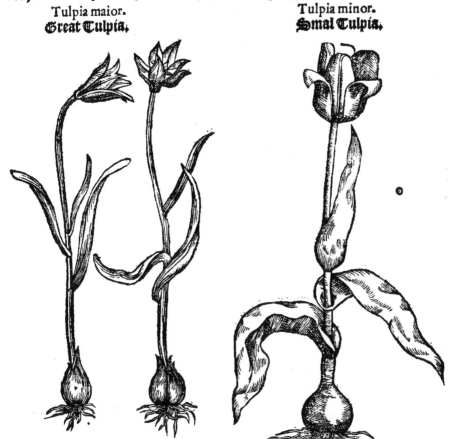

3   There is alſo placed with the Tulpia, a certayne ſtrange flower, whiche is called of ſome Fritillaria, whoſe tender ſtalkes are of a ſpanne long, with fiue oꝛ ſixe litle narrowe leaues growing at the ſame. There groweth alſo a flower at the toppe of the ſtalke with ſixe leaues, like to the leaues of Tulpia, but bending oꝛ hanging downewardes, of a purple violet colour, garniſhed and trimmed with certayne whitiſhe violet markes oꝛ ſpottes on the outſide, and with blacke ſpottes in the inſide. It hath alſo a bulbus oꝛ rounde roote.

❧ *The Place.*

1   The greater Tulpia is bꝛought from Grece, and the Countrie about Conſtantinople.
2   The leſſe is founde about Mounte-pelier in Fraunce.
3   Fritillaria is alſo founde about Aurelia in Fraunce.

❧ *The Tyme.*

They flower bytimes with the Narciſſis, oꝛ a litle after.

❧ *The Names.*

1   The greater is called both Tulpia, and Tulpian, and of ſome Tulipa, whiche is a Turkie name oꝛ woꝛde, we may call it Lillynarciſſus.
2   The ſmal is called Tulipa, oꝛ Tulpia minor, that is the ſmall Tulpian: and it is neither Hermodactylus, noꝛ Pſeudohermodactylus.

The

3 The third is called of the Grekes and Latines, Flos Meleagris, and Meleagris flos, as a difference from a kinde of birde called also Meleagris, whose feathers be speckled lyke vnto these flowers, but not with violet speckes, but with white & blacke spots, lyke to the feathers of the Turkie or Ginny hen, which is called Meleagris auis: some do also cal this flower Fritillaria. ✻ *The Nature and Vertues.*

The nature and vertues of these flowers, are yet vnknowen, neuerthelesse they are pleasant and beautifull to looke on.

Of bastarde Narcissus. Chap.liij.

✻ *The Description.*

This flower hath long narrowe leaues much lyke vnto Leeke blades, but not so long: amongst which springeth vp a round stalke bearing a faire yellowe flower diuided into sixe leaues like the flower of Narcissus, with a long rounde litle bell in the middle iagde about the edges, and of a deeper yellowe then the rest of the flower. After the flowers commeth the seede inclosed in round huskes or cods. The roote is round after the maner of bulbus, & like to Narcissus. ✻ *The Place.*

It groweth in moyst places in shadowy woods & in the borders of feeldes, as by Vuers, and Bornehem, & in the Parke wood by Louayne, where as it groweth abundantly, it is also planted in gardens. ✻ *The Tyme.*

This herbe bringeth foorth his leaues, stalkes, and flowers in Februarie, and is in flower somtimes vnder the snow. The sede is ripe in Marche. The herbe doth so perish in Aprill and May, that afterward it is no more seene. ✻ *The Names.*

This flower is called in high Douche, Geel Hornungsblumen, p is to say, the yellow flower of February, of some also Geel Tijdeloosen, & geel Sporckelbloemen: it is now called in Latine of some Narcissus lute', or Pseudonarcissus, bycause his flowers are somwhat like to Narcissus: in English, yellow Crow bels, yellow Narcissus, & bastarde Narcissus: in French, *Coquelourde*, and there is none other name to vs yet knowē.

✻ *The Nature.*

Yellow Narcissus is hoate & drie, much like in temperature to Narcissus.

✻ *The Vertues.*

Men haue proued this true and certayne by experience, that two drammes of this roote freshe and newly gathered, boyled in wine or water with a litle

Meleagris Flos, Fritillaria quorundam.

Pseudonarcissus.

the Historie of Plantes.

Anys or Fenell seede, and a litle Ginger and dronken, driueth foorth by siege tough and clammy flemme: wherfore the saide roote is good against al diseases, that happen by reason of tough and clammy flegme.

Of Theophrastus Violet, or the white Bulbus Violet. Chap.liiij.

❀ The Kindes.

There be three sortes of Leucoion, two small, and the thirde is bigger: wherof the flower of the first lesse kinde is three leaued: And the flower of the later kinde is sixe leaued.

Leucoium bulbosum triphillum &c.   Leucoium bulbosum hexaphillum. &c.

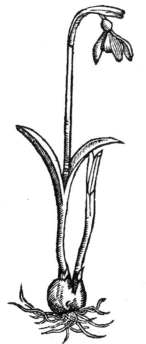 

❀ The Description.

1. THE first kinde of Leucoion bulbosum, beareth two or three narrowe leaues, a short stemme, and vpon it a litle faire and pleasant flower growing foorth of a little long huske vpon a smal stemme hanging downewards, with three white leaues, amongst which also there appeare three other litle greene leaues.

2. The second sort hath bigger leaues then the aforesayd, yet smaller & tenderer then Leeke blades, but otherwise they be alyke. The flowers be also white & compact or made of sixe little leaues somewhat straked or crested, in the middest of the flower are certayne hearie stemmes with small yellowe tippes: the seede is small and yellowe, conteyned in litle rounde huskes. The roote is white and Bulbus, and doth soone multiply and increase other.

3. The third kind is the greatest, & this sort beareth two or three or mo flowers togither vpon one stem, altogither lyke the flowers aforesayd: sauing that the stalkes and leaues are longer. ❀ The Place.

These kindes of Violets do growe in shadowy places, and lowe wooddes standing neare vnto waters in Italy and Germanie, they growe not in this Countrie, but in certayne gardens.

❊ *The Tyme.*

They begin to spring in Februarie, and yeelde their seede in Aprill, and in May the stalke with his leaues doth vanish cleane away, but the roote remayneth in the grounde like to yellowe Crowe belles or bastarde Narcissus.

But the thirde kinde flowreth not with the other twayne, but long after in Aprill.

❊ *The Names.*

These pleasant flowers are nowe accounted for a kinde of violettes, which Theophraste calleth in Greke Λευκόϊον, that is to say in Latine, Viola alba Therfore it is now called Leucoion, or Viola alba Theophrasti: we may call it in Englishe, White Bulbus violet, Narcissus violet, and Theophrastus white Violet: in Frenche, *Violette blanche*. in high Douch, Weitz hornungs blumen: in base Almaigne, Witte Sprockel bloemen, Soomersottekens, and Witte Tydeloosen.

❊ *The Nature.*

The temperament and vertues of these flowers are not yet knowen.

Of Saffron   Chap.lv.

❊ *The Description.*   Crocus.

Saffron hath long narrowe blades like grasse, the flowers grow vpon naked stemmes and are of a watcheth or pale blewe colour, diuided into sixe smal leaues (but somewhat long) from out of the middle whereof hange downe the Saffron blades or threddes of a redd colour. The roote is rounde lyke an Onyon, hauing sometimes foure or fiue small rootes adioyning.

❊ *The Place.*

Saffron, as Dioscorides and other the Auncientes haue written, groweth in the mount Coricus of Cilicia, and that was esteemed for the best, and in the mount Olympe of Lycia, neare about Aegis a towne of Aetolia: it groweth now in sundrie places of Douchland, especially about Vienne in Austriche, the which now is counted for the best: it groweth plentifully also in some places of England and Irelande.

❊ *The Tyme.*

Saffron flowreth (before his leaues are sprong out of the grounde) in September: and after that it bringeth foorth his grassie leaues, whiche do last vntill Maye: but in sommer a man shall not finde neither leaues nor flowers. The roote onely remayneth aliue, growing vnder grounde, and bringing foorth other small rootes.

❊ *The Names.*

Saffron is called in Greke κρόκος in latine, Crocus, of some (as witnesseth Dioscorides) Castor, Cynomorphos, or Herculis sanguis: in ÿ Arabian spech Zahafara,

from

from thence it was called in French & high Douch Saffran: in base Almaigne Sofferaen: and in English Saffron.

❊ The cause of the Name.

Saffron was named Crocus, after the name of a certayne Damsell called Crocus, (as Ouid writeth) from whence Galen borowed this Historie, who reciteth the same In nono de medicamentis secundùm loca, whereas it is writen in this sorte. A yong wenche called Crocus, went forth into the fieldes with Mercurie to throw the sledge, & whiles she tooke no heede, she was vnawares stroken in the head by Mercurie, and greeuously hurte, of whiche hurte shee died incontinent: than of hir bloud so shedde vpon the grounde, the Saffron sprang vp.

❊ The Temperament.

Saffron is hoate in the second degree, and dry in the first.

❊ The Vertues.

Saffron is good to be put into medicines, which are taken against the diseases of the Breast, the Lunges, the Liuer, and the Bladder: it is good also for the stomake to be taken in meates, for it comforteth the stomacke, and causeth good digestion, and dronken in sodden wine it preserueth from dronkennesse, and prouoketh bodily lust. [A]

Saffron taken in sweete wine causeth one to be long winded, & to fetch his breath easily, and it is good for them that are shorte winded, and Asthmatique. [B]

Saffron mengled with womans milke & layde to the eyes, preserueth them from the flowing downe of humors, and from the Masels, and small Pockes, and stoppeth the flurion or bloudshoting of the same, being layde therevpon. [C]

It is also good to be layde vpon inflammations, cholericke impostems, and wilde fire, and it is very good to be mengled with all medicines for the eares. [D]

Also it is very good to mollifie, and soften all hardnesse, & to rypen all rawe tumors, or swellings. [E]

The roote of Saffron dronken in wine prouoketh vrine, & it is good for the that haue the stone or grauell, and that cannot pisse but droppe after droppe. [F]

Of Standelworte, or Standergrasse. Chap.lvi.

❊ The Kyndes.

There are diuers sortes of Standergrasse called in Greeke Orchis, and in Douch Standelcruyt, wherof there were but two sortes described of the old & Auncient writers: but we haue ioyned to them certayne other, not knowē nor described of any other that haue traueled before vs in the searching out (or knowledge) of herbes: so that now we haue thought good to comprehend them all in fiue kindes. Wherof the first is Cynosorchin, siue canis testiculum: The second is Testiculum Morionis: the third is Tragorchin: the fourth is Orchin Serapian: the fifth is Testiculum odoratum, or Testiculum pumilionem.

❊ The first Kynde.

There be fiue sortes of the firste kinde of Orchios, whiche the Greekes call Cynosorchin.

1. Whereof the first hath foure or fiue great broade leaues, and thicke, almost like to the leaues of Lillies, but somwhat smaller: the stalke is of a foote & half long: at which groweth a great sort of floures tuffetwise, fayre & sweete, & of a carnation or fleshly colour like the colour of mans body, but speckled full of purple spots, the floures alone are but smal & like to an open hood or helmet, out of y inside wherof, there hägeth forth a certayne ragged thing, fashioned almost like y proportiō of a litle fourefooted beast. The rootes (ouer & bysides certaine small

The second Booke of

small hearie things growing aboute them) are round like to a payre of stones, or a couple of Oliue berries, one hanging somewhat shorter than the other, whereof the highmost is the smaller, fuller, and harder: and the nethermost is the greatest, the lightest, and most wrinckled or shriueled.

2 ¶ The second is somewhat like to the aforesayde, but his leaues be narrower and playner, whereof some do compasse or as it were embrace or clippe aboute the stalke: the spikie tuffte is short and thicke with a number of floures, of a bright or white purple colour, & speckled on the inside with a great many purple spottes, and small darke lynes: fashioned also like to an open Hoode or Helmet, out of whiche also there hang certayne thinges as it were small rabbets, or yong myse, or littell men without heades, with their armes & legges spread and cast abroade, in like manner as they were wonte to paynte little chyldren hanging out of Saturnes mouth: at the foote of the stalke are a couple of roūd buttons, as big as Nutmegges: with certayne great hearie strings or thredes annexed or growing by them.

Cynosorchios prima species.　　　Cynosorchios tertia species.
The first kinde of Standergrasse,　　The thirde kinde of Dogges
　　or Dogges Cullion.　　　　　　　　　　Cullion.

3 ¶ The thirde kinde of Cynosorchios putteth vp narrow straked leaues, narrower than the leaues of the seconde Cynosorchios, somewhat lyke the leaues of Rybworte Plantayne: a shorte stemme of nine inches long. The floures growe thicke togither in a shorte spykie busshe or tuffte of a Chestnutte, or darke purple colour without, and whitishe within: his rootes also are like to a payre of stones or Cullions, whereof also one is bigger than the other.

The

the Historie of Plantes. 219

4 The fourth Cynosorchios, in his floures is like to the thirde, but in leaues it is like the second kinde.

5 The fifth Cynosorchios his leaues be somewhat broade like to the second, and his spikelike bushe or toppe is meetely long, but his floures are of a faint, or pale greene colour, and that ragged thing which hangeth downe out of them is as it were foure square: the rootes are like to the rest.

The seconde Kinde.

The second kinde of Orchios, called Testiculus Morionis, is of twoo sortes Male and Female.

Testiculus Morionis mas.
Fooles Cullion the malekinde.

Testiculus Morionis foemina.
Fooles Cullion the female.

1 The male kind hath fiue or sixe long, broade, and smooth leaues, almost like to Lillie leaues, sauing they are full of blacke spottes: the small floures do likewise grow altogither in a spykie busshe or tufte, in proportion like to a Fooles hoode, or Coxcombe, that is to say, wide open or gaping before, and as it were crested aboue, hauing eares standing vp by euery side, and a tayle hanging downe behinde: of a violet colour, and pleasant sauour.

2 The female his leaues are likewise smooth. The floures also are somwhat like the Male, sauing they haue not such smal eares standing vp. Of these some be of a deepe violet colour, some white as Snow, and some Carnation or flesh colour.

3 Of this kinde there is also an other sorte, with narrow straked leaues, like to the leaues of narrow Plantayne (whiche some cal Ribworte): The floures of this kinde are of an orient redde purple colour.

T ij Testi-

220 The second Booke of

Testiculus morionis mas alter.　　Testiculi morionis feminæ species.
An other fooles cullion of ye male kind.　　A kinde of female, fooles Cullion.

There is yet an other muche smaller kinde, with fiue or
4 sixe small leaues : and a fewe small floures, thin set and standing farre a sonder one from another, of a sleight violet colour, turning towarde Azure or skie colour, and sometimes
white or of a decayed and darke purple colour : and of a grieuous vnpleasant sauour.

Tragorchis.

¶ *The thirde Kynde.*

The thirde kinde of Orchios, called in Latine Hirci testiculus, and Tragorchis, his leaues are like to the firste
leaues of the Lillie, sauing they be smaller, but yet they be
larger than any of the leaues of the other Orchios. The
stalke is of a foote long, and oftentimes wrapped aboute a-
lowe with some leaues : vpon the sayde stalke or stemme
groweth a greate many of small floures togither in a spikie
tuffte or bushe, of a very strong fashion or making, much like
to a Lezarde, bycause of the twisted or wrythen tayles, and
speckled heades. Euery one of the sayde floures alone, is at
the firste, as it were a small rounde close huske, of the bignesse or quantitie of a Pease : and whan it openeth, there
groweth out of it a little long and slender tayle, the whiche
is white aboue where as it is fastened to the stalke, and speckled with redde speckes, hauing vpon eache side a small thing
adioyned to it, like to a little legge or foote : the residue of the
sayde

sayde tayle is twisted about, & hangeth downewarde. The floure is of a ranke stinking sauour, like to the smell of a Goate, and prouoketh headache, if it be much and often smelled vnto. The rootes are like a couple of Nutmegges, or a payre of stones.

❧ The fourth Kynde.

THe fourth kinde of Orchios called Serapias, is of three sortes, one hauing a floure somewhat like a Butterflie: an other hath in his floure a certayne figure of a Dorre, or Drone Bee: the thirde hath in it the proportion of a certayne flie.

1 The firste Serapias Orchis hath two or three leaues somewhat long, broade, & smooth, yet not so large as the leaues of white Lillies: the stalke is of a foote long, on which groweth here and there in a spikie bushe or top certayne pleasant white floures, somwhat like Butterflies, with a little tayle hanging behind, in whiche is a certaine sweete iuyce or moisture, like hony in tast: and the sayde floures are ioyned to the stemme as it were with small twisted stalkes: the rootes are like to the other sortes of Orchis.

2 The secōd Serapias Orchis hath narrow leaues, & certaine of them are crokedly turned, and wrythed aboute next the groūd, the other grow about the stalke whiche is of a span or nine inches long, aboute the top whereof grow certayne floures, whose lowest or basest leafe, is like to a Dorre or Droone Bee, but the vpper parte and leaues of the floure are sometimes of a greenish colour, but most commonly of a light violet or skie colour.

3 The third, which is the least of al y̆ Serapias Orchis, hath small floures like to a kinde of Horseflies.

Orchis Serapias primus.   Orchis Serapias alter.

❧ The fifth Kynde.

THe fifth kind of Orchis is the least of all, and commonly it hath not aboue three smal leaues with veynes somewhat like Plantayne, but no bigger than fielde Sorrell, or the small leaues of the cōmon Daysies. The stalke is small and slender of a span long, aboute whiche growe little white floures, of a sweete sauour almost like to Lyllie Conuall, placed in a certayne order and winding aboute the stalke like to a kinde of Hatbande, or the rolling of a Cable Rope: the roote is like to a payre of Stones, or small long kernelles, wherof one is harde and firme, the other is lighte and Fungus, or spungie.

❧ The Place.

The Städelwurts, or Städergrasse, do grow most cōmōly in moyst places, & marrishes, marishes, woodes, and medowes: and some delite to grow in fatte clay groundes (as the kinde whiche is called Tragorchis) whiche lightly groweth in very good ground: some grow in barren ground. But the sweete Orchis, or Ladie traces are moste commoly to be found, in high, vntilled, & dry places, as vpon hilles and Downes.

Testiculus odoratus.

❧ *The Tyme.*

These herbes do all floure in May and Iune, sauing the final sweete Orchis, which floureth last of al in August and September.

❧ *The Names.*

1. The first kinde is called in Greeke ὄρχις, Orchis, & κυνὸς ὄρχις, Cynosorchis: in Latine Testiculus, & Testiculus canis, that is to say, Dogges Cullions, or Dogges coddes: in Shoppes Satyrion: in English some cal it also Orchis, Standelwort, Stadergrasse, Ragworte, Priest pintell, Ballock grasse, Adders grasse, and Bastard Satyrion: in French *Couillons de chien*, and *Satyrion à deux Couillons*: in Italian *Testiculo di cane*: in Spanish *Coyon de perro*: and in Douch Knabenkraut, and Standelkraut: in base Almaigne Standelcruyt.

2. This second kinde is called of some in Latine Testiculus Morionis: in English great Standelworte, and Fooles Balloxe.

3. The third kinde doubtlesse, is also of the kindes of Orchis, and bycause of his ranke sauour is called in Greeke τράγορχις, that is to say, Testiculus hirci, in Latine: in English Hares Balloxe, and Goates Cullions: in French *Couillon de bouq*: in Douch Bocxcullekens: they call it also in Latine Testiculus leporis: and in Shoppes Satyrion, wheras without iudgement it is vsed for the right Satyrion.

4. The fourth kinde is called in Greeke ὄρχις σεραπιας, Orchis Serapias: in Latine Testiculus serapias: in English Serapias stones, Priestes pintle, and Ragwurtz: in base Almaigne Ragwortel: some also call it in Greeke τρίορχις, Triorchis, that is to say, three Ballocks, or three Stones, wherfore Fuchsius sey ned Serapias Orchis to haue three Stones, or three Bulbus rootes, yet Plinie attributeth vnto it but twayne. We may call it in English properly flie Orchis, bycause al the kindes of Serapias Orchis, haue in all their floures the proportion and likenesse of one kinde of flie or other.

5. The fifth kinde is called Testiculus odoratus, Testiculus pumilio: that is to say, sauerie Standelwurte, or sweete Ballocke, and Dwarffe Orchis: in base Almaigne, welrieckende Standelcruyt, and cleyn Standelcruyt.

❧ *The Nature.*

All these kindes of herbes, are of complexion hoate and moyst.

❧ *The Vertues.*

The ful and sappie rootes of Standergrasses (but especially of Hares Balloxe, or Goates Orchis) eaten, or boyled in Goates milke and dronken prouoketh Venus, or bodily luste, doth norishe and strengthen the bodie, and is good for them that are fallen into a consumptiō or feuer Hectique, which haue great neede of nourishment.   A

The withered or shriueled roote is of a cleane contrary nature, for it restrayneth   B neth oʒ repʒeſſeth fleſhly luſt.

And it is written of this roote, that if men do eate of the greateſt and fulleſt rootes, (and eſpecially of the firſte kinde of Orchis) that they ſhall beget Sonnes: and if wemen do eate of the wythered rootes, they ſhall bring foorth Doughters.

The ſame rootes, but eſpecially of Serapias, oʒ flie Orchis boyled in wine and dronken ſtoppeth the laſke oʒ fluxe of the bellie.

The ſame roote, being yet freſh and greene, doth waſte and conſume all tumors, and mundifieth rotten vlcers, and cureth Fiſtulas, being layde thereto: and the ſame made into pouder, and caſt into fretting & deuouring vlcers and ſores: ſtayeth the ſame from any farder feſtering oʒ fretting.

The ſame roote (but eſpecially the roote of Dwarffe Orchis) boyled in wine with a little hony, cureth the rotten vlcers and ſores of the mouth.

Of Double leafe, and Gooſe neſte.   Chap.lvij.

✻ *The Kyndes.*

Bſides the kinds of Standergraſſe, deſcribed in the former Chapter, there are yet two other herbes alſo, which are ſomewhat like vnto the aforeſayde Standergraſſes, eſpecially in their ſtalkes & floures, and therfore are comprehended of ſome wrypters, vnder the title of the Standergraſſes. Whereof the one ſorte is called Double leafe oʒ Baſtarde Orchis : and the other Birdes-neſt: The whiche we haue thought good to place alone in a Chapiter by themſelues, bycauſe their rootes are much vnlike the rootes of Standergraſſe.

Bifolium.   Nid d'oyſeau.
Double leafe oʒ Twayblade.   Gooſeneſt.

The Description.

1. Double leaffe hath a rounde smothe stalke, and it beareth but two leaues onely, like to the leaues of great Plantayne. The stalke from the middle vp to the top, is compassed or beset round about with a great many of little smal floures, of a yellowishe greene colour, almost like to little yong Gooslings, or birdes lately hatched, and not much vnlike the floures of diuers sortes of Standergrasse. The roote is full of threddie strings.

2. Goosenest hath a bare naked stalke without leaues, bearing a floure at the top like to a spiky tufft or eare, of a browne colour like vnto wood. It is almost like the stalke of Orobanche or Broome Rape (wherof we shal write in the vj. parte of our Historie of Plantes) sauing it is tenderer, and not so thicke as the stalke of Orobanche. The roote is nought else but a sorte of threddy strings, as it were interlaced, snarled, or tangled one in an other.

The Place.

The Twayblade or Doubleleaf, delighteth best in moyst & waterie places. Goose neste is to be founde in moyst and sandie fieldes and pastures, and in grauely wooddes.

The Tyme.

These two herbes do spring in May, and June.

The Names.

1. The first of these herbes is called of the writers in our time, in Latine Bifolium: in English Twayblade, Dubble leafe, Bastard Orchis, & Eunuche Städergrasse: in high Douch Zueyblat: in base Almaigne Tweebladt: and it is thought of some to be Plinies Ophris, others thinke it to be a kinde of Perfoliatum, or Thorow waxe: & some thinke it to be Alisma, or water Plantayne: and of some it is taken for Helleborine, that is to say, the wilde white Hellebor, or Nieswort.

2. Herom Bouq calleth the second kinde Margendrehen: & some Herborists amongst (vs bicause that the rootes be so tangled & wrapped like to a nest) haue named it Goosenest: in French Nid d'oyseau: in base Almaigne Voghels nest.

The Nature and Vertues.

The nature & vertues of these herbes are not yet very wel knowen: howbeit the late wryters do take it to be good for woundes, ruptures or burstings: some do also say, that they be in nature like vnto Orchis, or Standergrasse.

Of the right Satyrion, or Dioscorides Satyrion. Cha. lviij.

The Kindes.

Besides the aforesayde Orchis, or Bastard Satyrions, which are also called Satyria of Apuleius & Plinie, Dioscorides also hath wryten of two kindes of Satyrion: one called in Greeke τρίφυλλον, and the other ἐρυθρονιον.

The Description.

1. The first of Dioscorides Satyrions, his leaues be somwhat broade like the leaues of Lillies, sauing they be smaller, and somewhat redde: the stalke is about the heigth of halfe a foote, bare, and naked, and it hath a white floure at the toppe, almost like vnto a Lillie: a Bulbus or rounde roote like to an apple, of a fyrie pellow or reddishe colour without, and white within, like the white of an egge, of a sweete and pleasant taste.

2. The other Satyrion, his seede is smooth and shyning, like vnto Lyne seede sauing it is bigger: and the rinde of the Bulbus roote is reddishe, but the roote it selfe is white, and sweete, and pleasant in tast, as Dioscorides writeth.

The Place.

It groweth in open sunnie places, vpon high mountaynes.

❊ *The Names.*

1. The firſt is called in Greeke σατύριον τρίφυλλον: in Latine Satyrium Triphyllum, or Trifolium: in Engliſh Satyrion, alſo right Satyrion, and three leaued Satyrion.

2. The other Satyrion is called in Greeke σατύριον ἐρυθρόνιον: in Latine Satyriũ erythronium: we may call it alſo Redde Satyrion, and Syrian Satyrion.

❊ *The Nature.*

Satyrion is hoate and moyſt of complexion.

❊ *The Vertues.*

The rootes of Satyrion prouoketh Venus, or bodily luſte, and they nouriſhe and ſtrengthen the body, as the auncient wryters ſay. 1

Of Royall Standergraſſe, or Palma Chriſti. Chap.lix.

❊ *The Kyndes.*

Beſides the two Satyrions, deſcribed of the auncient wryters, there is alſo at this day, an other ſorte found out of learned men.

❊ *The Deſcription.*

Satyrion Baſilicõ mas.   Satyrion Baſilicon fœm.
Satyrion Royall.   Satyrion Royall.

1. The greate Royall Satyrion which is alſo the male kinde, hath long thicke ſmooth leaues, ſmaller than Lillie leaues, without any apparant or manifeſt ſpots, and ſtalkes of a foote long or more, not without ſmal leaues growing by it: þ floures grow in a ſpiky buſhe or tuſſet, at the top of the ſtalke of a light purple colour, and ſweete ſauour: ſpeckled with ſmal ſpeckes of a deeper purple, like to Cuckow Orchis, or fooles ballockes, ſauing they lacke ſuche a come or coppe: vnder euery one of the ſayde floures, there groweth a ſmall ſharpe poynted leafe: the rootes be double, like to a payre of handes, and eache parted into iiii. or fiue ſmall rootes like fingers: whereof one is more withered, light, & ſpõgie: the other is full and ſounde, or firme, with a few ſmall rootes or ſtrings growing out, or faſtned thereto.

Of this ſorte there is alſo a kinde founde whiche is very ſmall, and it hath very narrow leaues, like to Saffron, or Leeke blades, and a ruſſhie ſtalke of nine inches long, with a ſharp pointed tufte, or ſpikie eare, at the top of þ ſtalke like the tuft, or ſpikie buſhe of floure Gentill, or Veluet floure, & of ſuch a bright crimſon, or purple colour. Of a very ſweete & fragrant ſauour like vnto muſke, whan they are freſh & new gathered: the rootes are like to the others, but not ſo large & greene.

2 The other great kinde whiche is the female of this royall Satyrion, hath leaues like to the leaues of ye male kind of royall Satyrion, sauing they be smaller, & dasshed full of blacke spottes: the floures be like vnto gaping hoodes or Cockescomes, & like to the floures of Fooles ballockes or Cuckowes Orchis: of colour sometimes white, & sometimes purple, or redde, or a light skie colour, alwayes speckled and garnished with more small spottes or speckes.

✤ *The Place.*

The royall Satyrions are found in certayne medowes and moyst woodes of England and Germanie. But that kinde whiche beareth the sweete spikie tufte or eare, is found vpon the high hilles and mountaynes of Sauoy.

✤ *The Tyme.*

Royall Satyrion floureth in May and June.

✤ *The Names.*

These plantes are now called σατύριον βασιλικον: in Latine Satyria Basilica siue regia, also Palmas Christi: we may call it in English Satyrion Royall, Palmas Christi, or noble Satyrion: in French *Satyrion royall*: in Douch Crutzblum: in base Almaigne Handekens cruyt.

✤ *The Nature.*

The rootes of Royall Satyrion, are in sent and tast like to Orchies, & therfore they are thought to be of the same complexion, whiche is hoate and moyst.

✤ *The Vertues.*

The roote of Royall Satyrion brused or stamped, & giuen to drinke in wine prouoketh vomit, & purgeth both the stomacke and bealy, by meanes wherof it cureth the old feuer Quartayne, after couenient purgation, if an inch or asmuch as ones thumbe of this roote be pounde, & ministred in wine before the accesse or comming of the fit: As Nicholas Nycols writeth Sermone secundo.

Of Hyssope. Chap.lx.

✤ *The Description.*

1 THe common Hyssop hath fouresquare, greene, harde, & wooddishe stemmes, or braches set with small narrow leaues, somewhat like the leaues of Lauander, but a great deale smaller and greener. The floures growe at the toppes of the branches in small tuftes, or nosegays almost like to a spikie eare, sauing that they growe by one side of the stalke. Whan the floures be past, there commeth seede which is blacke, and lieth in the smal huskes from whence the floures are fallen. The roote is blackishe, and of wooddie substance.

2 There is also an other kinde of Hyssope sowen and planted of the Herboristes: the whiche is somewhat like to the other in stalkes and leaues, sauing that his braches be shorter, & it groweth fast by the ground: the leaues be brouner & of a deeper greene, and thicker, and of a bitterer taste then the leaues

Hyssopus communis.

the Historie of Plantes.

leaues of common Hyſſope. The floures be well like the floures of the other Hyſſope, of a fayre deepe blew, and growing thicke togither at the toppe of the ſtalke, in proportion almoſt like to a ſhorte thicke & well ſet ſpikie tufte or eare. The roote is of a woddie ſubſtance, like to the roote of the other Hyſſope.

3   There is yet a thirde kinde like to the others in leaues and ſtalkes: but the floures of this kinde are milke white.   ✤ *The Place.*

Hyſſope groweth not of his owne kinde in this countrey, neuertheleſſe ye ſhall finde it commonly planted in all gardens.

✤ *The Tyme.*

Hyſſope floureth in June and July.

✤ *The Names.*

This herbe is now called in Shoppes Hyſſopus, and Yſopus, in Italian and Spaniſh *Hyſſopo*: in Engliſh Hyſope, in French *Hyſſope*: in Douch Hyſop, Hyſope, and Yſope: howbeit this herbe is not the right Hyſſope wherof Dioſcorides, Galen and the Auncients haue written, as it is ſufficiently declared by certaine of the beſt learned writers of theſe dayes.

✤ *The Nature.*

Hyſſope is hoate and dry in the thirde degree.

✤ *The Vertues.*

A   The Decoction of Hyſſope, with figges, Rue, and Hony boyled togither in water and dronken, is good for them whiche haue any obſtruction or ſtopping of the breaſt, with ſhortneſſe of breath, and for them that haue an olde difficult, or harde cough, and it is good alſo for the ſame purpoſe to be mengled with hony and often licked in, after the manner of Lohoc or Loch.

B   Hyſſope taken in with Syrupe Acetoſus (that is, of vineger) purgeth by ſtoole tough and clammy flegme, and killeth and driueth foorth wormes. It hath the like vertue eaten with figges.

C   Hyſſope boyled in water with figges, and gargled in the mouth and throte, ripeth and breaketh the tumors, and impoſtems of the mouth and throte.

D   Hyſſope ſodde in vineger, and holden in the mouth, ſwageth tooth ache.

E   The Decoction of Hyſſope, doth ſcatter & conſume the bloud that is congeled, clotted, & gathered togither vnder the ſkinne, and all blacke and blew markes that come of ſtripes or beating.

F   The ſame decoction cureth ye itche, ſcurffe, & foule mangines, if it be waſhed therewithal.

Satureia vulgaris.

Of common garden Sauorie. Chap.lxj.

✤ *The Deſcription.*

THe Sauorie is a tender ſommer herbe, of a foote long: the ſtalkes be ſlender, and blackiſhe, very full of branches, & ſet with ſmal narrow leaues, ſomwhat like the leaues of cõmon Hyſſope, but a great deale ſmaller. The floures grow betwixt the leaues, of carnation

228　The second Booke of

nation in white colour, of a pleasant sauour. The seede is browne or blackishe. The roote is tender and threedie.

❊ *The Place.*

This herbe is sowen in all gardens, and is muche vsed about meates.

❊ *The Tyme.*

This herbe floureth in Iune.

❊ *The Names.*

This herbe is now called in Latine Cunila, and Satureia: in Shoppes Saturegia: in Italian *Coniella Sauoregia*: in Spanish *Segurella*: in English somer Sauorie, and common garden Sauorie: in French *Sarriette*, & *Sauorie*: in Douch garten Hyssop, zwibel Hyssop, kunel, Saturey, & Sadaney: in base Almaigne Cuele, Saturepe, Lochtekol.

❊ *The Nature.*

Sommer or garden Sauorie, is hoate and dry in the thirde degree.

❊ *The Vertues.*

This Sauorie (as Dioscorides saith) is in operation like vnto Time, and is very good, and necessarie to be vsed in meates.

Of Tyme.　Chap.lxij.

❊ *The Kindes.*

There be two kindes of Tyme, the one called Thymum Creticum, that is to say, Tyme of Candie, the other is our common vsuall Tyme.

Thymum Creticum.　　　Thymum durius.
Tyme of Candie.　　　**Our common Tyme.**

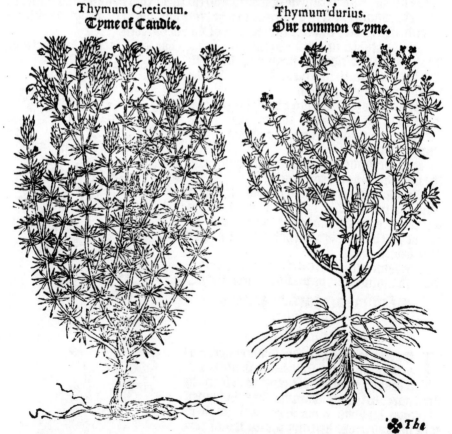

❧ The Description.

1 Tyme of Candie hath many smal wooddy stalkes, set round about with smal narrow leaues, at the top of the stalkes groweth certayne knoppie tuftes, like vnto small short eares, thrust togither, not much vnlike the flourie spike or knops of Stæcados, but much smaller, and bringing forth purple floures. The roote is brittle, and of wooddie substance.

2 The common tyme also hath many small, weake, and wooddie braches. The leaues be small, of sharpe and byting taste. The floures growe at the top of the stalkes of incarnate colour. The roote is small and wooddishe.

❧ The Place.

1 The first kinde of Tyme groweth in hoate countries, in dry & barren soyle, and stony mountaynes, and such like vntilled places. And it is found very plentifull in the countrie of Greece, but principally in Candie.

2 The second groweth also in hoate countries, vpon the stonie mountaynes, in leane & barren ground and such like places, as in many places of Spayne and Italie, and throughout all Lanquedoc, whereas it groweth very plentifully.

❧ The Tyme.

It floureth in May and June.

❧ The Names.

Tyme is called in Greeke θύμος, that is to say, in Latine Thymus: in Italian *Thymo*. in Spanish *Tomillo*.

1 The first kinde is called in Greeke κεφάλωτος θύμος: in Latine Thymū capitatum: of the later writers Thymum Creticum, that is to say, Tyme of Candie.

2 The second is called also in Greeke θύμος: of Dioscorides Thymum durius, the which is seldome foūd in season without his Epithymus: it is called in English Tyme, & the common garden Tyme: in French *Thym*: in Douch Thymus.

❧ The Nature.

Tyme is hoate and dry in the thirde degree.

❧ The Vertues.

A Tyme boyled in water & hony and dronken, is good against a hard & painefull cough and shortnesse of breath, it prouoketh vrine, & expulseth the Secondine and the dead fruite from the Matrix, it bringeth to women their naturall termes, and dissolueth clotted or congeled bloud in the bodie.

B The same made into pouder, and taken in waight of three drammes with honied vineger, whiche they call Oximel, and a little Salte, purgeth by stole tough and clammy flegme, and sharpe and cholerique humors, and all corruption of bloud.

C The same taken in like sorte is good against the Sciatica, the payne in the side, & the breast: also it is good against blastings and windinesse of the side and bellie, and of the stones or genitors, and it is profitable for those that are fearefull, melancholique, and troubled in spryte, or minde.

D Tyme eaten in the morning fasting, and in the euening before Supper is good for bleared and watering eyes, & the payne in the same. And it is also good for the same purpose to be often vsed in meates.

E It is also singuler against the Goute, taken in wine out of the time of the greefe, & with a dramme of Oximel, whan one is tormented with the same.

F Tyme mingled with honie after the māner of a Lohoc, to be often licked in, clenseth the breast, & ripeth flegme, causing it easily to be spet or cast out.

G Tyme stamped with vineger, consumeth and wasteth colde swellings, and taketh away Wartes being layde therevpon.

H The same pound with Barlie meale & wine, appeaseth ye payne of the hanch or hippe which some call the Sciatica Goute, being applied therevnto.

Time is also good to be giuen to them that haue the falling sicknesse, to smell vpon.

Of Thymbra, or winter Sauorie. Chap.lxiij.

❧ *The Description.* Thymbra.

Winter Sauorie hath many slender wooddie stalkes, set full of smal narrow leaues. The floures be small, incarnate or white, growing in littell huskes alongst the stalkes betwixte ye leaues, & floureth by little & littell, from the lowest parte of the stalke euen vp to the toppe of the branches, leauing after the floures be fallen away, as it were a greene spikie eare or tufte, côteyning the seede, whiche is very small. The roote is of wooddy substance.

❧ *The Place.*

This herbe groweth in certaine places of Fraunce, especially in Languedoc, & other hoate countreys, in vntilled places. It is found in this countrey in the gardens of suche as haue pleasure in herbes.

❧ *The Tyme.*

It floureth in this countrey in July & August, and somtimes later.

❧ *The Names.*

This herbe is called in Greeke θύμβρα: in Latine Thymbra, & Cunila: in English Tymbra, & Winter Sauorie, also Pepper Hyssope: in French *Thymbre*, and *Sarriette d'Angliterre*: of some Douch Herborists Tenderick. This is not Satureia, for Satureia is an herbe differing from Thymbra, as Columella and Plinie haue very well taught vs.

❧ *The Nature.*

Tymbra is hoate and dry like Tyme.

❧ *The Vertues.*

Winter Sauorie is good and profitable to be vsed in meates, like Tyme, Sauorie, and common Hyssope.

It hath power and vertue like Tyme, being taken in the like sorte, as Dioscorides sayth.

Of Wilde Tyme. Chap.lxiiij.

❧ *The Description.*

The running Tyme, hath diuers smal wooddie braches, somtimes trayling alongst the ground, & somtimes growing vpright of a foote & half long, set full of smal leaues, much like to the leaues of common garden Time, but much larger. The floures grow about the toppe of the stalkes like to crownes or garlands, after ye maner of Horehound floures, or knops, most commonly of a purple red colour, & somtimes (but very seldom) as white as snow.

The roote is harde, and of wooddie substance, with many threeddie strings.

❧ *The Place.*

This herbe groweth plentifully in all this countrie in places that are rude, rough, dry, vntilled, and stonie, by the high way sides, and in the borders of fieldes.

❧ *The Tyme.*

Running Tyme floureth from after May vntill the end of Sommer.

❧ *The Names.*

This herbe is now called in Shoppes Serpillum, and in some places Pulegium montanum: in Italian *Serpillo*: in Spanish *Sepollo Serpam*: in English wilde Tyme, Puliall mountayne, Pellamountayne, & running Time: in Frēch *Serpolet*: in high douch Quendel, and of some also Kumel, & Kieulin: in base Almaigne Quendel, & in Brabant Onser vrouwen bedstroo, & in some places wilden Thymus. Many iudge it to be that whiche the Greekes do call ἕρπυλλος κηπωτός, the Latines Serpyllum hortense, howbeit it should seeme rather to be a kind of Thymum durius, or that which is called of Dioscorides in Greeke σαξιφραγγον, in Latine Saxifranga, than Serpyllum.

Serpillum vulgare.

❧ *The Nature.*

Pellamountayne is hoate and dry in the thirde degree.

❧ *The Vertues.*

Wilde Tyme boyled in water or wine and dronken, prouoketh and bringeth to women the fluxe Menstruall, driueth out the stone and grauell, and prouoketh vomit. A

The same taken in the like manner, stoppeth the laske, and cureth gripings, or knawings, and is excellent against Crampes, and the drawing togither or shrinking of Synewes. B

This herbe taken in meates and drinkes, (or brothes,) is a soueraigne medicine against all poyson, and against the bytings and stingings of venemous beastes and Serpentes. C

The iuyce of Pellamountayne or Running Tyme, dronken to the quantitie of halfe an vnce with Vineger, is good agaynst the spetting and vomiting of bloud. D

Running Tyme mengled with Vineger and oyle of Roses, and applied to the forehead and temples, swageth head ache, & is very good against rauing, and frensie. E

The perfume of the same, driueth away all venemous beasts. F

Of Penny Royall or Podding grasse. Chap.lxv.

❧ *The Description.*

PEnny Royall hath smal brittle stalkes of a foote long & somtimes more, not vpright, but creeping alongst the ground, & taking new rootes, here & there in sundrie places. The leaues be somwhat round, almost like the leaues of Marierom, but they be greener, browner, and of a stronger sauour.

The floures growe here and there by certayne spaces aboute the stemmes like whorles or garlandes, and as the floures of Horehound, of a blewishe colour and sometimes very white. The roote is threeddie.

❧ *The Place.*

Pulegium.

Penny Royall loueth moyst, & vntilled places, whiche are dry in the Somer, and full of water in winter.

❧ *The Tyme.*

It floureth in Iune, and in August.

❧ *The Names.*

This herbe is called in Greeke γλήχων: in Latine & in Shoppes Pulegium: in Italian *Pulegio*: in Spanish *Poleios, Poleio*: in English Penny Royall, Pulioll Royall, Pudding grasse, and Organie: in French *Pouliot*: in high Douch Poley: in base Almaigne Poley, and Paley.

❧ *The Nature.*

Penny royall is hoate & dry in ye third degree, & of subtile partes, and cutting.

❧ *The Vertues*

Penny royall boyled in wine & dronken, prouoketh the monethly termes, bringeth foorth the Secondine, the dead frute, and the vnnaturall birth, it prouoketh vrine & breaketh the stone, especially the stone of kidneys.

Penny royal taken with hony clenseth the Lunges, & voydeth them & the breast from all grosse and thicke humors.

The same taken with Hony and Aloes, purgeth by stole the Melancholique C humor, & preuayleth much against crapes, & the drawing togither of sinewes.

The same taken with water and vineger, asswageth the inordinate desire D to vomit, and the gnawing paynes of the stomacke.

Penny royall taken in wine, helpeth the bitings of venemous beastes, and E with vineger it helpeth them that haue the falling sickenesse.

If at any time men be constrayned to drinke corrupt, naughtie, stinking, or F salte water, throw Penny royall into it, or strow the pouder thereof into it, and it shall not hurte any bodie.

A garlande made of Penny Royall, and worne about the head, is of great G force against the swymming paynes, and giddy turnings of the head.

The same pounde with Vineger, and giuen to smell vpon, to people that H are much giuen to sounding quickeneth their Senses, and causeth them to returne to them selues agayne, and is good for them that haue colde and moyst braynes.

The pouder or asshen of this herbe, doth fasten and sttengthen the gummes I that are rubbed therewith.

Penny royall pounde asswageth the payne of the Goute, and Sciatica, be- K ing rubbed vpon the greeued parte vntill it waxe redde.

The same mengled with vineger & hony cureth the crampes, and is profita- L ble for the diseases of the Splene or Melte, being layde therevnto.

The

the Historie of Plantes. 233

The Decoction thereof is very good against ventositie, windinesse, and blastings, also against the hardnesse and stopping of the Mother, whan one sitteth ouer the vapour or breath thereof in a stewe or bathe, whereas the sayde Decoction is. The same is also good against the itche and manginesse, to washe the scabbed parties therein.

The perfume of the floures of Penny royall (being yet fresshe and greene) driueth away fleas.

Xenocrates saith, that a braunch of Penny royall wrapped in a little wooll and giuen to smell vnto, or layde amongst the clothes of the bedde, cureth the feuer Tertian.

Of Poley. Chap.lxvi.

❧ *The Kyndes.*

Poley (as Dioscorides saith) is of two sortes, whereof one may be named great Poley, or as Dioscorides termeth it, Poley of the Mountaine: & the other may be called small Poley.

❧ *The Description.*

1 Poley of the Mountayne is a little, small, tender, base, and sweete smelling herbe, hauing small stemmes, and slender branches, of a spanne or halfe foote long. The leaues bee small, narrow, and grayish, whereof they that grow lowmoste are somewhat larger, and a little snipt or iagged aboute the edges: and they that growe aboue, are narrower and notso much iagged or snipte. The floures be white and do grow at the toppe of the branches. The roote is threedie.

2 The lesse Poley is not muche vnlike the other, sauing that his leaues are tenderer, smaller, narrower, and whiter than the other: it hath also a great many moe small, slender, and weake branches. But it hath not so great vertue, nor so strong a sauour as Poley of the Mountayne.

Polium.

❧ *The Place.*

It groweth not of him selfe in this countrie, and is not lightly found, sauing in the gardens of some Herboristes, who do plante and cherishe it with great diligence.

❧ *The Tyme.*

It floureth at the end of May and June, whereas it groweth of his owne kinde, and in this countrey in July.

❧ *The Names.*

It is called in Greeke πόλιον, πόλιον ὀρεινόν, καὶ τεύθριον: in Latine Polium, Polium montanum, and Theuthrium: in Italian *Polio*: in Spanish *Hierua vssa*: in English Poley, & Poley mountayne. It hath neither French nor Douch name that we know: for it is yet vnknowen of the Apothecaries them selues in the Shoppes of this countrey.

❧ *The Nature.*

Poley is hoate in the second degree, and dry in the thirde.

U iij ❧ The

The second Booke of

❧ *The Vertues.*

Poley boyled in water or wine prouoketh the floures, and vrine, and is very good against the Dropsies and Iaunders.

It profiteth much against the bytings of venemouse beastes, and against poyson taken in maner aforesayde, and it driueth away all venemous beasts from the place whereas it is strowen or burnte.

The same dronken with vineger, is good for the diseases of the Mylte and Splene.

Also it healeth, and closeth vp woundes, being yet fresh and greene, pound and applied, or layde therevpon.

Of Marierom. Chap.lxvij.

❧ *The Description.*
Maiorana vulgaris.

Marierom is a delicate and tender hearbe, of sweete sauour, very wel knowen in this countrie, hauing small weake and brittle stalkes, set with softe and tender leaues, somewhat round and of grayishe colour: it bareth about the toppe, and vpper parte of the braunches a great many of small buttons or knoppes, like to a little spike eare made of many scales, out of which groweth very smal white floures, yeelding a very small reddish seede. The roote is wooddish and very thready.

❧ *The Place.*

This Marierom is planted in gardens, and in pottes with earth, and it loueth fatte and well mainteyned ground.

❧ *The Tyme.*

It floureth in Iuly and August.

❧ *The Names.*

This noble and odoriferous plant, is now called in Shoppes Maiorana: in Italie *Persa:* in English Marierom, sweete Mariorom, and Marierom Gentle: in French *Mariolaine:* in high Douch Maioran, or Meyran: in base Almayne Marioleine, and Mageleyne. It is taken for the right σάμψυχον καὶ ἀμάρακον of the Greekes, and Amaracus & Maiorana in Latine: howbeit it trayleth not alongst the ground at all, as Dioscorides writeth that Sampsicon shoulde do: wherefore it shoulde rather be somewhat like that herbe whiche the Grecians call μάρον, and the Latines Marum, for this is an herbe of a most sweete and pleasant smell, bearing his floure almost like to Origanum Heracleoticum, whiche thing Dioscorides attributeth vnto Marum.

❧ *The Nature.*

Marierom is hoate and dry in the third degree.

❧ *The Vertues.*

Marierom boyled in white wine and a quantitie thereof dronken, is very good for such as begin to fall into the Hydropsie, & for such as cannot pisse but
Drop,

the Historie of Plantes. 235

drop, after droppe, and that with great difficultie, & it is good for them that are tormented with the gryping paynes, and wringings of the bellie.

The same taken in the like manner, prouoketh the floures: and so doth it being ministred beneath in manner of a Pessarie, or mother Suppositorie. B

Dried Marierom mengled with Hony, dissolueth Congeled bloud, and driueth away the blacke and blew markes after strypes and bruses, being applied thereto. C

The same with Salte and Vineger, is very good to be applied vnto the prickings and stingings of Scorpions. D

A playster made of Marierom with oyle & waxe, resolueth colde swellings or tumors, and is much profitable to be layde vpon places that be out of ioynt or wrenched. E

Marierom brused or rubbed betwixt the handes, & put into the Nosethrils, or the iuyce thereof snift vp into the nose, draweth downe humors from the head, mundifieth the brayne, causeth to sneese, and is very good for them that haue lost their smelling. F

And if Marierom be Marum of the Gretians, then is it also a very good herbe (as Galen saith) & fit to be put into all medicines, and compositions made against poyson: it is also good to be mengled with all odoriferous and sweete oyntments, as the oyntment called Vnguentum Amaricinum, and such like. G

Marum is also good to be layde vpon fretting & consuming vlcers, & is very profitable against all colde griefes and maladies, as Dioscorides writeth. H

Of Clinopodium, or Mastic. Chap.lxviij.

Clinopodium.

❧ *The Description.*

THis herbe hath smal, naked, roud, and woodish stemmes: the leaues be small and tender almoste like Marierom. The floures whiche are white and very small do grow like a Crowne or garland rounde about the stemme, in small rough or woolly huskes. The roote is of wooddishe substance. The whole herbe is of a very pleasant sweete sauour, almost like Marierom.

❧ *The Place.*

This herbe groweth not of him selfe in this countrie, but the Herboristes do plant it in their gardens.

❧ *The Tyme.*

It floureth in this countrie in August or there aboutes.

❧ *The Names.*

This herbe is taken of some Herboristes for Marum, (that is the English and French Mastic,) but seing that it floureth not like Organe or wild Marierom, it seemeth vnto me to be nothing like Marum, but rather to be like vnto the herbe which they call in Greeke κλινοπόδιον: in Latine also Clynopodium, for the whiche we haue described it: Turner calleth Clinopodium.

podium, for the whiche we haue described it: Turner calleth Clinopodium, Horse tyme, and so doth Cooper English Clinopodium, he calleth it also Puliall mountayne.

✻ *The Nature.*

It is in complexion very much like Marierom.

✻ *The Vertues.*

They vse to drinke the herbe Clinopodium in wine, and the Decoction or iuyce therof made in wine, against Crampes, burstings, difficultie of vrine, and the bitings of Serpents.

It prouoketh the floures, expelleth the dead fruite, and Secondine, if it be vsed as is aforesayde.

The same boyled vntill the thirde parte be consumed, stoppeth the bellie, but it muste be dronken with water in a feuer, and with wine without a feuer.

Menne wryte also of Clinopodium, that if it be taken with wine by the space of certayne dayes, it will cause the wartes that are vpon the body to fall away.

Of Origan, or wilde Margerom. Chap. lxix.

✻ *The Kindes.*

Origan is of three sortes, that is to say, garden Origan, wilde Origan, and that kinde whiche they call Origanum Onitis.

Origanum Heracleoticum.
Spanish Origan.
Bastard Margerom.

Origanum syluestre.
Wilde Origan.
Groue Margerom.

the Historie of Plantes.

❧ The Description.

Marum quibuſdam.
Engliſh Margerom.

1 THe firſt kinde hath harde, rounde, and ſometimes reddiſh ſtemes, wherevpon are round whitiſh leaues, ſmaller than the leaues of wild Origan, and nothing hearie, but otherwiſe ſomewhat like in faſſhion. The floures grow not in knoppie Crownets, but like vntoſmal ſpikie eares, growing vpon little fine ſtemmes, at the toppe of the ſtalke. And afterwarde it bringeth forth ſmall ſeede.

2 The ſecond kinde hath whiter leaues, and is not of ſo great vertue, but otherwiſe not much vnlike the firſt. It is not knowen in this countrie.

3 The thirde wilde kind, hath many round, browne, long, & hearie ſtalkes, the leaues be ſomewhat round, and ſofte heared, greater than the leaues of Penny Royall. The floures are reddiſh, and growing a great many togither in tuftes like Noſegayes. The ſeede is ſmall & reddiſhe. The roote is long harde & wooddiſh.

There is alſo a ſorte of this thirde kinde founde, bearing floures as white as ſnow, of ſtronger ſmell & ſauour, than the abouesayde wilde kinde, but in all things elſe lyke vnto it.

Yet there is found a thirde kinde, the which is cōmonly called Engliſh Marierom. This is a baſe or low herbe, not much vnlike to wilde Origan, with leaues ſomewhat rounde, and of a darke greene colour, ſmaller than the leaues of wild Origan, not hearie but plaine and ſmoth. The floures are purple in redde, and grow in crownelike tuftes. The roote is of wooddy ſubſtance.

❧ The Place.

Theſe herbes do grow in Candie, and other hoate countries, ſometimes alſo in Spayne: here they plant them in gardens.

❧ The Tyme.

The firſt kinde floureth very late in this countrey, and yet it floureth not at all ſauing whan the Sommer is very hoate. The wilde Origan & his kindes do floure at Mydſomer.

❧ The Names.

Theſe herbes be called in Greeke ὀριγάνοι, in Latine Origana.

1 The firſt is called ὀρίγανο ἡρακλεωτικὴ, Origanum heracleoticum, and of ſome Cunila: here in Shoppes it is called Origanum Hiſpanicum, bycauſe they bring it dry from Spayne to ſell at Antwerpe, and this is the cauſe that the Brabanders call it Origano as the Spaniards do call it Oreganos

2 The ſecond is called ὀρίγανο ὀνίτις, Origanum onitis, which is yet vnknowē in this countrie.

3 The thirde is called in Greeke ἀγριορίγανο: in Latine Origanum ſylueſtre, that is to ſay, wilde Origan: in Spaniſh Oregano campeſtre.

The firſt is commōly taken in the Shoppes of this countrie for Origanum, & is called in Engliſh wilde Origan, and Baſtarde Marierom: in French Origan ſauuage, and Mariolaine baſtarde: in high Douch Doſten wolgemut: in baſe Almaigne groue Marioleyne.

The ſecond may be called wilde Origan with the white floures.

The thirde is called Engliſh Marierom: in French Mariolaine d'Angleterre:

and

238 The second Booke of

and in base Almaigne Engelsche Marioleyne: and it is taken in some shoppes, and of some Herborists, for Marum. ✱ *The Nature.*

All the kindes of Origan are hoate and dry in the third degree, the one being stronger than the other. ✱ *The Vertues.*

Origan boyled in wine and dronken, is good against the bytinges of venemous beasts, or the stinginges of Scorpions and fielde spyders. And boyled in wine as is aforesayde, it is good for the that haue taken excessiuely of the iuyce of Homblocke, or Poppie, whiche men call Opium. A

The same dronken with water, is of great vertue against the paynes of the stomacke, and the stitches or griping tormentes aboute the harte, and causeth light digestion: and taken with Hydromel (or honied water) it loseth the bellie gentilly, and purgeth by stole aduste and Melancholique humors, and prouoketh the flure menstruall. B

The same eaten with figges, profiteth them much that haue the Hydropsie, and against the shrinking and drawing togither of members. C

It is profitably giuen to be licked vpon with Hony, against the Cough, the Pleurisie, and the stopping of the Lunges. D

The iuyce of Origan is of great force against the swelling of the Almondes or kernels of the throte, and cureth the vlcers of the mouth. E

The same iuyce drawen or snift vp into the Nose, purgeth the brayne, and taketh away from the eyes, the yellow colour remayning, after that one hath had the Iaunders. F

It appeaseth the paynes of the eares, being dropped in with Milke. G

It is good against all kinde of scuruinesse, roughnesse of the skinne, manginesse, and against the Iaunders, if one bathe in the Decoction thereof made in water, or if the body onely be washed with the same. H

The same herbe being mengled with vineger and Oyle, is good to be layde on with wool vpon squats or bruses, and blacke and blewe markes, ₳ to partes displaced or out of ioynt. I

The wilde Origan with the white floure, is of singuler vertue against all the abouesayde maladies or diseases, as Galen saith.

Of Tragorigan, or Goates Origan. Chap.lxx.

✱ *The Kyndes.*

There be two sortes of Tragoriganum, as Dioscorides hath left in wryting.

✱ *The Description.*

1 The first kinde is very much like Organū, sauing that his stalkes & leaues be tenderer.

2 The seconde kinde hath many browne woddish stēmes, the leaues be meetely large & of a swart greene colour, larger than the leaues of Pellamountayne or running time, and somwhat rough & ouer couered as

Tragoriganum alterum.

the Hiſtorie of Plantes.

as it were, with a certayne fine and ſofte hearie. The ſmall floures are purple, and grow like Crownes or whorles, at the toppe of the ſtemmes.

❧ The Place.

Theſe herbes are not common in this countrie, but are onely founde in the gardens of certayne diligent Herboriſts.

❧ The Tyme.

Tragoriganum floureth here in Auguſt.

❧ The Names.

1 This kinde of Origan, is called in Greeke τραγορίγανο: in Latine Tragoriganum, we may alſo call it in Engliſh Tragoriganum, or Goates Origan.

2 The ſecond kinde is called alſo Praſium: ⁊ of ſome of this countrie, it hath ben deemed or taken for Tyme.

❧ The Nature.

The Tragoriganum is hoate and dry like Origan: alſo it hath a certayne aſtringent vertue.

❧ The Vertues.

The decoction of Tragoriganum dronken maketh a good looſe bellie, and auoydeth the Choletique humors, and taken with vineger, it is good for the Melte or Splene. **A**

Tragoriganum is very good againſt the wambling of the ſtomacke, and the ſowre belkes whiche come from the ſame, and againſt the paine or deſire to vomit at the Sea. **B**

Tragoriganum mengled with Hony and oftentimes licked vpon, helpeth againſt the Cough and ſhortneſſe of breath. **C**

It prouoketh vrine ⁊ bringeth to wemen their monethly termes: the ſame layde on with the meale of Polenta, hath power to diſſolue colde tumors or ſwellings. **D**

Of Baſill.   Chap.lxxi.

❧ The Kyndes.

There be two ſortes of Baſill, the one of the Garden, ý other is wilde. Wherof the garden Baſill alſo is of two ſortes, one great, the other ſmall.

❧ The Deſcription.

1 The Baſill Royall or great Baſill hath round ſtalkes full of braunches, with leaues of a faynt or yellowiſhe greene colour, almoſt like to the leaues of Mercury. The floures are rounde about the ſtalkes, ſometimes purple, and ſometimes as white as ſnow. Whan they are gone there is founde a ſmall blacke ſeede. The roote is long with many ſtringes or threedes.

2 The ſecond kinde is not much vnlike to the aboueſayd. The ſtalkes be rolid with many littell collaterall or ſide braunches. The leaues be ſnipte or iagged round aboute, a great deale ſmaller than the leaues of Baſill Royall, or great Baſill. The floures are very much like to the others.

Theſe two kindes are of a maruelous ſweete ſauour, in ſtrength paſſing the ſmell of Marierom, ſo as in deede their ſent is ſo ſtrong, that they cauſe Headache, whan they are to much or to long ſmelde vpon.

The wilde Baſill hath ſquare hearie ſtemmes, beſet with ſmall leaues, much lyke to the leaues of Buſhe (or ſmall) Baſill, but a great deale ſmaller ⁊ hearie. The floures are purple or of a ſkie colour very like the floures of garden Baſill. The roote is full of hearie threedes, and creepeth alongſt the grounde, and ſpringeth vp yearely a new, the whiche the other two garden Baſils doth not, but muſt be newe ſowen yearely.

❧ The

The second Booke of

Ocimum maius.
Great Baſill gentle.

Ocimum minus.
Buſhe Baſill, oꝛ ſmall Baſill gentle.

❧ *The Place.*

Baſill gentill is ſowen in gardens.
The wilde Baſill groweth in ſandie groundes alongſt by the water ſide.

❧ *The Tyme.*

Theſe herbes do floure in Iune and Iuly.

❧ *The Names.*

1.2 The garden Baſill is called of the Auncients in Greeke ὤκιμον, ἢ ὄκιμον: in Latine *Ocimum*, and of ſome *Baſilicum*, that is to ſay, Royall, it is now called Ocimum gariophyllatum: in Engliſh Baſill Royall, Baſill gentle, oꝛ garden Baſill, and the ſmaller kinde is called buſhe Baſill: in French *Baſilicq*, oꝛ *Baſilic*: in high Douch Baſilgen, Baſilgram: in baſe Almaigne the great is called Groue Baſilicom, and the ſmall Edel Baſilicom.

3 The wilde Baſill is called in Greeke ἄκινος καὶ ἄκονος: in Latine *Acinus*: in French *Baſilic ſauuage*: in high Douch wilde Baſilgen: in baſe Almaigne wilde Baſilicom.

❧ *The Nature.*

1.2 Garden Baſill is of complexion hoate and moyſt.
3 The wilde Baſill is hoate and dꝛy in the ſecond degree.

❧ *The Vertues.*

The auncient Phiſitions are of contrary iudgements about the vertues of A Baſill. Galen ſaith that foꝛ his ſuperfluous moyſture, it is not good to be taken into the body. Dioſcorides ſaith that the ſame eaten is hurtfull to the ſight, and ingendꝛeth windineſſe and doth not lightly digeſt. Plinie wꝛiteth that the ſame eaten

the Hiſtorie of Plantes. 241

eaten is very good and conuenient for the ſtomacke, & that if it be dronken with Vineger it dryueth away ventoſities or windineſſe, ſtayeth y appetite or deſire to vomit, prouoketh vrine, beſides this he ſaith, it is good for the hydropſie, and for them that haue the Iaunders.

The later writers ſay that it doth fortefie & ſtrengthen the harte, & the brayne, and that it reioyceth and recreateth the ſpirites, & is good agaynſt Melancholie and ſadneſſe, & that if it be taken in wine, it cureth an olde cough.

B The ſame after the minde of Galen is good to be layde too outwardly, for it doth digeſt and ripe. Wherefore (as Dioſcorides ſaith) the ſame layde too with Barley meale, oyle of Roſes, and Vineger, is good for hoate ſwellings.

C Baſill pounde or ſtamped with wine, appeaſeth the payne of the eyes: And the iuyce of the ſame doth clenſe & mundifie the ſame, and putteth away all obſcuritie & dimneſſe, & drieth vp the Catarrhes or flowing humors that fall into the eyes, being diſtilled or often dropped into the ſame.

D The herbe bruſed with vineger, & holden to the noſe of ſuche as are faynt & falle into a ſound, bringeth them againe to themſelues. And the ſeede therof giue to be ſmelled vpon cauſeth the ſternutation or nieſing.

E The wilde Baſil (howſoeuer it be take) ſtoppeth the laſke, & the inordinate courſe of the Monethes.

Of Vaccaria / or Cow Baſill. Chap.lxxij.

✿ *The Deſcription.*

Acinos.

Forte Ocimoides.

THat herbe which men do now cal Vaccaria, hath rounde ſtalks full of ioyntes & branches: the branches haue vpon euery knot or ioynt two leaues ſomwhat broad, not much vnlike to y leaues of Baſill. At the top of y braches are ſmal red floures, after the whiche there cometh round huſkes, almoſt like y huſkes of Henebane, in whiche is conteined the ſeede, which is blacke like to the ſeede of Nigella.

✿ The

❈ *The Place.*

This herbe is found in certaine fruitefull fieldes or pastures, alongst by the riuer of Mense. In this countrey the Herboristes do plante it in their gardens.

❈ *The Tyme.*

It floureth from after Midsomer vntill September.

❈ *The Names.*

The Herboristes do call this herbe Vaccaria: and it seemeth to be the herbe whiche is called in Greeke ὀκιμοειδἐς: in Latine Ocimastrum, and after the opinion of some (as witnesseth Galen) Philitærium, whiche is a kinde of Echium in Nicander, it is called of Valerius Cordus Tamecnemum, we may call it fielde Basill, or Cowe Basill.

❈ *The Nature.*

The seede of Ocimastrum is hoate and dry.

❈ *The Vertues.*

The seede of Ocimastrum is good for such as are bitten of Serpentes, Vipers, and such other venemous beasts, if it be dronken with wine.

Of Oke of Hierusalem. Chap. lxxiij.

❈ *The Description.* Botrys.

This herbe at the first hath small leaues, deepely cut in, or iagged aboute, and somewhat rough or hearie, & vnderneath the leafe is of a red purple colour: afterward it putteth forth a straight or vpright stem of a foote long or more, with diuers braunches on the sides, so that it sheweth like a little tree: The leaues that groweth thereon, are long, and deepely cut, hearie, and wrinckled, fat or thicke in handling, in proportion like to the first leaues, sauing they be longer, and nothing at all redde or purple vnderneath. The seede groweth clustering about the braunches, like to the yong clusters or blowings of the grape or vine. The roote is tender, and hath hearie or threddie strings. The whole herbe is of an amiable and pleasant smell, and of a faynte yellow colour, and whan the seede is ripe the plante dryeth, and wareth all yellow, and of a more stronger sauour.

❈ *The Place.*

This herbe groweth in many places of Fraunce, by the waters or ryuer sides: but it groweth not of him selfe in this countrey: but whereas it hath ben sowen once, it springeth vp lightly euery yeare after.

❈ *The Tyme.*

It beareth his clustering seede in August, but it is beste gathering of it in September.

❈ *The Names.*

This herbe is called in Greeke Βότρυς, and in Latine Botrys, of some in Cappadocia (as Dioscorides writeth) Ambrosia: vnknowen in Shoppes: it is called in English Oke of Hierusalem, and of some Oke of Paradise: in French *Pyment*, and *Pyment Royall* in high Douch Traubekraut, and after the same in base Almaigne it is called Druyuencruyt, that is to say, Vine Blossom herbe.

❊ *The Nature.*

The Oke of Paradise is hoate and dry in the seconde degree, and of subtill partes.

❊ *The Vertues.*

Oke of Paradise boyled in wine, is good to be drōken of them whose breast is stopped, and are troubled with the shortnesse of winde or breath, and cannot fetche their breath easily, for it cutteth and wasteth grosse humors and tough flegme that is gathered togither about the Lunges, and in the breast.

It proudketh vrine and bringeth downe the termes, if it be taken as is aboue sayde.

The same dryed is also right good to be vsed in meates, as Hysope, Tyme, and other like hearbes, yeelding vnto meates a very good taste and sauour.

Oke of Hierusalem dryed, and layde in presses and Warderobes, giueth a pleasant smell vnto clothes, and preserueth them from mothes and vermin.

Of the kyndes of Mynte. Chap. lxriij.

❊ *The Kindes.*

THe Mynte is diuers, aswell in proportion, as in his manner of growing: whereof some be garden Myntes, and some be wilde Myntes. The garden Myntes are of foure sortes, that is to say, Curlde Mynte, Crispe Mynte, Spere Mynte, and Harte Mynte.

The wilde Mynte is of two sortes, that is, the Horse Mynte, and the Water Mynte.

❊ *The Description.*

1 The firste kinde of garden Myntes hath foure square, browne redde, and hearie stemmes, with leaues almost rounde, snipte, or dented rounde about, of a darke greene colour, and of sauour very good and pleasant. The floures are Crymsin or reddishe, and do grow in knops about the stalke lyke whorles, or like the floures of Pennyroyall. The roote hath threddie stringes, and creepeth alongst the ground, and putteth foorth yong shootes or springs yearely.

2 The second kind is very like vnto the first, in his round, swarte, and sweete sauouring leaues, also in his square stemmes, and the creeping rootes in the grounde, but his floures growe not in knoppes or whorles rounde aboute the stemmes, but at the toppe of the stalkes lyke to a small spike or busshie eare.

3 The thirde kinde hath long narrow leaues, almost like wythie leaues, but they be greater, whiter, softer and hearie. The floures grow at the top of the stalkes like spikie eares, as in the second kinde. The roote is tender with threddishe strings, and springeth foorth in diuers places, like to the others.

4 The fourth kinde is like to the abouesayde in his leaues, stalkes, and roote, but that his floures are not fasshioned like spykie eares growing at the toppes of the stalkes, but they compasse and grow round about the stalkes like whorles, or garlandes, like to the Curled, or Crispe Mynte.

5 The fifth kinde of Mynte, whiche is the first of the wilde kindes, and called Horse Mynte, hath square woollie stemmes, and his leaues be somwhat long, wrinkled and soft, and couered or ouerlayde with a fine downe, or soft cotton, both

Menta satiua prima.	Menta satiua secunda.	Menta satiua tertia.
Curlde Mynte.	Crispe Mynte, or Crispe Balme.	Spere Mynte, or right garden Mynte.

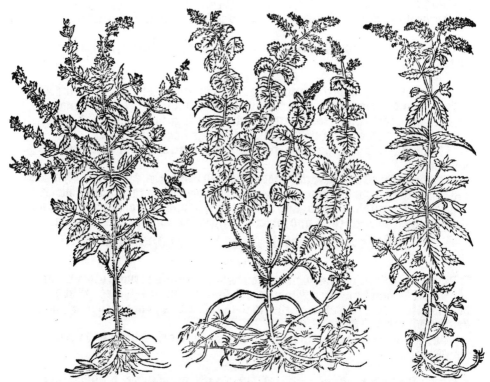

ouer and vnder. The floures grow at the toppe of the stalkes in spikie tuftes. The roote is tender with threddes or sucking strings.

6   The second wilde kinde, whiche is the sixth in number of the Myntes, and called water Mynte, is much like vnto ye Curlde Mynte, in his stalkes, leaues, and creeping rootes, sauing that his leaues & stalkes be greater, & of stronger sauour. The floures be purple growing at the top of the stalkes in small tuftes or knoppes like round bullets.

❧ *The Place.*

1   The garden Myntes are founde in this countrie in gardens, especially the Curlde Mynte, the which is most common and best knowen.

2   The wilde kindes do growe in lowe, moyst places as neare vnto springs, and on the brinkes of ditches.

❧ *The Tyme.*

All the sortes of Myntes do floure most commonly in August.

❧ *The Names.*

The garden Mynte is called in Greeke ἡδύοσμος, καὶ μίνθη: in Latine and in Shoppes *Mentha*: in Spanish *Yerua Ortelana, yerua buena*: in English Myntes: in French *Mente*: in high Douch Munte.

1   The first kind is called in high Douch Deyment, and Kraunszdyment, that is to say, Curlde Mynte: in French *Mente crespue*: in base Almaigne Bruyn heylighe.   The

the Historie of Plantes. 245

2 The second is also called of the high Douchmen krausmuntz, and Krausz-balsam, ÿ is to say, in French *Baulme crespu*: in English Crispe Baulme, or Crispe Mynte: also Crosse Mynte: in base Almaigne Cruysmunte, and of some also Heylighe.

3 The third kinde is called at this time in ÿ Shoppes of this countrey Menta Sarracenica, ⸿ Menta Romana: in English Spere Mynte, or the cōmon garden Mynte: also of some Baulme Mynte: in French *du Baulme*, and *Mente Romayne*: in high Douch Balsam Muntz, vnser frauwen Muntz, Spitz muntz, Spitz-balsam: in base Almaigne Roomsche munte, and Balsem munte.

Menta satiua quarta.	Mentastrum.	Sisymbrium.
Harte Mynte.	**Horse Mynte.**	**Water Mynte.**

4 The fourth kinde is called in high Douch Hertzkraut, that is to say Harte wurte, or Harte Mynte: in French *Herbe de cueur*: of the later wryters in Latine Menta Romana angustifolia, Flore coronata, siue Cardiaca Mentha.

5 The fifth wilde kinde, which is the fifth kinde of Mynte, is called in Greeke ἡδύοσμ(Ꙍ) ἄγρι(Ꙍ): in Latine Mentastrum, and of the newe writers Menta aquatica: in English Horse Mynte: in French *Mente Chevaline ou sauuage*: in high Douche katzenbalsam, Rosmuntz, wilder Balsam, wild Muntz: in base Almaigne witte water Munte.

6 The seconde wilde kynde whiche is the sixthe Mynte, is called in Greeke σισύμβριον, in Latine Sisymbrium, and of Damegeron Scimbron, as Constantine the Emperour witnesseth: in English Fisshe Mynte, Brooke Mynte, Water Mynte, and white water Mynte: in French *Mente Aquatique*: in high Douch Fischmuntz, Wassermuntz: in base Almaigne Roo munte, and Roo water munte.

✳ The Nature.

All the kindes of Myntes, whiles they are greene, are hoate and dry in the second degree: but dried they are hoate in the thirde degree, especially the wild kindes, whiche are hoater then the garden Myntes.

✳ The Vertues.

Garden Mynte taken in meate or drinke, is very good and profitable for the stomacke, for it warmeth and strengtheneth the same, and drieth vp all superfluous humors gathered in the same, it appeaseth and cureth all the paynes of the stomacke, and causeth good digestion.   A

Two or three branches of Myntes, dronken with the iuyce of soure Pomegranets do swage and appease the Hicquet or yeoxe, and vomiting, and it cureth the cholerique Passion, otherwise called the felonie, that is whan one doth vomit continually, and hath a laske withall.   B

The iuyce of Myntes dronken with vineger, stayeth the vomiting of bloud, and killeth the rounde wormes.   C

The same boyled in water and dronken by the space of three dayes togither, cureth the gryping payne and knawing in the belly, with the colique, and stoppeth the inordinate course of the menstruall issue.   D

Mynte boyled in wine and dronken, easeth women which are tomuch grieued with harde and perillous trauell in childebaring.   E

Mynte mengled with parched Barley meale, and layde vnto tumors and swellings doth wast and consume them. Also the same layde to the forehead, cureth headache.   F

It is very good to be applied vnto the breastes that are stretched foorth and swollen and full of milke, for it slaketh and softeneth the same, and keepeth the mylke from quarring, and crudding in the brest.   G

The same being very well pounde with Salte, is a speciall medicine to be applied vpon the biting of madde Dogges.   H

The iuyce of Mynte mengled with honied water, cureth the payne of the eares being dropped therein, and taketh away the asperitie, and roughnesse of the tongue, whan it is rubbed or washed therewith.   I

The sauour or sent of Mynte, reioyceth man: wherefore they sow & strow the wilde Mynte in this countrie in places whereas feastes are kepte, and in Churches.   K

5   The Horse Mynte called *Mentastrum*, hath not bene vsed of the Auncients in medicine.   L

6   The water Mynte is diuers wayes of the lyke operation vnto the garden Mynte, it cureth the trenches or gryping payne in the small of the bellie or bowels, it stayeth the yeoxe or hicket and vomyting, and appeaseth headache to be vsed for the same purpose as the garden Mynte.   M

It is also singuler against the grauell and stone of the kydneys, and against the strangury, whiche is whan one cannot pisse but droppe after droppe, to be boyled in wine and dronke.   N

They lay is with good successe vnto the stingings of Bees and Waspes.   O

Of Calampnt.   Chap.lxv.

✳ The Kyndes.

There be three sortes of Calampnt described of the Auncient Gretians, each of them hauing a seuerall name, and difference.

Cala-

the Historie of Plantes.

Calaminthæ alterum genus.
Corne Mynte, or wilde Pennyryall.

Calaminthæ tertium genus.
Catmynte.

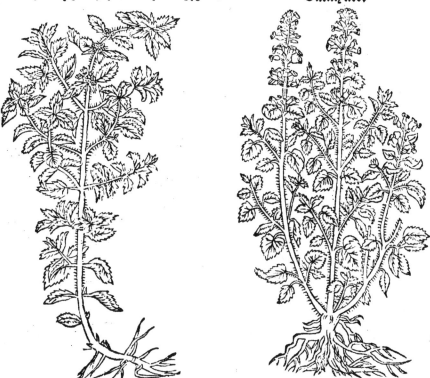

❧ The Description.

1. The first kinde, whiche may be called Mountayne Calamynte, hath harde square stalkes couered with a certayne hoare, or fine Cotton. The leaues be somwhat like ye leaues of Basill, but they are rougher. The floures grow onely by one side of the stalke amongst the leaues, somtimes three or foure vpon a stem, of a blewishe colour, the roote is threddy.

This herbe altogither is not much vnlike the secod kinde of Calamynte, sauing it is greater, the stalkes be harder, and the leaues be rougher and blacker, and it creepeth not alongst the grounde, but groweth vp from the yearth.

2. The second kinde which is called wild Pennyryall, hath also square stalkes couered with softe Cotton, & almost creeping by the ground, hauing euer two, and two leaues standing one against an other, small and softe, not much vnlike the leaues of Penny royall, sauing they are larger & whiter. The floures grow about the stalkes in knoppes like to whorles or garlandes, of a blewishe purple colour. The roote is small and threddie.

3. The thirde kinde whiche is called Catmynte, or Cattis herbe, is not much vnlike (as Dioscorides saith) vnto the whiter wilde Mynte. It hath square softe stalkes full of ioyntes, and at euery ioynt two leaues standing one against an other, and it hath also betwixt the sayde leaues & the stalkes, little branches. The leaues be not much vnlike to the leaues of Horse Mynte, sauing they are somewhat longer and dented, or natched rounde aboute, in proportion like to a Nettell leafe, but yet softe and gentill, and of a white hoore colour, especially in the vnderside of the leafe. The floures grow most cōmonly aboute the toppe of

The stalkes after the order of Crownets. The roote is tender and threddie.

4 There is yet an other kinde of Cattis herbe, a great deale smaller in all respects than the first, otherwise they be altogither alyke, and it hath a very good sauour.

❧ The Place.

1 The firste kinde, as Dioscorides saith, groweth in Mountaynes and hillie places. In this countrey it is planted in the gardens of Herboristes or louers of herbes.

2 The second kinde groweth in this countrie in rest fieldes, and vpon certaine small hilles or knappes.

3 The third kinde groweth in euery garden, and is very well knowen in this countrie.

❧ The Tyme.

All the sortes of these herbes, do for the most parte floure in June and July.

❧ The Names.

This kinde of Mynte, is called in Greeke καλαμίνθη: in Latine Calamintha: in Italian Nipotella: in Spanish Laueuada: in Shoppes Calamentum: of Plinie and Apuleius Mentastrum: in English Calamynte.

1 The firste kinde is called in Shoppes Calamentum montanum, that is to say, Calamynte mountayne: in English rough Calamynte: in high Douche Stein, oder berch Muntz.

2 The second kinde is called in Greeke γλήχων ἄγριον: in Latine Pulegium syluestre, and Nepita: in English wild Penny royall, and Corne mynte: in French Pouliot sauuage. in high Douch Kornmuntz, wilden Poley, in base Almayne wilde Poley, and velt Munte.

3 The third kinde is now called in Shoppes Nepita: in English Neppe, and Cat Mynte: in French Herbe de Chat: in high Douch Katzenmuntz: in base Almaigne Cattencruyt and Nepte.

❧ The Nature.

These herbes are hoate and dry in the third degree, especially the first kinde whiche is gathered vpon Mountaynes.

❧ The Vertues.

Calamynte (especially of the Mountayne) boyled and dronken, or layde too outwardly preuaileth much against the bitings of venemous beasts. The same dronken first or afore hande with wine, preserueth a bodie from all deadly poyson, and chaseth, & driueth away all venemous beasts, from that place whereas it is eyther strowen or burned. A

The same drōken with honied water warmeth the bodie, and cutteth or seuereth the grosse humors, and driueth away all cold shiuerings, and causeth to sweate. It hath the same power, if ye boyle it in oyle, and annoynt all the body therewith. B

Calamynte dronken in the same manner, is good for them that haue fallen from a lofte, and haue some bruse or squat, and bursting, for it digesteth the congeled and clotted bloud, and is good for the payne of the bowels, the shortnesse of breath, the oppillation or stopping of the breast, and against the Jaundice. C

The same boyled in wine and dronken, prouoketh vrine, and floures, and expelleth the dead childe, and so doth it also if it be applied vnder in manner of a Pessarie or Mother suppositorie. D

It is very good for Lazer people and Lepers if they vse to eate it, & drinke the whay of sweete milke after. E

The same eaten rawe or sodde with meates, or dronke with salte and hony, sleeth and driueth foorth al kindes of wormes, in what part of the bodie soeuer they F they be. The same vertue hath the iuyce dronken, & layde to any place whereas wormes are.

Also it taketh away scarres, and blacke and blewe markes, whan it is boyled in wine and the places often washed therewith, or else the herbe it self fresh gathered, pounde and layde vpon.

Of Costemary, or Balsampnte. Chap.lxvij.

The Kyndes.

Balsampnte is of two sortes, great and small, resembling one an other in sauour, leaues and seede.

Balsamita maior.
Costemary.

Balsamita minor.
Mawdelepn.

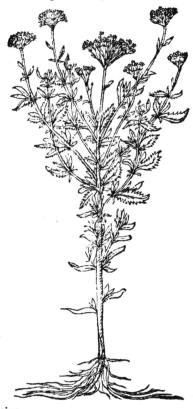

The Description.

1. The great Balsampnt hath slender stalkes, round and harde: the leaues be long and meetely large, of a white or light greene colour, very finely hackte or snipt about the edges. The floures grow in tuftes, or bundels like Nosegayes, and are nothing else like but to small yellow buttons, very like the floures of Tansie, sauing that they be smaller. The roote is threddy and beareth diuers stalkes and branches, and putteth vp yearely new springs. The whole herbe is of a strong sauour, but yet pleasant, and in tast bitter.

2. The small Balsampnte is much like to the first & great Balsampnte, aswell in stalkes, floures and seede, as in smell and sauour, but altogither smaller and not so high of groweth, his leaues be a great deale smaller and narrower, and much deeper snipt or cut about the edges. The roote also is threddie and putteth

teth vp yearely many new springs.

❧ The Place.

They are both planted in the gardens of this coūtrie, but especially the first, the whiche is very common in all gardens.

❧ The Tyme.

Balsamynte floureth in July and August.

❧ The Names.

1 The first kinde is called in Latine Balsamita maior: in the Shoppes of Brabant Balsamita, of some Menta Græca, Saluia Romana, Lassulata, and Herba diuæ Mariæ: in English Cooste Marie, and of some Balsampnte: in French Coq, or du Coq: in high Douch Frauwenkraut: in base Almaigne Balseme. It should seeme to be that Panax Chironia, whiche Theophrastus describeth in his ix. Booke.

2 The seconde kinde is called of some Balsamita minor: in Languedoc Herba diuæ Mariæ: in English Mawdelein, and of some small Balsampnte: in Italie Herba Giulia: And some take it for ἀχρατον, Ageratum of Dioscorides: others take it to be ἐλίχρυσον, Elichrysum: and others for Eupatorium Mesue. But in mine opinion it is none of them three, for I thinke it an herbe not described of any of the Aunctients vnlesse it be a kinde of Panaces Chironium Theophaasti.

❧ The Nature.

These two herbes be hoate and dry in the second degree, as their smell, and bitter taste doth declare.

❧ The Vertues.

The leaues of Costemarie alone, or with Parsenip seede boyled in wine & dronken cureth the trenches of the belly, that is a griping payne and torment in the guttes or bowels, and it cureth the bloudie flire.

The conserue made of the leaues of Costemarie and Suger, doth warme and dry the brayne, and openeth the stoppings of the same, and it is very good to stoppe all superfluous Catarrhes, Reumes, and distillations, to be taken in quantitie of a Beane.

This herbe is also vsed in meates as Sage and other herbes, especially in Salades and sawces, for whiche purpose it is excellent, for it yeeldeth a proper sent and taste.

As Mawdelein herbe or small Balsaminte, is like to Costemary or great Balsampnte in taste and sauour, so is it like in vertues and operations, & may be alwayes vsed in steede of the great Balsaminte.

Of Sage.   Chap.lxxvij.

❧ The Kyndes.

There be two sortes of Sage, the one is small & franke, & the other is great. The great Sage is of three sortes, that is to say, greene, white, and redde.

❧ The Description.

1 The franke Sage hath sundry wooddie branches, and leaues growing vpon long stemmes whiche leaues be long, narrow, vneuen, hoare, or of a grayishe white colour, by the sides of the sayde leaues at the lower ende, there groweth two other small leaues, like vnto a payre of little eares. The floures growe alongst the stalkes in proportion like the floures of Dead Nettell, but smaller and of colour blewe. The seede is blackishe, and the roote wooddie.

2 The great Sage is not much vnlike the small or franke Sage, sauing it is larger: the stalkes are square and browne. The leaues be rough, vneuen

euen and whitishe, like to the leaues of franke Sage, but a greate deale larger, rougher, and without eares. The floures, seede, and roote are like vnto the other.

Saluia minor.   Saluia maior.
Franke Sage, or small Sage.   Great Sage, or broade Sage.

There is found an other kinde of this great Sage, the which beareth leaues as white as snow, sometimes all white, and sometimes partie white, and this kinde is called white Sage.

Yet there is founde a thirde kinde of great Sage, called redde Sage, the stemmes whereof, with the synewes of the leaues, and the small late sprong vp leaues, are all redde: but in all things else it is like to the great Sage.

❧ *The Place.*

Sage, as Dioscorides saith, groweth in rough stonie places, both kindes of Sage, are planted almost in all the gardens of this countrie.

❧ *The Tyme.*

Sage floureth in June and July.

❧ *The Names.*

The Sage is called in Greeke ἐλελίσφακος: in Latine and in Shoppes Saluia: of some Corsaluium: in Spanish *Salua*: in English Sage: in French *Sauge*: in high Douch Salbey: in base Almaigne Sauie.

The first kinde is now called in Latine Saluia minor, Saluia nobilis, and of some Saluia vsualis: in English Smal Sage, Sage royall, and common Sage: in French *Sauge franche*: in high Douch Spitz Salbey, klein Salbey, edel Salbey, & Creutz Salbey: in base Almaigne Cruys sauie, and Oorkens sauie.

The

2 ❧ The second kinde is called in Latine Saluia maior, and of some Saluia agrestis: in English great Sage, or broade Sage: in French *grande Sauge:* in high Douch Grosz salbey, Breat salbey: in base Almaigne groue, & groote Sauie.

❀ *The Nature.*

Sage is hoate and dry in the thirde degree and somewhat astringent.

❀ *The Vertues.*

Sage boyled in wine & dronken, prouoketh vrine, breaketh the stone, comforteth the harte, and swageth head ache. A

It is good for weme with childe to eate of this herbe, for as Aëtius saith it closeth the Matrice, causeth the fruite to liue, and strengtheneth the same. B

Sage causeth wemen to be fertill, wherefore in times past the people of Egypt, after a great mortalitie and pestilence, constreyned their wemen to drinke the iuyce thereof, to cause them the sooner to conceyue and to bring foorth store of children. C

The iuyce of Sage dronken with hony in the quantitie of two glasse fulles, as saith Orpheus, is very good for those whiche spitte and vomit bloud, for it stoppeth the flure of bloud incontinent. Likewise Sage brused and layde too, stoppeth the bloud of woundes. D

The decoction thereof boyled in water and dronken cureth the cough, openeth the stoppings of the Liuer, and swageth the payne in the side: and boyled with wormewood it stoppeth the blouddy flire. E

Sage is good to be layde to the woundes and bitings of venimous beasts, for it doth both clense, and heale them. F

The wine wherein Sage hath boyled, helpeth the manginesse and itche of the priuie members, if they be wasshed in the same. G

Wild Sage. Chap.lxxviij.

❀ *The Description.*

Saluia agrestis.

WOode Sage is somewhat like garden Sage, in fashion & sauour, it hath square browne stalkes, set with a certaine kind of small heare, the leaues are not much vnlike the leaues of great Sage, but somewhat broader, shorter and softer. The floures are not much vnlike to the floures of Sage, growing onely vpon one side alongst the branches, euen vp to the very top of the same branches or stemmes, of a whitisshe colour, whan they are paste, there commeth a rounde blackish seede. The roote is threddie, & sendeth foorth new springs or branches euery yeare.

❀ *The Place.*

This kinde of Sage groweth in this countrey alongst the hedges, in woodes, and the bankes or borders of fieldes.

❀ *The Tyme.*

It floureth in June, and July.

❀ *The Names.*

This herbe is now called in Shops Saluia

the Historie of Plantes. 253

Saluia agrestis, and Ambrosiana: in high Douche wilde Salbey: in base Almaigne wilde Sauie. There are some that thinke it to be the seconde kynde of Scordium whiche Plinie describeth, bycause that whan it is bruised, it sauozeth of Garlike, and this is the cause why Cordus calleth it Scorodonia. It is called in English woodde Sage, wild Sage, and Ambros: in French *Sauge de Boys.*

❦ *The Nature.*

The woode Sage is hoate and dry, meetely agreable in complexion vnto garden Sage.

❦ *The Vertues.*

Woode Sage dissolueth congeled bloud in the body, and cureth inwarde woundes, moreouer it wonderfully helpeth those that haue taken falles, or haue bene sore bruised and beaten, if it be boyled in water or wine and dronken. A

Woodde Sage taken in manner aforesayde doth consume and disgest inwarde impostems and tumers, auoyding the matter and substance of them with the vrine. B

Of Clarey. Chap. lxxix.

Gallitricum.

❦ *The Description.*

CLarye hath square stalkes, with rough, grayish, hearie, & vneuen leaues, almost like to the leaues of great sage, but they are foure or fiue times larger: the floures be of a faynte or whitish colour, greater than the floures of Sage. Whan they are fallen of there groweth in huskes the seede, which is blacke. The roote is yellow & of woodie substance. The whole herbe is of a strong, and penetratiue sauour, in somuch that the sauour of it causeth headache.

❦ *The Place.*

In this countrie they sow it in gardēs.

❦ *The Tyme.*

Clary floureth in June & July a yeare after the first sowing thereof.

❦ *The Names.*

Clarie is now called in Latine and in Shoppes Gallitricum, Matrisaluia, Centrum galli, and Scarlea oruala: in English Clarye, or Cleare-eye quasi dicas, oculum clarificans: in French *oruale, & Toutebonne:* in high Douche Scharlach: in base Almaigne Scarleye. It seemeth to be a kind of Horminum, but yet it is not Alectorolophos as some men thinke.

❦ *The Nature.*

Clarey is hoate and dry, almost in the thirde degree.

❦ *The Vertues.*

In what sorte or maner soeuer ye take Clarey, it prouoketh the floures, it expulseth the Secondine, and stirreth vp bodely luste. A

Also it maketh men dronke, & causeth headache, & therefore some Brewers do boyle it with their Bier in steede of Hoppes. B

R This

This herbe also hath al the vertues and properties of Horminum, and may be vsed in steede of it.

Of Horminum, wilde Clarey, or Oculus Christi. Chap. lxxx.

❧ *The Kindes.*

There be two sortes of Horminum, as Dioscorides writeth, the garden and wilde Horminum.

Horminum satiuum.
Dubble Clarey.

Horminum syluestre.
Oculus Christi.

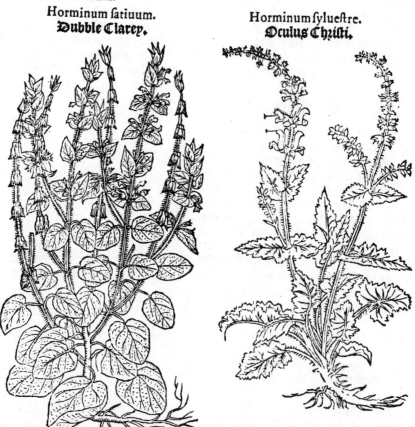

❧ *The Description.*

The garden Horminum hath leaues in a maner rounde, and somewhat ouerlayde with a softe Cotton, almost like Horehounde. The stalkes be square and hearie of the heigth of a foote, bearing all aboue at the top, fiue or sixe fayre small leaues of a blewish purple colour: the leaues stand at euery ioynte, one against an other, amongst the whiche there commeth forth little huskes, that bring forth purplish blew floures. The which whan the seede beginneth to waxe ripe, they turne towardes the grounde, and hang downewardes, hauing in them blacke seede and somewhat long, the whiche whan it is a little while soked or steeped in any licour, it waxeth clammy or slimie, almost like to the kernelles of Quinces.

The wilde Horminum beareth great, broade leaues, gasht, or natched rou̅d aboute.

aboute. The ſtalkes alſo be ſquare, and ſomewhat hearie, but yet they be longer and bigger than the ſtalkes of the garden Horminum. The floures be of a deepe blew colour, and do alſo grow by certayne ſpaces aboute the ſtemme like to whorles or Crownettes, out of little huſkes, whiche do alſo turne downewardes whan the ſeede is ripe, the ſeede is of a dunne or blackiſhe colour, round, & alſo ſlymie whan it is ſteeped or ſooked. The roote is of wooddie ſubſtance, and blacke.

These twoo herbes haue no ſpeciall ſauour, eſpecially the garden kinde: for the floures of the wilde kinde do ſauour ſomewhat like to Clarey.

❦ The Place.
Theſe two kindes are founde in this countrie, ſowen in the gardens of Herboriſtes.

❦ The Tyme.
They yeelde their floures in June, July and Auguſt, in the whiche ſeaſon their ſeede is alſo ripe.

❦ The Names.
This herbe is called in Greeke ὅρμινον: in Latine Horminum and Geminalis. The firſt is called Horminum ſatiuum, and Hortenſe. The ſeconde Horminum ſylueſtre: they may be both called wilde Clarie, ſome call thē Dubble Clarey, and ſome Oculi Chriſti.

❦ The Nature.
Horminum is of complexion hoate and dry.

❦ The Vertues.
The ſeede of Horminum mengled with Hony driueth away the dimneſſe of the ſight, and clarifieth the eyes. [A]

The ſame ſeede with water ſtamped and tempered togither, draweth out [B] thornes and ſplinters, and reſolueth or ſcattereth all ſortes of ſwellings, being layde or applied thereto. The ſame vertue hath the greene herbe whan it is ſtamped or bruſed and layde vpon.

The ſame ſeede dronken with wine ſtirreth vp bodely luſte, eſpecially the [C] ſeede of the wilde kinde, which is of greater efficacie, than the ſeede of garden Horminum.

Of Horehounde. Chap. lxxxi.

❦ The Kyndes.
Here be foure kindes of Horehounde, in faſhion one like to another. The whiche for all that in Latine haue their particular or ſeuerall names. The firſt kinde is our white Horehounde, the ſeconde is the blacke ſtinking Horehounde. The third is Stachys or field Horehounde. The fourth is water or Marriſhe Horehounde.

❦ The Deſcription.
The white Horehounde hath many ſquare & white hoare, or hearie ſtalkes, the leaues be rounde, crompled, hearie, aſhe coloured, and of no lotheſome ſauour. The floures be white, and growing forth of ſmall, ſharpe, and prickley huſkes, compaſſing the ſtalkes, like in faſhion to a ringe or garlande, in whiche (prickley huſkes) after that the floure is vaniſhed, there is founde a rough ſeede. The roote is blacke with many threddie ſtrings.

The blacke Horehounde, is ſomewhat like vnto the white. The ſtalkes be alſo ſquare and hearie, but yet they be blacke or ſwarte. The leaues be larger and longer than the leaues of white Horehounde, dented or ſnipte rounde aboute the edges almoſt lyke vnto Nettell leaues, they are blacke, and of a ſtrong vnpleaſant ſauour. The floures are purple lyke to the dead Nettell growing in whoꝛling knoppes rounde aboute the ſtalkes, like to white Hoꝛehounde.

Marrubium.	Ballote.
White Hoꝛehounde.	**Blacke Hoꝛehounde.**

3   Stachys oꝛ wilde Hoꝛehounde hath a round ſtemme, oꝛ ſtalke full of ioyntes couered with a fine white woolly downe oꝛ cotton: the leaues do euer grow by coupples, two and two at euery ioynte, and are white and woolly almoſt like the leaues of white Hoꝛehounde, ſauing they be longer and whiter. The floures grow like Crownets oꝛ garlandes compaſſing the ſtalke, of yellow colour, and ſometimes purple. The roote is harde and of a wooddy ſubſtance. All this herbe differeth nothing in ſmell oꝛ ſauour from white Hoꝛehounde.

Byſides theſe there is yet another herbe called ſweete ſmelling Hoꝛehoũd, oꝛ ſweete wilde Sage, the whiche beareth ſquare ſtalkes, thicke and woollie: The leaues be whitiſh and ſoft, and ſomewhat dented rounde about, but much longer, larger and bꝛoader, than the leaues of the other Hoꝛehoundes. The floures be reddiſh growing about the ſtalkes like to whoꝛles oꝛ garlãdes. The ſeede is blacke and rounde. The roote is yellowiſh.

4   The water Hoꝛehounde is much like to blacke Hoꝛehounde, aſwell in his ſtalkes and pꝛickle huſkes, as in his leaues and floures. The leaues be alſo of a ſwarte greene colour, but larger and moꝛe deeply indented, and not very hearie, but ſomewhat crompled, and wꝛinckled, like to the leaues of the Birche tree, whan they begin to ſpꝛing. The floures be white, and ſmaller than the floures

the Historie of Plantes.

floures of the other Horehoundes.

✻ The Place.

The white Horehound and the blacke do grow with vs in all rough and vnmanured places, by walles, hedges, wayes, and aboute the borders of fieldes. The third groweth on ye playnes of Almaigne and else where, it is not to be founde in this countrie, but in the gardens of Herborists. The water Horehounde is found very plenteously growing in this coūtrie by diches and watercourses, and in lowe moyst places.

✻ The Tyme.

All these herbes do moste commonly floure in July. The sauery Horehounde or wilde Sage doth floure in August.

✻ The Names.

Stachys.
Mountayne Horehounde.

1 The firste kinde is called in Greeke πράσιον: in Latine Marrubium: in Shops Prasium: in Italian *Marrabio*: in Spanish *Marruuios*: in English Horehounde, and white Horehounde: in French *Marrubin* and *Marochemin*, also *Marrube blanc*: in high Douche weitz Andorn, Marobel, Gottzsvergitz, and Andorn meunlin: in base Almaigne Malroue, Malruenie, Witte Andoren, and Andoren Manneken.

2 The second is called in Greeke βαλλωτή: in Latine Marrubium nigrum, Marrubiastrum: in Shoppes Prasium fœtidum: in Italian *Marrobio nero: Marrobio fendo*: in Spanishe *Marroios negros*: in English blacke Horehounde, and stinking Horehounde, & of some blacke Archangell: in French *Marrubin noir, Marrubin puant*: in high Douch schwartz Andorn, and Andorn weiblin: in base Almaigne stinckende and swerte Malruenie and Andoren, or Andoren wijfken.

3 The thirde is called in Greeke στάχυς: and in Latine Stachys: vnknowen in Shoppes, it may be also called in English Stachys or wilde Horehounde: in Frenche *Saulge sauuage*: in high Douch rieckende Andorn, fielde Andorn: in base Almaigne rieckende Andoren: in Italian *Herba odoraea*: in Spanish *Yerua olodera*, and *Yerua de souto*.

4 The fourth is now called in Latine Marrubium palustre: in English Marrishe or water Horehounde: in French *Marrubin d'eau*: in high Douche wasser Andorn, weiher Andorn: in Brabant water Andoren, and of some Egyptenaers cruyt, that is to say, the Egyptians herbe, bycause of the Rogues and runnegates whiche call themselues Egyptians, do colour themselues blacke with this herbe. Some men make it the first kinde of Sideritis.

The three first kyndes of Horehounde are hoate in the seconde degree, and dry in the thirde. The water Horehounde is also very dry, but without any manifest heate.

❧ *The Vertues*.

The white Horehounde boyled in water and dronken, doth open and comforte the Lyuer and the Melte, or Spleene, and is good against all the stoppings of the same, it clenseth the breast & the lunges, also it is profitable against an olde Cough, the payne of the side, and the olde spitting of bloud, & against the Tysike and vlceration of the lunges.

The same taken with the roote of Iris, causeth to spet out al grosse humors, and tough flegmes, that are gathered togither within the breast.

The same vertue also hath the iuyce thereof, to be boyled togither with the iuyce of Fenill vntill the thirde parte be consumed, and taken in quantitie of a spoonefull, and it is also profitable against an olde Cough.

The white Horehounde boyled in wine, openeth the Matrix or Mother, and is good for women that cannot haue their termes or desired sicknesse, it expulseth the Secondyne and dead children, and greatly helpeth women, which haue harde and perillous trauell, and is good for them that haue ben bitten of Serpentes, and venemous beastes.

The iuyce of white Horehounde mingled with wine and Hony, and dropped into the eyes, cleareth the sight. The same iuyce poured into the eares, asswageth the payne, and openeth the stoppings of the same. It is also good to be drawen or snifte vp into the nose, to take away the yellownesse of the eyes, whiche remayneth after the Iaundice.

The leaues tempered with Hony is good to be layde vnto olde vlcers, and corrupt vlcered nayles, or agnayles whiche is a paynefull swelling aboute the ioyntes and nayles. The same mengled with Hennes greace, resolueth and scattereth the swelling about the necke called Strumes. The dryed leaues mengled or tempered with vineger, do cure noughty virulent & spreading vlcers.

2 The blacke Horehounde pounde, is good to be applied and layde vpon the bytings of madde Dogges. The leaues of the same rosted in a Call leaffe, vnder the hoate immers or ashes, do stoppe and driue backe the harde lumpes or swellings whiche happen to arise aboute the siege or fundament, and layde to with hony, they cure and heale rotten vlcers.

3 Stachys or wilde Horehounde boyled and dronken, causeth women to haue their floures, & bringeth forth the Secondine or afterbirth, & the dead fruyte.

4 Water Horehounde is not vsed in Medicine.

❧ *The Daunger*.

The white Horehounde is hurtfull both to the bladder and kidneys, especially whan there is any hurte or exulceration in them.

Of Bawme. Chap. lxxxij.

❧ *The Kyndes*.

Vnder the title of Melissa, are comprehended both the right Bawme, and the Bastard Bawme, the whiche both are somewhat like to the Horehounde.

❧ *The Description*.

1 The right Bawme hath square stalkes, & blackish leaues like to blacke Horehounde, but a great deale larger, of a pleasant sauour, drawing towardes the smell of a Citron. The floures are of Carnation colour. The roote is single, harde, and of a wooddie substance.

2 The common Bawme is not much vnlike to the aforesayd, sauing that his sauour is not so pleasant and delectable, as the sauour of the right Bawme.

3 There is a certayne herbe bysides these, the whiche some take for the right Bawme (yet they are much deceyued that do so thinke) it hath a square stalke with

with leaues like to common Bawme, but larger and blacker, and of an euell sauour: the floures are white, and much greater than the floures of the common Bawme: the roote is harde, and of wooddie substance.

Melissa vulgaris. **Bawme.** Melissophylli species. Herba Iudaica.

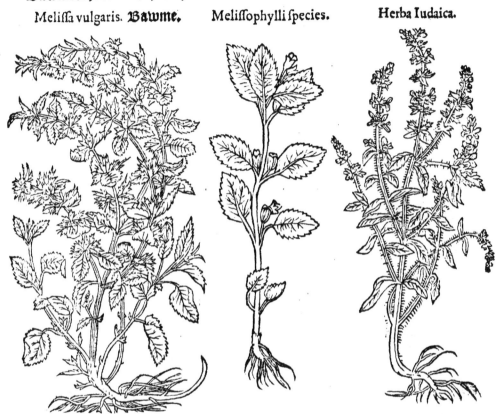

¶ A man may also place, amongst these sortes of Bawme, that herbe whiche ordinarily is called Herba Iudaica. It hath square hearie stalkes diuided or parted into many branches. The leaues be long and dented round about, and smaller then the leaues of Sage: alongst the toppes of the braunches groweth the floures, of a fainte blew or whitishe colour. The roote hath hearie strings. All the herbe draweth towardes the sauour of Bawme, or Melissa.

❧ *The Place.*

These herbes do grow in certaine countries in wooddes, and in some countries ye shall finde them growing about olde walles, & sometimes also ye shall haue it growing by the way sides: but now both sortes are plated in gardens.

Herba Iudaica groweth in Fraunce and Flaunders, in vntilled places, in vineyardes, and sometimes also alongst the hedges.

❧ *The Tyme.*

They floure in Iune and Iuly. The Iudaicall herbe floureth in Iuly and August.

❧ *The Names.*

Melisses is called in Greeke μελισόφυλλον, καὶ μελίφυλλον: in Latine Apiastrū, Melitæna, and Citrago: in Shoppes Melissa: in English Bawme: in Italian Cedronella, Herba rosa: in Spanish Torongil, yerua cidrera: in high Douch Melissenkraut,

kraut, and Mutterkraut: in base Almaigne Confilie de greyn and Melisse.

4 The fourth kinde is called of some in Latine Herba Iudaica: in English it may be called the Iudaicall herbe: in French *Tetrahil*, or *Tetrahit*: some count it to be the first kinde of Sideritis, called Sideritis Heraclea.

❧ *The Nature.*

These herbes are hoate and dry in the second degree, and somewhat like to Horehounde, but in vertue much feebler. ❧ *The Vertues.*

Bawme dronken in wine is good against the bitings, and stingings of ve- A
nemous beasts, it comforteth the harte, and driueth away all Melancholy and sadnes, as the learned in these dayes do write.

Bawme may be vsed to al purposes wherevnto Horehounde serueth, how- B
beit it is in all respects much weaker, so that according to the opiniōs of Galen, & Paulus Aegineta, it shoulde not be vsed for Horehounde in medicine, but for wante of Horehounde, in steede whereof Melissa may be alwayes vsed.

If a man put Bawme into Bee hyues, or else if the Hyues be rubbed there- C
wal, it keepeth Bees togither, & causeth other Bees to resorte to their cōpanie.

The cōmon Bawme is good for wemen whiche haue the strangling of the D
matrix or mother to be eyther eaten or smelled vnto. The iuyce thereof is good to be put into greene woundes, for it gleweth togither, sodereth and healeth the same.

Of Rue, or Herbe grace. Chap.lxxiij.

❧ *The Kyndes.*

There are two sortes of Rue, that is garden Rue, and wilde Rue.

Ruta hortensis.   Ruta sylueſtris minima.

Herbe grace, or garden Rue.   **The small wilde Rue.**

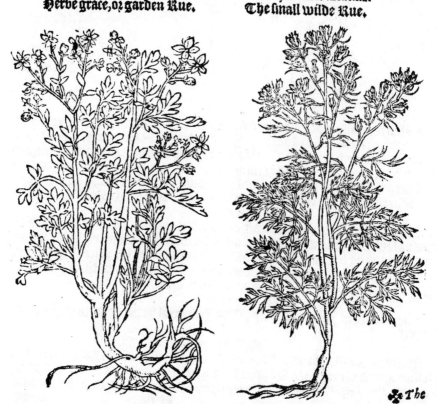

❧ *The*

❧ The Description.

THe garden Rue hath rounde harde stemmes, with leaues diuided into diuers other small roundish leaues, of a gray or blewish colour, and of a very ranke or strong sauour. The floures be yellow, growing at the top of the branches, after which there springeth vp square huskes contayning the seede whiche is blacke. The roote is of wooddie substance and yellow within. This Rue lasteth both winter and sommer, & dieth not lightly.

The wilde Rue, is much like to the other in his stalkes, leaues, floures, seede, colour, taste, and sauour: sauing that euery little leafe his cuttes are a great deale narrower.

But there is yet an other kinde whiche is the least of all, whose little leaues are very narrow and tender, and of colour somewhat whiter than the reste. Al this plante (as the other wilde Rue) is of a very grieuous sauour, and cannot abide the colde, but as the other wilde Rue, so doth this perishe with the firste colde or smallest froste.

❧ The Place.

The tame Rue is planted in gardens, and delighteth moste in dry groundes where as the Sonne shineth moste. The wilde Rue groweth vpon the mountaynes of Cappadocia, and Galatia, in the lesser Asia: in this coūtrie it is found sowen in the gardens of Herboristes.

❧ The Tyme.

They do all floure in this country in July and August, and the seede is ripe in September.

❧ The Names.

Rue is called in Greeke πήγανον: in Latine Ruta, and of Apuleius Eriphion.

The garden Rue is called in Greeke πήγανον κήπωτον: in Latine Ruta hortensis: in Shoppes Ruta: in English Rue of the garden, and Herbe grace: in Italian Rutta: in Spanish La arruda: in high Douch Zam Rauten, & wein Rauten: in base Almaigne Wijn ruyte.

The wilde Rue is called in Greeke πήγανον ἄγριον: in Latine Ruta syluestris, and in some places as Apuleius sayth Viperalis: in Shoppes Harmel: in high Douche wald Rauten: in base Almaigne wilde Ruyte.

❧ The Nature.

Rue is hoate and dry in the thirde degree: But the wilde Rue (& especially that which groweth in mountaynes) is a great deale strōger then gardē Rue.

❧ The Vertues.

The leaues of garden Rue boyled in water & dronken causeth one to make water, prouoketh the floures, and stoppeth the laske.

The leaues of Rue eaten alone with meates, or recepued with walnuttes, and dryed figges stamped togither, are good against all euil ayres, and against the Pestilence and all poyson, and against the bitings of vipers & Serpentes.

The same pounde and eaten or dronken in wine, helpeth them that are sicke with eating of venimous Tadstooles or Mousheroms.

The iuyce of Rue is good against the same mishappes, and against the bytings and stingings of Scorpions, Bees, Waspes, Hornettes, and madde Dogges, whan it is either dronken with wine, or whan that the leaues be stamped with hony and salte, and layde vnto the wounde.

The body that is annoynted with the iuyce of Rue, or that shall eate of Rue fasting, shalbe (as Plinie writeth) assured against all poyson, and safe from all venimous beastes, so that no poyson, or venimous beast shall haue powre to hurte him.

The

The same iuyce of Rue dronken with wine purgeth wemen after their deliuerance, & driueth forth the Secondine, the dead childe, & the vnnatural birth.

Rue eaten in meate or otherwise vsed by a certayne space of tyme, quencheth and dryeth vp nature, and naturall seede of man, and the milke in the breastes of wemen that giue sucke.

Rue boyled with Dyll and dronken, swageth the gnawing torment, or griping payne of the belly called the trenches, & is good for the paynes in the side and breast, the difficultie or hardnesse of breathing, the cough, the stopping of the lunges, the Sciatica, and against the rigour and violence of feuers.

Rue boyled in good wine vntill the halfe be sodden away, is very good to be dronken of such as begin to fall into the Dropsie.

Rue eaten rawe or condited with Salte, or otherwise vsed in meates, cleareth the sight, and quickeneth the same very much: so doth also the iuyce therof layde to the eyes, with hony, the iuyce of fenill, or by it selfe. The leaues of Rue mengled with Barley meale, asswageth the payne of the eyes being layde thereupon.

The iuyce of Rue warmed in the shell of a Pomgranete, and dropped into the eares swageth the paynes of the same. The same mengled with oyle of Roses, or oyle of Bayes & Hony, is good against the singing or ringing sounde of the eares, whan it is often dropped warme into them.

The leaues of Rue pounde with oyle of Roses and vineger, are good to be layde to the paynes of the head.

The same pounde with Baye leaues, and layde too, is good to dissolue and cure the swelling and blastings of the genitors.

The leaues of Rue mingled with wine, Pepper, and Nitre, do take away all spottes of the face, and clenseth the skinne: and mengled with Hony and Alom, it cureth the foule scabbe or naughtie Tetter. The same leaues poūd with Swines greace, doth cure all ruggednes of the skinne, and the scurffe or roome of the head, the kings euill or harde swellings about the throote, being applied and layde thereto.

Rue mengled with Hony, doth mitigate the paynes of the ioyntes, & with figges it taketh away the swelling of the Dropsie.

The iuyce of Rue with vineger giuen to smell vnto, doth reuiue and quicke such as haue the Lethargie, or the sleeping and forgetfull sicknesse.

The roote of Rue made into pouder and mengled with hony, scattereth & dissolueth congeled and clotted bloud, gathered betwixte the skinne and the flesh, and correcteth all blacke and blew markes, scarres, & spottes, that chaunce in the bodie, whan they are anoynted or rubbed therewith.

The oyle wherein Rue hath bene sodden or long infused & stieped, doth warme and chaafe all colde partes or members, and being annoynted or spread vpon the region of the bladder it prouoketh vrine, and is good for the stopping and swelling of the spleene or Melte: and giuen in glister, it dryueth forth windinesse, blastings, and the gryping payne in the bowels or guttes.

Some write also, that the leaues of Rue pounde, and layde to outwardly vpon the Nose, stancheth the bleeding of the same.

The iuyce of wilde Rue mengled with Hony, wine, the iuyce of fenill, & the gaule of a Henne, quickeneth the sight, & remoueth al clowdes & the pearles in the eyes. Also the wilde Rue hath the like vertue as the Rue of the garden, but it is of greater force, in somuch as the auncient Physitions would not vse it, bicause it was so strong, sauing about the diseases and webbes of the eyes in maner as is aboue writen.

Of Harmall, or wilde Rue. Chap.lxxiiij.

Harmala.

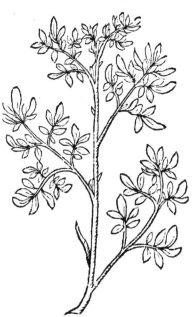

The Description.

This herbe hath three or foure stemmes growing vpright, and in them are small long narrow leaues, more tenderer, and diuided into smaller or narrower leaues than the common or garden Rue, the floures grow at the toppe of the stemmes or branches, of colour white, after whiche cōmeth triangled huskes coteyning the seede. And this plante is of a very strōg and grieuous smell, especially in hoate regions, or countries, where as it groweth of his owne kinde.

The Place.

Harmala groweth (as Dioscorides writeth) in Cappadocia and Galatia, in this countrie the Herborists do sowe it in their gardens.

The Names.

This herbe is called in Greeke πήγανον ἄγριον: in Latine Ruta syluestris: of some it is called Harmala: of the Arabian Physitions, and of the late wryters Harmel. The people of Syria in times past called it Besasa, and some Moly. We may also call it Harmala, or Harmel.

The Nature.

Galen writeth, ỹ this herbe is hoate in the third degree, & of subtill partes.

The Vertues.

[A] Bycause Harmala is of subtil partes, it cutteth asunder grosse and tough humors, it prouoketh vrine, and womens naturall fluxe.

[B] The seede of Harmala stamped with Hony, Wine, Saffron, the iuyce of Fenell, and the gaule of a Henne, doth quicken the sight, and cleareth dimme eyes.

Of Rosemary. Chap.lxxv.

The Description.

Rosemary is as it were a little tree or wooddish shrubbe, with many small branches and slender boughes, of harde and wooddie substance, couered and set full of little, smal, long, and tender leaues, white on the side next the ground, and greene aboue. The floures are whitishe, and mixte with a little blewe, the whiche past, there commeth forth smal seede. The roote and the stemme are likewise harde and wooddie. The leaues and the floures are of a very strong and pleasant sauour, and good smacke or taste.

The Place.

Rosemary groweth naturally, and plentifully, in diuers places of Spayne and France, as in Prouence and Languedoc. They plante it in this countrie in gardens, and mayntayne it with great diligence.

The Tyme.

The Rosemary floureth twise a yeare, once in the spring time of the yeare, and secondarily in August.

The

❧ The Names.

This herbe is called in Greeke λιβανωτίς στεφανωματικὴ: in Latine Rosmarinũ coronarium: in Shoppes Rosmarinus: in English Rosemary: in Italian *Rosmarino*: in Spanish *Romero*: in French *Rosmarin*: in Germany Rosmarein: in base Almaigne Rosmarijn. They call it in Latine Rosmarinum coronarium, that is to say, Rosemarie whereof they make Crownes & Garlandes, to put a difference from the other Libanotis which is of diuerse sortes, wherof wee shall intreate in Chapters following. The blossoms or floures of this Rosemarie is called in shoppes Anthos.

❧ The Nature.

This Rosemarie is hoate and dry in the second degree.

❧ The Vertues.

Dioscorides and Galen do write that this Rosemary boyled in water, and giuē to drinke in the morning fasting & before labor or exercise, cureth the Jaunders.

The Arrabians and their successours Physitions, do say that Rosemarie coforteth the brayne, the memory, and the inwarde Sences, & that it restoreth speach, especially the conserus made of the floures thereof with Sugar, to be receyued dayly fasting.

The ashes or axen of Rosemarie burnte, doth fasten loose teeth, and beautifieth the same if they be rubbed therewith.

Rosmarinum coronarium.

Of Lauender and Spyke.   Chap. lxxxvi.

❧ The Kyndes.

Lauender is of two sortes, male and female. The male hath his leaues, floures, spikie eares, and stemmes, broader, longer, higher, thicker, and of a stronger sauour. The female is smaller, shorter, lower, and of a pleasanter sauour.

❧ The Description.

Both kindes of Lauender haue square hollow stalkes, with ioyntes & knottes, vpon whiche groweth grayishe leaues, whiche be long, narrow & thicke. Yet larger and longer than the leaues of Rosemarie. The floures (whiche are most commonly blew) grow thicke set, and couched togither in knoppes or spiked eares, at the toppes of the stalkes. The roote is of wooddie substance with many thready strings.

❧ The Place.

Lauender groweth in certayne places of Italy, Spayne, and Fraunce, on the Mountaynes & rough stonie places, that lie against the Sunne: they plant it here in gardens especially the female Lauender, whiche is very common in all gardens, but the male kinde is not founde sauing amongst the Herboristes.

Lauandula

the Historie of Plantes. 265

Lauandula mas.
English Spike.

Lauandula fœmina.
Lauender.

❧ *The Tyme.*

Lauender floureth in Iune and Iuly.

❊ *The Names.*

It is called in Latine Lauandula: in Shops Lauandula: in English Spike and Lauender: in Italian *Spigo*, and *Lauanda*, in Spanish *Alhuzema*, & *Alfazema*: of some in Greeke Pseudonardus, and of others Hirculus, and of some also Rosmarinum coronarium. It seemeth to be the herbe that Virgil calleth Casia, and Theophrastus Cneorus albus.

1 The first kind is Lauandula mas: in English Lauender or Spike: in French *Lauande masle*: in high Douch Spica, and Spica nardi: in base Almaigne Lauender, and Lauender manneken.

2 The seconde kinde is called Lauendula, and Lauendula fœmina: in English Spike and Female Lauender: in French *Lauande femelle*: in high Douch Lasendel: in base Almaigne Lauender wijfken.

❊ *The Nature.*

Lauender is hoate and dry in the second degree.

❧ *The Vertues.*

Lauender boyled in wine and dronken prouoketh vrine, & bringeth downe A the floures, and driueth forth the Secondine, and the dead Childe.

The floures of Lauender alone, or with Cinnamome, Nutmegs, & Cloues B do cure the beating of the harte, and the Iaunders, and are singuler against the Apoplexie, and giddinesse, or turning of the head, they comforte the brayne

Z   and

and members taken or subiect to the Palsie.

The conserue made of the floures with Sugar, profiteth much against the sayde Diseases, to be taken in the morning fasting, in quantitie of a Beane. C

The Distilled water of the floures of Spike or Lauender healeth members of the Palsie if they be washed therewith. D

Of Stæchados, or French Lauender. Chap.lxxvij.
Stæchas.

❧ The Description.

This is a beautiful herbe, of a good & very pleasant smell, with diuers weake & tender branches, set full of long, small, & whitish leaues, but smaller, narrower & tenderer, & of a more amiable sauour than the leaues of Lauender. At the toppe of the stalkes there growe fayre thicke knoppes or spikie eares, with smal blew floures, thicke set and thrust togither. These knops or eares are solde euery where in Shops by the name of Stæcados Arabicum.

There is yet an other herbe which the Apothecaries do call Stichas citrina, the whiche we haue described in the lx. Chapter of the first Booke.

❧ The Place.

This herbe groweth in the Iles called Stæchades, standing directly ouer agaynst Marsiles, and in diuers places of Languedoc and Prouince, & in Arabia. In this countrie some Herboristes do sowe it, and mayntayne it with great diligence.

❧ The Tyme.

Stæcados floureth in May and June, somwhat before Lauender.

❧ The Names.

It is called in Greeke σιχας, χαι σοιχας: in Latine Stichas and Stæchas: in Shoppes Stichas Arabica, & Stęchados Arabicum: in the Arabian tongue Astochodos: in English Stæchados, French Lauender, Cassidonie, & of some Lauender gētle: in Italian Sticados: in Spanish Cantuesso, Rosmarinho: in Frēch Stachados.

❧ The Nature.

The complexion of Stæchados is hoate and dry.

❧ The Vertues.

The decoction of Stæchados with his floures, or else the floures alone, droken, do ope the stoppings of the Liuer, the lunges, the Melte, the Mother, the bladder, and of all other inward partes, clensing and driuing forth, all euill and corrupt humors. A

It is also very good against the paynes of the head, and diseases of the brest and lunges, and it bringeth forth the floures, if it be taken in maner as is aforesayde. B

They mengle the floures, with good successe in conterpoysons, & medicines that are made to expell poyson. C

The leaues and floures of Stæchados giuen often to smell vpon doth comforte the brayne, the memorie, and inwarde senses. D

the Historie of Plantes.

Of Dictam, or Dittani of Candie. Chap. lxxxviij.

❧ *The Kindes.*

Dioscorides that auncient Herborist, writeth of three sortes of Dictam, whereof the first onely is the right Dictam. The second is the Bastarde Dictam. The thirde is an other kinde bearing both floures and seede.

❧ *The Description.*

Pseudodictamnum.
Bastarde Dictam.

1. The first kinde, whiche is the right Dictam, is (as Dioscorides saith) a hoate and sharp herbe much like vnto Penniropall, sauing that his leaues be greater & somewhat hoare or mosly with a certaine fine downe, or wolly white Cotton: at the top of the stalkes or smal braches, there grow as it were certayne small spikie eares or tuffets, hanging by little smal stës, greater & thicker thã the eares or spikie tuffets, of wild Margerome, somwhat redde of colour, in which there grow little floures.

2. The second kinde whiche is called Pseudodictamnum, that is to say, Bastarde Dittam, is much like vnto the first as Dioscorides saith: sauing that it is not hoate, neyther doth it bite the tongue: wherof we haue here giuen you the figure, the whiche we haue caused to be cut according to the naturall & true proportion of the plant. Pseudodictamnum hath round soft wolly stalkes with knottes and ioyntes, at euery of whiche ioynts or knottes, there are two leaues somewhat rounde softe and wollie, not muche vnlike the leaues of Penniroyall, sauyng they bee greater, all hoare, or white, softe and woolly, like to the firste leaues of white Mollyn or Tapsus Barbatus, without sauour and not sharpe, but bitter in taste. The floures be of a light blewe, cõpassing the stalke by certaine spaces like to garlandes or whorrowes, and like the floures of Pennyropall and Horehounde. The roote is of wooddy substance.

3. The thirde kynde in figure is lyke to the seconde, sauing that his leaues are greener and more hearie, couered with a fine white softe heare, almost like to the leaues of Water Minte. All the herbe is of a very good and pleasant smell, as it were betwixt the sauour of Watermynte & Sage, as Dioscorides witnesseth.

❧ *The Place.*

1. The first kinde or the right Dictam commeth from Crete, whiche is an Ilande in the Sea Mediterrane, whiche Ilande we do now call Candie, and it is not founde else where, as all the Auncients do write. Therefore it is

no maruell that it is not founde in this countrie, otherwise than dry, and that in the Shoppes of certayne wise and diligent Apothecaries, who with great diligence get it from Candy to be vsed in Phisicke.

2.3 The two other kindes do not grow only in Candy, but also in diuers other hoate countries.

※ *The Names.*

1 The first kinde is called in Greeke Δίκταμ۞: in Latine Dictamnum, & Dictamnum Creticum, of some as Dioscorides writeth, Pulegium syluestre: in Shoppes Diptamū, yet notwithstanding the Apothecaries haue vsed an other herbe in steede of this, which is no kinde of Dictam at all, as shalbe declared in his place: it may be called in English as Turner writeth Dictam, or Dictamnū of Candie.

2 The second kinde is called in Greeke ψευδοδίκταμ۞, and Pseudodictamnum, that is to say, Bastarde Dictam.

3 The thirde kinde is called Δίκταμ۞: in Latine Dictamnum, and may be well called Dictamni tertium genus, or Dictamnum non Creticum.

※ *The Nature.*

1 The right Dictam is hoate and dry like Pennyroyall, but it is of subtiller partes.

2.3 The other twayne are also hoate & dry, but not so hoate as the right Dictam.

※ *The Vertues.*

The right Dictam is of like vertue with Pennyroyall, but yet it is better & A stroger: it bringeth downe the floures, it expulseth the afterbirth and the dead childe, whether it be dronken or eaten, or put in vnder as a Pessarie or mother Subpositorie. The like vertue hath the roote, whiche is very hoate and sharpe vpon the tongue.

The iuyce of Dictam is very good to be dronken against all venim, and a- B gainst the bitings of all venemous beasts and Serpents.

Dictam is of suche force against poyson, that by the onely sauour and smell C thereof, it driueth away all venimous and wicked beastes, and in manner killeth them, causing the same to be astonied, if they be but onely touched with the same.

The iuyce of the same is of soueraigne and singular force, against all kindes D of woundes made with Glayue, or other kinde of weapons, and against all bitings of venimous beasts, to be dropped or powred in, for it doth both mundifie, clense and cure the same.

Dictam qualifieth and swageth the payne of the Splene or melte, and wa- E steth or diminisheth the same, whan it is tomuch swollen, or blasted: if it be eyther taken inwardely, or applied and layde outwardely.

It draweth forth shiuers, splinters, and thornes, if it be bruised and layde F vpon the place.

We may see it lefte to vs written of the Aunciencts, that the Goates of Can- G die being shotte in and hurte by any shafte or Iaueline hanging or sticking fast in their fleshe: how that incontinent they seeke out Dictamnum and eate therof, by vertue whereof the arrowes fall of, and their woundes are cured.

2 The Bastarde Dictamnum is somewhat like the vertues of the first, but it H is not of so great a force.

3 The thirde kinde auayleth much to be put into Medicines, drenches, and I implaysters that are made against the byting of wicked and venemous beasts.

Of

the Historie of Plantes.

Of Fenell. Chap. lxxix.

❧ The Kyndes.

There are two sortes of Fenell. The one is the right Fenell called in Greke Marathron. The other is that which groweth very high, and is called Hippomarathron, that is to say, great Fenell.

❧ The Description.

Fœniculum. Fenell.

1 The right Fenell hath round knottie stalkes, as long as a man, and full of branches: the sayde stalkes are greene without & hollow within, filled with a certaine white pithe or light pulpe. The leaues are long and tender, and very much, and small cut (so that they seeme but as a tuffte or bushe of small threedes), yet greater and gentler, and of better sauour than the leaues of Dill. The floures be of pale yellow colour, and do growe in spokie tuffets or rundels at the top of the stalkes: the floure perisshed it turneth into long seedes, alwayes two growing togither. The roote is white, long, and single.

There is an other sorte of this kinde of Fenell, whose leaues waxe darke, with a certayne kinde of thicke or tawny redde colour, but otherwise in all things like the first.

2 The other kinde called the great Fenell hath round stemmes with knees & ioyntes, sometimes as great as ones arme, and of sixtene or eightene foote long, as writeth the learned Ruellius.

❧ The Place.

Fenell groweth in this countrie in gardens.

❧ The Tyme.

It floureth in Iune and Iuly, and the seede is ripe in August.

❧ The Names.

1 The first kynde is called in Greeke μάραθρον: and of Actuarius μάλαθρον: in Latine and in Shoppes Fœniculum: in Englishe Fenell: in Italian Finochio: in Spanish Finicho: in French Fenoil: in high Douch Fenchel: in base Almaigne Wenckel.

2 The seconde kinde is called in Greeke ὑππομάραθρον: in Latine Fœniculum erraticum, that is to say, wilde Fenell, and great Fenell: and of some Fenell Giant.

❧ The Nature.

Fenell is hoate in the thirde degree, and dry in the first.

❧ The Vertues.

A The greene leaues of Fenell eaten, or the seede thereof dronken with Ptisan, filleth wemens breastes or dugges with milke.

B The decoction of the crops of Fenel dronken, easeth the payne of the kidneys, causeth one to make water, & to auoyde the stone, & bringeth downe ỹ floures.

The roote doth the like, the which is not only good for the intentes aforesayd, but also against the Dropsie to be boyled in wine and dronken.

The leaues and seede of Fenell dronken with wine, is good agaynst the stingings of Scorpions and the bitings of other wicked & venimous beastes.

Fenell or the seede dronken with water, asswageth the payne of y stomacke, and the wamoling or desire to vomite, which such haue, as haue the Ague.

The herbe, the seede and the roote of Fenell, are very good for the Lunges, the Liuer and the kidneys, for it openeth the obstructions or stoppings of those partes, and comforteth them.

The rootes pounde and layde too with honie, are good against the bytings of madde Dogges.

The leaues pounde with vineger are good to be layde to the disease called the wilde fire, and all hoate swellings, and if they be stamped togither with waxe, it is good to be layde to bruses and stripes that are blacke and blewe.

Fenell boyled in wine, or pounde with oyle is very good for the yearde, or secrete parte of man, to be eyther bathed or stued, or cubbed and anoynted with the same.

The iuyce of Fenell dropped into the eares, killeth the wormes breeding in the same. And the sayde iuyce dryed in the Sunne, is good to be put into Collyres, and medicines prepared to quicken the sight.

Of Dill. Chap. xc.

Anethum.

✤ The Description.

Dill hath rounde knottie stalkes, full of bowghes & branches, of a foote & halfe, or two foote long. The leaues be all to iagged, or frenged with small threddes, not much vnlike to fenel leaues, but a great deale harder, and the strings or thredes therof are greater. The floures be yellow & grow in round spokie tuffets or rundels, at the toppe of the stalkes like Fenell: whan they are vanished, there cōmeth the seede, whiche is small and flat, the roote is white, and it dieth yearely.

✤ The Place.

They sowe Dill in al gardens, amōgst wortes, and Pot herbes.

✤ The Tyme.

It floureth in June and July.

✤ The Names.

This herbe is called in Greeke ἄνηθον: in Latine and in the Shoppes Anethum: in English Dil: in Italian Anetho: in Spanish Eueldo, Endros in French Aneth: in high Douch Dyllen, & Hochkraut: in base Almaigne Dille.

✤ The Nature.

Dill is almost hoate in the thirde degree, and dry in the second.

✤ The Vertues.

The decoction of the toppes and croppes of Dill, with the seede boyled in water

the Historie of Plantes.

water and dronken, causeth wemen to haue plentie of Milke.

It driueth away ventositie or windinesse, and swageth the blasting & griping torment of the belly, it stayeth vomiting and laskes, and prouoketh vrine to be taken as is aforesayde.

It is very profitable against the suffocation or strangling of the Matrix, if ye cause wemen to receyue the fume of the decoction of it, thorough a close stole, or hollow seate made for the purpose.

The seede thereof being well chauffed, and often smelled vnto, stayeth the pexe, or hiquet.

The same burned or parched, taketh away the swelling lumpes, and riftes or wrincles of the tuell, or fundement, if it be layde thereto. The herbe made into arsen doth restrayne, close vp, and heale moyste vlcers, especially those that are in the share or priuie partes, if it be strowed thereon.

Dill boyled in oyle, doth digest and resolue, and swageth payne, prouoketh carnall luste, and rypeth all rawe and vnripe tumors.

❧ *The Daunger.*

If one vse it to often, it diminisheth the sight, and the seede of generation.

Of Anyse. Chap. xci.

❧ *The Description.*

Anisum.

Anise hath leaues like to yong Persley, that is new sprong vp: his stalkes be rounde and hollow, his leaues at the first springing vp, are somewhat round, but afterwarde it hath other leaues cut and clouen like to the leaues of Persley, but a great deale smaller & whiter. At the toppe of the stalkes groweth diuers faire tuftes, or spokie rundels with white floures, like to the tuftes of the small Saxifrage, or of Coriandre. After the floures are past, there cōmeth vp seede, which is whitish, and in smell and taste, sweete and pleasant.

❧ *The Place.*

Anise groweth naturally in Syria, & Candie. Now one may find good store sowen in the gardens of Flauders, and Englande.

❧ *The Tyme.*

It floureth in June, and July.

❧ *The Names.*

Anise is called in Greeke ἄνισον, καὶ ἄννησον: in Latine and in Shoppes Anisum: in Italian *Semenze de Anisi*: in Spanish *Matahalua, yerua doce*. in high Douch Anitz: in base Almaigne Anijs.

❧ *The Nature.*

The Anise seede, the whiche onely is vsed in Medicine, is hoate and dry in the thirde degree.

❧ *The Vertues.*

Anise seede dissolueth the windinesse, and is good against belching, and vpbreaking and blasting of the stomacke and bowels: it swageth the paynes and griping torment of the belly: it stoppeth the laske: it causeth one to pisse, and to auoyde the stone, if it be taken dry, or with wine or water: and it remoueth the hicquet or yex, not onely when it is dronken and receyued

receyued inwardly, but also with the onely smell, and sauour.

It cureth the blouddie flire, and stoppeth the white issue of wemen, and it is B very profitably giuen to such as haue the dropsie: for it openeth the pypes and conduits of the Liuer, and stancheth thirst.

Annise seede plentifully eaten, stirreth vp fleshly lust, and causeth wemen to C haue plenty of Milke.

The seede chewed in the mouth, maketh a sweete mouth, and easie breath, & D amendeth the stench of the mouth.

The same dried by the fier, and taken with Hony, clenseth the breast from E flegmatique superfluities, and if one put therevnto bitter Amandes, it cureth the olde Cough.

The same dronken with wine, is very good against al poyson, and the sting- F ing of Scorpions, and biting of all other venimous beastes.

It is singuler to be giuen to infants or yong children to eate, that be in dan- G ger to haue the falling sicknesse, so that such as do but only hold it in their hāds (as saith Pythagoras) shall be no more in perill to fall into that euill.

It swageth the squināce, that is to say, the swelling of the throte, to be gar- H gled with Hony, Vineger and Hyssope.

The seede thereof bounde in a little bagge or handecarcheff, and kept at the I Nose to smell vnto, keepeth men from dreaming, and starting in their sleepe, & causeth them to rest quietly.

The perfume of it, taken vp into the Nose, cureth head ache. K

The same pounde with oyle of Roses, and put into the eares, cureth the in- L warde hurtes, or woundes of the same.

Of Ameos, or Ammi. Chap.xcij.

❧ *The Kindes.*

Ameos is of two sortes, according to the opinion of the Physitions of our time, that is the great Ameos, and the small.

❧ *The Description.*

1 The great Ameos, hath a rounde greene stalke, with diuers bowes & braunches, the leaues be large and long, parted into diuers other little long narrow leaues, and dented rounde aboute. At the top of the stalke there groweth white starlike floures in great rundels, or spokie tuftes, the whiche bringeth forth a small sharpe and bitter seede. The roote is white and threddie.

2 The small Ameos, is an herbe very small and tender, of a foote long or somwhat more. The stalke is small & tender. The first and oldest leaues are long, and very much cut and clouen round aboute. The vpper leaues draw towards the proportion of the leaues of Fenell or Dill, but yet for all that they are smaller. At the toppe of the stalke there groweth also in spoky littell tuffets or rundels, the small little white floures, the whiche afterwarde do turne into small gray seede, hoate and sharpe in the mouth. The roote is little and small.

❧ *The Place.*

These two herbes grow not in this countrie of themselues, without they be sowen in the gardens of Herborists. Neuerthelesse whereas they haue bene once sowen, they grow yearely of the seede whiche falleth of it selfe.

❧ *The Tyme.*

They floure in July and August, and shortely after they yeelde their seede.

❧ *The Names.*

1 The first kinde is called in Shoppes Ameos, by whiche name it is knowen in this

the Historie of Plantes. 273

in this countrie. The same as we thinke is the right ἄμμι, Ammi described by Dioscorides, who calleth it also Cuminum Aethiopicum, Cuminum regium, & as Ruellius saith, Cuminum Alexandrinum.

2 The small is taken of diuers of the learned writers in our dayes, for ἄμμι, Ammi, and therefore we haue placed it in this Chapter.

Ammi commune.　　　　　　　Ammi paruum.
Great Ameos.　　　　　　　**Small Ameos.**

❧ *The Nature.*

The seede of Ameos is hoate and dry in the third degree.

❧ *The Vertues.*

The seede of Ameos is very good against the griping payne and tormēt of A the belly, the hoatepisse, and the strangurie, if it be dronken in wine.

It bringeth to wemen their naturall termes, and the perfume thereof, toge- B ther with Rosin and the kernels of Rapsons, strowed vpō quicke coales, mundifieth and clenseth the Mother, if the same be taken in some hollow vessell or close stoole.

It is good to be dronken with wine, agaynst the bytings of all kindes of C venimous beastes: they vse to mingle it with Cantharides, to resist the venim of the same bycause they should not be so hurtefull vnto man, as they are whan they are taken alone.

Ameos breyed and mengled with Hony, scattereth congeled bloud, and put- D teth away blacke & blew markes, whiche happen by reason of stripes or falles, if it be layde too in manner of a playster. ❧ *The Daunger.*

The seede of Ameos taken into great a quantitie, taketh away the colour, and bringeth such a paalnesse, as is in dead bodies.

Of

Of Caruwayes. Chap. xciij.

Caros.

※ The Description.

Caruway hath a hollow, straked or crested stalke, with many knots or ioyntes, the leafe is very like to Carot leaues. The floures are white, and grow in tuffets or rundels, bearing a small seede, and sharpe vpon the tongue. The roote is meetely thicke, long and yellow, in taste almoste like vnto the Carot.

※ The Place.

Caruway groweth in Caria, as Dioscorides writeth. Now there is of it to be found in certayne dry medowes of Almaigne. In this countrie it is sowen in gardens.

※ The Tyme.

It floureth in May, a yeare after the sowing thereof, and deliuereth his seede in Iune and Iuly.

※ The Names.

This herbe is called in Greeke κάρος: in Latine Careum or Carum: in Shops and in Italian Carui: and it tooke his name of the countrie of Caria, whereas it groweth plentifully: in English it is called Caruway, and the seede Caruway seede: in French Carui, or Carotes: in Spanishe Alcaranea, Alcoronia: in high Douche Weitz kummel: in base Almaigne Witte Comijn.

※ The Nature.

Caruway seede is hoate and dry in the thirde degree.

※ The Vertues.

The Caruway seede, is very good and conuenient for the stomacke, and for the mouth, it helpeth digestion, and prouoketh vrine, and it swageth and dissolueth all kinde of windinesse and blastings of the inwardes partes. And to conclude, it is answereable to Annis seede in operation and vertue.

The rootes of Caruway boyled, are good to be eaten like Carottes.

Of Comijn. Chap. xciiij.

※ The Kyndes.

Comyn, as Dioscorides writeth, is of two sortes, tame and wilde.

Cuminum satiuum.	Cuminum syluestre.
Garden Comyn.	Wilde Comyn.

※ The Description.

1. The Garden Comyn hath a streight stem, with diuers branches: the leaues be all iagged and as it were threedes not much vnlike Fenell. The floures grow in rundels or spokie toppes, like to the toppes of Anyse, Fenell, and Dill. The seede is browne and long.

2. The wilde Comyn (as Dioscorides saith) hath a brittle stalke, of a span log, vpon whiche groweth foure or fiue leaues all iagged & snipt, or dented rounde about, and it is not yet knowen.

The

The other wilde kinde wherof Dioscorides writeth shalbe hereafter described in the lxxxvj. Chapter amongst the Nygelles, or Larke spurres.

✤ The Place.

The garden Comyn groweth in Ethiopia, Egypte, Galatia, the lesser Asia, Cilicia, and Terentina. They do also sowe it in certayne places of Almaigne, but it desireth a warme and moyst grounde.

✤ The Names.

1 The common & garden Comyn is called in Greeke κύμινον ἥμερον: in Latine Cuminum satiuum: in Shoppes Cyminum: in English Comyn or Comijn: in Italian *Cimino*: in Spanish *Cominos, Cominhos*. in French *Comyn*: in high Douch Romische Kummel, and zamer Kummel: in Brabante Comijn.

2 The wilde Comyn is called in Greeke κύμινον ἄγριον: in Latine Syluestre Cuminum, and Cuminum rusticum.

✤ The Nature.

The seede of Comyn is hoate and dry in the thirde degree.

✤ The Vertues.

Comyn scattereth and breaketh all the windinesse of the stomacke, the belly, the bowels and Matrix: also it is singuler against the griping torment, and knawings or frettings of the belly, not onely to be receyued at the mouth, but also to be powred into the bodie by clysters, or to be layde to outwardly with Barley meale. A

The same eaten or dronken is very profitable for suche as haue the Cough, and haue taken colde, and for those whose breastes are charged or stopped: and if it be dronken with wine, it is good for them that are hurte with any venimous beastes. B

It slaketh and dissolueth the blastings and swellings of the Coddes and Genitors being layde therevpon. C

The same mengled with Puray meale, and poulpe or substance of raysins, stoppeth the inordinate course of the floures, being applied to the belly in forme of a playster. D

Comyn seede pounde, and giuen to smell vnto with vineger, stoppeth the bleeding at the Nose. E

✤ The Daunger.

Comyn being to much vsed, decayeth the naturall complexion and liuely colour, causing one to looke wanne and paale.

✤ Of Coriander.   Chap.xcv.

✤ The Description.

Coriander is a very stinking herbe, smelling like to the stinking worme called in Latine Cimex, & in French *Punaise*, it beareth a round stalke full of branches of a foote and halfe long, the leaues are whitish all iagged and cut: the vnder leaues that spring vp first are almost like to the leaues of Charuell or Persele: and the vpper & last leaues are not much vnlike to the same, or rather like to Fumeterrie leaues, but a great deale tëderer, & more iagged. The floures be white & do grow in round tuffets. The seede is all rounde, and hollow within, & of a pleasant sent whan it is dry. The roote is harde and of wooddie substance.

✤ The Place.

Coriander is sowen in fieldes and gardens, and it loueth a good and frutefull grounde.

✤ The Tyme.

It floureth in July and August, and shortly after the seede is ripe.

❧ The Names.

This herbe is called in Greeke κόριον, ἢ κορίανον: in Latine & in Shoppes Coriandrum: in English Coriander, and of some Coliander: in Italian *Coriandro*: in Spanish *Culantro*, *Coentro*: in Frenche *Coriandre*: in Douch Coriander.

❧ The Nature.

The greene and stinking Coriander, is of complexion colde and dry, and hurtefull to the body: the dry and sweete sauoring seede is warme, and conuenient for many purposes.

❧ The Vertues.

Coriander seede prepared, and taken alone (or couered in Sugar) after meales, closeth vp the mouth of the stomacke, stayeth vomiting, and helpeth digestion.

The same rosted or parched and dronke with wine, killeth and bringeth foorth wormes of the body, and stoppeth the laske and the bloudy flire, and all other extraordinarie issues of bloud.

The seede of Coriander is prepared after this maner. Take of the seede of Coriander well dried, vpon whiche ye shall power or caste good strong wine and vineger mingled togither, and so leaue them to stiepe & soke by the space of xxiiij. houres: than take it forth of the liquor, and drye it, and so keepe it to serue for Medicine.

Ye must also note, that the Apothecaries ought not to sell to any person, of Coriander seede vnprepared, nor to couer it with Sugar, nor to put it in Medicine: for albeit it be wel dryed and of good taste, yet notwithstanding it may not be but a little vsed in medicine without great perill and danger.

The herbe Coriander being yet fresh and greene, & boyled with the crommes of white bread, or Barley meale, dryeth away & consumeth all hoate tumors, swellings and inflammations, and with Beane meale it dissolueth the kings euill, and wennes or harde lumpes.

The iuyce of Coriander layde to with Ceruse, Litharge, or skume of Siluer, vineger, and oyle of Roses, cureth S. Anthonies fire, and swageth and easeth all inflammations that chaunce on the skinne.

❧ The Daunger.

Greene Coriander taken into the bodie causeth one to waxe hoarse, and to fall into frensie, and doth so much dul the vnderstanding, that it seemeth as the partie were dronken. And the iuyce thereof dronken in quantitie of foure Drammes, killeth the bodie, as Serapio writeth.

Of Git or Nigella. Chap.xcvi.

❧ The Kyndes.

Nigella is of two sortes, tame and wilde, whereof the tame or garden Nigella, is agayne parted into two sortes, the one bearing blacke seede, the other

Coriandrum.

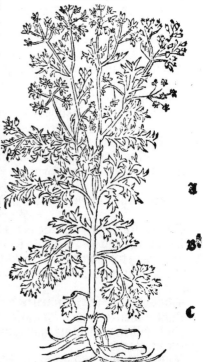

other a Citren colour or pale yellow seede, but otherwise like one to an other, as in stalkes, leaues, floures, and smell.

| Melanthium satiuum. | Melanthium sylueſtre. |
| Garden Nigella. | Wilde Nigella. |

❦ The Description.

1 The garden Nigella hath a weake and brittle stalke, full of braunches, and of a foote long. The leaues be all to cut and iagged, much like to the leaues of Fumeterrie, but much greener. The floures grow at the toppe of the braunches, and are white, turning towardes a whitish or light blewe, eache floure parted into fiue small leaues, after the maner of a little starre or rowell. After that the floures be past, there commeth vp small knops or heads, with fiue or sixe little sharpe hornes vpon them, eache knop is diuided in the inside into fiue or sixe celles, or little chambers, in whiche is contepned the seede, the whiche (as we haue before sayde) is sometimes blacke, and somtimes a bleeke or faynt yellow, and like to Onyon seede, in tast sharpe, and of a good pleasant strong sauour.

2 The wilde Nigella hath a straked, or crested stalke, of two spanes long, his leaues be ashe colour, and all to cut, more iagged than the leaues of garden Nigella, drawing towardes the leaues of Dill. The floures are like to the floures of garden Nigella, sauing that they be blewer: The heades or knoppes are also parted into fiue horned huskes, much like to Columbyne huskes, in whiche is contepned the sweete and pleasant seede.

3 There is yet an other Nigella, whiche is both fayre and pleasant, and is called Damaske Nigella, it is much like to the wilde Nigella in the small cut and iagge.

iagge of his leaues, but his stalke is longer. The floures are blewe and diuided into fiue partes lyke to the others, but a greate deale fayrer and blewer with fiue little leaues vnderneath them, very small cut and iagged, from the middle poynt or center whereof, the floure springeth. Whā the floures are gone, there appeareth the knoppes or horny heades, like as in the garden Nigella, in whiche also is conteyned the seede, and it is blacke like to the seede of the gardē Nigella, but it hath no sweete sauour.

Melanthium Damascenum.
Damaske Nigella.

❧ *The Place.*

1.3 These Nigellas are not found in this countrie, sauing in gardens whereas they be sowen.

2 The wild is found growing in fields, in certayne places of Fraunce and Almaigne.

3 The Damaske Nigella groweth plentifully through-out all Languedoc.

❧ *The Tyme.*

The Nigellas do floure in June and July.

❧ *The Names.*

Nigella is called in Greeke μελάνθιον: in Latine Melanthium, Nigella, and Papauer nigrum: in Shoppes Nigella, and of some Gith: in French *Nielle*.

1 The firste kinde is called Melanthium satiuum, and Nigella domestica, of some Salusandria: in English Garden Nigella: in Italian *Nigella ortelana*: in Spanish *Alipiure, Axenuz*: in high Douche Schwartz kumich, Schwartz kumel, in base Almaigne Nardus, and the seede is called Nardus saet: in French *Poyurette*, and of some *Barbue*.

2 The wilde Nigella is called Melanthium syluestre, and Nigella syluestris: in French *Nielle saunage, or Barbues*. in high Douch S. Catharinen blumen, that is to say, S. Catharines floure, of some Waldt schwartz kumich: some learned men thinke it to be wilde Comyn, whereof we haue written in the lxxxiiij. Chapter of this Booke.

3 The thirde kinde is now called Melanthium Damascenum, and Nigella Damascena, that is to say, Damaske Nigella: in French *Nielle de Damas*: in high Douch Schwartz Coriander.

❧ *The Nature.*

The seede of Nigella is hoate and dry in the thirde degree.

❧ *The Vertues.*

The seede of Nigella dronken with wine, is a remedie against the shortnesse of breath, it dissolueth, and scattereth all ventositie and windinesse in the body, it prouoketh vrine, & floures, it increaseth womans milke, if they drinke it often.

The same slayeth, and driueth out wormes, whether it be dronken with **B** wine or water, or else layde to the Nauell of the belly. The same vertue hath the oyle that is drawen forth of Nigella seede, to annoynt the region of the belly and nauell therewith.

The quantitie of a Dramme of it dronke with water, is very good against **C** all poyson, and the biting of venimous beasts.

The onely fume or smoake of Nigella tosted or burnt, driueth away Ser- **D** pents and other venimous beasts, and killeth Flies, Bees, and Waspes.

The same mingled with the oyle of Ireos, and layde to the forehead cureth **E** the head ache: and oftentimes put into the Nose, is good against the webbe, & bloudshotten of the eyes, in the beginning of the same.

The same well dried and pound, and wrapped in a piece of Sarsenet, or fine **F** linencloath, and often smelled vnto, cureth all Murres, Catharrhes, & poses, drieth the brayne, and restoreth the smelling being lost.

And boyled with water and vineger, and holden in the mouth, swageth the **G** tootheache, and if one chewe it (being well dried,) it cureth the vlcers and sores of the mouth.

It taketh out Lentils, Freckles, and other spottes of the face, and clenseth **H** foule scuruinesse and itche, and doth soften olde, colde, and harde swellings, being pounde with vineger and layde vpon.

The same stieped in olde wine, or stale pisse (as Plinie saith) causeth the **I** Cornes and Agnayles to fall of from the feete, if they be first scarified and scotched rounde aboute.

❧ *The Daunger.*

Take heede that ye take not to much of this herbe, for if ye go beyonde the measure, it bringeth death. Turner lib.secundo, fol.10.

Of Libanotis Rosmarie. Chap. xcvij.

❧ *The Kyndes.*

Libanotis, as Dioscorides writeth, is of twoo sortes, the one is frutefull, the other is barren. Of the frutefull sorte there is two or three kindes.

❧ *The Description.*

1 The first frutefull kinde, hath leaues (as Dioscorides saith) very much diuided and cut lyke vnto Fenell leaues, sauing they be greater and larger, moste commonly spread abroade vpon the grounde: amongst them groweth vp a stalke of a cubite, that is a foote and halfe long or more, vpon whiche grow the floures in spokie tuffets like Dill, and it beareth great, round, cornered seede, of a strong sauour, and sharpe taste. The roote is thicke, and hearie aboue, and sauoring like Rosin.

2 The seconde kinde hath a long stalke with ioyntes like the Fenell stalke, on whiche growe leaues almoste like Charuill, or Homlocke, sauing they be greater, broader, and thicker. At the toppe of the stalkes groweth spokie tuffets, bearing white floures, the whiche do turne into sweete smelling seede, flatte, and almost like to the seede of Angelica and Brank brsine. The roote is blacke without and white within, hearie aboue, and sauereth like to Rosin or Frankencence.

3 There is yet an other sorte of these fruteful kindes of Libanotis, the which is described by Theophrastus Lib.ix.Chapt.xij. It hath also a straight stalke with knottes and ioyntes, and leaues greater than Marche or Smallache. The

The floures grow in tuftes, like as in the two other kindes, & bringe foorth great long, and vneuen seede, which is sharpe in taste. The roote is long, great, thicke, and white, with a certayne kinde of great thicke heare aboue, and smelleth also of Frankencence or Rosin.

Libanotidis alterum genus.   Libanotis Theophrasti.

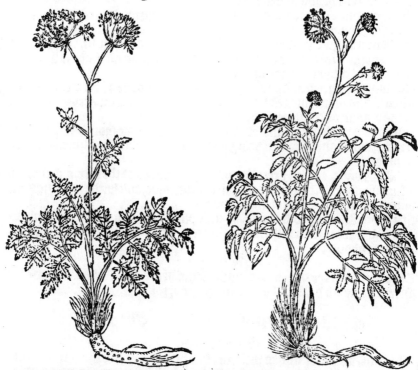

4   The barren Libanotides (as Dioscorides writeth) are like to the frutefull in leaues & rootes, sauing they beare neither stalkes, floures, nor seede.

5   The other kinde of Libanotis called Rosmarinum coronarium, in English Rosmarie, hath bene already described Chap. lxxb. of this Booke.

❀ *The Place.*

The frutefull Libanotides, are now founde vpon the high mountaynes, hilles, and desertes of Germany.

❀ *The Tyme.*

These herbes do floure most commonly in July.

❀ *The Names.*

This herbe is called in Greeke λιβανωτις, Libanotis, bicause that his roote sauoreth like þ Encens, which is called in Greke Libanos, in Latine Rosmarinus.

1   The first kind (as Dioscorides writeth) is called of some Zea, and Campsanema: in Shoppes Fœniculus porcinus: in high Douche Beerwurtz: in base Almaigne Beerwortel, that is to say, Beers roote.

The seede therof is called in Greeke κάγχευς, και κάχευς: in Latine Canchrys or Cachrys.

2   The second kinde is called in high Douch Schwartz hirtzwurtz, that is to say, blacke Harte roote.

3 ¶ The thirde is described of Theophrastus, wherefore we haue named it Libanotis Theophrasti: in high Douch weisz Hirtzwurtz, that is to say, white Hartes roote, the seede of this kinde is also called of Theophrastus, Canchrys, or Cachrys.

❀ *The Nature.*

These herbes with their seedes and rootes are hoate and dry in the second degree, and are proper to digest, dissolue, and mundifie.

❀ *The Vertues.*

The rootes of Libanotis dronken with wine, prouoketh vrine, & floures, & healeth the griping paynes and torment of the belly, and are very good against the bytings of Serpents, and other venimous beasts. A

The seede of Libanotis, is good for the purposes aforesayde: Moreouer it is singuler good against the falling sicknesse, and the olde and colde diseases of the breast. They vse to giue it to drinke with pepper against the Iaunders, especially the seede of the seconde kynde of Libanotis, for as touching the seede of the first kinde called Cachrys, it is not very good to be taken into the bodie, seing that by his great heate and sharpnesse, it causeth the throote to be rough and grieuouse. B

The leaues of al the Libanotides pounde, do stoppe the fluxe of the Hemorrhoides or Pyles, and do souple the swellings and inflammations of the tuell or fundement, and it mollifieth and ripeth all olde colde and harde swellings, being layde thereupon. C

The iuyce of the herbe and rootes put into the eyes with hony, doth quicken the sight, and cleareth the dimnesse of the same. D

The dry roote mengled with Hony, doth scoure and clense rotten vlcers, and doth consume and waste all tumors or swellings. E

The seede mengled with oyle, is good to annoynt them ẏ haue the Crampe, and it prouoketh sweate. F

The same mengled with Puray meale and vineger, swageth the payne of the goute when it is layde thereto. G

It doth also clense and heale the white dry scurffe, and manginesse, if it be layde on with good strong vineger. H

They lay to the forehead the seede called Cachrys, against the bloudshotten or watering eyes. I

Of Seseli.   Chap. xcviij.

❀ *The Kindes.*

SEseli, as Dioscorides writeth, is of three sortes. The first is called Seseli Massiliense. The second Seseli Aethiopicum. The third Seseli Peloponnense.

❀ *The Description.*

1 ¶ The first kinde of Seseli named Massiliense, his leaues are very much clouen and finely iagged, but yet they be greater and thicker than the leaues of Fenell, the stalke is long and high, with knottie ioyntes, and beareth tuffets at the toppe like to Dill, and seede somewhat long, & cornered, sharpe and biting. The roote is long like to the roote of the great Saxifrage, of a pleasant smell (as Dioscorides writeth) and sharpe taste.

2 The seconde Seseli (as Dioscorides saith) hath leaues like Iuye, but smaller and longer drawing neare to the proportion of Woodbine leaues. The stalke is blackishe, of three or foure foote long and ful of branches. The floures are yellow and grow in spokie rundels like Dill. The seede is as great as a wheate Corne, thicke, swarte, and bitter: And this is counted to be the Ethiopian Seseli, although in deede it is not the right Ethiopian Seseli.

Seseli Massiliense.   Seseli Aethiopicum.

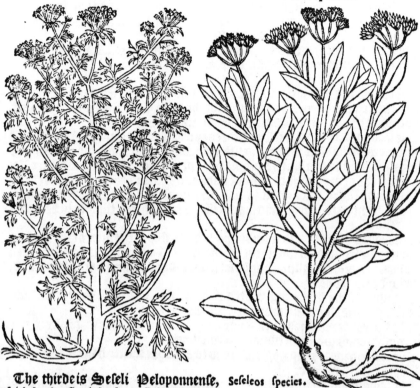

3 The thirde is Seseli Peloponnense, which hath a straight long stalke like Fenell, or longer, and groweth higher then Seseli of Marsiles. The leaues are all to cut and parted into diuers other small leaues, yet greater and larger than the leaues of Homlock. The seede groweth likewise in spokie toppes, and is broade and thicke.

Seseleos species.

4 Amongst the kindes of Seseli, we may place that strange herbe which is found in the gardens of certayne Herboristes. It hath at the first broade leaues spread vpon the grounde, very tender & finely iagged. The stalke is aboute foure or fiue foote long, with knottie ioynts, and round like to a Fenel stalke, but a great deale slenderer, and of a faynt greene colour changing towards yellow. The leaues that grow at the knots or ioyntes of the stalkes do bende and hang downewardes, but especially ye highest (except a few small leaues) whiche grow betwixt the others, & they grow vpward. The toppes of the stalkes and

and branches, are full of small spoky tuffets, bearing yellow floures, and afterwarde seede. The roote is long and lasteth many yeares.

❧ *The Place.*

1. The first kinde (as writeth Dioscorides) groweth in Prouence, and especially about Marsels, wherfore it is called Seseli of Marsels.
2. The second groweth, as witnesseth the sayd Dioscorides, in Ethiopia: and it groweth also meetely plentifully in Prouence, and Languedoc.
3. The third kinde groweth in Peloponneso, the whiche is now called Morea, and it lieth in Greece, and is now vnder the Empire and dominion of the Turcke.
4. The fourth is found vpō certayne Mountaynes of Lombardie, a man shall also finde, it as some say, in certayne places of Brabant.

❧ *The Tyme.*

1. The first floureth twise a yeare, in the spring, and Autumne.
2.3.4. The second, thirde, and fourth, do floure in Autumne.

❧ *The Names.*

1. The first kinde is called in Greeke σίσιλι μασσαλεωτικόν: in Latine Seseli Massiliense, of some πλατυκύμινον, that is Latum Cuminum, which is as much to say in English as large, and broade Comyn.
2. The second kind is called in Greeke σίσιλι αἰθιοπικόν: in Latine Seseli Aethiopicum, and of Egyptians κύον φρίκη, Cyonos phrice.
3. The third kinde is called σίσιλι πελοποννησιακόν: in Latine Seseli Peloponnése, that is to say, Seseli Peloponnense.
4. The fourth hath no speciall name, sauing that some take it for a kinde of Seseli, and some for Libanotis.

❧ *The Nature.*

The seede and roote of Seseli, are hoate and dry in the second degree, and of subtile partes.

❧ *The Vertues.*

The seede dronken with wine, comforteth and warmeth the stomacke, helpeth digestion, and driueth away the gnawing and griping of the belly, it cureth the shakings and brusing of a Feuer, and is very good against the shortnesse of breath & an old Cough, to be short it is good for al the inwarde partes. A

It prouoketh vrine, and is good against the strangurie, and hoate pisse, it prouoketh the menstruall Termes, expulseth the dead childe, and setteleth in his naturall place againe the Matrix or Mother that is risen out of his place. B

It is much worth vnto them that haue the falling sicknesse. C

The traueler that drinketh the seede of Seseli, with Pepper and wine, shal not complayne much of colde in his iourney. D

The same giuen vnto Goates, and other foure footed beasts to drinke, causeth them easilie to deliuer their yōg ones. The same propertie hath the leaues to be giuen to the cattell to eate. E

Of Seseli of Candie. Chap. xcix.

❧ *The Description.*

This is a tender herbe, about the length of a foote and halfe, his branches are tender and small, and set but with a fewe leaues, whiche be very small iagged and cut. At the toppe of the branches growethe little spokie tuffets or rundels, with white floures, the whiche being past, there commeth seede, whiche is redde, round, and flatte, garnished or compassed aboute with a white border, two seedes growing togither one against

Aa iiij an

284 The second Booke of

an other, eache of them hauing the shape and proportion of a Target or Buckler. The roote is small and tender, and dieth yearely, so that it muste be euery yeare new sowen againe.

❦ *The Place.*

This herbe (as Dioscorides writeth) groweth vpon the Mountayne Amanus in Cilicia: it is to be found in this countrie, in the gardēs of some diligent Herboristes.

❦ *The Tyme.*

It floureth in July and the seede is ripe in August.

❦ *The Names.*

This herbe is called in Greeke τορδύλιον, ἢ στοιλι κρητικὸν, of Paulus Egineta γορδύλιον: in Latine Tordylium, & Seseli Creticum: in English Seseli of Candy: vnknowen for the most parte in Shoppes.

❦ *The Nature.*

The seede of Seseli of Candie, is hoate and dry in the second degree.

❦ *The Vertues.*

The seede of Tordylion dronken in wine, prouoketh vrine, and is good against the strangurie, & causeth women to haue their moneths or termes.

The iuyce of it drōken (in the quātitie of a drāme, boyled with good wine,) by the space of tenne dayes, cureth the disease of the raynes or kidneys.

The roote thereof mengled with Hony and often-licked vpon, causeth to spitte out the tough and grosse Phlegmes, that are gathered aboute the breast and lunges.

Tordylion.

Of Daucus.   Chap. c.

❦ *The Kyndes.*

MEn do finde three sortes of herbes, comprehended vnder the name of Daucus, as Dioscorides & all the Auncients do write, whereof the third, is onely knowen at this day.

❦ *The Description.*

1   The firste kinde of Daucus is a tender herbe, with a stalke of a spanne long, set with leaues a great deale smaller and tenderer than Fenell leaues. At the toppe of the stalke groweth little spokie tuffets, with white floures, like to the tops of Coriander, yeelding a little long rough white seede, of a good sauour and a sharpe taste. The roote is of the thicknesse of ones finger, and of a spanne long.

2   The seconde kinde is like to wilde Persley, the seede whereof is of a very pleasant and Aromaticall sauour, and of a sharpe and byting taste: & both these kindes are yet vnknowen.

3   The thirde kinde (as Dioscorides writeth) hath leaues like Coriander, white floures, and a tufte or spokie bushe, like to wilde Carot, and long seede. For this kinde of Daucus, there is now taken, the herbe whiche some do call wilde Carrot, other call it burdes nest: for it hath leaues like Coriander, but greater, and not muche vnlike the leaues of the yellow Carrot. His floures
be

the Historie of Plantes.

be white growing vpon tuffets or rundels, like to the tuffets of ye yellow Carrot, in the middle whereof there is founde a little small floure or twayne of a broune redde colour, turning towardes blacke. The seede is long and hearie, and sticketh or cleaueth fast vnto garmēts. The roote is small and harde.

Dauci tertium genus.

❧ *The Place.*

1 The firste kinde groweth in stony places, that stād full in the Sunne, especially in Candy, as Dioscorides writeth.

3 The third kinde groweth euerywhere in this countrie, aboute the borders of fieldes, in stony places, & by the way sides.

❧ *The Tyme.*

The third kinde of Daucus floureth in July and August. ❧ *The Names.*

The Daucus is called in Greke δαῦκ[ον]: in Latine Daucum and Daucium.

1 The first kind is called Daucum Creticum, that is to say, Daucus of Candie.

3 The third kinde is called in Shoppes Daucus, & of some also Daucus Creticus: in English Daucus, and wilde Carrot: in Frēch *Carrotte sauuage.* in high Douch Vogelnest, that is to say, Birdes nest: in base Almaigne Croonkēs cruyt: & the same is but a certayne wilde Carrot.

❧ *The Nature.*

The seede of Daucus is hoate and dry, almost vnto the thirde degree.

❧ *The Vertues.*

The seede of Daucus dronken is good against the strangurie, and painefull A making of water, against the grauell & the stone: it prouoketh vrine, & floures, and expulseth the dead fruyte and Secondine.

It swageth the torment and griping payne of the bellie, dissolueth windinesse, cureth the Colique, and ripeth an old Cough. B

The same taken in wine, is very good against the bitings of venimous C beasts, especially against the stingings of Phalanges or fielde spiders.

The same pounde and layde to, dissolueth & scattereth colde softe swellings D and tumors.

The roote of Daucus of Candie dronken in wine, stoppeth the laske, and is E a soueraigne remedie against venim and poyson.

Of Sarifrage.   Chap.ri.

❧ *The Kyndes.*

The Sarifrage is of two sortes, great and small.

❧ *The Description.*

1 The great Sarifrage hath a long hollow stalke with ioyntes or knees, whereon groweth darke greene leaues, turning towardes blacke, made & fashioned of many small leaues growing vpon one stem, after the order of the garden (Carrot or) Parsenip, but much smaller, & each little leafe alone, is snipt round about the edges saw-fashiō, the floures are white, & grow in roūd

Cronets

Cronettes oꝛ ſpoky tuffets. The ſeede is like to common Parſelie ſeede, ſauing that it is hoater, and byting vpon the tongue. The roote is ſingle, white and long, like the Parſelie roote, but ſharpe and hoate in taſte like Ginger.

Saxifragia maior.	Saxifragia minor.
The great Sarifrage.	The ſmall Sarifrage.

2   The ſmall Sarifrage is altogither like the great, in ſtalkes, leaues, floures and ſeede, ſauing that it is a great deale ſmaller, and of a greater heate, and ſharpneſſe. The roote is alſo long and ſingle, of a very hoate and ſharpe taſte.

3   There is yet an other ſmall Sarifrage like to the aforeſayde in ſtalkes, floures, ſeede and roote, and in proportion, ſmacke and ſmell, ſauing his leaues are deeper cut, and of an other faſſhion, not much vnlike the leaues of Parſelie of the garden, oꝛ the wilde Parſelie.

❧ *The Place.*

1   The great Sarifrage groweth in high medowes, and good groundes.

2.3 The ſmal Sarifrages growe vnder hedges, and alongſt the graſie fieldes, in dꝛy paſtures, both theſe kindes are very common in this countrie.

❧ *The Tyme.*

Sarifrage floureth after June vnto the ende of Auguſt, and from that time foorth the ſeede is ripe.

❧ *The Names.*

The Sarifrage is called in Latine and in the Shoppes of this countrie Saxifragia, and Saxifraga, of Symon Janneuſis Petra ſindula, of ſome Bibinella: in high Douch Bibernell, and Feldmoꝛen: in baſe Almaigne Beuernaert, and Beuernelle. There be ſome alſo whiche call it Bipennula, Pimpinella, and Pampinula,

pinula, the whiche is the peculier or proper name of our Burnet described in the xcv. Chapter of the first booke: and doth not apperteyne vnto these herbes, as it appeareth by this olde Verse: Pimpinella pilos, Saxifraga non habet vllos: that is to say, Pimpinell or Burnet hath heares but Sarifrage hath none. Whereby it appeareth that our Pimpinell commonly called in English Burnet, (which) hath certayne fine heares appearing in the leaues whan they are broken) was called in times paste in Latine Pimpinella, and this whiche hath no heariness at all was called Saxifragia. Some learned men of our time, traueling to bring the small Sarifrage vnder certayne Chapiters of Dioscorides do call it Sison: and others Petroselinum Macedonicum: The third sorte wolde haue it a kinde of Daucus. But in my iudgement it is much like to Dioscorides βούνιον, Bunium.

The Nature.

Sarifrage with his leaues, seede and roote is hoate and dry euen to the thirde degree.

The Vertues.

The seede & roote of Sarifrage dronken with wine, or the decoction thereof made in wine, causeth to pisse well, breaketh the stone of the kidneys and bladder, and is singular against the strangurie, and the stoppings of the kidneys and bladder. A

The roote bringeth to women their termes, & driueth forth of the Matrix, the Secondine and the dead frupte, if it be taken in maner aforesayde. B

The roote dried and made into pouder, and taken with Sugar, comforteth and warmeth the stomacke, helpeth digestion, and cureth the gnawing and griping paynes in the belly, and the Colique, by dryuing away ventositie or windinesse. C

The same with the seede, are very good for them whiche are troubled with any Conuulsion or Crampe, and Apoplexie, and for such as are troubled with long colde Feuers, and for them that are bitten with any venimous beast, or haue taken any poyson. D

The same dronken with wine and vineger, cureth the Pestilence, and holden in the mouth preserueth a man from the sayde Disease, and purifieth the corrupt ayre. E

The same chewen vpon, maketh one to auoyde much flegme, and draweth from the brayne all grosse and clammy superfluities, it swageth toothache, and bringeth speach againe, to them that are taken with the Apoplexie. It hath the same vertue if it be boyled in vineger alone, or with some water put thereto, and afterwarde to holde it in the mouth. F

The iuyce of the leaues of Sarifrage, doth clense and take away all spots and freckles and beautifieth the face, and leaueth a good colour. G

It mundifieth corrupt and rotten vlcers, if it be put into them. The same vertue hath the leaues brused and layde vpon. H

The destilled water, alone or with vineger cleareth the sight, and taketh away all obscuritie and darkenesse, if it be put into the same. I

Of white Sarifrage or Stonebreake. Chap. cij.

The Description.

The white Sarifrage hath round leaues, cōmonly spread abrode vpon the ground & somwhat iagged about the borders, not much vnlike the leaues of groud Iuie, but softer & smaller, & of a more yellowish grene. The stalke riseth amongst the leaues, & is round and hearie, and of the length

The second Booke of

Saxifraga alba.
White Sarifrage.

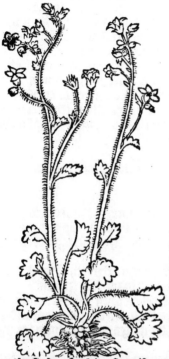

Saxifraga aurea.
Golden Sarifrage.

of a foote and halfe, it carieth at the toppe diuers white floures, almost like to stocke Gillofers. The roote is blackishe with many threddie strings, by whiche hangeth diuerse little rounde graynes, cornes, or berries, of a darke or reddish purple colour, greater than Coriander seede, sharpe and bitter, the which little graynes or berries they vse in medicine and do cal it Semen Saxifragæ albæ, that is to say, the seede of white Sarifrage or Stone breake.

2 There is yet an other called golden Sarifrage, which groweth to the length of a spanne and halfe, with compassed leaues, and iagges like to the other, at the toppe of the stalke growe two or three littel leaues togither, and out of the middle of them springeth small floures, of a golden colour, and after them little rounde huskes, full of small redde seede, and they open and disclose themselues whan the seede is ripe. The roote is tender creeping in the grounde, with longer threades and heares, and putteth foorth a great many stems or branches.

❧ *The Place.*

1 The white Sarifrage groweth in drye, rough, stony, places, as about the Colempnes, bysides Bathe in England: it groweth also in France and Almaigne. Ye shall also finde it planted in the gardens of Herborists.

2 The golden Sarifrage groweth in certayne moyst and watery places, in England, Normandie and Flaunders.

❧ *The Tyme.*

1 The white Sarifrage floureth in Maye, and in Iune the herbe with his floures perissheth, and are nomore to be seene, vntill the next yeare.

2 The golden Sarifrage floureth in March and Aprill.

❧ *The Names.*

1 This herbe is called in Latine Saxifraga alba: in English Stone breake, & white Sarifrage: in French *Rompierre, & Saxifrage blanche*: in high Douch weiß Steinbrech: in base Almaigne Wit Steenbreeck.

2 The secōd is called Saxifraga aurea: in English Goldē Sarifrage: in French *Rompierre, or Saxifrage dorée*: in high Douch Goldē Steinbrech: in base Almaigne
Gulden

the Hiſtorie of Plantes.

Gulden Steenbꝛeek, and this name is giuen it: becauſe it is like to the white Saxifrage, and beareth yellow oꝛ golden flowers.

※ *The Nature.*

This herbe eſpecially the roote with the ſeede, is of a warme oꝛ hoate complexion.

But the golden Saxifrage is of a colde nature, as the taſte doth manifeſtly declare.

※ *The Vertues.*

1 The roote of white Saxifrage with the graynes oꝛ berries of the ſame, boyled in wine, and dꝛonken, pꝛouoketh vꝛine, mundifieth and clenſeth the kidneyes and bladder, bꝛeaketh the Stone, and bꝛingeth it foorth, and is ſinguler againſt the Strangurie and all the imperfections, and griefes of the reynes.

2 What vertue the ſeconde hath, is to vs as yet vnknowen, bycauſe there is none hath yet pꝛoued it.

Of Gromell. Chap.ciij.

※ *The Kyndes.*

The Gromell is of two ſoꝛtes, one of the garden, the other wilde: and the garden Gromell alſo is of two ſoꝛtes, great and ſmall.

Lithoſpermum maius.   Lithoſpermum minus.
The great Gromell.   **The ſmall Gromell.**

The Description.

1. THE great Gromel hath long, slender, hearie stalkes, the whiche do most commonly trayle alongest the grounde, beset with long browne hearie leaues, betwixt the whiche leaues and the stalkes groweth certayne bearded huskes, bearing at the first a smal blewe floure, and afterwarde, a little harde, rounde, stonie seede, of a reasonable quantitie. The roote is harde of a wooddie substance.

2. The small garden Gromel hath straight rounde wooddie stalkes, and full of branches, his leaues be long, smal, sharpe, and of a swart greene colour, smaller then the leaues of the great Gromel. Betwixt the leaues and the stalkes groweth smal white floures, and they bring foorth faire rounde, white, harde, and stonie seede, lyke vnto Perles, and smaller than the seede of the aforesayd kinde.

3. The wilde Gromel is like vnto the small in stalkes, leaues and floures, sauing that the seede is not so white, neither so smooth & playne: but somwhat shriueled or wrinckled, like to the seede of the common langue de beufe, and the leaues be a little rougher.

4. Besides these two kindes there is yet founde a wilde kinde of Gromel, which is very small, of whiche kinde the learned Ierome Bocke hath treated in his herball, it groweth a span long, with his stalke set with small narrowe leaues, like to the leaues of lyne or flare, betwixt the whiche leaues & the stalke, it bringeth foorth a little smooth, blacke, harde seede, very lyke the seede of the small garden Gromel.

The Place.

1. The garden or tame Gromel groweth in some Countries in rough places: here they sowe it in gardens. The smaller garden Gromel groweth not often of him selfe, sauing alongest the Riuers and water sides.

2. The wilde is founde in rough and stonie places.

The Tyme.

Gromel floureth in Iune, Iuly, and August, in whiche season it doth also deliuer his seede.

The Names.

Gromel is called in Greke λιθόσπερμον: and in Latine Lithospermum, of some ρεγόνωρ: of the Arabians Milium Soler: in Shoppes Milium solis: in Englishe Gromel and Gremil: some name it also Pearle piante: in Frenche Gremil, or Herbe aux perles: in high Douch Meerhirsch, or Meerhirsen, and Steinsomen: in base Almaigne Peerlencrupt, and Steenslaet: in Italion Milium Solis.

The Nature.

The seede of Gremil is hoate and drie in the seconde degree.

The Vertues.

Gromel seede pounde and dronken in white wine, breaketh the Stone, driueth it foorth and prouoketh vrine: but especially the Stone in the bladder, as the Authors write. Turner.

Of Betony. Chap.ciiij.

The Description.

BEtony hath leaues somwhat long and broade, of a darke greene colour, bluntly iagged rounde about the edges like a sawe, and of a good sauour. Amongst the sayd leaues groweth vp a rough square stalke of a foote and halfe

the Historie of Plantes.

halfe long, decked with suche like leaues but a great deale smaller, and bearing at the top a short spykie eare, full of flowers, most commonly of a crymsin, or redde purple colour, and somtymes (but very seldome), as white as snowe: after whiche flowers there commeth in the sayd spykie tuffets, blacke seede, long and cornered. The roote hath threddie stringes.

2 Paulus Aegineta maketh mention of another Betony, called of the later writers Veronica, the which we haue described in the 17. Chapter of the first booke.

❧ The Place.

Betony groweth in meddowes, shadowy wooddes, and mountaynes: It is also commonly planted in gardens.

❧ The Tyme.

Betony flowreth commonly in July and August.

❧ The Names.

Betony or Betayne, is called in Greeke κέστρον, καὶ ψυχότροφον: In Latine and in shoppes Betonica and Vetonica: In Spanishe Bretonica: In frenche Betonie: In high Douch Braun Betonick: In base Almaigne Betonie.

❧ The Nature.

Betony is hoate and dry in the seconde degree.

❧ The Vertues.

The Decoction of Betonie dronken, prouoketh vrine, breaketh the Stone A of the kidneyes, doth clense and scoure the breast and lunges from flegme and slyme, and is very profitable for such as haue the Phthisik or consumptiō, and are vexed with the Cough.

The leaues of Bettayne dried, are good to be giuen the quantitie of a dram B with Hydromel, that is to say, Honied water, vnto such as are troubled with the Crampe, and also agaynst the diseases of the Mother or matrix.

The same taken in like manner, bringeth the flure menstruall. C

The dried leaues dronken in wine, are profitable against the biting of Serpentes, and so be they to be applyed or layde outwardly vpon the wounde: and it is good also for them that haue taken any poyson. And if it be taken before hande, it preserueth the people from all poyson.

Betany openeth and cureth the oppilation or stopping of the liuer, the melt, E and the kidneyes, and is good agaynst the Dropsie.

The same dronken with wine and water is good for them that spet blood, F and it cureth al inwarde and outwarde woundes.

The same taken with Hydromel or Meade, looseth the belly very gentilly, G and helpeth them that haue the falling sicknesse, madnesse, and head ache.

It comforteth the stomacke, helpeth digestion, swageth belching & the desire H to

Betonica. Betony.

Of Panax. Chap.cv.

✠ *The Kyndes.*

Dioſcorides that famous, and auncient writer of Plantes, hath deſcribed vnto vs three ſortes of Panaces: wherof the firſt is Panaces Heraclium: The ſeconde is Panaces Aſclepij: The thirde is Panaces Chironium.

✠ *The Deſcription.* Πάνακες ἡράκλειον. Panaces Heracleum.

1 THE firſt kinde of Panaces, hath great greene and rough leaues, layd & ſpread abroade vpon the ground, and parted into fine iagges and cuttes, almoſt lyke the leaues of the figge tree. Amongſt them ſpringeth vp, a long thicke ſtalke with ioyntes, white without and hearie, ſet here and there with the lyke leaues: but ſomwhat ſmaller, and bearing at the top a buſhe, or ſpokie tuffete lyke vnto Dyll, the floure or bloſſom of it pellowe, and the ſeede is of a pleaſant ſauour ſharpe & hoate. It hath diuers white rootes growing or comming foorth of one head, of a ſtrong ſauour, and couered with a thicke bitter barke. Out of the ſayde roote, and the ſtem, or ſtalke cut, and ſcarrified, floweth the gomme or liquor, called Opopanax, the whiche being freſh and newly drawen foorth of the plante is white: but beyng drie it wareth all pellowe without, as though it were coloured with Saffron.

2 The ſeconde kinde of Panaces, hath a ſlender ſtalke of a cubite long with knottes or iopntes, the leaues be greater, more hearie, and of a ſtronger ſauour than the leaues of Fenell. The floures growe alſo in tuffetes or rundels, and they are pellowe of an odiferous ſauour, and ſharpe taſte. The roote is ſmall and tender.

3 The thirde kinde as Dioſcorides and others do write, hath leaues like vnto Maricrom, floures of a golden colour, a ſmall roote, not goyng deepe in the grounde, and of a ſharpe taſte. But as Theophraſtus, and Plinie do deſcribe it, This thirde kinde of Panaces ſhoulde haue leaues lyke vnto Patience, or Sorrel, floures of a golden colour, and a long roote, ſo that amongſt the olde writers, is no perfit conſent touching this thirde kinde of Panax.

✠ The

the Historie of Plantes.   293

Quarta Panacis Species.
The fourth kinde of Panax.

❧ *The Description.*

4 Vnto these three kindes of Panaces, we may ioyne a certayne other strange plant, whose seede is founde amongst Opopanax. And this plante hath great large leaues, somwhat rough & hearie, largely spread abroade, and made of sundry leaues ioyned togither all in one, wherof eache collaterall (or by, leafe) is long and large almost like to ye leaues of Patience: The stalke or stem of this plante is full of ioyntes, and of fiue or sixe foote long, diuiding it selfe agayne into other stalkes and branches: The floures be yellow, growing in spokie tuffetes or rundels: The seede is playne, & the roote is long and white.

❧ *The Place.*

1 The first kinde groweth about Cyren in Lybia, and Macedonia: also in Bœotia, & in Phocis of Arcadia, whereas they vse to sowe it, and manure it diligently, for the gayne that is gotten of the sappe or iuyce thereof.

3 The thirde kinde groweth vppon the mount Pelius in Thessalie, & loueth good grounde.

❧ *The Tyme.*

The Opopanax is drawen, gathered in the time of harueft.

❧ *The Names.*

1 The first kinde is called in Greeke πάνακες ἡράκλειον, that is to say in Latine Panaces Herculeum, of Galien also Panax: vnknowen in the shoppes here.

The liquor that commeth from it, is called in Greeke ὀποπάναξ: in Latine also Opopanax: in shoppes Opopanacum.

2 The seconde kinde is called πάνακες ἀσκληπιοῦ, that is to say in Latine Panaces Asclepij, or Aesculapij Panaces.

3 The thirde is called πάνακες χειρώνιον, Panaces Chironium.

4 The fourth shoulde seeme to be Panaces Syriacum, wherof Theophraftus & Plinie haue mentioned: which differeth from the former kindes, as we haue els where, more largely written in Latine.

Panaces: in shoppes is called Siler montanum.

❧ *The Nature.*

1 The first Panaces is hoate in the thirde degree, and drye in the seconde.

The liquor thereof is also of the lyke temperament.

2. 3 The three other kindes are of the like temperature, but not so hoate, nor so
4 strong.

Bb iij   ❧ The

The Vertues.

1 ❧ The seede of the first Panaces dronken with wormewood, moueth womens flowres: And taken with Herbe Sarrasine, whiche is Aristolochia Clematitis, it is good agaynst the poyson of all venimous beastes. Being dronken with wine, it cureth the suffocation and strangling, or choking of the Matrix or Mother, and causeth the same to fall and returne agayne to his naturall place. A

The roote of Panaces chopped or hackt very small, and applied belowe to the Mother or Matrix, draweth foorth the dead Chylde, and the vnnaturall birth. B

The same roote mengled with Hony, and layde vppon, and also put into olde vlcers, cureth the same, and couereth bare, or naked bones with flesh agayne. C

2 The flowers and seede of the seconde kinde of Panaces, are very profitable agaynst the bytinges of Serpentes, to be dronken in wine, or layde vpon the wounde with Oyle. D

The same flowers and seede mingled with Hony, and layde thereunto, do cure olde malignant, corrupt, and fretting soares, and also knobbes or harde swellinges. E

3 The seede, the flowers, and also the roote of the thirde Panaces, are very good to be dronken, against the venom of Serpentes, and Vipers. F

Of Iouage. Chap.cvi.

❧ The Kyndes.

If men take that herbe whiche is commonly called in Shoppes Leuisticum, for one of the sortes of Ligusticum: Then there are two kindes of Ligusticum, the one whiche is the right Ligusticum, described by the Auncientes, And the other whiche may be a bastarde or wilde kinde of Ligusticum.

❧ The Description.

1 The right Ligusticum, described by Dioscorides, is in his roote lyke to the first kinde of Panax: it hath slender stalkes, with ioyntes like vnto Dill. The leaues are lyke to the leaues of Melilot, but they be softer and of a better sauour, whereof the vppermost leaues are tenderest, and more iagged or cut. At the top of the stalkes groweth the seede in spokie tuffetes, the whiche is harde and longe, almost like to Fenell seede, of an aromaticall or Spycie sauour, and in taste sharpe and byting. The roote is white, and odoriferous, much lyke to the roote of the first kinde of Panax. Neuerthelesse it is not yet knowen in this Countrie.

2 The other herbe, whiche is taken in this Countrie for Ligusticum, hath great, large, odoriferous leaues, muche iagged and cut, almost lyke to the leaues of Angelica: but a great deale larger, fayrer and of a deeper greene colour, deeper cut and more clouen. The stalke is smooth, rounde, holowe, and ioyntie, of the length of a man or more, with spokie rundels, or tuffetes, at the top of the stalkes: bearing a yelow flower, and a round, flat, broade, seede: larger then Dyll seede, and smaller then Angelica seede. The roote is long and thicke and bringeth foorth yerely newe Stemmes.

❧ The

the Historie of Plantes. 295

Ligusticum verum.
The right Louage.

Ligusticum vulgare.
The common Louage.

❋ *The Place.*

The right Ligusticum, groweth in Liguria, vppon the mount Apennian, neare to the Towne or Citie of Genues, and in other mountaynes there about.

The seconde kinde is planted in our gardens.

❋ *The Tyme.*

Louage flowreth most commonly in July and August.

❋ *The Names.*

1 The first and right kinde is called in Greeke λιγυστικὸν, and of Galien λιβυστικόν: in Latine Ligusticum: and of some also as Dioscorides writeth Panaces, by the which name it is yet knowen in the Shoppes of Genues: in the Shoppes of Flaunders they call it Siler Montanum: in Englishe Louage: in Frenche *Liuesche*: and in Douch Ligusticum.

2 The seconde kinde is called in Shoppes Leuisticum, and the Apothecaries vse it in steede of the right Ligusticum: in Englishe Louage: in Frenche *Leuesse*, or *Liuesche*: in Douche Liebstockel: in Brabant Lauetse, and Leuistock.

❋ *The Nature.*

Ligusticum is hoate and dry in the thirde degree.

Louage is also hoate and drye, and of qualitie muche like to Ligusticum.

❋ *The*

❦ The Vertues.

1　The roote of Ligusticum is very good for all inwarde diseases, driuing away all ventositie, or windinesse, especially the windinesse of the stomacke, and is good agaynst the byting of Serpentes, and al other venimous beastes.

　The same roote well dried and dronken with wine, prouoketh vrine, and the menstruall termes: it hath the same vertue, if it be applyed to the secrete place in a Pessarie or mother Suppositorie.

　The seede of Ligusticum warmeth the stomacke, helpeth digestion, and is pleasant to the mouth and taste, wherfore in times past the people of Genues dyd vse it in their meates in stede of Pepper, as some do yet, as witnesseth Antonius Musa.

2　The roote and seede of louage dryed and dronke in wine, doth drie vp and warme the stomacke, easeth trenches or griping payne of the belly, driuing away the blastinges and windinesse of the same.

　The same roote and seede do moue vrine, and the naturall sicknesse of women, whether they take it inwardly, or whether they bathe them selues with the decoction thereof, in some hollowe seate, or stue.

　To conclude, the louage in facultie and vertues, doth not differ much from Ligusticum, and it may be vsed without errour, in steede thereof.

　The distilled water of louage, cleareth the sight, and putteth away all spottes, lentiles, or frecles, and rednesse of the face, if it be often washed therewith.

❦ Of Angelica.　　Chap.cvij.

❦ The Kyndes.

ANGELICA is of two sortes, that is the garden and wilde Angelica.

❦ The Description.

1　THE garden Angelica hath great broade leaues, diuided agayne into other leaues, which are snipt and dented about, much like to the highest leaues of Spondilium, or Douch Branck vrsine, but they be tenderer, longer, greener, and of a stronger sauour. Amongst those leaues springeth vp the stalke, three yeeres after the sowing of the seede, the whiche stalke is thicke, and ioyntie, hollowe within, and smelleth almost like to Petroleum. At the top of the stalkes groweth certayne little felmes, puffed or bolne vp lyke to small bladders or bagges, out of which commeth the spokie toppes or rundels almost like vnto the tops of Fenell, bearing white floures, & afterward great, broade double seede, muche greater then Dill seede, and like to the seede of the thirde kinde of Sesely. The roote is great and thicke, blacke without & white within, out of which, when it is hurt or cut, there floweth a fat or oylie liquor, like gomme, of a strong smell or taste.

2　The wilde Angelica is like to that of the garden, sauing that his leaues are not so deepely cut or clouen, and they be narrower and blacker. The stalkes be muche slenderer and shorter and the floures be whiter. The roote is a great deale smaller, and hath more threddie stringes, and it is not by a great deale of so strong a sauour.

❦ The Place.

　The tame Angelica is sowen and planted in the gardens of this Countrie.

　The wilde groweth in darke shadowy places, alongest by water sides, and wooddes standing lowe.

❦ The Tyme.

　The two kindes of Angelica, do flower in July and August.

the Historie of Plantes. 297

Angelica Satiua.
Garden Angelica.

Angelica Syluestris.
Wilde Angelica.

❧ *The Names.*

This herbe is called in Englishe Angelica: in Frenche *Angelique*: in high Douch Angelick, des heylighen gheists wurtzel, oder Brustwurtz: in the shoppes of Brabante Angelica. There is yet none other name knowen to vs.

❧ *The Nature.*

Angelica especially that of the garden is hoate and dry, almost in the thirde degree.

❧ *The Vertues.*

The late writers say, that the rootes of Angelica are contrarie to all poyson, the Pestilence, and all naughtie corruption, of euill or infected ayre.

If any body be infected with the Pestilence or plague, or els is poysoned, they giue him straightwayes to drinke a Dram of the powder of this roote with wine in the winter, and in sommer with the distilled water of Scabiosa, Carduus Benedictus, or Rosewater, then they bring him to bedde, and couer him well vntill he haue swet well.

The same roote being taken fasting in the morning, or but only kept or holden in the mouth, doth keepe and preserue the body from the infection of the Pestilence, and from all euyll ayre and poyson.

They say also that the leaues of Angelica pounde with the leaues of Rue and honie, are very good to be layde vnto the bitinges of mad Dogges, Serpentes, and Vipers, if incontinent after his hurt, he drinke of the wine wherin the roote or leaues of Angelica haue boyled.

Of

The second Booke of
Of Horestrange or Sulphurwort.
Chap.cviij.

❧ *The Description.*   Peucedanus.

THIS herbe hath a weake slender stalke, with ioyntes or knottes, the leaues are greater than the leaues of Fenill, like to the leaues of Pine tree. At the top of ye stalkes groweth rounde spokie tuffetes full of little yellowe flowers, the whiche afterwarde do turne into broade seede. The roote is thicke and long, blacke without, and white within, of a strōg greeuous smell, and full of yellow sap or liquer smelling not muche vnlike to Sulphur, or Brymstone, and it beareth at the hyghest of the roote aboue the earth a certayne thicke or bushe of heare, like to the rootes of Libanotides, before described, amōgst whiche the leaues and stalke do spring vp.

❧ *The Place.*

This herbe groweth vppon the high mountaines of Almaigne, & in the woodes of Languedoc, & certayne other countries. Heare the Herboristes do sowe it in their their gardens, It is found in certayne places of Englande, and D. Turner sayth, he founde a roote of it at S. Vincentes rocke by Bristowe.

❧ *The Tyme.*

Peucedanum flowreth in July and August.

❧ *The Names.*

It is called in Greeke πευκίδανον: In Latine and in shoppes Peucedanum, of some also ἄγαθοδαίμων, id est, Bonus Genius, Pinastellum, Stataria, and Fœniculus Porcinus: In Englishe also Peucedanum, Horestrong, or Horestrange, Sowefenill, and of some Sulpherwurt: In Italion *Peucedano*: In Spanishe Heruatum: In Frenche Peucedanon, and Queuë de Pourceau: In high Douch Harstrang, & of some Schwebelwurtz, and Sewfenchel, that is to say, Sulpher roote, and Sowfenell: In base Almaigne Verckens Venckell.

❧ *The Nature.*

This herbe, but specially the sap or iuys of the roote, is hoate in the seconde degree, and drie almost in the beginning of the thirde degree.

❧ *The Vertues.*

A   The sappe of the roote of Peucedanum or Horestrange taken by it selfe, or with bitter Almondes and Rue (as Plinie sayth) is good agaynst the shortnesse of breath, swageth the griping paynes of the belly, dissolueth and driueth away ventositie, windinesse, and blastinges of the stomacke and of all inwarde partes, it wasteth the swelling of the Melte or Splene, It looseth the belly gentilly, and purgeth by siege both fleme and choler.

The

The same taken in manner aforesayde, prouoketh vrine, easeth the payne of the kidneyes, and bladder, it mooueth the fluxe menstrual, causeth easie deliuerance of childe, and expulseth the Secundyne and the deade childe.

The iuyce of Peucedanum is good agaynst the Cough, if it be taken with a reare egge.

The same giuen to smell vpon, doth greatly helpe such women as are greeued with vprising and strangling of the Mother, and stirreth vp agayne or waketh suche people as haue the lethargie, or the forgetfull and sleeping disease.

The same layde to the forehead with oyle of Roses and Vineger is good agaynst the madnesse called in Greeke Phrenitis, and the olde greeuous head aches, and giddinesse of the same, terrible dreames, and the falling sicknesse.

The same sappe applyed as is aforesayde, cureth the Paulsie, the Crampe, and drawing togyther of sinewes, and all olde, colde diseases, especially the Sciatica.

The perfume of Peucedanum burned vppon quicke coales, driueth away Serpentes and all other venemous beastes creeping vpon the grounde.

The iuyce of it put into the concauitie or hollownesse of a naughtie tooth, swageth toothache: and powred into the eares with oyle of Roses, cureth the payne of the same.

They lay it with good successe vnto the rupture or bursting of younge children, and vpon the Nauelles that stande out, or are to muche lifted vp.

The roote in vertue is lyke to the iuyce: but it is not althing so effectuall, yet men drinke the decoction thereof, agaynst all the diseases whereunto the the iuyce is good.

The roote dried and made into powder, doth mundifie and clense olde stincking and corrupt vlcers, and draweth foorth the splinters and peeces of boones, and bringeth to a scarre, and closeth vp vlcers, that be harde to heale.

They mingle it very profitably with al oyntmentes and Emplaisters, that are made to chafe and heate any part of the body, whatsoeuer.

The same dryed and mengled with the Oyle of Dill, causeth one to sweate if the body be annoynted and rubbed therwith.

Of great Pellitorie of Spayne, Imperatoria, or Masterwort. Chap.cix.

❧ The Kyndes.

Masterwort is of two sortes, tame & wilde, not much onlyke one another, aswel in leaues as in floures and rootes, & both kindes are wel knowen in this Countrie.

❧ The Description.

Imperatoria or Masterwort hath great broade leaues, almost like Alexander: but of deeper greene, and stronger sauour, euery leafe is diuided into three others, & which agayne hath two or three deepe cuttes or gashes, insomuch as euery leafe is diuided into seuē, or nine parts, and euery part is toothed or natched rounde about like a sawe. Amongst these leaues groweth the tender knottie stalkes, whiche be of a reddishe colour next the grounde, bearing at the top rounde spokie tuffets with white floures, after the whiche commeth the seede, whiche is large and lyke to Dyll seede. The roote is long of the thicknesse of ones finger, creeping alongst and putteth vp newe new leaues in sondrie places, somwhat blacke without and white within, hoate or byting vpon the tongue, & of a strong sauour.

2 The wylde Imperatoria, commonly called Herbe Gerarde, or Ashe Weede, is not much vnlyke ÿ abouesaide in leaues, flowers, & rootes, sauing that the leaues are smaller growing vpon longer Stemmes, and the roote is tenderer whiter and not so thicke. Also the whole plante with his roote is not althing so strong in in sauor, yet it is not altogither without a certayne strong smell or sauor.

✤ *The Place.*

1 Asterantium or Masterwort, is sometymes founde in wooddes and desertes vpon littel hylles or small mountaynes. They do also plante it meetely, plentifullye in the gardins of high and base Almayne, and Englande.

2 The seconde Imperatoria, or wylde Masterwort, groweth commonly in most gardens of his owne kinde, and this is surely a weede or vnprofitable plante. And wheras these herbes haue once taken roote, they wyll there remayne willingly, and do yearely increase & spreade abroade, getting more grounde dayly. For which cause as I thinke it was first called Imperatoria or Masterwoorts in Douch.

Asterantium, Ostrutium.

✤ *The Tyme.*

These herbes do flower here in June and July.

✤ *The Names.*

1 The first kinde is called of some Herboristes and Apothicaries, Osteritium, Ostrition, Ostrutium, or Asterantium: of some Imperatoria: In English also Imperatoria Masterworte, and Pellitorie of Spayne: In Italion *Imperatoria*: In Frenche Ostrutium, or Imperatoire, and Herbe du Benioin, but falsely: In high Douch Meysterwurtz: In base Almaigne Meesterwortell.

2 The second or wilde Imperatoria, is now called Herba Gerardi, ἑπτάφυλλον, and Septifolium, that is to say, Herbe Gerarde, and Setfoyle: In Englishe some call it Ashweede: In base Almayne Geraert, and Seuenblat.

✤ *The Nature.*

Asterantium, but chiefely the roote is hoate and dry in the thirde degree. The wilde is almost of the same nature and qualitie, but not so strong.

✤ *The Vertues.*

1 Masterworte is not onely good agaynst al Poyson, but also it is singuler agaynst all corrupt and noughtie ayre, and infection of the Pestilence, if it be dronken with wine and the same roote pounde by it selfe or with his leaues, doth dissolue and cure Pestilential Carboncles and Botches, and suche other apostumations and swellinges, being applyed therto.

2 The roote thereof dronken in wine, cureth the extreme and rigorous fittes

of olde feuers, and the Dropsie, and it prouoketh swet.

The same taken in manner aforesayde, comforteth and strengtheneth the stomacke, helpeth digestion, restoreth the appetite, and dissolueth the ventositie and blasting of the flankes and belly.

It helpeth greatly such as haue taken great squattes, brusis, or falles from aloft, and are sore hurt, and inwardly bursten, for it cureth the hurtes, and dissolueth and scattereth the blood that is astonyed, and clotted or congeled within the body.

The same roote pounde with his leaues, is very good to be layde to the bytinges of madde Dogges, and to all the bytinges and stinginges of Serpentes, and suche lyke venimous beastes.

The wilde Imperatoria, or herbe Gerarde, pounde and layde vppon suche members or partes of the body, as are troubled and vexed with the gowte, swageth the payne, and taketh away the swelling.

And as it hath ben proued in sundrie places, it cureth the Hemorrhoides, if the fundement or siege be fomented, or bathed with the decoction thereof.

Of Ferula. Chap.cx.

The Description.

Ferula.

1 THE leaues of Ferula are great and large, and spreade abroade, and cut into very small threddes or heares lyke Fenell, but a great deale bigger: The stalke or stem is thicke, ioyntie, and very long: in the toppes of the stalkes groweth great round spokie tuffetes, bearing first yellowe flowres, and afterward long, broade, and blacke seede, almost as large as the seede of Melones or Pepones. The roote is thicke and white, and groweth deepe in the grounde, or in the ioyntes or cliftes and Choppes of Cleeffes and Rockes.

2 There is also founde an other kinde of this Ferula, but his leaues are not so smally cut, and vnderneath they be white, or of a grayshe colour, but otherwyse they be as large as the other, the seede is also lesse, but in proportion lyke the other.

The Place.

These Ferulas do growe in Grece, and Italie, and other hoate regions, but they are strange in this Countrey, and Flaunders.

The Names.

1 The first is called in Greeke Νάρτηξ: in Latine Ferula.

2 The other is also a kinde of Ferula, and is counted of some to be a certayne Ferulago, The whiche of Theophrastus is called in Greeke Ναρθηκία.

The Nature.

There is no peculier or special vse of these Ferulas, sauing that the liquor or gummes

gummes that floweth out of them, as Sagapenum, Ammoniacum, and Galbanum, are vsed in medicine, wherefore their nature and vertue shalbe described in the Chapters folowing.

To the Reader.

Considering, welbeloued Reader, that we haue written in the Chapters going before of some herbes, out of the whiche flowe very costly sappes or gummes geathered, dried, and preserued, the which are greatly vsed in Medicines and Surgerie, especially as the sappe of Panax, the whiche is called Opopanax, and the sappe of Laserpitium, the whiche is named Laser, whiche in farre Countries do flowe out of the same herbes, and are brought into this Countrey, & into all partes of Christendome, of whose strength and vertue we haue not written: therefore haue we in the ende of this part for a conclusion & finishing of the same, written of the nature and vertue of the same gummes. And not onely of the gummes flowing out of the herbes aboue rehearsed: but also of gummes and sappes flowing out of herbes or thereof made, the whiche commonly we finde at the Apothecaries and are vsed in Medicines, although that the herbes (bicause they are not knowen in Christendome) are not written or spoken of by vs, omitting the sappes and gummes whiche flowe out of wooddes and trees, as Rosin, Pitche, Turpentine, and suche lyke, we wyll write of the historie of wooddes and trees. And in the description of these gummes and sappes we wyll folowe the learning of the Auncientes, as Dioscorides, Galen, Plinie, &c. Declaring their names as they are called by the sayd Auncientes in Greeke and in Latine, by the whiche they are nowe at this time knowen to the Apothecaries, like as we haue yet hitherto done and written in the historie of herbes.

Of Opopanax. Chap.cxi.

Popanax is the gumme or sappe of the first kinde of Panaces, called Heracleoticum, as Dioscorides writeth, & it floweth out of the roote and stalke of Panaces, as they shalbe hurt or cut, and the sappe when it is yet fresh, and first flowen out, is white, and when it is drie, it is altogyther yellowe lyke that which is coloured with Saffron. And the best of this sappe or gumme is that same whiche on the outsyde is yelowe and within whitish, for that is yet fresh.

❧ *The Names.*

The gumme is called in Greeke ὀποπάναξ: in Latine Opopanax: and of the Apothecaries Opopanacum: in Englishe Opopanax.

❧ *The Nature.*

Opopanax is hoate and drie in the thirde degree.

❧ *The Vertues.*

Opopanax is very good against the colde shiuerings, and brusing of Agues, the payne and griefe of the syde, the gnawing & griping payne of the bowelles or guttes, the Strangurie, and for them that are squatte or bruysed within, by occasion of falling, if it be dronken with Meade or Honied water. And to be taken in the same manner or with wine, it cureth the inwarde scuruinesse or hurt of the bladder.

Opopanax as Mesue writeth, taken the waight of two drammes or lesse, pourgeth by siege, the flegme and colde, tough, clammie, and slymie humours, drawing the same from partes farre of, as fro the head, the sinewes & ioyntes. Moreouer it is very good against al colde diseases, of the brayne and sinewes, as the Crampe and Paulsie, &c.

the Historie of Plantes.   303

The same taken in the like manner and quantitie, doth mundifie and scoure C
the breast, and is good for Asthmatique people, and for them that are troubled
with the shortnesse of winde or breath, and with an olde dangerous cough.

It cureth also the hardnesse, and other mishappes of the melt or splene, and D
Dropsie, if it be tempered or stieped in muste, and dronken.

Opopanax doth scatter, soften, & resolue, al hard, cold, swelling, or tumours, E
being stieped in vineger, and applyed or layde therto.

It is good to be layde to the Sciatica (whiche is the gowt in the hippe or F
huckle bone) and it easeth the payne of the gowt of the legges and feete, beyng
layde therevpon with the substance or pulpe of dried Raysons.

The same mingled with Hony, and put in vnder in manner of a Pessarie or G
mother suppositorie, prouoketh the flowres, driueth foorth the Secondine, and
dead fruite, dispatcheth the ventositie of the Matrix or mother, and cureth all
hardnesse of the same.

Opopanax being layde vpon Carbuncles, and Pestilentiall botches, and tu- H
mors, breaketh the same, especially after that it hath ben soked in vineger, and
mingled with leccayne.

It swageth tooth ache, being put into the hollownesse of perished teeth: or I
rather as Mesue sayth, to be boyled in vineger, and holdē or kept in the mouth.

Being layd to the eyes alone, or mingled with Collyries made for the pur- K
pose, it cleareth the sight.

With this gumme and Pitche they make a playster, the whiche is very sin- L
guler agaynst the bytinges of al wilde and mad beastes, being layd therevnto.

Of Laserpitium, and Laser.   Chap.cxij.

✻ *The Description.*

Laserpitium (by that we may gather of Theophrastus & Dioscorides)
is an herbe that dyeth yerely, his stalke is great and thicke lyke Ferula: the leaues be lyke Persley and of a pleasant sent: The seede is
broade as it were a little leafe, it hath a great many rootes growing
out of one head, which is thicke and couered with a blacke skinne.

From out of these rootes and stalkes being scarified and cut, floweth a certayne strong liquor, the which they drie, and is verie requisite in medicine, and
it is called Laser: but it is not all of a sorte, nor in al places alyke, for it chaun-
geth in taste, sauour, and fashion, according to the places where as the Laserpitium groweth.

1   The sappe or liquor that floweth out of the Laserpitium growing in Cyrene,
is of a pleasant sauour, and in tast not very grieuous: so as in tymes past, men
dyd not onely vse it in shoppes for Physick, but also in fine Cakes, Junkettes,
and other meates, as Plinie writeth.

2.3   That whiche floweth out of the Laserpitium, that groweth in Media, and
Syria, is of a very lothsome, and stinking sauour.

✻ *The Place.*

Laserpitium groweth on the high mountaynes and desertes of Cyrene and
Aphrica, and this is the best and chiefest, and it yeeldeth a liquor which is very
good and of a pleasant smell. It groweth also in Syria, Media, Armenia, and
Lybia, but the iuyce or liquor thereof is not so good, but is of a very lothsome
detestable, and abominable smell.

✻ *The Names.*

This plant is called in Greeke σιλφιον: In Latine Laser, and Laserpitium:
of some, as witnesseth Dioscorides, Magudaris, especially that whiche yeeldeth

no liquor, as in Lybia.

The stalkes of the right Laserpitium are called in Greeke σίλφιον: and in Latine Silphium.

The rootes are called μαγύδαρις, and Magudaris.

The first leaues ÿ spring vp out of the ground, are called μάσπετον, Maspetũ.

The iuyce or liquor of Laserpitium, is called in Latine Laser: and of the Arabian Physitions Asa, or Assa.

The iuyce whiche floweth from the stalkes is called of Plinie Caulias, and of Gaza the interpreter of Theophrastus, Scaparium Laser.

That whiche floweth from the rootes, is called Rhizias, of Gaza Radicarium Laser.

¹ The sweete sauering gumme or liquor is called in Greeke ὀπὸς κυρηναικός: in Latine Succus Cyrenaicus, or Laser Cyrenaicum, of some Asa Adorata: vnknowen in Shoppes: for that whiche they take for Laser (as all the learned men of our tyme thinke) is called of the Apothecaries Gummi benzui, or Belzui, or Assa dulcis: in Englishe Belzoin, or Benzoin: in Frenche Benioin, and it is not Laser: but the gumme or liquor of a certayne great tree to vs vnknowen, as the trauelers do affirme, and as it doth manifestly appeare by the thicke peeces of barke and wood, which is often found in and amongst the Benzoin, that it cannot be the gumme or liquor of an herbe that perisheth yercly.

² That Laser whiche commeth from Media, is called in Greeke ὀπὸς μηδικός: in Latine Laser Medicum, or Succus Medicus.

³ That whiche commeth from Syria is called ὀπὸς συριακός: in Latine Laser Syriacum.

These two last recited kindes of Laser that come from Syria, and Media, bycause of their lothsome sauour, are called of the Arabian Physitions and Apothecaries Assa fœtida: in Englishe also Assa fetida: in high Douche Teufels dreck, that is to say Deuilles durt: it is called in Brabant by a very strange name Fierilonfonsa.

❧ The Nature.

Laserpitium, especially the roote, is hoate and drie in the thirde degree.

Laser is also hoate and drie in the thirde degree, but it exceedeth muche the heate of the leaues, stalkes, and rootes of Laserpitium.

❧ The Vertues.

The rootes of Laserpitium are very good (as Dioscorides and Galen writeth) to be dronken against al poyson: and a little of the same eaten with meat, or taken with salte, causeth one to haue a good and sweete breath. **A**

The leaues of this plante (as Plinie writeth) boyled in wine and dronken, mundifieth the Matrix, and driueth foorth the Secondine, and the dead fruit. **B**

The rootes well pounde or stamped with Oyle, scattereth clotted blood, taketh away blacke and blewe markes that come of bruses or stripes, cureth and dissolueth the kinges euill, and all harde swellinges and Botches, the places being annoynted or playstered therewith. **C**

The same roote made into powder, and made into a playster with the Oyle of Ireos and waxe, doth both swage and cure the Sciatica or gowte of the hippe or huckle bone. **D**

The same boyled with the pilles of pome Granattes and vineger, doth cure the Hemorrhoides, and taketh away the great wartes, & all other superfluous outgrowinges about the fundement. It hath the same vertue, if one foment or bathe the fundement with the Decoction of the same rootes boyled in water. **E**

They do also mundifie and clense the breast, & it dissolueth and ripeth tough **F**
flegme,

flegme, and it is very profitable against an olde cough comming of colde, to be taken with hony in maner of a Lohoc, or electuarie.

They prouoke vrine, they mundifie and clense the kidneyes and bladder, they breake and driue foorth the Stone, they moue the flowres, and expulse the Secondine, and the dead fruit. G

If they be holden in the mouth and chewed vpon, they swage tooth ache, and drawe from the brayne a great quantitie of humours. H

The liquor or gumme of Laserpitium, especially of Cyrene, broken and dissolued in water and dronken, taketh away and cureth the hoarsenesse that cometh sodenly: and being supt vp with a reare Egge, it cureth the cough, and taken with some good broth or supping, it is good against an olde Pleurisie. I

Laser cureth the Jaunders and Dropsie taken with dryed figges. K

It is very good agaynst Crampes, and the drawing togyther or shrincking of sinewes, and other members, to be taken the quantitie of a scruple, and take with Pepper & Myrrhe, it prouoketh the flowres, and driueth foorth the Secondine and dead fruit. L

To be taken with Hony and vineger, or with Syrupus Acetosus, it is singuler agaynst the falling sicknesse. M

It is good against the flire of the belly comming of the debilitie and weakenesse of the stomacke (which disease is called in Latine Cœliacus morbus) with the skinne, or rather the kernelles of raysons. N

It driueth away the shakinges & shiueringes of agues, to be dronken with Wine, Pepper and Franckencense. And they make thereof an Electuarie with Pepper, Ginger, and the leaues of Rue pounde togyther with hony, the which is called Antidotum ex succo Cyreniaco, the whiche is a singuler medicine against feuer Quartaynes. O

It is good against the bytinges of al venimous beastes, and venimous shot of dartes and arrowes, to be taken inwardly, and applied outwardly vpon the woundes. It is also very profitable layde to all woundes, and bytinges of Dogges and other madde beastes, and vpon the stinging Scorpions. P

It quickeneth the sight, and taketh away the hawe or webbe in the eyes, at the first comming of the same, if it be straked vpon them with hony. Q

Dioscorides saith, that if it be put into the hollownesse of corrupt & noughty teeth, it taketh away the ache and payne of them: but Plinie bringeth agaynst the same the experience of a certayne man who hauing tried the same, for the extreame rigour & anguishe he felt after that medicine, threw him selfe downe headlong from aloft. Neuerthelesse if it be wrapped with Frankencense in a fine linnen cloute and holden vpon the teeth, it cureth the ache of the same, or els the Decoction thereof with figges and hysope boyled togyther in water, and holden or kept in the mouth. R

Being layde to with hony it stayeth the vuula, and cureth the Squinance, if it be gargled with Hydromell or Mede: and if it be gargled with vineger and kept in the mouth, it will cause the Horseleaches, or Loughleaches, to fall of, which happen to cleaue fast in the throote or wesande of any man. S

It breaketh Pestilentiall Impostemes and Carboncles, being layd thereto with Rue, Niter, & hony: after the same manner it taketh away Cornes, when that they haue ben scarrified rounde about with a fine knife. T

Being layd to with Copperous & Uerdigris, it taketh away al superfluous outgrowinges of flesh, and the Polypus growing in the Nosthrilles, and all scuruie manginesse: and layde to with vineger Pepper and wine, it cureth the noughtie scurffe of the head and the falling of, of heare. U

Cc iij It

If it be boyled in vineger with the pil of the Pomegarnet, it taketh away al outgrowinges, which chaunce in the fundement.

Against kybed heeles, they first bathe the heeles or feete with wine, & than they annoynt the kybes with this gumme boyled in oyle.

The stinking gumme called Assa foetida, is good for al purposes aforesayde, howbeit, it is not so good as the Laser of Cyrene: yet it is very good to smell vnto, or to be layd vpon the Nauell, against the choking or rising vp of the mother.

They vse Benzoin in steede of Laser Cyrenaicum, for all the purposes aforesayde that be attributed vnto sweete Laser.

✤ *The Choyse.*

The best Laser is that which is reddish, cleare and bright, and sauering like Myrrhe, not greenish, and of a good and pleasant smel, the which being dissolued waxeth white.

Of Sagapenum. Chap.criij.

Sagapenum ye is sap or gumme of a kinde of Ferula or Fit, like vnto Panax growing in Media, altogyther vnprofitable, sauing for ye gumme or liquor that is drawen out of it. And the best is that, which (as Mesue sayth) doth melt, by and by, in the water, and sauereth like garlike: or betwixt Laser, and Galbanum, as Dioscorides saith: whiche is sharpe and cleare, of a yellowishe colour without, and white within. ✤ *The Names.*

This gumme is called in Greeke σαγάπηνον: in Latine Sagapenum, and Sagapeniū, of Plinie Sacopenium, of Galen, ὀπὸς σαγαπηνῦ, that is, Sagapeni Succus: They call it in shoppes Serapinum. ✤ *The Nature.*

Sagapenum is hoate in the thirde degree, and drye in the seconde.

✤ *The Vertues.*

Sagapenum taken the waight of a dram, purgeth by siege, tough & slymie humours, and al grosse flegme and choler. Also it is good against al olde & cold diseases that are harde to cure: it purgeth the brayne, and is very good against all the diseases of the head, and against the Apoplexie, and Epilepsie.

To be taken in the same sorte, it is good against Crampes, Paulsies, shrinkinges, and paynes of the sinewes.

It is good against the shortnesse of breath, the colde long and olde cough, the paynes in the side and breast, for it doth mundifie and clense the breast of al cold mentes or flegme.

It doth also cure the hardnesse, stoppinges, and windinesse of the melte, or splene, not onely taken inwardly, but also to be applyed, outwardly in oyntplaysters.

It is good against the shakinges and brusinges of olde and colde feuers.

If Sagapenum be dronken with honyed water, it prouoketh the flowers, and deliuereth the dead Childe. And to be taken with wine, it is of great force against the bytinges and stinginges of all venimous beastes.

The sente or sauour of this gumme, is very good against the strangling or vprising of the mother.

Sagapenum soked or stieped in vineger, scattereth, dissolueth, and putteth cleane away all harde, olde colde swellinges, tumoures, Botches, and harde lumpes growing about the ioyntes: And it is good to be be mingled amongst all oyntmentes and emplaysters that are made to mollifie and soften.

It cleareth the sight, & at the beginning it taketh away the hawe or webbe in the eye & al spottes or blottes in the same, if it be dropped into the eyes with the

the iuyce of Rue: it is also good agaynst the bloodshoting and dimnesse of the same, which commeth by the occasion of grosse humors.

Of Galbanum. Chap.cxiiij.

Galbanum is also a gumme or liquor, drawen foorth of a kinde of Ferula in Syria called Metopium. And the best is gristel, or betwixt hard and soft, very pure, fat, close and firme, without any stickes or splinters of wood amongst the same, sauing a fewe seedes of Ferula, of a strong sauour, not moyst, nor to drye.

❧ *The Place.*

The plant out of which Galbanum floweth, groweth vpon the mountayne Amanus in Syria.

❧ *The Names.*

Plinie calleth ye plant out of which Galbanū floweth, in Latine Stagonitis. The liquor or gumme is called in Greeke χαλβάνη: in Latine and in shoppes Galbanum : of some also Metopium.

❧ *The Nature.*

Galbanum is hoate almost in the third degree, & drie almost in the seconde.

❧ *The Vertues.*

Galbanum is good against an olde cough, and for such as are short winded, and cannot easily drawe their breath, but are alwayes panting and breathing. It is very good for such as are broken, and brused within, & against Crampes and shrinking of sinewes. **A**

The same dronken in wine with Myrrhe, is good against al venome dronken, or shot into the body with veninous Dartes, Shaftes, or Arrowes. **B**

To be taken in the same manner, it prouoketh the termes, and deliuereth the dead childe. It hath the same vertue if it be conueyed into the secrete place, or if a perfume therof be receiued at the place couenient: and if the quantitie of a beane thereof be taken in a glasse of wine, it helpeth against the payneful traueil of women, as Plinie sayth. **C**

The parfume or sent thereof driueth away Serpentes, frō the place where as it is burned, & no veninous beastes haue power to hurt such as be annoynted with Galbanum, and those veninous beastes or Serpētes as be touched with Galbanum, mingled with oyle, and the seede or roote or Spondilium, or Angelica, it will cause them to dye. **D**

The parfume of Galbanum doth also helpe wemen that are greeued with the rising or strangling of the mother, and them that haue the falling sicknesse: and being layde to the nauel, it causeth the Matrix or mother that is remoued from his naturall place, to settel agayne. **E**

Galbanum doth mollifie and soften, and draweth foorth thornes, splinters, or shiuers, and colde humours : and it is good to be layd vpon al colde tumors and swellinges, and it is mingled with all oyntmentes, oyles and emplaysters, that haue power or vertue to warme, to digest, to dissolue, to ripe and breake impostemes, and to drawe out thornes and splinters. **F**

It is good to be layde vpon the stoppinges and hardnesse of the melte, and against the payne of the syde. **G**

The same layde to with vineger and Nitrum, taketh away the spottes and freckles of the face, and from other partes of body. **H**

If it be put into the holowe and naughtie tooth, it taketh away the ache of the same. **I**

It is good to be poured into the eares with the oyle of roses, or Nardus, agaynst the corrupt filth and matter of the same. **K**

Of

Of Ammoniacum. Chap.clv.

Mmoniacum is the gumme or liquor of a kinde of Ferula, whiche is called Agasyllis, as Dioscorides saith, growing in the Countrie of Cyrene in Aphrica, nigh to the Oracle of Ammon in Lybia, whereof it is called Ammoniacum, as some thinke. The best Ammoniacum, as Dioscorides writeth, is that whiche is close or firme, pure, and without shardes, splinters, or stonie gristels or grauell, and without any other baggage intermeddled with the same, of a bitter taste, & drawing towardes the sauour of Castoreum, and it is almost lyke the right Frankenseuce, in small peeces and gobbetes.

✻ *The Names.*

This gumme is called in Greeke after the name of the Temple of Ammon, ἀμμωνιακόν: in Latine Ammoniacum: in Shoppes Armoniacum, and Gummi Armoniacum.

The best and purest of this gumme or liquor, is called Thrausma, as Dioscorides saith, that is to say, Friatura in Latine.

That which is full of earth and grauell, is called Phyrama.

✻ *The Nature.*

Ammoniacum is hoate in the second degree, & almost drie in the same degree.

✻ *The Vertues.*

Ammoniacum taken the waight of a Dram, loseth the belly, and driueth foorth colde slymie flegme, drawing the same to it from partes a farre of: also it is good against the shortnesse of breath, and for such as are Asmatique and alwayes panting and breathing, and against the stoppinges of the breast, the falling sicknesse, the gowt, the payne of the hanche or hucklebone, called the Sciatica, against the olde head ache, and diseases of the brayne, the sinewes, and extreame partes.

It doth mundifie and clense the breast, it rypeth flegme, & causeth the same to be easily spet out, to be mingled with hony and lickt as a Lohoc, or taken with the decoction of hulled Barley.

It is good against the hardnesse and stopping of the Spleene or Milte, it deliuereth the dead Childe, and prouoketh vrine: but there must be but a little of it taken at once: for if it be taken in to great a quantitie or to ofte, it wil cause one to pisse blood.

It cureth all swellinges and hardnesse, it slaketh the payne of the liuer and Splene being stieped in vineger, and spread or layde vpon the place.

If it be mingled with hony or pitch and layd to, it dissolueth harde lumpes or swellinges, and taketh away Tophi, whiche be harde tumoures engendred of the gowte in the ioyntes and extreme partes: it consumeth also all colde tumours and Scirrhus matter being layde vpon: And it is very good to be put into al oyntmentes and playsters that are made to chafe and warme, to swage payne, to soften and drawe.

It is good to be layde to the Sciatica or gowt of the hippe, and vppon all payne and wearinesse of any parte, with the oyle of Cyprus and Nitrum.

Ammoniacum is good to be put into Colyria and all Medicines that are made to cleare the sight, & medicines that are made to take away the dimnesse and webbe of the eyes.

Of Euphorbium. Chap.xvi.

Vphorbium is the gumme or teare of a certayne strange plante growing in Lybia on the mount Athlante, or Athlas, next to the Countrie of Mauritania, nowe called Morisco, or of the Moores. And it was
first

first founde out in the tyme of Iuba king of Lybia: the leafe of this plant is long and rounde, almost lyke to the fruit of Cucumer, but the endes or corners be sharper, & set about with many prickles, which are somtimes foūd in the gumme it selfe: one of those leaues set in the grounde, doth increase and multiply diuers. The sappe or liquor that commeth foorth of the sayde leaues, burneth or scaldeth, and straightwayes it congeleth and becommeth thicke, and that is the Euphorbium. The first Euphorbium is yellowish, cleare, brittle, very sharpe and burning in the mouth and throte, freshe and newe, not muche elder then a yere: for this gomme doth soone lose much of his heate and vertue by age, as Galen and Mesue saith.

Euphorbium.

❧ The Place.

The Euphorbium described of the Auncientes groweth vppon the mount Athlas in the Countrie of Lybia, bordering vpon Mauritania: it groweth also in Africa and Iudea, from whence it hath ben conueyed into certayne places of Spayne, Fraunce, & Italie, where as it bringeth foorth neyther floures nor fruit. Pena hath seene it growing at Marselles and Monspellier in France, where as he saw the floures and tasted of the fruite.

❧ The Tyme.

It putteth vp his leaues in the spring time, whereof the first, the second, and the thirde, is the stalke or stem, and the rest growe foorth as branches, and whan the plant is seuen or eyght yeeres olde, it bringeth foorth yellow floures, like in proportion to Balaustia, and in Autumne the fruit is ripe, of colour red and prickley. &c.

❧ The Names.

This gumme is called in Greke ἐυφορβιον: in Latine Euphorbium: in shoppes Euforbium: some call it Carduus Indicus, and Ficus Indica, that is to say, the Thistell, or figge of India, some take it to be Opuntia Plinij: This Euphorbiū should seeme to be that whereof Solinus hath made mention in the xxvij. Chap. of his Historie, wheras he saith, Proficere ad oculorum claritatem, Et multiplex sanitatis præsidium fore, ac non mediocriter percellere vim venenorum. It is also the Euphorbium described by John Leo in his African historie.

✱ The cause of the Name.

Iuba king of Lybia, was the first finder out of this herbe: and named it after the name of his Physition, the brother of Musa who was also a Physition to the Emperour Auguste.

❧ The Nature.

Euphorbium is very hoate and drie almost in the fourth degree.

The second Booke of

❧ The Vertues.

Euphorbium prepared in manner as shalbe vnder written, purgeth and driueth foorth by siege (as Mesue saith) tough, colde, and slymie flegmes, and draweth vnto it, from the sinewes and partes a farre of, and also purgeth choler. Moreouer it is very good against the olde head ache, the Paulsie, the Crampe, the weakenesse that foloweth after the Frenche pockes, the payne of the sinewes and extreme partes, that are of continuance, & against the Jaunders. It is also good against the Pestilence, and suche lyke contagious sicknesses, as one Gentilis writeth.

They make a playster with Euphorbium, and twelue times so much Oyle, and a little waxe, very singuler against all paynes and aches of the ioyntes, the Takinges, Lamenesse, Paulsies, Crampes, and shrinking of sinewes, and against all aches, paynes, & disorder of the same, as Galien in his fourth booke de Medicamentis secundùm genera, declareth more at large, shewing how and whan the quantitie of Euphorbium, is to be augmented or diminished, whiche shoulde be to long to recite in this place.

Euphorbiū mingled with Oyle of Bay, Beares grease, or Woolfes grease, or such like, cureth the scurffe and scales of the head, and pyldenesse, causing the heare to renewe and growe againe, not only vpon the head and other bare places, but it will also cause the bearde to growe that is slacke in comming, if it be annoynted therwithal.

The same mingled with Oyle, and straked or layd vpon the temples of such as are very sleepie, or troubled with the lethargie, and raging, doth awaken and quicken their sprites agayne. And if it be applied to the nuque, or nape of the necke, it restoreth the speach agayne vnto them that haue lost it by reason of the Apoplexie.

Euphorbium mingled with vineger, and straked vpon the place, taketh away al fowle, & euilfauoured spots from the body, especially the white scurffe and scales of the skinne.   ❧ The Daunger.

Euphorbium by reason of his extreame heate, is very hurtfull to the liuer and stomacke, and all the inwarde partes, when it is receiued into the body, for it chafeth and inflameth the same out of measure.

❧ The correction and preparation therof.

1. The malice and violence of Euphorbium is corrected many waies: and first ye must annoynt it with Oyle of sweete Almondes, after put it into the midle of a Citron, and wrap it, or close it vp in leauened paste, and so bake it, & when the paste is readie, ye may take the Euphorbium out of it, to vse in medicine.

2. Maynardus taketh Mastick & gumme Dragagante, as much as the Euphorbium commeth to, and mingling them well togeather, putteth it into the midle of an vnbackte loafe, so letting it bake vntil the bread be wel backte: then taketh he of the crumbe or pulpe of that loafe, and maketh small pilles thereof, whiche be very singuler against the weakenesse or debilitie comming of the Frenche pockes, and al anguish and payne of the outwarde partes.

3. An other mingleth with Euphorbium, the lyke quantitie of Masticke, and maketh pilles with the iuyce of Citrons or Orenges, the whiche are muche praysed against the Pestilence.

Of Sarcocolla. Chap. cxbij.

Sarcocolla is the gumme of a certaine thornie plant growing in Persia. And the best is that which is yellowish, bitter in taste and like to the fragmentes or small peeces of Frankensence: yet Plinie in the xiij. Chap. of the xj. booke
of

of his historie preferreth the white before the other, and so doth he also in the xxiiij. booke, the xiiij. Chap.

❧ The Names.

This gumme is called in Greeke σαρκοκόλλα : in Latine and in Shoppes Sarcocolla : in Englishe Sarcocoll : in Frenche Sarcocolle : in Douche Sarcocolla.

❧ The cause of the Name.

The Greekes called this gumme or teare Sarcocolla, bycause it sodereth and gleweth togyther woundes and cuttes of the flesh, euen as glewe doth ioyne togyther timber.

❧ The Temperament or Nature.

Sarcocolla is hoate in the second degree, and drie almost in the same degree, and it drieth without any byting sharpnesse, as Galen saith.

Sarcocolla, as Mesue writeth, purgeth rawe and grosse fleame, and the tough flymie humours, that are in the ioyntes and extreame partes: It mundifieth the brayne, the sinewes, the breast, and the lunges: and is very good against an olde cough that hath continued long, and for suche as are flegmatique and Reumatique, to be taken the quantitie of a Dram or somwhat more.

It is very consolidatiue or healing, wherefore it closeth vp woundes and vlcers, and it mundifieth and clenseth malignant and corrupt vlcers, and filleth the same with newe flesh, especially being reduced and brought into a powder, and strowed thereon, or applied or layde therebnto with honie.

This gumme is very conuenient to bloodshotten eyes, the spottes, darkenesse, scarres, and such lyke impedimentes or defaultes of the same: especially if it be stieped in Asses milke by the space of foure or fiue dayes (as Mesue writeth) but the milke must be euery day renewed, and the stale or olde milke cast away.

❧ The daunger and correction of the same.

They that vse it muche waxe balde: it is slowe in operation, and it troubleth them that haue Cholerique stomackes: wherefore heede must be taken, that it be not giuen to suche.

One may augmente and increase his vertue to loose the belly, by putting thereto some ginger and Cardamome.

The ende of the seconde part.

*Twise corrected and augmented
by the Aucthor.*

312 The thirde Booke of

¶ The thirde part of the Historie of
Plantes, intreating of Medicinal rootes, and herbes, that purge the body, also of noysome weedes, and dangerous Plantes, Their sundrie fashions, Names, and Natures, their vertuous Operations and dangers.

Compiled by the learned D. Rembert Dodoens, nowe Phisition to the Emperour.

Of Aristolochia.   Chap.j.

❧ *The Kyndes.*

ARistolochia, as Dioscorides writeth, is of three sortes, that is to say long Aristolochia, rounde Aristolochia, and the Aristolochia called clematitis. Whereunto Plinie hath added a fourth kinde, called Pistolochia, and the later writers haue ioyned to them a fifth kinde, called Sarrasines herbe or Astroloche.

| 1. Aristolochia longa. | 2. Aristolochia rotunda. |
| Long Aristoloche. | Rounde Aristoloche. |

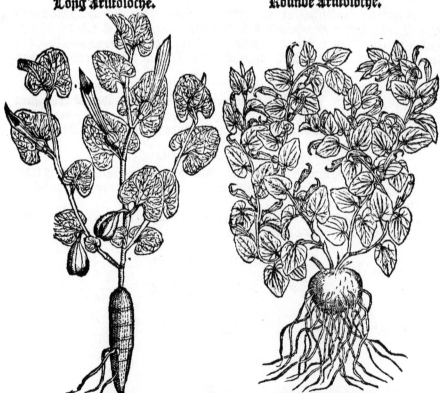

❧ *The Description.*

THE long Aristolochia, hath diuers square slender branches of a span long or more, growing vp from the roote, about which groweth here and there certayne broade leaues like Iuy leaues. The floures be purple and most commonly pale, of a strong greeuous sauour, they growe

the Hiſtorie of Plantes.

growe faſt by leaues, and are in proportion long and holowe, yet longer by one ſyde than by another: whan they are paſt, there foloweth a certayne fruit like vnto ſmall peares, ſauing they be ridged alongeſt the ſydes, or creſted and clouen lyke garlike heades: the which do alſo chop and cleeue a ſunder whan the ſeede is rype, and the ſeede that than appeareth is triangled, and of blackiſh colour. The roote is halfe a foote long or more, and as thicke as ones thombe or finger, of a yellowiſh colour like Boxe, of a ſharpe bitter taſte, and ſtrong ſauor.

2 ¶ The rounde Ariſtolochia in his ſtalkes and leaues is like to the firſt, but his leaues be ſomewhat rounder. The flowres differ onelye in this, that they be ſomewhat longer and narrower, and of a faynte yellowiſhe colour: ſhorter by one ſide than another, and of a blackiſhe purple colour vpon that ſyde that turneth backe agayne: The fruit of this Ariſtolochia is alſo ſharpe faſhioned lyke to a top, or peare, ſauing it is rounder and fuller, and ſtraked or ribbed like the other. The ſeede is like to the ſeede of the log Ariſtolochia. The rootes be round and ſwollen like to a Puffe or Turnep, in taſte and ſauour like to the long.

3. Ariſtolochia Clematitis.   4. Piſtolochia.   5. Ariſtolochia Sarracenica.
Branched Ariſtolochia.   Smal Ariſtolochia.   Saraſins Ariſtolochia.

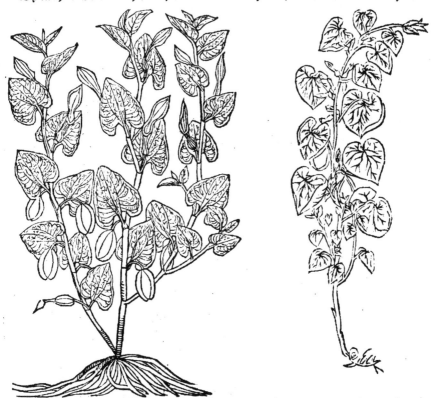

3 ¶ The thirde kinde of Ariſtolochia his ſtalkes and branches are ſmal and tender, his leaues be like to the others, but the little ſtemmes or footſtalkes of the leaues are ſomwhat longer. The flowers alſo be long and holow, of a yellow or deepe violet colour: The rootes be ſmall and ſlender, diſperſed or growing here and there.   Dd   The

The thirde Booke of

4 　The fourth Aristolochia in his leaues and stalkes, is like to the long and rounde Aristolochias, sauing it is smaller and finer or tenderer, his leaues be also broade lyke Iuy leaues. The flowres be also long and hollowe, and blackishe about the toppes or endes. The fruit is also round and like to the others, his rootes be long and small as russhes, or threddes.

5 　The fifth kinde which is called Sarasins wurt, or Sarasins Aristolochia, hath longer and higher stalkes than any of the kindes aforesayd: his leaues be also larger, but otherwise they differ not, for they be also lyke Iuy leaues. The small flowres growe betwixt the leaues, in proportion also long and hollowe of a yellowish colour. The fruit also is fashioned lyke to a peare. The rootes be long, and sometimes thicke, and couered with a thicke rinde or barke, in sauour and taste lyke the others. ❧ *The Place.*

1.2 　The long and rounde Aristolochias, growe plentifully in Spayne, and in many places of Italie, and certayne places of Fraunce, it delighteth muche in fertile grounde and good pastures.

3 　Aristolochia Clematitis (as Peter Bellon writeth) groweth vppon the mountayne Ida in Crete or Candie. Carolus Clusius saith it groweth about Hispalis a Citie in Spayne nowe called Ciuill, and that he hath founde it amongst the busshes and briers there.

4 　The Pistolochia also groweth in certayne places of Fraunce and Spayne.

5 　The Sarasines Aristolochia, delighteth muche in vineyardes, and high deserte places, and wildernesses, and is founde in sundrie places of Germanie, and Brabant. ❧ *The Tyme.*

　The Aristolochias do flowre in May & July, & timelier in hoate Countries.
❧ *The Names.*

　They are called in Greeke ἀριϛολόχια: in Latine Aristolochia: in English Aristologia, and of some Byrthwort, & Hartwort; in Shoppes also Aristolochia.

1 　The first is called in Greeke ἀριϛολοχίας μακρὰν: in Latine Aristolochiam longam, bycause of the fashion of the roote: it is also called δακτυλίτην, μηλοκάρπον καὶ τεύξινον, Dactilis Melocarpon, and Teuxinon, & Aristolochiam marem : In Englishe long Aristolochia.

2 　The seconde is called in Greeke ἀριϛολόχια ϛρογγύλη, Aristolochia rotunda, and Aristolochia foemina: of some χαμαιμῆλον, and Malum terrę: in Englishe Aristolochia rotunda, and rounde Aristologia.

3 　The thirde is called ἀριϛολόχια κλημάτιτις, Aristolochia Clematitis: Branched Aristologia.

4 　The fourth kinde called of Plinie in the eyght Chapter of his xxv. booke πιϛολόχια καὶ πολύριζον, Pistolochia and Polyrhizon.

5 　The fifth Aristolochia is nowe called of some Herba Sarracenica: in frenche *Sarrasine*: in Douche Zarasijn cruyt: in Shoppes Aristolochia longa, which is in Douche lange Osterlucey: in English long Aristolochia, in steede whereof it may be vsed. We may also name it in Englishe Sarasines herbe, & Sarasines Aristolochia. ❧ *The Nature.*

　The rootes of Aristolochia, are all hoate and dry in the extremitie of the seconde degree. ❧ *The Vertues.*

1 　The rootes of Aristolochia are excellent against al poyson, and agaynst the bitinges & stinginges of venimous beastes, if it be taken in wine, or layd vpon the woundes, or bitinges.

2 　The long Aristolochia moueth the menstrual termes, and prouoketh vrine: And if it be dronken with Pepper and Myrrhe, it expelleth the Secondine, & dead childe, & al other superfluities gathered togyther in the Matrix. It worketh

keth the same effect, to be ministred in a Pessarie or mother Suppositorie.

2 The rounde Aristolochia is lykewise good for the same purpose: and it is also very good for them that are short winded, and troubled with the peor or hyquet, it is profitable against the payne of the syde, the hardnesse of the melt or splene, the crampe, or conuultion, or drawing togyther of the sinewes, the falling sicknesse, the gowt, and the shakinges or shiueringes of Agues: and for al such as are hurt or bursten inwardly, if it be giuen them to drinke with water.

The same draweth foorth splinters of broken bones, Shaftes and Dartes, thornes, and shiuers, if it be layde to the place with Pitche or Rosen, as Plinie writeth.

It mundifieth and scoureth all corrupt and filthy sores, fistulas, and virulent holowe vlcers: and filleth them vp agayne with newe flesh (if it be mixt with Ireos and hony) & especially it cureth the faultes, & vlcers of the secret partes, if ye wash the same with the Decoction of this Aristolochia made in wine.

Aristolochia rotunda, doth beautifie, clense, and fasten the teeth, if they be often frotted or rubde with the powder thereof.

3 The thirde kinde is much like to the other in vertue, sauing it is not so strog as Dioscorides writeth: and Galen saith, that this kinde is of the sweetest, and pleasantest sauor, and therfore is much vsed in oyntmentes: but it is weaker in operation than the aforesayde.

4 Pistolochia or smal Aristolochia, is also of the same vertues and operatios, but not so strong as the others.

5 Sarrasines or braunched Aristolochia is also lyke ye others, it is very hoate and bitter: and not inferior to Aristolochia longa, wherfore in all compositios one may be vsed in steede of the other, without errour.

The Choice.

2 The rounde Aristolochia, is of fine and subtile partes, and of stronger operatio than the rest, it mundifieth and clenseth mightily, and it soupleth and maketh thinne, grosse humours.

1 The long Aristolochia is not of suche subtile partes, neither doth it clense so mightily, but is better to incarnate, and ingender flesh in vlcers.

3 Aristolochia Clematitis hath the best sauour, wherfore it is best to make Oyntmentes.

Of Holeworte. Chap.ij.
The Kyndes.

Holeworte is of two sortes, the one hath a rounde roote, which is not holowe within: And the roote of the other is holowe within: but otherwayes they are like one another, in their stalkes, leaues, floures, & seede.

The Description.

Holeworte hath smal tender stalkes of a span long: his leaues be also small and iagde lyke Rue or Coriander, of a light greene or rather a grayishe colour. At the top of the stalke it beareth flowers after the proportion of larkes spurre, but muche smaller, and of Carnation or a light redde purple colour, and oftentimes white, and growing meetly thicke togyther. After the flowers there cometh certaine huskes or coddes, in which is the seede, whiche is rounde and blacke. The roote of one of these kindes, is all rounde, and firme, yellowe within and couered ouer with a blackishe pyll or skinne. The roote of the other is most commonly long, & growen like a peare, holowe both vnderneath and within.

The Place.

These rootes growe by olde quicke set hedges, and bushes in the borders of feeldes, and in the pendant and hanging of hilles and mountaynes. The smaller roote whiche is not hollowe is founde in certayne places of Brabant by Louaigne. The greater whiche is also holow, groweth in Germany: & wheras the one groweth, the other groweth not at all, so that ye shal neuer finde ye full roote growing with the holowe roote, nor the holowe roote growing by the full roote.

Radix caua maior.
The great Holewurt.

Radix caua minor.
The small Holewurt.

❧ *The Tyme.*

This herbe springeth betimes, and bringeth foorth his stalkes and leaues in February, and flowreth in Marche, and deliuereth his seede in April, & afterwardes the herbe fadeth so, that nothing of him remayneth sauing the roote vnder grounde.

❧ *The Names.*

The roote whiche is holowe within is called in Germanie Holwurtz, that is to say in English Holowe roote, or Holewurt: in Frenche *Racine creuse*: in Brabant Hoolwortele: that is to say in Latine Radix caua.

The other whiche is full, close, and firme, is called in Brabant Boonkēs Holwortel.

This roote especially that whiche is holowe, hath ben of long time vsed in the Shoppes of this Countrey for rounde Aristolochia, & it is so taken yet of some ignorant Apothecaries. Some of the learned do thinke this herbe to be the Pistolochia described of Plinie, Others woulde haue it to be a kinde of fumetorie, caled Capnos Phragmites: and some thinke it to be θήσιον Thesium Theophrasti. Some also thinke it to be ἠρίφαι Eriphiam Plinij: and it seemeth to be somewhat lyke Eriphya (that is written with y) bycause it is founde in the spring time onely: and therefore it may be well called ἠριφύα, that is in Latine Planta veris.

❧ *The Nature.*

Holewort is hoate and dry in the seconde degree.

❧ *The Vertues.*

Holewort cureth the Squinancie, and olde tumoures or swelling of the throte, or kernelles and Almondes of the same, if one gargle or wash his mouth with the decoction of the same roote boyled in water onely or vineger, for it hath power to cut and consume grosse humours.

It is also good agaynst the tumoures, and inflammations of the vuula, to be kept in the mouth and chewed vppon, or the powder of the same layde thereto.

The same mingled with Vnguentum Populion nigrum, or with some other of the same nature, is good to waste and consume the Hęmęroydes, or piles, and to swage the paynes of the same.

Of Swallowurte or Vincetoxicum. Chap.iij.

❧ The Description.

Asclepias is somewhat lyke the third kinde of Aristolochia, in stalkes and leaues, his stalkes be smothe, rounde, and small, about two foote long, with blackish leaues, not much vnlyke Iuye leaues, sauing they be longer & sharper poynted. The flowers growe vpon small stemmes betwixt the leaues, of a pale or bleake white colour, and sometime yellowish, and also blacke, of a certayne strong sweetish sauour: after them commeth long sharpe-poynted huskes or coddes, the which do open of themselues whan they are ripe, and within them is conteined seede, lapped as it were in a certaine white wooll, the whiche seede is reddish and broade, not muche vnlyke the seede of Gentian. The rootes be long & round, as it were small round threddie stringes or laces, enterlaced one with another, almost lyke the rootes of blacke Hellebor, or Oxe heele, and of a rancke sauour.

❧ The Place.

Asclepias groweth in rough, high, grauely, and Stonie mountaynes.

❧ The Tyme.

It flowreth in June, and his seede is ripe in August.

❧ The Names.

This herbe is called in Greke ἀσκληπιάς: and in Latine Asclepias, of some it is called in Greeke κίσσιον, Hederuncula, & κισσόφυλλον, that is, Hederæ folium, and nowe it is called Hirundinaria, and Vincetoxicū: in Germanie Schwalben wurtzel: in Brabant Swaluwe wortele: we may call it in English Asclepias, Vincetoxicū, & Swallowurt.

❧ The cause of his first Name.

This herbe tooke his name of the Ancient father Esculapius, which was called in Greke ἀσκληπιός, whom both the Greekes and Gentils say, that he was the first that found out Physicke, wherefore they honoured him as a God.

❧ The Nature.

The rootes of Asclepias are hoate and drie, and resist poyson.

❧ The Vertues.

The roote of this Herbe boyled in water and dronken, slaketh the grypping paynes of the belly, & is very good for suche as are bitten of venimous beastes, and madde Dogges, not onely to be giuen to drinke inwardly with wine, but also if the leaues be applyed outwardly.

The leaues of Asclepias pounde and layde to, are good agaynst the malignant vlcers, and corrupt sores both of the breastes and Matrix, or mother.

Of Periploca. Chap.iiij.

❧ The Kyndes.

There are two sortes of Periploca: wherof one hath no surname, the other is called Periploca repens.

Asclepias.

The thirde Booke of

Periploca prior.　　　　　Periploca altera.
The firſt Periploca.　　**The ſeconde Periploca.**

❊ *The Deſcription.*

1   The firſt Periploca is many wayes like vnto Swallowurt o2 Aſclepiaſ, but his leaues be ſomewhat larger and greater, his little ſtalkes o2 b2anches are longer, his huſkes o2 coddes alſo are longer and thicker, and his rootes are like th2eddie ſtringes creeping on the grounde.

2   The other hath longer and larger leaues, his ſtalkes and b2aunches are thicker and harder, & they periſhe not in winter as the firſt do: and his huſkes o2 coddes are alſo greater.

Both theſe herbes (beyng ſcarrified o2 hurt) do giue foo2th a milkie iuyce, o2 liquo2, and ſpecially the laſt: fo2 the iuyce of the firſt is oftentimes yellowiſh.

❊ *The Place.*

Theſe plantes growe in Syria, and ſuche lyke hoate regions, they do not lightly beare their huſkes in B2abant.

❊ *The Names.*

They are both called Periplocæ: and the ſecond is called Periplocca repens: both are thought to be ἀπόκυνον, Apocynon of Dioſco2ides, the whiche is alſo called κυνοκράμβη, and Braſſica Canina, yet there is another Braſſica canina, a kind of wilde Mercurie.   ❊ *The Nature and Vertues.*

Apocynon is a deadly and hurtfull plant not onely to man, but alſo to cattel: his leaues mixt with meale, and tempered o2 made into b2ead, it deſt2oyeth Dogges, Wolues, and Foxes, and other ſuche beaſtes that eate thereof.

Of

the Historie of Plantes.

Of Asarabacca. Chap.v.

※ *The Description.*

Asarabacca hath swart greene, rounde, shining leaues, lyke Iuye, but a great deale rounder, and tenderer: in and amongst those leaues (next the grounde) growe the flowers vppon short stemmes, which be of a fayre browne purple colour, and of a good sauour somwhat like Nardus, & fashioned like the flower of a Granat tree, called Balaustia or Cytinus which is the buddes of Balaustia, and somewhat lyke the cuppes or huskes of Henbane. The rootes be small, long, and crookedly layd, ouerthwart, here and there, with diuers small hearie stringes, of a pleasant sharpe sauor and taste byting the tongue.

Asarum.

※ *The Place.*

It delighteth in shadowy places, and rough dry groundes, especially in the pendent or hanging of hilles & mountaynes, in thicke darke wooddes, and commonly vnder the Haselles (as Cordus sayth.)

It is alwayes greene, and springeth anew and floureth in the spring time, and it floureth agayne at the ende of Sommer.

※ *The Names.*

This herbe is called in Greeke ἄσαρον: in Latine & in shoppes Asarum : of some Nardus rustica, & Perpenia, Macer calleth it Vulgago: it is called in English Asarabacca, and folefoote, it may also be called Haselworte : in Frenche *Cabaret:* in Germanie Haselwurtz: in Brabant Haselwortel, and of some Mansooren.

※ *The Nature.*

Asarabacca is hoate and drie in the thirde Degree, especially the roote whiche is most vsed in Physicke.

※ *The Vertues.*

The roote of Asarabacca boyled in wine and dronken, prouoketh vrine, and is good against the strangurie, the cough, the shortnesse of breath, and difficultie of breathing, Conuulsions and Crampes, and the shrinking togyther of members. A

The same taken in lyke manner, is profitable against venome, and agaynst the bitinges and stinginges of Serpentes, and all venemous beastes. B

The same boyled in wine, is good for them that haue the Dropsie, and the Sciatica. C

The same dronken with honied wine, bringeth downe the menstrual fluxe, expelleth the Secondine and other superfluities of the mother. D

The leaues of Asarabacca stamped with wine, and strayned, and the iuyce thereof dronken, causeth to vomite, and purgeth by vomiting, tough flegme, and choler. E

The same leaues stamped are good to be applyed or layde to the ache and dolors of the head, to the inflammation of the eyes, and to womens breastes that are to full of milke, whan they list to drie vp the same, and it is good to be layde to the disease called the wilde fire, especially at the beginning. F

Of Dragons. Chap. vi.

❧ *The Kyndes.*

There are three sortes of Dragons, as Plinie writeth, that is to say, the great and the smal, and a certayne third kinde growing in waterie places.

1. Dracunculus maior.   2. Dracunculus minor.   3. Dracunculus palustris.
The great Dragonwurt.   The smaller Dragonwurt.   Water Dragonwurt.

❧ *The Description.*

The first kinde called the great Dragon or Serpentarie, beareth an vpright stalke of a cubit long or more, thicke, rounde, smothe, and speckled with diuers colours and spottes lyke to an Adder or Snakes skinne. The leaues be great and large, compackt or made of fire, seuen, or moe leaues: whereof eache single leafe is long & lyke to a Sorrell or Docke leafe, sauing they be very smothe and playne. At the top of the stalke groweth a long hoose or huske, lyke to the hoose or codde of Aron, or Wake Robin, of a greenish colour without, and of a darke red or purple colour within, and so is the clapper or pestill that groweth vp within the sayde huske, the whiche is long and thicke, and sharpe poynted peeked lyke to a horne: whose fruit by increase wareth so, as it streatcheth, and at length breaketh out of a certayne skin or belme, the sayde fruit appeareth like to a bunche or cluster of grapes, first greene, and afterwarde red as fier, the berries or grapes whereof are full of iuyce or liquor, in which is a certayne smal harde seede. The roote of this Dragon is lasting, thicke and white, and growen lyke to a Bulbus Onyon, couered with a thin pil, and of the quantitie of a prettie apple, and bearded with diuers little white heares or stringes, and oftentymes there is ioyning to it, other small rootes, whereby it is multiplyed.

The

the Historie of Plantes.

4. Dracunculus Matthioli.
Matthiolus Dragonwurte.

2 ¶ The smaller Dragon in his leaues, his huske or codde, his pestill or clapper, his berry and grape is like vnto Iron or Cockowpint: sauing that his leaues are not marked with blacke but with white spottes. Neyther do they perish so soone as Iron, but they growe togyther with their berries, euen vntyl winter. Their berries also are not fully so redde, but are of a certaine yellowish red. The roote is not muche vnlike Iron white, and rounde lyke an Onyon, and hath certayne hearie threddes, hanging by it, with certayne small rootes, or buddes of newe plantes.

3 ¶ The roote of water Dragon is not round after the order of Bulbus, but it is a long creeping roote, full of ioyntes, and of a reasonable thicknesse, out of whose ioyntes, spryngeth vp the stalkes of the leaues, whiche are smoth without, and spungie within: but downewardes towardes the grounde the sayd rootes sendeth out of their said ioyntes, certaine smal hearie rootes. The fruit groweth aboue, vppon a shorte stem, and commeth foorth with one of the leaues, compassed about with small white thromes or threddes, at the first, (which is the blowing) and afterward it groweth foorth into a cluster, which is greene at the first, and waxeth red whan it is rype, smaller than the grape or cluster of Irons berries, but as sharpe or byting. The leaues be large, greene, fine, smoth, & fashioned like Iuy leaues, yet smaller then the leaues of Cockowpint, or Iron. But that leafe in which ẏ cluster of berries groweth, is smallest of al, & on the vpper part or syde next the fruit, it is white.

4 ¶ Besides the aforesayde Dragons, there is an other kinde placed of Matthiolus, with great large leaues, growing folden and lapped one within another, with an vpright stalke, and beareth at the toppe a certayne blossome or flower lyke to a spyke eare. The roote is also round lyke the others, as ye may perceiue by ẏ figure. Surely this kinde of Drago (if any such be to be found) is rather a kinde of Bistort: howbeit there be that thinketh this figure to be false and fayned.

❦ *The Place.*

1 ¶ The first Dragonwort groweth well in shadowie places, and in this Countrie, they plante it in gardens.

2 ¶ The seconde also delighteth in shadowie places vnder hedges, and is found plentifully growing in the Ilandes called Maiorque, and Minorque.

3 ¶ This thirde kinde groweth in moyst waterish places, in ẏ brinkes of diches, and floting waters, and also alongst the running streames and riuers.

❦ *The Tyme.*

They flowre in Iuly, and in August the fruit is rype.

❦ *The Names.*

1 ¶ The first kinde is called in Greeke δρακοντία μεγάλη: In Latine Dracunculus maior, of some Serpentaria, and Colubrina: in Shoppes Serpentaria maior: of Serapio

Serapio Luf. in English Dragons, and Dragons wurte: in French Serpentaire, or Serpentyne: in Germanie Schlangekraut, Drachenwurtz: in Brabāt Speerwortele and Drakenwortele.

2 The seconde kinde is called in Greeke δρακόντιον μικρὸν: in Latine Dracunculus minor: and of some late writers Arum maculatum: in Englishe small Dragonwurte, and speckled Aron.

3 The thirde is nowe called Dracunculus palustris, siue aquatilis: in Englishe water Dragon, or Marshe Dragon: in Frenche Serpentaire d'eau, or aquatique: in high Douche Wasser Schlangenkraut, wasser Drachenwurtz: in base Almaigne, water Draken wortele.

4 The fourth set downe of Mathiolus for the great Dragonworte, in my iudgement is none of the Dragonwurtes, but that is the right great Dragonwurt, the which we haue described and set in the first place: & it is thought there is no such herbe to be founde, as Mathiolus figure doth represent.

❧ The Nature.

These herbes, but especially their rootes and fruit, are hoate and drye in the thirde degree. ❧ The Vertues.

The rootes of these herbes eyther boyled or rosted, & mingled with hony, A and afterward licked, is good for them that can not fetche their breath, and for those that are vered with dangerous Coughes and Catarrhes, that is to say, the distillation and falling downe of humours from the brayne to the breast, and agaynst conuulsions or Crampes: for they diuide, ripe, and consume, all grosse and tough humours, and they of scoure and clense al inwarde partes.

They haue the like power, whan they are three or foure times boyled, vntyl B they haue lost their acrimonye or sharpnesse, to be afterwarde eaten in meates, as Galen saith.

The same dried and mingled with hony, scoureth malignant, and fretting C vlcers, that are harde to cure, especially if it be mingled with the roote of Brionye, and it taketh away all white spottes, and scuruinesse, from any parte of the body that is rubbed therewithall.

The iuyce of the roote of the same, putteth away all webbes & spottes from D the eyes, and it is good to be put into Collyres, and Medicines that are made for the eyes.

The same dropped into the eares with oyle, taketh away the paine & greefe E of the same.

The fruit of Dragons cureth virulent and malignant vlcers, & consumeth F and eateth away the superfluous flesh (called Polyppus) that groweth in the Nose, and it is good to be layde vnto Cankers, and suche like fretting and consuming vlcers.

The freshe and greene leaues, are good to be layde vnto freshe and greene G woundes, but they are not profitable whan they be dryed.

It is thought of some, that if cheese be laid amongst Dragon leaues, it will H preserue the same from perishing and rotting.

Dioscorides writeth, that it is thought of some, that those whiche carrie I about them the leaues or rootes of great Dragonwurtes, cannot be hurt nor stong, of Vipers and Serpentes.

Of Aron, Calfes foote or Cockowpynt. Chap. vij.

❧ The Description.

Cockowpynt hath great, large, smoth, shining, sharpe poynted leaues, much larger than Iuy leaues, & spotted with blackish markes of blacke and blew: amongst them riseth a stalke of a spanne long, spotted here & there

there with certaine purple speckles, and it carieth a certayne long codde, huske, or hose: open by one syde like the proportion of a haares eare, in the middle of the sayd huske, there groweth vp a certayne thing lyke to a pestel or clapper, of a darke murry, or wanne purple colour: the whiche after the opening of the belme or huske doth appeare, whan this is gone, the bunche or cluster of beries also or grapes, doth at length appeere, which are greene at the first, and afterwarde of a cleare or shining yellowish red colour, lyke Corall, and full of iuyce in eache of the sayde berries, is a smal harde seede or twaine. The roote is swelling rounde lyke to a great Olife, or smal bulbus Onion, white and full of Pith or substaunce, and it is not without certayne hearie stringes by it: with much increase of small yong rootes or heades.

❧ The Place.

Aron groweth vnder hedgis, and cold shadowie places.

❧ The Tyme.

The leaues of Aron do spring foorth in Marche and Aprill: and they perishe and vanishe in June and July, so as nothing remayneth sauing onely the stalke and naked fruit in July, in August and after the fruit waxeth rype.

❧ The Names.

This plant is called in Greeke ἄρον: in Latine Arum: in Shoppes Iaron, and Barba Aron: of some Pes vituli: of the Assyrians Lupha: of the Cyprians Colocasia: (as amongst the bastardes and counterfet names) where as it is also called ὄλιμος, and δρακοντια. Plinie affirmeth in the xbj. Chapter of his xviij. booke, that there is much controuersie about Aron and Dragonwortes, and some affirme it to be the same, and so call it Serpentariam minorem: in Englishe also it is commonly called Aron, Priestes pyntill, Cockowpintell: also Rampe, and Wake Robyn: in Frenche Pied de veau, and Vit de Prestre: in Italian Gigaro: in Spanishe Yaro: in Germanie Pfaffen pint, and Teutschen iugbeer: in Brabant Papecullekens, and Calfsboet.

❧ The Nature.

Aron is of complexion hoate and drie, and as Galen sayth, it is hoater in one region than in an other, for that which groweth in Italie, is only hoate in the first degree, or almost in the seconde degree, but that which groweth in this Countrie, is hoate in the thirde degree.

❧ The Vertues.

The rootes, leaues and fruit of Aron, are in power and facultie much lyke vnto Serpentaria, or that kinde of Dragonwortes that groweth in this Countrie, the whiche is very hoate, as we haue sayde.

The second Booke of

Of Arisarom. Chap.viij.

✣ The Kyndes.

There is nowe founde two kindes of Arisarom, whereof one hath broade leaues, and the other narrowe.

Arisarum latifolium. Arisarum angustifolium.
Broadleaued Arisaron. **Narrowleaued Arisaron.**

✣ The Description.

THE first and right Arisarom, hath leaues fashioned like Aron, sauing they be muche smaller sharpepoynted & somwhat fashioned like Iuy-leaues, his stalke is smal and slender, his huskie couering, is but litle, and his pestill or clapper small: of a blackishe purple colour, his grape or berie whan it is ripe is red. The kernelles are smal. The roote is also white and fashioned like Aron, sauing it is smaller.

The seconde Arisaron hath fiue or sixe, or mo: long, narrowe, smothe, and shining leaues, his huskie bagge or hose is long and narrowe, the long tayle or slender pestill that groweth out of the sayde huske, is somewhat bigger than a rushe, and of a blackish purple, & so is part of the lining, or inside of the huske: to the which at the last there groweth, a lowe euen by the ground, and somtimes deeper, a certayne small number of kernelles or berries, growing togyther in a little bunche or cluster like grapes: which are greene at the first as the others be, and afterwarde red. The roote is also rounde and white lyke the other.

✣ The

the Historie of Plantes.   325

✿ *The Place.*

Both of these plantes are strangers in Germanie, and this Countrie. But the first kinde groweth in Italy, specially in certayne places of Tuscane: the other groweth about Rome, and in Dalmatia, as Aloisius Anguillara witnesseth.

✿ *The Tyme.*

Both of these plantes do beare their flowres and seede at suche tymes and seasons as Aron and Dragons do.

✿ *The Names.*

The first of these plantes is called of Dioscorides ἀρίσαρον: in Latine Arisarū, we may also call it in English Arisaron: Plinie in his xxiiij. booke and xvj. Chap. calleth it ἀρίς, saying, there is an Aris growing in Egypt, like vnto Aron, but it is smaller both in leaues and roote, and yet the roote is as bigge as an Olife. But the other Arisaron was vnknowen of the olde writers. Yet, that it is also a kinde of Arisaron, it is manifest aswel in the flowers, fruit, & rootes, as also in the qualities.

✿ *The Nature.*

Arisaron is of a hoater and dryer complexion than Aron, as Galen writeth.

✿ *The Vertues.*

Arisaron also in vertue and operation is lyke to Dragonwortes, and the roote thereof is proper to cure hollowe vlcers and paynefull sores, as Dioscorides writeth: they also make of it Collyria and playsters good agaynst Fistulas. It rotteth and corrupteth the priuie members of all liuing thinges being put therein, as Dioscorides writeth.

Of Centorie.   Chap.ix.

✿ *The Kyndes.*

CEntorie (as Dioscorides writeth) is of two sortes, that is to say, the great and the smal, the whiche in proportion and quantitie, are muche differing one from the other.

✿ *The Description.*

1 THE great Centorie hath rounde stemmes of two or three Cubites long: it hath long leaues, diuided into sundry partes, lyke vnto the walnut tree leaues, sauing ý these leaues are snipt, & dented about the edges lyke a Sawe. The flowers be of small hearie threddes or thrommes, of a lyght blewe purple colour, and they growe out of the scalye knoppes at the toppes of the braunches, the whiche knoppes or heades are rounde and somewhat swollen in the neather parte, lyke to a peare, or small Hartichock, in whiche knoppes (togyther with a certayne kinde of Downe or Cotton) are founde the long, rounde, smoth, and shining seede, like the seede of Cartamus or Bastarde Saffron, and our Ladyes Thistel. The roote is long, grosse, thicke, and brickle: of a blackish colour without, and reddish within, full of iuyce of sanguin colour, with sweetnesse and a certayne byting Astriction.

Of this great Centorie there is an other kinde, whose leafe is not diuided or iagde into partes, or peeces, but after the manner of a Docke leafe, it is long and broade, single, and not cut into partes: yet it is nickt & snipt rounde about the edges, Sawe fashion. The stalke is shorter than the other: The flowers, seede and roote, is lyke the other.

2 The small Centorie is a little herbe, it springeth vp with a smal, square, cornered stalke, of halfe a foote or nine inches long: with small leaues in fashion lyke Maricrom, or rather lyke the leaues of S. Johns worte. The pleasant flowers growe at the top of the little braunches, of a fayre carnation, or light

Ee   purple

The thirde Booke of

Centaurium magnum. The great Centozie.

Centaurium minus. The smal Centozie.

purple red colour, lyke the rose campine, but smaller: whiche by day tyme and after the Sunne rysing do open, and do close vp agayne in the euening. There commeth after the flowers little long huskes, or sharpe poynted coddes, somewhat lyke wheate cornes, in which is conteyned a very small seede. The roote is small, harde, and of wooddy substance, and serueth not to any purpose in medicine.

❧ *The Place.*

1   The great Centozie delighteth in a good and fruitfull grounde, and grasie hilles & playnes. Dioscorides sayth, it groweth in Lycia, Peloponneso, Arcadia, Helide, Messenie, and in diuers places of Pholoen, & Smyrna, that stande high and well agaynst the Sunne. It is also founde vpon the mounte Garganus or Idea, in the Countrie of Apuleia, and in the feelde Baldus vppon the mountaynes nere Verona: but that which groweth in the mount Baldus, is not so good as that of Apuleia, as Matthiolus writeth.

The single, or whole leaued great Centozie groweth in Spayne, and the rootes being brought to Antwarpe, and hyther, do sometime grow being planted in our gardens.

2   The small Centozie groweth in vntopled feeldes and pastures, but especially in dry groundes, and it is common in the most places of Englande, and also in Italie and Germanie.

❧ *The Tyme.*

1   The great Centozies do flower in sommer, and their rootes must be gathered in Autumne.

The

2 The small Centorie is gathered in July and August, with his flowers and seede.

❧ *The Names.*

1 The great Centorie is called in Greeke κενταύριον τὸ μέγα: in Latine Centauriũ magnum: Theophrastus also calleth it Centaurida: in Shoppes it is wrong named of some Rha Ponticum: for Rha Ponticum is that kinde of Rha which groweth in the Countrie of Pontus, and it is a plant muche differing from the great Centaurie. There be also other names ascribed vnto the great Centorie, which are fayned and counterfayted, as Apuleius writeth, wherof some seeme to apparteine to the lesser Cētorie, as ναρκῆ, μαρώνη ἢ μαρώνιον, νέαςιον, λιμνήςιον, λιμνης ις, πλεκτρονία ἢ πηλκτρόνιον, χειρωνία, αἷμα, ἠρακλίες, that is in Latine Herculis sanguis, Vnefera, Fel terræ, Polyhydion ἡμερῶτον.

2 The smal Centorie is called in Greeke κενταύριον τὸ μικρὸν: and of Theophrastus κενταυρίς: in Latine Centaurium paruum, and Centaurium minus: of some Febrifuga, Fel terræ, and Multiradix: of the Apothecaries Centauria minor: in Italie and Hetruria *Biondella*: in Spanish Cintoria: in Germanie Tausenguldenkraut: in Brabant Santorie, and cleyn Santorie: in French *Petite Centaure*.

❧ *The cause of the Name.*

Centorie was called in Greeke Centaurion, and Chironion, after the name of Chiron the Centaure, who first of all founde out these two herbes, & taught thē to Aesculapius, as Apuleius writeth. And as some other write they were so named, bycause Chiron was cured with these herbes, of a certayne wounde whiche he tooke (being receiued as a ghest or straunger in Hercules house or lodging) by letting fall on his foote, one of Hercules shaftes or arrowes, as he was handling and vewing of the sayde Hercules weapon and armour.

❧ *The Nature.*

1 The great Centorie is hoate and dry in the thirde degree, & also astringent.
2 The lesse or small Centorie, is of complexion hoate, and drie in the seconde degree.

❧ *The Vertues.*

The roote of great Centorie, in quantitie of two Drammes, taken with A water if there be a feuer, & in wine if there be no feuer: is good for them that are bursten, and for them that spet blood, and agaynst the Crampe & shrinking of any member, the shortnesse of winde, and difficultie of breathing, the olde cough, and griping paynes or knawinges of the belly.

The same dronken in wine, bringeth downe the monethes or womens natural termes, and expulseth the dead fruit, as it doth also being conueyed in at the naturall place, as a Pessarie or mother Suppositorie. B

The greene roote of great Centorie stamped, or the drie roote soked in water and brused, doth ioyne togyther and heale, al greene and fresh woundes being layde and applyed therevnto. C

The iuyce of the roote, the which they gather and keepe in some countries, hath the lyke vertue as the roote it selfe. D

The roote of the small, or lesse Centorie, is to no purpose for Medicine, but the leaues, flowers, and iuyce of the same, are very necessarie. E

The smal Centorie boyled in water or wine, purgeth downewardes Cholerique, flegmatique, & grosse humours, and therefore it is good for such as are greeued with the Sciatica, if they be purged with the same vntyll the blood come. F

It is very good agaynst the stoppinges of the liuer, against the Jaundise, and agaynst the hardnesse of the Melte or Splene. G

The decoction of Centorie the lesse dronken, killeth wormes, and driueth them foorth by siege. It is also very good against conuulsions and Crampes, and al the diseases of the sinewes. H

Ee ij The

The iuyce therof taken & applied vnder in a Pessarie, prouoketh the flowers, and expulseth the dead childe.

The same with hony cleareth the sight, and taketh away the cloudes and spottes of the same being dropped or distilled into the same, and it is very good to be mingled with all Collyries, and medicines that are made for the eyes.

The small Centorie, greene pounde and layde to, doth cure and heale freshe and newe woundes, and closeth vp, and sodereth olde malignant vlcers, that are harde to cure.

The same dried & reduced into powder, is profitable to be mingled amongst oyntmentes, playsters, powders, and suche lyke medicines as are ordayned to fyll vp with flesh, fistulas and holowe vlcers, and to mollifie and soupple all hardnesse.

Of Reubarbe or Rhabarba. Chap.x.

The Kyndes.

There be diuers sortes of Rha, or as it is nowe called Reubarbe, not so muche differing in proportion, but their diuersitie is altogyther in the places wher as they are found growing. For one kind of it groweth in Pontus, and is called Rha Ponticum: The seconde groweth in Barbaria, and is therefore called Rhabarbarum, and it is the common Reubarbe: The third commeth from beyonde the Indians, out of the regions of China, and it is that whiche the Arabians call Raued Seni.

The Description.

Rha. Reubarbe.

Rha (as it is thought) hath great broade leaues, lyke to the leaues of Tapsus Barbatus, or white Mollin: or lyke to the leaues of of Clot Burre: snipt and dented rounde about the edges like to a saw, greene and smothe aboue, and white and fryzed vnderneath. Amongst them springeth vp a round straight stalke of a cubite long, and at the top thereof groweth a fayre scaly knop or head, the which whan it bloweth and openeth, sheweth foorth a fayre purple flower, and afterwardes it beareth seede, not muche vnlyke the seede of the great Centorie, sauing it is somewhat longer. The roote is long, thicke, and spungie or open: and being chewed, it yeldeth a yellowish colour lyke Ocre, or Saffron.

The Place.

Rha groweth in the Regions about Bosphorus, and Pontus, by the riuer Rha, and in Barbaria, & in the Countrie of China. We haue found here in the gardens of certaine diligent Herboristes that strange plant whiche is thought of some to be Rha, or Rhabarbarum.

the Historie of Plantes.

❧ The Tyme.

It flowreth in June.

❧ The Names.

This herbe, & specially the roote, is called in Greeke ρᾶ ἢ ῥίον: in the Arabian speeche Rheu, and Raued, or Rauet, of Plinie in Latine Rhacoma, & Rhecoma.

1. That whiche groweth about Bosphorus is called in Greeke ρᾶ ποντικόν: in Latine Rha Ponticum, or Rheon Poticum: of Mesue Raued Turcicum, that is to say, Rha of Turkie.

2. The seconde which groweth in Barbarie, is called Rha Barbarum: of Mesue and the Apothecaries Rheubarbarum.

3. The third kinde (called Chinarum) is called also Rha, or Rheum Seniticum: and Rheum Indicum, and of the Arabians Raued Seni.

❧ The Nature.

Rha is hoate in the first degree, and dry in the second, and of an astringent or binding nature.

❧ The Vertues.

The roote of Rhaponticum, as saith Dioscorides, is good against the blastinges, wamblinges, and the debilitie or weakenesse of the stomacke, and all the paynes of the same. Moreouer it is singuler agaynst conuulsions and Crampes, or agaynst the diseases of the liuer and splene, agaynst the gnawing or griping tormentes of the belly, the kidneyes, and bladder. Also agaynst the aking paynes of breastes and Mother, and for suche as are troubled with the Sciatica, the spitting of blood, sobbing, yeering: it is good also agaynst the blooddie flixe and the laske, and against the fittes of feuers, and the bitinges and stinginges of all sortes of venimous beastes. A

For the same purpose, it is giuen y quantitie of a Dragme with Hydromel or honied water in a feuer: & with syrupe Acetosus against the diseases of y splene or melt: with honied wine it is good against y diseases of y breast: & it is taken drie without any moysture, agaynst the weakenesse or loosenesse of y stomacke. B

The roote of Rha Pontike stamped and mingled with vineger, cureth the vile white scurffe or manginesse, & clenseth the body from pale or wan spottes (or the Morphew) being straked or annoynted with the same. C

Reubarbe and Raued Seni (as Mesue writeth) taken in quantitie of a Dramme, purgeth downewardes cholerique humours, wherefore they are good against all hoate feuers, inflammations, and stoppinges of the liuer, and the Iaunders, especially to be giuen or ministred with whaye or any other refreshing or cooling drinke or potion. D

Reubarbe of him selfe, or of his owne proper nature, is also good against al manner of issue of blood, eyther aboue or belowe, and is good for them that are hurt or burste inwardly, and against greeuous falles and beatinges, & against Crampes, and the drawing together of any part or shrinking of sinewes. E

Also it cureth the blooddy flixe, & al manner laskes, being first a litle tosted, or dried agaynst the fire, and dronken with some astringent liquor, as the iuyce of Plantayne, or grosse and thicke redde wine. F

❧ The Choice.

The best Rha, as Mesue writeth, is y which is brought fro beyond India, & groweth in y Countrie of Chinæ, called Raued Seni. The next to that is the Reubarbe of Barbarie, & that which is of the least vertue is the Rha Potike.

Of Sowbread. Chap. ri.

❧ The Kyndes.

There be two sortes of Cyclamen, as Dioscorides writeth. The one is a lowe plant with a round roote, and is called Cyclamen Orbiculatum. The other

other groweth high, and wrappeth it selfe about shrubbes and plantes, and it hath no notable roote, and it is called Cyclaminus altera.

❧ The Description.

1. Cyclaminon (which we may cal round Sowbread) hath broade leaues spread vpon the grounde with peaked corners lyke to Iuy leaues, and slightly dented round about the edges: and of a swart or darke greene colour aboue, yet powdered or garnished with white speckes or spots, and the middle part of the sayde leafe is somewhat white: but that syde of the leafe whiche is next the grounde, is purple colour, but sometimes deeper and sometimes lighter. The flowers hang vppon tender stalkes, nodding or beckning downewardes, and their leaues turning vpwardes or backwardes, in colour lyke to the purple violet, but not so faire: and of but a little or no sauour. There folowe small knoppes with seede, growing vpon small stalkes that are winded or turned two or three tymes about. The roote is turned rounde lyke to a Turnep, or Bulbus roote, and somewhat flat or pressed downe, with diuers hearie stringes by it, and it is blacke without, and white within, & in withering it gathereth wrinckles.

Cyclaminus orbicularis.
Sowbread.

2. The second Cyclaminon, or Sowbread, his leaues be also broade and nothing peaked or angled, but in a manner rounde, and nothing speckled vppon, or at least wayes very harde to be perceiued: they be also of a sadde or blackish greene colour, but vnderneath of a red purple colour. The flowers are lyke to the first, but of a better sauour. The roote is somewhat smaller.

3. The third kinde also hath leaues without corners, but they be somwhat dented or snip rounde about the edges: these leaues also are speckled, and blackish in the middle. The flower is of a deeper purple, and of a most pleasant sauor. But the roote is smaller than any of the rest.

❧ The Place.
Sowbread groweth in moyst and stony shadowy places, vnderneath trees, hedges, and bushes, and in certayne wooddes, but not euerywhere. It groweth about Artoys and Vermandoys in Fraunce, & in the forest of Arden, and in Brabant. It is also common in Germanie and other Countries. But the thirde kinde is the dayntiest, and yet not strange in Italie.

❧ The Tyme.
The kindes of Sowbread do flower in Autumne about September, afterwardes springeth vp the leaues, which are greene all the winter. The seede waxeth ripe about sommer next folowing.

❧ The Names.
1. The first is called in Greeke κυκλάμινῷ, ἢ ἰχθυόθηρον: in Latine Cyclaminus, Rapum terræ, Tuber terræ, and Vmbilicus terræ: of Apuleius Orbicularis, Palalia, Malum terre, Rapum porcinum, and Panis porcinus: in shoppes Cyclamen, and Arthanita: in English Sowbread: in frenche Pain de pourceau: in Italian

Pan

the Historie of Plantes.

Pan porcino: in Spanish some call it Mazam de porco: in Germanie Schweinbrot, Erdtapssel, Erdtwurtz, and Seuwbrot: in Brabant Uerckens broot, and Sueghen broot.

Plinie calleth the colour of this flower in Latine Colossinum, or Colossinus color.

2　The second kinde is called in Greeke κυκλάμινⓄ ἑτέρα: in Latine Cyclaminus altera: of some κίσιον κιατάνθεμον καὶ κισόφυλλον, and we take that to be Vitalba, the which shalbe described hereafter in the xlviij. Chapter of this booke.

❧ *The Nature.*

Sowbread is hoate and drye in the thirde degree.

❧ *The Vertues.*

The roote of Sowbread dryed, and made into powder, & taken in the quantitie of a dragme, or a dragme and a halfe with Hydromell called also honyed water, purgeth downewardes grosse & tough flegme, & other sharpe humours. A

The same taken in wine is profitable against al poyson, and agaynst the bytinges and stinginges of venimous beastes, to be applyed & layd to outwardly vpon the wounded or hurt place. B

The same dronken with wine or Hydromel, cureth the Iaundise & stopping of the liuer, & taketh away the yellow colour of the body, if after the taking of the same in manner aforesayd, one be so wel couered that he may sweat. C

The same prouoketh the menstrual termes, & expulseth the dead fruit, either dronken or conueyed into the body by a Pessarie or mother Suppositorie. D

The iuyce thereof straked vpō ye nauel or belly, loseth the belly very gētly. And it hath the same vertue being applied to wool to ye fundement as a suppository. E

The same iuyce with vineger, setleth the fundement that is loose and fallen downe out of his naturall place, if it be annoynted therewithall. F

The same mingled with hony, and dropped into the eyes cleareth the sight, & taketh away al spots, as the web, the pearle, & haw, & al impedimēts of ye sight. G

The same snift vp into the nose, clenseth the braynes, and purgeth at the nose grosse and colde flegmes. H

The roote of Sowbread maketh the skinne faire and cleane, and cureth all mangie scuruinesse and the falling of the heare, and taketh away the markes and spottes that remayne after the small pockes and measelles, and all other blemishes of the face. I

The same layde to the melt, or rather the iuyce thereof mingled with oyntmentes and Oyles for the purpose, wasteth and consumeth the hardnesse, and stopping of the Splene or melt. K

It also healeth woundes, being mingled with oyle and vineger, and layde vpon them, as Dioscorides sayth. L

The broth or decoction of the same roote, is good to bathe & stue such partes of the body as be out of ioynt: the gowt in the feete, and kybed heeles, and the scuruie sores of the head. M

The Oyle wherin this roote hath ben boyled, closeth vp olde vlcers, & with the same also & a litle waxe, they make an oyntment very good for kibed heeles and feete that are hurt with colde. N

The roote hanged vpon weimen, in trauayle with chylde, causeth them to be deliuered incontinent. O

❧ *The Daunger.*

In what sorte soeuer this roote be taken, it is very daungerous to women with childe: wherfore let thē take heede, not only how they receiue it inwardly, but also let them be aduised in any wise not to applye it outwardly: nor to carrie

Of Felwort or Gentian. Chap. xij.

The Description. Gentiana. Gentian.

1. THE first leaues of Gentian, are great and large, layd and spread abroade vpon the ground with sinewes or ribbes lyke Plantayne, but greater and more lyke to the leaues of white Hellebor, amongst which springeth vp a rounde, smothe, holowe stalke, as thicke as ones finger, full of ioyntes, and somtimes as long as a man, with smaller leaues growing by couples at euery ioynt, and sometymes somwhat snipt round about the edges, with yellow flowers growing round about the stalke at the sayde ioyntes lyke to Crownes or garlandes, whereof eache flower beyng spread abroade, shineth with sixe narrow leaues like a starre, and they grow out of little long huskes, in which afterward is found the seede, which is light, flat, & thin, like ye seede of Garnesey violets, or stocke-gillofers, or a darke euilfauoured red colour. The roote is long, rounde & thicke, sometymes forked or double, of the color of the earth without, & yellowish within lyke to Bore or Ocre, and exceeding bitter in taste.

Bysides the Gentian aforesayd there are two other sortes of herbes, which are also at this tyme taken for Gentian.

2. The one is altogyther lyke Gentian, sauing it is smaller and beareth blew flowers, & in taste it is farre bitterer, wherefore Tragus saith, it is of greater efficacie and vertue.

3. The other hath rounde stalkes, and smothe, set with greene smothe long narrow leaues, alwayes growing by couples, one agaynst another: at the top of the stalke groweth the flowers like little belles of a light blew colour, somewhat smaller than the flowers of ye second kinde of Ranunculus. The roote is yellow, long & bitter, and this is that plant the which we call Autumne violettes or Belflowers: & is described in the xxj. Chap. of the second part of this historie.

The Place.

Gentian groweth vpon high mountaynes, and in certayne Coomes or balleyes amongst ferne or brake, as in sundrie places of Germanie & Burgundie.

The Tyme.

It flowreth in June, and the seede is rype in July and August.

The Names.

Gentian is called in Greeke γεντιανή: in Latine and in Shoppes Gentiana: of Apuleius Aloe gallica, νάρκη, Narce, χειρώνιον, Chironion, Basilica, Cyminalis: in English Felworte: in Frenche *Gentiane*: in high Douche, Entzian, and Bitterwurtz: in base Almaigne, Gentiaen. It is also called Genuane in Italian and Spanishe.

The

※ *The cause of the Name.*

Gentius king of Illyria was the first founder out of this herbe, and the first that vsed it in medicine, and therefore it was called Gentian after the sayde kinges name. ※ *The Nature.*

The roote of Gentian is hoate and drie in thirde degree.

※ *The Vertues.*

The roote of Gentian made into powder, and taken in quantitie of a dram with wine, a little peper & Rue, is profitable for them that are bitten or stong of any venimous or madde beastes, and is also good for them that haue taken any poyson.

The same dronke with water, is good against the diseases of the liuer & stomacke, it helpeth digestion, and keepeth the meate in the stomacke, and the vse of it is very good agaynst all colde diseases of the interior or inner partes.

The iuyce of the same roote cureth the payne and ache of the syde, & helpeth them that haue taken great falles, and bruses, and are bursten, for it dissolueth and scattereth congeled blood, and cureth the sayde hurtes.

The roote of Gentian also cureth deepe festered, and fretting sores and woundes, whan the iuyce thereof is stilled or dropped into them.

The same iuyce applyed or layde to with fine linte or lynnen, doth swage and mitigate the payne and burning heate of the eyes: and scoureth away and clenseth the skinne of the body from all foule and euilfauoured spottes, beyng annoynted or straked therewith.

The roote of Gentian being applyed vnder in manner of a Pessarie or mother Suppositorie, prouoketh the flowers, and draweth foorth the dead fruit.

Of Cruciata or Dwarf Gentian and Alisma. Chap. xiij.

※ *The Description.*

Dwarf Gentian hath rounde stalkes of a spanne long or somewhat more, they be also holowe, & spaced with certayne knottie ioyntes, the leaues be long narrowe and thicke, and growe also by couples one agaynst another, and falling somewhat backwardes lyke the other Gentian, the flowers be blew, long and holowe within lyke belles, growing foorth of greene huskes, standing rounde togyther at the top of the stalkes and about the stem at certaine spaces. The roote is white, round, and long, and pearsed or thrust through in certayne places crossewise, which is ye cause it is called Cruciata, as some say: but it is rather so called of the fashion of the flowers, as Pena saith.

Some men also take the herbe Alisma or Saponaria for a kinde of Cruciata, it hath rounde stalkes with ioyntes or knottes: it is of a cubite or a foote and a halfe long, or more, the leaues be large with veynes or ribbes, lyke the leaues of broade plantayne, sauing they be smaller, & most commonly growing by couples at euery ioynt, and bending or falling backwardes, especially those which grow next the roote. The flowers grow in the top of the stalkes, & also about the vpper ioyntes in tuffets, of sweet sauour, & colour somtimes red as a rose, and somtimes of a light purple or white colour, growing out of long rounde huskes, & are made of fiue leaues set togyther, in the midle wherof are certaine small hearie threddes. The rootes be long & thicke, & grow or creepe crookedly, by whiche there hang certayne small hearie threddes lyke to the rootes of Beares foote or Setterworte. ※ *The Place.*

It groweth in certayne gardens of Brabant: and els where it groweth by suddes, brookes, & riuers, & in moyst places that are open against the Sunne. It continueth a long time in gardens.

The thirde Booke of

Cruciata. **Dwarfe Gentian.**　　Alisma siue Saponaria. **Sopewort Gentian.**

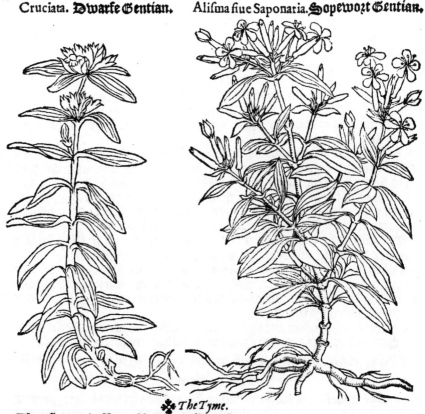

❊ *The Tyme.*

They flower in June, July, and August.

❊ *The Names.*

1 ⁋ The first is called in Germanie Modelgheer, and Speerenstich: in base Almaigne Madelgheer: of some in Latine Cruciata, that is to say, Crossed. Both in forme and facultie, it seemeth to be a kinde of Gentian, & Conrade Gesnere iudgeth it so to be, and therefore it may be called Gentiana minor, that is to say in English, the smal or Dwarf Gētian. For there is another Cruciata, so called bycause ẏ leaues are set togyther, standing like to a Burgonion Crosse, wherof shalbe spoken amongst the kindes of Madder. Some would haue it to be called Chiliodynamin: but Polemonia is called Chiliodynamis of the Cappadocions, as Dioscorides writeth, but with this Polemonia the Dwarfe Gentian hath no lykelyhode.

2 ⁋ The seconde is commonly called Saponariam, bycause of the clensing or scouring propertie that is in his leaues: for whan they are brused, they yeelde a certayne iuyce which wil scoure almost lyke soope. But Ruellius describeth another Soopeworte. Some call it Herbam tunicam: yet it is none of the cloue Gillofers, and muche lesse any of the kindes of Polimonij, which are taken for Sweete Williams or Tolmeyners, as we haue written in the Chap. of Gillofers.　It shoulde rather seeme to be Alisma or Damasonium, sauing that the stalke for the most part is not single, but most commonly groweth foorth into certayne branches or diuisions: & the rootes should be greater or thicker than the rootes of Bearefoote: But the leaues are agreable with the description of Alisma.

Alisma, and so is the tuft or bundle of flowers at the top. But the stalke of Alisma is single and slender, and the rootes shoulde be also slender: whiche declare the diuersitie betwixt this Saponaria, and Alisma. Some do also take it for Struthion, but it is nothing lyke: we may call it in English Soopewort: some call it Mocke Gilofer.

❧ *The Nature.*

The bitternesse of these herbes, doth manifestly declare, that they be hoate and drie, in qualitie not much vnlyke Gentian.

❧ *The Vertues.*

The decoction of the leaues or roote of Crossewort Gentian, or Dwarffe Gentian dronken, doth clense and scoure the breast, from all superfluities, and grosse flegmes, that are gathered togyther in the same, and it is good agaynst the falling sicknesse. [A]

If it be taken as is aforesayde, or taken in powder, it is good agaynst al venome and poyson and against the infection of the ayre, and the pestilence. [B]

It is good to washe woundes & corrupt vlcers, in the wine wherein it hath ben boyled, or to strawe the powder of it into the same: for it clenseth & healeth the same. [C]

The swine heardes of Germanie, do geue it chopt very smal to their hogges and swine to eate, and by this meanes do keepe them from the Murren, and suche lyke contagious diseases, as chaunce to their cattell in corrupt season. [D]

Of Elecampane. Chap.xiiij.

❧ *The Description.*   Helenium.

1 Elecampane hath great, broade, soft leaues, immediatly springing vp fro the roote, not much differing fro the leaues of white Mullin, but greater and larger, amongst which springeth vp a thicke hearie long stalke, commonly longer than a man, beset with leaues of y' same sorte, but smaller, of a light greene colour aboue, but whitish vnderneth, at the top of the stalke there growe fayre, large, yellowe, shining flowers lyke starres, and in figure lyke to Chrysanthemon or golden flower, but a great deale larger, & almost as large as y' palme of ones hande: The which whan they fade or perish, do chãge into a fine downe or soft Cotton, wherunto the seede is ioyned, & is carried away with the winde, like Thistell seede. The roote is great and thicke, with many other smal rootes & buddes vneuely adioyning, and couered with a thicke rinde or barke, of a browne earthly colour without, but most commonly white within, & is not very strõg or ranke of sauor whan it is fresh and greene: but whan it is drye, it is very aromaticall, and hath in it a certayne fat and Oylie moysture or substance.

2 The seconde Helenium whereof Dioscorides writeth, is vnknowen to vs: it hath tender branches creeping alongst the grounde, beset with many leaues like the pulse lentilles. The roote is whitish & thicke as ones little finger, large aboue and narrow downewardes.

❧ *The*

✿ The Place.

1 Elecampane delighteth in good fertill soyle, as in valleyes and medowes, it is also founde in hilles and shadowie wooddes, but not commonly in drye groundes. It is very common in England, Flaunders, and Brabant, and very well knowen in all places.

2 The second groweth in places adioyning to the Sea, and vpon litle hilles.

✿ The Tyme.

Elecampane flowreth in Iune and Iuly, the seede is ripe in August. The best time to gather the roote, is at the ende of September, whan it hath lost his stalkes and leaues.

✿ The Names.

This herbe is called ἑλένιον: in Latine Inula, and Enula: in Shoppes Enula Campana: of some Panaces chironion, or Panaces centaurion: in Englishe Elecampane, Scabworte, and Horseheele: in Frenche *Enula Campana*: in Germanie Alantwurtz: in base Almaigne Alantwortel, and Galantwortel: in Italian *Enoa* and *Enola*: in Spanishe Raiz del alla.

2 The seconde kinde is called Helenium Aegyptiacum, but yet vnknowen to men of this tyme.

✿ The Nature.

Elecampane being yet greene, hath a superfluous moysture whiche ought first to be consumed before it be occupied. But that moysture being dryed vp, it is hoate in the thirde degree, and dry in the seconde.

✿ The Vertues.

The decoction of Elecampane dronken, prouoketh vrine and womens A flowers, and is good for them that are greeued with inwarde burstinges, or haue any member drawen togyther or shronke.

The roote taken with hony in an Electuarie, clenseth the brest, ripeth tough B fleme, and maketh it easie to be spet out, and is good for the cough and shortnesse of breath.

The same made in powder and dronke, is good agaynst the bytinges and C stinginges of veninous beastes, and agaynst windinesse and blastinges of inwarde partes.

A Confiture made of the sayde roote, is very wholesome for the stomacke, D and helpeth digestion.

The leaues boyled in wine, and layde to the place of the Sciatica, swageth E the payne of the same.

Of Spicknel Mewe, or Meon. Chap.rv.

Matthiolus figure is almost lyke the first kinde of Libanotidis, & as Turner and he writeth, is called in Douche Bearewortes, or Hartes wortes.

✿ The Description.

Meon of Dioscorides is described amongst the rootes, wherefore we haue none other knowledge of the fashion of the same, but as our Auncientes haue left it vs in writing. This haue I sayde, to the intent that men may knowe, that those herbes which the Apothecaries and others do vse at this day in Physike, are not the true Meon, whiche we shoulde not tell howe to knowe, if that men coulde not finde the fashion, and nature of the right Meon described.

Meon according to Dioscorides, is lyke to Dyll in stalkes and leaues, but it is thicker and of the heigth of two cubites or three foote. The rootes are long, small, well smelling and chafing or heating the tongue, and they are scattering here and there, some right and some awry. ✿ The

the Historie of Plantes.

※ *The Place.*

Mew groweth plenteously in in Macedonia and Spayne.

※ *The Names.*

This herbe is called in Greeke μῆον: in Latine Meū: in shoppes Mew, which do but only keepe ye name, for the true Meon is yet vnknowē, but the Apothecaries do vse in the steede therof, a kinde of wilde Parcelie, the which is described in the fifth part of our history of plantes, & it hath no agreement or lykenesse with the description of Meon, wherfore it can not be Meon.

Meum. Meon.

※ *The Nature.*

The roote of Meon is hoate in the thirde degree, and dry in the seconde.

※ *The Vertues.*

A The rootes of Meum boyled in water, or onely soked in water and dronke, doth mightily open the stoppinges of the kidneyes & bladder, they prouoke vrine, ease and helpe the strangurie, and they consume all windinesse and blastinges of the stomacke.

B The same takē with hony, do appease the paynes and gripinges of the belly, are good for the affections of the mother, podagres and aches of ioyntes, and against al Catarrhes & Phlegmes falling down vpon the breast.

C If wemen sit ouer the decoction therof, it bringeth downe their sicknesse.

D The same layde vpon the lowest part of the belly of young children, wyll cause them to pisse and make water.

※ *The Daunger.*

If to muche of the roote of this herbe be dronken, it causeth head ache.

Of Peonie. Chap.xvi.

※ *The Kyndes.*

There be two sortes of Peonie, as Dioscorides and the Auncientes write, that is to say, the male and female.

※ *The Description.*

1 Ale Pœonie hath thicke redde stalkes of a Cubite long: the leaues be great and large, made of diuers leaues growing or ioyned togither, not muche vnlyke the Walnut tree leafe in fashion and greatnesse: at the hyghest of the stalke there groweth fayre large red flowers, very well lyke red roses, hauing also in the middes yellow threddes or heares. After the falling away of the leaues, there groweth vp great coddes or huskes three or foure togyther, the whiche do open whan they be ripe, in the opening whereof there is to be seene, a faire red coloured lining, and a pollished blacke shining seede, full of white substance. The rootes be white, long, small, and well smelling.

ff The

The thirde Booke of

2   The female Peonie at his first springing vp, hath also his stalkes redde and thicke: the leaues be also large and great, but diuided into more partes, almost like the leaues of Angelica, louage, or Marche. The flowers in like manner be great and red, but yet lesser and paler then the flowers of the male kinde. The coddes and seede are like the other. In these rootes are diuers knobbes or knottes as great as Acornes.

3   Yet haue you another kinde of Peonie, the which is like the second kinde, but his flowers and leaues are much smaller, and the stalkes shorter, the whiche some call Mayden or Virgin Peonie: although it beareth red flowers and seede lyke the other.

Pæonia mas. **Male Peonie.**

Pæonia fœmina. **Female Peonie.**

❧ *The Place.*
The kindes of Peonies are founde planted in the gardens of this Countrie.

❧ *The Tyme.*
Peonie flowreth at the beginning of May, and deliuereth his seede in Iune.

❧ *The Names.*
Peonie is called in Greeke παιονία: and in Latine Pœonia: of some πεντόροβ⊙, γλυκυσίς, Dulcisida, and Idæus Dactylus, of Apuleius Aglaophotis, σελήνιον, ῥιχομήνιον, θεολόνιον, πελινόγονον, and Herba casta: in shoppes Pionia: in high Douche Peonien blum, Peonie rosen, Gichtwurtz, Kunigzblum, Pfingstrosen: in base Almaigne Pioene, and Pioenbloemen, and in some places of Flaunders Mastbloemen.

❧ *The cause of the Name.*
Peonie tooke his name first of that good old man Pæon, a very ancient Physition, who first taught the knowledge of of this herbe.   ❧ *The Vertues.*

A   The roote of Peonie dried, and the quantitie of a Beane of the same dronken with Meade called Hydromel, bringeth downe womēs flowers, scoureth the mother of women brought a bed, and appeaseth the griping paynes, and tormentes of the belly.

B   The same openeth the stopping of the liuer, and the kidneyes, and sod with red wine stoppeth the belly.

C   The roote of the male Peonie hanged about the necke healeth, the falling sicknesse (as Galen and many other haue proued) especially in young children.

D   Ten or twelue of the red seedes, dronken with thicke and rough red wine, doth stop the red issues of women.

E   Fiftene or sixtene of the blacke cornes or seedes dronkē in wine or Meade, helpeth the strangling and paynes of the Matrix or mother, and is a speciall good remedie for them that are troubled with the night Mare (which is a disease wherin men seeme to be oppressed in the night as with some great burthē and sometimes to be ouercome with their enimies) and it is good against melancholique dreames.

the Historie of Plantes.

❀Of Valerian, Phu or Setwal.   Chap.lvij.

✠ *The Kyndes.*

THere be two sortes of Valerian, the garden and wilde: and the wilde Valerian is of two kindes, the great and small: Besides all these there is yet a strange kinde, the which is nowe called Greeke Valerian.

1. Valeriana hortensis.
Setwall or garden Valerian.

2. Valeriana sylueſtris maior.
The greater wilde Valerian.

✠ *The Description.*

1. Setwall or garden Valerian, at the first hath broade leaues of a whitish greene colour, amongst which there commeth vp a round holow, plaine, and a knottie stalke. Uppon the whiche stalkes there groweth leaues spread abroade and cut, lyke leaues of the roote called garden Parsenep: at the highest of þ stalke groweth tuffets of Corones with white flowers, of a light blew or carnation colour at the beginning and afterwarde white. The roote is as thicke as a finger, with little rootes and threddes adioyning therevnto.

2. The great wilde Valerian, is almost lyke to the garden Valerian, it hath also playne, round, holow stalkes, diuided with knottes. The leaues are lyke desplayed winges, made of many small leaues set one against another, lyke the leaues of Setwall or garden Valerian, whiche growe at the vpper part of the stalke, but much greater and more cloue or cut. The flowers grow and are like to the garden kinde, of a colour drawing towardes a light blew or skye colour. The roote is tender winding and trayling here and there, and putting foorth euery yere newe plantes or springes in sundrie places.

3. The little wilde Valerian, is very well like the right great Valerian, but it is

Ff ij   alwayes

alwayes lesse. The first and neathermost leaues are like the litle leaues of Plantaine, the rest which grow about the stalke, are very much and deepely cut, very wel lyke to the leaues of wilde Valerian, or like the leaues which grow about the stalkes of gardē Valerian. The stalkes be round with ioyntes, about the length of a hande. The flowers be like to ye flouers of the aforesaid kindes. The rootes be smal, & creeping alongst ye grounde.

The Greekish Valerian hath two or three holow stalkes, or moe: vpon ye which groweth spread leaues almost lyke the leaues of wilde Valerian, but longer, narrower, and more finely cut, lyke the leaues of the wylde Fetche, but somewhat bigger. The flowers grow thicke clustering togither at the top of the stalke of a light Azure or blew color, parted into fiue litle leaues, hauing in the midle smal white threddes pointed with a litle yellow at the tops. The seede is small growing in round huskes. The rootes are nothing els like, but smal threds.

※ *The Place.*

The garden Valerian and Greeke Valerian are sowen & planted in gardens. The other two kindes grow here in moyst places, and in watery medowes lying low. ※ *The Tyme.*

The three first kindes of Valerian do flower from May to August. The Greeke Valerian doth flower most commonly in June and July.

※ *The Names.*

1. The first kinde of these herbes is called in Greeke φȣ, Phu: in Latine Valeriana, and Nardus syluestris, or Nardus rustica: in shoppes Valeriana domestica, or Valeriana hortēsis, of some in these dayes Marinella, Genicularis, and Herba benedicta: in Frenche *Valeriane:* in high Douche Groß Baldrian: in base Almaigne, tame or groote Valeriā, & of some S. Joris crupt, or Speercrupt, that is to say, Spearwurte, or Speare herbe, bycause his first leaues at their first comming vp, in making are lyke to the Iron or head of a Speare: in English Setwal, or Sydwall.

2. The second kinde is called Valeriana syluestris, Phu sylueſtre, and Valeriana syluestris maior: in Frenche *grande Valerian sauuage:* in high Douch wilde Baldriā, Katzenwurtzel, Augenwurtz, Wendwurtz, & Dennenmarcke: in base Almaigne, wilde Valeriane: in English the great wilde Valerian.

3. The third is a kinde of wilde Valerian, and therefore we do call it, Valeriana syluestris minor, that is to say, the small wilde Valerian, and also Phu paruum, and Valeriana minor.

3. Phu paruum. Valeriana syluestris minor. The smal wild Valerian.

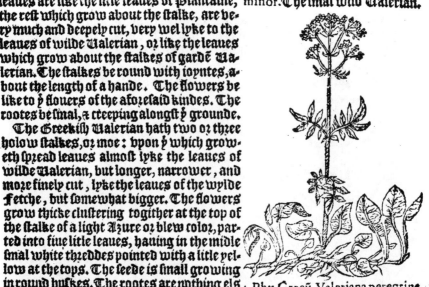

4. Phu Græcū. Valeriana peregrina. Greekish Valerian.

the Historie of Plantes. 341

4 ¶ The fourth is called of the Herboristes of our time Phu Græcum, & Valeriana Græca, that is to say, Greekish, or Greke Valerian, & it may be wel called Valeriana peregrina, or Pseudophu, for this is no Valerian, but some other strange herbe, the which we cannot compare to any of the herbes described by Dioscorides, except it be the right Auricula muris, for the which it is taken of some.

❧ *The Nature.*

The roote of Valerian is hoate and drie in the seconde degree.

❧ *The Vertues.*

The decoction of the rootes of Setwal dronken, prouoketh vrine, bringeth A downe womens flowers, and helpeth the ache and paynes of the side, and stomacke. They be of like vertue being made in powder and dronke in wine. And they be put into preseruatiues and medicines made agaynst poyson, and the pestilence, as Tryacles and Mithridats.

The leaues & rootes of the great wilde Valerian boyled in water, do heale B the vlceration and blistring of the mouth, especially the roughnesse, and inflammation of the throte, if one washe his mouth or gargarize therewith.

Men do vse to giue it with great profit in dreches, to such as are burste win.

The two other Valerians be not vsed in medicine.

3.4 English men vse Greeke Valerian, against cuttes and woundes.

Rosesenting ⎫ Roote.   **Of Rose wurte or Rhodia.**   Chap. xviij.
Rosesmelling ⎭   Rhodia radix. Rosewurt.

❧ *The Description.*

Rosewurte hath three or foure stalkes growing fro the roote, set ful of thicke leaues, lyke the leaues of Lyblong or Crassula maior, but they are more narrower, & cut or hackt at the top. The roote is thicke, hauing many smal hearie threddes, & whan it is eyther bruysed or bursten, it doth sente and sauor like the Rose, & of that it tooke his name.

❧ *The Place.*

Rosewurt or the roote sauering like the Rose, groweth in Macedonia and Hungarie: in this Countrie the Herboristes do plante it in their gardens.

❧ *The Tyme.*

It floureth in May, but it beareth flower very seldome.

❧ *The Names.*

This herbe is called in Greke ῥοδία ῥίζα: in Latine Radix Rhodia, & Radix rosata: in Frenche *Racine sentant les roses*: in high Douche Rosenwurtz: in base Almaigne Rosenwortel: in Englishe Rosewurt or the roote sauouring of the Rose.

❧ *The Nature.*

The roote which smelleth like the rose, especially of that sorte whiche groweth in Macedonia, is hoate in the second degree, and of subtile and fine partes.

❧ *The Vertues.*

The roote Rhodia layde to the temples of the forehead with oyle of roses, slayeth head ache.

ff iiij   Of

Of baſtard Pelitory or Bartram. Chap.xix.

Pyrethrum. Bartram.

The Description.

Pelitory hath leaues muche lyke to fenil, al finely cut or hackt. The flowers are yellow in þ midle, set round about with little white leaues somewhat blew vnder, like þ flowers of Camomil, or lyke the flowers of the great Dasie. The roote is long and straight, somtimes as byg as a finger, hoate and burning the tongue.

The Place.

This herbe is not founde growing of him selfe in this Countrie, but it is found planted in the gardens of certayne Herboristes.

The Tyme.

Pelitory flowreth after May vntyl the end of somer, in which season the seede is rype.

The Names.

This herbe is called in Greeke πύρεθρον: in Latine Saliuaris: in shops Pyrethrũ, of some also in Greke πύρινον, πύρωτον, & πύριτυς: in Frenche Pyrethre, or Pied d'Alexandre: in high Douch Bertrã. Albeit mine Author setteth foorth this herbe for Pyrethro, yet it is not aunswerable vnto Dioscorides Pyrethrum, or Saliuarem, wherfore I thinke we may wel cal it baſtard Pelitory or Bertram.

The Nature.

The roote of Pyrethre is hoate and dry in the thirde degree.

The Vertues.

The roote of Pelitory taken with hony, is good agaynst the falling sicknes, the Apoplexie, the long and olde diseases of the head, and against all colde diseases of the brayne. [A]

The same holden in the mouth & chewed, draweth foorth great quantitie of waterish fleme. [B]

The same sodden in vineger, & kept warme in the mouth, doth mitigate and alay the tooth ache. [C]

The Oyle wherein Pellitory hath ben boyled, is good to annoynt the body to cause a man swet, and is excellent good for any place of the body that is brused and shaken for colde, and for members that are benummed or foundered: and for such as are striken with the Palsie. [D]

Of wilde Pelitory. Chap.xx.

The Description.

Wilde Pellitory hath round brittle branches: the leaues be long & narrow hackt round about like a Saw, at the highest of the stalke grow flowers like the flowers of Camomil, yellow in the midle, & set round about with smal white leaues: the roote is tender & ful of threds: the whole herbe is sharpe & biting, almost in tast like Pellitory of Spayne, & for þ cause men cal it also wild Pellitory.

The

the Historie of Plantes.

Pyrethrum sylueſtre. **Wilde Pellitory.**

❧ The Place.

Wilde Pellitory is founde about the borders of feeldes, in high medowes and ſhadowy places, & ſomtimes vpon mountaynes and ſtony places.

❧ The Tyme.

This Pellitory flowreth from Maye vntyll September.

❧ The Names.

This herbe is nowe called in Latine Pyrethrum sylueſtre, that is to ſay, Wylde Pellitory: in Frenche Pyrethre ſauuage: in Douche Wilden Bertram: of ſome Weitʒ Reinfahrn, that is to ſay, White Tanſie. This is not πταρμικη, Ptarmice, or Sternumentaria, but another herbe vnknowen of the Auncientes.

❧ The Nature.

This herbe is hoate and drye.

❧ The Vertues.

This herbe holden in the mouth and chewed, bringeth lykewiſe frō the brayne ſlymie fleme, almoſt as mightily as Pelitory of Spayne: & it is very good againſt the tooth ache.

It is alſo good in Sallades, as Tarragon and Roquet, whereof ſhalbe written in the fifth booke.

Of falſe Dictam. Chap.xxi.

❧ The Deſcription.

THis herbe is lyke to Lentiſcus, or Lycoras in branches and leaues, it beareth rounde blackiſh and rough ſtalkes, and leaues diſplayed and ſpread lyke Lycoras, at the top of the ſtalkes growe fayre flowers, ſomewhat turning towarde blew, the whiche on the vpper part, or halfe deale hath foure or fyue leaues, and in the lower, or neather of the ſame flower it hath ſmall long threddes crooking and hanging downe almoſt lyke a bearde. The flowers periſhed, there commeth in the place of eche flower foure or fiue coddes, ſomething rough without, and ſlymie to be handled, and of a ſtrong ſauour almoſt ſmelling lyke a Goate: in the which is conteined a blacke, playne, ſhining ſeede. The rootes be long and white, ſometyme as thicke as a finger, and do growe a thwart one another.

❧ The Place.

It groweth in the Ile of Candie, as Dioſcorides wryteth, in this Countrie it is founde in the gardens of certayne Herboriſtes.

❧ The Tyme.

It floureth in this Countrie in June and July, and ſometymes the ſeede commeth to rypeneſſe.

❧ The Names.

This herbe is called in Greeke τράγιον: in Latine Tragium: and is the firſt kind of Tragium deſcribed by Dioſcorides. Some herboriſtes cal it Fraxinella:

344 The thirde Booke of

and some Apothecaries do vse the roote of it in steede of Dyctam, and do call it Dyptamum, not without great errour, and therfore it is called of some Pseudodictamum nothum, that is to say, Bastarde or false Dictam.

❧ *The Nature.*

Tragium is almost hoate in the third degree, and of subtil partes.

❧ *The Vertues.*

The seede of Tragium taken to the quantitie of a dragme, is good agaynst the strangurie, it prouoketh vrine, breaketh the stone in the bladder, & bringeth it foorth: and it moueth the termes or flowers of women.

The lyke vertue hath the leaues and iuyce to be taken after the same sorte: and being layde to outwardly, it draweth out thornes and splinters.

The roote taken with a little Rheubarbe, killeth, & driueth forth wormes, & is very singuler & of excellent vertue agaynst the same, as men in these dayes haue proued by experience.

It is sayde also (as recordeth Dioscorides) that the wilde Goates whan they be stroken with darts or arrowes, by the eating of this herbe do cause the same to fall from out of their bodyes, aswell as if they had eaten of the ryght Dyctam. And it is possible, that for the same cause this herbe was first taken in shoppes in steede of the right Dyctam.

Tragium.

Of Polemonium. Chap.xrij.

❧ *The Description.*

1. Polemonium hath tender stalkes, with ioyntes: the leaues are meetely brode, alwaies two set at euery ioynt one against another, at the highest of the stalkes groweth white flowers, hanging downewarde and ioyning one to another lyke a tuttay, or little nosegaye, after whiche flowers there commeth blacke seede, inclosed in rounde huskes. The roote is white, playne and long.

2. Yet there is an other herbe taken for Polemonium, whiche doth also bring foorth long stalkes, with knottes or ioyntes: it is muche longer than the aforesayd kinde, hauing long leaues, narrow at the top, and broade beneath where as they be ioyned to the stalke. The flowers of this kinde be of an orient or cleare red de colour, and do growe in tuffetes almost lyke Valerian. The roote is long, white, and thicke, and wel sauouring.

The Polemonium wherof Absyrtus speaketh, is the Horse minte described in the second de booke.

the Historie of Plantes.

Polemonium. Been album.   Polemonij altera species.

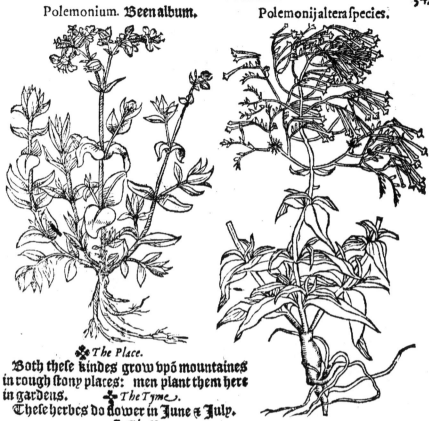

✣ *The Place.*
Both these kindes grow vpō mountaines in rough stony places: men plant them here in gardens.   ✣ *The Tyme.*
These herbes do flower in June & July.
✣ *The Names.*

1 The first is called in Greeke πολεμώνιον, καὶ φιλαιτερία· in Latine Polemonium, & Polemonia, of some, χιλιοδύναμις, Chiliodynamis, ῆ is to say, a hundred vertues, or properties: in shops, as witnesseth Bernardus of Gondonio, Herba tunica: of Herboristes at this day Behen, or Beën album. Yet notwithstanding this is not that kinde of Behen, wherof Serapio writeth in his CCxriij. Chapter.

2 The seconde is also taken for Polemonium, & is called of Herboristes Behen rubrum, this herbe should seeme to be Narcissus wherof Virgil in his Georgiques, and Columella in hortis maketh mention.   ✣ *The Nature.*
Polemonium is of complexion dry in the seconde degree.
✣ *The Vertues.*

Te roote of Polemonia dronken in wine, is good agaynst the blooddy flyxe, **A** and agaynst the bitinges and stinginges of venimous beastes.

The same dronken in water, prouoketh vrine, and helpeth the strangury and **B** paynes about the huckle bone or hanche.

Men vse it with vineger, against the hardnesse and stoppinges of the melt or **C** splene, and to all such as are by any meanes greeued about the melt.

The same holden in the mouth and chewed, taketh away tooth ache.   **D**

The same pounde & layde to, cureth the stinginges of Scorpions: & in deede **E** it hath so great strength against Scorpions, that whosoeuer do but holde the same in his hande, cannot be stong, or hurt by any Scorpion.

All these last recited vertues from B. to E. are not found in the last Douch copy. Yet they be all in my French copy, the which I haue and is in diuers places newly corrected and amended by the Author him selfe.

Of English Galangall. Chap. xxiij.

❦ *The Description.* Cyperus. English Gallangal.

1 Cyperus leaues are long, narrow, and hard. The stalke is triangled of a cubite long, in the top wherof groweth litle leaues white seede springing out. The roote is long interlaced one within an other, hauing many threddes, of a browne colour and sweete sauour.

2 Besides this there is found another kinde like to the aforesayd in leaues and stemmes: but it hath no lóg rootes, but diuers round little rootes of ẏ bignesse of an Olyue ioyning togither: And of this sort Dioscorides hath written.

3 One may wel place amongst ẏ kindes of Cyperus, the litle rootes called Trasi (of ẏ Italians) for their leaues be somewhat like the leaues of Cyperus, but they be smaller & narrower, the rootes be almost like to smal nuttes, or like the silke wormes wrapped rounde in their silke, before they turne into Mothes or Butterflyes, and hang togither plenteously, by little smal threds, these rootes be sweet in taste almost like Chestnuts.

❦ *The Place.*

Cyperus as witnesseth Dioscorides, groweth in low moyst places, & is not commonly founde in this Countrey, but in the gardens of some Herboristes. ❦ *The Tyme.*

This herbe bringeth foorth his spikie top, & seede with leaues, in June and July. ❦ *The Names.*

It is called in Greeke κύπερ⊙: in Latine Cyperus, Cypirus, and Cyperis, of some Aspalathum, & Erysisceptrum: in shoppes Cyperus: of Cornelius Celsus, Iuncus quadratus, of Plinie Iunculus angulosus, and Triangularis: in Frenche Souchet: in Douche wylden Galgan: in English Galangal.

The rootes called Trasos, are also named of them ẏ write now Dulcichimū: in Spayne *Auellanada*, & of the commons of Italy (as is aforesaid) Trasi, & Trasci. Some learned men thinke ẏ this is μαμίρας, Mamiras, wherof Paulus Aegineta writeth, which Auicen calleth Memirem, or rather ὁλωκονιτις, Holoconitis, of Hippocrates. ❦ *The Nature.*

The roote of Cyperus or English Galangal, is hoate and dry in the third degree. ❦ *The Vertues.*

A The rootes of Cyperus boyled and dronke prouoketh vrine, bringeth downe womens naturall sicknesse, driueth foorth the stone, and is a helpe to them that haue the Dropsi.

B The same taken after the same manner is a remedy against the stinging and poysons of Scorpions, and agaynst the cough.

C It is also good agaynst the coldnesse and stoppinges of the mother if the belly be bathed warme ther withall.

The same made into powder closeth vp and healeth the olde running sores D of the mouth and secrete partes (although they eate, and waste the flesh) if it be strowed therein, or layde thereupon with wyne.

It is customably, and also with great profite put into hoate oyntments and E playsters maturatiue.

The seede of Cyperus dronken with water, as Plinie sayth, stoppeth the F flure of the belly, and all the superfluous running foorth of womens flowers: but if to much thereof be taken, it engendreth headache.

Of white Hellebor or Nesewurte. Chap.xciiij.

❧ *The Description.* Veratrum album.

The white Ellebor hath great broad leaues, with ribbes or sinewes like the leaues of the great Plantayne or Gentian. The stalke is rounde two or three foote high, at the vpmost part whereof groweth alongest and rounde about the top, the flowers one aboue another, pale of color, diuided into sixe little leaues, the which haue a greene line ouerthwart. The same being passed, there commeth in their places smal huskes wherin is conteyned the seede, the roote is rounde, as thicke as a mans finger or thombe, white both without and within hauing many thicke laces or threddy stringes.

❧ *The Place.*

White Hellebor groweth in Anticyra, neare about the mountayne Oeta, and in Capadocia & Syria, but the best groweth in Cyrene. The Herboristes of this Countrie do set it in their gardens.

❧ *The Tyme.*

White Hellebor flowreth in this Countrie in June and July.

❧ *The Names.*

This kind of Hellebor is called in Greke ἑλλέβορος λευκός: in Latine Veratrum album: in shoppes Helleborus albus: of some Pignatoxaris & Sanguis Herculis: in French *Ellebore blanc*: in high Douche Weiß Nieswurtz: in base Almaigne Witte Nieswortel or wit Nieskruyt: in English White Hellebor, Nesewort, and Lingwort.

❧ *The Nature.*

The roote of Ellebor is hoate and drye in the thirde degree.

❧ *The Vertues.*

The roote of white Ellebor causeth one to vomit vp mightily and with A great force, all superfluous, slymie, venemous and naughtie humours. Likewise it is good agaynst the falling sicknesse, Phrensies, olde payne of the head, madnesse, sadnesse, the gowt, and Sciatica, all sortes of dropsies, poyson, and
agaynst

agaynst all colde diseases, that be harde to cure, and suche as wyll not yeelde to any medicine. But as concerning the preparation thereof before it be ministred to any, and also in what sort the body that shall receiue it ought to be prepared, it hath ben very well and largely described by diuers olde Doctors, wherof I minde not to intreate, bycause the rules to be obserued be so long, that they cannot be comprehended in fewe wordes, for they may well fyll a booke, and bycause Galen teacheth, that one ought not to minister this vehement and strong roote in inwarde medicines, but onely to apply the same outwardly.

Therefore it is good to be vsed agaynst all roughnesse of the skinne, wylde scurffe, knobbes, foule spottes, and the leprey, if it be layde thereto with Oyle or Oyntmentes.

The same cut into gobbins or slices, and put into fistulas, taketh away the hardnesse of them.

The same put vnder in manner of a Pessarie, bringeth downe flowers, and expelleth the dead childe.

The powder thereof put into the nose, or snift vp into the same, causeth snesing, warmeth and purgeth the brayne from grosse slymie humours, & causeth them to come out at the nose.

The same boyled in vineger and holden in the mouth, swageth toothache, and mingled with eye midicines, doth cleare and sharpen the sight.

The roote of Hellebor pounde with meale and hony, is good to kill Myse and Rattes and suche lyke beastes, and to driue them away: lykewyse if it be boyled with mylke, and Waspes and Flyes do eate thereof, it killeth them, for whatsoeuer doth eate of it, doth swell and breake: and by this we may iudge howe perilous this roote is.

❧ The Daunger.

White Ellebor vnprepared, and taken out of time and place, or to muche in quantitie, is very hurtfull to the body: for it choketh, and troubleth all the inwarde partes, draweth togyther and shrinketh al the sinewes of mans body, and in fine it sleaeth the partie. Therfore it ought not to be taken vnprepared, neyther than without good heede and great aduisement. For such people as be either to yong or to old, or feeble, or spit blood, or be greeued in their stomackes, whose breastes are straight and narrowe, and their neckes long, suche feeble people may by no meanes deale with it, without ieobardie and danger. Wherfore these landleapers, Roges, and ignorant Asses, which take vpō them without learning and practise, do very euill, for they giue it without discretion to al people, whether they be young or olde, strong or feeble, and sometimes they kil their patientes, or at the least they put them in perill or great daunger of their lyues.

Of wilde white Ellebor or Nesewurte. Chap.xxv.

❧ The Description.

This herbe is lyke vnto the white Ellebor abouesayd, but in al partes it is smaller: it hath a straight stalke with Sinowey leaues, like the leaues of Plantaine or white Ellebor, but smaller. The flowers hang downe from the stalke of a white colour, holowe in the middle, with small yellowe and incarnate spottes, of a very strange fashion, & whan they are gone, there cōmeth vp smal seede like sande closed in thicke huskes. The rootes are spread here and there full of sappe, with a thicke barke, of a bitter taste.

❧ The Place.

This herbe groweth in Brabant in certayne moyst medowes, and darke shadowie places.

❧ The

the Hiſtorie of Plantes. 349

❧ *The Tyme.*

This herbe flowreth in June and July.

❧ *The Names.*

This herbe is called in Greeke ἐπιπακτὶς, bicauſe it is lyke in faſhion to White Hellebor : in Latine Helleborine, and Epipactis : in high Douche Wildt wit Nieſcruyt, that is to ſay, Wilde white Ellebor. Some thynke, that Eleborine is an herbe lyke to Elleborus onely in vertues, and not in faſhion. Theſe fellowes wyl not receiue this herbe for Helleborine: but by this they may know their errour, bycauſe neyther Galen nor Dioſcorides do attribute any of the properties of Ellebor to Helleborine.

❧ *The Nature.*

This herbe is of hoate and drie complexion.

❧ *The Vertues.*

A The decoction of Helleborine dronke, openeth the ſtoppinges of the liuer, and is very good for ſuch as are by any kinde of meanes diſeaſed in their liuers, or haue receiued any poyſon, or are bitten by any manner venemous beaſt.

Of blacke Hellebor. Chap.xxvi.

Veratrum nigrum Dioſcorides.

Blacke Hellebor.

Helleborine,

Planta Leonis.
Chriſtwort.

The thirde Booke of

The Kyndes.

VNder the name of Helleborus niger, that is to say, blacke Ellebor, are comprehended (by the Herboristes of our time) three sortes of herbes, wherof ẏ first is muche lyke in description to Helleborus niger, of Dioscorides: The seconde is a strange herbe not muche differing in vertue from the true blacke Hellebor, and is called Christes herbe, and is much lyke in description to Helleborus niger, that Theophrastus speaketh of: The thirde is commonly called of the lowe Douchmen Viercruyt, that is to say, Fierwurte.

Pseudohelleborus. **Bastard Hellebor the blacke.**   **Louswurt.**

The Description.

1. THE true blacke Hellebor hath rough blackish leaues, parted with foure or fiue deepe cuttes, like the fashion of the vine leafe, or as Dioscorides saith, like the leaues of ẏ Plane tree, but much lesser, the stalkes be euen and playne, at the top whereof grow flowers in little tuffetes, thicke set like to scabeous, of a light blew colour. After the falling of whiche flowers commeth the seede whiche is not muche vnlike to wheate. The rootes are many small blacke long threddes comming altogither from one head.

2. Christes herbe hath great thicke greene leaues, cut into seuen or eyght parts, whereof eache part is long and sharpe at the top, and one halfe thereof is cut and snipt about like a sawe, the other halfe leafe next to the stalke is plaine and not cut. The flowers grow amongst the leaues vpon short stemmes comming from the roote, and are of the bignesse of a grote, or shilling, of a faire colour as white as snow, hauing in the middle many short, tender, & fine threddes, tipts with yellow. After the flowers haue staide a long time, whan they begin to perish, they become blew, & afterward greene. After the flowers it bringeth forth
foure

the Historie of Plantes.

foure or fiue cods or huskes ioyning togither almost like ye huske of Columbine wherein is conteyned the seede. It hath in the steede of a roote many thicke blacke stringes.

3 The leaues of Bastard Hellebor are somewhat like ye leaues of the aforesaid Christes herbe, but muche smaller, parted likewise & cut into diuers other narrow leaues, which are cut round about on euery side like a sawe. The flowers come not from the roote, but grow vpon the stemmes wher as the leaues take hold, & are much lesse then ye flowers of Christes herbe, of a greene or herbelike colour. After the passing away of which flowers, commeth vp also foure or fiue litle huskes or cods ioyning one in another, wherin is seede, which is blacke & round. The rootes are many blacke threddes wouen, or interlaced togither.

4 Louswurt which Fuchsius counteth for a kinde of blacke Hellebor, ye shall finde it hereafter amongst the Aconites, whereof it is a kinde.

5 The other which Hierom Bock setteth out for blacke Hellebor, the which also of the Apothecaries hath ben so taken, is described in the second part of this Historie, where as it is also declared, ye it is no kinde of blacke Hellebor, but the right Buphthalmum, or Oxe eye, and therefore neither hurtful nor dangerous as it hath ben more largely declared.

¶ *The Place.*

1 Blacke Hellebor groweth in Aetolia, vpon the mountayne Helicon in Beotia, and vpon Parnassus mount in Phocidia: and in this Countrie it is found in the gardens of certayne Herboristes.

2 Christes wurtes likewise, is not common in this Countrie, but is only found in the gardens of some Herboristes.

3 The bastarde blacke Hellebor groweth in certayne woods of this Countrie, as in the wood Soenie in Brabant, & it is set or planted in diuers gardens.

¶ *The Tyme.*

1 The blacke Hellebor in this Countrie flowreth in June, and shortly after the seede is rype.

2 Christes wurte flowreth al bytimes about Christmas, in Januarie, & almost vntill March, in Februarie the old leaues fall of, and they spring foorth againe in Marche.

3 The blacke Bastarde Hellebor flowreth also bytimes, but most commonly in Februarie, sometimes also vntill April. ❈ *The Names.*

1 Blacke Hellebor is called in Greke ἑλλέβορος μέλας: in Latine Veratrum nigrũ, and Helleborus niger, of some Melampodium, Prætium, Polyrhyzon, Melanorhizon, & of some writers now, Luparia, & Pulsatilla: in high Douch Schwartz Niesewurtz: in base Almaigne Swert Niesewortel.

2 The second should seeme to be ἑλλέβορος μέλας, Helleborus niger, which Theophrast describeth: and is called of learned men that write now, Planta Leonis, that is to say, Lions foote, & it is taken for that herbe which Alexander Trallian, and Paulus Aegineta call in Greke κορονοπόδιον: in Latine Coronopodiũ, and Pes cornicis. It is called in Brabant Heylichkerstcrupt, that is to say, the herbe of Christ or Christmas herbe, bycause it flowreth most commonly about Christmas, especially whan the winter is milde.

3 The thirde is now called Pseudohelleborus niger, Veratrum adulterinum nigrum, and it is taken of some for the herbe whiche Plinie calleth Consiligo: in high Douch it is called Christwurtz, that is to say, Christes roote: in Brabant Viercrupt, that is to say, Fier herbe, bycause with this herbe alone men cure a disease in cattel named in Frenche *Le feu*: of some it is called Wranckcrupt, as of the learned and famous Doctor in his time Spierinck resident at Louaigne: and some call it Uaencrupt.

The occasion of the Name.

This herbe was called Melampodium, bycause a shepheard called Melampus in Arcadia cured with this herbe the daughters of Prœtus, whiche were distract of their memories, and become mad: so that afterwarde the herbe was knowen.

The Nature.

Blacke Hellebor is hoate and dry in the thirde degree.

Christeswurt, and the blacke bastarde Hellebor are in complexion, very lyke to blacke Hellebor.

The Vertues.

Blacke Hellebor taken inwardly, prouoketh the siege or stoole vehemently, and purgeth the neather part of the belly from grosse and thicke fleme, and cholerique humours: also it is good for them that ware mad or fall beside them selues, and for suche as be dull heauy and melancholique: also it is good for them that haue the gowte and Sciatica. A

Lyke vertues it hath to be taken in potages, or to be sodden with boyled meate, for so it doth open the belly, and putteth forth al superfluous humours. B

The same layd to in manner of a Pessarie or mother Suppositorie, bringeth downe womens sicknesse, and deliuereth the dead childe. C

The same put into Fistulas and holowe vlcers, by the space of three dayes, clenseth them, and scoureth away the hardnesse and knobbes of the same. D

The roote therof put into the eares of them that be harde of hearing, two or three dayes togither helpeth them very muche. E

It swageth tooth ache, if one washe his teeth with vineger wherein it hath ben boyled. F

An emplayster made of this roote with barly meale and wine, is very good to be layde vpon the bellyes of them that haue the Dropsie. G

The same pounde with Frankencense, Rosom, & Oyle, healeth al roughnesse, & hardnesse of the skinne, scuruinesse, spots, & scarres, if it be rubbed therewith. H

Planta Leonis or Christeswurtes, is not much differing in properties from blacke Hellebor: for it doth also purge and driueth forth by siege mightily, both melancholy and other superfluous humours. I

The roote of bastarde Hellebor stieped in wine and dronken, doth also loose the belly like blacke Hellebor, and is very good against al those diseases, where vnto blacke Hellebor serueth. K

It doth his operation with more force and might, if it be made into powder, and a dram thereof be receiued in wine. L

The same boyled in water with Rue & Egrimony, or bastard Eupatory, healeth the Iaundise, and purgeth yellow superfluities by the siege. M

The same thrust into the eares of Oxen, Sheepe, or other cattel, helpeth the same agaynst the disease of the longes, as Plinie and Columella writeth, for it draweth all the corruption and greefe of the longes into the eares. N

And in the time of Pestilence, if one put this roote into the bodyes of any, it draweth to that part al the corruption & benomous infection of the body. Therefore assone as any strange or sodayne greefe taketh the cattell, the people of the Countrie do put it straightwaies into some part of a beast, where as it may do least hurt, and within short space all the greefe will come to that place, and by that meanes the beast is saued. O

The Danger.

Although blacke Hellebor is not so vehement as the white, yet it can not be giuen without danger, & especially to people that haue their health, for as Hippocrates saith, Carnes habentibus sanas, Helleborus periculosus, facit enim Cō-
uulsionem,

the Historie of Plantes.

uulsionem, that is to say, to suche as be whole, Hellebor is very perilous, for it causeth shrinking of sinewes: therefore Hellebor may not be ministred, except in desperate causes, and that to yong and strong people, and not at al tymes, but in the spring time only: yet ought it not to be geuen before it be prepared and corrected.

✻ The correction.

Whan Hellebor is giuen with long Pepper, Hysope, Daucus, and Annys seede, it worketh better & with lesse danger: also if it be boyled in the broth of a Capon, or of any other meate, and then the brothe giuen to drinke, it worketh with lesse danger.

Of herbe Aloë. Chap.lxvij.

✻ The Description.

Aloë hath very great long leaues, two fingers thicke, hauing rounde about short pointes or Crestes standing wide one from another. The roote is thicke and long. The flowers stalke and seede, are much like y flowers stalke & seede of Affodyll, as Dioscorides sayth, but in these partes they haue not ben yet seene. All the herbe is of strong sauour and bitter tast. And out of this herbe which groweth in India is drawen a iuyce, y which is dryed and is also named Aloë, and it is carryed into all partes of the worlde for to be vsed in medicine.

✻ The Place.

Aloë groweth very plenteously in India, and from thence commeth the best iuyce, it groweth also in other places of Asia and Arabia, adioyning to the sea, but the iuyce thereof is not commonly founde so good. It is to be seene also in this Countrie in the gardens of some Herboristes.

✻ The Names.

This herbe is called in Greeke ἀλόη: and from thēce sprang the Latine name, and is called Aloë in al other speeches of Christendome, & so is the sappe or iuyce thereof named. The Frenchmen call it *Perroquet*: bycause of his greenesse, we may call it in English Aloë, herbe Aloë, or Sea Aygreene.

✻ The Nature.

The iuyce of this herbe called Aloë, whiche only is vsed in medicine, is hoate almost in the seconde degree, and drye in the thirde.

✻ The Vertues.

The iuyce of Aloë whiche is of a browne colour, like to the colour of a liuer, whiche is cleare and cleane, openeth the belly, in purging colde, flegmatike, and cholerique humours, especially suche wherewithal the stomacke is burdened, and is the cheefest of all other purging medicines (which most commonly

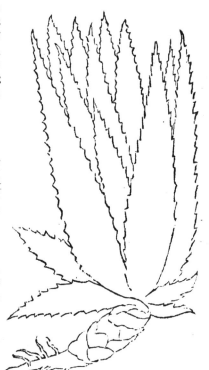

Aloë.

do hurt the stomacke) but this is a soueraigne medicine for the stomacke, for it comforteth, cleuseth, drieth vp, and driueth foorth all superfluous humours, if it be taken with water the quantitie of two drammes.

Men take it with Cynamome, Ginger, Mace, Cubibes, Galangal, Annys seede, and such spices to aswage and driue away the payne of the stomacke: by which meanes they comfort & heate the stomacke & cause fleme to be expulsed.

The same is also good agaynst the Iaunders, as Dioscorides writeth, and taken a litle at a time profiteth much against the spetting, and all other issues of blood, except that of the Hemorrhoides.

Aloë made into powder & strawen vpon newe bloody woundes, stoppeth the blood and healeth the wounde. Lykewyse layd vpon old sores closeth them vp, and it is a soueraigne medicine for vlcers about the secrete partes, and fundement.

The same boyled with wine and hony, healeth the outgrowinges & riftes of the fundement, & stoppeth the abounding fluxe of the Hemorrhoides, being layde vppon: for being receiued into the body, it causeth the Hemorrhoides to breake out, and to bleed.

The same with hony dispatcheth abroade al standing of blood, and bruses, with blacke spottes that come of stripes.

It is also good agaynst all inflammation, hurtes, and scabbes of the eyes, and agaynst the running and darkenesse of the same.

Aloë mixt with Oyle of roses and vineger, and layde to the forehead and temples, swageth headache.

If one do often rubbe his head with Aloës mingled with wine, it wil keepe the heare from falling.

The same layd to with wine, cureth the sores and pustules of the gummes, the mouth, the throte, and kernelles vnder the tongue.

To conclude, the same layd to outwardly, is a very good consolidatiue medicine, it stoppeth bleeding, and doth mundifie and clense all corruption.

Of Palma Christi. Chap. xxviij.

The Description.

Palma Christi hath a great, round, holow stalke, higher then a good long man, with great broade leaues, parted into seuen or nine diuisions, larger, and more cut in, then the leaues of the figge tree, lyke some byrdes foote, or lyke to a spread hande. At the highest groweth a bunche of flowers, clustering togither lyke grapes, whereof the lowest be yellow, & wither without bearing fruit, and the highest are red, bringing forth threecornered huskes, in which is founde three gray seedes somewhat smaller then kidney Beane.

The Place.

This herbe groweth not of hit selfe in this Countrie, but the Herboristes plante it in their gardens.

The Tyme.

It is sowen in Aprill, and his seede is rype in August and September, and as soone as the colde commeth, al the herbe perisheth.

The Names.

This herbe is called in Greeke κίκι καὶ κρότον: in Latine Ricinus: in Shoppes and of the Arabians Cherua: of some Cataputia maior, Peutadactylon, & Palma Christi: in English Palma Christi: in Frenche Paulme de Christ: in high Douche Wunderbaum, and Creutzbaum, & of some Zecken korner: in base Almaigne wonderboom, Cruysboom, and Mollencruyt.

the Historie of Plantes.

Ricinus.

❦ *The Nature.*

The seede of Palma Christi is hoate & drie in the thirde degree.

❋ *The Vertues.*

A The seede of Palma Christi taken inwardly, openeth the belly, causeth one to vomite, and to cast out slymie flegme, drawing the same from farre, and sometymes cholerique humours with waterish superfluities.

B The broth of meate, in whiche this seede hath ben sod, dronke, is good for the cholike (that is to say, payne in the belly) against the gowte & payne in the hippe, called the Sciatique.

C The same pounde and taken with whaye or new milke, driueth foorth waterish superfluities and cholerique humours, also it is good agaynst the Dropsie and Iaunders.

D The oyle which is drawen foorth of this seede is called Oleum Cicinum, in Shoppes Oleum de Cherua. It heateth and drieth, and is very good to annoynt and rubbe all rough hardnesse, and scuruie roughnesse, or itche.

E The greene leaues of Palma Christi pound with parched Barley meale, do mitigate and asswage the inflammatiō and swelling sorenesse of the eyes, and pounde with vineger, they cure the greeuous inflammation, called S. Antonies fire.

❋ *The Danger.*

The seede of Palma Christi turneth vp the stomacke, and doth his operation with much payne and greefe to the partie.

❋ *The Remedie.*

But if you take with it eyther Fenill or Annys seede, and some spices of Cynamome, and Ginger, &c. it will not ouertturne nor torment the stomacke, but will worke his effecte with more ease and gentlenesse.

Of the kindes of Tithymale or Spourge. Chap.xrir.

❋ *The Kyndes.*

THere are, as Dioscorides writeth, seuen sortes of Tithymal: whereof some at this time are welknowen, & some shalbe now by our endeuour brought agayne to light, and some are yet vnknowen.

❋ *The Description.*

THe first kinde of Tithymal, called þ male kind, hath round red stalkes, of the heigth of a cubite, þ is a foote & a halfe high. The leaues are lōg and narrowe, somewhat longer and narrower then the leaues of the Olyue tree, wherof the highest leaues, before they be throughly growen, shew rough or cottonlike. The seede groweth at the highest of the stalke in pretie

Gg iiij rounde

356   **The thirde Booke of**

round holow leaues, like as it were basons, or litle disshes, through which the stalke groweth. The seede is inclosed in threecornered huskes like the seede of Palma Christi, as well in growing as in shape or fashion, but much smaller: the roote is of a wooddie substance with many hearie stringes.

Tithymalus Characias.   Tithymalus Characiæ species.
Wood Spourge.

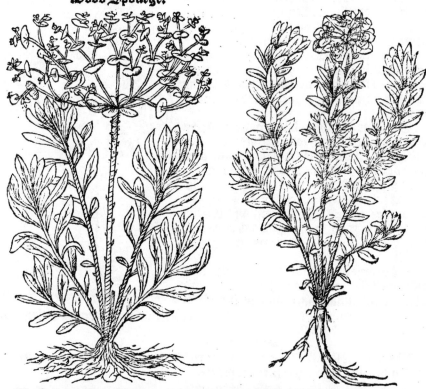

2 ❦ The second kinde of Tithymal hath straight stalkes of a span long, about the which growe many leaues, set a sunder without order, thicke, whitish, & sharpe poynted, not much vnlyke the leaues of Kne holme or Ruscus, but greater and thicker, not blacke but al white of þ colour of sea Spourge. And whan they be brused or bursten, there cometh forth milke as out of þ other kinds of spourge. The flowers are yellow, growing in tuffetes like Rosewurte or Rhodia, the fruit is triangled, like the fruit of the other Spurges or Tithymales.

3 The third kind, which may be wel called Tithymal of þ sea, or sea Spourge, hath sixe or seuen fayre red stemmes or moe, comming from one roote. The leaues are smal, almost lyke the leaues of flaxe, or lyneseede, growing rounde about the stalke, being thicke toothed, the flowers are yellow and grow out of litle dishes or sawsers, like the first kind of Spourge, after commeth the triangled seede as in the other Tithymales: the roote is long, & plaine, & of wooddie substance. This kinde of Tithymal, his leaues, dishes, and flowers are much thicker then any other kinde of Spourge.

4 The fourth kinde called Sonne Spourge, after his Greeke name, or Tithymal turning with the Sonne, hath three or foure stalkes somewhat reddish, about the length of a foote, & his leaues are not so thicke as garden Porcelane.

The

The flowers are yellowe growing in tuffetes. The roote is lyke the other Tithymales.

5. The fifth kinde callled Cypres Tithymal, hath rounde reddish stalkes of the length of a foote: The leaues are very small, greene, narrow, lyke the leaues of the Firre tree, but farre smaller & tenderer. The seede is smal but in al thinges els lyke the other, and it commeth in little blewish Cuppes or Sawsers, in the middest of the syde branches. This herbe hath leaues much narrower then Ezula minor.

Of this sort there is founde another kinde very smal, the which may be wel called Cypres Tithymal. It hath very small stalkes, both little and tender, about the heyght of a spanne, and vpon them small tuffetes, with flowers of a faynt yellow or pale colour, after cōmeth the seede lyke to the other, but a great deale smaller.

Tithymalus Myrsinites.
Myrtell Spourge.

Tithymalus Paralios. Sea Spourge.

Yet there is a thirde sort of this kinde, whereof the leaues be all white, but otherwayes it is lyke to Cypres Tithymale, as the great and diligent Herborist Jan the Ureckom hath declared vnto vs, who before this hath had suche Tithymale growing in his garden: neuerthelesse, I did neuer see it, and therefore I do not set out a larger description.

6. The sixth kinde is great, of eight or niene cubites high, growing like a little tree, the stalke is somtimes as bigge as ones legge (as Peter Belon writeth) and bringeth foorth many branches spred abroade, reddishe, and set with small leaues, like the leaues of litle the Myrtel tree, the fruit is like the fruit of the other Tithymales.

Tithymalus Helioscopius.	Tithymalis Cyparissias.
Wartewort or Son Spourge. Sonturner.	Cypresse Spourge.

7 The seuenth kinde hath soft leaues like Molin or Higtaper, but it is yet vnknowen.

All these kindes are full of white liquor or sappe like milke, the which commeth foorth whan they be broken or hurt, and it is sharpe and bitter vpon the tongue. ❦ *The Place*.

1 The first kinde of Spourge groweth not of his owne kinde in this Countrie, and is seldome founde, but in the gardens of diligent Herborістes.

2 The seconde, as saith Dioscorides, groweth in places that lye waste.

3 The thirde groweth about the Sea, and is founde in Zelande vpon trenches and drie sandie bankes and in wast places adioyning to the sea.

4 The fourth groweth about townes in plaine fieldes, and in some gardens: it is very common in this Countrie.

5 The fifth called Cypres Tithymale is not founde in this Countrie, but in the gardens of Herborістes.

But the litle of the same kinde groweth about Malines, in borders of some fieldes, yet it is not founde euery where.

6 The sixth kinde groweth in stonie places.

❦ *The Tyme*.

All the kindes of Tithymal or Spourge, are most commonly in flower in June and July, and their seede is ripe in August.

❦ *The Names*.

All kindes of this herbe are called in Greeke τιθυμάλοι: in Latine Lactaris: in French

the Historie of Plantes.

French *Tithymales* or *Herbe à laict*: in high Douch Wolfzmilch: in base Almaigne Wolfsmelck: in English Spourge.

1. The first kind is called in Greeke αιθυμάλ◌ χαρακίας, καὶ ἀμιγδαλοιδ῀ς: in Latine Tithymalus mas, or Lactaria mascula, that is to say in Frenche, *Tithymale masle*: in English Wood Spourge.

2. The seconde kinde is called in Greeke τιθυμάλ◌ μυρσινίτης, of some *Caryites*: in Latine Tithymalus foemina, that is to say, in Frenche *Tithymale femelle*: in English Femall Tithymall, of Theodor Gaza *Myrtaria*, it may be named in English Myrtell Spourge.

3. The thirde kinde is called in Greeke παράλι◌, Paralios, and Tithymalus, or Mecon, of Theophraste κόκκ◌, Coccos. This kind may be wel called in French *Tithymale marin*: in English Sea Spourge: in Douche Zee Wolfsmelck.

4. The fourth is called in Greeke τιθυμάλ◌ ἡλιοσκοπί◌, that is to say in Latine Tithymalus solsequius, or Lactaria solsequia: in Frenche *Tithymale suyuant le soleil*, and *Reueille matin*: in Almaigne Sonnewend, Wolfsmelck: and in Brabant Croonkens cruyt: in English Sonne Spourge, or Wartwurt.

5. The fifth is called in Greeke τιθυμάλ◌ κυπαρισσίας, that is to say, Tithymale lyke Cypres.

6. The sixth is called in Greeke δενδροειδὴς, and of some Leptophyllos: in Latine Tithymalus arborescens, that is to say, Tithymal growing lyke a tree: or Tree Tithymall.

7. The seuenth kinde is called in Greeke τιθυμάλ◌ πλατυφύλλ◌, and of some, as Hermolaus Barbarus writeth, Corymbites, & Amigdalites: in Latine Tithymalus latifolius, or Lactaria latifolia, that is to say, Large leaued Tithymall or Spourge. ❧ *The Nature.*

All the Tithymales are hoate and drie almost in the fourth degree, of a very sharpe, and biting qualitie, fretting and consuming, first of al the milke or sappe, then the fruit and leaues. The roote is of least strength. And amongst all the Tithymales as Galen sayth, the male is the strongest, then the female, thirdly the sixth kinde, and the Tithymale with broade leaues. The fifth in strength is that, which is lyke Cypres, the syxth is Sea Tithymall, the seuenth and of least force is the Sonne Spourge, or Tithymall folowing the Sonne.

❧ *The Vertues.*

The iuyce of Tithymal is a very strong medicine opening the belly, and somtimes causing vomit, bringing tough flegme & cholerique humours: like vertue is in the seede and roote, especially the barke therof, and are very good for such as fall into the Dropsie, whan it is ministred with discretion and wel corrected or prepared. A

The same mixt with hony, causeth heare to fall from the place that hath ben annoynted therewithall in the Sonne. B

The same put into the holes of corrupt & noughtie teeth, swageth the tooth ache, but ye must beware, ye put not the iuyce vpon any sounde tooth, or whole place, but first ye must couer them with waxe to preserue them from the sayde iuyce. The roote of Tithymal boyled in vineger and holden in the mouth, is good for the same intent. C

The same doth also cure all roughnesse of the skinne, manginesse, lepzie, wild scurffe, and spreading scabbes, the white scurffe of the head, and it taketh away and causeth to fall of all kindes of wartes, it taketh away the knobbes & hardnesse of fistulas, corrupt and fretting vlcers, and is good agaynst hoate swellinges and Carboncles. D

It kylleth fishe, if it be mixt with any bayte, and giuen them to eate. E

*The

The thirde Booke of

❧ The Danger.

The iuyce, the seede, and rootes of Tithymales, do worke their effect with violēce, and are hurtful to the nature of man, troubling the body, and ouerturning the stomacke, burning and parching the throte, and making it rough and sore, insomuch that Galen writeth, that these herbes ought not to be ministred or taken into the body, much lesse the iuyce ought to be dealt with, but onely it must be applied outwardly, and that with great discretion.

❧ The correction or remedie.

If one lay the barke of the rootes of Tithymales, to soke or stiepe in vineger by the space of a whole day, then if it be dryed and made into powder, putting to it of Annys or Fenell seede, gumme Tragagante and Masticke, and so ministred altogither with some refreshing or cooling liquor, as of Endiue, Cicorie, or Orenges, it wil do his operation, without great trouble or payne, and will neither chafe nor inflame the throte, nor the inwarde partes.

Of Ezula. Chap.xxx.

❧ The Kyndes.

EZula is of two sortes (as Mesue saith) the great and small, whereunto Dioscorides doth agree, where as he writeth, that Pityusa is small in one place, and great in another.

Pityusa maior. Great Ezula.
    Spourge Giant.

Pityusa minor. Smal Ezula.
    Pyne Spourge.

❧ The Description

1. THe great Ezula hath straight high stalkes, vpon ẏ which grow great brode leaues, greater then the leaues of male Tithymale. The floures and seede growe at the highest of the stalke, and sometimes they come

foorth

foorth at the sides of the stalkes, like the seede of Tithymale, the roote is great and thicke, couered ouer with a thicke barke.

2 The small Ezula in stalkes and leaues is much lesse, the leaues are narrow lyke the leaues of wilde flare, the flowers and seede are lyke the first kind, but smaller. The rootes be small couered with a smooth or fine barke. These two kindes be lyke the Tithymales: therfore they haue ben reckened of some Auncientes for kindes of Tithymale (as Dioscorides writeth) and as they be now counted, and they do also peelde a white sappe or liquor like milke, whan they be either brused or broken, the which liquor is sharpe and biting.

¶ *The Place.*
The great Ezula in some Countries groweth in wooddes and wildernes, and in this Countrie in the gardens of Herboristes.

The lesser groweth in rough stony places, and is found in this Countrie in arable fieldes and bankes, but not euerywhere.

¶ *The Tyme.*
These herbes do flower about Midsomer, like the Tithymales.

✻ *The Names.*
These herbes are called in Greeke πιτυούσα: in Latine Pityusa: in the Arabian speache of Mesue Alscebran. in Shoppes Ezula, and Esula, and it should seeme that this name Esula, was borowed of Pityusa: for in leauing out the first two syllables Pity, there remaineth usa, wherof commeth the diminutiue vsula, the whiche is quickly turned into Ezula, or Esula.

✻ *The Nature.*
Ezula is hoate and drie in the thirde degree, sharpe, byting, and burning inwardly, of nature much like Tithymale.

✻ *The Vertues.*
The iuyce, seede, and roote of Ezula, openeth the belly, and driueth foorth tough flegme and grosse humours, also it pourgeth Cholerique and sharpe humours like the Tithymales.

To be short, both kindes of Ezula are in al thinges like to the Tithymales, in facultie and operation agreable to all that, wherevnto the others are profitable.

✻ *The Danger.*
As Ezula is like the Tithymales in nature and working, so it is of hurtful qualitie agreable to the same.

✻ *The Correction.*
The euill qualitie of Ezula is amended, in lyke maner as Tithymale.

Of Spourge.   Chap.xxxi.

✻ *The Description.*

Spourge hath a browne stalke, of two foote high or more, of the bignesse of ones finger. The leaues be long and narrow, like the leaues of a withie or Almonde tree, the stalke breaketh abroade at the top into many other little branches, set with little rounde leaues, vpon the same little branches groweth the triangled fruit, like the fruit of Palma Christi, but smaller, where in is conteyned little round seedes, the which by force of the heate of the Sonne, do skip out of their huskes whan the fruite is ripe. The roote is of a wooddy substance and not very thicke.

All the herbe with his stalkes and leaues do peelde a white milke lyke the Tithymales being butsten or hurt.

✻ *The Place.*
It is planted in many gardens of this Countrie.

The thirde Booke of

Lathyris.

❧ *The Tyme.*

It hath flowers and seede in Iuly and August.

❧ *The Names.*

This herbe is called in Greeke λαθυρις: and in Latine Lathyris: in Shoppes Cataputia minor: of some, as Dioscorides saith, Tithymalus: in Frenche Espurge: in high Douche Springkraut, Springkorner, and Treikorner: in base Almaigne Springcrupt, & in some places of Flaunders Spurgie: in English Spurge.

❧ *The Nature.*

This herbe is hoate and drie in the thirde degree, and in facultie lyke Tithymale.

❧ *The Vertues.*

A ❧ If one take syxe or seuen seedes of Spurge, it openeth the belly mightily, & driueth foorth choler, fleme, and waterish humours. Like vertue hath the iuyce, but it is of stronger operation.

B ❧ To be briefe, Spurge and the iuyce thereof, are of facultie lyke to the Tithymales.

❧ *The Danger.*

Spurge is as hurtfull to mans body as the Tithymales.

❧ *The Correction.*

If one take the seede of Spurge with Dates, Figges, or gumme Tragagante, Mastik, Annys seede, or any cooling or refreshing herbe, or if one drinke water straightwayes after the taking of the same seede, it wil not stirre vp the inflammation of the inward partes, nor much trouble the partie receiuing the same, and it shall not be much hurtfull to mans body.

Of Pety Surge. Chap. xxxi.

❧ *The Description.*

Artwurt or rather Peplos is a plante fashioned like a little tree, not much vnlike the Tithymale that foloweth the Sonne, but farre smaller, growing of the higth of halfe a spanne with diuers branches, set ful of very smal leaues. The sede is smal growing in triangled huskes lyke Spurge. The roote is long and somwhat thready, all the herbe is full of milke like the Tithymales.

Bysides this there is yet founde an other kinde described by Hyppocrates and Dioscorides, called Peplis, the which hath many rounde leaues like the leaues of garden Porcelane, red vnderneath, the seede groweth amongst the leaues, like the seede of Peplos. The roote is smal and very tender, this herbe is also full of white liquor neither more nor lesse, but as the aforesayde.

❧ *The Place.*

Peplos groweth in this Countrie in gardens amongst pot herbes & beanes, and in some places amongst bines.

Peplis

the Historie of Plantes.

Peplos.   Sea Wartwurt or wilde Porcelayne.
                                     Peplis.

Peplis, as Dioscorides reporteth, groweth in salt grounde by the Sea syde.

❧ *The Tyme.*

Peplos flowreth, and deliuereth his seede at Midsomer, lyke the Tithymales.

❧ *The Names.*

Peplos is called in Greeke πἐπλ☉: in Latine Peplus: in Shoppes Ezula rotunda: in high Douche Teufels Milch: in base Almaigne Duyuels Melck: in frenche *Reueille matin des vignes*: in English of some Wartwurt, & Spurge time, we may cal it after the Greke Peplos, or folowing the Douche, Dyuels milke, also Pety Spurge, and Spurge time.

The other is called in Greeke πἐπλἰς: in Latine Peplis: Hippocrates calleth it πἐπλιον, Peplion, some call it Portulaca syluestris. Turner nameth this Sea Wartwurt.

❧ *The Nature.*

Peplos is hoate and dry in the thirde degree, lyke the Tithymales: and Peplis is of the lyke temperament.

❧ *The Vertues.*

A   The seede and iuyce of Peplos are both of lyke qualitie with the iuyce and seede of Spurge and Tithymal, and serueth to all ententes and purposes, as Tithymal doth, wherefore they lose the belly, and driue foorth tough flegme, with water and cholerique humours.

B   This herbe kept in brine and eaten, dissolueth windinesse in the bowels and Matrix, and cureth the hardnesse of the melt.

C   Of the lyke vertue is Peplis, as Dioscorides writeth.

The thirde Booke of

❧ The Danger and Remedie.

This herbe is also hurtfull vnto man, neyther more nor lesse, but euen lyke Spurge, and is corrected and amended in the same sorte, as is declared in the former Chapter.

Of Serapions Turbith. Chap. xxxiij.

❧ The Description.

This herbe hath long leaues, large, greene, playne, and shining lyke in fashion to the leaues of wade, amongst which cōmeth forth a straight rounde stalke, of the heyght of a foote and a halfe or there about, set with the lyke leaues but smaller, it parteth at the top into many braunches, vpon the which grow faire flowers, blew before their opening, and when they are open they haue within a crowne of yellow, compassed about with small azured leaues, lyke to the flowers of Camomyll in figure. After when they fade they turne into a rough or downie white seede, the whiche flyeth away with the winde. The roote is long and thicke, and couered with a barke somwhat thicke also.

Tripolium.

❧ The Place.

This herbe groweth alongst the sea coast, where as the tide and waues do ebbe and flowe, in suche sorte, that sometimes it is couered with the Sea, and sometimes it is drie. And it is founde in abundance in Zealande.

❧ The Tyme.

This herbe flowreth in July and August.

❧ The Names.

Some cal it in Greeke τριπόλιον: in Latine Tripolium: in the Arabian speeche of Serapio, Chap. CCCxxx. Turbith: but this is not the Turbith of Mesue or Auicenne. It hath no name in our vulgar speech, that I know, but that some call it blew Camomil or blew Dasies, the which name belongeth not properly vnto it, seing that it is not of the kinde of Camomil or Dasies: we may very well call it, Serapio his Turbith.

❧ The Nature.

The nature of Tripolium is hoate in the thirde degree.

❧ The Vertues.

The quantitie of two dragmes of the roote of Tripolium taken with wine, driueth foorth by siege waterie humours: Moreouer, it is very profitable for suche as haue the Dropsie.

The same is very profitable mixte in medicines, that serue agaynst poyson.

the Historie of Plantes.

¶ The leaues of this herbe, as some writers do now affirme, haue a singuler vertue agaynst all woundes, so that they heale and close them vp incontinent, if the iuyce thereof be powred in, or if the brused leaues be layde vppon the woundes.

Of Mesues Turbith Thapsia. Chap.xxiiij.

❧ The Description.

Thapsia, as Dioscorides writeth, is lyke Ferula, but his stalkes be smaller, and his leaues lyke Fenil. The flowers be yellow growing in tuffetes lyke Dyll. The seede is broade, but not so broade as Ferula. The roote is long and thicke, blacke without, & white within, hauing a thicke barke full of white liquor and sharpe in taste.

¶ The Place.

Thapsia groweth in the Ile of Thapsus by Sicilia, and it is to be founde at this day vpon the mount Garganus in Apulia, and in many other places of Italie.

¶ The Names.

This herbe is called in Greeke θαψία: in Latine Thapsia, Ferulago, and Ferula syluestris: of Mesue in the Arabique tongue Turbith. And this is that Turbith which ought to be vsed in Shoppes, in the composition of such medicines, as Mesue hath described.

❧ The Nature.

Thapsia, but chiefely the barke of the roote, is almost hoate in the thirde degree, hauing therevnto adioyning a superfluous moysture, whiche is the cause it doth so quickly putrifie, and cannot be kept long.

❧ The Vertues.

A The barke of the roote of Thapsia, taken in quantitie of a Dram or somewhat lesse, openeth the belly, and driueth foorth clammie fleme, and thicke humours, and sometimes cholerique humours. For it draweth them with it not onely from the stomacke (the which it doth throughly scoure and clense) but also from partes farre of. Moreouer it is good agaynst the shortnesse of breath, the stoppinges of the brest, the Cholique, and payne in the side, drawing togither of sinewes, the gowt and greefe or ache of the ioyntes with the extreme partes.

B It is good to be layde with oyle to the noughtie scurffe of the head, which causeth the heare to fall of, for it causeth the heare to growe agayne.

C The same layd to with Frankencense and waxe dispearseth congeled blood, and taketh away blacke and blew markes which come of bruses and stripes.

D The iuyce of the roote with honie, taketh away all lentils and other spots of the face, and scurffe.

E The same mingled with sulfre, dissolueth al swellinges being layd vpon.

F With the same roote Oyle and waxe, men make an oyntment very good agaynst the olde payne of the head, the ache in the syde, and outwarde partes.

❧ The Danger.

In the gathering and drawing foorth of the iuyce of this roote, or the pith of the same, there chanceth great inflammation in the face of him that draweth it foorth, and his handes will rise full of blisters. And being receiued into the body, it rayseth vp great windinesse, blastinges, tormoyling, & ouerturning the whole body: and being to largely taken, it hurteth the bowelles and inwarde partes.

❧ The Remedie.

When one wyll gather the iuyce of Thapsia, or strip the barke of the roote, he must annoynt his face and naked partes with an oyntment made with oyle of Roses and waxe.

And when one wyl minister it inwardly to open the belly, he must put therto Ginger or long Pepper, and a litle Sugar, and so to geue it. For prepared after this sort, it shall not be very hurtfull to mans nature.

Of Hermodactil or Mede Saffron. Chap. xxxv.

Colchicum cum floribus.	Colchici folia & Semen.
Wild Saffron with the flowers.	Wild Saffron with his leaues & seede.

❧ The Description.

HErmodactil hath great brode leaues lyke the Lilly, three or foure comming foorth of one roote, amongst which groweth the stalke about the heyght of a foote, bearing triangled huskes lyke to the Marsh flague or false Acorus, but alway smaller, the which being rype do open them selues into three partes: within that is inclosed a rounde seede, blacke, and harde. The flowers growe vp after the leaues and stalke are perished, vpō short stemmes or stalkes, lyke the flowers of Saffron. The roote is round, broade aboue, and narrow beneath, white & sweete, couered with many coates or felmes, hauing by one syde right in the midle as it were a clift or parting, where as the stalke bearing the flowre groweth. The roote being dryed becommeth blacke.

There is also to be seene in Shoppes litle white rounde rootes, the whiche they call Hermodactils in fashion partly lyke the aforesayde, but that they be more flatte, and haue no diuision in the middle, as the abouesayde, but what flowers and leaues they haue, Mesue hath not left vs in writing.

❧ The Place.

Medowe Saffron, as Dioscorides sayth, groweth in Messenia, and in the Ile of Colchis, whereas it tooke his first name. It is also found in this Countrie in fat medowes, and great store of it is found about Uiluorde, and about Bath in Englande.

❧ The Tyme.

The leaues of Medow Saffron, come foorth in March and April, the seede is rype in June, in July the leaues and stalke do perishe. And in September the pleasant flowers come forth of the grounde.

❧ The Names.

1. The kinde of Hermodactil here figured, is called in Greke κολχικὸν ἢ ἐφήμερον: of some in Latine Agrestis Bulbus: in Frenche Tue chien, or Mort aux chiens: in high Almaigne Zeitlosen, & Wilsen Zeitlosen: in base Almaigne of the Herboristes Hermodactilen: Turner nameth it, Mede Saffron, & wild Saffron.

2. The seconde kinde which is found in Shoppes, is called of Paulus Aegineta, Mesue, Serapio, and certayne other auncient Greeke Physitions ἑρμοδάκτυλος: in Latine Hermodactilus, and by this name it is knowen in Shoppes.

❧ The Nature.

Medow or wilde Saffron is corrupt and venemous, therefore not vsed in medicine.

The seconde Hermodactill is hoate and drie in the seconde degree.

❧ The Vertues.

That Hermodactil which is vsed in shoppes, driueth foorth by siege slymie fleme, drawing the same from farre partes, and is very good to be vsed against the gowte, the Sciatica, and all paynes in the ioyntes.

❧ The Danger.

Medow Saffron taken into the body stirreth vp knawing and fretting in all the body, as though all the body were rubbed with nettles, inflameth the stomacke, and hurteth the inwarde partes, so that in fine it causeth blooddy excrementes, and within the space of one day death.

The other Hermodactil vsed in Shoppes, stirreth vp tossinges, wamlings, windinesse and vomiting, and subuerteth and ouerturneth the stomacke.

❧ The Remedie.

If any man by chaunce haue eaten of wild Saffron, the remedie is to drinke a great draught of Cowe mylke, as maister Turner hath written.

If one put to that Hermodactill which is vsed in Shoppes, eyther Ginger, long Pepper, Annys seede, or Comin, and a little Mastik: so taken it doth not ouerturne the stomacke, neyther stirreth vp windinesse.

Of Lauriel or Lowrye. Chap.xxxvi.

❧ The Description.

Lauriel groweth of the heigth of a foote and a halfe or more, it hath many tough branches which will not easily breake with wresting or playing, couered with a thicke rinde or barke: round about the sayde branches, but most comonly at the top grow many leaues clustering togither, thicke and of a blackish colour, like in fashion to Baye leaues, but not so great, the which being chewed in the mouth, do chafe and burne the mouth, tongue, & throte exceedingly. The flowers grow vpon short stemmes, ioyning and vppon the leaues, well clustering togither about the stalke, of a white greene or herby colour. The fruit in the beginning is greene, and after being ripe, it is blacke almost lyke a Baye berie, but lesser. The roote is long and of a wooddy substance.

The thirde Booke of

Daphnoides.

❧ *The Place.*

Lauriel groweth in rough mountaines, amongst wood, and is found in the Countrie of Liege and Namure, alongest the riuer Meuse, & in some places of Almaigne. It groweth also in many places of Englande.

❧ *The Tyme.*

It flowreth all bytimes in February: the seede is ripe in May.

❧ *The Names.*

This plant is called in Greke δαφνοειδὴς: Daphnoides: in Shoppes Laureola: in Frenche and base Almaigne Laureole: in high Almaigne Zeilandt: in Englishe Lauriell.

❧ *The Nature.*

It is hoate and dry in the third degree, drawing neare to the fourth.

❧ *The Vertues.*

A   The leaues of Lauriel open the belly, and purge slymie fleme, and waterie superfluities, & are good for suche as haue the Dropsie. Like vertue haue fouretene or fiftene of the Beries taken at once for a purgation.

B   The leaues of the same holden in the mouth and chewed, drawe foorth muche water and fleme from the brayne, and put into the nose they cause sneesing.

❧ *The Danger.*

Lauriel doth vexe and ouerturne the stomacke very muche, and inflameth, hurteth, and burneth the inwarde partes.

❧ *The Remedie.*

The leaues of Lauriell are corrected and made more apt to be receiued, in like manner as Chamelæa.

Of Mezereon.   Chap. xxxvij.

❧ *The Kyndes.*

MEzereon, as Auicenne, Mesue, and Serapio do write, is of two sortes, whereof one hath broade leaues, the other narrowe. And is set foorth by the Auncient Greeke Physitions vnder these two names Chamelæa, and Thymelæa.

❧ *The Description.*

CHamelæa is but a lowe plante, about the heigth of a foote and a halfe, or two foote. The stalkes be of a wooddy substance, ful of branches: the leaues be long, narrow and blackish, much lyke the leaues of the Olyue tree, but smaller. At the highest of the stalkes growe little pale or yellowishe flowers, and afterward the three-cornered fruite, like the Tithymales and Spourge, greene at the beginning, and red when it is ripe: after, blackish or browne whan it is drye: whereof

the Historie of Plantes. 369

Chamelæa.
widowayle.

Thymelæa.

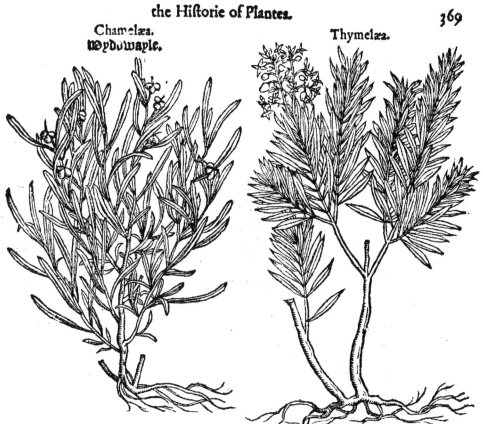

whereof eache seede is rounde almost lyke a Pepper corne, harde and bitter in the beginning, and after hoate burning the mouth.

Thymelea hath many smal springes or branches of the length of a cubite, or a cubite and a halfe, the leaues are smal, lesser and narrower then the leaues of Chamelea, and thicker. The flowers be small and white, growing at the toppe of the springes or twigges. The fruit is greene at the beginning and after red like the Haw, or white thorne fruit, hauing within it a white kernell couered with a litle blacke skinne, very hoate, and burning the tong. These two plantes do neuer lose their leaues, but are alwayes greene, both in winter and somer.

❧ The Place.

These plantes do grow in rough vntoyled places, about high wayes, and are found in some places of France as in Languedock, and about Mompelier, great store and abundance. ❧ The Tyme.

Chamelea flowreth at ye beginning of somer, & yeldeth his sede in Autumne Thymelea flowreth also in somer, and his fruit is rype in August.

❧ The Names.

The Arabian Physitions do call both these plantes by the name of Mezereon, and some call it Rapiens vitam, Et faciens Viduas.

1 The first kinde is called in Greeke χαμελαία: in Latine Chamelæa, Oleago, Oleastellus. of some Citocacium, and it may be well called Chamelæa tricoccos, to put a difference betwixt it and Chamelea Germanica.

2 The seconde kinde is called in Greeke θυμελαία: in Latine Thymelæa: of some κνέωρον ἢ κέστρον, Cneoron, Cestron, & also Chamelæa: in the Assyrian speech Apolinum,

linum, they are both vnknowen in the Shoppes of this Countrie.

The seede of Thymelæa, is called in Greeke κόκκ‍‌ κνίδι‍‌: in Latine Granũ Gnidium. vnknowen also in Shoppes: for in steede thereof the Apothecaries of this Countrie do vse the seede of common Mezereon, of the whiche we will speake in the Chapter folowing. And others take another blacke rounde seede or fruit, named Cuculus Indus, the which name should seme to come of Coccus Gnidius.

❧ The Nature.

Both kindes of these herbes are hoate & drie in the thirde degree, drawing very neare to the fourth degree: they be very hoate and sharpe, making great heate in the throte when one doth chew thereon.

❧ The Vertues.

The leaues of these two kindes of Mezereon purge downewarde with A great force and violence, fleme and Cholerique humours, especially heauy wa‐terishe humours, also they preuayle much against the Dropsie, if it be ministred with good iudgement and discretion.

To the same purpose serueth the seede of Thymelæa, when one doth take the B pulpe of twentie graynes.

The leaues of Chamelæa pounde with hony, doth mundifie & clense corrupt C vlcers.

❧ The Danger.

The qualitie of these herbes approcheth very neare to the nature of venome, being diuers wayes very euil and hurtfull to mankind. It bringeth great hurt to the stomacke, the liuer, & to al the noble and principal partes of man, chafing, hurting, and searching, causing vlcers in the entrayles, and in fine purging the belly vntill bleeding.

❧ The Remedie.

The greene leaues of Chamelæa must be stieped a day and a night in good strong vineger, then dried and kept to occupie. If first ye lay to soke in the saide vineger, Quinces, or the seede of Barberies, it shalbe the more apt for to pre‐pare the sayde leaues of Chamelæa. And when ye wyll occupie of your leaues so prepared ye must make them into pouder, and geue it with Annys seede and Mastik, or ye must boyle them in whaye of sweete milke, & specially of Goates milke, or in the broth of a Capon, and then minister the sayde whaye or broth.

Of Douch Mezereon.   Chap.xxxviij.

❧ The Description.

That Mezereon, the whiche is called in Douche Seidelbast, is a little shrubbe, or tree of three or foure foote high, with short branches which will not easily breake, vpon the same are long leaues like Priuet, but whiter and tenderer. The flowers growe alongst the branches of a purple colour and sweete sauour, after which commeth the berries, whiche are first greene, and red when they be ripe: after whan they be drye, they become blacke and wrinkled: and are lyke Hempseede when one hath taken from them their withered Skinne, but they are a little rounder, and bigger.   And whan they be chewed, they are founde very hoate and strongly burning in the mouth and throte. The whiche the seede only doth not, but also the leaues, barke, and roote.

❧ The Place.

Mezereon groweth in diuers places of Almaigne in moyst darke woods, and in rough vntoyled places.

❧ The Tyme.

It flowreth bytime in Februarie and Marche, before it beareth leaues, and the

the fruit becommeth red and ripe in August and September.

Camelæa Germanica.

❧ *The Names.*

This plant is called in Shoppes of Almaigne Mezereon, of some Piper montanū, and it hath ben taken a long tyme for the right Chamelea, wherefore it may be well called Chamelęa Germanica, in high Douch Seidelbast, Leuzkraut, and Ketterhals: in base Almaigne Zeelbast, & most commonly Mezereon.

The seede of this plante is wrongfully taken of the Apothecaries of this Countrie for Coccos Gnidios, & is called of the common people Dronkaerts besiekens, that is to say, Drunkards berries, bycause that after one hath eaten of these berries, he cannot easily swallow or get downe drinke.

❧ *The Nature.*

The leaues, barke, roote, & fruit of this plante, are hoate & dry, almost in the fourth degree, and of qualitie like the roote of Thymelea.

❧ *The Vertues.*

A The leaues of Mezereon do purge downewardes with violence & might, both fleme and cholerique humours. Likewise it purgeth waterish humours, and men do vse it in the Shoppes of this Countrie, in steede of the leaues of Chamelea.

B Lyke vertue haue the berries, the whiche being chewed, do leaue in the throte such a heate and burning, that it may hardly be quenched by meanes of drinke.

❧ *The Danger.*

This plant is without doubt hurtful vnto the body, bycause it is very hoate, and of strong and vehement working, wherefore it doth hurt and greeue the inwarde partes.

❧ *The Remedie.*

The leaues of this Mezereon are prepared euen as the leaues of Chamelæa, and in lyke manner ought the fruit and barke to be ordered, when one wil giue them to be taken with any medicine.

Of Stauisaker. Chap. xxxix.

❧ *The Description.*

Taphis-acre hath straight stalkes of a browne colour, with leaues clouen or cut into fiue, sixe, or seuen cliftes, almost lyke the leaues of the wild vine. The flowers grow vpō short stemmes of a fayre blewe or skie colour, parted into fiue or sixe litle leaues: when they are gone there commeth vp close huskes, wherein is conteined a triangled seede, blacke, sharpe, and burning the mouth, the roote is of a wooddy substance, and single.

❧ *The*

Staphis agria.

❀ *The Place.*

The Herboristes of this Countrie do sowe it in their gardens, and it groweth prosperously in shadowy places.

❀ *The Tyme.*

Staphis-acre flowreth at Midsomer.

❀ *The Names.*

This herbe is called in Greeke ϛαφὶς ἀγρία: in Latine Herba pedicularis, or Pituitaria, of some in Greke φθειροκτόνον, that is to say, Lousebane, or φθείριον: in shops Staphis agria: in Frenche *Staphisaigre*, or *Herbe aux pouilleux*: in high Douche Leuſzkraut, and Speichelkraut: in base Almaigne Luyſcruyt, and the seede made into powder Luyſepouder, that is to say, Louſepowder.

❀ *The Nature.*

Staphisacre, especially the seede, is hoate almost in the fourth degree.

❀ *The Vertues.*

A Fifteene seedes of Staphisacre taken with honied water, will cause one to vomit grosse fleme and slymie matter, with violence.

B The seede of Stafisacre mingled with oyle driueth away lise from the head and from all other places of the body, and cureth all scuruie itche, and mangines.

C The same boyled in vineger and holden in the mouth, swageth tooth ache.

D The same chewed in the mouth, draweth foorth much moysture from the head, and mundifieth the brayne.

E The same tempered with vineger, is good to rubbe vpon lousie apparell, to kill and driue away lise.

❀ *The Danger.*

The seede of Stafisaker to be taken inwardly, is very hurtfull to nature, for it chafeth and inflameth all inwarde partes, and ouerturneth the stomacke, if one holde it in his mouth, it causeth inflammation in the mouth and throte: wherefore one ought not rashly to vse this seede, except it be giuen outwardly.

❀ *The Remedie.*

Before ye occupie the seede of Stafisakre, ye must stipe it in vineger and drie it, and whan it is drie, ye may giue it to drinke with Meade or watered honie. Meade is honie and water boyled togither, and whosoeuer hath receiued of this seede, must walke without staying, and should drinke Hidromel very ofte, when he feeleth any kinde of choking, and in this dooing it shall perfourme his operation without any great danger.

Of the wilde spirting Cucumbre. Chap.cl.

❀ *The Description.*

WIlde Cucumbre hath leaues somewhat rounde and rough, but lesser and rougher then the leaues of common Cucumber. The stalkes be rounde and rough, creeping alongst the grounde without any claspers or holders, vpon whiche out of the holownesse of the collaterall branches

the Historie of Plantes. 373

ches or winges, amongſt ẏ leaues grow ſhorte ſtemmes bearing a flower of a faynte yellow colour, after the flowers there commeth little rough Cucumbers of the bigneſſe & length of ones thombe, full of ſappe with a browne kernell, the which being ripe, ſkippeth forth aſſoone as one touche ẏ Cucumbers. The roote is white, thicke, and great, with many other ſmall rootes hanging by. All the herbe is of a very bitter taſte, but eſpecially the fruite, whereof men vſe to gather the iuyce and drye it, the whiche is vſed in medicine.

Cucumis ſylueſtris.

❧ *The Place.*

This herbe is found in the gardens of Herboriſtes of this Countrie: and where as it hath ben once ſo wen, it commeth eaſily agayne euery yere.

❧ *The Tyme.*

Theſe Cucumbers do flower in Auguſt, & their ſeede is ripe in September.

❧ *The Names.*

This Cucumber is called in Greeke σίκυς ἄγριος: in Latine Cucumis Agreſtis, ſylueſtris, & erraticus: of ſome Cucumis anguinus: in ſhoppes Cucumis aſininus: in Engliſh Wylde Cucumbre: in French *Concombre ſauuage*: in high Douche Wilde Cucumer, or Eſels Cucumer: in baſe Almaigne Wilde Concommeren, or Eſels Concommeren: in Engliſhe Wilde Cucumber, or leaping Cucumber.

The iuyce of the roote being dry, is called Elaterium: in ſhoppes Elacterium.

❧ *The Nature.*

The iuyce of wilde Cucumbre is hoate and drie in the ſecond degree, and of a reſoluing and clenſing nature. The roote is of the ſame working, but not ſo ſtrong as the iuyce.

❧ *The Vertues.*

Elaterium (whiche is the iuyce of wilde Cucumbers dryed) taken in quantitie of halfe a ſcruple, driueth foorth by ſiege groſſe fleme, cholerique, and eſpecially wateriſhe humours. Moreouer it is good againſt the Dropſie, and for them that be troubled with ſhortneſſe of breath.

The ſame delayed with ſweete milke, and powred into the noſe, putteth away from the eyes the euyl colour whiche remayneth after the Jaundiſe, ſwageth headache and clenſeth the brayne.

The ſame put into the place of conception ſodden with honied wine, helpeth women to their naturall ſickneſſe, and deliuereth the dead childe.

Elaterium layd to outwardely with olde Oyle, or honie, or with the gall of an Oxe, or Bull, healeth the Squinancie, and the ſwellinges in the throte.

The iuyce of the barke and roote of wilde Cucumber, doth alſo purge fleme, and cholerique, and wateriſh humours, & is good for ſuch as haue the Dropſie, but not of ſo ſtrong operation as Elaterium.

Ii The

The roote of wilde Cucumber made soft or soked in vineger and layde to, swageth the payne, and taketh away the swelling of the gowte. The vineger wherein it hath ben boyled, holden in the mouth, swageth the tooth ache.

The same layde to with parched barlie meale, dissolueth cold tumours, and layde to with Turpentine, it breaketh and openeth impostemes.

The same made into powder, and layd to with honie, clenseth, scoureth, and taketh away foule scuruines, spreading tetters, manginesse, pushes or wheales, red spottes, and all other blemishes, and scarres of mans body.

The iupce of the leaues dropped into þ eares, taketh away the payne of the same.

The Danger.

Elaterium taken into the body, hurteth the inward partes, and openeth the smal vaynes, prouoketh gripinges and tormentes in the belly in doing his operation.

The Remedie.

To cause that it shal do no hurt, it must be geuen with Mede, or with swete mylke, a litle salt and Annys seede, or geue it in powder with gumme Tragagante, a litle Annys seede and salt.

Of Coloquintida. Chap. xli.

The Description.

Colocynthis.

1 Coloquintida creepeth with his branches alongst by the ground, with rough hearie leaues of a grayish colour, muche clouen or cut almost like the leaues of þ Citron Cucumber. The flowers are bleake or pale. The fruit round, of a greene colour at the beginning, and after yellowe, the barke thereof is neither thicke nor hard, the inner part or pulpe, is open & spogie, full of gray seede, in taste very bitter, the which men dry & kepe to vse in medicine.

2 There is yet founde another kind of Coloquintida, nothing lyke the first: for this hath long rough stalkes, mounting somewhat high, and taking holde with his claspers euerywhere, like Goordes. The leaues be like the leaues of wilde Cucumber. The fruite in all thinges is like the Goorde, but farre smaller, onely of the quantitie of a peare. These wilde Goordes haue a very hard vpper barke, or pille of a wooddy substance & greene, the inside is full of iuyce, and of a very bitter taste.

The Place.

1 The first kind groweth in Italie and Spayne, from which places the dryed fruite is brought vnto vs.

2 The seconde kinde we haue sometime seene in the gardens of certayne Herboristes.

The Tyme.

Coloquintida bringeth foorth his fruite in September.

the Historie of Plantes.

❧ *The Names.*

1 Coloquintida is called in Greeke κολοκύνθις: in Latine Colocynthis, of Paulus Aegineta *Sicyonia:* in shoppes Coloquintida: in Douche Coloquint opffelin, and Coloquint appel.

2 The seconde kinde may be called in Greeke κολοκύνθα ἄγρια: in Latine Cucurbita sylueſtris: in French *Courge sauuage:* in Douch Wilde Cauwoorden, for this is a kinde of the right Goorde.

❧ *The Nature.*

Coloquintida is hoate and drie in the thirde degree.

❧ *The Vertues.*

The white and inwarde pith or poulpe of Coloquintida, taken about the weight of a scruple, openeth the belly mightily, and purgeth grosse flemes, and cholerique humours, and the flymie filthinesse, and stinking corruption or scrapinges of the guttes, yea sometimes it causeth blood to come foorth, if it be taken in to great quantitie.   A

Like vertue it hath, if it be boyled, or layde to soke in honied water or any other liquor, and after geuen to be dronken: it profiteth muche against all colde dangerous sicknesses, as the Apoplexie, falling sickenes, giddinesse of the head, payne to fetche breath, the cholique, looseneſſe of the sinewes, and places out of ioynt.   B

For the same purposes, it may be put into Clisters and Suppositories, that are put into the fundement.   C

The Oyle wherein Coloquintida hath ben boyled, or whiche hath ben boyled in the Coloquintida, dropped into the eares, taketh away the noyse and singing of the same.   D

❧ *The Danger.*

Coloquintida is exceeding hurtfull to the hart, the stomacke and liuer, and troubleth and hurteth the bowelles, and other partes of the entrayles.

❧ *The Remedie.*

Ye muſt put to the pulpe or pithe of Coloquintida gumme Tragant and Maſticke, and after make it into trochiſques or balles with hony: for of this they vſe to make medicine.

Of Gratia Dei.   Chap.clij.

❧ *The Description.*

Gratiola is a lowe herbe, about a spanne long, something lyke to commō hysope, with many square stalkes or branches, the leaues are somwhat large, broader then the leaues of hysope, and longer then the leaues of the lesser Centaurie. The flowers growe betwixt the leaues vpon short stemmes, of a white colour mixt with a litle blewe. All the herbe in taſte is bitter, almoſt like the lesser Centaurie.

❧ *The Place.*

This herbe delighteth to growe in lowe and moyſt places, and is found in medowes: in this Countrie the Herboriſtes do plant it in their gardens.

❧ *The Tyme.*

This herbe is in flower in July and Auguſt.

❧ *The Names.*

This herbe is called of men in these dayes in Latine Gratiola, and of some also Gratia Dei, that is to say, the grace of God: and Limneſion: in Italian Stanca cauallo: and to the eye it sheweth to be a kinde of Centaurium minus, and

Ii ij   therefore

therefore of some it is called Centauris.

¶ The Nature.

Gratiola without doubt is of nature hoate and dry, and in dede it is more dry then hoate, in qualitie very like vnto the lesse Centorie.

❊ The Vertues.

A Gratiola boyled and dronke, or eaten with any kind of meat, openeth the belly freely, & causeth one to scoure muche, & by that meanes it purgeth grosse flemes, and cholerique humours.

B The same dried and made into powder, & strowed vpon wounds, doth heale and make sounde them that are newe or greene, and clenseth the old and rotten woundes. And therefore it is very necessarily put into Oyles & Oyntmentes that are made to clense and heale woundes.

Gratiola.

Of Sene. Chap. xliij.

❊ The Description.

Sena is but a litle lowe plante, with smal tender branches, the leaues are soft and tender, and somewhat rounde or hooked, not muche differing from the leaues of fenugrek. The flowers be of a pale or faynt yellowe colour: the whiche fallen or faded away, there commeth small coddes or huskes flatte and crooked, hauing a flatte seede, and somewhat browne.

❊ The Place.

Sena groweth in Alexandria, and in many places of Italie and Prouence, but the best is that of Alexandria.

❊ The Tyme.

Men do sowe it in the spring time, it floureth at Midsomer, and bringeth foorth his coddes, sodaynely after men gather and drie it.

❊ The Names.

Sena is called of Actuarius in Greke, and of the Arabian Physitions in their

Sena.

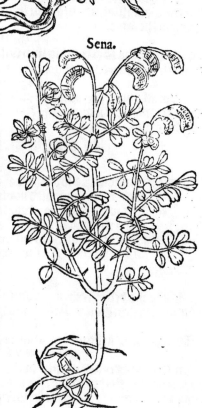

language

language Sena: and by that name it is knowen of the Apothecaries in France, Flaunders, and Englande.

❧ The Nature.

The coddes and leaues of Sena are hoat in the seconde degree, and drie in the first.

❧ The Vertues.

The coddes and leaues of Sena taken in the quantitie of a Dram, do lose and purge the belly, scoure away fleme and choler, especially blacke choler and Melancholie.

For the same purpose men geue it to drinke with the broth of a chicken, or with Perrie made of Pease, or some other lyke liquor.

The leaues of Sena taken in this sort, are good for people that are geuen to be sadde, and pensiue, heauie, dul, and careful, and that are sodainely afrayd for litle or nothing. They are good to be geuen to al melancholique people, and which are subiect to the falling sicknesse. Also they are good agaynst all stoppinges of the liuer, the splene, against the paynes of the head, the scurffe, manginesse, itche, and leprie. In fewe wordes, the purgation made with the leaues of Sena, is good agaynst all disease, springing of melancholique, adust, and salt humours.

❧ The Choise.

The coddes after the opinion of Mesue, are best to be vsed in medicine, and next the leaues, but the stalkes and branches are vnprofitable.

❧ The Danger.

Sena prouoketh windinesse, and gripinges in the belly, & is of a very slacke operation.

❧ The Correction or Remedie.

You must put to Sena Annys seede, Ginger, and some Sal Gemme. Or you must boyle it with Annys seede, Raysons, and a litle Ginger: for being so prepared and drest, it maketh his operation quickly, and without any greefe. H. Fuchius lib. primo, De Comp. s. medic. biddeth in the correction of Sena, to vse Mastick and Cloues. Cynnamome is excellent for the same purpose, as you may see in Matthiolus vpon Dioscorides.

Of Elder or Bourtre. Chap. xliiij.

❧ The Kyndes.

SUche as do trauell at this day in the knowledge of Simples, do finde that there be two kindes of Elder: wherof one is very common & wel knowen. The other is geason, and not very well knowen, and therefore it is called wilde or strange Elder.

❧ The Description.

THE common Elder doth oftentimes growe to the heyght of a tree, hauing a great tronke or body, strong, and of a wooddy substance, from whence grow forth many long branches or springes very straight, and ful of ioyntes, holow within, and ful of white soft pith, and couered without, or outwardly with a gray or ashcolour barke, vnder the whiche is also another barke or rinde, whiche is named the median or middle barke or pill: from euery knot or ioynt growe two leaues of a darke greene colour, and strong sauour, and parted or diuided into diuers other small leaues, whereof euery leafe is a litle snipt or tagged rounde about. At the highest of the branches growe white flowers, cluste-

Te thirde Booke of

clustering togither in tuftes, lke flowers of Parsenep. And when those bwers be fallen, there come little pret: rounde beries, first greene, and after blcke, out of the whiche they wring a reddiuyce, or winelike liquor. In the said beris is conteyned the seede whiche is small nd flat.

Of this kinde of Elder, ther is yet founde another sort, the beries wherof are white turning towardes yellow, in al thinges els like to the other: & this kinde is strange, and but seldome seene.

2 The seconde kinde, that is to say the wilde Elder is lyke to the first kinde in springes & knottie branches, full of wite pithe or substance, also in the sauour of he leafe. But it differeth muche in flowes and fruite: for the flowers of this wille kinde do not growe in flat & brode tufte: like the flowers of the first cōmon Eldren but clustering togither like the flowers of Medowe sweete or Medewort, or rather like the flowers of Priuet. And when the flowers of changeable colour betwixt pellowe and white, are fallen of: the beries grow after the same fashion, clustering together almost lyke a cluster of grapes. They re rounde and red, of a noughtie and strange sent, or sauour.

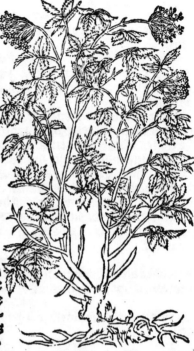
Sambucus.

❊ The Place.
1 The common Elder is found growing abundantly in the Countrie, about hedges, and it loueth shadowe and moyst places.
2 The wilde and strange kinde of Elder, doth growe likewise in darke and moyst places, but it is very seldome seene or founde.

❊ The Tyme.
1 The common Elder flowreth in May or somewhat after.
2 The wilde flowreth in April: and the fruite of them both is ripe in September.

❊ The Names.
1 The common Elder is called in Greeke ἀκτῆ: in Latine and in the Apotecaries shoppes Sambucus: in Frenche Suyn, or Hus: in high Douch Holder: in base Almaigne Ulier.
2 The wilde is nowe called Sambucus syluestris, and Sambucus ceruinus: in high Douche Waldt Holder: in base Almaigne Wilde Ulier.

❊ The Nature.
1 Common Elder is hoate and drie in the third degree, especially in the barke, the leaues, and young buddes.

❊ The Vertues.
The leaues and tender croppes of common Elder, taken in some broth or A potage, doth open the belly, purging by the same both slymie fleme, and choletique humours.

The greene median barke of the branches of Elder, do not much vary from B the

the leaues and tender croppes, but that it is of a stronger operation, purging the sayde humours with payne and violence.

The seedes, especially the litle flatte seede, dried, is profitable for suche as haue the Dropsie, and for suche as are to fatte, and woulde fayne be leaner, if it be taken in the morning the quantitie of a Dramme with wine, so that dyet be vsed for a certayne space.

The greene leaues pound, are very good to be layd vpon hoate swellinges and tumours, and being layde to playsterwise, with Dearesuet, or Bulles tallow, they aswage the payne of the gowte.

The nature and vertues of the wilde Eldren, are as yet vnknowen.

❧ *The Danger.*

Elder of his owne nature is very euyll for man, for it stirreth vp a great desire to vomit with great tossing and troubles to the stomacke, in the bowelles, and belly. It maketh all the body weake and feeble, and wasteth the strength and health of the liuer.

Of Walwort or Dane wort. Chap. xlv.

❧ *The Description.*

Ebulus.

Albeit Walwort is no tree, nor plant of a wooddy substance, but an herbe that springeth vp, euery yere a newe from his roote: yet notwithstanding it lyketh vs best in this place to set out his description, not onely bycause he is like vnto Elder, but also, bycause the auntientes haue alwayes set and described Elder & Walwort togither, the which I thought good to imitate in this matter. Therefore Walwort is no wooddy plante, but an herbe hauing long stalkes, great, straight, and cornered, parted by knottes, and ioyntes, as the branches of Elder, vppon whiche groweth the leaues of a darke greene colour, parted into diuers other leaues, muche like to the leaues of Elder, both in figure and smell. At the highest of the stalkes, it bringeth foorth his flowers in tuftes, and afterwarde it hath seede and beries like Elder. The roote is as bigge as a mans finger, of a reasonable good length, fitter to be vsed in medicine then the roote of Elder, the which is hard, and therefore not so fit as Walwort.

❧ *The Place.*

Walwort groweth in places vntopled, neare vnto high wayes, and sometimes in the feeldes, specially there where as is any moysture or good ground and fruitefull.

❊ The Tyme.

It flowreth in Iune and Iuly, his fruit is ripe in August.

❊ The Names.

This herbe is called in Greeke χαμαιάκτη, that is to say in Latine, Humilis Sambucus: and in Frenche *Suseau bas & humile:* it is called in Latine Ebulus, and Ebulum: in Frenche *Hyeble*: in high Douche Attich: in base Almaigne Hadick, Adick, and Wilden Vlier: in Englishe Walwort, Danewort, and Bloodwort.

❊ The Nature.

Walwort is hoate and drie like Elder, also it openeth and dissolueth, and is of subtill partes.

❊ The Vertues.

The leaues and newe buddes of Walwort, haue the same vertue, that the leaues and croppes of Elder haue, if they be taken after the same manner.   A

The leaues do also appease, and heale the tumours, and swellinges of the secrete partes or members, being boyled and layde thereupon.   B

The rootes boyled in wine and dronke, are good agaynst the Dropsie, for they purge downewardes the waterie humours.   C

The same do soften and vnstop the Matrix or Mother that is harde & stopped, and it doth dissolue the swelling paynes and blastinges of the belly, if women receiue the fume of the decoction thereof, through a holow chaire or stoole meete for the same purpose.   D

The iuyce of the fruite of Walwort, doth make the heares blacke.   E

The fume of Walwort burned, driueth away Serpentes, and other venemous beastes.   F

❊ The hurt or Danger.

Walwort is as noysome to the stomacke and inwarde partes of man, as is the Elder.

Of Brionie.   Chap. xlvi.

❊ The Kyndes.

There be two sortes of Brionie, as Dioscorides writeth, the white is common and well knowen in most places. The blacke is yet vnknowen to vs, and is not seene in this Countrie.

❊ The Description.

White Brionie is somthing like vnto the common Vine in his leaues and Claspers, sauing that it is both rougher and whiter: it hath smal tender branches or spruytinges, the which lifteth them selues very high, and are wrapped and entangled, about hedges & trees like Hoppes, taking holde vpon euery thing, with their sayde claspers. The leaues be great, parted into foure or fiue depe cuttinges, very like vnto the leaues of the mauted Vine, but whiter, rougher, and more hearie. The flowers do growe many togither, in colour white, after them commeth rounde Beries, in the beginning greene, but afterwarde all redde. The roote is very great, long and thicke, bitter, and of a very strange taste.

The blacke Vine (as Dioscorides sayth) hath leaues lyke vnto Iuye, but muche greater, and almost lyke the leaues of Bindeweede, or Withywinde, called Smilax. The stalkes or branches be also lyke wrapping themselues about he hedges and trees, and taking holde and cleauing to euery thing with their Claspers: the fruite clustereth togyther lyke to smal grapes, which in the beginning is greene, and afterwarde when it is ripe, al blacke. The roote is

blacke

the Historie of Plantes. 381

blacke without, and yellow within like Boxe. To this description of Dioscorides approcheth that herbe, whose figure we do here set before you (the which of some men is taken to be the blacke Vine, and the wilde blacke Brionie) sauing that his branches do not mount so high, neyther do they wrap themselues nor cleaue vnto hedges and trees, as Discorides writeth that the blacke Brionie doth: wherefore you must haue regarde to these Latine wordes, Caules etiam cognatos, capreolis suis arbores quasi adminicula comprehendit, whether they be spoken in vayne: for if those wordes be superfluous, whiche are alleaged in the translation of Dioscorides, in his description of Vitis nigra:

Brionia alba.	Christophoriana. Brionia nigra fortè.
White Brionie.	Christophorin. Grapewort, or peraduenture blacke Brionie.

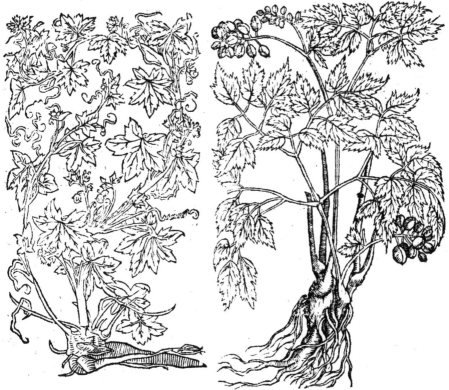

Then this wilde herbe must be without doubt the right Vitis nigra of Dioscorides. This herbe hath great and large leaues of a grayish colour, parted into diuers other leaues, of which each leafe is ranke toothed or snipt round about, in proportion almost like to the leaues of the Vine, or the flowers of the blew Bindeweede or Withywinde. The flowers be white, and do growe clustering togither at the top, or ende of the stalkes: after it beareth a fruite, which is nothing els but round berries, greene at the beginning, and blacke when they are ripe, clustering lyke grapes. The roote is blacke without and yellowe within, abyding alwayes in the grounde, and bringing foorth euery yere both newe leaues and branches: for the olde do perish in winter, euen lyke as doth both the leaues and branches of the white Brionie.

The thirde Booke of

❋ *The Place.*

1 Bryonie or the white Uinde do grow in most places of this Countrie in the feeldes, wrapping it selfe, and creeping about hedges and ditches.

2 The herbe whiche is taken for the blacke Bryonie, is founde in certayne woodes, on the hanging of hilles, in good ground, as in the Countrie of Fauquemont, and rounde about Coloygne, whereas of some it is accounted for a kinde of Naplus, whereunto it hath no kinde of lykenesse.

❋ *The Tyme.*

White Bryonie beginneth to flowre in May, and the fruite is ripe in September.

❋ *The Names.*

1 White Bryonie is called in Greeke ἄμπελος λευκὴ, ψίλωθρον, ἢ βρυωνία: in Latine Vitis alba: in the Arabian tongue Alphesera, of Mattheus Syluaticus, Viticella: in Shoppes Bryonia: in Frenche *Couleureé blanche*: in high Almaigne Stichwurtz, and Hunds kurbs: in base Almaigne Bryonie.

2 The other blacke kinde is called in Greeke ἄμπελος μέλαινα, ἢ βρυωνία μέλαινα: in Latine Vitis nigra, and Bryonia nigra, of some χειρώνιον, that is, Chironia vitis. And it may be well called in French *Couleureé noire*: in high Douche Schwartz Stickwurtz: in base Almaigne Swerte Bryonie.

The herbe, which some thinke to be the blacke Bryonie, is called of some Christophariana, and of others Costus niger, albeit it is nothing like the right Costus.

❋ *The Nature.*

1 The roote of white Bryonie is hoate and drie, euen vnto the third degree.
2 The blacke Bryonie is of the same complexion, but not altogither so strong.

❋ *The Vertues.*

A The roote of white Bryonie, especially the iuyce thereof doth mightily prouoke to the stoole, causing tough flemes to come foorth, and prouoking vrine, and is very good to mundifie and clense the braine, the brest, & inward partes from flemes, grosse and slimie humours.

B The roote of Brionie taken daily the quantitie of a Dragme by the space of one whole yere, healeth the falling euill.

C It doth also helpe them that are troubled with the Apoplexie, & turninges or swimminges of the head. Moreouer men do with great profite mingle it in medicines which they make agaynst the bitinges of Serpentes.

D The quantitie of halfe a dragme of the roote of Brionie, dronke with vineger, by the space of thirtie dayes, healeth the Melt or Splene that is waxen harde and stopped. It is good for the same entent, if it be pounde with figges, and layd outwardly vpon the place of the Splene.

E Of the same they make an Electuarie with honie, the whiche is very good for them that are short breathed, and whiche are troubled with an olde cough, and with payne in the sides, and for them that are hurt and bursten inwardly, for it dissolueth and dispatcheth congeled blood.

F Being ministred below in a Pessari or Mother suppositorie, it moueth womens flowers, and deliuereth the Secondine, and the dead childe.

G The like vertue hath a bath made of the Decoction thereof: bysides that it purgeth and clenseth the Matrix or Mother from al filthy vncleannesse, if they do sit ouer it.

H The same pound with salt, is good to be layd vpō noughtie spreading sores, that do freat, and are corrupt and running, especially about the legges.

And the leaues and fruit are as profitable for the same intent, if it be layde thrin like maner.

It

It clenseth the skinne, and taketh away the shriueled wrinckles, & freckles made with the Sonne, and all kindes of spottes and scarres: if it be mingled with the meale of Orobus, and Fenugrec. So doth the oyle wherin the roote of Brionie hath ben boyled.

The same pounde & mingled with wine, dissolueth the blood that is astonde or fixed, it dispatcheth al scarres, and blewe markes of bruysed places, and dissolueth newe swellinges, it bringeth to ripenesse and breaketh old Apostemes. It draweth foorth splinters and broken bones, and appeaseth noughtie vlcers and agnailes, that grow vp about the rootes of the nayles.

The fruit of Brionie is good against the itche, leprie, or noughtie scabbe.

The first springes or sprutinges are very good to be eaten in Salade, for the stomacke: they do also open the belly and prouoke vrine.

The roote of blacke Brionie is as good for al the greeues abouesaid, as the white Brionie, but not so strong: yet it preuayleth muche against the falling euill, and the giddinesse or turninges of the head, to prouoke vrine, the natural sicknesse of women, to waste and open the Splene or Melt that is swollen or stopped.

The tender springes of this kinde of Brionie, are also very good to be eaten in Salade, for to purge waterie superfluities, and for to open the belly, neither more nor lesse then the white Brionie.

※ *The Danger.*

The roote of Brionie by his violence doth trouble & ouerturne the stomacke, and other of the inner partes. Moreouer the same with his leaues, fruite, stalkes, and rootes, is altogither contrarie and euill to women with childe, whether it be prepared or not, or whether it be mingled with other medicines: insomuche that one cannot geue of the sayde roote, or any other medicine compounded of the same, without great daunger and perill.

※ *The Correction.*

The malice or noughtie qualitie thereof is taken away, by putting thereto Masticke, Ginger, Cinamome, and to take it with hony, or with the decoction of Raysons.

Of the Wilde Vine, Brionie, or Our Ladies Seale. Chap.xlvij.

※ *The Description.*

OUR Ladies Seale hath long branches, flexible, of a wooddishe substance, couered with a gaping or clouen barke, growing very high, and winding about trees and hedges, lyke the branches of the Vine. The leaues are lyke the leaues of Morelle or garden Night shade, but much greater, not much varying from the leaues of the greater Wythie winde or Bindeweede: the flowers be white, smal, and mossie, after the fading of whiche flowers, the fruite commeth clustering togither like little grapes or Raysons, red when it is ripe, hanging within three or foure kernelles or seedes. The roote is very great and thicke, and sometimes parted or diuided at the ende, into three or foure partes, of a brownishe colour without, and white within, and clammie like the roote of Comferie.

※ *The Place.*

In this Countrie, this herbe groweth in low and moyst woods, that are shadowed and waterie.

384   The thirde Booke of

¶ *The Tyme.*   Vitis syluestris.

It flowreth in Maye and June, and the fruite is ripe in September.

✼ *The Names.*

It is called in Greeke ἄμπελος ἀγρία: in Latine Vitis syluestris, that is to say, the Wild Vine, yet this is not that kind of wild Vine, the which men cal Labrusca, for that resembleth altogither the garden and manured Vine, but this (as is aforesayde) is a plant or herbe of the kindes of Bryonie, the which is also called in Greeke Ampelos, that is to say, a vine, bycause that it groweth high, winding it self about trees & hedges like the vine. And of this I haue thought good to geue warning, lest any hereafter happen to fal into errour, with Auicen, Serapion, & other of ye Arabian Phisitions, thinking that Labrusca and Vitis syluestris, shoulde be any other then one selfe plante. Columel calleth this plante Tamus, by folowing, of whom Plinie called the fruite Vua Taminea, & this plant is called in some places Salicastrum. It is called in Shoppes of some Apothecaries Sigillum beatæ Mariæ, that is to say, the Seale or Signet of our Ladye: in Italian Tamaro: it may be called in Frenche *Couleuree sauuage*: in Douche Wilde Bryonie, bycause it is a kinde of Bryonie, as a difference from the right wilde Vine.

Some take this herbe for Cyclaminus altera, but their opinion may be easily reproued, and founde false, bycause this herbe hath a very great roote, and as Dioscorides writeth, Cyclaminus altera hath an vnprofitable and vaine roote, that is to say, very small and of no substance.

❧ *The Nature.*

Wilde Bryonie is hoate and drie, good to mundifie, purge, and dissolue.

❧ *The Vertues.*

The roote of this herbe boyled in water & wine, tempered with a litle Sea water and dronke, purgeth downewarde waterie humours, and is very good for suche as haue the Dropsie.   A

The fruite of this plant dissolueth all congeled blood, and putteth away the markes of blacke and blewe stripes that remayne after beatinges or bruses, freckles, and other spottes of the skinne.   B

Like vertue hath the roote, if it be scrapte or grated very small, and afterwarde layde vpon with a cloth as a playster, as we our selues haue proued by experience.   C

The newe springes at their first comming vp, are also good to be eaten in Sallade, as the other two kindes of Bryonie are.   D

the Historie of Plantes.

Of Clematis altera. Chap.xlviij.

❧ *The Kindes.*

OF this kinde of plante or Withywinde, the whiche for a difference from Pereuincle (which is named Clematis in Latine) and therfore men call this kind Clematis altera, there be founde two kindes, ouer and bysides that plant whiche is nowe called in Latine Vitalba, and in Frenche *Viorne*, the whiche some do also iudge to be a kinde of Clematis altera.

Clematis altera.   Clematis alterius altera species.
Biting Pereuincle.   **Bushe Pereuincle.**

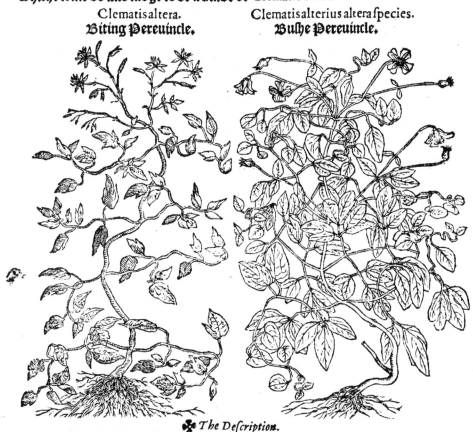

❧ *The Description.*

1. THE first kinde which is the right Clematis altera, hath smal branches, flexible, and tender, by the whiche it standeth and climbeth vp. The leaues be long & large, growing three or foure togither, very sharpe and byting the tongue. The litle flowers be white. The roote is litle and smal, and ful of heary threddes or stringes.

2. The seconde kinde is much like to the aforesayde in branches & leaues, sauing that his leaues be greater, & his stalkes or branches stronger, and in growing it is higher. The flowers are large & parted in foure leaues, fashioned like a crosse, of a blew or purple colour, and nothing lyke the flowers of the other.

3. Vitalba, or as the Frenchmen terme it *Viorne*, which some take for Clematis altera, hath long branches ful of ioyntes, easie to ploy, bigger, longer, & thicker then the branches of the aforesayde, not muche differing from the branches of the vine, by the which it climbeth vpon, and about trees and hedges: vpon the

sayde branches growe the leaues, whiche for the most part are made and do consist of fiue leaues: whereof eche leafe is of a reasonable breadth, and not muche vnlike to the leaues of Iuye, but smaller. The flowers do growe as it were by tuftes, and many togyther, of a white colour, and well smelling, after whiche flowers past, commeth the seede whiche is smal and somwhat browne, bearing smal, crooked, and downie stemmes: the roote is very full of small stringes, or hearie threddes.

Vitalba. Viorne, or Wilde Vine.

❧ *The Place.*

1 Clematis altera, is a strange herbe, and not found in this Countrie, except in the gardens of some Herboristes.

2 The seconde is also a stranger in this Countrie, but in Englande it groweth abundantly about the hedges, in the borders of feeldes, & alongst by high wayes sides.

3 Vitalba is common in this Countrie, and is to be founde in wooddes, hedges, and about the borders of feeldes.

❧ *The Tyme.*

1.2 The two first kindes do flower in this Countrie very late, in August and September.

3 But Vitalba floweth in June.

❧ *The Names.*

1 The first is called in Greeke κληματίς ἑτέρα, of some ἐπιγῆτις: in Latine Clematis altera, Ambuxum, Epigetis, and of some of our tyme Flammula.

2 The seconde is also accounted to be Clematis altera, bycause of the lykenesse it hath with the other, albeit his leaues do not muche bite vpon the tongue.

3 The thirde is nowe called Vitalba: in Frenche *Viorne*: in high Douch Lijnen or Lenen, and of some Waldreben. Some learned men take this herbe for a kinde of Clematis altera, although his leaues lykewise haue no very great byting sharpnesse, vpon the tongue. Wherefore it shoulde be rather iudged of me, to be more lyke the herbe whiche men call in Greeke κυκλάμινος ἑτέρα: in Latine Cyclaminus altera, of some Cissanthemon, and Cissophilon, whereof we haue written before in the eleuenth Chapter of this thirde booke.

❧ *The Nature.*

The leaues of Clematis altera, are hoate in the beginning of the fourth degree.

❧ *The Vertues.*

The seede of Clematis altera, taken with water, or Mede made with water **A** and honie, purgeth downewarde cholerique humours, with grosse and tough fleme, as sayth Dioscorides.

The leaues being layde vppon, doth take away, and heale the scurffe and **B** lepre.

The fruite of Cyclaminus altera, dronken with white wine fourtie dayes **C** togither,

togither, doth heale the stoppinges and hardnesse of the Melte or Splene, purging the same both by siege, and vrine. And is profitable for them that are short winded, to be taken into the body.

Of Juye. Chap.xlix.

❧ The Kyndes.

There be three kindes of Juye, as Dioscorides writeth. The first hath a white fruite and is vnknowen vnto vs. The seconde beareth a blacke or yellowish fruite, and of this kinde there groweth great plentie in this Countrie. The thirde kinde is small, and creepeth along vpon the grounde, and this kinde bringeth no fruite.

Hedera nigra. Hedera Helix.
Blacke Juye. **Smal Juye. Barren Juye.**

❧ The Description.

2. THE blacke Juye hath harde wooddy branches, couered with a graye thicke barke, whereby it embraceth and taketh holde vpon walles, old houses, and buildinges, also about trees and hedges, and all thinges els that it meeteth withal. The leaues be harde & playne, of a browne greene colour, triangled at the beginning, and after when they be more elder, they waxe somthing rounder. The flowers grow at the top or highest part of the branches, vpon long straight stemmes, many togither, like a round nosegay, of a pale color: after they turne into round beries, about the quantitie of a pease, clustering togither, greene at the beginning, but afterwarde when they be ripe, they waxe blacke.

Kk ij The

3 ❧ The thirde kinde is not muche vnlyke the Iuie abouesayde, but that his branches are both smaller and tenderer, not lifting or bearing it selfe vpwarde (as the other kinde) but creeping alongst by the grounde. The leaues are most commonly three square, of a blackish greene, and at the ende of sommer about Autumne, they are betwixt browne and red vpon one side: this Iuie hath neyther flowers nor fruite.

¶ *The Place.*

2 The blacke Iuie groweth in all partes of this Countrie, vppon olde buyldinges, houses, walles, tyles, or coueringes of houses, and vppon trees and hedges, about the which it embraceth, and taketh holdefast.

3 The small Iuie groweth in woodes, and creepeth alongst the grounde, amongst the mosse.

✤ *The Tyme.*

The blacke Iuie flowreth in sommer, and the fruite is rype in winter.

✤ *The Names.*

Iuie is called in Greeke κισσός, and of some κίτταρος: in Latine Hedera: in high Douche Ephew, or Eppich: in base Almaigne Ueyl.

1 The first kinde, whiche is vnto vs vnknowen, is called Hedera alba, and of Plinie Hedera fœmina.

2 The seconde kinde is called Hedera nigra, and Διονυσία, Dionysia, of Plinie Hedera mas, and that kinde whiche embraceth trees, is called (of men in these dayes) Hedera arborea, and that which groweth vpon walles, Hedera muralis: in French *Lyarre noir*: in high Douch Schwartzer Eppich, and Maur Ephew, or Baum Ephew: in base Almaigne Ueyl, and Boom Ueyl, or Muer Ueyl.

3 The third kinde is called in Greeke ἕλιξ: in Latine Clauicula, and Hederula: in French *Petit Lyarre*: in high Douche Klein Ephew: in base Almaigne Cleyne Ueyl.

✤ *The cause of the Name.*

Iuie is called in Greeke Cissos, bycause of a certaine Mayden or Damsell, whose name was Cissus, the whiche at a feast or banquet (wherevnto the Goddes were al bidden) so daunced before Bacchus, and kissed him often, making suche mirth and ioy, that being ouercome with the same fel to the ground, and killed her selfe. But as soone as the earth knew therof, she brought foorth immediatly the Iuie bushe, bearing still the name of the yong Damosel Cissus, the which as soone as it groweth vp a litle, commeth to embrace the Vine, in remembrance that the Damosell Cissus was wonte so to loue and embrace Bacchus the God of wine.

✤ *The Nature.*

The Iuie is partly colde, drie, and astringent, and partly hoate and sharpe. Moreouer being greene, it hath a certayne superfluous moystnesse and humiditie, the which vanisheth when it is drie.

✤ *The Vertues.*

The leaues of Iuie boyled in wine, do cure great woundes and vlcers, and do stay corrupt vlcers, and fretting sores. A

The same ordered as is aforesaid, & wel stampt or pound, & layd to, healeth burninges and scaldinges, that chaunce eyther by hoate water or fier. B

The same boyled in vineger, healeth the hardnesse and stopping of the melt or splene, if it be layde thereuppon. C

The iuyce of the leaues and fruite drawen, or snift vp into the nose, purgeth the brayne, and causeth slymie or tough fleme, and other cold humours, wherewithall the brayne is charged, to issue foorth. D

The same put into the eares, stayeth the running humours of the same, and healeth E

the Historie of Plantes. 389

healeth vlcers, and the corrupt sores happening in the same, and it doth the lyke to the sores and vlcers in the nose.

The same layd to by it selfe, or with oyle of roses, is very profitable against F the olde greeues of the head.

The flowers of Iuye layde to, in manner of a playster with oyle and waxe, G healeth all burninges.

The decoction of the same flowers made in wine, and dronke twise a day, H healeth the dangerous flixe called Dysenterie.

Fiue Iuie beries boyled with oyle of roses in the pille of a Pomgarnet: I This oyle doth cure and helpe the toothach, being put into the eare, on the contrarie syde where the payne of the teeth is.

The gumme of Iuy kylleth Lyce and Nittes. And being layde to, it taketh K away heare from the place you lay it vpon.

❧ The Danger.

The fruite of Iuye taken in to great a quantitie, weakeneth the hart, and troubleth the sense and vnderstanding. The vse therof is also very dangerous for women, especially for women with childe, and such as are newly deliuered.

Of grounde Iuye. Chap.I.

❧ The Description.   Hedera terrestris officinarum.

Grounde Iuye hath many square tender stalkes growing foorth from a roote full of threddes or stringes, vppon whiche growe leaues somewhat rounde, vneuen, and indented rounde about, of a strong smell and bitter taste, smaller, rounder, and tenderer then the leaues of Iuye. The flowers do growe amongst the leaues, in taste bitter, and of a purple colour.

❧ The Place.

Grounde Iuye is very common in all this Countrie, and groweth in many gardens, and shadowie moyst places.

❧ The Tyme.

It floureth from Apzill, vnto the ende of sommer, and continueth greene the most part of all the yere.

❧ The Names.

This herbe is called of men in these dayes, in Latine Hedera terrestris, and Corona terræ: and by this name it is knowen of the Apothecaries. It is called in Frenche Lyarre, or Lierre terrestre: in high Douche Gundelreb, and Grundreb: in base Almaigne Onderhaue. And this herbe hath ben long tyme taken, for that, which is called in Greke, χαμαικίσσος, Chamæcissus, but as I do thinke, it is better like ἑλατίνη, for whiche it is taken of some.

❧ The Nature.

Grounde Iuye is hoate and drie.

Kk iiij   ❧ The

The thirde Booke of

✥ *The Vertues.*

Grounde Iuie bruised and put into the eares, taketh away the humming noyse or ringing sounde of the same. And is good for suche as are harde of hearing.

Of Woodbine or Honysuckle. Chap.li.

| Periclymenum. | Periclymeni tertia species. |
| Woodbine or Honysucke. | The thirde kinde of Periclymenum. |

✥ *The Description.*

1. Woodbine or Honysuckle hath many small branches, whereby it windeth and wrappeth it selfe about trees and hedges: vpon the sayde branches grow long leaues and tender, white vpon the one syde, & on the other side, of a bleake or faint colour, betwixt white and greene, at the end of the branches grow the flowers in tuftes lyke nosegayes, of a pleasant colour and sweete sauour, betwixt white and yellow, or pale and purple, long & holow, almost like the little bags of Colombine. After the flowers come rounde beries, which are as red as Corall when they be ripe. The roote is of a wooddy substance.

2. There is yet another kinde, the whiche bringeth foorth leaues standing directly one agaynst the other, and so closed or ioyned togither, that the stalkes passe through them: but in all other poyntes, meetely well like to the aforesayd kinde.

Byside

3 Bylide thele two loꝛtes of Honyluckle oꝛ Woodbine, there is yet another, in leaues lyke the firſt, the whiche kinde doth not wꝛap noꝛ winde it ſelfe about trees and hedges, as the other ſoꝛtes do, but groweth and ſtandeth vpꝛight of it ſelf, without ẏ helpe of winding bꝛanches oꝛ clinging claſpers. The flowers are white, muche ſmaller then the other ſoꝛte of flowers, in figure ſomewhat long, conteyning within them many ſmall thꝛeddes, and they growe euer two and two togither by couples, and no moe, vpon a ſtemme, amongſt the leaues and bꝛanches: the whiche being gone & paſt, there grow vp two round beries, eyther red oꝛ bꝛowne when they be ripe.

✣ The Place.

Woodbine groweth in all this Countrie in hedges, about incloſed feeldes, and amongſt bꝛoome oꝛ firres. It is founde alſo in woodes, eſpecially the two laſt recited kindes. The third kind groweth in many places of Sauoye, and in the Countrie of the Swyſers.

✣ The Tyme.

Woodbine flowꝛeth in June, and July: the ſeede is rype in Auguſt and September.

✣ The Names.

1.2 This herbe oꝛ kinde of Bindeweede, is called in Greeke περικλύμενον, of ſome ἀιγίνη, κάρπαθον, σπλήνον, ἐπαιτιτίς, κληματίτις, καλυκάνθεμον, Aegina, Carpathon, Splenio, Epætitis, Clematitis, and Calycanthemon: in Latine Volucrum maius, Periclymenum, and Syluæ mater: of the Apothecaries Caprifolium, and Mater Sylua, and of ſome Lilium inter ſpinas: in French *Cheurefueille*: in high Almaigne Geiſzbladt, Speckgilgen, Zeunling, and Waldgilgen: in baſe Almaigne Gheytenbladt, and Mammekens cruyt: in Engliſhe Honyſuckle, oꝛ Woodbine, and of ſome Capꝛifoyle.

3 The thirde kinde is called in high Douchlande Hundtſzkirſchen, that is to ſay, Dogges Cherries.

❡ The Nature.

Woodbine is hoate and dꝛie, almoſt in the thirde degree.

✣ The Vertues.

The fruit of Honyſuckle dꝛonken in wine by the ſpace of fourtie dayes, doth heale the ſtopping and hardeneſſe of the Melt oꝛ Splene, by conſuming of the ſame, and making it leſſe. And purgeth by vꝛine the coꝛrupt and euil humours, ſo ſtrongly, that after the dayly vſe thereof, by the ſpace of ſixe, oꝛ ten dayes togither, it will cauſe the vꝛine to be red and blooddy.

It is good foꝛ ſuch as be troubled with ſhoꝛtnes of bꝛeath: & foꝛ them that haue any dangerous cough: moꝛeouer, it helpeth women that are in trauell of child, and dꝛieth vp the natural ſeede of man to be taken in manner aboueſayd.

The leaues haue the lyke vertue, as the fruite hath, as Dioſcoꝛides ſayth. ❡ Moꝛeouer, it keepeth backe the bꝛuſinges which are wonte to come at the beginning of Agues, when the ſayd leaues are ſodden in oyle, and pound oꝛ ſtamped very ſmal, and the backe oꝛ ridge be annoynted therewithal befoꝛe oꝛ at the firſt comming of the fittes of the Ague.

The ſame healeth woundes and coꝛrupt moyſt vlcers, and taketh away the ſpottes and ſcarres of the body and face.

✣ The Danger.

The leaues and fruit of Woodbine, are very hurtfull to women with child, and altogither contrarie.

Of ſmothe Bindweede, oꝛ Withiwinde. Chap.lij.

✣ The Kyndes.

There be two ſoꝛtes of Bindeweede oꝛ withywinde, the one bearing a blewe flower, the other a white, whereof one is great, the other ſmall.

The thirde Booke of

The greater kind windeth it selfe about hedges and trees, the lesser most commonly trayleth vpon the grounde.

Smilax lenis maior.　　　　Smilax lenis minor. Chamæcissus.
Gentle Withiwinde the great.　　**Gentle Withiwinde the smal.**

❧ *The Description.*

1 THE blewe Withiwinde hath slender branches and small, by whiche it clymbeth vp, and wrappeth or windeth it selfe about trees and poles. The leaues be large and cornered, lyke to the olde leaues of Iuye, sauing that they be not so harde. The flowers are fashioned like belles, blewe and holowe, the seede is blacke, and almost three square, lying in knoppes or huskes, after the same manner, as the seede of the white Bindeweede.

2 The great white Bindeweede or soft withiwinde hath lykewise stalkes and branches, small and tender, whereby it windeth it selfe about trees and hedges lyke the hoppe. Vpon the same branches, grow tender and soft leaues, greene, and smothe, almost like the leaues of Iuye, but muche smaller and tenderer. The flowers be great, white, and hollowe, in proportion like to a Bell. And when they are gone, there come in their steede little close knoppes or buttons, which haue in them a blacke & cornered or angled seede. The roote is smal and white, like to a sort of thicke heares, creping alongst vnder the earth, growing out or sending forth new shutes in sundrie places, of taste somewhat bitter, and full of white iuyce or sappe.

3 The lesser white Withywinde, is muche lyke to the aforesayd, in stalkes, leaues, flowers, seede, and rootes, sauing that in all these thinges, it is muche smaller,

smaller, and most commonly it creepeth alongst vppon the grounde. The branches are small and smooth: the little leaues are tender and soft: the flowers are like to litle belles of a purple or flesh colour: the seede is cornered or angled, as the seede of the others. ✤ *The Place.*

1 The blew groweth not in this Countrie, but in the gardens of Herboristes, whereas it is sowen.

2 The great white Withywinde groweth in most places of this Countrie, in euery garden, and about hedges, and inclosures.

3 The litle white Withiwinde groweth in feeldes, especially amongst the stubble and sometimes amongst the Barley, Otes, and other grayne.

✤ *The Tyme.*

1 The blew flowreth very late in this Countrie.

2.3 The white kindes do flower in June and July.

✤ *The Names.*

The Withiwinde or Bindeweede is called in Greeke σμίλαξ λεία, of Galen μίλαξ, Milax: in Latine Smilax lenis, of Marcus Cato Coniugulum: in shoppes Volubilis, of some Campanula, and Funis arborum: in frenche *Liset*, or *Liseron*: in Douche Winde, and Wranghe.

1 The kinde which beareth blewe flowers, is called Coniugulum nigrum: and after the opinion of some learned men in these dayes, of Columella in hortis, Ligustrum nigrum: of Herboristes Campana Lazura.

2 The great white smothe Withiwinde, is called of the Apothecaries Volubilis maior: in high Douche Grosz Windenkraut, and Groszweisz glocken: in base Almaigne Groote Winde. This kinde is taken of some to be Ligustrum album, whereof Virgil treateth.

3 The smal Withiwinde or Bindeweede is called Volubilis minor: in french *Campanette*, or *Vitreole*: in high Douchlande Klein Windenkraut: in Neather Douchlande Cleyne clocxkens Winde. And it seemeth to be much like to that which the Greekes cal χαμαίκισσος in Latine Chamæcissus, & Hedera terrestris.

✤ *The Nature.*

Bindeweede or Withiwinde, is of a hoate and drie qualitie or nature.

✤ *The Vertues.*

Withiwinde or Bindeweede, is not fit to be put in medicine, as Galen and Plinie witnesseth.

Of blacke Withiwinde, or Bindeweede. Chap.liij.

✤ *The Description.*

Blacke Bindeweede hath smothe red branches, very small lyke great threddes, wherewithal it wrappeth and windeth it selfe about trees, hedges, stakes, and about al herbes that it may catch or take holde vpon. The leaues are lyke to Iuie, but smaller and tenderer, much resembling the leaues of the white Bindeweede. The flowers be white and very small. The seede is blacke and triangled, or three square, like to the seede of Bockweyde or Bolymong, but smaller and blacker, growing thicke togither. Euery seede is inclosed and couered with a litle skinne. The roote is also small and tender as a thred.

✤ *The Place.*

Blacke Bindweede groweth in Vineyardes, and in the borders of feeldes, and gardens, about hedges and ditches, and amongst herbes.

✤ *The Tyme.*

It deliuereth his seede in August and September, & afterward it perisheth.

✱ *The Names.*

This kinde of Bindeweede is called in Greke ἑλξίνη κισσάμπελ‍ος, and of the Emperour Constantine μαλακοκίσσ‍ος, Malacocissos, hoc est, Mollis Hedera. Some call it in Latine Conuoluolus, of some Vitealis, that is to say, Bindeweede of the Vineyardes, or belonging to the Vine: in Shoppes Volubilis media, that is to say, The meane Bindeweede: in high Douche Swertewinde, and Middelwinde: in English Weedewinde, and Windweede, or Iuybindweede.

❡ *The Nature.*

Swerte Bindeweede is of a hoate nature, and hath power to dissolue.

✱ *The Vertues.*

A The iuyce of the leaues of this Bindeweede dronken, doth lose and open the bellye.

B The leaues pounde, and layde to the greeued place, dissolueth, wasteth, and consumeth swellinges, as Galen sayth.

Of Soldanella or Sea Cawle. Chap. liiij.

✱ *The Description.*

Soldanella hath many small branches, somwhat red, by the whiche it trayleth or creepeth alongst the grounde, casting or spreading it self here and there, couered or decked here & there with litle, rounde, greene leaues, more rounder and smaller, then the leaues Asarabacca; or lyke to the leaues of the round Aristolochia, or Birthworte, but smaller. The flowers are lyke them of the lesser Bindeweede, of a bright red, or incarnate colour. The seede is blacke, and groweth in huskes or rounde coddes, like the Bindeweedes. The roote is small and long. But to conclude, this kinde of Bindeweede is muche like the lesser Withiwinde, sauing that the leaues are muche rounder and thicker, and of a saltish taste.

✱ *The Place.*

This herbe groweth abundantly in Zealande vpon the Sea bankes, and

Helxine Cissampelos.

Brassica Marina.

alongst

the Historie of Plantes. 395

alongst the coast, or Sea side in Flaunders, and in all salt grounde standing neare the Sea.

❧ *The Tyme.*

This herbe flowreth in June, after which time men may gather it, to keepe to serue in medicine.

❧ *The Names.*

This herbe is called in Greeke κράμβη θαλασσία: in Latine Brassica Marina: in Shoppes of the Apothecaries and common Herbaries, Soldanella: in high Douche Zeewinde.

❧ *The Nature.*

Soldanella, is hoate and drie in the seconde degree.

❧ *The Vertues.*

Soldanella purgeth downe mightily all kindes of waterie humours, and openeth the stoppinges of the liuer, and is geuen with great profite vnto suche as haue the Dropsie: but it must be boyled with the brothe of some fatte meate or fleshe, and dronken: or els it must be dried and taken in powder.

❧ *The Danger.*

Soldanella, especially if it be taken in powder, hurteth and troubleth the stomacke very muche.

❧ *The Correction.*

Men take to it Annys seede, Cynamome, Ginger, and a great quantitie of Sugar, and it must be so receiued, in powder altogither.

Of Rough Bindeweede. Chap.lv.

❧ *The Description.*

Rough or prikeley Bindeweede hath tender stalkes and branches, garnished, or set round about, with many sharpe prickes or thornes, winding and wrapping it selfe about trees, hedges and bushes lyke to the other kindes of Bindeweede, taking holde with their clasping branches vppon euery thing standing agaynst it. The leaues be very well lyke Juye, but they are longer and sharper at the poynt. The flowers are white, and for his fruite, it hath round beries clustering togither lyke grapes, the whiche are red when they be ripe. The roote is thicke and harde.

❧ *The Place.*

Rough Bindeweede, as witnesseth Plinie, groweth in vntoyled waterie places, and in lowe and shadowie valleyes. It is not founde in this Countrie, but in the gardens of some diligent Herboristes.

❧ *The*

The Tyme.

Rough Bindeweede flowreth in the spring time, but in hoate Countries it flowreth agayne in Autumne.

The Names.

This Bindeweede is called in Greeke σμίλαξ τραχεᾶα, ἢ μίλαξ τραχεᾶα. in Latine Smilax aspera, of some Volubilis acuta, or Pungens: in Frenche Smilax aspre, or Liset piquant: in high Douch Stechend windt: in base Almaigne Stekēde winde. And the roote of this plant is the Zarsa parella, or as some do write Sparta parilla. The whiche some of our tune commende very muche for diuers diseases, albeit very small effecte commeth thereof.

The Nature.

This herbe is hoate and drye.

The Vertues.

The leaues and fruite of sharpe Windeweede, are very profitable against all venome and poyson, and it doth not serue onely for the venome receiued beforehande, but also agaynst all poyson taken after that a man hath eaten of the leaues or fruite of this plant. In somuch that whosoeuer eateth hereof dayly, no venome may hurt him.

Men do also write of this herbe, that if ye geue to a childe newly borne, the iuyce of this herbe, that no venom shall after hurt him.

Of Scammonie. Chap.lvi.

Scammonea. Diadrygium.

The Description.

Scammonie is a kinde of Windeweede, whiche bringeth foorth many branches from one roote, of the length of foure or fiue foote, meetely great and thicke, hauing leaues triangled and rough, not much varying from the leaues of the blacke Bindeweede, almost like the leaues of Iuye, but more softer. The flowers be white and rounde, fashioned like a Cup or Bell, of a strong and noughtie sauour. The roote is long, very thicke, and of a strong sauour, ful of sappe or iuyce, the whiche men do gather and drie calling it Scammonium, and is of great vse in Physicke.

The Place.

It groweth in Asia, Mysia, Syria, and Judea, but the best commeth from Asia, and Mysia.

The Names.

This Bindeweede or Windeweede, is called in Greeke σκαμμωνία, and of some also, as Dioscorides writeth, σκαμβωνίας ῥίζα, of the Auncient Romains in Latine Colophonium.

The iuyce of the roote dryed, is called in Greeke σκαμμώνιον: in Latine Scammonium: in Shoppes when it is yet vnprepared, Scommonea, and whan it is prepared, Diagredium, or Diagridium.

The Nature.

Scammonie is hoate and drie in the thirde degree.

The Vertues.

The iuyce of Scammonie dried, the whiche is called Scammonium, as is abouesaid, taken to the weight of sixe wheate cornes, doth purge downeward vehemently cholerique humours. Moreouer, it is good against the Jaundise, Pleuresie, Frensie, hoate feuers, and agaynst all diseases, the which take their originall beginning of hoate and cholerique humours.

The same layde to with hony and Oyle, dissolueth all colde swellinges, and with

the Historie of Plantes. 397

with vineger, it healeth all spreading scabbes, scuruinesse, and hardnesse of the skinne.

Scammonie layde to with oyle of Roses & vineger, healeth the olde paynes of the head.

The same with wooll, put into the naturall places of women, as a Pessus, or mother suppositorie, prouoketh the flowers, and expelleth the secondine and dead childe.

The Danger.

Scammonie, that is the iuyce of Scammonium, is a very strong & violent medicine, bringing a number of inconueniences, and dangerous euils, if it be eyther taken vnprepared, or out of due time and place.

First, it ouercommeth and tormenteth the stomacke very muche, causing wambling and windinesse in the same.

Secondarily, it doth by heate so chafe the liuer & blood, that it engendreth feuers, in suche as be of a hoate complexion.

Thirdly, it openeth the veynes, and hurteth the bowels and inward parts, euen to the prouoking of bloody excrementes. And therefore without doubt, Scammonie is very hurtfull to the liuer, the hart, and other inwarde partes.

The Correction.

The first danger is corrected, by putting the Scammonie to boyle, or digest in a Quince, or in the paste of Quinces, vntyl the sayd Quinces be very tender, and perfectly boyled. When the Scammonie is thus prepared, it is called Diagredium.

The second danger is preuented, by mixing your Scammonie, with some cold iuyce, as of roses, Psylium, or with the substance or pulpe of prunes.

The third is amended, by putting to the Diagredium, some Masticke, or the iuyce of Quinces.

Of Dulcamara. Chap.lvij.

The Description.

This plant hath his stalkes and branches, smal and tender, of a wooddy substance, by ý which it climeth vp, by trees, hedges, & bushes. The leaues be long & greene, not muche differing from the leaues of Iuie, but somwhat lesser, hauing sometime two eares, or two little leaues adioyning to the lowest part of the same leaues, like vnto franke Sage. The flouers be blew growing togither, euery flower diuided or parted, vnto fiue little narrow leaues, hauing in ý midle, a small yellowe pricke or poynt. The flowers being past, there come in their steede long beries, red, and very playne or smoth, of a strange sauour, clustering togither lyke the beries of Iuie. The roote is smal and threddy.

Dulcamera. Wood Nightshade.

❧ *The Place.*

This herbe groweth in moyst places, about ditches and pondes, in quick-settes and hedges.   ❧ *The Tyme.*

It flowreth in July, and his seede is ripe in August.
❧ *The Names.*

The learned men of our age, do cal this herbe in Greke γλυκύπικρον, ἢ κλυκυπικρίς: in Latine Dulcamara, and Amara Dulcis: some Herboristes of Fraunce, do cal it Solanum lignosum, that is to say, Wooddy Nightshade: in high Douche it is called, Ie lenger ie lieber, and Hynschkraut: in Neather Douchelande Alfsraucke.   ❧ *The Nature.*

Dulcamara is of complexion hoate and drie.
❧ *The Vertues.*

The decoction of this herbe in wine dronken, openeth all the stoppinges of the liuer. Moreouer, it is good agaynst the Iaunders comming of obstructions or stoppinges.

The same decoction taken as is aforesayde, is very good for such as are fallen from high places, agaynst brusinges, and dislocations, burstinges and hurtes of the inward partes: for it dissolueth congeled and fixed blood, causing the same to come foorth by the vrine, and doth cure and heale woundes and stripes.

Of Doder or Cuscuta. Chap.lviij.

❧ *The Description.*

Cassytha.

DOder is a strange herbe, without leaues, & without roote, lyke vnto a threed, muche snarled and wrapped togither, confusely winding it selfe about hedges and bushes, and other herbes. The thredes be sometimes red, sometimes white, vpon the said thredes are fastened, here and there little rounde heades or knoppes, bringing foorth at the first, small white flowers, and afterwarde a littie seede.

❧ *The Place.*

This herbe groweth muche in this Countrie vpon Brambles, Hoppes, and vpon Line or Flaxe, and sometimes it is also founde growing vpon other herbes, especially in hoate Countries, as vpon Thyme, Winter Sauerie, Tithymale, Germander, Sea Holme, but it is very little and smal, and in drie places of this Countrie it groweth vpon Wodwaxen, and vpō wormwood, as I haue seene in my garden.

❧ *The Tyme.*

Most commonly, this herbe is founde in July and August, and after that, it beareth his flowre and seede.

❧ *The Names.*

This herbe is called in Greke κασύθα: in Latine Cassytha: in shoppes Cuscuta: of some Podagra lini, and Angina lini: in French Goute, or Agoure de lin: in high Douch Filtzkraut, Flachſzseiden, and Todtern: in Neather Douchlande Scorfte, and of some Wrange, and Wildtcrupt. The Doder whiche groweth vpon Thyme, is named of the Auncient Greke Physitions & of the Arabians Epithymū: & in like maner you may call by diuers names ye Doder growing vpon & about other herbes, according to ye diuersitie of ye same, as

Epi-

the Historie of Plantes.

Epichamædris, that whiche groweth vpon Germander.
Epitithymalos, that whiche groweth vpon Tithymale.
Eperingium, whiche groweth about Sea Holme.
Epigeniston, whiche groweth about Broome.
Epibaton, whiche wrappeth about Brambles.
Epilinum, whiche groweth vpon Flaxe.
Epibryon, whiche windeth about Hoppes.
Epapsinthion, whiche groweth about Wormwood. Et sic de alijs.

❧ The Nature.

The nature of this herbe changeth, according to the nature and qualitie of the herbes, whereon it groweth, insomuche that, that whiche groweth vppon hoate herbes, as Thyme, Sauerie, & Tithymale, is likewise very hoate. That which groweth vpon other herbes, is not so feruent hoate. Neuerthelesse of it selfe, it is somwhat hoate and drie.

❧ The Vertues.

Doder or Cuscuta, boyled in water or wine, and dronke, openeth the stoppinges of the liuer, the bladder, the galle, the melt, the kidneyes, & the veynes: and purgeth both by siege and vrine, the Cholerique humours. A

It is good agaynst olde Agues, and agaynst the Iaunders, especially that B kinde whiche groweth vpon the Hoppes and vpon Brambles.

The other sortes haue propertie, according to the herbes whereuppon they C growe

Of Hoppes. Chap. lix.

❧ The Kyndes.

THere be two sortes of Hoppes, the manured or toyled Hop, and the wilde hedge Hoppe. The husbanded Hoppe, beareth his flowers or knoppes ful of scales or litle leaues growing one ouer another, & clustering or hanging downe togither like belles. The wilde is not fruitefull, but if by chance they happen to beare, it is but little and small.

❧ The Description.

1. THE tame Hoppe hath rough branches, beset with small sharpe prickels, it groweth very high, and windeth it selfe about poles and perches standing neare wheras they be planted. The leaues be rough almost like the leaues of Briony, but lesser, and nothing so muche, nor so deepely cut, of a deeper or browne colour. About the top of the stalkes amongst the leaues, grow rounde and long knoppes or heades of a whitish colour, whiche are nothing els, but many small leaues, betwixt white and yellow, or pale growing togither. Under the sayde small leaues or scales, is hidden the seede, which is flat. The belles or knoppes be of a very strong smell when they be ripe: The brewers of Ale and Bier, do heape and gather them togither, to giue a good relish, and pleasant tast vnto their drinke. The roote creepeth along in the earth, & is enterlaced or tangled, putting foorth in sundrie places newe shutes and springes.

2. The hedge or wilde Hoppe is very much like the manured and tame Hoppe in leaues & stalkes, but it beareth no knoppes or flouers: and if they beare any, they be very small and to no purpose. The roote of the same doth also trayle or creepe alongst in the grounde, and at diuers places, putteth foorth also newe shutes, and tender springes, the whiche are vsed to be eaten in Salades before they bring foorth leaues, and are a good and holesome meate.

Ll ij ❧ The

The Place.

1 The tame Hoppe is planted in gardens and places fit for the same purpose, & is also found in the borders of feeldes and about hedges.

2 The wild Hoppe groweth in hedges and bushes in the borders of feeldes, and herbe gardens.

The Tyme.

The bell knoppes and heades of Hoppes come foorth in August, and are rype in September.

The Names.

Some of our tyme do cal the Hoppe in Greeke βρύον: in Latine Lupulus Salictarius, or Lupus Salictarius: in shoppes Lupulus: in high Douche Hopffen: in Neather Douchlande Hoppe, and Hoppecruyt.

The Nature.

The Hoppe, but especially his flowers, are hoate and drie in the second degree.

The Vertues.

Lupus Salictarius.

The Decoction of Hoppes dronken doth open the stoppinges of the liuer, the splene or melte, and kidneyes, and purgeth the blood from all corrupt humours, causing the same to come foorth with the vrine. Also it is good for them that be troubled with scabbes and scuruinesse and suche lyke infirmities, whose blood is grosse and corrupted. A

For the same purpose serueth the young springes and tender croppes, at their first comming foorth of the grounde in Marche and Aprill, to be eaten in Salade. B

The iuyce of Hoppes openeth the belly, and driueth foorth the yellowe cholerique humours, and purgeth the blood from all filthynesse. C

The same dropped into the eares, clenseth them from their filth, and taketh away the stinking of the same. D

Of Ferne or Brake. Chap.lx.

The Kyndes.

There be two kindes of Fernes (as Dioscorides writeth) the male and female, the whiche in leaues are very well lyke one another.

The Description.

1 THE male Ferne hath great long leaues, sometimes of two foote in length, spread abrode vpon eche side like winges cut in euen to the middle ribbe or sinew, and snipt or toothed round about like a sawe: vnder whiche leaues ye may see many little spottes or markes, the whiche in continuance of time become blacke, and after they fall of, the roote is thicke and blacke without, putting foorth many leaues, and small dodkins or springes, whiche are the beginning of leaues.

This kinde of Ferne beareth neither flowers nor seede, except we shal take for seede the blacke spottes growing on the backside of the leaues, the whiche some do gather thinking to worke wonders, but to say the trueth, it is nothing els but trumperie and superstition.

Filix mas. Osmunde Royall.   Filix fœmina.
                                  Brake or common Ferne.

2 The female Ferne also, hath neyther flowers nor seede, but it hath long, greene, bare stemmes, vpon the whiche growe many leaues on euery syde, cut in, and toothed rounde about, very like to the leaues of male Ferne, but somewhat lesse. The roote of this Ferne is long and smal, blacke without, and creeping along in the grounde. ❦ *The Place.*

1 Male Ferne groweth almost in al rough and vneuen places, in moyst sandy groundes, and alongst the borders of feeldes, standing lowe or in vallies.
2 The female kinde is founde in woods, and mountaynes.
❦ *The Tyme.*
The leaues spring foorth in April, and wither or fade in September.
❦ *The Names.*

1 The firste kinde of Ferne, is called in Greeke πτέρις, ἤ πτέριον : in Latine Filix mas, that is to say, The Male Ferne: in Frenche *Feuchiere masle*: in high Douch Waldtfarn mennle: in neather Douchland Uaren manneken, of Matthœolus and Ruellius, it is called Osmunde Royall.

2 The seconde kinde is called in Greeke θκλυ πτέρις, and of some νυμφαία πτέρις: in Latine Filix fœmina: in Frenche *Feuchiere femelle*: in Englishe female Ferne: in high Almaigne Waldtfarn Weiblin, and of some Grotz farnkraut: in base Almaigne Uaren wijfken: in English Brake, Common Ferne and female Ferne.

The thirde Booke of

❧ *The Nature.*

Both kindes of Ferne are of like temperament or qualitie, that is hoate and drie in the seconde degree.

❧ *The Vertues.*

The roote of male Ferne taken with Mede or honied water, to the weight of halfe an ounce, driueth foorth, and killeth brode wormes. A

The same sodden in wine, is very good agaynst the hardnesse and stopping of the Melt or Splene. B

The roote of the female Ferne, taken in lyke manner as you take the male, bringeth foorth the brode and rounde wormes. C

The leaues of both kindes of Ferne put into the bedstrowe, driueth away the stinking punayses, and al other suche wormes. D

❧ *The Danger.*

The vse of Ferne is very dangerous for women, especially those that are with childe.

Of Osmunde or Water Ferne. Chap.lxi.

❧ *The Description.*

This kinde of Ferne is almost lyke the female ferne, sauing that the leaues be not dented or toothed: it hath a triangled, straight, and small stemme, about a cubite and a halfe long, hauing vppon eche side large leaues, spread abrode like winges, and cut in, like Polipodie. At the top of some of the branches grow round about small, rough and round graynes, which are lyke vnto seede. The Roote is great and thicke, folded, and couered ouer with many small enterlacing rootes, hauing in the middle a litle white, the whiche men call the Harte of Osmunde.

Filix aquatica Osmunda.

❧ *The Place.*

This kinde of Ferne groweth in woods, and moyst shadowie places.

❧ *The Tyme.*

It springeth vp in Aprill with the other fernes, and fadeth at the comming of winter: yet the roote abideth stil in the grounde.

❧ *The Names.*

This herbe is called in Latine of the Herboristes or Herbaries of our tyme, Osmunda, Filix aquatica, and of some Filicastrum: of the Alcumistes Lunaria maior: in Frenche *Osmonde*, or *Feuchiere aquatique*: in Douche Water Varen, or Wildt Varen, and of some Sinte Christoffels cruyt. We may cal it in English Osmonde the Waterman, Waterferne, and Saint Christophers herbe.

❧ *The Nature.*

Osmunde in hoate is the first degree, and drie in the seconde.

❧ *The*

❧ The Vertues.

The Hart or middle of the roote of Osmonde, is good against squattes and bruses, heauie and greeuous falles, burstinges aswel outwarde, as inwarde: or what hurt or dislocation soeuer it be. And for this purpose, many practisers, at this day: do put it into their brothes and drinkes whiche they make for woundes, causing it to boyle with other herbes: some do also put it in their Consolidatiue, or healing playsters.

Of Polypodie, Wall Ferne, or Oke Ferne. Chap.lxij.

❧ The Description. Polypodium.

Polypodie hath leaues of a spanne long, diuided into many cuttes or slittes, rent and torne, euen harde to the middle ribbe or sinewe, and yet not snipt about the litle leaues. The roote is almost as bigge as a mans finger, and very long, creeping hard by the ground, bringing foorth many litle leaues, browne without, hauing many small heares, and within of a greene herbelike colour. It hath neither branche nor flower, nor seede.

❧ The Place.

Polypodie groweth in the borders of feeldes, standing somewhat high, & about the rootes of trees, especially of Okes. Sometimes also ye shall finde it growing vppon olde wythiese, houses, and olde walles.

❧ The Tyme.

Polypodie keepeth his leaues bothe sommer and winter, but his newe leaues come foorth in April.

❧ The Names.

This herbe is called in Greeke πολυπόδιον: in Latine Filicula, and Polypodium: in frenche Polypode: in high Douche Engelsuz, Baumfarn, and Dropffwurtz: in base Almaigne Boomvaren, and of some Eyckenvaren: in Englishe Polypodie, Wall Ferne, and Oke Ferne.

❧ The Nature.

The roote of Polypodie, is drie in the seconde degree.

❧ The Vertues.

The roote of Oke Ferne openeth the belly, and purgeth Melancholique grosse, and flegmatique humours. Moreouer, it is very good agaynst the Colique, that is the payne or griping in the belly, agaynst the hardnesse and stopping of the Splene or Melt, and agaynst quartayne agues, especially if you ioyne to it Epithymum.

You must boyle it in mutton brothe, or the brothe of a Cocke or Capon, or the decoction of Mallowes or Beetes, and a little Annys, and after drinke thereof: or els you may make it in powder and drinke it with honied water or Mede.

The powder of Polipody often put into the nose, healeth and taketh cleane away the superfluous flesh growing in the nosethrilles, whiche men call Polypus.

❧ *The Choise.*

The roote of Polypody which groweth at the foote of the Oke, is the best and most fitte to be vsed in medicine, and is called in Latine Polypodium quercinum.

Of Oke Ferne, Petie Ferne, or Pilde Osmunde. Chap. lxiij.

There is now a dayes found two kindes of Dryopteris, or Oke Ferne, the one is white, the other swarte, the which are not much vnlike one another.

Dryopteris candida. White Oke Ferne.   Dryopteris nigra, Blacke Oke Ferne.

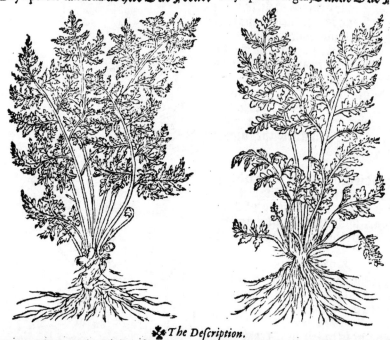

❧ *The Description.*

1. THE white kinde of Dryopteris, and the male, are not muche vnlyke, but it is much smaller, and not exceeding a spanne in height, and lykewise it beareth neither stalke, flowers, nor sede. The leaues be white, with great and deepe incisions and cuttes, snipt rounde about with smaller and thicker snips or iagges, then the leaues of male Ferne, and it hath also smal spottes or markes vnderneath the leafe. The roote is thicke and blackish, with many litle rootes, twisted, pressed, and enterlaced one with another.

The blacke Dryopteris, hath the stalke or stemme of his leaues blacke, the leaues brownish, the whiche are neyther so large, nor yet so long, neyther so muche creuished or snipt, as the leaues of the white Dryopteris, but in al other partes like, and it is beset also with litle markes or spottes vnderneath. The leaues of this kind do not perish nor fade in winter, but continue greene all the yere.

❧ *The Place.*

2. Both kindes of Dryopteris grow in holowe wayes, in shadowy and couered places, in the foote or rootes of Okes, that be aged, and of many yeres continuance: but yet they are not to be founde in all places.

❧ *The*

The Tyme.

1 The white Dryopteris springeth vp in Aprill as Ferne doth.
2 The blacke bringeth foorth his leaues at the same time.

The Names.

This kinde of Ferne is called in Greeke δρυοπτερίς, Dryopteris: In Latine Filix querna: that is in English Oke Ferne: Mathiolus, and Ruellius, both men of great knowledge, do call it in Latine Osmunda, and Osmunda Arborea. Wherefore we considering the propertie of this herbe in taking away heare, as also for a difference from the other Oke Fernes, and Osmundes, do thinke good to name this herbe in our language Osmunde Baldepate, or Pylde Osmunde.

1 The white is called in shoppes Adianthum, and to the great perill and danger of such as be sicke, is vsed for Adianthum.
2 The blacke is not very well knowen of the Apothecaries, but whereas it is knowen, they do lykewise call it Adianthum. This may be very wel called in our tongue, Small Osmunde, or Petie Ferne.

The Nature.

The white Dryopteris, is hoate, sharpe, and very abstersiue, or clensing.
The blacke agreeth with the nature or facultie of Saluia vita, or Stone Rue.

The Vertues.

White Oke Ferne, whiche is the right Dryopteris, is of such strong power or vertue, that it causeth the heare to fal of, and maketh the skinne balde. But for the doing of the same, the roote must be pounde very small, and layde vpon the place whiles a man is in the stoue or hoate house, vntill he sweate well: then it must be taken away, and newe layde on, two or three times, as witnesseth both Dioscorides and Galen.

The blacke may be vsed for Adianthum, that is to say, Uenus or Mayden heare. Phyllitis.

Of Stone Hartes tongue.
Chap.lxiiij.

The Description.

Hartes tong hath long narrow leaues, about ye length of a spanne, playne, and smothe vppon one side, and vpon ye side next the ground, it is straked ouerthwart, with certayne long rough markes, as it were small wormes, hanging vppon the backside of the leafe. The roote is blacke, hearie and twisted, or growing as it were wounden togither. And it bringeth foorth neyther stalke, flower, nor seede.

The Place.

Hartes tongue loueth shadowie places, and moyst stonie vallies, about welles, fountaynes, and olde moyst walles.

❧ The Tyme.

It beginneth to bring foorth newe leaues in April.

❧ The Names.

This herbe is called in Greeke φυλλίτις: and in Latine Phyllitis : in shoppes Scolopendria, and Lingua Ceruina : in Frenche *Langue de cerf* : in high Douche Hirtzung: in base Almaigne Hertstonge, and for a diuersitie betwixt it and Bistorte, the whiche they do likewise cal Hertstonge, Steenhertstonghe, this is not Hemionitis, as some do thinke.

❧ The Nature.

Hartstong is of complexion very drie, and astringent.

❧ The Vertues.

The decoction of the leaues of Hertstong dronke, is very good agaynst the bitinges of Serpentes, it stoppeth the laske, and the blooddy flixe.

Of brode or large Splenewort, or Miltwast. Chap.lxv.

❧ The Description.

Hemionitis is also an herbe without fruite, as the abouesayde Fernes, and Hartes tong, without stalke, without flowers and seede, bearing leaues somewhat great, large beneth, and somewhat sharpe at the top, not muche differing (as witnesseth Dioscorides) from the leaues of the seconde Dracunculus, the whiche leaues are playne by one side, & of the other side they haue also strakes or rough markes, euen as Hartes tong, his roote is compact of many stringes.

❧ The Place.

This herbe groweth in shadowy, moyst, stony, and fresshe places, and is nowe found about the decayed places and ruines of Rome, & in some other places of Italie, especially planted and set in the gardens of Herboristes. In this Countrie it is yet a stranger.

❧ The Names.

It is callled in Greeke ἡμιονῖτις, καὶ σπλήνιον: in Latine Hemionitis, Splenium, and of Gaza Mula herba : not knowen of the Apothecaries : we may call it Broade Splenewort, or large Splenewort.

❧ The Nature.

Hemionitis is meetely warme, and drie of Complexion.

❧ The Vertues.

Hemionitis taken with vineger, doth open and helpe the hardnesse and stopping of the splene, and is a soueraigne medicine for the most part of accidentes, and greeues comming or proceeding from the Rate or Spleene.

Hemionitis.

Of wild or rough Splenewort. Chap.lxvi.

✽ *The Description.*

Onchitis aspera, is partly lyke the other Fernes, for it beareth neyther stalke nor seede. The leaues be long, about the length of a spanne or foote, not muche differing from the leaues of Polypodie: but muche narrower, creuised, and cut, into more diuisions. The roote is browne and thicke, like to the roote of Dryopteris.

✽ *The Place.*

It groweth vppon the brinkes of ditches, in wooddes and low moyst places, of drie Countries.

✽ *The Tyme.*

It abideth al the winter, and bringeth forth newe leaues in April.

✽ *The Names.*

This kind of Ferne is called in Greke λογχῖτις τραχεῖα: in Latine Lonchitis aspera: of some Longina, and Calabrum, of our later writers Asplenium magnum, & Asplenium syluestre. in high Douche Spicant, & Grosz Miltzkraut: in Neather Douchlande Grachtvaren: We may name it in Englishe, Great Splenewort, or Wilde Splenewort.

Lonchitis aspera.

✽ *The Nature.*

Lonchitis is hoate in the first degree, and drie in the seconde.

✽ *The Vertues.*

Lonchitis is very good agaynst the hardnesse, stoppinges, and swellinges of the Splene or Melt: when it is dronken, or layde vpon with vineger, vpon the place of the Splene outwardly.

This herbe is also good for to be layde vnto woundes, for it keepeth them from inflammation and apostumation.

Of Ceterach, or the right Scolopendria. Chap.lxvii.

✽ *The Description.*

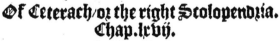Eterach hath little leaues, almost of the length of a mans fingar, cut and iagged vpon both sides euen harde to the ribbe or middle sinewe (with cuttes halfe rounde or compassed, standing not directly, but contrarie one to another) fat and greene vpon one side: but on the other syde, it is rough and somewhat hearie, reddishe, or of a browne colour. The roote is small, blacke and rough, muche platted or enterlaced. And this herbe hath neither stalke, flower, nor seede.

✽ *The Place.*

This herbe groweth in shadowy and stony places, and it is muche founde about Welles, neare vnto Namur, and the quarters thereabout.

✽ *The*

Asplenum.

The Tyme.

This herbe continueth greene al the winter, and putteth foorth newe leaues in April.

The Names.

This herbe is called in Greke ἀσπλήνον, and of some σκολοπένδριον, ἡμιόνιον, ἢ πτέρις: in Latine Asplenum, or Asplenium: in Shoppes Ceterach: in Frenche *Scolopendrie vraye*: in high Douch Steinfarn, and Miltzkraut: in base Almaigne Steenvaren: in English Right Scolopendria, Scaleferne, Fingerferne, Stone ferne, Ceterach, and Myltewaste.

The Nature.

Ceterach is temperate in heate and cold, of subtil partes, & somwhat drying.

The Vertues.

A The leaues of Ceterach, taken with vineger, by the space of fourtie dayes, healeth the Melt that is hard and stopt, and is very good agaynst Quarteyne Agues, like vertue they haue, boyled in wine, and playstered vpon the left side.

B The same is also very good against the strangurie, the hoate pisse, the stone in the bladder: it stayeth peoxsing, or yexing: it openeth the stoppinges of the liuer, and it is giuen with great profite, to suche as haue the Jaunders.

Of Venus heare, or lumbardie Maydenheare. Chap. lxviij.

The Kyndes.

Vnder ŷ name of Capillus Veneris, at this day, is set before vs two kindes of herbes not a litle lyke one ŷ other: wherof one, who is the stranger, is ŷ right Adiantum. True Maydenheare. Ladies heare. Venus heare.

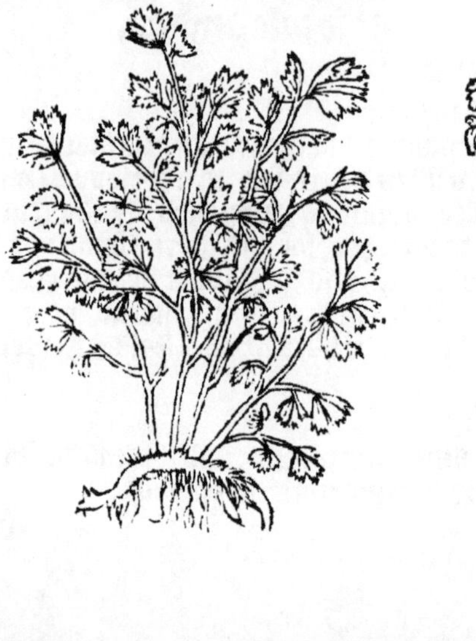

Ruta Muraria.

Stone Rue, or Wall Rue.

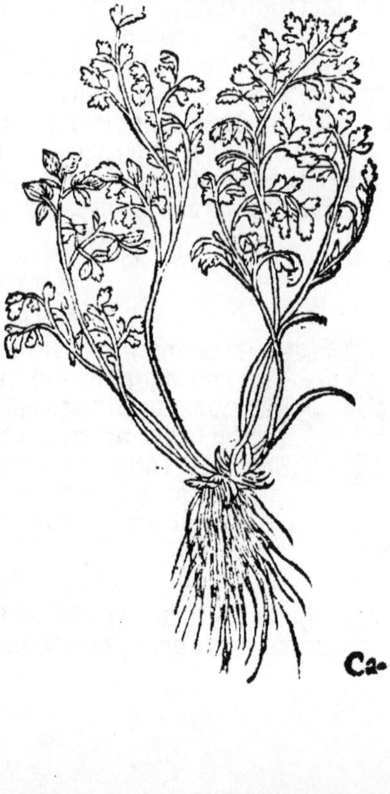

Ca-

Capillus Veneris, named of the Auncientes Adiantum. The other is very common, and hath bene vsed here for Capillus Veneris, the whiche some men call Ruta Muraria in Latine, and of others it is called Saluia vita.

❧ The Description.

1 THE right Venus heare hath the footestalkes of his leaues very smal, blackishe, and glistering with a certayne brightnesse. The leaues are smal & tender, hackt or snipped round about, like vnto the first leaues of Coriander, but muche smaller. The roote is tender.

2 The second kinde called Wall Rue, hath likewise his leaues set vpon shorte and smal stemmes, the which do somwhat resemble the leaues of garden Rue, but lesser, and something dented about, playne and smothe vppon one side, but the other side is laden, or charged with small prickes or spottes. The roote is tender and hearie. And both these herbes be without eyther flowers or seede like to the Ferne.

❧ The Place.

1 Venus heare groweth in walles, and in stony shadowy places, neare about waters and welspringes, and there is great plenty thereof found in Italie, and Prouence. It groweth not in this Countrie, but it is brought drie to vs from Italie.

2 Rew of the wal is very common in this Countrie: for it is to to be found almost vpon all olde walles that are moyst, and not comforted or lightned with the shining of the Sonne, as are the walles of Temples or Churches.

❧ The Tyme.

They remayne all the yeere, and renewe their leaues in Aprill.

❧ The Names.

1 The first kinde is called in Greeke ἀδίαντον, πολύτριχον, καλλίτριχον, ἐβενότριχον: in Latine Adiantum, Polytrichum, Callitrichū, Cincinnalis, Terræ capillus, Supercilium terræ. Apuleius calleth it Capillus Veneris, Capillaris, & Crinita: in the Shoppes of Fraunce and Italie Capillus Veneris: it is for the more part vnknowen in the Shoppes of this Countrie: in French Cheueux de Venus: in high Almaigne frauwenhar: in base Almaigne Vrouwen hayr.

2 The seconde kinde is called in the Shoppes of this Countrie Capillus Veneris, and of some it is taken for Adiantum: in the Shoppes of Fraunce Saluia vita: of the learned at this time Ruta Muraria, that is to say, Rue of the wall: in high Douche Maurrauten, and Steinrauten: in base Almaigne Steencruyt.

❧ The Nature.

Both these herbes be drie, and temperate in heate and colde.

❧ The Vertues.

The decoction of Capillus Veneris, made in wine and dronke, helpeth them A that are short breathed, and cannot fetch winde, also it helpeth such as are troubled with an harde or vnesie cough, for it ripeth tough fleme, and auoydeth it by spetting.

It prouoketh vrine, breaketh the stone, moueth the flowers, deliuereth the B secondine, and vnstoppeth the liuer, and the melt, and is very good agaynst the diseases of the Melt and the Iaunders.

Capillus Veneris stoppeth ye flixe of the belly, & stayeth the spitting of blood: C and is profitable against the flurions and moystnesse of the stomacke, & against the bitinges and stinginges of venemous beastes.

Capillus Veneris as yet greene, pounde and layde to the bitinges of vene- D mous beastes, and mad Dogges, preuayleth very muche, and layde vppon the head, causeth heare to come agayne in places that are pilde or balde.

The thirde Booke of

It dispatcheth also the swellinges of the throte called Strumes, especially in young children, when it is pound greene, and layde thereupon.

The lye wherein the same hath ben stieped and boyled, is very good to washe the scurffe of the heade: for it healeth the same, causing the rome and scales to fall of.

A cap or garlande of Maydenheare worne vpon the head, healeth the ache and payne of the same, as Plinie affirmeth.

The leaues of Adiantum mixed togither with a little Saltpeter, and the brine of a young child, taketh away the shreueled wrinckles that appeare vpon the bellies of women lately deliuered of child, if the belly be washed therwithall after their deliuerance.

Men vse in this Countrie, to put Rue of the wall in steede of Capillus Veneris, in all their medicines: and haue founde it to profite muche, in the colde passions or diseases of the breast.

Of English or common Maydenheare. Chap.lxix.

The Description. Trichomanes.

Trichomanes is a litle herbe, of the length of a span, without flowers and seede, and hath the stalkes of his leaues, very small and leane, browne, shining, and smoth, beset on both sides with many little pretie round leaues, euery leafe of the bignesse of a Lentill, straked and dashed on that side whiche is next the grounde, with many small markes and strakes, lyke Rue of the wall. The roote is small and blackishe.

The Place.

It loueth moyst and shadowie places, and groweth about waters, especially vpō moyst rockes, and olde walles, and great store thereof is found in this Countrie.

The Tyme.

It abydeth alwayes greene, like Uenus heare, and Rue of the wall.

The Names.

This herbe is called in Greeke τριχομανὲς: in Latine Fidicula capillaris, and also Trichomanes: in the Shoppes Polytrichon: in high Douche Widertodt, Abthon, and of some Roter Steinbrecke: in neather Douchland Wederdoot: in Englishe Maydenheare, and Common Maydenheare.

The Nature.

This herbe is drie and temperate betwixt hoate and colde, and of the same nature that Uenus heare is.

The Vertues.

Trichomanes after the minde of Dioscorides and Galen, hath the same faculties in operation, that Capillus Veneris hath.

the Hiſtorie of Plantes.

Of Stone Liuerwort. Chap.lxx.

✿ *The Deſcription.*

Stone Liuerwort ſpreadeth it ſelfe abroade vpon the ground, hauing wrinckled, or crimpled leaues layde one vpon another as the ſcales of fiſhe, and are greene on the vpper part, and browne on that ſide which is next the ground: amongſt the leaues there grow vp ſmal ſtemmes or twigges, in the toppes wherof are certayne knappes or thinges like ſtarres. The rootes are like ſmal threddes, growing vnder the leaues, wherby it cleaueth, and ſticketh faſt vpon the ground, and vpon moyſt or ſweating rockes.

✿ *The Place.*

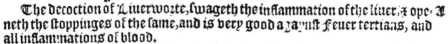

This herbe (if a man may ſo cal it) groweth in moyſt groundes, and ſtonie places, and ſhadowie, where as the Sonne ſhineth ſeldome.

✿ *The Tyme.*

It bringeth foorth his ſtarres in June and July.

✿ *The Names.*

This herbe is called in Greeke λειχὴν: in Latine Lichen: in Shoppes Hepatica: in French *Hepatique*: in high Almaigne Brunnenlebercraut, or Steinlebercraut: in baſe Almaigne Steenleuercruyt, and Leuercruyt: in Engliſhe Liuerwurt and Stone Liuerwort.

✿ *The Nature.*

Liuerwort is colde and drie of complexion.

✿ *The Vertues.*

The decoction of Liuerworte, ſwageth the inflammation of the liuer, & openeth the ſtoppinges of the ſame, and is very good agaynſt Feuer tertians, and all inflammations of blood. A

This herbe (as Dioſcorides and Plinie writeth) bruſed when it is yet greene, and layd vpon woundes, ſtoppeth the ſuperfluous bleeding of the ſame and preſerueth them both from inflammation and Apoſtemation. B

The ſame doth alſo heale all foule ſcurffes and ſpreading ſcabbes, as the Pockes, and wilde-fire, and taketh away the markes and ſcarres made with hoate irons, if it be pounde with hony and layde therevpon. C

The ſame boyled in wine, and holden in the mouth, ſtoppeth the Catarrhes, that is, a diſtilling or falling downe of Reume, or water and flegme from the the brayne to the throte. D

Of Moſſe. Chap.lxxi.

✿ *The Kyndes.*

There be many ſortes of Moſſe, whereof ſome growe in the feeldes, ſome vpon trees trees, and ſome in ſhadowie and moyſt woods, and ſome in the rockes of the ſea.

✿ *The Deſcription.*

The firſt kinde of Moſſe, which groweth vpon trees, and is moſt properly called Moſſe, is nothing els but a ſorte of ſmall white leaues, all iagged, hackte, or finely kerued, twiſted, and enterlaced one in another, without roote, without flower or ſeede, hanging and growing vpon trees.

M m ij   The

The thirde Booke of

1. Muſcus. **Moſſe.**   2. Pulmonaria. **Lungwurt.**

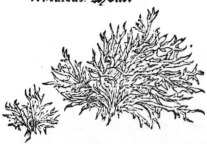

2 ❧ The ſeconde kinde groweth alſo about trees, the whiche is called Lungwurt, and it doth ſomwhat reſemble Liuerwurt, but that it is greater and larger, with great ſcales layd one vpõ another, metely greene vpon one ſide, and whitiſhe vpon the other ſide.

3. **Goldylockes,** Polytrichon, oʒ **Golden Maydenheare.**

4. Ros Solis. **Sonne Dewe.**

The third kind which ſome call Golden Polytrichon, hath very ſmall ſlender ſtalkes, nothing ſo lõg as a mans hand, couered with ſhoʒt heares, of a bʒowne greene colour changing vppon yellow, the which doth ſometymes put foozth other little bare ſtemes, with ſmall graynes oʒ ſeedes at the top.

Of this ſoʒt is founde another ſmal kinde, B like vnto the afoʒeſayd, ſauing that it is much leſſe.

The fourth kinde called Roſa Solis, hath reddiſhe leaues, ſomewhat rounde, hollowe, rough, with long ſtemmes, almoſt faſhioned lyke little ſpoones, amongſt the whiche commeth vp a ſhoʒt ſtalke, crooked at the toppe, and carrying little white flowers. This herbe is of a very ſtrange nature and maruelous: foʒ although that the Sonne do ſhine hoate, and a long time thereon, yet you ſhall finde it alwayes moyſt and bedewed, and the ſmall heares thereof alwayes full of little dʒoppes of water: and the hoater the Sonne ſhineth vpon this herbe, ſo muche the moyſtier it is, and the moʒe bedewed, and foʒ that cauſe it was called Ros Solis in Latine, whiche is to ſay in Engliſhe, The dewe of the Sonne, oʒ Sonnedewe.

5 ❧ The fifth kinde of Moſſe, called Wolfes clawe, creepeth and ſpʒeadeth with his bʒanches abʒoad, wel and thickly couered with a certaine heare of changeable colour, betwixt greene and yellowe, cleauing faſt, and taking holde in certayne places with his ſmal rootes. Theſe bʒanches agayne do put foʒth o-
thers

the Historie of Plantes. 413

5. Lycopodium.
Woolfs Clawe.

6. Muscus Marinus.
Coralin, or Sea Mosse.

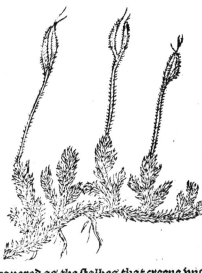

thers, parted into three or foure, hauing at their extremities or endes certayne whites fashioned like gripes, or clawes, almost lyke the clawes of Wolfe: And sometymes they bring foorth straight stalkes, small, whitishe, whiche are not couered as the stalkes that creepe vpon the ground: and they carie smal white eares, full of little leaues, whiche are lyke to small white flowers.

6 The sixth kinde of Mosse, called Mosse of the Sea, hath many smal stalkes, harde, and of a stony substance, diuided into many ioyntes, and many branches growe foorth togither from one hat, or litle stony head, by the whiche it is fastened vnto rockes.

7. Muscus Marinus, Theophrasti, & Fuci species Dioscoridis.
Slanke, Wrake, or Lauer.

7 The seueth kind of Mosse, whereof Theophrastus speaketh, is a plante without stalke or stemme, bearing greene leaues, crimpled,

8. Fuci marini species.
Wrake, or Sea girdell.

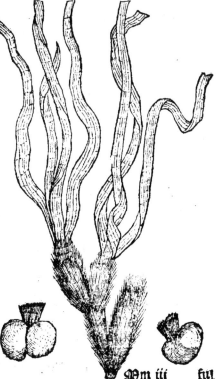

Mm iij ful

full of wrinckles, and broade, not muche differing in fashion from the leaues of some Lettise, but muche more wrinckled and drawen togither, the whiche leaues come vp many togither, growing vpon rockes.

8 The eight which is a kinde of Fucus Marinus, hath long narrow leaues, almost as narrow as a Leeke: the roote is thicke, ful of branches and rough heared, taking holde vpon rockes.

※ *The Place.*

1 The first kinde of Mosse groweth vpon trees, especially there where as the grounde is nought.

2 Lungwurt groweth vpon Mossie trees, in moyst, sandie, and shadowie places.

3.4.5 Golden Maydenheare, or Goldylockes Polytrichon, the Ros Solis, and Woolfes clawe, do growe in drie waterie Countries, and also in feeldes that lye vnmanured, or toyled, and in some shadowy wooddes. The Golden Polytrichon is very common.

6.7.8 The Sea Mosse groweth vpon stones and rockes in the Sea.

※ *The Names.*

1 The first kind of these plantes is called in Greeke βρύον, & of some σπλάγχνον: in Latine Muscus, of Serapio and in Shoppes Vinea: of Aetius Dorcadias: in Frenche *Mousse*: in high Douche Mosz: in base Almaigne Mosch. The best and most fittest for medicine is that whiche groweth vpon the Cedar tree, and next to that, is that whiche groweth vpon the Popler.

2 The seconde kinde is nowe called Pulmonaria in Latine, in English Lungwort: in high Douch Lungenkraut: in base Almaigne Longencruyt: in Frēch *Herbe aux Poulmons*.

3 The thirde is called in Douche Gulden Widdertodt: in base Almaigne, Gulden Wederdoot, that is to say, Golden Polytrichon, of some Jungfraw hare. Some thinke it to be Polytrichon Apuleanum, albeit there is but small similitude betwixt the one and the other: for Apuleius his Polytrichon, is the true Trichomanes of Dioscorides: we may cal it in English Goldylockes Polytrichon: in Frenche *Polytrichon doré*.

4 The fourth kinde is called in this Countrie Ros Solis: in Frenche *Rosee de Soleil*: in Douch Sondaw, and of some Sindaw, and Loopich cruyt.

5 The fifth is called in high Douche Beerlap, Gurtelkraut, Seilkraut, Harschar, Teuffels clawen: in Brabant Wolfs clawen, and of some Wincruyt: in some Shoppes Spica Celtica: and is taken for the same of the vnlearned, to the great detriment, dammage, and hurt of the sicke and diseased people. What the Greeke or Latine name is, I know not, and therefore after the common name I do call it in Greeke λυκοπόδιον, Lycopodion, that is, Pes Lupi, in Latine, and *Pied de Loup*, in Frenche: in Englishe, Woolfes clawe.

6 The sixth kind is called in Greke βρύον θαλάσσιον: in Latine Muscus marinus, that is to say, Mosse of the Sea: in Frenche *Mousse Marine*: in Douche Zee Mosch: in Shoppes it is called Corallina, that is to say, Herbe Corall, and of the vnlearned Soldanella, vnto whiche it beareth no kinde of lykenesse.

7 The seuenth is called also of Theophrastus βρύον θαλάσσιον: in Latine Muscus marinus, that is to say, Sea Mosse with the large leaues: in Frenche *Mousse marine a larges fueilles*: in Brabant Zee Mosch: it is to be thought, that this is the first kinde of φύκος, that is to say, Fucus, or Alga, whereof Dioscorides treateth in his fourth booke.

8 The eight is called in Greeke φῦκος, in Latine Fucus, and Alga: this is the second kind of Fucus in Dioscorides, the which Theophrastus nameth also in Greke

the Historie of Plantes.

Greke πρασον, that is to say in Latine Porrum, bycause the leaues are lyke vnto Lecke blades.

※ *The Nature.*

1. The Mosse is drie and astringent, or of a binding qualitie, without any manyfest heate or colde.
2. Lungworte is lyke to the aforesayde, sauing that it cooleth more.
3.5 Golden Maydenheare, and Woolfes Clawe, are drie and temperate in heate and colde.
4. The Ros Solis is hoate and drie almost in the fourth degree.
5. The Sea Mosse, is colde, drie, and astringent.

※ *The Vertues.*

The decoction of Mosse in water, is good for women to washe them selues in, whiche haue to muche of their naturall sicknesse: and put into the nose, it stayeth bleeding: to conclude, it is very well, and profitably put into all oyntmentes and oyles that be astringent. A

The Physitions of our time do muche commend this Pulmonaria, or Lungwort, for the diseases of the lunges, especially for the inflammations, and vlcers of the same, if it be made into powder and dronke with water. B

They say also that the same boyled in wine and dronke, stoppeth spitting of blood, pissing of blood, the flowers of women, and the laske or fluxe of the belly. C

The same made into powder, and cast into woundes, stoppeth the bleeding, and cureth them. D

Ros Solis brused with Salt, and bounde vppon the flesshe or bare skinne, maketh blisters and holes, euen as Cantharides, as you may proue by experience. E

The common sort of people do esteeme this herbe (but especially the yellow water) distilled of the same, to be a singuler and special remedie for such as begin to drie away, or are fallen into consumptions, and for them that are troubled with the disease, called Asthma, whiche is a straightnesse in drawing of breath, or with any vlceration in their lunges: thinking that it is very consolidatiue, and that it hath a special vertue to strengthen and nourish the body: but that whiche we haue recited before concerning the vertue of this herbe, declareth sufficiently, that their opinion is false. F

Men vse not Golden Maydenheare, nor Woolfes Clawe in medicine. G

Sea Mosse is af a very astringent and preseruing qualitie. Therefore men lay it to the beginning of hoate tumours or swellinges, and vpon all kindes of gowtes that require refreshing or cooling. H

The same also is very good agaynst wormes, to be made in powder, and giuen to take: for it stayeth them, and driueth them foorth mightily. I

The two other sortes of Mosse of the Sea, is also good against flegmons or hoate tumours, and the hoate gowte if they be vsed, as the first kinde of Sea Mosse commonly called Corallina. K

Of Crowfoote. Chap.lxriij.

※ *The Kyndes.*

There be foure kindes of Ranunculus, or Crowfoote, as Dioscorides and Galen do affirme, whereof the first is of many sortes. The one hath great thicke leaues, the whiche is called Water Crowfoote. The seconde hath white leaues, and is called White Crowfoote. The thirde hath blacke leaues, the whiche is called Leopardes Clawes. And these be comprised of Apuleius

M m iiij   vnder

The thirde Booke of

vnder the first kinde. The seconde kinde hath rough stalkes and leaues: the thirde is small with yellowe flowers. The fourth hath white flowers.

Bysides these there be yet other Crowfootes, the whiche growe commonly in gardens whiche are called Butterflowers, the whiche are set foorth in the the lxxiiij. Chapter. And yet there be other, as hereafter is declared.

✿ *The Description.*

Ranunculorum primum genus quadruplex.

Ranunculus palustris.	Ranunculus albus, siue echinatus.
Water Crowfoote.	**White, Vrchin Crowfoote.**

THE water Crowfoote hath white greene stalkes, hollowe, and smooth, vppon the whiche growe leaues deepely cut or clouen, almost lyke the leaues of Parsely, or Smalache, but muche whiter, softer, and thicker, very hoate and burning in the mouth. The flowers be pale, in fashion lyke Golde cuppes. The which being faded, there come vp in their places little heades or knoppes almost lyke the first buddes of Asparagus. The roote is compact of a number of white threddes.

The white or Vrching Crowfoote, hath also playne whitishe stalkes, vpon the whiche grow leaues also of a whitishe colour very deepely cut and clouen, especially the vppermost, almost lyke the leaues of Coriander. The flowers be lyke them aforesayde: when they be fallen away, in place of euery flower commeth foure or fiue round graines or beries, plat, rough like vrching. The roote is threddie lyke the other.

Golden

the Historie of Plantes. 417

Ranunculus auricomus.	Pulsatilla.
Golden Crowfoote.	**Mischieuous Passeflower.**

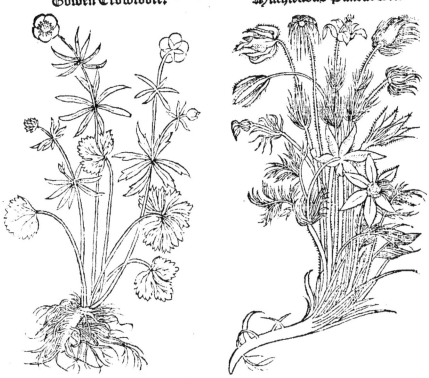

Golden Crowfoote hath his first leaues somewhat round, but afterwarde very muche cut and diuided, of a browne greene colour, & speckled in the midle with brode spottes, somwhat blacke or blackish, drawing toward the colour of fire. The flowers be of a fayre yellowe colour as golde, and shining: after the flowers there come vp rounde heades or buttons, more rougher then the knoppes of water Crowfoote.

Of this sorte there is yet one kinde founde (as Dioscorides, and Apuleius are witnesses) which beareth a purple flower, and the same is yet vnknowen.

The second kind of Ranunculus, that is called Illyricus, hath thinne stalkes, and thereon grow cut leaues, and with white, small, soft heares, the flowers be of a pale yellowe, the seede is as the other: but the rootes are otherwise, and be as many, and somwhat more then the wheate or barlie cornes ioyning togither, out of the whiche some threddes sprout, with the whiche it setteth foorth and multiplieth.

Of this kind there is yet also another strange Ranunculus, and it hath long narrow leaues, as grasse, of colour after white and blew, drawing it out of the greene. The flowers & seedes are as the aforesaid, but the rootes are threddie.

To this kind of Ranuculus is drawen another herbe which is called Passeflower, and it hath rough hearie stemmes, all iagged, and small cut, or splitte, sometimes thicke maned, and lying for the most part vpon the grounde: at the highest of the stalkes growe flowers, almost after the fashion of little Cymballes, hauing in the insyde smal yellowe threedes, as in the middle of a Rose,

of

of colour most purple browne, sometimes white, and in some places red or yellow, and whan the flowers be fallen, there commeth vp a round head, couered ouer with a certayne gray and browne heare.

Ranunculus Illyricus.   Ranunculus Lusitanicus.

3 ¶ The thirde kinde of Ranunculus, is lesser and lower then the aforesayde, his leaues be broade and vndiuided, and slipperie: betweene these two there groweth a stalke, and one flower therevpon lyke vnto the other, of a fayre yellow colour lyke vnto golde, and of a very pleasant smel. The rootes are of many cornes gathered, the whiche be longer then the rootes of Ranunculus Illyricus.

4 ¶ The fourth kinde groweth high, and hath brode leaues like vnto the Leopardes clawes, but bigger, the flowers are fashioned as the other, of colour white. The rootes are muche threeddie.

5 ¶ Byside these kindes of Ranunculus, is yet another stange kind reckoned, the whiche is called Troll flowers, and it hath great leaues diuided into many partes, and cutte rounde about: the flowers growe vppermoste of the stalke, and are yellow lyke vnto gold, fashioned lyke the flowers of Ranunculus: but bigger, and not whole open, but abiding halfe shut: thereafter folowe many small coddes togither, in the whiche the seede lyeth. The rootes are muche threddie.

✱ The

the Historie of Plantes. 419

Ranunculus albo flore.
Crowfoote with white flowers.

Ranunculus flore globoso.
Trol flowers.

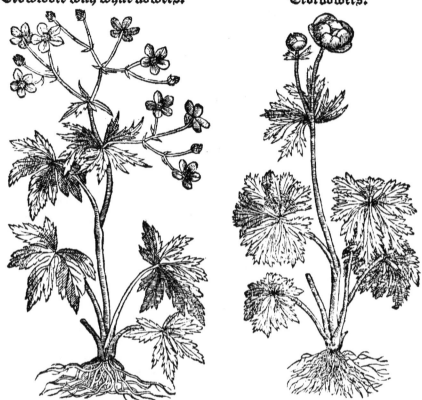

✱ *The Place.*

Crowfoote of the water, groweth in ditches and standing waters, sometimes also in medowes, and lowe sandy groundes, especially in moyst yeres.

B   The white and golden Crowfoote, groweth in medowes, & moyst feldes. C
These three kindes be common in this Countrie.

2   The seconde kinde of Crowfoote groweth in the Countrie of Illyria and Sardine, and loueth sandy and drie ground that is vntoyled, and is founde in many places of Fraunce and Almaigne. In this Countrie the Herboristes do plant it in their gardens.

3   The third Ranunculus is found vpon certaine mountaines in the Countrie of Portingal, and of Ciuil.

4   The fourth is here in this Countrie very strange.

5   The Trol flowers grow vpon the mountaynes of Switserlande.

✱ *The Tyme.*

The kindes of Crowfootes flower from April til June, & sometimes later.

✱ *The Names.*

Crowfoote is called in Greke βατράχιον: in Latine Ranunculus, of Apuleius Herba scelerata: in high Douche Hanenfusz: in base Almaigne Hanenboet, that is to say, Cockes foote: in French *Bassinet*: in Spanish *Yerua belida*: in Italian *Pie Coruino*.

The first of the first kinde is called of some in Greeke σίλινον ἄγριον, καὶ σίλινον ὑδάτιον:

The thirde Booke of

ἰλάτιον· in Latine Apium palustre, and Ranunculus palustris: in Frenche *Grenoillette aquatique*, or *Baßinet d'eaue*: in high Douch Waſſer epffs, & Waſſer hanenfutz: in neather Douchlande Water hanenboet. It may be called in Englisch Water or Marrish Crowfoote.

The seconde is nowe called in Latine Ranunculus echinatus: in Frenche *Grenoillette Heriſſonnée*: in high Douche Weiſz Hanenfutz, Ackerhanenfutz: in Brabant Witte Hanenboet: in French *Baſſinet blanc*: in Englisch White Crowfoote, or Urchin Crowfoote.

The thirde is called Ranunculus auricomus: in Frenche *Grenoillette dorée*: in high Douch Wiſen Hanenfutz, Schwartz Hanenfutz, and Geelb Hanenfutz: in base Almaigne Lupaerts clawen, and according to the same it is called in Latine Pes Leopardi, that is to say, Leopardes foote, Crowfoote and Golden Crowfoote.

2 The second kinde is called Herba Sardoa, Apium sylueſtre, Apium ruſticum, Apiaſtrum, and Apium riſus, & Ranunculus Illyricus, after the Countrie where it is founde. Paſſe flower, is called in Latine Pulſatilla, and of some Apium riſus: in Frenche *Paſſe fleur*: in high Douche Kuchenſchelle: in base Almaigne Cueckenſcelle.

3 The thirde kinde of the Emperour Constantine, is called Chriſanthemum: in Englisch Golden flower, and nowe Ranunculus Luſitanicus.

4 The fourth is called Ranunculus albus: in french *Grenoillette petite, ou de Bois*: in high Almaigne Weiſz Hanenfutz: in base Almaigne Witte Hanevoet: in Englisch White Crowfoote.

✤ The Nature.

All the Crowfootes be hoate and drie, in the fourth degreee.

✤ The Vertues.

The leaues or rootes of Crowfoote pounde, and layde to any part of the body, causeth the skinne and flesh to blister, and rayseth vp wheales, bladders, scarres, cruſtes, and vlcers. Therefore it is layd vpon corrupt and euil nayles, and vpon wartes, to cause them to fall away.

The leaues of Crowfoote may be also vsed agaynst the foule scurffe or tetter, wheales, gaules, scabbes, if it be layde to wel pound or brayed: ye may not let it lye long, but it muſt be taken of immediatly.

Also the roote of Crowfoote dried, and made into powder, and put into the nose, prouoketh sneſing.

✤ The Danger.

Al the Crowfootes are dangerous, and hurtfull, yea they kyl and ſlay, eſpecially the second, & Apium riſus, the whiche taken inwardly ſpoyleth the senſes, and vnderſtanding, and doth so drawe togither the ſinewes of the face, that such as haue eaten therof do seeme to laugh, and so they dye laughing, without some present remedie.

Of Rape Crowfoote. Chap. lxriij.

✤ The Description.

This herbe is also a kinde of Crowfoote, it groweth to the length of a span or twayne, the leaues are very muche parted and cut, lyke to the leaues of Goldknap or Goldecup, the flowers be faire and yellow, the seede groweth in rounde heades or knoppes, as the seede of Goldcuppes, the roote is white and round as a litle Turnep, sometime of the quantitie of a Nut with a beard, or threddes vnderneath.

The

the Historie of Plantes.

❧ *The Place.*

This herbe groweth in drie sandy medowes, and in suche lyke grassie places.

❧ *The Tyme.*

It floureth in Apriil and May.

❧ *The Names.*

Apuleius calleth this herbe in Greeke βατραχιον, and separateth it from the kinds of Crowfoote called Ranunculus. It is called in Neather Douchlande, Sint Anthuenis Raepken, that is to say, Saint Anthonies Turnep: we may call it, Rape Crowfoote, Goldknappe, Yellow Craw.

❧ *The Nature, Vertue, and Danger.*

This herbe is of like qualitie, and complexion, as the Crowfootes are, and is as dangerous & hurtfull to be taken inwardly.

Of Golde Cuppes, or Golde Knoppes. Chap.lxxiij.

❧ *The Kindes.*

Gold knoppe is of two sortes, ÿ single and double, or els the garden Goldecuppe, and the wild. The single is the wild kind, ÿ double is planted in gardens.

Batrachion Apulei.

Polyanthemū simplex. The single Goldcuppe, or Butter flower.

Polyanthemū multiplex. The double Goldcup, Batchelers Buttons.

The thirde Booke of

❧ *The Description.*

THE Goldeknop hath bare slender stemmes, the leaues are blackish, slit and clouen, not much differing from the leaues of Crowfoote, but more large, and not so muche cut. The flowers be yellow as fine gold, altogither like to goldē Crowfoote. The roote is threddie or hearie.

The double Goldcup is like to the single, in his leaues, stalkes, and rootes, but the flower is very double. To cōclude, Goldknop is very much like Crowfoote, and especially to the golden kind (which I thinke to be Chrysanthemon Constantini Imperatoris) sauing that it hath no blacke spottes in the leaues, as golden Crowfoote hath, neyther is it burning vpon the tongue, as Crowfoote is. ❧ *The Place.*

Goldeknoppes do grow vpon grassie downes or playnes, and in gardens, wher as it is planted.

❧ *The Tyme.*

It flowreth from April, almost al the sommer.

❧ *The Names.*

Plinie calleth this herbe in Greke πολυάνθεμον: and in Latine Polyanthemū, and it is described lib. xxvij. Chap. xij. Some do also name it Batrachion, that is to say, Ranunculus, bycause it is lyke the sayde herbe: in Frenche *Bassinet*: in high Almaigne Schmalzblum: in base Almaigne Booterbloeme: in Englishe Goldcuppes, Goldknoppes, and Butterflowers. The double Goldcuppes, are now called in English, Bachelers Buttons.

❧ *The Nature.*

The Goldknop is of complexion hoate and drie, and yet not so hoate as Ranunculus, or Crowfoote.

❧ *The Vertues.*

This herbe is not vsed in Physicke, yet in some places of Almaigne (as Hierom Bock writeth) they do mingle it amongst other herbes, in rounde salades, and Junkettes with egges.

Of Anemone. Chap. lxxv.

❧ *The Kindes.*

Dioscorides describeth two kindes of Anemone. The one is tame, and the other wilde: of the tame are founde many sortes.

❧ *The Description.*

PAsseflower or the first Anemone, hath leaues like Coriander, as witnesseth Dioscorides, or almost like the leaues of Ranunculus, but muche lesser. The flowers be sometimes red, sometimes white, and sometimes purple. The roote is thicke & rounde, greater then an Olyue, in some places not very euē, but as though it had certayne knottes and ioyntes.

Anémone Passe Huer. Rose persley, or Winde Passeflower.

The

the Historie of Plantes. 423

2. Anemone.   3. Anemone.

The seconde Anemone hath leaues lyke Goldcuppe, but lesser. The flowers be for the most part blewe, sometimes also white, being beset rounde about the middle with xiij. or xiiij. narrowe leaues. The roote is thicke, knottie, and lyeth ouerthwart.

The third Anemone hath leaues very much snipt or indented, & flowers of seuen or eight litle leaues, of a purple violet colour, or red, or white. The roote is muche lyke to the seconde Anemone.

The fourth Anemone, is lyke to the thirde in leaues and rootes, but the flowers are thicke, and very double, and red of Colour.

The fifth Anemone in leaues is like the aforesayde, but commonly greater. The flowers are some purple red, some white, and some yellowe. The rootes be very hearie..   ✲ The Place.

The first Anemone groweth in some places of Almaigne alongst by the riuer Reyn.

The fifth groweth alongst by fieldes, and in wooddes, in lowe places and grassie: and is very common in this Countrey.
    ✲ The Tyme.

It flowreth in Marche and in Aprill.
    ✲ The Names.

Anemone is also called in Greeke ἀνεμώνη, and in shoppes likewise : of some Flos Adonis: and of some Herboristes, Herbaventi, although this name is common vnto other herbes : for as Antonie Musa writeth, Cotyledon is likewise

An ij   called

The thirde Booke of

called Herba venti, and also diuers others.

4. Anemone.   5. Anemone.

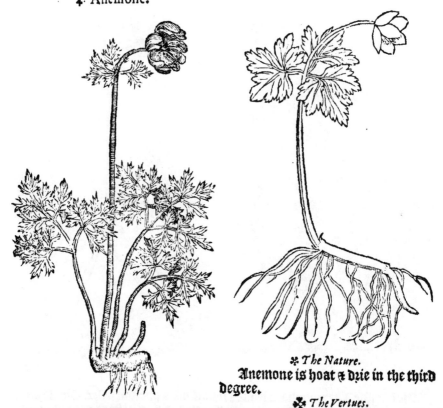

❧ *The Nature.*

Anemone is hoat & drie in the third degree.

❧ *The Vertues.*

The roote of Anemone chewed in the mouth, draweth vnto it selfe flemes, and causeth muche moysture to be auoyded out of the mouth.

The same boyled in wine prepared (called in Latine Passum) and after layd vpon the eyes, cleareth the sight, and taketh away webbes and spottes.

The leaues and yong branches boyled with cleane husked barley, causeth Nurses (that drinke thereof) to haue abundance of milke.

The same put vnder, as a Pessarie or mother Suppositorie, stirreth vp the menstruall flowers of women.

Of Spereworte or Baneworte. Chap.lxxvi.

❧ *The Description.*

This herbe hath reddish stalkes, holow, ful of knees or ioyntes, vpon the whiche growe long narrowe leaues, almost lyke to the Wythie leaues, but yet longer, and a litle snipt or toothed round about, especially those that growe lowest. The flowers are yellowe as golde, somwhat rough in the middle, in fashion and color altogither like the flowers of golden Crowfoote. Those being past, it hath knops or heades like the seedie knoppes of golden Crowfoote. The roote is ful of threddes or laces.

¶ *The Place.*

It groweth in moyst medowes, waterie places, and standing puddelles.

❧ *The Tyme.*

It flowreth in May, and soone after yeeldeth his seede.

❧ *The*

the Historie of Plantes.

Flammula.

❧ *The Names.*

This herbe is nowe called in Latine Flammula, that is to say, Flame, or the fierie herbe, bycause it is very hoate, and burning like fire. The Douchemen call it Egelcoolen, bycause ẏ sheepe that haue eaten of this herbe, haue the disease whiche they call Egel, that is to say, the inflammation and blistering of the liuer. I know not by what name the olde writers haue called this herbe, except this be that herbe, whiche Octauius Horatianus doth name Cleoma, the whiche groweth also in moyst places, and is of a very hoate temperament or complexion. It is called in some places of Englande Sperworte, it may be also called Baneworte.

¶ *The Nature and operation.*

It is hoate and drie in the fourth degree, and burneth, and blistereth the body, as Ranunculus, vnto which it is partly lyke in complexion and operation.

❧ *The Danger.*

This herbe is hurtful both vnto man and beast: for it slayeth both the one and the other. The sheepe whiche do happen to eate of it, are vexed with a maruelous inflammation, and they dye therewith, bycause their liuers are inflamed and consumed.

Of Herbe Paris, or One Berie. Chap.lxxvij.

Herba Paris.

❧ *The Description.*

Herbe Paris hath a smoth round stalke, about a span long, vppon the whiche growe foure leaues, set directly one agaynst another crossewise, or like a Crosse: amongst the sayde leaues groweth a faire starrelike flower, in the middle whereof there commeth foorth a bud or knop, growing harde by, and square, the which turneth into a browne berie. The roote is long and small, casting it selfe hither and thither.

❧ *The Place.*

This herbe groweth in darke shadowed wooddes, as in the wood Soignie by Brussels, where as it groweth abundantly.

❧ *The Tyme.*

This herbe flowreth in April, and the sede is ripe in May.

¶ *The Names.*

This herbe is now called in Latine Herba Paris, and of some Vua Lupina, and Vua versa:

in Frenche Raisin de Renard: in high Douche Wolfsbeer, Einbeer: in Neather Douchlande Wolfsbesie: in English, Herbe Paris and One berrie.

The Nature, and Vertues.

The fruite and seede of this herbe, are very good agaynst al poyson, especially for suche as by taking of poyson, are become peeuishe or without vnderstanding: insomuche that it healeth them, if it be giuen euery morning by the space of twentie dayes, as Baptista Sardus hath first written, and after him the excellent learned man Andreas Matthiolus.

Of Aconitum. Chap.lxxviij.

The Kindes.

Aconit is of two sortes (as Dioscorides writeth) the one is named Aconitum Pardalianches, that is to say, Aconite that baneth, or killeth Panthers. The other is Aconitum Lycoctonū, that is to say, Aconit that killeth Woolfs, whereof shalbe spoken in the next Chapter.

The Description.

Aconitum Pardalianches.
Panther, or Leopardes bane.

THE first kinde of Aconite, called Pardalianches, hath three or foure leaues, partly rounde, and somewhat rough heared, the whiche do resemble the leaues of Sowebread, or lyke the wilde Cowcumber, but they be smaller. The stemme groweth of the height of spanne. And thereupon grow yellow flowers, which when they perishe, they change into wooll hearie threddes, which are caried away with the wind. By them hangeth blacke seede. The roote is not vnlyke to a Scorpion, or Tortese, and is white, shining like Alablaster.

Of this kinde there is also found another whiche is somewhat greater. The roote also is somwhat longer, and more lyke to a freshwater Creauis. The whiche roote is most commonly solde of the Apothecaries, for Doronicum.

The Place.

This herbe loueth shadowie, and rude or wilde places, and is not founde in this Countrie.

The Names.

This kinde of Aconit, is called in Greeke ἀκόνιτον παρδαλιαγχές, μυοκτόνον, θηλύφονον καὶ κάμμορον: in Latine Aconitum Pardalianches, Myoctonū, Theliphonum, Cammorum: in the Apothecaries shoppes, is this roote vsed for Doronicū: but it is very vnlike to the Doronicum of the Arabian maisters.

The Nature.

Aconit is hoate and drie in the fourth degree, very hurtful to mans nature, and

the Historie of Plantes. 427

and killeth out of hande.

※ *The Vertues.*

The report goeth, that if this herbe or the roote thereof, be layde by the Scorpion, that he shall lose his force, and be astonied, untill suche time, as he shall happen agayne to touche, or be touched, with the leaues of white Elebor, or Nieswort, by vertue whereof he cometh to him selfe agayne.

※ *The Danger.*

Aconit taken into the body, killeth Wolues, Swine, and all beastes both wilde and tame.

Of Woolfes bane, or Leopardes bane.   Chap.lxix.

※ *The Kindes.*

VVoolfes bane is of two sortes. The one beareth blewe flowers, and the other yellowe. And of both those kindes are diuers other.

Lycoctonum cæruleum maius, Napellus verus. Blew Woolfs bane, or Monkes Hoode.

Lycoctonum cæruleum minus.

※ *The Description.*

THE first kinde of blewe Woolfes bane is small, the leaues be splitte and somewhat parted, as Leopardes bane. The flowers be as litle hoodes, like to the leaues of the greater Woolfes bane, with three coddes folowing the same commonly togither. On the hearie roote groweth as it were a litle knoppe, wherewith it spreadeth it selfe abroade and multiplyeth.

Lycoctonum Ponticum.	Lycoctonum flore Delphini.
Pontike Leopardes bayne, or yellow Woolfes bayne.	

 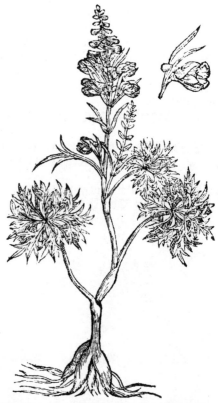

The great blewe Woolfs bayne, hath lykewise large leaues, and clouen or cut, not much differing from the aforesayde, but muche greater and more finely iagged and cut, and in colour likewise browne. The stalkes grow to the height of three or foure foote, and they beare at the toppe faire blewe flowers, rough within, and made like a Hoode or Helmet, of fiue leaues: whereof the two neathermost are narrowe and straight, they of eche side a little larger, and the leafe that is all vpmost is great and holowe, as a Cappe or Hoode, couering the leaues that are by the sides. In the holownesse of the sayde flower, growe two small crooked heares, somewhat great at the ende, fashioned like a fooles bable: in the middle of the sayde flowers are many smal hearie threddes, at the ende of the sayde small threddes, are litle prickes or poyntes, turning vpon yellowe. When the flowers doo fall, there come in their steede, three or foure huskes togither, hauing a harde, blacke, and cornered seede. The roote is thicke and blacke, fashioned like a peare, and hath many hearie stringes or strappes.

To these kindes of blewe Woolfes bayne, is lykewise accounted another purple flower, the leaues be much cut, the flowers grow along the stalkes, and are of a violet colour, of fashion like vnto a knights Spurre, with a litle taile hanging behinde the flower. The coddes are as the aforesayde. The rootes are ioyning three or foure togither.

The yellow Woolfes bayne, is likewise of two sortes, the one great, and the

the Historie of Plantes. 429

the other small. The great yellowe Woolfes bane, hath large blackishe leaues, sitte and clouen, almost lyke to Crowfoote, but farre greater. The stemmes be roundishe, about two foote high or more: at the toppe of those stemmes or braches grow pale flowers, almost like the flowers of wilde Lineseede, after which flowers there folow small coddes, conteyning a blacke and cornered seede. The roote is blacke and very threddy.

The litle yellowe Woolfes bane is a lower herbe, his leaues come foorth of the roote, the whiche are deepely cut rounde about: the flowers growe vpon some of the leaues, and they be of a yellowe colour, fashioned lyke vnto one of the Ranunculus flowers. Thereafter folowe coddes like vnto the Woolfes bane. The rootes be thicke and knotty as on the Anemone.

Lycoctonum luteum minus.

✻ *The Place.*

These venemous & noughtie herbes are founde in this Countrie planted in the gardens of certayne Herboristes, and the blew is very common in diuers gardens. The two laste kindes are founde in Almaigne & other Countries, in lowe valleyes, and darke wooddes or wilde forestes. The yellowe are also founde vpon wilde mountaynes in darke places.

✻ *The Tyme.*

These herbes do flower in April, May, and June.

✻ *The Names.*

This herbe is in called in Greke ἀκόνιτον λυκοκτόνον, ἢ κυνοκτόνον: in Latine Aconitum lycoctonum, and of some Luparia: in Frenche *Tueloup*.

1 The first is called in high Douchelande Blowolfwurtz, Psenhut, and Blopsenhutlin: in Neather Douchelande Blauw Wolfs wortele, and of some Munckes capkens, and therefore they call it in Latine Cucullus Monachi, or Cappa Monachi, that is to say, The Cape or Hoode of the Monke: And the second is counted of many learned men to be the right or true Napellus described of Auicen, & he calleth it Napellus, quasi paruus Napus, bycause the roote is like to a litle Rape or Nauew, called in Latine Napus.

2 The yellowe is called of Dioscorides, ἀκόνιτον ποντικόν, ἢ λυκοκτόνον ποντικόν: in Latine Lycoctonum Ponticum: in Frenche *Tueloup iaulne*: in high Douche Wolffwurtz, and Gelbwolffwurtz: in Neather Douchlande Geel Wolfe wortell: in Englishe Yellowe Woolfes bane: playne Woolfes bane, and Heath Crowfoote.

The litle yellowe seemeth well to be that Aconitum, the whiche Theophrastus hath spoken of, and is nowe called of some Aconitum hyemale: bycause it is preserued in the gardens of this Countrie, and in the winter it flowreth.

❧ The

The thirde Booke of

✱ The Nature.

All these Leopardes or Woolfes bane, are hoate and drie in the fourth degree, and of a venemous qualitie.

✱ The Danger.

Woolfes bane taken into the body, inflameth the hart, burneth the inwarde partes, and killeth the body, as it hath ben seene not long sithens, in Anwarpe, where as some did eate in Salade the roote of blewe Woolfes bane, in steede of some other good herbe, and died incontinent. The kindes of Woolfes bane, do not onely kill men, but also Woolfes, Dogges, and suche other beastes, if it be giuen them to eate with flesh.

Of Oleander, or Rose Baye. Chap.lxxx.

Nerium.

✱ The Description.

Oleander is a little tree or shrub, bearing leaues greater, thicker, and rougher, then the leaues of the Almonde tree, the flowers be of a fayre red colour, diuided into fiue leaues, and not much vnlike a litle Rose. The fruite is as long as a finger, full of rough hearie seede, like the coddes or huskes of Asclepias, called in Englishe Swallowe wort.

✱ The Place.

Oleander groweth in some Countries by riuers, and the sea syde, in pleasant places (as Dioscorides writeth) in this Countrie in the gardens of some Herboristes.

✱ The Tyme.

In this Countrie it bringeth foorth his flower in June.

✱ The Names.

This plante is called in Greeke νύριον, ῥοδοδάφνη, ἤ ῥοδόδευδρον: Nicander calleth it also Neris: in Latine Laurus rosea, and Rosea arbor: that is to say, Rose tree: in Shoppes Oleander, in frenche Rosagine, or Rosage: in Douche Oleander boom: in Englishe Rose tree, or Rose Baye tree, Oleander, and Nerium.

✱ The Nature.

Oleander is also very hoate and drie of Complexion.

✱ The Vertues.

It hath scarse one good propertie. It may be compared to a Pharisee, who maketh a glorious and beautifull shewe, but inwardly is of a corrupt and poysoned nature. God graunt all true Christians and Christian Realmes, whereas this tree, or any branche thereof, beginneth to spread and florishe, to put to their helping handes to destroy it, and all the branches thereof: as dissimulation,

the Historie of Plantes. 431

mulation, Couetousnesse, Briberie, or Symonie, and maister Usurie. It is high tyme, if it be the wyl of God, to supplant it. For it hath alredy flowred, so that I feare it wil shortly seede, & fil this holsome soyle ful of wicked Nerium.

The Danger.

Oleander or Nerium, is very hurtfull to man, but most of all to Sheepe, Goates, kine, Dogges, Asses, Mules, Horses, and al foure footed beastes: for it is deadly, and killeth them. Yea if they do but drinke the water, wherein Oleander hath ben stieped or soked, it causeth them to dye sodaynly, as Dioscorides, Plinie, and Galen do write.

Of Poppie. Chap.lxxxi.

The Kindes.

There be three sortes of Poppie, as Dioscorides sayth, wherof the first kind is white, and of the garden, the two other are blacke and wilde.

| Papauer satiuum. | Papauer syluestre. |
| Garden Poppie. | Wilde Poppie. |

The Description.

THe garden white Poppie beareth a straight stem, or straight smoth stalke, about ý height of foure or fiue foote in length, with long leaues therevpon, large and white, vneuenly iagged and toothed about: at the highest of the sayde stemmes, groweth a round bud or button, the whiche openeth into a large white flower, made of foure leaues, the whiche flower hath in the middle many small hearie threddes, with little tippes at the endes, and a round head, the which head waxeth great and long, wherin is the seede, which is white, and very necessarie in medicine. Of

Of this kinde there is yet another, whose flowrie leaues be iagged or frenged, in all thinges els lyke to the aforesayde.

2 The seconde kinde of Poppie, hath his stalkes and leaues much lyke to the white, but the flowers be of a fayre red colour, and the heades are more rounder, and not long. The seede is blackish.

Of this sorte there is found another kind, whose flowers be snipt & iagged, the whiche sometimes be very double, lyke to the other.

3 The thirde kinde of Poppie, is lyke to the two other sortes in leaues, and stemmes, sauing that it is smaller, and beareth moe flowers, and headdes. The flowers be of a colour betwixt white and red, changing towarde blacke, hauing blacke spottes, at the lower part of euery flowers leaues. The heades be somewhat long, much smaller then the heades of the others, wherein there is also blacke seede, and when the sede is ripe, the heades do open aboue, vnder the shel or scale whiche couereth the sayde heades. And afterwarde the seede falleth out easily, whiche happeneth not to the other two Poppies, whose heades remayne alwayes close.

There droppeth or runneth out of Poppie, a liquor as white as milke, when the heades be pearced or hurt, the whiche is called Opium, and men gather and drie it, and is kept of the Apothecaries in their shoppes to serue in medicine.

※ *The Place.*

Al these kindes of Poppie are sowen in this Countrie in gardens. The third kind is very common, insomuch as it is sowen in many feelds for the commoditie and profite which commeth of the seede. In Apulia and Spayne, and other hoate Countries, they gather the iuyce, whiche is the Opium, that men of this Countrie put in medicines.

※ *The Tyme.*

It flowreth most commonly in Iune.

※ *The Names.*

Poppie is called in Greeke μήκων: in Latine and in shoppes Papauer, of some Oxytonon, Prosopon, Lethe, Lethusa, and Onitron: in high Almaigne, Magsamen, Moen, Magle, and Olmag: in base Almaigne Huel, & of some Mancop.

The iuyce of Poppie is called in Greke ὄπιον: in Latine, & in shoppes Opium.

1 The first kind is called in Greke μήκων ἥμερος: in Latine Papauer satiuum, of some Thylacitis: in shoppes Papauer album: in Frenche *Pauot cultiué & blanc*: in Almaigne Witten Huel, and Tammen Huel: in Englishe White Poppie, and Garden Poppie.

2 The seconde kind Dioscorides calleth μήκων ἄγριος, and Papauer syluestre, & erraticum. some also cal it Pithitis: in Shoppes Papauer nigrum, magnum, of the vnlearned Papauer rubrum, and according to the same, the Frenchmen call it *Pauot rouge*: in Douche Rooden Huel: in Englishe, Blacke Poppie, and Wilde Poppie.

3 The thirde sorte is also taken for a kinde of wilde Poppie, and is called in Shoppes Papauer commune, and Papauer nigrum, that is to say, Common Poppie, and blacke Poppie: in Douche Huel. This should seeme to be Poppie Rhœas, that is to say, flowing and falling, bycause the seede thereof floweth out when it is ripe, whiche chanceth to none of the other kindes, as is abouesayde.

※ *The Nature.*

Al the Poppies be colde and drie, almost euen harde to the fourth degree. Opium is colde and drie, almost harde to the fourth degree.

※ *The*

❦ *The Vertues.*

The decoction of the leaues and heades of Poppie, made in water & dronke causeth sleepe. It hath the lyke vertue, if the head and handes be washed therwith.

Of the heades boyled in water, is made a Syrupe, whiche doth also cause sleepe, and is very good agaynst the subtil Rheumes, and Catharrhes, that distill and fal downe from the brayne vpon the lunges, and against the cough, taking his beginning of such subtil humours.

The seede of blacke Poppie dronke in wine, stoppeth the fluxe of the belly, and the vnreasonable course of womens issues: & if it be mingled with water, and layde to the forehead, it will cause sleepe also.

A playster is made with the greene knoppes or heades of Poppie (before it is ripe) & parched barley meale, the which is good to be layde vpon the disease, named in Latine Ignis sacer, and hoate tumours, which haue neede of cooling.

Opium, that is the iuyce of Poppie dried, taken in quantitie of a fetche, swageth all inwarde paynes, causeth sleepe, cureth the cough, and stoppeth the flixe.

The same layde to with Oyle of Roses, swageth headache: and with Oyle of sweete Almondes, Myrrhe, and Saffron, it healeth ache, or payne of the eares.

With vineger it is good to be layde to the disease, called Erysipelas, or wild fire, and all other inflammations, and with womans milke and Saffron, it swageth the payne of the gowte.

The same put into the fundement, as a Suppositorie, bringeth or causeth sleepe.

To conclude, in what manner soeuer Opium be taken, eyther inwardly or outwardly, it causeth sleepe, and taketh away paynes. Yet ye must take heede, to vse it euer with discretion.

❦ *The Danger.*

The vse of Poppie is very euill and dangerous, and especially Opium, the which taken excessiuely, or to often applyed vpon the flesh outwardly, or otherwise without good consideration and aduisement, it wyll cause a man to sleepe to muche, as though he had the Lethargie, which is the forgetful sicknesse, and bringeth foolish and doting fansies, it corrupteth the sense and vnderstanding, bringeth the Palsie, and in fine it killeth the body.

❦ *The Correction.*

Whan by great necessitie ye are forced to vse Opium, mixe Saffron with it, for it shall let, and somewhat hinder the euill qualitie of Opium, in suche sort as it shall not so easily do harme, as it woulde, if Saffron were not mingled with it. See Turners Herbal for the remedie against Opium lib. 2 fol. 76.

Of Red Poppie, or Cornerose. Chap. lxxxij.

❦ *The Kindes.*

There be two sortes of red Poppie, or Cornerose, the great and the small, differing onely in leaues, but the flowers are lyke one another.

❦ *The Description.*

THE smal Cornerose, or wild Poppie, hath smal rough branches, the leaues be somewhat long, toothed rounde about, not muche differing from the leaues of the other Poppie, sauing that they be muche smaller, and not smothe, but rough. The flowers be of a faire red colour, not differing in figure from the flowers of the

Oo other

other Poppie with blacke thids in the midle. After the falling of the flowers, there rise heades muche smaller then the heades of Poppie, and in proportion longer, wherein is conteyned blacke seede. The roote is long and yelowish.

Papauer Rhœas.
Shadowie Poppie, or red Poppie.

Papauer Rhœas alterum.
Cornerose or shadding Poppie.

2   The great Cornerose hath large leaues, very muche iagged, or rather rent, lyke to the leaues of white Senuey, but alwayes longer and rougher. The stalkes, flowers, and knoppes, or heades, are lyke to the smaller Cornerose. The roote is great, and whiter then the roote of the lesser Cornerose.

❧ *The Place.*

The Cornerose groweth amongst the Wheate, Rye, Otes, and Barley. The least is most common.

❧ *The Tyme.*

Cornerose flowreth in May, and from that time foorth, vntill the ende of sommer.

❧ *The Names.*

This kinde of wilde Poppie is called in Greke μήκων ῥοιάς: in Latine Papauer erraticum, Papauer fluidum, and Papauer Rhœas: in some Shoppes Papauer rubrum: in frenche *Coquelicoc*, or *Ponceau*: in high Douch Klapperrosen, Kornrosen: In base Almaigne Clapperroosen, and Rooden Huel, or wilden Huel. And it is not without cause to be doubted, whether the second Cornerose be a kinde of Ἀργεμόνη, Argemone, or no.

❧ *The Nature.*

Corneroses do coole and refreshe also, and are of complexion muche lyke Poppie.

❧ *The*

❧ *The Vertues.*

Fiue or sixe heades of wilde Poppie, or Cornerose boyled in wine & dronke, causeth sleepe. Like vertue hath the seede taken with hony.

The leaues with the greene heades brused togither, are very good to be layde vpon all euil hoate swellinges, and vlcers, and vpon Erysipeles, or wild fire, as the other Poppies are.

Of Horned or Codded Poppie. Chap.lxxxij.

And Hypocoum forte.

Papauer Corniculatum.   Papaueris corniculati alia species.
Horned Poppie, yellow Poppie.   **Horned Poppie, an other kind.**

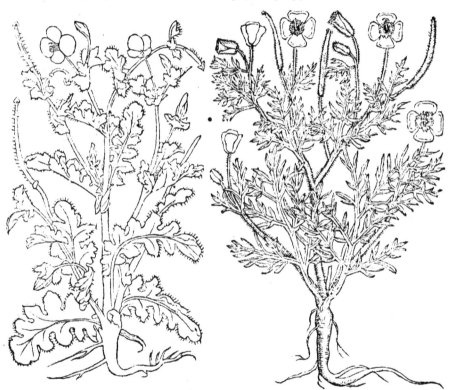

❧ *The Description.*

1. The Horned Poppie his leaues be very muche cut and clouen, not much vnlyke the leaues of the other Poppie, but more rough, and heary, lyke the leaues of Cornerose. The stalkes be round & somwhat rough also, wherevpon growe yellow flowers, made of foure leaues, the whiche falling away, they bring foorth long, narrow huskes, or coddes, something crooked, wherein the seede is conteyned. The roote is great and thicke, and abideth winter, bringing foorth euery yere newe leaues, and stemmes.

There is yet two other sortes of this Poppie, as some men of good knowledge do testifie, the whiche are very common in Spayne.

2. The one hath his leaues, stalkes, & coddes, altogither like to the aforesayd,

Do ij   sauing

The thirde Booke of

sauing that his flowers be not yellowe, but shining red: but for the rest, the flower is agreable with the proportion of the yellow.

3 The seconde is lyke to the others, sauing that it is muche lesse in leaues, stalkes, flowers, and coddes. And the flowers be neyther yellow, nor red, but of a faire blewe violet colour, parted lykewise into foure leaues.

4 There is founde in some places of Fraunce, a kinde of herbe very fayre, the whiche may be very wel brought vnder this Chapter, bycause it is lyke to the herbes described in the same. First it hath large leaues finely iagged & white, lyke the leaues of Rue, the whiche do partly lye vpon the grounde, and partly are lifted vp from the earth: amongst the which cometh vp a stalke or twayne, set by certayne spaces, with the lyke leaues, but smaller, and diuided towardes the toppe, into other smal branches, whiche bring foorth a yellowe flower with two leaues onely, in the midle whereof, ye may see a thing like to a little clipper, the which is nothing els, but the huske or codde, and afterwarde it waxeth long, & hath within a reddish sede. The roote is white and tender, hauing a number of threddes.

Corniculati Papaueris peregrina species. A strange codded Poppie.

❧ *The Place.*

1 Horned Poppie groweth of his owne kinde, by the sea side in rough places (as Dioscorides sayth) in this Countrie the Herboristes do set and sowe it in their gardens.

2 The other two kindes are founde in Spayne by the Sea coaste, amongst Corne, and by the high wayes.

3 The thirde groweth about Monpellier, amongst the wheate and Otes.

4 The fourth is founde in some places of Languedoc, as neare about Vouer, where as there is great store in the feeldes, that are by the high wayes.

❧ *The Tyme.*

Horned Poppie flowreth in July and August.

Hypecoum flowreth in April, and the seede is rype in June.

❧ *The Names.*

This kinde of Poppie is called in Greeke μήκων κερατῖτις: in Latine Papauer cornutum, and of some Apothecaries that are ignorant Memitha, whereunto it is nothing lyke: in Frenche *Pauot cornu*: in high Douche Gehornter Magsamen, and Geel Olmagen: in base Almagne Geelen Huel: in English Horned Poppie.

Some of the learned sort do thinke, that this herbe is a kinde of Papauer Corniculatum, that is to say, Horned Poppie, described by Dioscorides in his fourth booke. Some woulde haue it Papauer spumeum, described of the same Dioscorides in the same place. But if it may be lawfull for me to giue a iudgement

the Historie of Plantes. 437

ment aswel as the rest, it shalbe neither of those herbes: but rather that Hypecoum of Dioscorides, named in Greke ὑπήκοον, and ὑποφεωρ: for all the signes and tokens do agree very well with the same.

❧ *The Nature.*

Horned Poppie is hoate and drie in the thirde degree.

If the fourth kinde be Hypecoum, it shoulde be colde and drie in the thirde degree, not muche differing from Poppie, as Galen sayth.

❧ *The Vertues.*

The roote of Horned Poppie boyled in water vntil halfe be consumed, provoketh vrine, vnstoppeth the liuer, and it is giuen to drinke with great profite to such as make grosse and thicke vrine, and to such as are diseased in the liuer, and that haue any greefe in their raynes, their lining, or hanche. 𝔄

The seede of this Poppie, taken in quantitie of a spooneful, looseth the belly very gently, and purgeth fleme. 𝔅

The leaues and flowers brused or pound, and afterward layd to old sores, and rotten vlcers, clenseth them wel. 𝔆

Of Mandrake, or Mandrage. Chap. lxxxiij.

❧ *The Kyndes.*

MAndrake (as Dioscorides writeth) is of two sortes, that is to say, The white and the blacke. The white is called ỹ male Mandrake, the whiche is very well knowen. The blacke is called the female Mandrage, the whiche is not yet muche knowen.

❧ *The Description.*

THe white Mandrake hath great large leaues, of a whitish greene colour, thicke, and playne, spread vpon the ground, not much differing from the leaues of Beetes, amongst the whiche there commeth vp, vpon short small and smooth stemmes, fayre, yellowe, round apples, and of a strong sauour, but yet not vnpleasant. The roote is great and white, not muche vnlyke a Radishe roote, diuided into two or three partes, and sometimes growing one vpon another, almost lyke the thighes and legges of a man.

The blacke or female Mandrake, hath likewise no vpright stemme; his leaues be in lyke manner spread abroade vppon the grounde, narrower and smaller then the leaues of lettise, of an vnpleasant smel or sauour. The apples be pale, in figure lyke the Sorbappel or Corne, by halfe lesse then the apples of the Malemandrage. The roote is blacke without, and white within, clouen beneth into two or three diuisions or cliftes folding one vpon another. It is smaller then the roote of the male.

¶ *The Place.*

Mandrage groweth willingly in darke

Mandragora mas.
The male Mandrake.
Mandragora fœmina.
The female Mandrake.

The thirde Booke of

and shadowie places. It groweth not of him selfe in this Countrie, but ye shall finde it in the gardens of some Herboristes, the whiche do set it in the sonne.

✿ *The Tyme*.

The Apples of Mandrage, in this Countrie be ripe in August.

✿ *The Names*.

Mandrage is called in Greeke μανδραγόρας: in Latine Mandragoras, of some Circæa, and Antimalum, and of Pythagoras also Anthropomorphos, bycause that the rootes of this herbe are lyke to the lower partes of man.

The first kinde is called Mandragoras mas, of some (as Dioscorides saith) Morion: in Frenche *Mandragore masle*: in high Douch Alraun mennlin: in Neather Douchlande Mandragora manneken, or Alruyn manneken: in Englishe White Mandrake, and Male Mandrage.

The other is called Mandragoras fœmina, of some Thridacias: in Frenche *Mandragore femelle*: in Almaigne Alraun weibling, and Mandragora wijfken: in Englishe Blacke Mandrake, and Female Mandrage.

✿ *The Nature*.

The roote of Mandrake, and especially the barke, is colde and drie euen harde to the fourth degree, the fruite is not so colde, and it hath some moysture adioyning.

✿ *The Vertues*.

The iuyce drawen foorth of the rootes of fresh Mandrake, dried, and taken in a very small quantitie, purgeth the belly vehemently from fleme, and blacke melancholique humours, euen lyke the roote of blacke Hellebor.

It is good also to be put in Collyres, and medicines, that do mitigate the paynes of the Eyes: and being put vnder as a Pessarie, it draweth foorth the Secondine, and the dead childe.

A suppositorie made of the same, and put into the fundement, causeth sleepe.

The greene and fresh leaues of Mandragoras, pound with parched barley meale, are good to be layde vnto al hoate swellinges and vlcers, and they haue vertue to dissolue, and consume al swellinges and impostemes, if they be brused and layde thereupon.

It is also good to put of the roote vpon hoate vlcers & tumours: and with Oyle and hony, it is good, it is good to be layde to the bitinges of venemous beastes.

The wine wherein the roote of Mandrage hath ben stieped or boyled, causeth sleepe, and swageth all payne, wherefore men do geue it (very wel) to such as they intende to cut, sawe, or burne, in any part of their bodies, bycause they shal feele no payne.

The smel of the apples causeth sleepe, but the iuyce of the same taken into the body doth better.

✿ *The Danger*.

It is most dangerous to receiue into the body, the iuyce of the roote of this herbe, for if one take neuer so little more in quantitie, then the iust proportion which he ought to take, it killeth the body. The leaues and fruit, be also dangerous, for they cause deadly sleepe, and peeuish drowsines like Opium. *See Turners remedie agaynst this euill in the Chapter of Mandrage*.

Of Madde Apples, or Rage Apples. Chap. lxxxv.

✿ *The Kindes*.

There be two kindes of Amorus, or Raging loue apples. The one beareth apples of a purple colour, the other pale or whitishe, in all thinges els one lyke to the other, as in making, fashion, stalkes, leaues, and flowers.

the Historie of Plantes. 439

✤ *The Description.*   Mala insana.

This plante hath a round stalke or stemme of a foote high, bearing broade browne greene leaues, almost lyke to the leaues of Dwale or deadly Nightshade, but a litle more rougher, amongst the whiche growe the flowers vpon short stemmes, whiche do turne afterward into a great, rounde, long fruite, almost like an apple, full of seede within as the Cowcumber, & of colour outwardly somtimes browne as a Chesnut, somtimes white, or yellow. The roote is full of laces like threddes.

¶ *The Place.*

Apples of loue, grow not of their owne kinde in this Countrie: but the Herboristes do set and mainteyne them in their gardens, as Cowcumbers & Gourdes, with the which they do spring, and vanish perely.

✤ *The Tyme.*

This plant flowreth in August, and his fruite is ripe in September.

¶ *The Names.*

They be called nowe in Latine Mala insana: in French Pommes D'amours: in base Almaigne, Verangenes: in high Douche, according to the Latine name Melantzan, and Doll opffel, that is to say, Raging or mad Apples, also they be called in English Amorus Apples, and Apples of loue.

✤ *The Nature.*

These apples be of complexion colde and moyst lyke Gourdes.

✤ *The Vertues.*

They be not vsed in medicine, but some do prepare and trim them with oyle, pepper, salt, and vineger, for to eate. But it is an vnholsome meate, ingendring the body full of euill humours.

Of Amorus Apples or Golden Apples.
Chap. lxxxvi.

✤ *The Kyndes.*

These strange Apples be also of two sortes, one red, and the other yellowe, but in all other poyntes they be lyke as in stalkes, leaues, and growing.

✤ *The Description.*

These apples haue rounde stalkes of a gray or ashe colour, and hearie: three or foure foote long full of branches. The leaues be great, broade, and long, spread abroade vpon euery side, and deepely cut, almost lyke the leaues of Agremonie, but muche greater and whiter. The flowers are yellowishe, growing vpon short stemmes, fiue or sixe togither, and when they are fallen, there come in their places great flatte apples, bollen or by certayne

Do iiij   spaces

Poma Amoris.

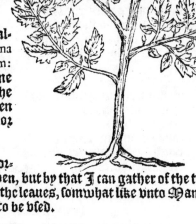

ſpaces bunched out, on the ſides, and of colour ſometines red, ſometimes white, and ſometimes yellowe, lyke Oreges, or Mandrake apples, wherin is conteined the ſeede. Al the herbe is of a ſtrange ſtinking ſauour, and it muſt be ſowen euery yere as the Cowcombers be.

❧ *The Place.*

This is a ſtrange plante, and not founde in this Countrie, except in the gardens of ſome Herboriſtes, where as it is ſowen.

❧ *The Tyme.*

This herbe floweth in July and Auguſt, his apples be ripe in Auguſt and September.

❧ *The Names.*

This ſtrange plante, is nowe called in Latine Pomum Amoris, Poma Amoris, and of ſome Pomum aureum: in Frenche *Pommes dorées*, and of ſome alſo *Pommes D'amours*: in high Douche Golt oﬀel: in baſe Almaigne Gulden appelen: in Engliſh Apples of loue, or Golden Apples.

❧ *The Nature and Vertue.*

The complexion, nature, and working of this plante, is not yet knowen, but by that I can gather of the taſte, it ſhould be colde of nature, eſpecially the leaues, ſomwhat like vnto Mandrake, and therefore alſo it is dangerous to be vſed.

Of Apples of Perowe. Chap. lxxxvij.

❧ *The Deſcription.*

THE apple of Perow hath a rounde ſtalke, about two foote long, the leaues be grayiſhe, almoſt lyke the leaues of Solanum, or Nightſhade, but greater, eſpecially ÿ lowmoſt next the roote, the flowers be white, rounde, and holowe as a bell, of a pleaſant ſauour like the white Lilie, and when they are fallen, there commeth fruite, rounde as an apple, of a greene colour, beſet rounde about with many prickley thornes, and therefore they call it Thorne apple, ful of ſeede within lyke the Apples of loue. The roote is ful of threddie ſtringes, interlaced, wouen, and winded one in another.

❧ *The Place.*

The apple of Perow, is a ſtranger alſo, the whiche is not to be found except in the gardens of the Herboriſtes, and yet not often.

❧ *The Tyme.*

Theſe apples are in flower, in May and June.

❧ *The*

the Historie of Plantes. 441

Stramonia.

✻ *The Names.*

This strange plant is called of the Italians, Stramonia, and Pomum spinosum, of some Corona regia: at Venize Melospinus, and Paracoculi: in Frenche *Pomme de Perou, or Pomme espineuse*: in high Douche Stech opffel, Rauch opffel, & Stecheud opffel: in base Almaigne Doren appel: we may call it in Englishe, The apple of Perrow, Thornie apples, Prickle apples, and Stramonia.

✻ *The Nature.*

The complexion, vertue, and facultie of this plant, is not yet knowen.

Of the Balme Apple or Momordica. Chap.lxxxviij.

✻ *The Kindes.*

By the name of Balsamine, you must now vnderstand two sorts of apples, or fruites, varying muche one from another, both in figure and growing. The one is called the Male Balsem, or Balme apple. The other is called the female Balsem apple.

✻ *The Description.*

1. THE first kinde of these Maruelous Apples, hath long branches and smal, with litle claspers or tendrelles, wherewithal it taketh holdefast vpon hedges, trees, poles, and rayles, agaynst whiche it is planted. The leaues be large and round, cut in round about with certayne deepe cuttes, almost like the vine leaues, but smaller. The flowers be pale, the fruite round, sharpe poynted, and rough without, like the fruit of the wild Cowcumber, greene at the beginning and afterwarde red. In these apples are founde broade, rough, and blackishe seede. The roote putteth foorth many branches, or moores, spread abrode here and there.

2. The seconde kinde hath a thicke stalke or stemme, of a reddishe colour lyke Purselane, about a foote high or somewhat more. The leaues be long and narrowe, and not muche vnlyke the leaues of Wythie, a little toothed or creuised about. The flowers be fayre, of an incarnate or liuely colour changing vpon blewe, with a little tayle turned agayne, not muche differing from the flowers of Larkes Spurre. The fruite or apple is rounde, sharpe at the point, and rough without, greene at the beginning, but after yellowishe pale, the whiche openeth it selfe whan it is ripe, and the seede falleth out, the which is very well lyke vnto a Fetche. The roote is lyke the abouesayde.

✻ *The*

Charantia. **Balſam apple, the male.**
Maruelous apples.

Balſaminum.
Balſam the Femal.

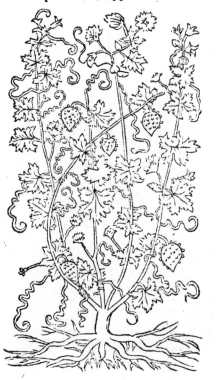 

❊ *The Place.*

Theſe two ſtrange herbes, are founde in this Countrie, in the gardens of certayne Herboriſtes.

❊ *The Tyme.*

Theſe plantes do flower in July and Auguſt, and their fruite is ripe in Auguſt and September.

❊ *The Names.*

The firſt kind of theſe herbes is called in Italian Charantia, Balſamina, Momordica, and in ſome places, Pomum Hieroſolymitanum: in Frenche *Pomum mirabile*, *Pomme de merueille*, *& Merueille maſle*: in high Douch Balſam opffel mennlin: in baſe Almaigne Balſam appel manneken: in Engliſhe we may call it Momordica, and the Male Balſam apple.

The other kinde is called Balſaminum, and is not Charantia, Balſamina, or Momordica, as ſome do thinke. The high Douchmen do call it Balſam opffel weiblin, that is to ſay, *Merueille femelle*: and in baſe Almaigne, Balſem appel wijfken: in Engliſh, the female Balſam apple.

❊ *The Nature.*

The complexion of theſe apples, according to the iudgement of ſome, is hoate in the firſt degree, and drie in the ſeconde.

❊ *The Vertues.*

A man ſhal find in writing, that the Maruelous apples, are named Charantia, for the vertues folowing.

The leaues of Charantia taken in wine, are a present remedie for al paines, aswell within the body as without, and doth comfort the strength of suche as take it, in suche sort, that no griefe may happen to them.

The same made into powder and dronke in wine, doth cure and heale all inward woundes, that is to say, of the bowels or entrayles, and are very profitable agaynst the Colique.

The onely iuyce of the leaues, put vpon the teeth, healeth the ache of the same.

The Oyle whiche men drawe foorth of the fruit of the same in the Sonne, closeth vp al woundes, aswageth all paynes, helpeth Crampes, and the drawing togither, or shrinking of sinewes, being layde to the places hurt & greeued.

The same is also good agaynst the vlcers of the breast, and paynes of the Matrix: causing women to be easily deliuered and without great payne, if it be layde to or annoynted vpon their bellyes.

The same cureth al vlcers, hurtes, impostumes, and gatherings togither of euil humours in the Matrix, being cast into the same, with an instrument made for that purpose.

The same with Cotton layde to the fundement, healeth the Hemorrhoides, and swageth all paynes of the same.

Burstinges be also holpen, when the diseased place is annoynted with the Oyle aforesaid, but for the same purpose, ye must giue the powder of the leaues to drinke in wine.

The Oyle of Momordica, or Maruelous Apples, made as is aforesayd, putteth away al scarres and blemishes, if it be applyed thereto.

There is nothing founde written of the properties of the female Balsam, bycause they be not knowen.

Of Nightshade or Morelle. Chap. lxxix.

❦ The Description.

Nightshade hath rounde stalkes of a foote long, full of branches. The leaues are blackishe, large, soft, and full of iuyce, lyke to the leaues of Basil, but muche greater: the litle flowers be white, hanging three or foure one by another. After that they be passed, there come in their places, berries hanging togither like the fruit of Iuie, of colour most commonly blacke, whan they be ripe sometimes red, and somtimes also yellow. The roote is white and ful of heary threddes.

❦ The Place.

Nightshade is very common in this Countrie, about olde walles, vnder hedges, about pathes, and hollow wayes, and al about the borders of feeldes, and in the gardens of pot herbes.

❦ The Tyme.

This herbe floureth from the moneth of Iune, during all the sommer, and in this space deliuereth his seede.

❦ The Names.

This herbe is called in Greke στρύχνο, ή τρύχνο, και στρύχνο κηπαίο: in Latine Solanum, and Solanum hortense: In Shoppes Solatrum, and of some Morella, Vua lupina, and Vua Vulpis: in Frenche *Morelle*: in high Douche Nachtschat: in base Almaigne Nascaye, and Nachtscade: in Englishe Nightshade, Petimorel, and Morel.

❦ The Nature.

Morell is colde and drie in the seconde degree.

The

Solanum hortense.

❧ *The Vertues:*

A The greene leaues of Petiemorel, or Nightshade, pounde with parched barley meale, is maruelous profitable to be applied, or layd to Saint Antonies fire, to corrupt and running vlcers, and all hoate inflammatiós. And for the same purposes men make an oyntment of the iuyce of the same, with Oyle of Roses, Ceruse, and Littarge.

B The same pound by it selfe and layd to, is good against paynes in the head, and is very profitable against a hoate stomacke, and all hoate distemperature of the eyes, the eares, the liuer, the melte, or Splene, and the bladder, to be layde to outwardly vpon the places of the same.

C The same with Salt, dissolueth the aposteines and swellinges behinde & about the eares, named Parotidas, if it be layde therevnto after the forme of a playster.

D The iuyce of Nightshade, mingled with the white of an egge, is good to be layde vpon the forehead, against inflammatiós, rednesse, reumes, fluxions, and all other hoate diseases of the eyes.

E The same dropped into the eares, swageth the paynes of the same, & laid to with Cotton, in the manner of a mother Suppositorie, stayeth the inorditate course of womens issues.

F To conclude, Galen affirmeth, that Nightshade or Morell is very good agaynst al diseases and Accidents, wherein is any neede of cooling and restrayning.

Of red Nightshade, Winter Cherrie and Alcakengy. Chap. xc.

❧ *The Description.*

THE common Alcakengie, beareth slender stalkes, leaues lyke Petie Morel, but muche larger and greater. The flowers be pale, greater, but not so white as the flowers of Nightshade or Petimorel, & whan they perish, they bring foorth rounde balles, or blasted bladders, hollow, close, greene at the beginning, but afterward red: in the sayd bladders be rounde red beries, full of seede, flat, and yellowish. The roote is smal, creeping along, and casting foorth new euery yere, and in sundry places it putteth foorth newe shutes, and tender stalkes.

Bysides this there is founde a strange kinde, which is also taken for Alcakengie, the which hath smal and tender stalkes, the leaues be somewhat long, creuised & deepely cut round about. The flowers be white as snowe, bringing foorth also bladders, or rounde blasted balles, at the beginning greene, but afterwarde blackishe: wherein groweth blacke beries, about the quantitie of a pease. The roote is small and threddie.

❡ *The*

the Historie of Plantes. 445

Veſicaria vulgaris.
Alkakengie oȝ winter Cherie.

Veſicaria peregrina.
Blacke winter Cherie.

❉ *The Place.*

Alkakengie groweth in ſome wooddes of this Countrie, about hedges and lowe moyſt places, and is much planted in gardens.

❉ *The Tyme.*

The little bladders, and the fruite of this plante are rype in Auguſt and September.

❉ *The Names.*

1. Alkakengie is called in Greeke ςρύχνο ἀλκάκαβο, καὶ φυσαλίς: in Latine Veſicaria, of Plinie Veſicula, of ſome Callion, in Shoppes Alkakengie: in French *Alquequanges*, and *des Coquerelles*: in high Douche Schlutten, Boberellen, Juden Kirtzen, Teuffels Kirſen, Juden Hutlin, and Rot Nachtſchad: in baſe Almaigne, Criecken van ouer ſee, That is to ſay, Beyondſea Cheries: in Engliſh it is called Nightſhade, Alkakengie, and Winter Cheries.

2. The other ſtrange kinde is called of men of this time, Veſicaria peregrina, and Veſicaria nigra: in French *Pois de merueilles*: in high Douch Munchs kopffin, Schwarte Schlutten, and Welſch Schlutten: in baſe Almaigne Vremde Criecken van ouer zea, and Swerte Criecken van ouer zea.

❉ *The Nature.*

The leaues of Alkakengie are colde lyke Petimozell. The fruite is not ſo colde. Moreouer, it is of ſubtill partes.

The complexion of the ſtrange Alkakengie, is yet vnknowen.

The Vertues.

The leaues of Alkakengie are good for all suche thinges, as the leaues of a Petimorel serue for, but not to be eaten.

The Cheries or fruite of Alkakengie, openeth the stoppinges of the liuer, & the kidneyes, clenseth the bladder, and prouoketh vrine. Therfore they be very good against the Jaunders, the ache & greefe of the raines and bladder, against the difficultie & sharpnesse of making water, and against the stone and grauell.

Of great Nightshade, or Dwale. Chap.xci.

The Description.

Mandragoras Theophrasti.
Solanum lethale.

This noughtie and deadly plant is taken for a kinde of Solanum, bycause it doth somewhat resemble it. It hath round blackish stalkes of two or three foote high, or more, vppon the whiche growe great broade leaues, somewhat rough, greater and larger, yea & blacker then the leaues of Morel, the flowers be of a browne colour, fashioned lyke to litle holow belles, after the whiche there comme vp great round beries, euery one vpon a stalke by him selfe, about the bignesse of a Cherie, greene at the beginning, but afterwarde when they ware towarde rypenesse, they be of a faire blacke shining colour, within the sayd beries is coteyned a litle browne seede. The roote is great, putting foorth newe euery yere, and bringing foorth a number of newe stalkes.

The Place.

This herbe is founde in some places of this Countrie, in woods, and hedges, and in the gardens of some Herboristes.

The Tyme.

The fruit or beries of this venemous Solanum, are ripe in August.

The Names.

This herbe is nowe called Solanum lethale: in Shoppes Solatrum mortale: in Frenche Solanum mortel: in high Douche Dollkraut, Seukraut: in base Almaigne Groote Nascaye, and Dulcruyt, or Dulle besien. This is not Solanum Manicum, neither Solanum Somniferum, neither yet Mandragoras Morion, the whiche Dioscorides describeth. But it shoulde rather seeme to be that kinde of Mandrage, whereof Theophrast speaketh in his sixth booke the second Chapter. And for that cause it may be well called Mandragoras Theophrasti.

The Nature.

The leaues and fruit of this herbe are very cold, euen in the fourth degree.

The working.

The greene and fresh leaues of this deadly Nightshade, may be applyed outwardly as the leaues of Petimorel to S. Antonies fire, and the lyke hoate inflammations

flammations, but it must be done by great aduise, seing that this Solanum cooleth agayne more strongly than the common Nightshade.

❧ The mischeuous Danger.

The fruite of this Solanum is deadly, and bringeth such as haue eaten therof into a deepe sleepe, with rage and anger, the which passion leaueth them not, vntill they die, as it hath ben seene by experience, as well in Almaigne, as at Mechlen, vpon some children who haue eaten of this fruite, thincking that it was not hurtful. Wherfore eche man ought to take heede, that they plant not, neyther yet suffer in their gardens, any suche venemous herbes, especially of suche sortes whiche beare a faire and pleasant fruite, as this last recited kinde doth : or if they wil haue it in their gardens, then at the least way, they ought to be carefull, to see to it, & to close it in, that no body enter into the place where it groweth, that wilbe entised with the beautie of the fruite to eate thereof, as it commeth very oftentimes to passe vnto wemen and young children.

Of Solanum Somniferum, & Manicum.   Chap. xcij.
Sleeping Nightshade.   Furious Nightshade.

❧ The Kindes.

THE deady Nightshade, whereof I haue written in the former Chapter, causeth me yet to remember two other kinds of Solanum, or Morel, described of the Ancientes, and of Dioscorides. Whereof one is called Solanum Somniferum, that is to say, Sleeping Nightshade: The other is called Solanū Manicum, that is to say, Mad, or Raging Nightshade.

❧ The Description.

Solanum Somniferum, that is, Sleeping Nightshade, hath grosse and harde stalkes, vpon the whiche groweth great broade leaues, almost like to the leaues of the Quince tree. The flower is great and red, the fruite as yellowe as Saffron, conteyned in puffed balles or coddes. The roote is long and wooddy, and on the outside browne.

The other Solanum called Manicum, that is to say, Madde or Raging, hath leaues like Senuie or Mostarde, but greater, and somewhat like to the leaues of the right Branke Vrsine, called in Latine Acanthus, the which shalbe described in the fifth booke. It bringeth foorth from one roote ten or twelue stalkes of the height of two or three foote, at the toppe of the sayd stalkes or branches groweth a rounde head of the bignesse of an Olyue, and rough like the fruit of the Plane tree, but smaller & longer. The flower is blacke, & when it perisheth, it bringeth foorth a littie grape, with ten or twelue beries, like the fruite of Iuie, but playner, and smother like the berries of grapes. The roote is white and thicke of a cubite long, and holow within. To this Description agreeth that kinde of strange Mallowe, whiche is called Malua Theophrasti, and Alcea Veneta, the whiche shalbe described in the xxvij. Chapter of the fifth part of this Historie.

❧ The Place.

Solanum Somniferum, according to the opinion of Dioscorides, groweth in stony places, lying not farre from the Sea.

Solanum Manicum, groweth vpon high hilles, whose situation or standing is agaynst the Sonne.

❧ The Names.

The first kinde of these two herbes, is called in Greeke ςρύχνος ὑπνωτικός: in Latine Solanum Somniferum, that is to say in English, Sleeping Nightshade, of some Halicacabon, Dircion, Apollinaris minor, Vlticana herba, and Opsago.

The second kind is called in Greeke ςρύχνΘ μανικός: in Latine Solanum Manicum: that is to say, Furious or raging Solanum, or Nightshade, of some Persion, Thryon, Anydron, Pentadryon, and Enoron.

❋ The Nature.

The sleeping Nightshade or Solanum, is colde in the thirde degree, approching very neare vnto the nature or complexion of Opium, but muche weaker.

The roote of ye mad or furious Solanum or Nightshade, especially the barke thereof, is drie in the thirde degree, and colde in the seconde, as Galen writeth.

❋ The Vertues.

The fruit of Solanum Somniferum, causeth one to make water, and is very good agaynst the Dropsie, but ye may not take aboue twelue of the beries at once: for if you take moe, they will do harme.   A

The iuyce of the fruit is good to be mixed with medicines, that do asswage and take away payne.   B

The same boyled in wine, and holden in the mouth, swageth tooth ache.   C

The roote of raging Solanum, especially the barke thereof, is very good to be rubbed and layd to Saint Antonies fier, in forme of a playster, and vpon vlcers that be corrupt and filthy.   D

It is not good to take this kinde of Solanum inwardly.   E

❋ The Danger.

If you giue more then twelue of the beries or grapes of Solanum Somniferum, it will cause suche as you do giue it vnto, to raue, and waxe distracte or furious, almost as muche as Opium.

The roote of Solanum Manicum, taken in wine to the quantitie of a Dram, causeth idle and vayne imaginations: & taken to the quantitie of two Drams, it bringeth frensie and madnesse, whiche lasteth by the space of three or foure dayes: and if foure Drammes thereof be taken, it killeth.

Of Henbane. Chap. xciij.

❋ The Kyndes.

Of Henbane are three kindes (as Dioscorides and others haue written) that is, the blacke, the yellowe, and the white.

❋ The Description.

The blacke Henbane hath great stalkes and softe, the leaues be great broade, soft, gentle, woolly, grayishe, cut and iagged, especially those at the lowest part of the stalke, and neare the roote: for they that grow vpon branches, are smaller, narrower and sharper. The flowers be browne-blewe within, and lyke to little belles, and when they fall of, there folowe round huskes, like litle pottes, couered with smal couers, inclosed within with small rough velmes or skinnes, open aboue, and hauing fiue or sixe sharpe pointes. These pottes or cuppes are set in a rewe, one after another, alongst the stalkes. Within the sayd pottes is contepned a browne sede. The roote is long, sometimes as great, as a finger.

The yellowe Henbane hath broade whitishe and soft, or gentle leaues, neither carued nor cut, almost like the leaues of Mortal Nightshade, but greater, whiter, and softer. The flowers be of a feynt or pale yellow colour, and round, the whiche being past, there come in their steede rounde huskes, almost like litle cuppes, not much differing from the cuppes or huskes of blacke Henbane, wherein is the seede, which is like to the seede of other Henbanes. These small pottes do growe and are inclosed in a rounde skinne, but the same is gentle and pricketh not. The roote is tender. This kinde of Henbane, hauing once borne his seede, dyeth before winter, and it must be sowen yerely.

the Historie of Plantes.

Hyoscyamus niger. **Blacke Henbane.**   Hyoscyamus luteus.   **Yellowe Henbane.**

Hyoscyamus albus. **White Henbane.**

3 ❧ The thirde kinde of Henbane, called the white Henbane, is not much vnlike to the blacke, sauing that his leaues be gentler, whiter, more woolly, and much smaller. The flowers be also whiter, & the seede which is inclosed in litle cups, is lyke the seede of blacke Henbane, but the shel or skin that couereth the huskes is gentle and pricketh not. The roote of this kind is not very great. It dieth also before winter, and it must be likewise newe sowen euery yere.

❦ *The Place.*

The Henbane doth growe very plenteously in this Countrie, about wayes & pathes, and in rough & sandy places.

The two other kindes, þ Herboristes do set in their gardens, whereof þ white sort groweth of his owne kind, as Dioscorides saith, vppon dunge heapes, or mixens by the sea coast. In Languedoc they haue scarse any other, sauing the white kind.   Pp iij   ❦ *The*

The thirde Booke of

❧ *The Tyme.*

These three kindes of Henbane do flower in July and August.

❧ *The Names.*

This herbe is called in Greeke ὑοσκύαμος, καὶ ἀπολλινάρις: in Latine Hyoscyamus, Apollinaris, and Faba suilla, of some Dioscyamos, that is, Iouis faba, Fabulonia: of Apuleia Symphoniaca, Cicularis, Remenia, Faba Lupina, Mania: of the auncient Romaines, and Hetruscians, or Tuscans, Fabulum: of the Arabian Physitions Altercum, and Altercangenum: of Mattheus Spluaticus, Deus Caballinus, and Cassilago: of Jacobus Manlius Herba Pinula: of some others Canicularis, and Caniculata: in French *Iusquiame*, or *Hanebane*: in high Almaigne Bilsamkraut, Sewbon, and Dolkraut: in neather Douchlande Bilsen, and Bilsencrupt.

The first kind is called bycause of his darkish browne flowers, Hyoscyamus niger, that is to say, blacke Henbane.

The seconde is called Hyoscyamus luteus, that is to say, Yellowe Henbane, bycause it beareth yellow flowers.

The thirde whiche hath white flowers, is called Hyoscyamus albus, that is to say, White Henbane.

❧ *The Choise.*

The white Henbane is best to be vsed in medicine. The two other be not so good, especially the blacke whiche is most hurtfull.

❧ *The Nature.*

The seede of the white Henbane, and the leaues are cold in the third degree. The two other kindes are yet more colder, almost in the fourth degree, very hurtfull to the nature of mankinde.

❧ *The Vertues.*

The iuyce drawen foorth of the leaues and greene stalkes of Henbane, and afterwarde dryed in the Sonne is very good to be mingled with Colyries, that are made agaynst the heates, rheumes, and humours of the eyes, and the payne in the same, in the eares, and mother.

The same layde to with wheaten meale, or with parched Barley meale, is most profitable against all hoate swellinges of the eyes, the feete, and other partes of the body.

The sede of Henbane is good for the cough, the falling downe of Catarrhes, and subtill humours into the eyes, or vpon the breast against great paynes, the inordinate fluxe of womens issues and al other issue of blood to be taken in the waight of an halfe pennie, or ten graynes with Hydromel, that is to say honied water.

The same swageth the payne of the gowt, healeth the swelling of the genitors or stones, asswageth the swelling of wemens pappes after their deliuerance. If it be bruised with wine, and layde vpon. It may be also put into al emplaysters anodins, that is suche as are made to swage payne.

The leaues alone, or by them selues, pound with parched Barley meale, or mingled with other oyntmentes, emplaisters and medicines swage also all paynes.

If one do washe his feete with the Decoction of Henbane, or if it be giuen in glister, it will cause sleepe. The same vertue hath the seede to be layd to with oyle, or any other liquer vppon the forehead, or if one do but smell often to the herbe and his flowers.

The roote of Henbane boyled in vineger, and afterwarde holden in the mouth, appeaseth the tooth ache.

the Historie of Plantes.

To conclude, the leaues, stalkes, flowers, seede, roote, and iuyce of Henbane, do coole al inflammations, causeth sleepe, and swageth al payne: yet notwithstanding this mitigation of payne doth not continually helpe or remayne: for by suche remedies as consist of thinges that are extreme colde as Opium, Henbane, Hemlocke, and suche other, the disease or paine is not cleane taken away, but the body and greeued place is but onely astonied, or made asleepe for a season, and by this meanes it feeleth no payne. But when they come agayne to their feeling, the payne is most commonly more greeuous then before, and the disease more harder to be cured, by the extreme cooling of the sayde herbes, whiche bring to the sicke (especially to such as be of a colde nature) intolerable Crampes and retractions of sinewes. Therefore these herbes ought not to be vsed for the appeasing of payne, except in time of great neede when the greefe is great and intolerable.

The Danger.

The leaues, seede, and iuyce of Henbane, but especially of the blacke kinde, the which is very common in this Countrie, taken either alone or with wine, causeth raging, and long sleepe, almost like vnto dronkennesse, whiche remayneth a long space, and afterwarde killeth the partie.

The leaues or iuyce taken in to great quantitie, or to often, or layde to any member or part of the bodie hauing no neede, quencheth the naturall heate of the same, and doth mortifie and cause the sayde member to looke blacke, and at last doth putrifie and rot the same, and cause it to fall away.

Of Hemlocke. Chap. xciiij.

The Description.

Cicuta.

Hemlocke hath a high long stalke, of fiue or sixe foote long, great and hollowe, full of ioyntes like the stalkes of fenil, of an herbelike colour, poudered with small redde spottes, almost like the stemme of Dragon, or the greater Serpentarie. The leaues be great, thicke, and small cut, almost like the leaues of Cheruil, but much greater, and of a strong vnpleasant sauour. The flowers be white, growing by tuftes, or spokie toppes, the whiche do change and turne into a white flatte seede. The roote is short, and somewhat holowe within.

The Place.

This noughtie and dangerous herbe, groweth in places not toyled, vnder hedges, and about pales, and in the fresh, cold shadowe.

The Tyme.

Hemlocke flowreth most commonly in July.

The Names.

This herbe is called in greke κώνειον: in Latine Cicuta: in English Hemlocke:

Pp iiij

in Frenche *Cigne*: in high Almaigne Shirling, Wutzerling, wundtscherling, and Weterich: in base Almaigne, Scheerlinck, and Dulle keruel, or Dulle Peterselie: of some vnlearned Apothecaries Harmel, the whiche albeit they haue bene sundrie times warned of their errours by many learned, as Leonicenus, Manardus, and diuers others, yet wil they not leaue, but continue obstinate in in their ignorance, vsing yet dayly in steede of the seede of Rue called in Greke Harmel, the seede of Hemlocke (the whiche they take peruersly for Harmel) and do put it dayly into their Medicines.

The Nature.

Hemlocke is very colde, almost in the fourth degree.

The Vertues.

Hemlocke layd vpon the stones of young children, causeth them to continue in one estate, without waxing bigger. Likewise layde to the brestes of young maydens, do cause them to continue small: neuerthelesse, it causeth suche as do vse it, to be sicke and weake, all the dayes of their liues.

The same layde to and applyed in manner of a playster vpon wilde fire and hoate inflammations, swageth the payne and taketh away the heate, euen as Henbane and Opium doth.

The Danger.

Hemlocke is very euyl, dangerous, hurtful, and venemous, in so much that whosoeuer taketh of it, dyeth, except he drinke good olde wine after it: for the drinking of suche wine, after the receiuing of Hemlocke, doth surmount and ouercome the poyson, and healeth the person: but if one take the wine and Hemlocke togither, the strength of the poyson is augmented, and then it killeth out of hande, insomuche that he is no kinde of wayes to be holpen, that hath taken Hemlocke with wine.

The ende of the thirde parte of the
Historie of Plantes.

the Historie of Plantes. 453

¶ The fourth part of the Historie of
Plantes, treating of the sundrie kindes, fashions, names, vertues, and operations, of Corne or Grayne, Pulse, Thistelles, and suche lyke.

By Rembertus Dodonæus.

Of Wheate. Chap.i.
❧ *The Kindes.*

THE Auncient writers haue described diuers sortes of wheate, according to the places and Countries, from whence it hath ben brought to Rome and other suche great Cities. But suche as make no account of so many kindes, as Columella and Plinie, haue diuided wheate but onely into three kindes: whereof the one is called Robus, the other Siligo. The whiche twayne are winter corne or fruites, and the third Setanium, which is a sommer wheate or grayne. Yet to say the trueth, this is as it were but one sort or kinde, and the diuersitie consisteth but onely in this point, that the one kinde is browner or blacker, and the other sort is whiter and fairer, & the one is to be sowen before winter, and the other after.

Triticum. Wheate.

❧ *The Description.*

Uery kinde of wheate hath a rounde high stemme, strawe, or reede, most commonly many strawes growing frō one roote, euery one hauing three or foure ioyntes, or knottes, greater and longer then barley strawe, couered with two or three narrowe leaues, or grayishe blades, at the highest of the sayd stemme or straw, a good way from the said leaues, or blades, groweth the eare, in which the graine or corne is set, without order, very thicke, and not bearded.

❧ *The Place.*

The wheate groweth in this Countrie, in the beast and fruitful feeldes. ❧ *The Tyme.*

Men sow their winter corne in September, or October, & the sommer corne in March, but they are ripe altogither in July. ¶ *The Names.*

Wheate is called in Grecke πυρὸς: in Latine Triticum: in high Douche Weyssen, & Weytzen: in neather Douchlande Terwe.

1 The first kinde, whiche of Columella is iudged the best, & groweth not in this Countrey, is called Robus, & of Plinie Triticū: in English Red Wheat.

2 The second kinde, which is more light, and whiter, is called in Latine Siligo, & that is our common wheate growing in this Countrie, as we haue euidently declared in Latine, in Historia Frugū, wheras we haue also declared, that our common Rye is not Siligo, whereof Columel and Plinie haue written.

3 The third kind is called in Greke τρίμηνον καὶ σητάνιας: in Latine Setanium, and Trimestre Triticum: in French *Blé de Mars*: in base Almaigne, Zoomer Terwe: in English March, or sommer Wheate. ❧ The

The Nature.

Wheate layde to outwardly as a medicine, is hoate in the firſt degree, without any manifeſt moyſture. But the bread that is made therof, is warmer, and hath a greater force, to ripe, drawe, and digeſt.

The Amylum made of wheate, is colde and drie, and ſomewhat aſtringent.

The Vertues.

Raw wheate chewed in the mouth, is good to be layd to agaynſt the biting **A** of mad Dogges.

The whole wheate is very profitable againſt the paynes of the gowt, whē **B** a man plongeth him ſelfe therein, euen vp to the knees, as ye ſhall reade in Plinie of Sextus Pompeius, who being ſo vſed, was cured of the gowte.

Wheaten meale mingled with the iuyce of Henbane, & layde to the ſinewes, **C** is good againſt the rheumes and ſubtill humours falling downe vppon the ſame.

The ſame layde vpon with vineger and hony (called Oximel) doth clenſe **D** and take away all ſpottes and lentilles from of the face.

The meale of Marche or Sommer wheate, layd to with Vineger, is very **E** good againſt the bitinges of venemous beaſtes.

The ſame boyled lyke to a paſte or pappe, and licked, is very good agaynſt **F** the ſpetting of blood: and boyled with Butter, & Mintes, it is of great power againſt the cough, and roughneſſe of the throte.

The flower of wheaten meale boyled with hony and water, or with Oyle **G** and water, diſſolueth all tumours, or ſwellinges.

The Branne boyled in vineger, is good againſt the ſcuruie itche, and ſprea- **H** ding ſcabbe, and diſſolueth the beginninges of hoate ſwellinges.

The ſayde Branne boyled in the decoction of Rue, doth ſlake & ſwage the **I** harde ſwellinges of womans breaſtes.

The leauen made of wheaten meale, draweth foorth ſhiuers, ſplinters, and **K** thornes, eſpecially from the ſoles of the feete. And it doth open, ripe, and breake al ſwellinges and impoſtumes, if it be layde to with Salt.

Wheaten bread boyled in honied water, doth ſwage and appeaſe all hoate **L** ſwellinges, eſpecially in putting thereto other good herbes and iuyces.

Wheaten bread newe baked, tempered or ſoked in brine or pickle, doth cure **M** and remoue all olde and white ſcuruineſſe, and the foule creeping or ſpreading ſcabbe.

The Amylum or Starche, that is made of wheate, is good againſt the fal- **N** ling downe of rheumes and humours into the eyes, if it be layd therevnto, and it cureth and filleth agayne with fleſh, woundes and holow vlcers.

Amylum dronken ſtoppeth the ſpetting of blood, and mingled with milke, it **O** ſwageth the roughneſſe, or ſoreneſſe of the throte and breſt, and cauſeth to ſpet out eaſyly.

Of the Corne called Spelt or Seia.
Chap. ij.

The Kindes.

Spelt is of two ſortes. The one hath commonly two cornes or ſeedes ioyned togither, whereof eche grayne is in his owne ſkinne, or chaffie couering. The other is ſingle, and hath but one grayne.

The

the Historie of Plantes.

Zea. Far. Spelt.

❧ *The Description.*

Spelt, hath straw, ioyntes, and eares, much lyke to wheate, sauing that the corne therof is not bare as the wheate corne is, but is inclosed in a litle skinne or chaffie huske, from whiche it can not be easily purged, or clensed, except in the myll, or some other deuise made for the same purpose, and whan it is so pylde and made cleane from the chaffe, it is very well lyke to a wheat corne, both in proportion and Nature: in so muche that at the ende of three yeres, the Spelt being so purged, changeth it selfe into faire wheate, whan it is sowen, as Plinie, Theophraste, and diuers other of the Auncientes haue written.

¶ *The Place.*

Spelt requireth a fat and fruitfull grounde well laboured, and groweth in high & open feeldes. In times past, it was founde onely in Grece, but at this day, it groweth in many places of Italie, Fraunce, and Flaunders.

❧ *The Tyme.*

It is sowen in September, and October, lyke vnto wheate, and is ripe in July.

❧ *The Names.*

This grayne is called in Greeke ζἑια: in Latine Zea: of the Auncient Romaynes Semen, and Far, and at this day Spelta: in Frenche *Espeautre*: in high Douche Speltz, and Dinckelkorne: in base Almaigne Spelte: and amongst the kindes of Far, it shoulde seeme to be Venniculum album.

❧ *The Nature.*

Spelt is of Nature like vnto wheate, but somwhat colder, drawing neare to the complexion of barley, and somewhat drying.

❧ *The Vertues.*

The meale of Spelt, with red wine, is very profitable against ý stinginges of Scorpions, and for suche as spet blood.

The same with sweete Butter vnsalted, or with newe Goates suet, doth souple and mitigate the roughnesse of the throte, and appeaseth the cough.

The same boyled with wine and Saltpeter, cureth corrupt and running sores, and the white scurffe of all the body, the payne of the stomacke, the feete, and womens brestes.

To conclude, Spelt in qualitie is very like wheate, and is a good nourishment both for man and beast, as Theophrastus writeth.

The bread thereof is not muche inferior to that is made of wheate, but it nourisheth lesse. Turner lib.2.fol.131.

The fourth Booke of
Of Amilcorne. Chap. iij.

❊ *The Description.*   Amyleum frumentum.

THis grayne is also lyke vnto wheate in the strawe, ioyntes, and growing, but that the eares be not bare or not like wheate, but rough with many sharpe pointed eares or beardes, like the eares of Barley: & the cornes grow by ranges, like to the cornes or graynes of Barley. The seede is also inclosed in little huskes or coueringes, like to spelt, and being clensed and purged from his chaffie huske, it is much lyke to wheate.

❊ *The Place.*

This Corne groweth in many places of Almaigne. ❊ *The Tyme.*

Men do also sow it before winter, and it is cut downe in July.

❊ *The Names.*

This Corne is called in high Douch Ammelkorne, That is to say, in base Almaigne Amelcorne, and in Latine Amyleum frumentum: and is a kinde of Zea, and Far: and it shoulde seeme to be Halicastrum. It may be englished, Amelcorne, or bearded Wheate.

The Nature, and Vertue.

As this grayne is a kinde of spelt, euen so it is very muche lyke vnto it in complexion and working, beyng in the middle betwixt Wheate and Barley, agreeable to all purposes whereunto Spelt is good.

The bread that is made of it, is also somewhat lyke the bread of wheate.

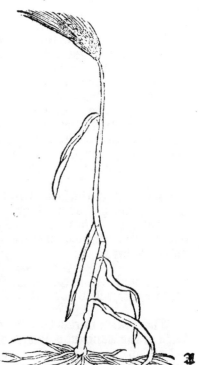

Of Typhewheate, called in Latine Triticum Romanorum. Chap. iiij.

❊ *The Description.*

1   **R**Omayne Wheate, is like common Wheate in his blades and knottie strawes, but the eares are more rounde and playne, and better compact, very muche bearded, the grayne is like the Wheate.

2   There is another kinde lyke vnto this, whose strawe and eares are smaller, the eares be also pointed and bearded. The seede is lyke vnto wheate, sauing that it is smaller, and blacker then our common Wheate is.

❊ *The Place.*

1   This Corne groweth not in all places, nor is not very common, but it is founde in some partes of Almaigne, as in Iussois, about the mountaynes and forestes, where as wilde Bores, and Swine do commonly haunt. And the husbandmen of the Countrie do sowe it for the same purpose, bycause of the Swine, whiche do ordinarily destroy the other Corne, but they come very seldome to feede vppon this kinde of grayne, bycause of the rough and prickely beardes which do hinder and let them, as Hierome Bocke writeth.

2   The seconde kinde groweth in the Iles of Canarie, and in certayne places of Spayne.

❊*The*

the Historie of Plantes. 457

Triticum Romanorū, aut Triticum Barbatum. **Romaine Wheat.**   Triticum Typhinum. **Typh wheate.**

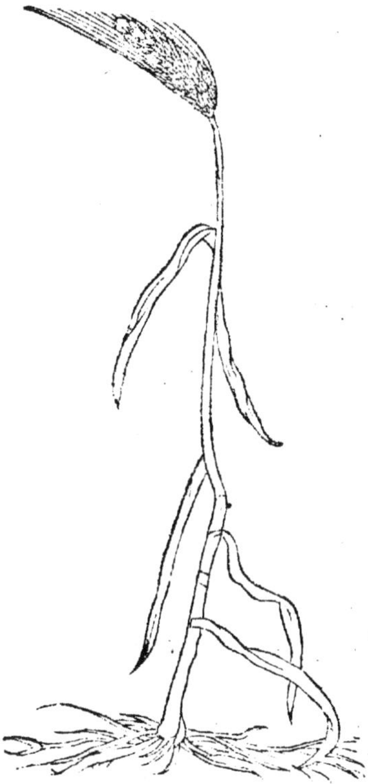

✣ *The Names.*
1 The first kind is called in French *Meteil*: in high Douch Welsche Weyssen, that is to say in Latine, Triticum Romanū: in base Almaigne, Romsche terwe: of some it is called in Greeke τύφκ, in Latine Typha, and also Typha cerealis, as a token of knowledge from another Typha, which is called Typha palustris: we may call it Typhe wheate or Bearded wheate, and Romaine wheat.

2 The second might also be a kinde of Typha, if the seede were inclosed in litle chaffie skinnes like vnto Spelt, but seing it is naked lyke wheate, therefore it cannot be Typha, although in other respectes it is very muche like Typha, of Theophrastus and Galen, therefore it may be wel called Triticum Tiphinum.

✣ *The Nature.*
This Corne is of temperature, somewhat lyke to the other, but not so good.

Of Spelt corne / Spelt wheate. Chap.v.
✣ *The Description.*
Zeopyron is a strange grayne, very muche like Spelt, in the strawe, knottes, and eares: yet the seede or grayne is better like wheate, for it is not closed vp in the huske like Spelt, but it commeth foorth easyly in threshing like wheat: & it hath a browne yellowish colour like wheat.

¶ *The Place.*
This kinde of grayne, doth also growe in some places of Almaigne.

✣ *The Tyme.*
Men sowe, and cut it downe like other corne.

Qq ¶ The

The fourth Booke of

✤ The Names.

This corne as Galen writeth in his first booke, De alimentis, hath ben called in the Countrie of Bithynia in Greece ζωπύρον, Zeopyron. The whiche is a compounde name, of Zea (that is to say, Spelt) and Pyros, that is to say, Wheate, the whiche name is very agreable vnto this Corne, bycause it is like to them both, or as a meane betwixt them both. The Almaignes call it Kern, Drinkelkern, and Kernsamen, that is to say in base Almaigne Keerensact.

The Nature, and operation.

A Zeopyron is of temperature, not muche differing from Spelt.

B The bread of Zeopyron is better then the bread made of Briza, and is as it were a meane or middle cast bread, betwixt wheaten bread, and the bread made of Briza, as witnesseth Galien.

Of single Spelt. Chap. vi.

✤ The Description.

Riza is also something like to Spelt, sauing that it hath the eares, motes, and strawes, lesser, smaller, and shorter, the eares be bearded, and the beardes are sharpe, like the beardes of Barley. The seede is couered with a huske lyke to Spelt. The whole plante with his strawe, eares, and grayne is of a browne redde colour, and it maketh browne bread, of a very strange and vnpleasant taste.

✤ The Place.

This corne loueth rough and rude places, and hath not to do with the champion ground. It hath ben founde in times past of Galen, in Macedonia, & Thracia: but now it is growen in some places of Douchlande, being brought first thither out of Thracia, as it is easie to coniecture. The whiche Countrie the Turkes do nowe cal Romaine, the chiefest citie whereof is Constantinoble.

✤ The Tyme.

Men sowe it in September, and cut it in Sommer, as other fruites of the lyke kindes.

❧ The Names.

This grayne is nowe called in Douche Blicken, Sant Peters Corne, and Einkorn: in Neather Douchlande Eencoren. It shoulde seeme to be a kinde of Zea Monococcos, and

Zea

Zeopyron.

Briza.

the Historie of Plantes.

Zea simplex, of Dioscorides, and the Zea of Mnesitheus, the which Galen in his booke, De aliment.facult. thinketh to be that grayne, whiche in his tyme was called in Thracia and Macedonia, βρίζα, Briza. It shoulde also seeme to be the kinde of Far whiche Columella nameth Far Venniculum rutilum.

❧ *The Nature, and operation.*

To what purpose this corne serueth in Physicke, hath not yet bene written of, nor proued to my knowledge. But the bread made thereof is very heauie, nourisheth euill, and is vnholsome.

Of Rye. Chap.vij.

❧ *The Kyndes.*

AS the wheate described in the first Chapter, is diuers, according to the times or seasons of sowing, euen in like manner is the Rye: for the one kinde is sowen before winter, and the other after winter.

❧ *The Description.*

1 Rye bringeth foorth of one roote, sixe or seuen and somtime moe, long, slender, and leane strawes with foure or fiue ioyntes, the whiche in good and fertill grounde groweth to the length of sixe foote or more, lyke to the strawe or reede of wheate, but softer, smaller, and longer. At the hiest of the sayde strawes, grow long eares, bearded with sharpe yles, like Barley eares, but nothing so rough or sharpe. The whiche when the corne is ripe do hang or turne downewardes, within the sayde eares is the grayne or corne, smaller, and muche blacker then wheate, and lesse then Barley, and is not enclosed in small huskes, but commeth foorth lightly. Of this kinde is made a very browne bread.

2 The other Rye is lyke to the aforesayde, in al respectes, sauing that the strawes and eares are smaller.

❧ *The Place.*

Rye groweth in all the lowe Countrie of Flaunders, and in many other Regions, it loueth the barren soyle, that is dry & sandy, where as none other corne or grayne may grow, as in the Countrie of Brabant, the whiche is called Kempene, and other like drie soyles. Yet for al that, the best Rye groweth in good and fertill soyles.

❧ *The Tyme.*

The first kinde is sowen in September, and the other in Marche, and are both ripe in July.

❧ *The Names.*

This grayne is called of Plinie in Latine Secale : in Englishe Rye : in Frenche Seigle : in high Almaigne Rocken : in base Alemaigne Rogghe : in Italian Segala : of some Asia, of others Farrago : although this is not the true Farrago, for Farrago is none other

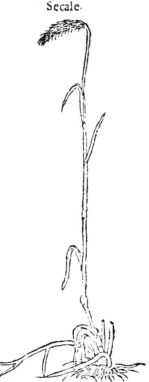

Secale.

ther thing, but Barley, Otes, and suche lyke graynes mingled togither, and sowen for forrage or prouender for Cattell: the whiche men do mowe and cut before it is ripe, to feede their Oxen, Kyen, Horses, and other lyke cattell.

And that this grayne is not Siligo, it is sufficiently declared in our fourth booke of the Historie of Plantes Chapt.j.

※ *The Nature.*

Rye layde outwardly to the body, is hoate and drie in the seconde degree.

※ *The Vertues.*

Rye meale put into a litle bagge, and layed vppon the head, cureth the olde ❡ and inueterate paynes of the head, and drieth the brayne.

The leuen made of the same, draweth foorth thornes, & splinters, or sheuers, ❡ and it ripeth al swelling and impostumations, insomuch that for this purpose, it wil worke better and is of more vertue, then the leuen made of wheate meale.

Rye bread with butter is of ye like vertue, but yet not so strong as the leuen. ❡

Rye bread is heauie and hard to digest, most meetest for labourers, and such ❡ as worke or trauell much, and for suche as haue good stomackes.

Of Barley. Chap.viij.

※ *The Kindes.*

Barley is of two sortes, great and small, to the whiche they haue nowe put two other kindes, that is to say, a kinde without huske: and another kinde called Douche Barley or Rice.

1. Hordeum Polystichū. Winter Barley. Beare Barley.
2. Hordeum Dystichum. Sommer Barley. Common Barley.
3. Hordeum Nudum. Naked or bare Barley. Wheate Barley.
4. Oriza. Rice.

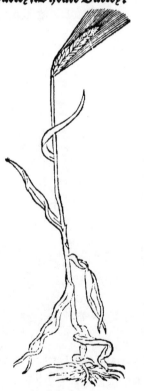 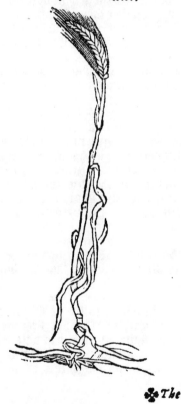

The Description.

1 Barley hath helme or strawe, lyke wheaten strawe, but it is shorter and more britle with fixe or moe ioyntes, and knottes. The eares be long and very rough, couered & set ful of long bearded sharpe ayles, where as the grayne or cornes are placed in order or rewes, sometimes in foure ranges or moe lines. The seede is lyke to wheate, and is closed vp fast in a chaffie couering or skinne, like Spelt.

2 The small common Barley is very well like the other, sauing that his spike or eare hath but two rewes or orders of Cornes.

3 Besides these two sortes of Barley, there is yet another kinde the whiche hath the Barley in strawe and eare, but the grayne is not so closed vp in the huske as the other Barley, but is naked bare, and cleane, and commeth foorth easily from his eare like wheate and Rye.

4 Yet there is another kinde, whiche some do call Douche Rysz, the same in his straw, ioyntes, and in his long bearded eares doth much resemble Barley. It hath also his graynes or cornes inclosed in chaffie huskes, lyke to Barley, but it is whiter then Barley.

The Place.

1.2. Barley is common in all Countries, and it loueth good grounde and fertile soyle.

3 The naked or hulled Barley groweth in some places of Fraunce, as about Paris.

4 That which is called Douch Rysz, is sowen in some places of Almaigne, as in Westerich.

The Tyme.

Men do sowe the great Barley in September, and they mowe or cut it in July, and sometime in June.

The lesser or common Barley is sowen in the spring time, and is ripe in August.

The Names.

Barley is called in Greeke κριθή: in Latine Hordeum: in Frenche Orge: in Douche Gerst.

1 The great Barley is called in Greke πολυστιχή: in Latine Hordeum Cantherinum: in high Douche Grosz Gerst: in base Almaigne Groote Gerste. I take this for Beare Barley,

2 The lesser Barley is called δυστιχή, and Galatinum Hordeum: in high Douch Fuder Gerst: in base Almaigne Voeder Gerste.

3 The third kinde (as witnesseth Ruelius) is called Hordeum mundum, and may be wel called in Greeke γυμνοκριθον, Gymnocrithon, that is to say in Latine Hordeum nudum, as Galen setteth foorth in his booke De aliment. faculta.

4 Hierome Bock nameth the fourth kinde Teutsch Rysz, that is to say in Latine Oriza Germanica. It should seeme to the eye, to be a kinde of Far, especially that Far Clusinum, which resembleth muche Santalum Plinij. It shoulde seeme also to be ὄλυρα, Olyra, of Dioscorides, whiche is called in Latine not Siligo, but Arinca: in Englishe Rise.

The Nature.

Barley is colde and drie in the first degree.

The Vertues.

Barley meale boyled with figges in honied water dissolueth hoate and cold A tumours, and it doth soften and rype all hard swellinges with Pitche, Rosen, and Pigeons donge.

The same mingled with Tarre, Oyle, Waxe, & the Urine of a young childe B doth digest, soften, and ripe the harde swellinges of the Necke, called in Latine Strumæ.

462   The fourth Booke of

 The same with Melilote and the heades of Poppie swageth the ache of the ☧
side, and with Lineseede, Fenugreck, and Rue, it is good to be layd vpon the
belly against the paynes and windinesse of the guttes.

1 Barley giuen with Mirtels, or wine, or wilde tarte peares, or with Bram- ☙
bles, or with the barke of Pomgarnet, stoppeth the running of the belly.

2 They make a playster with Barley meale against the scurffe and leprie. ☧

 The same mingled with vineger or Quinces swageth the hoate inflamma- ☙
tions of the gowt, and if it be boyled with vineger and Pitche, and layd about
the ioyntes, it stayeth the humours from falling into them.

 It is also vsed in meates, and bread is made of it, the which doth not nou- ☙
rish so wel as the bread made of wheate or spelt.

Of Mill or Millet. Chap. ix.

Milium.   Lachryma Iob.

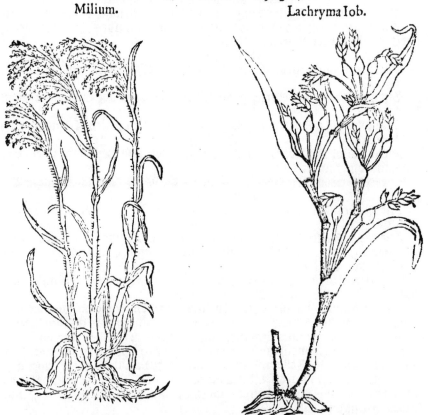

✣ *The Description.*

MIllet hath a hearie stalke, with seuen or eight knottes or ioyntes.
The leaues be long and like the leaues or blades of Polereede: at
the highest of the stemmes come foorth the bushie eares, very
muche seuered and parted, like the plume or feather of the Cane or
Polereede, almost lyke a brushe or besome to sweepe withall, in
whiche groweth the seede, very rounde and playne almost like to lineseede, but
that it is not so blacke.

For

the Historie of Plantes.   463

For one kinde of Milium is likewise taken of some, that which is named Lachrima Iob, and it hath many knottie stalkes, about a foote and a halfe high, and thereon broade reede leaues, betweene the whiche commeth foorth round fruite vpon thinne stalkes, about the bignesse of a pease, thereof come foorth small eares. The rootes haue strong threddie stringes.

※ *The Place.*

Mill loueth a moyst and claye ground, it groweth abundantly in Italy and Spayne.

Lachryma Iob is in this Countrie strange, and is found only in the gardens of some diligent Herboristes.

※ *The Tyme.*

They sowe it in the spring time, and it is ripe in somer, it may be kept a long time, euen a hundred yeres, so it be kept from the winde.

※ *The Names.*

This plant is called in Greke κέγχος: in Latine of the Apothecaries Miliũ: in English Mill, Millet, and Hirse: in Frenche Mil, or Millet: in high Douche Hirsen: in neather Douchlande Hirs, or Milie. What other name Lachryma Iob hath, is vnknowen vnto vs.

※ *The Nature.*

Millet is colde in the first degree, and drie almost in the third degree, and of subtill partes.

※ *The Vertues.*

Millet parched in a frying panne, and well heated and made warme, and A put vp into a bagge, and so layde to the belly, doth helpe the gripinges & gnawing paynes of the same: and swageth all paynes, and aches, especially of the sinewes: and is good to dry vp that which requireth to be dried, being most conuenient to drie, and comfort the brayne.

For want of other corne men may make bread of Millet, the which bindeth B the belly, and prouoketh vrine: but it nourisheth litle, and is very leane or slender.

Of Turkie Corne, or Indian Wheate.   Chap.x.

※ *The Kindes.*

TUrkish wheate is of one, and of many sortes. A man shall not finde in this Countrie (in fashion and growing) more then one kind, but in collour the seede or grayne doth muche differ: for one beareth a browne grayne or Corne, the other a red, the thirde a yellowe, and the fourth a white Corne or grayne. The which colour doth likewise remayne both in the eares & flowers.

※ *The Description.*

THis Corne is a marueilous strange plante, nothing resembling any other kinde of grayne: for it bringeth foorth his seede cleane contrarie from the place where as the flowers growe, which is agaynst the nature and kindes of all other plantes, whiche bring foorth their fruite there, where as they haue borne their flower.

This corne beareth a high helme or stemme, & very long, rounde, thicke, firme, and belowe towardes the roote of a brownishe colour, with sundrie knottes and ioyntes, from the whiche dependeth long, and large leaues, like the leaues of spiere or Polerrede : at the highest of the stalkes, growe idle and barren eares, whiche bring foorth nothing but the flowers or blossomes, which are sometimes browne, sometimes redde, sometimes yellow, and sometimes white, agreable with the colour of the fruit which commeth foorth afterwarde.

Qq iiii

464　The fourth Booke of

warde. The fruitefull eares do growe vppon the sides of the stemmes amongst the leaues, the which eares be great and thicke and couered with many leaues, so that one cannot see the sayde eares, vpon the vppermost part of y sayde eares there grow many long hearie threddes, which issue foorth at the endes or pointes of the leaues couering the eare, and do shewe them selues, about the time that the fruit or eare waxeth ripe. The grayne or seede which groweth in the eares, is about the quantitie or bignesse of a pease of colour in the outside sometimes browne, sometime red, and sometime white, and in the inside it is in colour white, and in taste sweet, growing orderly about the eares, in niene or tenne ranges or rewes.

Frumentum Turcicum.
Turkish or Indian wheate.

❊ *The Place.*

This grayne groweth in Turkie wher as it is vsed in the time of dearth.

❊ *The Tyme.*

It is sowen in Aprill, and ripe in August.

❊ *The Names.*

They do nowe call this grayne Frumentum Turcicum, and Frumētum Asiaticum: in Frenche Blé de Turquie, or Blé Sarrazin: in high Douche Turkie Korn: in base Almaigne Torckschcoren: in Englishe Turkish Corne, or Indian wheate.

❊ *The Nature and Vertues.*

There is as yet no certaine experience of the natural vertues of this corne. The bread that is made thereof is drie and harde, hauing very small fatnesse or moysture, wherefore men may easily iudge, that it nourisheth but litle, and is euill of digestion, nothing comparable to the bread made of wheate, as some haue falsly affirmed.

Of petie Panick, Phalaris grise, grasse corne. Chap.xi.

❊ *The Description.*

Phalaris hath a rounde strawe or helme, with three or foure ioyntes, the leaues be narrowe and grassie, lyke the blades of Spelt or wheate but smaller and shorter, vppon the sayde strawe groweth a short thicke eare, and clustered or gathered togither. it bringeth foorth a seede lyke vnto Mill, and in fashion lyke to Line seede.

❊ *The Place.*

This seede groweth in Spayne, and in the Iles of Canarie. And is onely sowen in this Countrie of the Herboristes.

❊ *The Tyme.*

It is ripe in this Countrie in July and August.

The

❧ The Names.

This seede is called in Greeke Φαλάρις: & likewise in Latine Phalaris: of some Douchmē Spaensch saet, and Saet van Canarien, that is to say, Spanishe or Canarie seede, some Apothecaries do sell it for Millet. Turner calleth it Petie Panicke.

❧ The Nature.

In complexion, it is much like to Millet.

❧ The Vertues.

A The iuyce of Phalaris dronken with water, is good agaynst the payne or greefe of the bladder.

And a spoenfull of the seede made into powder is good to be taken for the same purpose.

Phalaris.

Of Panicke. Chap.lij.

❧ The Description.

1 Panicke commeth vp lyke Millet, but his leaues are sharper & rougher. It hath a rounde stemme or straw ful of knottie ioyntes, for the most part sire, or seuen knottes vppon one stemme, and at euery knot a large narrow leafe. The eares be round, and hanging somwhat downewardes, in the which groweth smal seede, not muche vnlike the seede of Millet, of colour sometimes yellowe, and sometimes white.

2 There is also founde another plant like vnto Panicke, the which some hold for a kind of Panicke, the Italians do cal it Sorghi. This strange grayne hath foure or fiue high stemmes, which are thicke, knottie, and somwhat brownish, beset with long sharpe leaues, not muche vnlike the leaues of Spier or Poole reede, at the vppermost part of the stalkes, ther grow thicke browncred eares, greater & thicker then the eares of Panick, the which at the first do bring forth a yellow flower, & afterward a round reddish sede, of the quantitie of a lentil, & somwhat sharpe or pointed.   ❧ The Place.

1 Panick is not much knowen in this Countrie, it groweth in some places of Italie and France, and it loueth grauel and sandy ground, it desyreth not much raine or moysture: for when it rayneth muche, it maketh the leaues to loll and hang downewarde, as Theophrastus wryteth.

2 The Indian Panick is also a strange sede, & is not found in this Countrie, but in the gardens of Herboristes.   ❧ The Tyme.

1 Men do sow Panick in the spring of the yere, and it is cut downe againe (in hoate Countries) fourtie dayes after. The Gascons do sowe it after they haue sowē their other corne, yet for al that, it is ripe before winter, as Ruelius saith. In this Countrie when it is sowen in April, it is ripe in July.

Also

Panicum.
Panik.

Sorghi. Melica.
Indian Panick.

2 Also the Indian Panicke is sowen in the spring time, and ripe at the ende of sommer.

❋ *The Names.*

1 Panick is called in Greeke ἔλυμ☉: of Theophraste also μελίν☉: in Latine Panicum: and nowe a dayes in Italian *Melica*: in high Douche Feuch, Fenich, and Heydelpfenich: in base Almaigne Panickoren.

2 The Indian Panick is nowe called of some Italians *Melegua*, or *Melega*, of some others, *Saggina*, and *Sorgho*: in Latine Melica Sorghi, Milium Saburrum, and of some Panicum peregrinum: of the Almaignes Sorgsamen: of the Brabanders Sorgsaet. It is very lyke that this is Milium Indicum, whiche as Plinie writeth, was first knowen in the time of the Emperour Nero.

❋ *The Nature.*

Panick is colde and drie of complexion.

❋ *The Vertues.*

The seede of Panick dronke with wine, cureth the dangerous and bloody flire, and taken twise a day boyled in Goates milke, it stoppeth the laske, and the gnawinges or gripings of the belly.

They make bread of Panick, as of Millet, but it nourisheth, and bindeth lesse then the bread of Millet.

The Indian Panick is like the other Panick in operation and vertue.

the Historie of Plantes. 467

Of Otes. Chap.xiij.

❧ The Description.

1. Otes (as Dioscorides saith) in grassie leaues, and knottie straw or inotes, are somwhat like to wheate: at ẏ vpper part of the strawes growe the eares, diuided into many small sprinnges or stemmes, displayed and spread abroade farre one from another, vppon the which stemmes or small branches the grayne hangeth sharpe pointed alwayes togither, well couered with his huske.

2. There is an other kinde of Otes, whiche is not so inclosed in his huskes as ẏ other is, but is bare, and without huske whan it is threshhed.

3. Also there is a barren Ote, of some called the purre Otes, of others wilde Otes.

Auena.

❧ The Place.

1. Otes are very common in this Countrie, and are sowen in al places in the feeldes.

2. The pilde Otes are sowen in the gardens of Herboristes. *Turner saith they growe in Sussex.*

3. The Purwottes or wilde Otes, commeth vp in many places amongst wheate and without sowing.

❧ The Tyme.

Otes are sowen in the spring time, and are ripe in August.

❧ The Names.

1. Otes are called in Greeke βρόμος : in Latine Auena: in high Douche Habern: in base Almaigne Hauer: in Frenche *Auoyne*.

2. The seconde kinde may be called in Englishe, Pilcorne, or pylde Otes.

3. Turner calleth the thirde kinde by the Greeke name αἰγίλωψ: and in Latine Auena sterilis: whiche you may see described in the xvj. Chapter of this fourth booke.

❧ The Nature.

Otes do drie much, and are of complexion somwhat colde, as Galien saith.

❧ The Vertues.

A. Otes are good to be put in playsters and Cataplasmes wherein Barley is vsed, men may also vse the meale of Otes in steede of Barley meale, forasmuche as Otes (as Galen saith) do drie and digest without any biting acrimonie.

B. Oten meale tempered with vineger, driueth away the Lentiles and spots of the face.

C. The same taken in meate stoppeth the belly.

D. Oten bread nourisheth but litle, and is not very agreable or meete for mankinde.

The fourth Booke of
Of Bockwheate. Chap.xiiij.

The Description. Fegopyron. Tragopyron.

Ockwheate hath round stalkes chauellured and fluted (or forowed and crested)of a reddishe colour, about the height of two foote or more: The leaues are broade and sharpe at the endes, not muche vnlyke the leaues of Iuie or common Wythiwinde. It putteth foorth shorte stemmes, aswell on the sides as on the top of the stalkes, vpon the said short stemmes there growe many white flowers in tuftes or clusters, after the said flowers commeth the sede, which is triangled and gray, enclosed in a little felme or skinne, lyke the seede of blacke bindeweede, described in the third part of the historie of plantes.

The Place.

They sow it in leane and drie ground, and is very common in the landes of Brabant called Kempene.

The Tyme.

It is sowen in the spring tyme, & in somer after the cutting downe of Corne, and is ripe niene or ten weekes after.

The Names.

This kind of grayne and plant is called in Frenche *Dragée aux cheueaux*, in high Douche Heydenkorne: in base Almaigne Bockweydt, after whiche name it may be englished Bockwheat, The Authour of this worke calleth it Tragopyró, certaine others do call it in Greeke φαγοπυρον, and in Latine Fagotriticum, whiche is not Ocymum, described by Columel, as we haue sufficiently declared in the fourth booke of our Historie of Plantes, where as we haue in lyke manner declared howe it was vnknowen of the Auncientes. *I thinke this to be the grayne called in some places of Englande Bolimonge.*

The Nature.

This seede without fayle is indued with no heate, and is not very drie.

The Vertues.

The meale of Bockewheate is vsed with water to make pappe, whitpottes and great cakes of light digestion, whiche do lightly lose the belly, and prouoke vrine, yet they be but of small nourishment.

The bread which men do make of this grayne is moyst, & sharpe or sower, without any great nourishing.

It hath none other vertue that I knowe, sauing that they giue the greene herbe as fodder and fourrage for cattell, and they feede hennes and chickens with the seede, which doth make them fat in short space.

Of

the Historie of Plantes.

Of Juray or Darnell. Chap.xv.

The Description. Lolium.

Uray is a vitious grayne that combereth or anoyeth corne, especially wheat, and in his knottie Strawe, blades, or leaues is like vnto wheate, but his eares do differ both from wheat and Rye eares, for they are diuided into many small eares growing vppon the sides at the toppe of the straw, in the whiche small eares the seede is conteyned, in proportiō almost lyke wheate cornes, but muche smaller.

The Place.

Juray for the most part groweth amongst wheate, and sometimes it is also founde amongst Barley, especially in good lande, where as wheate hath growen before.

The Tyme.

It wareth ripe with þ wheate and other corne.

The Names.

This plant is called in Greke ἄιρα, καὶ θύαρος: in Latine Lolium: of the Arabians Zizania: in Frenche Yuraye, or Gasse. in Englishe it is also called Juraye, Darnell, and Raye.

The Nature.

Juray is hoate euen almost in the thirde degree, and drye in the seconde.

The Vertues.

The meale of Juray layde on with Salt and Radish rootes, doth stay and A keepe backe wilde Scurffes, and corrupt and fretting sores.

The same with sulfer and vineger, cureth the spreading skabbe, and leprie, B or noughtie scurffe, when it is layde thereon.

The same with Pigeons donge, oyle, and lineseede, boyled & layde plap- C sterwise vpon wennes, and such harde tumours, doth dissolue and heale them.

It draweth foorth also al splinters, thornes, and shiuers, and doth ripe and D open tumours and impostemes.

If it be sodden with Mede, or as Plinie saith Orimel, it is good to be layd E to, to swage the payne of the gowte Sciatique.

They lay it to the forehead with birdes greast, to remoue and cure the head- F ache.

It is also founde by experience, that Juray put into Ale or Bier causeth G dronkennesse and troubleth the brayne.

The fourth Booke of Pour Otes Festuca and Melampyrum. Chap.rbi.

Aegilops.
Pour Otes.

Festuca altera.
Drauick wilde Otes.

❧ The Description.

1. Pour Otes or wilde Otes, are in leaues and knottie strawes like vnto common Otes, the eares be also spread abroade, like to the common Otes. The graine is blackishe & rough heared, inclosed in hearie huskes, eche one hauing a long bearde or barbe. This is a hurtfull plant as well to the Rye as other corne.

2. Festuca, or as the Douchmen call it Drauick, is also a hurtfull plant, hauing his leaues and strawe not much vnlyke Rye, at the top whereof growe spreading eares, wherein is conteyned a small seede of grayishe colour, inclosed in litle skinnes or small huskes, muche lesse and smaller then any other kinde of corne or grayne.

3. We may wel place with these, that herbe or plant which of the Brabanders is called Peertsbloemen, that is to say, Horse flower, whose description you may see in the second booke Chapter riiii. placed with those wild flowers, that growe amongst corne: for his seede is lyke to wheat, and a hurtful or noysome weede to corne, especially to wheate, as Galen saith.

❧ The Place.

You shal finde much of this geare amongst Rye, and oftentimes amongst wheate and Barley.

❧ The Names.

The first is called in Greeke αἰγίλωψ: in Latine Aegilops, and according to Plinie

the Historie of Plantes. 471

Plinie Festuca: in English Wilde Otes, or Poure Otes.

2 The seconde is called in high Douche Dort: in Neather Douchlande Drauick: it may be also very well called in Latine Festuca, or Festucaaltera: in Englishe Wilde Otes, or Drauick.

※ *The Nature.*

Poure Otes are hoate, as Galen testifieth.

※ *The Vertues.*

A The greene leaues layde to, with the meale of th the seede of Poure Otes (if it be Aegilops) is good to heale hollowe vlcers called Fistulas, especially those whiche are in the corners of the Eyes, called Aegilopes.

B The same sodden with Ale or Bier, causeth the head to be dul and heauy, after a dronken sort or manner, like to Iuraye, and the seede of the same grayne which the Brabanders call Peertsbloemen.

Of Blight or Brantcorne. Chap. xvij.

Melampyrū Blacke wheat. Cow wheat or Horse flower.

※ *The Description.*

Stilago is a certayne disease, or infirmitie, that happeneth vnto suche fruits as eare eares, but especially vnto Otes. This kinde of plante, before it shuteth out in eare is very lyke vnto Otes, but when it beginneth to put foorth his eare, insteede of a good eare, there comneth vp a blacke burnt eare, ful of blacke dust or powder.

※ *The Place.*

It groweth most commonly (as is beforesaid) amongst Otes, and sometimes amongst wheate.

※ *The Tyme.*

It is founde most commonly in Aprill, when as the Sonne shineth very hoate, & after a rayne folowing.

※ *The Names.*

This barren and vnfruitefull herbe is nowe called Vstilago, that is to say, Burned, or Blighted: in French *Brulure:* in high and base Almaigne Brant.

※ *The Nature, and faculties.*

Vstilago hath no good propertie in Phisicke, and serueth to no manner of good purpose, but is rather a hurt or maladye to all Corne.

Vstilago.

The fourth Booke of
Of Beanes. Chap.xviij.

❧ *The Kyndes.*

There be two sortes of beanes. The one sort is commonly sowen, the other is wild. The comon or manured beane, is diuided againe into two sortes, that is: great, and smal.

Phaselus satiuus.
Sowen Beanes.

Phaselus syluestris.
Wilde Beanes.

❧ *The Description.*

1. THE great sowen Beane hath a square stalke, vpright, and hollowe. The leaues growe vpon short stemmes standing vpon both sides of the stalkes one against another, and are long & thicke. The flowers grow vpon the sides of the stalke, and are white with a great blacke spot in them and somtimes a browne. After which flowers there come vp long coddes, great and round, soft within, & frised, or cottonlike. In the sayd coddes the beanes are inclosed, of colour most commonly white, sometimes redde or browne, in fashion flat, almost lyke to the nayle of a mans finger or toe.

2. The lesser beane that is vsed to be sowen, is like to the aforesayd, in stalkes, leaues, flowers, and woolly coddes, sauing that in all pointes it is lesser. The fruite also is nothing so flat, but rounder and smaller.

3. The wilde beane hath also a square holow stalke, as the garden and sowen beanes haue. The leaues be also like to the common beane leaues, but the litle stemmes, whereon the leaues do growe, haue at the very ende tendrelles and claspers, as the pease leaues haue. The flowers be purple. The coddes are flat,

and

and woolly within, as it were laid with a soft Downe or Cotton, but nothing so much as the coddes of the common sowen beanes. The fruite is all rounde and very blacke and no bigger then a good pease, of a strong vnpleasant sauor, and when it is chewed, it filleth the mouth full of stinking matter.

❧ The Place.

The domesticall, or husbandly beanes, do growe in feeldes and gardens where as they be sowen or planted. The wilde is to be founde amongst the Herboristes: and groweth of his owne kinde in Languedoc.

❧ The Tyme.

They are planted and sowen in Nouember, January, February and April, and are ripe in June and July.

❧ The Names.

Beanes are called in Greeke φασηλοι, of Dioscorides also Phasioli: in Latine Faseli: nowe a dayes they be called in Shoppes, and commonly Fabæ: in high and base Almaigne Bonen.

1. The great kinde is called in Latine Phaselus maior, or Faselus satiuus maior: in Douche, Groote Boonen: in English, Great Beanes, and garden Beanes.

2. The other may be well called Faseli minores, that is to say, The smaller Beane, in Brabant Zeeusche Boonkens, and Peerde Boonkens. That the common Beane is not that kind of pulse called of the Auncientes Cyamos, and Faba, hath ben sufficiently declared, In Historia nostra.

3. The wilde kinde may well be called in Latine Faselus syluestris, and Faselus niger, that is to say, The wilde Beane, and the blacke Beane: in Douch, Wilde Boonen, and Swerte Boonen, or Moorkens, as some do cal them. This may well be that Pulse whiche is called Cyamos, and Faba.

❧ The Nature.

Greene beanes before they be rype, are colde, and moyst: but when they be drie they haue power to binde and restrayne.

❧ The Vertues.

The greene and vnripe Beanes eaten, do loose & open the belly very gently, but they be windy, & engender ventosities (as Dioscorides saith.) The which is well knowen of the common sortes of people, and therefore they vse to eate their beanes with Commine.

Drie Beanes do stop & binde the belly, especially when they be eaten without their huskes or skinnes: and they nourishe but litle, as Galen saith.

Beane meale layde to outwardly in manner of a Cataplasme or plaister, dissolueth tumours and swellinges. And is very good for the vlcers and inflammation of womens pappes, and against the mishappes and blastings of the genitors.

The wilde Beane serueth to no vse, neither for meate nor medicine, that I do knowe.

Of Kidney Beane, or garden Smilar. Chap. xix.

❧ The Description.

Arden Smilax hath long and small branches, growing very high, griping, and taking holdfast when they be succoured with rises or long poles, about the whiche, they wrappe and winde them selues, as the Hoppe, otherwise they lye flat and creepe on the ground, & beare no fruite at all. The leaues be broade almost like Juie, growing three and three togither as the Trefoil or three leaued grasse. The flowers be somtimes white and sometimes red, after the flowers there come in their places long coddes, which

which be somtimes crooked, and in them lye the sedes or fruit, smaller then the common beane, and flat fashioned lyke to a kidney of colour somtimes red, somtimes yellow, somtimes white, somtimes blacke, & sometimes gray, & speckled with sundrie colours. This fruit is good and pleasant to eate, in so much that men gather and boyle them before they be ripe, and do eate them coddes and all.

Phaseolus.

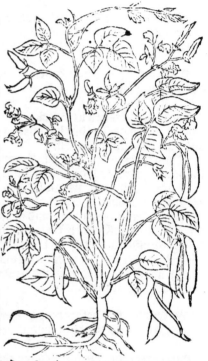

❧ The Place.
In this Countrie men plant this kind of Beanes in gardens, & they loue good grounde and places that stande well in the Sonne.

❧ The Tyme.
They are planted in Aprill after that the colde and frostes be past: for at their first comming vp, they can by no meanes at all indure colde. They are ripe in August and September.

❧ The Names.
This kinde of Beanes are called in Greeke φάσκολ⊙, δολιχός, καὶ σμίλαξ κηπαία: in Latine Faseolus, Dolichus, and Smilax hortensis. The coddes or fruite are called λόβοι, that is in Latine Siliquæ, and Lobi: of Serapio Lubia: in Frenche Phaseoles: in high Douch Welsch Bonen: in base Almaigne Roomsche Boonen: in Englishe of Turner it is called kidney beane, and Sperage, of some they are called Faselles, or Long Peason, it may be also named Garden Smilax, or Romaine Beanes.

❧ The Nature.
Kidney beanes are somewhat hoate and moyst of Complexion, after the opinion of the Arabian Physitions.

❧ The Vertues.
Kidneybeanes do nourishe meetely well, and without engendring windinesse, as some other pulses do: also they do gently loose and open the belly, as Hipocrates and Diocles do write. A

The fruite and Coddes boyled and eaten before they be ripe, do prouoke vrine, and cause dreames, as Dioscorides sayth. B

Of Pease or Peason. Chap. xx.

❧ The Kyndes.
There be three kindes of Peason, the great, the meane, & the smal, the which are lyke one another, in stalkes, leaues, flowers and coddes, but not in fruite, as ye may perceaue hereafter.

❧ The Description.
1 THe great branched Peason, are not muche knowen in this Countrey. They growe when they be stayed vp, by rises, stakes, or other helpes to the length of a man, or higher. The stalkes be rounde and holowe, and somewhat cornered, as big as a finger, vpon the which at

the Historie of Plantes. 475

at euery knot there growe two leaues, very well closed and ioyned togither, as if it were but one leafe: amongst the leaues growe small stemmes, the which haue foure or fiue grosse or fatte leaues set directly opposite, one against another, hauing at the ende foure or fiue griping or ramping claspers, whereby the Pease doth take holde, and is stayed vp, and fastened to such thinges as it standeth by. Adioyning harde to the stemmes of the leaues aforesayd, there growe other naked and bare stemmes, vpō the which grow pleasant flowers of blew or purple colour. After the sayde flowers there come vp long coddes, rounde, playne, and shining, hanging downewardes, in which the Peason are couched and layde, the whiche beyng yet but greene, are rounde and whitishe, but when they be drie, they are gray and cornered.

Pisum.	Ochros. Eruilia.
Great Peason. Branche Peason.	Middle Peason.

2 The seconde sorte whiche are the common pease, is muche like to the aforesayde, but that his leaues and branchie stalkes, are smaller, neyther do they growe so high, although they be stayed vp and succoured with bowes or branches. The flowers are most commonly white, the fruite is rounde and white, and remayneth rounde whan it is drie.

3 The thirde, whiche is the least kinde, is lyke vnto the seconde, sauing that it is much smaller in leaues, stalkes, coddes and fruite. It is suffered for the most part to lye vppon the grounde in the feeldes, without any stay or helpe of branches or bowes. The fruite thereof is lykewise rounde, of colour, sometimes white, sometimes greene, and sometimes gray or blackish.

Rr iiij   Besides

The fourth Booke of

Eruilia syluestris. Wilde Peason.

4 Besides the aforesayde kindes, there is yet a certayne kind of Pease lyke vnto the wilde or least kinde. It hath flat stalkes, the leaues are long with clasping tendrels at the endes, whole beneath next to the stemme, but at the toppe of the branches, the leaues are clouen and diuided into two or three small narrowe leaues, almost lyke the leaues of Cicercula, (whiche Turner calleth Cicheling.) The flowers are white, after the which flowers there come vp round coddes or huskes, lesser then pease-coddes, within them groweth the fruite, which is rounde lyke vnto Pease, sauing it is lesser, and in taste bitter, while it is yet greene, & very harde when it is drie.

❧ *The Place.*

1 The great and branched Peason are planted in gardens: but the midle and least kind are sowen and planted in fruitful feeldes, and are very common in this Countrie.

2 The Herboristes do sow the wild kinde in their gardens.

❧ *The Tyme.*

Men plant them in Marche and Aprill, and they be rype in August.

❧ *The Names.*

1 The branche Peason are called in Greke πίσον: in Latine Pisum: in Brabant Groote Erweten, Roomsche Erweten, and of some Stock Erweten. This kinde is the right Pisum, described of Plinie and the Auncientes: in Englishe Great Peason, Garden Peason, and Branche Peason, bycause, as I thinke, they must be holpen or stayed vp with branches.

2.3. The two other kindes are called in Greke ἄχερ: in Latine of Plinie Eruilia: in French *Pois*: in high Douch Erweyssen: in base Almaigne Erweten: in Italian *Rouiglione*: at Venice *Pisareli*: in Englishe, Common Peason.

4 The fourth is very wel like to be a kinde of wilde pease, and especially that kinde whereof Hermolaus Barbarus writeth, calling it Eruilia syluestris, that is to sayde wilde Peason.

❧ *The Nature, and Vertues.*

Branche Peason being eaten do nourish meanely, engender windinesse, but not so muche as the pulse whiche the Auncientes call Faba.

The other rounde and common Pease are beter, and do nourish better then the great or branched Peason, and they do lose and open the belly gently.

Of the Cicheling or flat Peason. Chap.xxi.

❧ *The Kindes.*

There be two sortes of Cichelinges, the great and the small, or garden and wilde Cichelinges.

❧ The

the Historie of Plantes. 477

Lathyrus Cicercula.
Cicheling or brode Peason.

Lathyrus sylueſtris.
Wilde Cicheling.

⚜ The Description.

1. The Cicheling or flat peaſe, hath flat and creſted ſtalkes: the leaues be long and narrow, ſtanding vpward, almoſt like the two eares of a Hare, with Claſping tendrelles, by the which they take holde by poles and branches that are ſet by thē. The flowers be white lyke the flowers of branche Peaſon, after them come flat coddes, & large, wherein is a white fruite, large, flat, & vneuenly cornered hauing almoſt the ſent or ſmell of the peaſe. The roote is tender and threddy.

2. The leſſer Cicheling is like to the aforeſaid, in ſtalkes leaues & coddes. The flowers are reddiſh. The fruit is alſo flat, vneuenly cornered as the great kind, but it is ſmaller, harder, and of a more browne colour, drawing towardes blacke.

3. There is alſo founde a wilde kinde of this pulſe much lyke to the aforeſayd in the flatneſſe of the ſtalkes, and in his long and narrowe leaues. The whiche in like manner bringeth foorth reddiſh flowers, and afterward narrow coddes, wherin is contepned a ſmall browne ſeede, round and hard. The roote is great and thicke, of a wooddy ſubſtance, and dieth not, but putteth foorth new euery yere.

¶ The Place.

1.2. Theſe pulſes are found in this Countrie, amongſt ſome diligent Herboriſtes.

3. The wild groweth in hedges, and in the borders of feeldes, in good & fertill ground, and is found in great plentie about Louayne and Bruſſels.

⚜ The

The fourth Booke of

❦ *The Tyme.*

These Pulses do flower in June, and are ripe in July and August.

❦ *The Names.*

1 The first and greatest kinde is called in Greeke λάθυϱ⁀, Lathyrus: of Columella and Paladius Cicercula. Turner calleth it a Cichelng.

2 The seconde is called in Greeke ἄϱακ⁀, Aracus: in Latine Cicera They are both called in Frenche *Des Sars* but they haue no Douche name that I knowe, yet the Authour of this booke in the last Douche copie by him corrected, calleth the first kinde in Douche Platte Erwten, that is to say in English, Broade or Flat Pease: not knowen of the Apothecaries.

❦ *The Nature, and Vertues.*

A The first kinde is of nature and qualitie like vnto Pease, and doth meanely nourishe the body, as Galen saith.

B The seconde is like to the first, as witnesseth the same Galen, sauing that it is harder, for whiche cause it ought to be longer boyled.

Of Chiche Peason.   Chap. xxij.

❦ *The Kyndes.*

There be three kindes of Ciche Peason (as Dioscorides writeth) the domesticall or tame kinde, the square or cornered kinde, the which some do cal Arietinum, and the wilde Ciche, and there be two sortes of that kinde whiche is called Arietinum, white, and blacke.

Cicer satiuum. **Tame Ciches.**   Cicer Arietinum. **Sheepes Ciches.**

the Historie of Plantes.

❧ *The Description.*

1 The tame Ciche Peason is a small kinde of pulse, almost like to a lentil, it hath foure or fiue branches, and thereupon small, narrowe, diuided leaues, not muche vnlyke the leaues of lentilles. The flowers grow vpon short stemmes, small and somewhat whitishe, after the whiche there come vp small rounde huskes or coddes, wherein is commonly founde three or foure round Peason, hauing a certayne bunch, hillock, or outgrowing by one side, not muche vnlike Sheepes Ciche Peason, but a great deale smaller, and not so harde, and of a better taste.

2 Sheepes Ciches haue slender stalkes, and harde with many branches, and rounde leaues iagged about the brimmes, like the lentil or feche: growing directly or opposite one against the other, the flowers be either white or purple, and bring foorth shorte rounde Coddes or huskes, bollen or swelling vp like small bladders, wherein growe two or three Peason cornered, and fashioned almost lyke a sheepes head, in colour sometymes white, and sometimes blacke.

3 The wild Ciche pease, in leaues are lyke to the tame, but they are of a ranke and strong sauour, and the fruite of another fashion (as Dioscorides saith) vnlike the tame Ciches.

❧ *The Place.*

These Ciche Peason, are founde planted in the gardens of Herboristes.

❧ *The Tyme.*

All the Ciches are ripe in August, like to the other sortes of pulse.

❧ *The Names.*

Ciche Peason are called in Greeke ἐρέβινθος: in Latine Cicer: in Frenche *Cices*, or *Pois Cice*.

1 The first kind is called Cicer satiuum, Columbinum Venereũ: and in Greke ὀροβιαῖον, Orobiæon, that is to say in Latine Cicer eruillum: vnknowen in Shoppes. This is not Eruum, as many at this time do thinke, and for that purpose they put it into their triacles and other suche medicines.

2 The seconde kinde is called in Greeke ἐρέβινθος κριός: in Latine Cicer Arietinum, that is to say, Sheepes Ciche pease: in Shoppes Cicer: in Englishe Sheepes Cichpeason, in French *Pois Ciches*: in high Douch Zysern erweyssen: in base Almaigne Ciceren.

3 The thirde kinde is called Cicer syluestre, that is to say, Wilde Ciches.

❧ *The Nature.*

The Ciche pease is hoate and drie in the first degree.

❧ *The Vertues.*

1 The domestical or tame Ciches, prouoke vrine, and cause milke to encrease in womens brestes, it taketh away the euyll colour, and causeth good colour to ensue. A

The same boyled with Orobus (called in Englishe the bitter fiche) doth asswage and heale the blastinges or swellinges of the yearde or priuie members, if it be layde thereon: also men vse with great profite, to applie it to running sores, and vlcers of the head, and the scurffe. B

The same mingled with Barley meale and honie, is good against corrupt and festred sores, and Canckers, being layde thereuppon. C

2 Sheepes Ciches do prouoke vrine, and vnstoppe the Melt, the Liuer, and the kidneyes: and the decoction thereof drunken, breaketh the stone, and grauell. D

To conclude, the Ciche peason do wast, clense, and make thinne, all cold and grosse humours, and are good agaynst all spreading sores, and the inflammations E

480 The fourth Booke of

tions and swellinges behinde the eares.

¶ They do likewise nourish sufficiently, but they engender muche windinesse.

☙ *The Danger.*

The vse of Ciches is not very good for them whiche haue any vlceration, in the kidneyes or bladder, for they be to much scouring, and do cause the vrine to be sharpe.

Of Lupines. Chap. lxiij.

☙ *The Kindes.*

THere be two sortes of Lupines, the white or garden Lupine, and the wild Lupine. The wild kind agayne is of sundrie sortes for somtimes you shall see some of them with a yellowe flower, sometymes with a blewe flower, and sometimes with a reddishe flower.

Lupinus Satiuus.   Lupinus syluestris. Lupines.

☙ *The Description.*

1. THE tame or garden Lupine hath round harde stemmes, standing vpright of him selfe, without any succour stay or helpe, eyther of bowes, or branches: and after it hath brought foorth his first flowers, then it parteth it selfe aboue, into three branches, which when they haue also brought foorth their flowers, euery of the sayde branches doth part and diuide them selues agayne into three branches, continuing so in flowers & parted branches vntill they be hindered by frostes. The leaues are cut and slit downe into fiue, sixe, or seuen partes. The flowers do grow many togither at the end, or parting of the stalkes, after whiche flowers there come in their places long coddes, somwhat rough without. The fruit is white and flat like a cake, in taste very bitter.

The

the Historie of Plantes.

2 The wilde Lupine hath yellow flowers, and is very like to the aforesayd, sauing that his leaues and stalkes are much lesse, & his flowers are not white, but yellow, and the seede or fruite is not white, but spotted.

3 The wilde Lupines, with the blewe and red flowers, are yet lesser then the yellow, the fruite is also marked or spotted, and it is the least of the Lupines.

❧ The Place.

The Herboristes do plante Lupines in their gardens. The wilde with the blewe, do growe amongst the corne about Monpellier.

❧ The Tyme.

In warme Countries and hoate seasons, the Lupine floweth three times a yere. The first flower commeth foorth about the end of May, afterward the three first collaterall branches do spring out, the whiche three branches do likewise flower about the beginning of July. The sayd collaterall bowes or branches, do agayne bring foorth three other branches, & they do flower in August, where as they be well placed in the Sonne. The fruite of the first and seconde blowing doth come to perfect ripenes in this Countrie, but the thirde blowing doth hardly come to ripenesse, except it chaunce in a very hoate sommer.

❧ The Names.

This kinde of pulse is called in Greeke θέρμος: in Latine and in Shoppes, Lupinus: in Frenche *Lupin* in English Lupines: in high Douche Feigbonen: in base Almaigne Lupinen, and Wijchboonen.

The first kinde is called Lupinus satiuus, that is to say, The manured or garden Lupine.

The three other sortes are called Wilde Lupines, in Latine Lupini syluestres: and these be not vsed in medicine.

❧ The Nature.

The garden Lupine is hoate and drie in the seconde degree, it hath vertue to digest, make subtil, and to clense.

❧ The Vertues.

The meale of Lupines taken with hony, or els with water and vineger, doth kill and driue foorth by siege al kindes of wormes. The same vertue hath the decoction of Lupines, when it is dronken. And for the same purpose men vse to lay Lupines stamped vpon the nauel of young children fasting. **A**

Men giue the decoction of Lupins, boyled with Rue and Pepper, to drinke to open the stoppinges of the liuer and melt. **B**

A pessarie made of Lupins, Mirrhe, and Hony mingled togither, mooueth womens natural sicknesse or flowers, and expelleth or deliuereth the dead birth. **C**

The decoction of Lupines doth beautifie the colour of the face, and driueth away all frekles, and spottes like lentils. The meale thereof is of the like vertue, mingled with water and layde therto. **D**

The flower or meale of Lupines, with the meale of parched barley & water swageth all impostumations and swellinges. **E**

The same with vineger, or boyled in vineger, swageth the payne of the Sciatica, it digesteth, consumeth and dissolueth the kinges euill or swelling in the throte, it openeth and bursteth wennes, botches, boyles, and pestilential or plague sores. **F**

Lupins may be eaten, when as by long soking in water they are become sweete, and haue lost their bitternesse: for when they be so prepared, they take away the lothsomnesse of the stomacke, and the desyre to vomit, and do cause good appetite. Yet for all that this kinde of foode or nourishment, engendreth **G**

Ss grosse

grosse blood, and grosse humours. For Lupins are harde to digest, and vneasie to descende, as Galen saith.

The wilde Lupins haue the lyke vertue, but more strong.

Of the bitter Vetche called in Greeke Orobos, and in Latine Eruum. Chap.xciiij.

The Description.

ERuum or the bitter Fetche is nowe vnknowen, and therefore we can geue none other description, but so much as is written in Dioscorides and Galen. They say that Orobos, or Eruilia is a small plant, bearing his fruit in coddes, round, of a white or yellowish colour, of a strange and vnpleasant taste, so that they serue not to be eaten, but of cattel, neither wil cattell feede vpon them, before that with long soking or stieping in water, their vnpleasant taste be gone and lost: wherefore it is very easie to iudge, that the flat Pease called in Greeke Lathyri, and described in the xxj. Chapter of this booke, are not Ers or Eruilia, as some haue thought: for those flat Peason are in taste lyke the common Peason, as we haue before declared.

The Names.

This pulse is called in Greeke ὄροβος: in Latine Eruum: and the Frenchmen folowing the Latine name, do cal it Ers: in Douche Eruen: in Englishe Bitter Vetche, or Ers.

The Nature.

Ers are hoate in the first degree, and drie in the seconde.

The Vertues.

The meale of Eruum often licked in with hony in maner of a Lohoch, clenseth the breast, and cutteth and ripeth grosse and tough humours, falling vpon the lunges. [A]

It loseth the belly, prouoketh vrine, maketh a man to haue a good colour, if it be taken in reasonable quantitie: for to muche thereof is hurtfull. [B]

With honie it scoureth away lentiles or freckles from the face, and all other spottes and scarres from the bodie. It stayeth spreading vlcers: it doth soften the hardnesse of womens breastes, it breaketh Carboncles and impostumes. [C]

Being kneded or tempered with wine, it is layde very profitably vnto the bitinges of dogges, of men, and wilde beastes. [D]

The decoction of the same, helpeth the itche, and kibed heeles, if they be washed therein. [E]

Ers are neuer taken in meate, but it fatteth oxen well. [F]

The Danger.

Ers or Orobos being vsed often, and in to great a quantitie causeth headache, and heauie dulnesse, it bringeth foorth blood, both by the vrine, and excrementes of the belly.

Of the Vetche. Chap.xrv.

THe Vetche hath stalkes of a sufficient thicknesse, and square about the height of three foote, with leaues displayed & spread abroad, compassed about with many small leaues, set opposite one directly against another: at the ende of whiche leaues, ye haue tendrelles or claspers wherby it taketh hold and is stayed vp. The flowers are purple and fashioned like the Beane flowers, afterward there come vp long flat coddes, wherein are Vetches, which are flat and of a blackish colour.

the Historie of Plantes. Vicia. 483

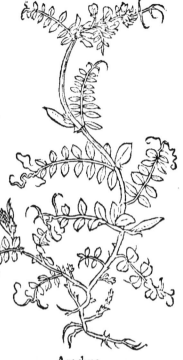

❊ *The Place.*

They sowe Uetches in this Countrie, in the feeldes, for fourrage or prouender for Horses.

❊ *The Tyme.*

They be rype in July and August.

❊ *The Names.*

This pulse is called in Greke βίκιον, and of some σαρακον: in Latine Vicia: of some Osmūdi: in English a Uetche, or Uetches: in Frenche Vesse: in high Douche Wicken: in base Almaigne Witsen. And that this is not Eruum, it appeareth euidently by that which is described in the former Chapter. This shoulde seeme to be Theophrastus Aphace or Taare.

❊ *The cause of the Name.*

The Uetche, as Uarro sayth, is called in Latine Vicia, bycause it bindeth it selfe about other plantes, and ouercometh them, and is deriued from this Latine worde (Vincire) whiche signifieth to binde sure, to ouercome and to restrayne from libertie.

❊ *The Nature and Vertues.*

The Uetche is not vsed in medicine, neyther vsed to be eaten of men, but to be giuen to Horses and other catttel, and this Galen doth also witnesse.

Arachus.

Of Arachus or wilde Fitche. Chap. xxvi.

❊ *The Description.*

Rachus is muche lyke to the common Uetche, in stalkes, leaues, and coddes, but in all these muche lesse. The stalkes be tēder, weake and slēder with cornered traples or square crested edges. The leaues are spread abroade like the other Uetche, but clouen and parted aboue at the endes, into two or three clasping tendrelles. The flowers be smal, of a light purple, or incarnate colour, and do growe vppon the stalke scife, as the flowers of beanes or common Uetches do, without any foote stalkes. The coddes be small, long, and narrowe, wherein is couched fiue or seuen seedes, of a blackishe colour, harde, and smaller than Uetches.

Ss ij Of

The fourth Booke of

2 ¶ Of this sorte there is found an other kinde, the which is very wel like to the abouesaid, in leaues and stalkes, but it is smaller. It hath smal white flowers growing clusterwise at the ende vpon long stems, almost like the wild Vetche, the whiche do turne into litle short huskes, clustering togither, smaller then the lentil huskes, in the whiche is founde, but onely two graynes, harde, rounde, gray speckled, blackish, in making and taste like to the Vetche.

❧ The Place.

These two kindes of Vetches do grow in the feeldes, amongst Rye & Otes, and other lyke graynes.

❧ The Tyme.

They are both ripe in June.

❧ The Names.

This plant is called of Galen, Lib.1.de alimentorum facultatibus, ἄραχος, A-rachus, the whiche name is written by ch in the last syllabe, as a difference from the other ἄρακος Aracus, written with a c, wherof we haue alredy treated. They cal it in French Vesseron: in Brabant, Crock: in Englishe, Wilde Vetche.

❧ The Nature, and Vertues.

Arachus or the wild Vetche, is not fit for man, but serueth only for prouender or fourrage, for Beues and horses, vnto whom the whole herbe is giuen.

Of smal wild Fetihelinges. Chap. xcvij.

Galega altera. **Vetcheling or smal wild Fitches.**

Onobrichis forte, Medica Ruellij. Saint Foin. **Medick Vetcheling.**

1 ¶ THE wild Vetche is much like Arachus, described in the former Chapt. in stalkes leaues and clasping tendrelles, but that his flowers grow not in the like order, but do grow

in tuffed clusters about long stemmes, almost like to spiked eares, of colour purple in blew, the which past & gone, there come vp litle flat huskes, wherin lieth the seede, like to the seede of Arachus.

2 You may set by this wild Vetche, a certaine plant not much vnlike the aforesayd in leaues and growing, the whiche beareth rough and prickie buttons, or bosses. It doth not commonly growe in this Countrie, but planted in the gardens of some Herboristes. This kind of plant hath leaues like to the other, but somwhat narrower, whiter & smother. His flowers do likewise growe thicke vpon long stemmes, commonly of a cleare red or Crymsen colour. After whiche there come flat prickley round huskes, bossed or bunched, and somtimes fashioned like a smal Hedgehogge, which is nothing els but the seede.

❧ *The Place.*

1 The wilde Vetche groweth in the borders of feeldes, in medowes, & oftentimes in moyst places, and about water courses, and running streames.

2 The other kind (for daintines sake) is planted in the gardens of the Herboristes of this Countrie. They say it groweth plentifully by the Sea side, vpon bankes or trenches made with mans handes, and such like places. They vse to sow it in medowes about Paris, and otherwhiles: it is found growing there of his owne accorde. ❧ *The Tyme.*

The wilde Vetche flowreth most commonly in June, and soone after it deliuereth his coddes and grayne.

The other flowreth in July, and for the most part deliuereth his seede foorthwith. ❧ *The Names.*

1 The first should seeme to be Galega altera, & a kind of litle Vetches, & may also be wel called Arachus, & taken for a kinde of Arachus: in Frenche *Vesce sauuage*: in Douch Wilder Wicken: in neather Douchlande, Wilde Vitsen: in English Small wilde Vetches or Vitchelinges.

2 The other is counted of some to be ὀνόβρυχις, Onobrychis, of Ruelius for Medica, they name it in Frenche *Saint Foin*: we may call it, Yellow Fitcheling, and Medick fitche. ❧ *The Nature and Vertues.*

The wilde Vetche is no better than Arachus, and therefore it serueth onely but for pasture, and feeding for cattell, as other like herbes do.

If the other be Onobrychis, you shall finde his properties described in the B Chapter of Onobrychis.

Of Tares. Chap. xxviij.

❧ *The Description.*

The Tare hath long, tender, square stalkes, longer and higher then the stalkes of the lentil, growing almost as high as the wheat or corne, or the other plantes whereamongst it groweth. The leaues be smal and tender (triangled like a scuchion) somwhat round, growing alwaies two togither, one against another at the ioyntes, betwixt the said leaues there grow vp clasping tendrels, & other smal stems or shutes, wherevpon growe flowers, of a yellowish colour. The flowers past there rise coddes somewhat large, & longer then the coddes or huskes of the Lentiles, in whiche is conteyned fiue or sixe blacke seedes, harde, flat, and shining, lesser then the seedes of lentiles.

❧ *The Place.*

The Tare groweth in feeldes, & is found growing in this Countrie, in fertil groundes amongst wheat & Rye. ❧ *The Tyme.*

In this Countrie it flowreth in May, and in June and July the seede with the coddes is ripe. ❧ *The Names.*

This kinde of Pulse is called in Greke ἀφάκη: in Latine Aphaca: in English,

Ss iij   Tares:

Tares: vnknowen in shoppes, this is the Aphace of Dioscorides & Galen: for it should seeme, that the Uetche is the Aphace of Theophrastus.

❦ *The Nature.*

The Tare is temperate in heate, & of like nature to the Lentil: but drier.

❦ *The Vertues.*

A The Tare seede is of a restringent vertue like y Lentil, but more astringent, for it stoppeth the fluxe of the belly, and drieth vp the moysture of the stomacke.

B The Tare in vertue is lyke to the Lentil.

C Men in tyme past dyd vse to eate this pulse (as witnesseth Galen) neuerthelesse it is harder of concoction or digestion, then the Lentil.

Of Birdes foote. Chap. xxix.

Ornithopodium.

❦ *The Description.*

BIrdes foote is lyke to Arachus, & to the wilde Uetche, but far smaller. It hath very slender and small stalkes or branches, soft and tender, the leaues be small and rounde, fashioned like to a small fether. The flowers be yellowishe and small, growing close togither vpon huskes or stems, the which being withered, there commeth vp in their places small crooked huskes or coddes, growing fiue or sixe togither, y which in their standing do shewe almost like the closing foote of a smal bird. Within the sayde litle crooked coddes the seede is inclosed, in fashion not much vnlike Turnep seede.

❦ *The Place.*

Birdes foote groweth in certaine fields, and is likewise found in high medowes, & in drie grassie wayes & Countries. That which groweth in medowes, and grassie wayes, is a great deale smaller, then that which groweth amongst the corne.

❦ *The Tyme.*

Birdes foote floureth from after the moneth of Iune, vntill September, and within this space it deliuereth his seede.

❦ *The*

the Historie of Plantes. 487

❧ *The Names.*

This wild herbe is called in Brabant Voghelvoet, that is to say in English, Birdes foote, or Fowle foote, bycause his huskes or cods are lyke to a birdes foote, & for that cause men may wel cal it ὀρνιθοπόδιον, Ornithopodion, for it hath none other Greke nor Latine name (that I know) except it be that Polygala of Dioscorides, as it may be called, whervnto it is very like.

❧ *The Nature and Vertues.*

This herbe is not vsed in medicine, nor receiued any wayes for mans vse, but is a very good foode both for horses and cattel.

Of Lentilles. Chap.xxx.

❧ *The Description.* Lens.

THe Lentil hath small tender and plyant braunches, about a cubite high. The leaues be very smal, the which are placed two and two vppon litle stems, or small footestalkes, and do sometimes ende with clasping tendrelles, wherby it hitcheth fast and taketh sure hold. The flowers be smal, of a brownishe colour, intermixt with white. The huskes or shelles are flat. The fruite is round and flat, of colour now blacke, now white, and sometimes browne.

❧ *The Place.*

The Lentil is not very wel knowen in this Countrie, but is founde sowen in the gardens of Herboristes.

❧ *The Tyme.*

The Lentil doth both flower and waxe ripe in July and August.

❧ *The Names.*

This Pulse is called in Greeke φακός, & φακῆ: in Latine Lens, and Lenticula, by whiche name it is knowen in Shoppes: in Englishe Lentilles: in Frenche Lentille: in high Douche Linsen.

¶ *The Nature.*

The Lentil is drie in the seconde degree, the residue is temperate.

❧ *The Vertues.*

The first decoction of Lentilles doth lose the belly. A

If after the first boyling you cast away the broth wherein they were sodden, and then boyle them agayne in a freshe water: then they binde togither and drie, and are good to stop the belly, and agaynst the blooddy flixe or dangerous laske. also they stoppe the inordinate course of womens termes, but it wyll make their operation more effectual in stopping, if you put vineger vnto them, or Cichorie, or Purselayn, or redde Beetes, or Myrtilles, or the pill of Pomegarnates, or dried Roses, or Medlers, or Seruices, or vnripe binding Peares, or Quinces, or Plantayne, and whole Gawles, or the berries of Sumach. B

Ss iiij The

The pill or shel of Lentiles hath the like propertie, and in operation, is of C more force then the whole Lentil.

The meate that they vse to make of the huſked or vnſhelled lentil, drieth the D ſtomacke, but it ſtoppeth not, and is of harde digeſtion, and engendreth groſſe and noughtie blood.

They vſe to ſwallow downe thirtie graines of Lentilles ſhelled, or ſpoyled E from their huſkes, againſt the weakeneſſe, and ouercaſting of the ſtomacke.

The lentil boyled with parched barley meale, & laid to, ſwageth the paynes F or ache of the gowte.

The meale of Lentiles, mixed with hony, doth mundifie and clenſe corrupt G vlcers and rotten ſores, and filleth them againe with newe fleſh.

The ſame boyled in vineger, doth diſſolue and driue away wennes, and H harde ſwelling ſtrumes.

With Melilot, a Quince & oyle of Roſes, they helpe the inflammation of the I eyes and fundement, and with ſea water it is good againſt the hoate inflammation called Eriſipilas, S. Antonies fier, and ſuche lyke maladies.

The lentil boyled in ſalt (or ſea) water, ſerueth as a remedie againſt clotted K cluſtered milke in womens breaſtes, & conſumeth the abundant flowing of the ſame. ⁂ *The Danger enſuing the vſe of this pulſe.*

The Lentil is of hard digeſtion, it engendreth windineſſe, and blaſtinges in the ſtomacke, & ſubuerteth the ſame, they cauſe doting madneſſe & fooliſh toyes, and terrible dreames: it hurteth the lunges, the ſinewes and the braine. And if one eate to muche thereof, it dulleth the ſight, and bringeth the people that vſe thereof, in danger of Cankers, and the Lepzie.

Of Hatchet Fitche, Axſede or Axwurt. Chap.xxxi.

Hedyſaron. Securidaca. Axeſiche, or Axwurt.   Securidaca altera.

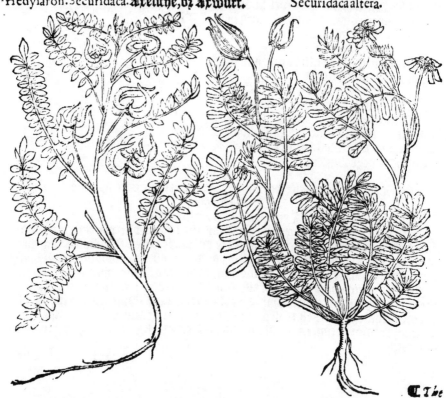

the Historie of Plantes. 489

❋ *The Description*

1 SEcuridaca hath small slender stemmes, wherevpon grow small leaues spread abroade lyke to the leaues of Arachus, or the wilde Vetche, but smaller and rounder. The flowers growe thicke togither, after the sayde flowers, there come long coddes, round and small, growing one agaynst another, bowing inward, & crooking or crompling lyke vnto hornes, within whiche crooked coddes, is conteyned a flat seede with flower corners, like to a litle wedge. The roote is smal and tender.

2 The other Securidaca set foorth by Matthiolus, in leaues is very well lyke to the aforesayd, neuerthelesse his coddes be longer, and not so much crooked: and for that consideration not very well approching to the description of Securidaca, set foorth by Dioscorides.

Ferrum equinum. Horse shoe.

3 There is found another herbe with many stalkes, trayling vpon the ground, hauing at euery ioynt a leafe, or rather a branche of leaues, very well like to the leaues of the Lentil or Securidata, but yet somewhat lesser, betwixt those leaues, & the trayling branches it beareth smal yellow flowers, in fashion lyke to the flowers of the Vetche or Lentill: the which afterward do change into flat huskes or coddes, the whiche are vpon one side full of deepe Chinkes, or Cliftes, and the graine or seede waxing ripe, the saide coddes do turne crooked vpon one side, so that they beare the forme and fashion of a horse shoe. The sede also is crooked, and turned rounde like a Croissant or newe Moone.

¶ *The Place.*

1.2 Securidaca, as Galen saith, groweth in some places amongst Lentiles: or according to Dioscorides, amongest Barley and wheate, vnknowen in this Countrie, and is not knowen to growe but in the gardens of Herboristes.

3 The thirde kinde groweth in some places of Italie, and of Languedoc, alongst the wayes, and like vntoyled places.

❋ *The Tyme.*

1.2. Securidaca flowreth in this Countrie, in July and August, and afterwarde the seede is ripe.

3 The Horse shoe flowreth in June, and July, and the seede is ripe in August.

❋ *The Names.*

1.2. The Pulse called in Greke ἡδύσαρον: in Latine Securidaca, of some also Pelecinon: in Douch and French it hath no name that I knowe. Turner calleth it in Englishe Arsuch, or Axeworte, bycause Dioscorides saith the seede is lyke a two edged Axe. The

490   The fourth Booke of

3   The thirde kinde is called in Italian *Fer di Cauallo*, that is to say in Latine, Ferrum equinum: and in English Horse shoe: in Frenche *Fer de Cheual*: in Brabant Peerts ysere. It shoulde seeme that this is a kinde of Securidaca: and therefore we haue placed it in this Chapter.

※ *The Nature.*

The seede of Securidaca is hoate and drie of complexion.

The Horse shoe is in qualitie and vertue lyke to Securidaca, as you may knowe by his bitternesse.

※ *The Vertues.*

The seede of Securidaca, openeth the stoppinges of the liuer, the Spleue, and all the inwarde partes, and is very good for the stomacke, bycause of his bitternesse.

Of the like vertue are the newe leaues and tender croppes of the same.

Of Italian Fitche, or Goates Rue.   Chap.xxxij.

※ *The Description.*   Galega.

This herbe is not muche onlyke Arachus or the wild Vetche in stalkes and leaues: it hath round hard stalkes, and therupon displayed leaues, made of diuers small leaues lyke to the leaues of Vesseron or Arachus, but muche greater and longer. The flowers be eyther cleare blewe, or white, and do grow clustering togither spikewise, and like to the wild Vetche, after come long, small, and round coddes, wherein is the seede. The roote is meetely great, and doth not lightly die.

※ *The Place.*

Galega in some Countries (as in Italy) groweth in the borders of feeldes, it groweth also in the wood called Madrill by Paris. Ye shall not lightly finde it in this Countrie, but sowen in the gardens of Herboristes.

※ *The Tyme.*

Galega flowreth in July and August, and foorthwith the seede is ripe.

※ *The Names.*

This herbe is called of the Herboristes of these dayes, in Latine Galega, Ruta Capraria, and of some Fœnogræcum syluestre. And some do also count it to be Glaux, or Polygala, but as I thinke it is nothing lyke any of them: it is called in English, Italian Fetche, and Goates Rue.

¶ *The Nature.*

Galega is of nature hoate and drie.

※ *The Vertues.*

Galega, as Baptista Sardus wryteth, is a singuler herbe against al venome and

the Historie of Plantes.

and poyson, and against wormes to kill and driue them foorth, if the iuyce of it be giuen to little children to drinke.

It is of like vertue fried in Oyle of Lineseede, and bounde vpon the nauel B of the childe.

They giue a sponefull of the iuyce of this herbe euery morning to drinke, to C young children against the falling sicknesse.

It is counted of great vertue, to be boyled in vineger, and dronken with a D litle Treacle, to heale the plague, if it be taken within twelue houres.

Of the Pease Earthnut.   Chap.xxiij.

❧ *The Description.*   Chamæbalanus.

1 THE Earthnut hath three or foure little stalkes or tender branches, somewhat reddishe belowe next the grounde, with clasping tendrelles, whereby it taketh holde vpon hedges, and al other thinges that it may come by. The leaues be small and narrowe. The flowers be of a fayre red colour, and of an indifferent good smell. After the fading of those flowers there come in their steede small coddes, in which is conteined a small seede. The rootes be long and small, whereunto is hanging here and there certeyne nuttes or kernels like Turneps, of an earthlike colour without, and inwardly white, sweete in taste, almost lyke the Chesnut.

2 The other kinde of Earthnut, called in some places, the litle Earthnut, shalbe described in the fifth part of this history, in the xxiij.Chapter.

❧ *The Place.*

The Pease Earthnut, groweth abundantly in Hollande and other places, as in Brabant, neare Barrow, by the riuer Zoom, amongst the Corne, and vppon, or vnder the hedges. It groweth in Richmonde heath, and Coome parke, as Turner saith.

❧ *The Tyme.*

This herbe flowreth in June, and afterwarde the seede is ripe. In some places they drawe or plucke vp the rootes in May, and do eate of them.

❧ *The Names.*

This herbe is called in high Douche Erdnusz, Erckelen, Erdfeigen, Erdamandel, Acker Eychel, and Grund Eychel: in Brabāt Eerdtnoten, and of some Muysen met steerten: of the writers in these dayes in Greeke, χαμαιβάλανΘ. Chamæbalanos: but this is not that Chamæbalanus, whiche is called ίσχας, and άπιΘ.: in Latine Glandes terrestres, that is to say, Earthnuttes. Some of the learned do count it to be Astragalus described by Dioscorides, and some hold it for Apios. But that it is not Apios, it is manifest ynough by the third Chapter of the thirde parte of this booke, where as we haue playnely set foorth the right Apios.

❧ *The*

492 The fourth Booke of

❧ *The Nature.*

The peafe Earthnut is drie in the seconde degree.

❧ *The Vertues.*

The rootes of peafe Earthnut, are boyled in many places of Hollande and Brabant, and eaten as the rootes of Turneps and Parsneps, and they nourish aswel: yet for all that they be harder of digestion then Turnep rootes, and do stop the belly, and running of the laske.

If these herbes be the right Astragalus, his roote wil prouoke vrine, and stop all fluxes of the belly, being boyled in wine and dronke.

The same receiued in the same manner, stoppeth also the inordinate course of womens flowers, and all vnnaturall fluxe of blood.

The same roote of Astragalus dried and made into powder, is very good to be strowen vpon olde sores, and vpon freshe newe woundes, to stop the blood of them.

Of Fenugrec. Chap.xxxiiij.

❧ *The Description.*

Fenugreck hath tender stalkes, rounde, blackishe, hollow, and ful of branches, the leaues are diuided into three partes, lyke the leaues of Trifoil, or the threeleaued grasse. The flowers be pale, whitish, and smaller then the flowers of Lupins. After the fading of those flowers, there come vp long coddes or huskes, crooked and sharpe pointed, wherein is a yellowe seede, the roote is ful of small hanging heares.

❧ *The Place.*

The Herboristes of this Countrie, do sowe it in their gardens.

❧ *The Tyme.*

It flowreth in July, and the seede is ripe in August.

❧ *The Names.*

This herbe is called in Grecke τῆλις: in Latine and in Shoppes Fœnum Græcum, of Columella Siliqua: in Frenche *Fenugrec, or Fenegrec*: in high Douche Bockshorn, or kuhorne: in base Alemaigne Fenigriek: in English Fenegreck.

❧ *The Nature.*

The seede of Fenugreck, is hoate in the second degree, and dry in the first, and hath vertue to soften and dissolue.

❧ *The Vertues.*

A The decoction or broth of the seede of Fenugreck, dronken with a litle vineger, expelleth al euil humours, that sticke fast to the bowels.

B The same decoction first made with Dates, and afterward with a litle Hony, vntil it haue gotten the substance or thicknesse of a Syrupe, doth mundifie & clense the breast, and is very good for greeues

Fœnum Græcum.

and

the Historie of Plantes.

and diseases of the breast, so that the patient be not vexed with a feuer or the head ache: for such a syrupe is hurtfull to the head, & to them that haue agues.

The meale of Fenugreck, boyled in Meade or Honied water doth consume, soften, and dissolue colde harde impostumes and swellinges. The same tempered or kneded, with Saltpeter and vineger, doth soften & waste the hardnesse, and blasting of the Melt.

It is good for women that haue either impostume, vlcer, or stopping of the Matrix, to bathe and sit in the decoction thereof.

The strayning or iuyce of Fenugreck mingled with Goose grease, & put vp, vnder, in the place conuenient, after the maner of a mother Suppositorie, doth mollifie and soften all hardnesse, and paynes of the necke of the Matrix, or the naturall place of conception.

It is good also to washe the head with the decoction of Fenugreck: for it healeth p scurffe, and taketh away both nittes, and scales, or brand of the head.

The same layd to with Sulphur (that is, brimstone) & hony driueth away pusshes or little pimples, wheales, and spottes of the face: and healeth al manginesse and scuruie itche, and amendeth the stinking smell of the armepittes.

Greene Fenugreck bruised, or pounde with a litle vineger, is good agaynst weake and feeble partes, that are without skinne, vlcerated, and rawe.

The seede of Fenugreck may be eaten, being prepared as the Lupines, and is then of vertue like, and looseth the belly gently.

Men do also vse to eate of the young buddes and tender croppes in salades with oyle & vineger (as Galen saith) but such meate is not very holesome, for it ouerturneth the stomacke & causeth headache, to be vsed to much, or to often.

Of Cameline. Chap. xxxv.

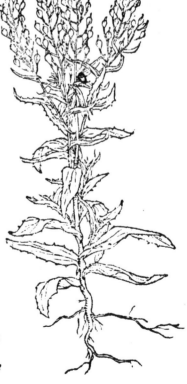

Myagrum.

The Description.

Myagrū or Cameline, hath straight rounde stalkes, of p height of two foote or more, diuiding it self into many branches or boughes. The leaues be long and narrowe, almost lyke to the leaues of Madder, at the highest of p stalkes, alogest by thē grow smal flowers, and afterward smal cuppes, or sede vessels, almost lyke the knoppes of Lineseede, but flatter, within the which is inclosed a small yellowish seede, of the whiche an oyle is made, by pounding, and pressing foorth of the same.

The Place.

This herbe groweth in many places amongst wheat, and flaxe, and the most part of mē do deeme it as an vnprofitable herbe: yet for all that it is sowen in many places, bycause of the oyle which the sede yeeldeth, as in this Countrie, Zeelande, and the Countrie of Liege, or Luke beyonde Brabant.

The Tyme.

It is sowen in Marche and Aprill, and ripe about August.

Tt   The

The fourth Booke of

❧ The Names.

This herbe is called in Greeke μύαγρον, of some also μελάμπυρον: in Latine Myagrum, Linum triticeum, and Melampyrum: in Frenche and base Almaigne, Cameline and of some Camemine: in high Douche Flaschdotter, and Leindotter: in English Myagrū, or Cameline: It should seeme that this herbe is Erysimon, wherof Theophrast & Galen, Lib.1. de alimentorum facultatibus, haue written.

❧ The Nature.

Myagrum is of a hoate complexion.

❧ The Vertues.

The oyle of the seede of Cameline, or Myagrum straked, or annoynted vpon the body, doth cleare and polish the skinne from all roughnesse. **A**

It is vsed at this day to prepare and dresse meate withal, as Rape oyle, and it may be vsed to burne in lampes. **B**

Of medow Trifoyl, or three leaued grasse. Chap.xxxvi.

❧ The Kindes.

There be diuers sortes of Trefoyles, the which for the more parte of them shalbe set foorth in diuers Chapters. But that kinde wherof we shal now intreate, is the common Trefoyl, growing in medowes. The whiche is of two sortes, the one with redde flowers, the other with white, but for the rest there is no great difference in these two herbes.

❧ The Description.

Trifolium pratense.

1. Three leaued grasse of the meddowe, hath a rounde tender stalke: and leaues somewhat rounde, alwaies standing togither vpon a stemme, the flowers do grow at the top of the branches or stemmes, in tuftes or knoppes tuffed, and set full of small flowers, of a red purple colour: lyke to a short tuffed eare, the whiche flowers once vanished, there commeth vp rounde seede inclosed in small huskes. The roote is long, and of a wooddy substance.

2. The Trefoyl with the white flowers, is muche lyke to the aforesayde, but that his stalkes are somwhat rough and heatic, and the leaues be longer and narrower, and in the middle of euery leafe is sometimes a white spotte, or marke, lyke to the new Moone. The flower is white, in all thinges els lyke to the other, and groweth after the selfe same fashion.

❧ The Place.

These two kindes of Trefoyles, doo growe in all places of this Countrie, in medowes, especially suche as stande somwhat high.

❧ The Tyme.

The three leaued grasse flowreth in May and June, and sometimes all the Sommer.

The Names.

This kinde of Trefoyl is called in Greeke τρίφυλλον ἐν χορτοκοπείοις γινόμενον: in Latine Trifolium pratense: in Frenche Treffle de pres, or Triolet : in high Douche Wisen klee, and Fleyschblum: in base Almaigne Claueren, and Ghemeyn Claueren: in English, Medow Trefoyle, or Common Trefoyle.

The Nature.

The Trefoyl is colde and drie, as one may easily know by the taste thereof.

The Vertues.

Trefoyl with his flowers, or by him selfe, boyled in Meade, or honied water, or wine, and dronken, doth slake and swage the hoate burning and fretting of the bowels and inwarde partes. Of the like vertue is the decoction therof, made in water, and powred into the body by glister.

The same decoction dronken in due time, and season, stoppeth the white flowers in women.

The flowers or leaues of Trefoyle sodde in Oyle, and layd to in manner of a plaister, doth ripe hoate inflammations and swellinges, & other like tumors, and breaketh them, yea sometimes they do scatter and dissolue them cleane.

Of sweete Trefoyl. Chap.xxxvij.

The Description

Lotus satiua, or Vrbana.

Sweete Trefoyl hath a round holow stalke, of two or three foote long or more, full of bowes and branches.

The leaues do alwaies grow three and three togither, euen as the common medow Trefoyl, but somwhat longer, & iagged round about like a Sawe. At the top of the branches grow flowers, clustering togither in knoppes, like the flowers of the medowe or common Trefoyl, sauing that the tuftes or knappes, are not so great as the knappes of the other: after the fading of those flowers there come little huskes, or sharpe poynted heades, wherin the seede lyeth. Al the herbe, especially when it is in flower, is of a very good smel or sauour, the whiche as some say, looseth his sent or smell seuen times a day, & recouereth it againe as long as it is growing, but being withered and dried, it keepeth still his sauour, the whiche is stronger in a moyst and cloudy darke season, then when the wether is fayre and cleare.

The Place.

In this Countrie men sowe the sweete Trefoyl in gardens, & where as it hath bene once sowen, it groweth lightly euery yere of the seede which falleth, *In Maister Riches garden.*

The Tyme.

This herbe flowreth in July & August, during which time, the sede is ripe.

The Names.

This herbe is called in Greke λωτὸς ἥμερος: in Latine Lotus vrbana, & Lotus satiua, of some Trifolium, and now a dayes Trifolium odoratum: in French Trefle odoriferant:

odoriferant: in high Douch Sibengezeyt: and in base Almaigne, Seuen getijde cruyt, bycause that seuentymes a day it looseth his sweete sente and smell, and recouereth it againe. Turner calleth Lotus vrbana in English, Garden or Hallet Clauer: we may call it sweete Trefoyl, or three leaued grasse.

✤ The Nature.

Sweete Trefoyl, is temperate in heate and cold, & taking part of some litle drynesse.

✤ The Vertues.

The sweet Trefoyl doth swage & ripe, al cold swellings, being laid therto.

The iuyce of the same taketh away the spot or white perle of the eyes, called in Latine Argema.

The Oyle wherin the flowers of the sweet Trefoyl haue ben soked, cureth all new woundes, and burstinges, as some affirme.

Of Wilde Lotus. Chap. xxxviij.

✤ The Kindes.

There is commonly founde in this countrie, two sortes of wilde Lotus or Trefoyl, with yellow flowers, one hauing Coddes, and the other none.

Lotus syluestris.	Lotus syluestris minor.
Wild Trefoyl. Yellow stone Clauer.	Petie Clauer or stone Trefoyl.

✤ The Description.

1. The first kinde of wilde Lotus is a litle low herbe, creeping alongst the grounde. The leaues be somewhat lyke to the leaues of the common three leaued grasse, or medow Trefoyl, almost of an asshe colour. The flowers be faire and yellow, fashioned like to the flowers of peason, but muche smaller: the whiche decayed and fallen away, there come vp three or foure round coddes, standing togither one by another, wherein is conteyned a round sede. The roote is long & reddish.

The

2 The seconde kinde hath rounde stalkes, and very small. The leaues be like to medow Trefoyl. The flowers be yellow, growing thicke togither in round knopped heades, the which do chaunge into a rounde crooked blacke seede, couered with a blacke hushe or skinne. The whiche seede groweth rounde about the knoppes, orderly compassing the same.

❧ *The Place.*

These two kindes of wilde Lotus, or Trefoyl do grow in this Countrie in drie places, alongst the feeldes and high wayes.

❧ *The Tyme.*

These Tresoyles, are in flower from after the moneth of June, al the rest of the Sommer, and in the meane season they yeelde their seede.

❧ *The Names.*

These Tresoyles, are nowe called Loti sylvestres, yet they be not the Lotus sylvestris of Dioscorides, the which groweth very high, and hath seede lyke to Fenugreck.

1 The first kind is called in high Douch, Wilden klee, Steenklee, Edelsteenklee, Vogels wicken, Unser Frawen schuchlin: in Frenche *Tresle sauuage iaulne:* in Brabant Steenclaueren, and Geelsteenclaueren, Wilde Claueren, and of some Vogels Vitsen. Some take it for a kinde of Melilotus, and therefore it is called in Latine Melilotus sylvestris, or Melilotus Germanica: in Englishe, The wilde yellowe Lotus, the Germaines Melilot, or the wilde yellowe Trefoyl.

2 The second is called in Frenche *Petit Tresle iaulne:* in high Douche Geelklee, Kleiner, Steenkle, and Geel wisen klee: in base Almaigne, Cleyn steenclaueren, and Cleyn geelclauere. This shoulde seeme to be a kind of Medica, wherof we shall speake hereafter.

❧ *The Nature and Vertues.*

These herbes are colde, drie, and astringent, especially the first: therfore they may be vsed aswell within the body, as without, in al greefes that require to be cooled and dried.

Of Melilot. Chap.xcix.

❧ *The Kyndes.*

There is nowe founde two sortes of Melilot, the one whiche is the right Melilot, and the other whiche is the common Melilot.

❧ *The Description.*

1 THe true and right Melilot, hath rounde stalkes, the leaues iagged rounde about, not muche vnlyke the leaues of Fenugreck, alwayes growing three and three togither like to the Trefoyl. The flowers be yellow and smal, growing thicke togither in a tuft, the which past there come in their places, a many of smal crooked huskes or coddes, wherin the seede is conteyned. The roote is tender, and full of small hearie threddes.

2 The common Melilot hath rounde stalkes, about two or three foote long, & full of branches. The leaues do alwayes grow by three and three, lyke to Trefoyl, hacked rounde about lyke the leaues of Fenugreck, or the right Melilot. The flowers be yellow, clustering togither, after the fashion or order of spike, the whiche vanished, there come vp small huskes, whiche conteyne the seede. The roote is long, al the herbe with his flowers, is of a right good sauour, specially whan it is drie. ❧ *The Place.*

1 The right Melilot groweth plentifully in Italy, especially in the Countrie of Campania, neare the Towne of Nola. In this Countrie the Herboristes do sowe it in their gardens.

2 The common Melilot groweth in this Countrie in the edges and borders of fieldes, and medowes, alongst by diches, and trenches.

Tt iij ❧ The

The fourth Booke of

Melilotus Italica.
The right Melilot.

Melilotus Germanica.
The common Melilot.

❈ *The Tyme.*

These two kindes of Melilot do flower in July and August, during which time they yeelde their coddes and seedes.

❈ *The Names.*

Melilot is called in Greke μελίλωτ⊙: in Latine Melilotus, and Sertula Campana.

The first kinde of these herbes, is taken at Rome and in Italy for Melilot, & therfore is called Melilotus Italica that is to say, Italian Melilot: in French *Melilot d'Italie*: and in Douche, Italiansche, or Roomsche Melilote.

The other kinde is called in Shoppes of this Countrie, and of Almaigne Melilotus, and is vsed for the same, and hereof it commeth to passe that men cal it Melilotus Germanica: in Frenche *Melilot vulgaire*: in base Alemaigne, Ghemeyne, or Douche Melilote. Some do also call it Saxifraga lutea, that is saye, Yellow Sarifrage: and in high Douche, Grosse steinklee: in Englishe, The common and best knowen Melilot.

❈ *The Nature.*

Melilot is hoate, and partly of an astringent nature, and hath part of a digesting, consuming, dissoluing, and riping power.

❈ *The Vertues.*

Melilote boyled by it selfe in sweete wine, or with the yolke of a rosted egge, or the meale of Fenugreck, or Lineseed, or with the fine flower of meale, or with Cichorie, doth smage and soften all kindes of hoate swellinges, especially those that chaunce in the eyes, the matrix, or mother, the fundement, and geni-

the Historie of Plantes. 499

genitors or coddes, being layde thereto.

If it be layde to with Gawles or Chalke, or with good wine, it healeth B the scurffe, and suche sores, as yeelde corrupt matter or filthe.

The same rawe, and pounde, or sodde in wine, swageth the payne of the C stomacke, and dissolueth the impostumes and swellinges of the same, being layde thereto.

The iuyce of the same dropped into the eares, taketh away the payne of D them, and layde to the forehead with oyle of roses and vineger, cureth the head ache.

The common Melilote is vsed and found good for all suche thinges as the E other serueth: it is most vsed to swage and slake payne, as the flower of Camomill is.

The same boyled in wine and dronke, prouoketh vrine, breaketh the stone, F and swageth the payne of the kidneyes, the bladder and belly: and ripeth fleme, causing it to be easily cast foorth.

The iuyce therof dropped into the eyes, cleareth the sight, and doth consume, G dissolue, and take away the web, pearle, or spot of the eye.

Of Horned Clauer, or Medic fother. Chap.xl.

Medica.
Spanish Clauer.

Italian or Spanishe Clauer.

❧ *The Kindes.*

THere be three sortes of Medica, the which we haue seene in this Countrie. The first kind hath flat huskes, and turned or folded rounde togither. The other hath long, rough, & sharpe poynted huskes, turning in also togither lyke a Rammes horne, or Snaple (as Turner writeth) otherwise one muche lyke to the other. There is also a thirde kinde, wherof both Turner and this Author do write.

Tt iiij ❧ *The*

✣ The Description.

1. The first kind of Medica, hath many rounde tender stalkes, which grow not vpright, but are spread abrode vpon the grounde, like the common medow Trefoyl. The leaues be like them of the commō Trefoyl. The flowers be small, of a pale yellowish colour, & for the most part they grow three and three togither. The which once past, there grow vp flat huskes or coddes, turned round togither, like a water snayle, wherein the seede is conteyned, the whiche is flat. The roote is leane or slender, and withereth or perisheth in this Countrie, after that it hath once borne seede.

2. The second kind of Medica, is much like ye other in stalkes & leaues. The cods only be not so flat, but longer, & sharpe pointed, wherin is a sede like to ye other.

3. The third kind hath many stalkes, growing almost right vp, & theron leaues like vnto the other. The flowers grow in tuftes almost like to the cōmon Trefoyl, of color faire purple blew, somtimes yellow, & therafter folow many roūd flat cods turned togither, of ye which eche asunder about the bignes of a Lentil. The roote of this is long, and continueth many yeres, especially in Spayne.

4. Bysides these there is yet another kind of Medica or strange Trefoyl, ye which lieth not alōgst the ground, but standeth vpright, a foote & a halfe or two foote long. It hath hard round stalkes, diuided into diuers branches, vpō the which grow meetly large leaues, gray & thicke, three vpō one stemme, almost like the leaues of Trefoyl or Fenugreck, but muche lesse. The flowers be white mixt with Crymsen or Carnation color. Al the herbe, aswel the stalkes as leaues, is whitish, and couered with a soft and gentle cotton, or woolly roughnesse.

✣ The Place.

These kindes of Trefoyl growe in Spayne. They growe not of their owne kinde in this Countrie, but are sowen in the gardens of Herboristes.

✣ The Tyme.

Medica flowreth in this Countrie in July, and within short space after commeth foorth his crooked or crompled huskes.

4. The fourth kind flowreth in this Countrie at the ende of Sommer.

✣ The Names.

1.2. The first two haue no certaine name which is knowē vnto vs, therfore haue we named them in Latine Trifolia cochleata: in Douche, Gedrayde Claueren: in French Tresle au limaſon: in English Horned Trefoyl or Ciauer, bycause their coddes be turned as water snayles, wherein the seede is conteyned.

3. The third is called in Greke μηδική: in Latine Medica: in Spanish Alfafa, after the Arabian name Fasfasa, or Alfasfasa: with the whiche Medica of Auicenna is named: in Douch Spaensche Claueren: in Englishe Spanish Trefoyl.

4. The fourth kind is counted of some to be Glaux, of some to be Anthyllis, of others it is taken for Polygala.

✣ The Nature.

Medica is of a colde nature.

✣ The Vertues.

A Medica is good against al hoate diseases, & impostumes that require cooling (& drying.

B This is also an excellent fodder for Oxen and kine, and for the same purpose it was vsed to be sowen of the Auncient Romynes in olde tune.

Of the right Trefoyle, or Treacle Clauer. Chap. xli.

✣ The Description.

Amongst al the sortes of Trefoyles, ye same here is the largest in leaues that we haue yet seene, it hath great round stalkes of a foote & a halfe or two foote long, ful of branches, vpō the which there grow alwaies three leaues togither, vpon one footestalke or stemme, of a blackish colour, and muche greater then the leaues of the common Trefoyl. The flowers growe grow from the sydes of the stalkes vpon long stemmes, thicke tufting and clustering togither, almost like the flowers of Scabiouse, of a deepe blew or skye colour. The seede is broade and rough, or a little hearie, and sharpe at the ende. The roote is smal and slender. ❀ *The Place.*   Trifolium.

The Herboristes of this Countrie, do also sowe this kinde of Trefoyl in their gardens.   ❀ *The Tyme.*

This Trefoyl flowreth in this Countrie in August.   ❧ *The Names.*

This kinde of Trefoyl is called in Greke τρίφυλλον, ὀξυτρύφυλλον, μηνιανδές, ἀσφάλτιον, καὶ κνίκιον: in Latine Trifolium, & Trifolium odoratum, at this time they cal it Tritolium fœtidum, Trifolium bituminosum, in Frenche *Vray Trefle*, and *Trefle puant*: in base Almaigne, Groote Claueren: in Englishe, The right Trefoyl, stinking Trefoyl, Smelling Clauer, Treacle Clauer, Clauer gentle, and Pitche Trefoyl. And this is that Oxytriphyllon, of the which Scribonius Largus hath written.

❀ *The Nature.*

This Trefoyl is hoate and drie in the thirde degree.   ❀ *The Vertues.*

A   The leaues and flowers, or seede of this Trefoyl, dronken in water, is good for the payne of the syde, the strangurie, the falling sicknesse, the dropsie, and for women that are sicke of the mother, or stuffing of the matrix: for taken in suche sort, it prouoketh vrine, and the menstrual termes or flowers.

B   The same leaues taken in the syrupe Oximel, helpeth against the bitinges of venemous beastes. The decoction of this Trefoyl, with his rootes is very good for the same, if the bitinges and stinginges of suche hurtful beastes be washed therewithall.

C   Moreouer they do with great profite mingle the said leaues or rootes, with Treacles and Mithridates, and suche lyke preseruatiue medicines, whiche are vsed to be made agaynst poyson.

D   Also they say, that three leaues of this Trefoyl dronke a little before the comming of the fit of the feuer tertian, with wine, do cure the same, & foure leaues so taken, do helpe agaynst the Quartayne.

Of Hares foote, or rough Clauer.   Chap.xlij.

❀ *The Kyndes.*

THere be two sortes of Hares foote, the great & the smal, but in leaues and figure one is lyke to the other.   ❀ *The Description.*

Hares foote hath a round stalke, & rough : the leaues are very like the leaues of Trefoyl or Trinitie grasse. The flowers grow at the top of ye stalkes, in a rough spikie knap or eare very like to Hares foote. The roote is small and harde.   ❀ *The Place.*

Hares foote, especially the lesser, is very common, throughout all the feeldes of this Countrie.   ❧ *The*

Lagopus.

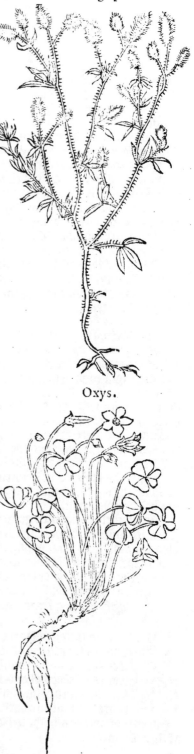

Oxys.

❊ *The Tyme.*

Hares foote is most commonly in flower in July and August.

❊ *The Names.*

This herbe is called in Greeke λαγόπȣς: in Latine Lagopus: of some Leporis Cuminum, now Pes Leporis, and Trifolium humile: that is to say in english, Hares foote, Rough Clauer, & base Trefoyl: in French *Pied de Lieure*, and *Trefle bas*: in high Douche Hasenfusz, Katzenklee, Katzle: in base Almaigne Hasenpootkens, Hasen voetkens.

❡ *The Nature.*

Hares foote is drie in the third degree, and indifferent colde.

❊ *The Vertues.*

A Hares foote boyled in wine and dronke, stoppeth the laske, and the bloody flixe.

Of wood Sorrel, or Sorrel de boys. Chap. xliij.

❊ *The Description.*

1 Wode Sorrel is a lowe or base herbe, without stalkes: the leaues do growe from the roote vpõ short stemmes, and at their first comming foorth are folden togither, but afterwarde they spread abroade, and are of a faire greene colour, and fashioned almost like the Tresoyl, sauing that eche leafe hath a deepe clift in the middle. Amongst the leaues, there growe also vppon shorte stemmes comming from the roote, little smal flowers, almost made like litle belles, of a white colour with purple veynes, all alongst, sometimes of a yellowishe colour: when they be fallen, there rise vp in their places sharpe huskes or cuppes, full of yellowishe seede. The roote is browne, somewhat red, and long.

2 Of this is founde yet another kind, the which beareth yellow flowers, and afterwarde small coddes.

❊ *The Place.*

This herbe groweth in this Countrie in shadowie wooddes, vpon the rootes of great olde trees, sometimes also vpon the brinkes and borders of ditches.

❊ *The Tyme.*

This herbe flowreth in Aprill, and at the beginning of May.

❊ The

the Historie of Plantes. 503

❋ The Names.

This herbe is called in Greeke ὀξύς: in Latine Oxys: in Shoppes Alleluya, of some Trifolium acetosum and Panis Cuculi Alimonia: in French Pain de Cocu: in high Douche Saurerklee, Buchklee, Buchamffers, Buchbrot, Gauchklee, and Gauchgauchklee: in base Almaigne Coeckoecks broot: in English Wood-forel, Sorel du bois, Alleluya, Cockowes meate, Sower Trifoly, Stubwurt, and Woodsower.

❋ The Nature.

This herbe is colde and drie lyke Sorrell.

❋ The Vertues.

Sorel du bois is good for them that haue sicke & feeble stomackes, for it drieth and strengthneth the stomacke, and stirreth vp appetite.

It is good for corrupt sores, and stinking mouthes, if one wasshe with the decoction thereof.

Of Grasse. Chap.xliiij.

❋ The Kindes.

A Man shal finde many sortes of grasse, one lyke another in stemme, and leaues, but not in the knoppes or eares: for one hath an eare like Barley, the other lyke Millet, another like Panick, another lyke Juray, and such vnprofitable weedes that growe amongst corne. Some haue rough prickley eares, and some are soft and gentle, others are rough & mossie lyke fine downe or cotton, so that there are many sortes and kindes of grasse: whereof we will make no larger discourse, but of suche kindes onely, as haue bene vsed of the Auncient Physitions, and are particularly named Agrostis and Gramen.

❋ The Description.

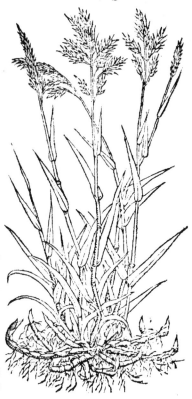

Gramen Couche grasse.

THE grasse whereof we shall nowe speake, hath long rough leaues almost lyke the Cane, or Pole reede, but a great deale lesser, yet muche greater & broder then the leaues of that grasse which groweth comonly in medowes. The helme or stemmes are small, a foote or two long, with fiue or sixe ioyntes, at the vppermost of ÿ stalkes there grow soft & gentle eares, almost like ÿ bushy eares of ÿ Cane or Pole reede, but smaller and slenderer. The roote is long and white, full of ioyntes, creeping hither & thither, & platted or wrapped one with another, & putting forth new springs in sundry places, & by the meanes hereof it doth multiplie and increase exceedinly in leaues and stalkes.

❋ The Place.

This grasse groweth not in medowes & lowe places, lyke the other, but in the corne feldes, & the borders therof, & is a noughty & hurtful weede to corne, the which the husbandmen would not willingly haue in their lande, or feeldes: & therfore they take much payne to weede and plucke vp the same.

❋ The Names.

This grasse is called in Greeke ἄγρωστις, Agrostis, bycause it groweth in the corne

corne feeldes, whiche are called in Greeke ἄγροι, Agroi, therfore men may easily iudge, that the common grasse is not Agrostis.

This grasse is called in Greeke ἄγρωστις: in Latine Gramen: in French Grame, or Dent au chien: in base Almaigne Ledtgras, and knoopgras: and of the Countrie or husbandmen Pœen: in Englishe Couche, and Couche grasse.

The Nature.

Couche grasse is colde and drie of complexion.

The Vertues.

The roott of Couche grasse boyled in wine and dronken: doth swage and heale the gnawing paynes of the belly, prouoketh vrine, bringeth forth grauel, and is very profitable against the strangurie.

The same with his leaues newe brused, healeth greene woundes, and stoppeth blood, if it be layde thereto.

Of wall Barley or way Bennet. Chap.xlv.

The Description.

Phœnix.

1 Phœnix is a kind of vnprofitable Grasse, in eare and leaues almost like Iuray, or Darnel, but smaller & shorter.

It hath leaues meetely long and large, almost like Barley, but smaller. The litter or steyns is short, full of ioyntes, and reddish. The eares growe in fashion like Iuray, but the litle knoppes or eares, stande not so farre asunder one from an other.

2 There is yet another grasse much like to ye aforesaid, ye which groweth almost throughout al medowes and gardens. Neuerthelesse his leaues be narrower, & the stalkes smaller, and are neuer red, but alwayes of a sad greene colour, and so is all the residue of the plant, whereby it may be very wel discerned frō the other.

¶ The Place.

Phœnix groweth in the borders or edges of feeldes, and is founde in great quantitie, in the Countrie of Liege or Luke. And as Dioscorides writeth, groweth vpon houses.

The Tyme.

Phœnix is ripe in July and August, as other grayne is.

The Names.

This herbe is called in Greke φοῖνιξ: in Latine Phœnix, and of some Lolium rubrum: in Englishe Wall Barley, or Way Bennet: it may be called Red-Ray, or Darnell.

The Nature.

Phœnix drieth without sharpnesse, as Galen writeth.

The Vertues.

Phœnix taken with red wine stoppeth the flure of the belly, and the abundant

the Historie of Plantes.   505

dant running of womens flowers, and also the inuoluntarie running of vrine.

Some do write, that this herbe wrapped in a Crymson skinne, or peece of leather, and bounde fast to a mans body, stoppeth bleeding.

Of Hauer Grasse. Chap.xlvi.

❊ *The Description.*

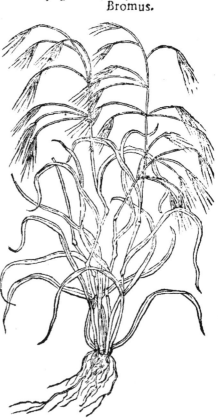

Bromus.

BRomus or Hauergrasse, is also an vnprofitable grasse, much like to Otes, in leaues, stemmes, and eares: sauing that the grasse or leaues be smaller, the stalkes or motes be both shorter and smaller, and the eares are longer, rougher, & more bristeled or bearded, standing farther asunder one from the other.

❊ *The Place.*

It groweth in ỹ borders of feeldes, vpon bankes and Rampers, & alongst by way sides.

❊ *The Tyme.*

It is to be found, in eare, wel neare all the sommer.

❊ *The Names.*

This herbe is called in Greke Βρόμος: in Latine Bromus: and as a difference from Otes (whose Greeke name is Bromus) they put to this addition, Βρόμου πόα, Bromus herba, and Auena herba. It had this name first, bycause of the likenesse it hath with Otes: it is called in Frenche *Aueron*, or *Aueneron*, we may call it in Englishe, Hauer, or Ote grasse.

❊ *The Nature.*

Bromus is of a drie complexion.

❊ *The Vertues.*

This herbe and his roote boyled in water vntil the third part be consumed, and afterwarde the same decoction boyled agayne with hony, vntill it waxe thicke, is good to take away the smel or stenche of the sores in the nose, if it be put in with a weeke or matche, but especially if you put to it Aloes.

The same also boyled in wine with dried roses, amendeth the corrupt smell of the mouthe, if it be washed throughly therwithall.

Of Stitchwurt. Chap.xlvii.

❊ *The Description.*

THis herbe hath round tender stalkes, ful of knots or ioyntes creeping by the ground, at euery ioynt grow two leaues one against another, hard, brode, and sharpe at the endes. The flowers be white, diuided into fiue small leaues, when they be fallen away there growe vp litle round heades or knoppes, not much vnlike the knops or heades of Line, wherin the seede is. The rootes be small and knottie, creeping hither, and thither. The

The Place.

It groweth in this Countrie alongst the fieldes, and vnder hedges and busshes.

The Tyme.

A man may finde it in flowers in Apzill and May.

The Names.

This herbe hath the likenesse of the herbe called in Greke κραταιόγονον, κραταίονον, καὶ κραταίο: in Latine Crataeogonum, Crataeonum, and Crataeus: it is called in high Douche Augentroostgras: and the Brabanders folowing the same call it Oogentroostgras, that is to say, Grasse comforting the eyes. And may wel be named Gramen Leucanthemum.

The Nature.

The seede of Crataeogonum, heateth and dryeth.

The Vertues.

Men haue written, that if a woman drinke the seede of Crataeogonum three daies togither fasting after the purging of her flowers, that the childe which she may happen to conceiue within fourtie dayes after, shalbe a man childe.

Gramen Leucanthemum.

Of Bupleuros. Chap. xlviij.

The Description.

1. Bvpleuron hath long narrowe leaues, longer & larger then the blades of grasse: otherwise not muche vnlyke. The stalkes be of a three or foure foote long or more, rounde, vpright, thicke, full of ioyntes, the whiche do part and diuide agayne, into many branches, at the toppe whereof there growe yellow flowers in round tuftes or heades, & afterward the seede, whiche is somewhat long.

2. There is another herbe much like to the aforesayd, in fashion and growing, sauing that his leaues which are next the grounde, are somewhat larger, the stemme or stalke is shorter, and the roote is bigger, and of a wooddy substance: in al thinges els lyke to the aforesayde.

The Place.

1. This herbe groweth not of it selfe in this Countrie, but the Herboristes do sowe it in their gardens.
2. The seconde is founde in the borders of Languedoc.

The Tyme.

It flowreth and bringeth foorth seede in July and August.

The Names.

1. The first is called in Greke βούπλευρον, in Latine Bupleurum: we know none other name.
2. The seconde sort is called of the Herboristes of Prouince Auricula Leporis. It is very lyke that which Valerius Cordus nameth Isophyllon.

The

the Hiftorie of Plantes. 507

Bupleuri prima fpecies.
The firſt kind of Bupleures.

Bupleuri altera fpecies.
The ſecond kind of Bupleures.

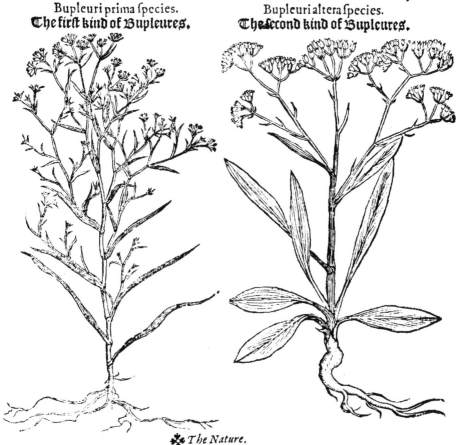

❧ *The Nature.*
Buplerum is temperate in heate and dryneſſe.
❧ *The Vertues.*

This herbe in time paſt was vſed as pot herbe, and counted of Hippocra- A
tes, as a conuenient food, as Plinie writeth.

The leaues of the ſame pounde with ſalt and wine, and layde to, doth con- B
ſume and driue away the ſwelling in the necke, called the Kinges euill.

It is alſo a ſpeciall remedie againſt the bitinges of Serpentes, if the partie C
that is ſo hurt, doth drinke the ſeede of the ſame in wine, and waſh the wound
with the decoction of the leaues of the ſame.

They that vſe it much do take Auricula Leporis, againſt the ſtone & grauell. D

Of Catanance. Chap.xlix.

❧ *The Deſcription.*

1 THis herbe is alſo like vnto graſſe, it hath narrow leaues & ſmoth,
like to the blades of graſſe, but ſmaller, the whiche afterward being dried, do turne crooked or bend round towardes the ground.
The ſtalkes be tender, ſmal, and ſhort, vpon the which grow litle
Crymſen flowers, and afterwarde long ſmal rounde coddes, in
the whiche is conteyned a ſeede, ſomewhat reddiſhe.

2 Of this ſort there is yet another kinde, the whiche hath no rounde coddes,
but large and ſomwhat broade, in all thinges els lyke to the other.

Uv ij ❧ *The*

Catanance.

¶ *The Place.*

This herbe groweth in Copses that be seuerall, and in pastures, but that with the broade coddes is found most commonly by the sea coast.

❊ *The Tyme.*

Catanance bringeth forth his flowers and coddes, in July and August, and sometimes sooner.

❊ *The Names.*

This grasse is muche lyke to that which the Greekes call κατανάγκη: and the Latinistes Catanance, and it should seeme to be the first kind of Catanance, described by Dioscorides.

❊ *The Nature and Vertues.*

Catanance was not vsed for medicine, in times past: neyther yet is vsed that I can tel of.

Of Moly. Chap.I.

❊ *The Description.*

1. Moly according as Dioscorides writeth hath leaues like grasse, but broader, and spreaden or laid vpon the ground. The flowers be white, in fashion like the stocke or wall Geleflowers, but smaller. The stalke is white of foure cubites long, at the top wherof there groweth a certayne thing fashioned like Garlike. The roote is small and rounde as an Onyon.

2. Plinie in the fourth Chapter of his xxb. Booke writeth of another Moly, whose roote is not bolefashion, or like an Onyon, but long and slender. His leaues be also lyke vnto grasse, and layd flat vpon the ground, amongst which springeth vp, a rounde, small, and playne stalke diuided aboue into many branches, wherevpon grow white flowers, not muche vnlyke the flowers of stocke Gelleflowers, but muche smaller. The rootes be long and small, and very threddie.

3. You may also recken amongst the kindes of Moly, a sort of grasse growing alongst the sea coast which is very tender and smal, bearing smal, short, narrow leaues, and most commonly lying flat and thicke vppon the grounde, amongst whiche commeth vp small short and tender stalkes bearing flowers at the top tuft fashion, of a white purple, or skie colour. The rootes of the same kinde be likewise long, smal, and tender.

❊ *The Place.*

2. The second Moly, as Plinie writeth, groweth in Italie in stonie places, the Herboristes of this Countrie do plant it in their gardens.

3. The grasse that groweth by the sea coast, is founde in some places of Zealande, in lowe moyst places or groundes.

❊ *The Tyme.*

Plinies Moly, floureth in this Countrie in July.

the Historie of Plantes. 509

Liliago.
Phalangium.

Pſeudo Moly.
Sea graſſe.

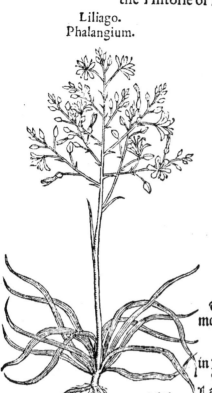

The baſtarde Moly flowreth moſt commonly all the ſommer.

❧ *The Names.*

The firſt is called in Greeke μῶλυ, and in Latine Moly.

The ſeconde is called Plinies Moly, in Latine Moly Plinij, and is taken to be the right Phalangium, or Spiders worte of Dioſcorides, and that in Greke φαλαγγιον, is of Valerius Cordus named Liliago.

That kinde of graſſe whiche groweth by the ſea ſyde, is called in Greeke ψευδομᾶλυ, Pſeudomoly, that is to ſay, Baſtarde Moly. Neuertheleſſe it is no kinde of Moly, but rather a kinde of graſſe, the whiche you may well name Gramen marinum: ſome call it in Engliſhe our Ladies quſhion.

❧ *The Nature.*

The true Moly, which is the firſt kinde, is hoate in the third degree, and of ſubtill partes. ✻ *The Vertues.*

The roote of Moly, eaten or dronken, prouoketh vrine, and applyed as a peſſarie or mother ſuppoſitorie, openeth the ſtoppings of the matrix or mother.

Moly is alſo excellent againſt enchauntementes, as Plinie and Homer do teſtifie, ſaying, That Mercurie reuealed or ſhewed it to Vlyſſes, whereby he eſcaped all the enchauntments of Circe, the Magicien.

Of the graſſe of Parnaſus. Chap.li.

✻ *The Deſcription.*

THis herbe hath litle rounde leaues, in faſhion much vnlike the leaues of Iuie or Aſarabacca, but farre ſmaller, and not of ſo darke a colour : amongſt the which ſpring vp two or three ſmall ſtalkes, of a foote high, and of a reddiſh colour belowe, and bearing faire white flowers at the top, the which being paſt, there come vp round knops or heads, wherin is conteyned a reddiſh ſeede. The roote is ſomwhat thicke, with many thredy ſtringes thervnto annexed.

510　The fourth Booke of

Hepatica alba.
Gramen Parnasi.

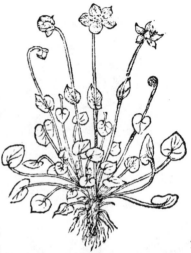

❀ *The Place.*

This herbe groweth in moyst places, and is founde in certayne places of Brabant.

❀ *The Tyme.*

This herbe flowreth in July, and soone after yeeldeth his seede.

❀ *The Names.*

Bycause of the lykenesse that this herbe hath with the grasse called in Greeke ἄγρωςις ἐν τῷ παρνασῷ γενωμένη: in Latine Gramen Parnasium: they call it in this Countrie, The grasse of Parnasus: in base Almaigne Gras van Parnasus: Valerius Cordus nameth it Hepatica alba.

❀ *The Nature.*

The seede of Parnasus grasse is drie, and of subtill partes.

❀ *The Description.*

The decoction of Parnasus grasse dronken, doth drie and strengthen the feeble and moyst stomacke and the moyst bowels, stoppeth the belly, and taketh away the desire to vomit.

The same boyled in wine or water, prouoketh vrine, especially the seede B thereof, the which doth not only prouoke vrine, but also breaketh the stone, and driueth it foorth, if it be dronken.

The young leaues brused, and layd to fresh woundes, stoppeth the bleeding C of the same, and healeth the woundes.

Of the iuyce of this herbe is made a singuler Collyrium, or medicine for the D eyes, the whiche comforteth the sight, and cleareth the eyes, if you put vnto it asmuch wine as you haue of the iuyce, and halfe as much Myrrhe, with a litle Pepper and Frankensence. And for to keepe the sayd Collyrium a long time in his goodnesse, it must be put into a copper Boxe.

Of Rushe. Chap.lij.

❀ *The Kindes.*

There are founde in this Countrie, foure or fiue kindes of vsual or common rushes.　❀ *The Description.*

All Rushes seeme nothing els, but lōg twigs, shutes, or springs, that are slender, smooth, rounde, and without leaues, & the roote from whence they grow and spring vp, is large and enterlaced. The flowers & seede grow vpon one side, almost at the top of the shutes or rushes in tuftes or tassels.

1　The first kinde is full of white substance or pith, the whiche being drawen out, sheweth like long white softe or gentle threds, and serueth for Matches to burne in lampes, and of the same is made many plesant deuises.

2　The seconde kind is somwhat rough and harder in handling then the first, and hath but litle pith within, and the the same not thicke nor close, so that in drawing it foorth, it yeeldeth small substance, wherefore the sayde pith is vnprofitable: but the Rushe being dried, is more plyant, and better to binde any thing withall, then any of the other sortes.

3　The third kind also hath not much pith, and groweth not farre apart from the rootes, but many togither, as the flagge or gladen leaues, so that one rushe groweth out of another.

The

4 The fourth kinde is great, of eight or nienc foote long, of the bignesse of ones finger, spongie within, as the Flagge or water Laver, whereof they vse to make Mattes: and of this kinde they do likewise make Mattes, which are called Rushe Mattes.

5 Bysides these sortes of common rushes, there is also a strange, aromaticall, or sweete smelling rushe, the whiche is not to be founde in this Countrie, but onely in Apothecaries shoppes, vnder the name of Squinantum.

⁋ *The Place.*

The Rushes grow in low moyst sugges, or waterie places. The small kinde groweth onely in drye leane and sandy groundes, & barren Countries, as is aforesayde: but the sweete rushe groweth in Arabia, Africa, and India.

❊ *The Names.*

Rushes are called in Greke χοίνοι: in Latine Iunci: in Frenche *Ioncs*: in Douche Bintzen: in base Almaigne Biesen.

1 The first kinde is called in Greke χοῖνΘ- λυα: in Latine Iuncus læuis, of Plinie Mariscus: in base Almaigne Merch biesen, that is to say, The pith, or pithy Rushe: and in English, the Rush candle, or Candle rushe: Camels strawe.

2.3 The seconde is called in Douche Pferen Biesen, and the third Strop Biesen, the which are like a kinde of ὀξυχοῖνΘ-: in Latine Iuncus acutus: they be our common harde Rushes: in Frenche *Ionc agu*.

The small Rushe seemeth to be a kinde of ὀξυχοῖνΘ-, and Iuncus acutus, especially that kinde which is Sterile, or barren without flowers.

4 The fourth is called in Greke ὁλοχοῖνΘ-, and folowing the Greke Holoschœnus: in English, the pole Rushe, or bull Rushe, or Mat Rushe: in Frenche *Ionc a cabas*, that is to say, The frayle Rushe or panier Rushe, bycause they vse to make figge frayles and paniers therwithall: in base Almaigne Matten biesen: bycause they vse to make Mattes therewith.

5 The strange Rushe is called in Greke χοῖνΘ-: in Latine Iuncus odoratus, & Iuncus angulosus, the flower whereof is called in Greke χοίνυ ἄνθΘ-: in Latine Iunci flos, and Schœnu anthos: and from hence came that name Squinan un, whiche is the name whereby this kinde of Rushe is knowen in Shoppes: in Englishe Squinant. ❊ *The Nature.*

The common Rushe is of a drie complexion.

❊ *The Vertues.*

The sede of the common Rush parched, & stieped in wine: stoppeth the laske, A and the redde flowers of women, and prouoketh vrine. But to be taken in to great a quantitie, it causeth headache.

You must search farther for the vertues of Squinant, which are not descri- B bed in this place.

Of Typha palustris. Chap.liij.

❧ The Description.

This herbe hath long, rough, thicke, and almost threesquare leaues, within filled with soft marow. Amongst the leaues sometimes groweth vp a long smoth naked stalke, without knottes or ioyntes, not hollowe within, hauing at the top a gray, or russet long knap or eare whiche is soft, thicke, and smooth, and seemeth to be nothing els but a throm of gray wooll or flockes, thicke set and thronge togither. The whiche at length when as the sayd eare or knap wareth ripe, is turned into a Downe, and caried away with the wind. This downe or cotton is so fine, that in some Countries they fill quishions and beddes with it, as Leonardus Fuchsius writeth. The rootes be harde, thicke, and white, with many hanging threddes ouerthwart one another, and when these rootes are drie, then they burne very well.

Typha palustris. Reede Mace, Cattes tayle, or Water torche.   Typha abíque caule. The water flagge or Liuer.

❧ The Place.
This Typha groweth in this Countrey in shadowe pooles, and standing waters, and in the brinkes or edges of great riuers, and commonly amongst Reedes.

❧ The Tyme.
This Mace or torche is founde in July and August.

❧ The Names.
This herbe is called in Greeke τύφη: in Latine Typha, and of the writers in these

these dayes Typha palustris, as a difference from the other Typha, called Typha cerealis, whiche is a kinde of grayne or corne, the whiche hath bene already described in the fourth Chap. of this booke, of some it is also called Typha aquatica, and Celtrum morionis: in Frenche *Marteau, Masses*: in high Douche Narrenkolben, and Lieszknospen: in base Almaigne, Lisch Dodden, and Donsen. Turner calleth it in Englishe, Reede Mace, and Cattes tayle: to the which we may ioyne others, as Water Torche, Marche Betill, or Pestill, and Dunche Downe, bycause the Downe of this herbe will cause one to be deafe, if it happen to fall into the eares, as Matthiolus writeth. The leaues are called, Matte reede, bycause they make mattes therewith, to the whiche they onely serue when it bringeth foorth neither stalkes nor cattes tayle. like as this plant yeeldeth his cattes tayles, so likewise be the leaues not necessarie to make any thing thereof.

※ *The Nature.*

This herbe is colde and drie of complexion.

※ *The Vertues.*

The Downe of this herbe mingled with Swynes grease well washed, healeth burninges and scaldinges with fire or water. A

Men haue also experimented and proued, that this cotten is very profitable B to heale broken or holowe kibes, if it be layde vpon.

Of Pole Reede, or Canes. Chap. liiij.

※ *The Kindes.*

THERE are diuers kindes of Reedes, as Dioscorides and Plinie do write, whereof the sixth kinde is very common and well knowen in this Countrie.

Harundo Vallatoria.
The common Pole Reede.

※ *The Description*

1 THE common Reede or Cane hath a long stalke or strawe full of knottie ioyntes, whereuppon grow many long rough blades or leaues, and at the top large tufts, or eares spread abrode, the whiche do change into a fine downe or cotton, and is carried away with the winde, almost like the eares of Mill or Millet, but farre bigger. The roote is long & white, growing outwardly in the bottome of the water.

7 The Cane of Inde, or y Indian Cane, is of the kind of Reedes, very high, long, great, and strong, the which is vsed in temples & Churches to put out y light of candels, whiche they vse to burne before their Images.

8 To these we may ioyne that Cane, whereof they make Sugar, in the Ilandes of Canare, and els where.

9 Besides these sortes, there is another aromatical, and sweete smelling kind, vnknowen in this Countrie.

※ *The*

❧ *The Place.*

The common Reede or Spier groweth in standing waters, and on the edges and borders of riuers.

❧ *The Names.*

This plante is called in Greke κάλαμος: in Latine Harundo, or Arundo, and Calamus: in Frenche Canne, or Roseau: in high Douche Rohr: in base Almaigne Riet: in English, Common Pole Reede, Spier, or Cane Reede.

1. The first kinde is called νάστος, Nastus, of this kinde in times past they made arrowes and dartes.

2. The seconde is called κάλαμος θῆλυς, Arundo fœmina, this kinde dyd serue to make tongues for pipes, shaulmes, or trumpettes.

3. The thirde is called συριγγίας, Syringias, Fistularis, of whiche they make pipes and flutes.

4. With the fourth men did write in times past, as they do now vse to do with pennes and quilles of certayne birdes, the whiche for the same purpose were named Calami.

5. The fifth kinde is called δόναξ, Donax, κάλαμος κύπριος, Arundo Cypria.

6. The sixth, which is our commō Canereede, is called in Greke κάλαμος φραγμίτης, that is to say in Latine, Arundo vallatoria, and Arundo vallaris, and Arundo sepicularis: in Englishe Cane Reede, Pole Reede, Spier, and the Reede or Cane of the vally.

7. The seuenth is called κάλαμος ἰνδικός, and Arundo Indica: in Frenche Canne: in base Almaigne, Riet van Indien, and of some also, Riet van Spaengien: in English Spanish Canes, or Indian Reede.

8. The Sugar Cane, hath none other particuler name, but as men do nowe cal it Arundo saccharata, or Arundo sacchari, that is to say in Englishe, Sugar Reede, or Sugar Cane: in Frenche Canne de succré: in Douche Suycker Riet.

9. The Aromaticall and sweete Cane, is called in Greke κάλαμος ἀρωματικός: in Latine Calamus odoratus, Calamus Aromaticus, Arundo odorata, altogither vnknowen in shoppes, for that whiche they vse to sel for Calamus Aromaticus, is no reede nor roote of a reede, but is the roote of a certayne herbe lyke vnto the yellow Flagge or bastard Acorus, the whiche roote is nowe taken for the right Acorus.

❧ *The Nature.*

The Cane Reede is hoate and drie, as Galen sayth.

9. The Aromatical and sweete Cane, is also hoate & dry in the second degree.

❧ *The Vertues.*

The roote of Cane Reede or Spier, pound smal and layd to, draweth forth thornes and splinters, and mingled with vineger it swageth the paine of members out of ioynt.

The greene tender leaues finely stamped and layde to, healeth cholerique inflammations or wilde fier, also hoate swellinges and impostumes.

The ashes of the Pole Reede mingled with vineger and layde to, healeth the roome and scales of the head, whiche do cause the heare to fal of.

The Aromatical or sweete Calamus being dronken, prouoketh vrine, and boyled with Parsley seede, is good agaynst the strangurie, the payne of the raynes, the bladder and dropsie.

The same taken in any kind of wayes, is very agreable to al ÿ inner partes, as the stomacke, the liuer, the spleene, the matrix, & agaynst burstinges or ruptures.

❧ *The Danger.*

The downe that is in the top of the Cane reede, or in the tufting tassels therof, if it chaunce to fal into the eares, bringeth such a deafenesse, as is hard to be cured.

❧ The

the Historie of Plantes.

Of Reede Grasse. Chap.lv.

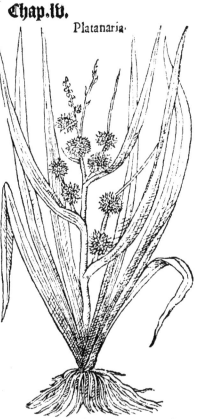
Platanaria.

❀ *The Description.*

Eede grasse hath long narrow leaues, two edged or sharpe on both sides, with a sharpe crest or backe, raysed vp, so that they seeme almost triangled or three square. The stalkes growe amongst the leaues, to the height of two or three foote or more, and do beare about the vpper part of the stalkes rounde prickley knoppes, or boullettes, as bigge as a Nut. The roote is ful of hearie stringes.

❀ *The Place.*

It groweth in this Countrie in moyst medowes, & in the borders, or brinkes of ditches & riuers.

❀ *The Tyme.*

It bringeth foorth his boullettes, or prickley knoppes in August.

❀ *The Names.*

This herbe is called in base Almaigne Rietgras, and therefore some take it for a kinde of grasse which Dioscorides calleth in Greeke καλαμάγρωσις, Calamagrostis: in Latine Gramen Arundinaceum: in Englishe, Reede grasse. With the which it hath no likenesse, and therefore it serueth better to be named Platanaria, and lykewise it is not lyke vnto Spargamum, but it is more lyke that Butomon of Theophrastus, that likewise in Greeke is called βούτομον. ❀ *The Nature.*

It is of a colde and drie complexion.

❀ *The Vertues.*

Some write, that the knoppes or rough buttons of this herbe boyled in wine, are good agaynst the bitinges of venemous beastes, if it be either dronken, or the wounde be washed therewith.

Of Kattel grasse. Chap.lvi.

❀ *The Kindes.*

There be two kindes of this grasse, one which beareth redde flowers, and leaues finely iagged or snipt, the other hath pale yellow flowers, and long narrowe leaues snipt like a sawe rounde about the edges.

❀ *The Description.*

The first kind hath leaues very smal iagged, or dented, spread abrode vpō the ground: The stalkes be weake & smal, whereof some lye along trayling vpon the ground, & do beare the litle leaues: the rest do growe vpright, as high as a mans hand, & vpon them grow the flowers from the midle of the stemme round about, euen hard vp to the top, of a browne red or purple color, somwhat like to ẏ flower of the red nettle. The which being fallē away, there grow in their place litle flat powches or huskes, wherein the seede is conteined, which is flat, & blackish. The roote is smal & tender. The

Fiſtularia. Reede Rattel.   Criſta gallinacea. Yellow Rattel.

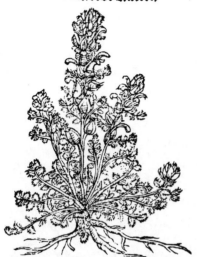

2 The ſeconde kind hath a ſtraight ſtemme, ſet about with narrowe leaues, ſnipt rounde about vpon the edges. The flowers growe rounde about the ſtemme, at the higheſt of the ſtalke, faſhioned like the flowers of the firſt kinde, ſauing that they be of a fainte or pale yellowe colour, or whitiſhe, after the whiche there come vp litle flat powches or purſes, couered as it were with a little bladder, or flat ſkin, open before like the mouth of a bladder. Within the litle purſes is the ſeede, the whiche is flat, yellowe or browniſh. The roote is ſmall and ſhort.

✻ *The Place.*

1 That with the red flowers groweth in moyſt medowes, and is very noyſome to the ſame.
2 That with the pale or yellowe flowers, groweth in drie medowes, and in the feeldes alſo, and is to them very euill and hurtfull.

✻ *The Tyme.*

1 That with the red flowers, flowreth in May, and his ſeede is ripe in June.
2 The other flowreth in June and July, and almoſt all the ſommer.

✻ *The Names.*

1 The firſt is called of the writers in theſe dayes, Fiſtularia, and Criſta, and of ſome in Greeke φθίριον, Phthirion: in Latine Pedicularis, that is to ſay, Louſe herbe: in high Douche Braun Leuſzkraut, bycauſe the cattell that paſture where plentie of this graſſe groweth, become full of lice. They call it alſo in high Douch Rodel, and Browne Rodel: in baſe Almaigne, Roode Ratelen: ſome take it for ἀλεκτοροόλοφος, Alectorolophos of Plinie: in Engliſhe, Redde Rattel.

2 The other kind is called of the writers in theſe dayes Criſta gallinacea, and Criſta galli, bycauſe that in proportion of flowers and pouches, it is like to Red Rattel: in high Douch Geel Rodel: in baſe Almaigne, Geel and witte Ratelē, & of ſome Hanekammekens, that is to ſay, Hennes Commes, or Corecombes: alſo yellowe or white Rattel. This may wel be that herbe, whereof Plinie writeth in his xviii. booke, the xxviii. Chapter, and there is called Nimmulus, the which is very hurtfull to medowes.

✻ *The*

the Historie of Plantes. 517

⸿ *The Nature.*

Both of these herbes are colde, drie, and astringent.

❊ *The Vertues.*

Redde Rattel is taken of the Physitions in these dayes, against the flure A
menstruall, and all other issue of blood, boyled in wine and drunken.

The other kinde hath no peculier vertue that I knowe. B

Of the Thistel Chameleon. Chap.lvij.

❊ *The Kindes.*

Chamæleon is of two sortes, as Dioscorides writeth, the white and the blacke.

The figures which my Author attributeth to Leucacantha, *wil agree well with this description, and they be so placed of Matthiolus.*

❊ *The Description.*

1. The great Chameleon, hath great brode prickley leaues, not much
vnlike the wilde Thistle, but rougher & sharper, the which leaues
are greater, stronger & grosser, then the leaues of the blacke Chameleon: amongst which leaues there riseth immediatly from the
roote, a prickley head or bowle, almost like the heades of Hartichokes, and beareth a purple thrommed flower like veluet. The
seede is almost lyke the seede of bastarde saffron. The roote is long, and white
within, of a sweete taste, and aromaticall smell.

2. The blacke Chameleons leaues, are also almost lyke to the leaues of the
wilde Thistel, but smaller and finer, and sprinckled or spotted with red spottes.
The stemme is reddish or browne red, of the bignesse of a finger, & groweth to
the height of a foote, wherupon grow round heades with smal prickley flowers
of diuers coloures, whereof eche flower is not much vnlyke the flowers of the
Hyacynthe. The roote is great & firme, or strong, blacke without and yellowish
within, sharpe and biting the tongue, the whiche for the most part is found, as
it were already tasted or bitten.

❊ *The Place.*

1. The white Chameleon groweth vpon hilles and mountaynes, & such lyke
vntoyled places. Yet for all that it desyreth good grounde. It is founde in
Spayne in the region of Arragon by the high way sides.

2. The blacke groweth in drye soyles, and places neare the Sea.

❊ *The Names.*

1. The first kinde is called in Greeke χαμαιλέων ὁ λευκός: in Latine Chamæleo albus, of Apuleius Carduus syluaticus, of some also Erisisceptrum, Ixia, Carduus
varinus, Carduus irinus, Carduus lacteus. Matthiolus sayth, that the Italians
cal this herbe Carlina in Spayne Cardo pinto: in Frenche Carline.

2. The seconde kinde is called in Greeke χαμαιλέων μέλας: in Latine Chamæleo
niger, of some Pancarpon, Vlophonon, Cynomazon, Cynoxylon, Ocymoides,
Cnidos coccos, Carduus niger, Veruilago, Vstilago, &c. Both these kindes are
vnknowen in this Countrie.

❊ *The Nature.*

Chameleon is temperate in heate and drynes, specially the blacke, the which
is almost hoate in the seconde degree, and altogither drie in the thirde degree.

❊ *The Vertues.*

1. The roote of the white Chameleon dronken with redde wine wherin Origanum hath bene sodden, killeth and bringeth foorth large or brode wormes. A

The same boyled in good wine, is very good for such as haue the Dropsie, B

Xx and

and ſtrangurie, for it delayeth the ſwelling of them that haue the Dropſie, and diſpatcheth vrine.

It is good againſt al kindes of venome or poyſon that may be giuen.

2 The roote of the blacke Chameleon, is not very meete to be receiued inwardly: for it is indued with a certayne hidden euill qualitie, as Galen ſayth, and therefore it was neuer miniſtred of the Auncientes, but in outward medicines.

The ſame with a litle Copperoſe and Swines greaſe, healeth the ſcabbe, and if you put thereto of Brimſtone and Roſen or Tarre, it wil heale the hoate running or creeping ſcabbe or ſcurffe, foule tetters, and all noughtie itche or manginesse.

The ſame layde to in the Sonne but onely with Brimſtone, putteth away the creeping ſcabbe and tetter, white ſpottes, ſonne burning, and other ſuche deformities of the face.

The decoction of this roote boyled in water or vineger being holden in the mouth, healeth the tooth ache. The lyke propertie hath the roote, broken or bruſed with Pepper and Salt, to be applyed and layde vppon the noughtie tooth.

Of Sea Holly. Chap.lviij.

❧ *The Kindes.*

IN this Countrie is founde two kindes of Eryngium, the one called the great Eryngium, or Eryngium of the Sea, and the other is called but Eryngium onely.

Eryngium marinum.
Sea Holly.

Eryngium vulgare.
The hundred headed Thiſtel.

❧ The Description.

1 THE great Eryngium hath great, large, whitishe, somewhat rounde and thicke leaues, a litle crompled or cronkeled about the edges, set here and there with certayne prickles rounde about vppon the edges, the sayde leaues be of an aromaticall or spicelyke taste. The stalkes be rounde, and growe about the height of a foote, of a reddishe colour belowe neare the grounde, vpon the toppes of the branches come foorth round knoppie and sharpe prickley heades, about the quantitie of a nut, set rounde about full of small flowers, most commonly of a Celestiall or skie colour, and in this Countrie they haue small tippes or white markes. And harde ioynung vnder the flowers grow fiue or sixe small prickley leaues, set in compasse round about the stalke like a starre, the whiche with the vppermost part of the stemme are altogither of a skie colour in this Countrey. The roote is often or twelue foote long, and oftentimes so long, that you cannot drawe it vp whole, as bigge as ones finger, full of ioyntes by spaces, and of a pleasant taste.

2 The seconde kinde hath broade crompled leaues, al to pounced and iagged, whitish, & set rounde about with sharpe prickles. The stalke is of a foote long, with many branches, at the toppe whereof growe rounde, rough, and prickie bullettes or knoppes, like to the heades of Sea Holly or Huluer, but muche smaller, vnderneath which knoppes grow also fiue or sixe small narrow sharpe leaues, set rounde about the stemme after the fashion of Starres. The roote is long and playne or single, as bigge as a mans finger, blackishe without, and white within.

❧ The Place.

1 Erynge, as Dioscorides writeth, groweth in rough vntoyled feeldes: it is founde in this Countrie in Zealand, & Flaunders, vpon banckes, and alongst by the Sea coast.

2 The common Erynge groweth also in this Countrie in the like places: it groweth also in Almaigne alongst by the riuer Rhene, and in drie Countries by the high wayes. There is plentie growing about Strasbourge.

❧ The Tyme.

Both these kindes do bring foorth their flowers in this Countrie, in June and July.

❧ The Names.

1 The first kinde of these Thistels is called in Greeke ἠρύγγιον: in Latine Eryngium: Plinie calleth it also Erynge: the writers of our time cal it Eryngium marinum: the Arabians with the Apothecaries cal it Iringus: in Almaigne Cruyswortele, and Endeloos, and in some places of Flaunders, Meere wortele: in Englishe, Sea Holme, or Huluer, and Sea Holly.

2 The other kinde is called in English, the Hundred headed Thistel: in French Chardon a cent testes: in high Douch Manstrew, Brachen distel, and Rad distel: in base Almaigne Cruysdistel: in the Shoppes also it is nowe called Iringus. This without doubt is a kinde of Eringium, the whiche may also very be well called Centumcapita.

❧ The Nature.

1 Sea Holly is temperate of heate and colde, yet of drie and subtil partes.
2 The hundred headed Thistell, is hoate and drie as one may easyly gather by the taste.

❧ The Vertues.

1 The first leaues of Eryngium are good to be eaten in Salade, and was for that purpose so vsed of the Auncientes, as Dioscorides writeth.

Xx ij   The

¶ The rootes of the same boyled in wine and dronken, are good for them that are troubled with the Colique and gripings of the belly, for it cureth them, and driueth foorth windinesse.

¶ The same taken in the same manner, bringeth foorth womens natural sicknesse.

¶ It is good to drinke the wine wherein Sea Holly hath boyled, against the stone and grauel, and against the payne to make water, for it prouoketh vrine, driueth foorth the stone, & cureth the infirmities, that chaunce to the kidneyes, if it be dronken fifticne dayes togither one after another.

¶ The same rootes taken in the same manner, are good for suche as be liuer sicke, and for those that are bitten of any benemous beastes, or haue receiued or dronke poyson, especially if it be dronken with the seede of wilde Carrot.

¶ It doth also helpe those that are troubled with the Crampe, and the falling sicknesse.

¶ The greene herbe is good to be pounde, and layde to the bytinges of benemous beastes, especially to the bitinges of Frogges.

¶ The Apothecaries of this Countrie do vse to preserue and comfit the roote of Eringium, to be giuen to the aged, and olde people, and others that are consumed or withered, to nourishe and restore them againe.

2 ¶ The roote of Centumcapita, or the Thistel of a hundred heades, is likewise comfited, to restore, nourishe, and strengthen, albeit it commeth not neare by a great way, to the goodnesse of the other.

Of Starre Thistel, or Caltrop. Chap. lix.

❧ The Description.   Carduus stellatus.

Starre Thistell hath softe frised leaues, deepely cutte or gayste, the stalkes grow of a foote and a halfe high, full of branches, whereuppon growe small knappes or heades like to other Thistelles, but muche smaller, and set rounde about with sharpe thornie prickles, fashioned lyke a Starre at ye beginning, either greene or browne redde, but afterwarde pale or white: when those heades do opē, they bring foorth a purple flower, & afterwarde a small flat and round seede, the roote is long and somewhat browne without.

¶ The Place.

This Thistell groweth in rude vntoyled places, & alongst the waies, & is founde in great quaintitie, about the Marte Towne of Anwarpe, nere to the riuer Scelde, and alongst by the newe walles of the Towne.

❧ The Tyme.

This Thistell flowreth from the moneth of July, vntill August.

❧ The

the Historie of Plantes.

❧ *The Names.*

This herbe is nowe called in Latine Carduus ſtellatus, and Stellaria, alſo Calcitrapa : and ſome take it for πολυάκανϑ⁃, Polyacanthus of Theophraſt, the which Gaza calleth in Latine Aculeoſa, they call it in Frenche *Chauſſetrape* : in high Douch, Wallen Diſtell, and Raden Diſtel : in baſe Almaigne, Sterre Diſtel: in Engliſh, Starre Thiſtel, or Caltrop.

❧ *The Nature.*

This Thiſtel alſo is of a hoate nature, as the taſte of the roote doth ſhewe.

❧ *The Vertues.*

They vſe greatly to take the powder of the ſeede of this Thiſtel in wine to drinke, to prouoke vrine, and to driue foorth grauel, and againſt the ſtrangury.

Of the Teaſel. Chap.lx.

❧ *The Kindes.*

The Cardthiſtel or Teaſel is of two ſortes, the tame & the wild. The tame Teaſel is ſowen of Fullers and clothworkers to ſerue their purpoſes, the wilde groweth without huſbanding of it ſelfe, & ſerueth to ſmal purpoſe.

Dipſacum ſatiuum.
Fullers Teaſel.

Dipſacum ſylueſtre.
Wilde Teaſel.

1 ❡ The Cardthiſtel his firſt leaues be long, and large, hackt round about with natches, lyke the teeth of a ſawe, betwixt thoſe leaues riſeth a holowe ſtalke of three foote long or more, with many branches, ſet here and there with diuers hooked ſharpe prickles, and ſpaced or ſe-
uered

uered by ioyntes, & at euery of the sayd ioyntes, grow two great long leaues, the which at the lower endes be so closely ioyned and fastened togither, round about the stalke, that it holdeth the water, falling either by rayne or dewe, so sure, as a dishe or bason. At the top of the branches growe long, rough, and prickle heades, set full of hookes: out of the same knops or heades, grow smal white flowers placed in Celles and Cabbins, like the honie Combe, in whiche Chambers or Celles (after the falling away of the flower) is found a sede like Fenil, but bitter in taste. The knoppes or heades are holow within, and for the most part hauing wormes in them, the whiche you shall finde in cleauing the heades. The roote is long, playne, and white.

2 The wild Teasel is much like to the other, but his leaues be narrower, and his flowers purple, the hookes of this Teasel be nothing so harde, nor sharpe as the other.

3 There is yet another wilde kinde of these Carde Thistels, the which grow highest of al the other sortes, whose knopped heades are no bigger then a nut, in all thinges els lyke to the other wilde kindes.

✠ *The Place.*

1 The tame Teasel is sowen in this Countrie, and in other places of Flaunders, to serue Fullers and Clothworkers.

2 The wild groweth in moyst places, by brookes, riuers, & such other places.

✠ *The Tyme.*

Carde Thistel flowreth for the most part in June and July.

❡ *The Names.*

This kinde of Thistel is called in Greeke Δίψακος, in Latine Dipsacum, and Labrū Veneris, of some also Chamæleon, Crocodilion, Onocardion, Cneoron, Meleta, Cinara rustica, Moraria, Carduus Veneris, Veneris lauacrum, & Sciaria: in Shoppes Virga Pastoris, and Carduus Fullonum. in French *Verge de berger, Cardon a Foulon*, or *A Carder*: in high Douche Karten distel, Bubenstrel, Weberkarten: in base Almaigne, Caerden, and Volders Caerden: in Englishe, Fullers Teasel, Carde Thistell, and Venus bath or Bason.

1 The tame Teasel is called Dipsacum satiuum, and Dipsacum album.

2 The wilde Teasel is called Dipsaca sylucstris, or Purpurea.

✠ *The Nature.*

The roote of Carde Thistell (as Galen saith) is drie in the seconde degree, and somwhat scouring. ✠ *The Vertues.*

The roote of Teasell boyled in wine, and afterwarde pounde vntill it come A to the substance or thicknesse of an oyntment, healeth the chappes, riftes, and fistulas of the fundement. But to preserue this oyntment, ye must keepe it in a boxe of Copper.

The small wormes that are founde within the knoppes or heades of Tea- B selles, do cure and heale the Quartayne ague, to be worne or tyed about the necke or arme, as Dioscorides writeth.

Of Artechokes. Chap.lxi.

✠ *The Kyndes.*

1 There is now found two kindes of Artechokes, the one with brode leaues, and nothing prickley, which is called the right Artechoke, the other whose leaues be all to gashed full of sharpe prickles and deepe cuttes, which may be called the Thistell, or prickley Artechoke. ✠ *The Description.*

The right Artechoke hath great long broade leaues, like the leaues of our Ladyes Thistel, but blacker, greater, & without prickles, amongst the whiche springeth vp a stalke garnished or set here and there with
the

the Historie of Plantes.

Cynara. **Artechokes.** Cynaræ aliud genus. **Prickley Artechokes.**

the like leaues, but smaller, bearing at the top great rounde scalie heades, the whiche at their opening beare a purple flower or blossom, and after it yeeldeth seede, like to the seede of our Ladies Thistel, but greater. The roote is long and grosse.

2 The Thistell or prickley Artechoke, hath great long leaues, very much and deepely cut vpon both sides (euen to the very sinewes which depart the leaues) and full of sharpe cruel prickles. The stalke is long, vpon the which grow scaly heades, almost like the others.

They are both of one kinde, & not otherwise to be accounted: for oftentimes of the seede of one springeth the other, especially the Thistell Artechoke commeth of the seede of the right Artechoke. Which thing was very well knowen of Palladius, who commaunded to breake the point of the seede, for bycause it shoulde not bring foorth the prickley kinde.

¶ *The Place.*

These two kindes growe not in this Countrie of their owne accorde, but are sowen and planted in gardens.

❦ *The Tyme.*

The right & prickley Artechokes, bring foorth their great heades in August.

❦ *The Names.*

1 This kinde of Thistell, especially the first sorte, is called of Galen in Greeke κύναρα of Math. Σκόλιμος: in Latine Cynara, Cinara, Carduus, & Carduus satiuus, of writers of our time, Arocum, Alcoralum and Articocalus: in Frenche Artichaut: in Italian Articoca: in high Douch Strobildorn: in Brabāt Artichauts,

X x iiij folowing

folowing the Frenche: the heades be called in Greeke σπονδύλαι, Spondyli: in Englishe, the great and right Artechok.

2 The other is called Cinara acuta: in French Chardonnerette: in Brabant, Chardons. It may be wel Englished, the Thistel or prickley Artichoke.

❧ The Nature and Vertues.

The heades of Artechokes are vnholesome to be eaten, as Galen writeth in his boooke, De Alimentis, and of harde digestion, wherefore they engender noughtie humours, especially being eaten rawe and vnprepared. Therefore they must be boyled after the order of Asparagus, in some good broth of beefe or other flesh, then serue them with a sauce of butter or oyle, salt and vineger: some vse them rawe with pepper and salt, and the powder of Coriander, and so they yeelde a natural pleasant and kindly sweetenesse in taste. They are not vsed in medicine, as my Aucthor in folowing Dioscorides and Galen writeth.

Some write, if the young and tender shelles or Nuttes of the Artechok (being first stieped or soked in strong wine) be eaten, that they prouoke vrine, and stirre vp the lust of the body.

Also they write, that the roote is good agaynst the rancke smel of the arme-pittes, if after the taking cleane away of the pith, the same roote be boyled in wine and dronken. For it sendeth foorth plentie of stinking vrine, whereby the ranke and rammishe sauour of al the body is amended.

The same boyled in water and dronken, doth strengthen the stomacke, and so confirme the place of naturall conception in women, that it maketh them apt to conceaue male Children.

The first springes or tender impes of the Artechok sodden in good broth with Butter, doth mightely stirre vp the lust of the body both in men and women, it causeth sluggishe men to be diligent in Sommer, and wil not suffer women to be slowe at winter. It stayeth the inuoluntarie course of the naturall seede in man or woman.

Of our Ladyes Thistell.
Chap. lxij.

❧ *The Description.*

Our Ladies Thistel hath great, broade, white, greene leaues, speckled w many white spots, set rounde about with sharpe prickles. The stalkes be long, as bigge as ones finger, at the top whereof grow rounde knapped heades with sharpe prickles, out of the same knappes come foorth fayre purple flowers, and after them within the same heades groweth the seede inclosed or wrapt in a certayne cotton or downe. The which is not much vnlyke the seede of wilde Carthamus, but lesser, rounder, and blacker. The roote is long, thicke and white.

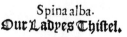

Spina alba.
Our Ladyes Thistel.

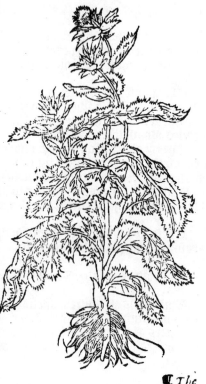

The

the Historie of Plantes. 525

✽ *The Place.*

Our Ladyes Thistel groweth of his owne kinde in this Countrie, almost in euery garden of pot herbes, and is also founde in rough vntoyled places.

✽ *The Tyme.*

It floweth in June and July, the same yere it is first sowen, and when it hath brought foorth his seede, it decayeth and starueth.

✽ *The Names.*

This Thistell is called in Greeke ἄκανθα λευκή: in Latine Spina alba, of some also Agriocinara, Donacitis, Erysiceptrum, Spina regia, and Carduus Ramptarius: of the Arabian Physitions, Bedeguar: in Englishe, Our Ladies Thistell: in Frenche, *Chardon nostre Dame*: in high Douche, Marien Distel, and Frauwen Distel: in base Almaigne, Onser Vrouwen Distel: in shoppes, Carduus Mariæ.

✽ *The Nature.*

The roote of our Ladies Thistel, is dry and astringent. The seede is hoate, and of subtill partes.

✽ *The Vertues.*

The roote of our Ladyes Thistel dronke in wine, is good for them that spit blood, and for those that haue feeble stomackes, and lose bellyes. A

Taken in the same sort, it prouoketh vrine, and driueth it foorth. B

It consumeth colde and soft swellinges, being layde thereunto. C

The wine wherein it hath bene boyled, swageth the tooth ache. D

The seede is giuen with great profite, to children that be troubled with the E cramp, or the drawing awry of any member, and to suche as are bitten with Serpentes, or other venemous beastes.

Spina peregrina.

Of the Globe Thistel. Chap.lxiij.

✽ *The Description.*

This thistel hath also great broade leaues, of a sadde greene colour aboue, or in the vpper side, and next the grounde they are rough, & of a grayish colour, deepely iagged and hackt rounde about, the indented edges are full of sharpe & prickley pointes. The stalke is rounde, and blackishe, as bigge as ones finger, and of foure or fiue foote long, wherupon grow faire round heades, and rough, bearing rounde about a great many of smal whitish flowers mixt with blew. The roote is browne without.

✽ *The Place.*

This Thistell is a stranger in this Countrie, and is not founde but in the gardens of Herboristes, and such as loue herbes.

❊ *The Tyme.*

It flowreth in June and July, a yere after it hath bene sowen.

❊ *The Names.*

The Thistel is called of the writers in these dayes, in Latine, Spina peregrina: ꝫ of Valerius Cordus, Carduus Sphærocephalus: in high Douche Welsch Distel, or Romisch distel: in base Almaigne, Roomsche distel, ꝫ Vremde distel, that is to say, the Romaynes Thistel, or the strange Thistel. How this Thistel was called of the Auncientes, we knowe not, except it be Acanthus sylvestris, wherewithall it seemes to be much like. Turner calleth it Ote Thistel, or Cotton Thistel: in folowing Valerius Cordus, we may also call it Globe Thistell, bycause the heades be of a rounde forme lyke to a Globe or bowle.

❊ *The Nature.*

This strange Thistel is hoate and drie, the whiche may be perceived by the strong smell, in rubbing it betweene your handes: also it may be discerned by the sharpe taste thereof.

❊ *The Vertues.*

This Thistell is not in use that I knowe, except as some do write, that in Italy they boyle the round heades with flesh, and eate them like Hartechokes.

Of white Cotton Thistel. Chap.lxiiij.

❊ *The Description.*   Acanthium.

Acanthium is not muche unlyke our Ladies Thistell, it beareth great large leaues al to mangled and cut by the edges, and set full of sharpe prickles, couered and layd ouer with a fine Cotton or soft Downe. The stalke is great ꝫ thicke set full of prickley stings, at the top of the stalkes are rough heades, in fashion like to the heades of our Ladies Thistel. The roote is great and thicke.

❊ *The Place.*

This Thistell groweth here by the high wayes and borders of feeldes, and in sandy untoyled places.

❊ *The Tyme.*

It flowreth from the moneth of June, unto the ende of August, and sometimes longer.

❊ *The Names.*

This Thistel is called in Greke ἀκάνθιον: in Latine Acanthium: in high Douche weisz wege distel: in neather Douche lande, Witte wech Distel, and Wilde or Groote witte Distel: in Frenche Chardon argentin, or Chardon saluage: in Englishe White Cotton Thistell, Wilde white Thistell, and Argentine, or Siluer Thistel.

❊ *The Nature.*

This Thistel is hoate of complexion.

❧ The

the Historie of Plantes. 527

❧ *The Vertues.*

Dioscorides and Galen write, that the leaues or rootes of Acanthium dronken, are good for such as are troubled with the cricke or shrinking of sinewes, by meanes of the Crampe.

Of Branke Vrsine. Chap.lxv.

❧ *The Kyndes.*

BRanke Vrsine called Acanthos in Greeke, is of two sortes, as Dioscorides sayth, to wit, the garden and wilde Branke Vrsine.

❧ *The Description.*

Acanthus satiuus.
Branke Vrsine.

1 THE tame Acanthus hath great large leaues, of a sadde greene color, thicke and grosse, smooth, & deepely cut in, rent, or iagged by the sydes or borders, lyke the leaues of white Senuie, or Roquet. The stalke is long, of the bignesse of ones finger, couered with long, little, and sharpe poynted leaues, euen all alongst vp to the toppe: amongst the leaues doo growe fayre white flowers, and after them broade huskes, wherein is founde a yellowish seede. The rootes be long and slymie.

2 The wild Acanthus is lyke to the wild Thistell, rough and prickley, but smaller then the aforesayde, as Dioscorides writeth. It is of leaues, flowers, and seede, growing vpwarde, lyke vnto the tame.

❧ *The Place.*

Branke Vrsine groweth in gardens, and in moyst stonie places, as Dioscorides sayth. In this Countrie it is founde but onely in the gardens of Herboristes.

❧ *The Tyme.*

The garden Branke Vrsine, flowreth in this Countrie in July and August, and sometimes later.

❧ *The Names.*

1 The tame or garden Branke Vrsine, is called in Greeke ἄκανθꝏ καὶ ἄκανθα: in Latine Acanthus, and Acantha, of some Pæderota, Herpacantha, Melamphyllon, Topiaria, Marmoraria, and Cræpula: in the Shoppes of Italy and France, *Branca Vrsina*: in Englishe, Branke Vrsine: in Frenche, *Branche Vrsine*. in high Douche, Bernklaw: in base Almaigne, Beerenclauw. It is knowen in the Shoppes of this Countrie, for they vse in stede of the same, the herbe described in the next Chapt. Cooper in his Dictionarie, calleth it Branke Vrsine, Beare Briche, and not Bearefoote, as some haue taken it.

2 The wylde is called of Dioscorides, Acanthus sylueſtris, that is to say, the wilde Acanthus.

❧ *The*

The Nature.

The roote of Acanthus is drie, and temperate in heate.

The Vertues.

The rootes of Acanthus taken in drinke, do prouoke vrine and stoppe the belly. They be excellent for suche as be troubled with crampes or drawing togither of sinewes, and for such as be broken, and those that haue the Ptysike or consumption, or consuming feuer.

The same greene is good against burning, and members out of ioynt, and with the same is made very good playsters agaynst the gowte of the handes and feete.

Dioscorides saith, that the wilde Acanthus hath the same vertue.

Of Douch Branck vrsine. Chap.lxvi.

The Description.

THE wilde Carrot, or Cow Parsnep, hath great rough blacke leaues, much clouen & diuided, into fiue or sixe lesser leaues. The stalke is long, round, and holowe within, full of ioyntes, and sometimes of an inche thicke, at the top of the stalkes growe spokie flowers, which are white, & after commeth the seede whiche is broade and flatte. The roote is white and long.

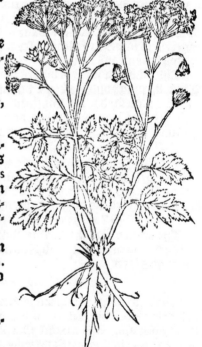

Branca vrsina Germanica.
Wild Carrot, or Cow Parsnep.

The Place.

The wilde Carrot groweth alongst the borders of feeldes, and in lowe grassie places and medowes.

The Tyme.

This herbe flowreth in June and July, and in this space the seede is ripe.

The Names.

This herbe is called in high and base Almaigne, Branca vrsina, and of some writers of our time Pseudacanthus, or Acanthus Germanica: in Frenche Panaiz sauuage: in Douche Bernclaw, or Berntailz: in Brabant, Beerenclauw: in English, Wild Carrot, or Douche Brank vrsine.

Some take it to be the herbe called in Greeke σφονδύλιον: in Latine Spondylium. Turner calleth it Cowe Parsnep, or Medo Parsnep.

The Nature.

Medow or Cow Parsnep, is of a manifest warme complexion.

The Vertues.

Douche Branck vrsine doth consume and dissolue colde swellinges, if it be brused and layde thereupon.

The people of Polonia, and Lituania, vse to make drinke with the decoction of this herbe and leauen, or some suche lyke thing, the whiche they vse in steede of Bier, or other ordinarie drinke.

Turner ascribeth moe Vertues to his Spondilion.

the Hiſtorie of Plantes.

Of Carline Thiſtel.　Chap.lxvij.

❧ *The Kindes.*

Of this kind of thiſtel there be two ſorts. The one beareth white flowers vpon a ſtalke of a handful and a halfe long, or ſomwhat more. The other beareth a red flower without ſtemme.

Leucacantha. Carlina.	Spina Arabica. Carlina minor.
White Caroline Thiſtel.	The Arabian thiſtel, or the leſſer Caroline.

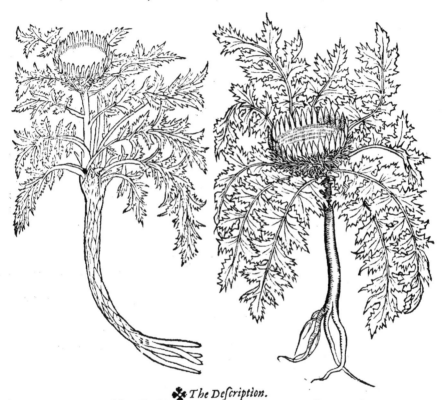

❧ *The Deſcription.*

1　The white Carline, hath long, narrow, rough, and prickley leaues, deepely cut and mingled vpon both ſides or edges, and they haue red ſinewes or ribbes in the middeſt of the leaues, from amongſt thoſe leaues ſpringeth vp a ſtemme or ſtalke of a handfull and a halfe long, or ſomewhat more, bearing ſuche leaues as aforeſayd, vpon whiche ſtemme groweth a round flat head, ſet round about with ſharpe prickles, lyke the ſhelles or huſkes of the Cheſtnut, the which head or knappe is open, & wide aboue in the middle, and thromdelyke Veluet, and rounde about that Veluet, throm, or Crowne, ſtandeth a pale or incloſure, of proper ſmall white leaues, whiche is the flower: the flowers being paſt, you ſhall finde a narrowe gray ſeede amongſt the fine heare or downe. The roote is long and rounde, moſt commonly ſplit, & diuided through the middeſt, of a pleaſant ſmell, and ſharpe bitter taſte.

2　The ſeconde kinde is lyke the other in leaues and rootes, but it is ſmaller. The flowers be of a fayre redde colour lyke the roſe, and growe harde by the

Y p　　　　leaues,

leaues, immediatly from the roote without stalke, almost lying harde by the grounde. The roote is reddish, and of a strong smell.

❦ The Place.

1 The white Carline groweth in many places of Italy, and Douchlande vpon high rough hilles. The Herboristes of this Countrie do sowe it in their gardens.

2 The other lykewise groweth in many places of Italy and Douchland, and in Fraunce, as Ruellius writeth, it is yet vnknowen to vs.

❦ The Tyme.

These two kindes of Carline do flower in July and August.

❦ The Names.

1 The first of these Thistelles is called in Greeke λευκάκανθα (the whiche name is distinct, and separated from Acantha leuce, as Dioscorides writeth) of some it is called Polygonatum, Phyllon, and Ischias, of the Auncient Romaynes Spina alba: nowe they call it Carlina, or Carolina, bycause of Charlemaigne Emperour of the Romaynes, vnto whom an Angel first shewed this Thistel, as they say when his armie was striken with the pestilence: some call it also Cardopatium: in Frenche, Carline: in high Douche, Eberwurtz, Grosz Eberwurtz, and Waisz Eberwurtz: in base Almaigne, Euerwortele, Witte Euerwortele, and Carlina.

2 The other is also a kinde of Carline, and is called in Frenche, Petite Carline: in high Douch, Klein Eberwurtz: and in base Almaigne according to the same it is called Euerwortele, and Cleyne Carlina. Some learned Fryers of Rome do thinke it to be that Thistel, whiche is called in Greke ἄκανθα ἀραβική: in Latine Spina Arabica, of some Acanthis, and of the Arabian Physitions Suchaha.

❦ The Nature.

The roote of Carline is hoate in the first degree, and drie in the thirde.

❦ The Vertues.

A The roote of Carline boyled in wine, is very good for the olde greefes of the side, and against the Sciatica, if you drinke three little cupfulles of wine wherein it hath bene sodden.

B The same taken in lyke manner, is good for them that are bursten, and troubled with the Crampe, or drawing togither of the sinewes.

C The same made into powder and taken to the quantitie of a Dramme, is of singuler vertue against the Pestilence, for as we may reade, al the hoast of the Emperour Charlemaigne, was by the helpe of this roote preserued from the Pestilence.

D The same roote holden in the mouth, is good against the tooth ache.

E The same layde to with vineger, healeth the scurffe and noughtie itche.

F The lesser Carline is the Thistel, which Dioscorides calleth Spina Arabica, and of the Arabian Physitions Suchaha, it stoppeth all issue of blood, the inordinate course of womens flowers, and the falling downe of Rheumes and Catarrhes vpon the lunges and inwarde partes, so that it be eaten. Cooper saith that Leucacantha is a kinde of Thistel with white prickle leaues, called in English, Saint Marie Thistel. Wherein he hath folowed Matthiolus, if their allegations be true, this place is to be amended. Seeke for Matthiolus Carlina in the Chapter Chameleon, where as he reciteth the tale of the Emperour Charlemaigne. The figures here expressed, Matthiolus vseth to *Chameleon*, and to *Leucacantha*, he hath giuen the figure of Saint Marie, or our Ladyes Thistel, whereof we haue before written. Chapt. 63.

Of wilde Caroline. Chap.lxviij.

❀ *The Description.*   Carlina fylueftris.

This Thiftel hath lōg narrow leaues, deeply cut vpon both edges or sides, and prickley, much lyke to the leaues of Carlina: from the middest of which leaues groweth vp a ſtraight rounde ſmall ſtemme, about a foote high, ſet ful of ſuch leaues as are before deſcribed, at the toppe whereof growe three or foure round heades or moe, ſet full of ſharpe prickles lyke the huſkes of the Cheſtnut, the which at their opening do ſpreade very brode in the middle, and about the roundneſſe therof it beareth litle pale yellowiſh leaues whiche is the flower. To conclude, the knoppes with their prickles, flowers, and ſeede, do much reſemble the heades or knoppes of Caroline, ſauing they be ſmaller & paler turning towardes yellowe. The roote is ſmall and hoate vppon the tongue.

❀ *The Place.*

This Thiſtel groweth in this Conntrie, in rude vntoyled places, about the high wayes.

❀ *The Tyme.*

It flowreth in July and Auguſt.

❀ *The Names.*

This Thiſtel is called in high Douch, Dreydiſtel, Frauwen Diſtel, and Seuw Diſtel, and in baſe Almaigne likewiſe, Dryediſtel. It ſhoulde ſeeme, that this is a ſorte or kinde of Carline, and therefore we call it Carlina ſyueſtris, that is to ſay, wilde Carline. It may be ἄκορνα, Acorna of Theophraſte.

❀ *The Nature.*

This Thiſtel is hoate of complexion. But what vertue or working it is of, is yet vnknowen.

Of wilde baſtarde Saffron. Chap.lxix.

❀ *The Description.*

This Thiſtell is not muche vnlyke Carthamus, that is to ſaye, the right Baſtarde Saffron. The leaues be rough and prickley, the little heades or knoppes are deckte, with many ſmall narrow leaues, ſharpe pointed and pricking out, of which growe thready or thrommed flowers, lyke as in Carthamus, of a faynt yellowiſhe colour, but much paler, than the flowers of Carthamus. The flowers paſt, there is founde within the knoppie heades, a ſeede lyke the ſeede of Carthamus, but browner.

¶ *The Place.*

This Thiſtel groweth not of it ſelfe in this Countrie, but is ſowen in the gardens of Herboriſtes.

❀ *The Tyme.*

This Thiſtel flowreth very late in Auguſt and September.

Yy ij   ¶ The

The Names.

This herbe is called in Greeke ἀτρακτυλίς, καὶ κνίκ☉ ἄγρια: in Latine, Atractilis, Syluestris Cnecus, Fusus agrestis, Colus rustica, of some also Amyron, Aspidion, Aphedron, and Presepium: they call it nowe a dayes, Syluestris Carthamus: in French, Quenoille rustique, Saffran bastard sauuage in Douch wilde Carthamus: vnknowen in Shoppes: in Englishe, wilde Carthamus, or wilde bastarde Saffron.

The Nature.

Wild bastard Saffron hath a drying qualitie, and partly digestiue.

The Vertues.

A The tender Croppes, leaues, and seede of this Thistel, wel brayed with Pepper and wine, is very good to be layde to the bitinges of Scorpions.

B Men say also (as Dioscorides hath written) that such as be stongue with the Scorpion, do feele no payne nor greefe so long as they beare this herbe in their handes, but as soone as they let it goe, the ache and payne taketh them agayne.

Of Blessed Thistel. Chap. lxx.

The Description.

Blessed Thistell hath long rough hoare leaues, deepely cut, and parted on both sides or edges. The stalkes be also rough & hearie, creeping or rather lying vpon the ground, and set full of smal leaues, but lyke the other, it beareth rough knoppes or heades, beset rounde about with long and sharpe poynted, little prickley leaues, out of whiche growe the flowers, of a faint yellowish colour. The whiche being past and gone, there is founde in the knoppes, a long gray seede (bearded with bristelles at the vpper ende) laid and wrapped in a soft downe or Cotton. The roote is long and tender full hearie threddes.

The Place.

This Blessed Thistell is sowen in gardens.

The

Atractilis hirsutior.

❧ The Tyme.

It flowreth in June, and July.

❧ The Names.

This herbe is also taken of Plinie, and Theophrast, for a kind of Atractilis, and they call it Atractilis hirsutior. It is nowe called in Shoppes Carduus benedictus, and Cardo benedictus, and accordingly in Frenche they call it Chardon benist: in high Douch Cardobenedict, and Besegneter Distel: in base Almaigne Cardobenedictus: in Englishe, Blessed Thistel, and Carduus benedictus.

❧ The Nature.

Blessed Thistel is hoate and drie of complexion.

❧ The Vertues.

The Blessed Thistel taken in meate or drinke, is good agaynst the great payne, and swimming giddinesse of the head, it doth strengthen memorie, and is a singuler remedie against deafenesse. [A]

The same boyled in wine and dronken hoate, healeth the griping paynes of the belly, causeth sweate, prouoketh vrine, driueth out grauel, and moueth womens flowers. [B]

The wine, wherein it hath bene boyled, doth cleanse and mundifie the infected stomacke, and is very good to be dronken against feuer quartaynes. [C]

The powder thereof dronken in wine, doth ripe and digest cold fleme in the stomacke, and purgeth, and bringeth vp that which is in the breast, scouring the same, and causeth to fetche breath more easily. [D]

To be taken in like manner, it is good for such as begin to haue the Ptysick or consumption. [E]

A Nut shell full of the powder of Carduus benedictus, is giuen with great profite against the pestilence: so that suche as be infected with the sayde disease, do receiue of the powder, as is abouesayde, within the space of xxiiij. houres, and afterward sweate, they shalbe deliuered incontinent. The like vertue hath the wine of the decoction of the same herbe, dronken within xxiiij. houres after the taking of the sayde sicknesse. [F]

The Blessed Thistel, or the iuyce thereof, taken in what sorte soeuer it be, is singuler good agaynst al poyson, so that whatsoeuer he be that hath taken poyson, he shall not be hurt therewithall, if immediatly he take of Carduus benedictus into his body, as was proued by two young folke, whiche when they could not be holpē with treacle, yet were they made whole by the vse of blessed Thistel, as Hierome Bock writeth. [G]

The iuyce of the same dropped into the eyes, taketh away the rednesse, and dropping of the eyes. [H]

The greene herbe pound and layd to, is good agaynst al hoate swellinges, Erysipilas, and sores or botches that be harde to be cured, especially for them of the pestilence, and it is good to be layde vpon the bitinges of Serpentes, and other venemous beastes. [I]

Of Scolymus, or the Wilde Thistel.
Chap. lxxi.

❧ The Kindes.

In this Countrie there is founde three sortes of wilde Thistelles, commonly growing by the way sydes, and in the borders of feeldes, and in wooddes, the whiche are all comprised vnder the name of wilde Thistelles.

The fourth Booke of

Scolymus.
Carduus syluestris. **Wild Thistel.**

Cardui syluestris tertium genus.
The third kind of wild Thistel.

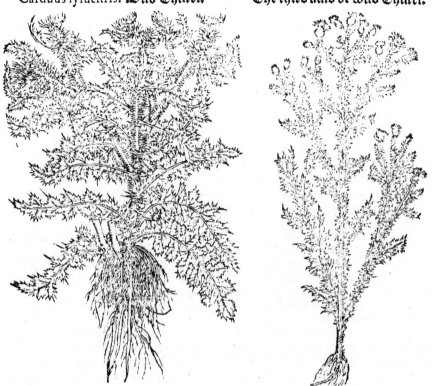

❧ *The Description.*

1. The first kinde of these Thistels groweth about a foote & a halfe high, it hath a round stem ful of branches, and set with prickley leaues, like the leaues of Acanthium, but smaller, and nothing at all frized or Cottonie, & of a browner colour, at the top of the stalke grow round rough knops, set round about full of sharpe prickles, in fashion lyke to a Hedge Hogge, the whiche being open, do shewe foorth a faire purple flower, within the whiche groweth the seede lyke to the seede of the other Thistelles, but smaller. The roote is long and browne, and very full of threddes, or sucking stringes.

2. The second kinde groweth three or foure foote high, and beareth a rounde naked stemme, with a few branches. The leaues be like to the leaues aforesaid, set on euery syde with sharpe prickles, but they be smaller, and not so large as the leaues of the other. The knoppes smal and somewhat long, not very sharpe or pricking: the whiche when it openeth, putteth foorth a purple flower. The roote is blacke and of a foote long.

3. The third kind of wild Thistel groweth also to the length of three or foure foote, hauing a straight stemme, without many branches, but set full of cruell prickles, the leaues are lyke to them of the seconde kinde. The knoppes of this Thistel, are smaller then the knoppes of the seconde. The flowers are purple. The seede is white and very smal. And for his roote, it is nothing els but smal hearie sucking stringes. ❧ *The Place.*

These Thistels grow in all places of this Countrie, by the way sides, & in the

the Historie of Plantes.   535

the feeldes. The second and the third sort are lykewise founde in medowes.

❧ *The Tyme.*

The Thistels flower in Iuly and August.

❧ *The Names.*

1.2. These Thistels be called Cardui syluestres, that is to say, Wild Thistels, & the two first sortes are of that kind of wild Thistels, called in Greke σκόλιμ۞, & of Plinie in Latin Carduus syluestris, & also Limoniũ, of some φέρυσα καὶ πυράκανθα, Pherusa, and Pyracantha. Cooper calleth this, wild Artichoke and Cowthistel.

3 The third is also a kinde of wild Thistel, yet it is not Scolymus, but it may be wel called Carduus Asininus, that is to say, Asse Thistel.

❧ *The Nature.*

The wilde Thistel is hoate and dry in the second degree, as Galen writeth.

❧ *The Vertues.*

The roote of the wilde Thistel, especially that of the second kinde, which is blacke and long, boyled in wine & dronke, purgeth by vrine, and driueth forth al superfluities of the blood, & causeth the vrine to stincke, & to be of a strong smel: also it amendeth the stenche of the armepittes, and of all the rest of the body.   A

The same layd to with vineger, healeth the wild scurffe, & noughty scabbe.   B

Plinie writeth, that in some places men do vse to eate this roote, & the first buddes or tender croppes of the same, as Galen reporteth, but it nourisheth but little, and the nourishment that it yeeldeth, is waterie and nought.   C

Of Tribulus.   Chap.lxxij.

❧ *The Kyndes.*

THeophrast and Dioscorides haue described two kindes of Tribulus, the one of the lande, whiche is also of two kindes. The other of the water, called Saligot.

❧ *The Description.*

Tribulus terrestris Theophrasti prior.

1 THe first kind of Tribulus terrestris, hath long branches, ful of ioyntes, spread abroade vpon the ground, garnished with many leaues, set about with a sort of litle round leaues, stãding in order one by another, all fastened and hanged by one sinewe or ribbe, lyke the leaues of ẏ Cichepease, amongst whiche growe small yellowe flowers, made & fashioned of fiue small leaues, almost like the leaues of Tormẽtil, or white Tansey called in Latine Potentilla, the whiche doo turne to a square fruit, ful of sharpe prickles, wherein is a Nut or kernel, the roote is white & ful of threedy stringes.

Yy iiij   The

The fourth Booke of

Tribulus aquaticus. Saligot.

2　The Saligot or water Tribulus, hath long slender stalkes growing vp, and rising from the bottom of the water, and mounting aboue the same, weake and slender, beneath vnder the water, hauing here and there certaine tuftes or tassels, full of small stringes and fine threddie heares, but the sayde stalke is big or great in the vpper part, where as the leaues grow foorth vpon long stemmes: the said leaues be large and somewhat round, a litle creauesed and toothed rounde about, amongst, & vnder the leaues groweth the fruite, which is triangled, harde, sharpe pointed, and prickley. Within the whiche is conteined a white kernel or nut, in tast almost lyke to the Chestnut.

❧ The Place.

1　The first groweth by the way sides, and neare vnto waters, in vntoyled places. It is founde in Italy and some places of Fraunce. It groweth abundantly in Thracia.

2　Saligot is found in certayne places of this Countrie, as in stues & pondes of cleare water.

❧ The Tyme.

Grounde Tribulus flowreth in June, and after that it bringeth foorth his prickley seede.

❧ The Names.

1　The first of these plantes is called in Greeke τρίβολος, καὶ τρίβολος: in Latine, Tribulus, and Tribulus terrestris. This is the first kinde of Tribulus terrestris, or grounde Tribulus described of Theophrastus. for he setteth foorth two sortes as we haue before sayde, that is to say, one bearing leaues lyke Ciche peason, whereof we haue nowe geuen you the figure to beholde, and the other hauing prickley leaues, for which cause it is called in Greke φυλλάκανθος, Phyllacanthus, that is to say, the prickley leafe. The seconde kinde seemeth to be that kinde of Grounde Tribulus which Dioscorides speaketh of in his fourth booke, whiche kinde is yet to vs vnknowen.

2　That whiche groweth in the water, is called in Greeke τρίβολος ἐνυδρος: in Latine Tribulus aquaticus: in French Chastaignes d'eau, and Saligot: in high Douch Wassernutz, Weihernutz, Stachelnutz, Spitz nutz: in base Almaigne, Water Noten, and of some Minckysers: in English, Water Nuttes, and Saligot.

❧ The Nature.

Grounde Tribulus is colde and astringent, is Galen writeth.

3　The Saligot is also of the same complexion, but moyster.

❧ The Vertues.

The greene Nuttes or fruite being dronken, is good for them whiche are troubled with the stone and grauell.

The same dronken or layde to outwardly, helpeth those that are bitten of Vipers. And dronken in wine, it resisteth all venome and poyson.

The

The leaues of Saligot or water Tribulus, are very good to be laide plaster- C
wise vpon all vlcers, and hoate swellinges.

They be good also agaynst the inflammations and vlcers of the mouth, the D
putrefaction, and corruption of the Jawes or gummes, and against the kings
euill, and swellinges of the throte.

The iuyce of them is good to be put into collyries, & medicines for the eyes. E

They vse to giue the powder of the Nuttes to be dronken in wine, to suche F
as pisse blood and are troubled with grauell.

Also in time of scarsitie they vse to eate them as foode, but they nourish but G
litle, and do stoppe the belly very muche.

Of Madder. Chap.lxxiij.

✤ *The Kindes.*

THere be two sortes of Madder, the tame Madder, the whiche they vse to plant and sowe, and the wild Madder, which groweth of his owne kinde.

Rubia satiua. Garden Madder.   Rubia sylueſtris. Wilde Madder.

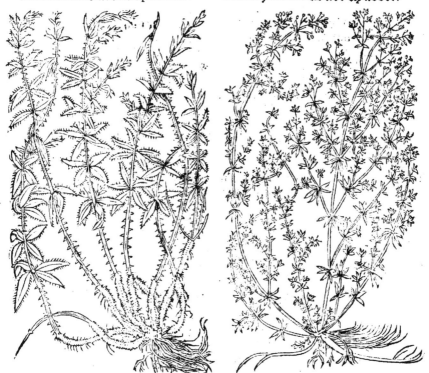

✤ *The Description.*

1. THE husbanded or garden Madder hath long stalkes or branches, square, rough, and full of ioyntes, and at euery ioynt set rounde with greene narrowe leaues fashioned lyke a Starre, the flowers growe about the top of the branches lyke as in the wilde Madder, of a faynt colour changing vpon yellow, after which commeth a rounde seede, at the first greene, then red, and at the last blacke. The roote is very long, small, and red.

2. The wilde Madder is lyke to that of the garden, but it is smaller, and not so rough. The flowers are white. The roote very smal & tender, and somtimes also reddish.

✤ *The*

The fourth Booke of

✤ *The Place.*

1 The husbanded Madder is planted in Zeelande and Flaunders, and in some places of Brabant, by Berrow, in good and fertill grounde.

2 The wilde groweth generally of it selfe, alongst the feeldes vnder hedges and bushes.

✤ *The Tyme.*

They do both flower in July and August.

✤ *The Names.*

1 Madder is called in Greke ἐρυθρόδανον: in Latine Rubia: in Shoppes, Rubia tinctorum: in high Douche, Rodte: in base Almaigne, Rotte, & most commonly Mee, and they call the powder of the Rotte, Meetrappen: in Frenche, Garance: in English, Madder.

2 The wilde is called Rubia syluestris, and of some learned men is thought to be Alysson, of Plinie it is named Mollugo.

✤ *The Nature.*

Garden Madder is drie of Complexion.

✤ *The Vertues.*

The roote of garden Madder, boyled in Meade or honied water and dronken, openeth the stopping of the liuer, the melt, the kidneyes, and matrix: it is good against the Iaunders, and bringeth to women their desyred sicknesse. A

The same taken in the like manner, prouoketh vrine vehemently, in somuch that the often vse thereof causeth one to pisse blood. B

The decoction of the same dronken, or the powder therof dronken in wine, dissolueth clotted or congeled blood in the body, and is good for such as are fallen from high, and are brused or bursten within. C

Men giue the iuyce of the roote to such as be hurt with venemous beastes: and also the wine wherein the rootes and leaues haue boyled. D

The seede thereof taken with Oximel, or honied vineger, doth swage and make lesse the Melt, and healeth the hardnesse thereof. E

The roote put vp vnder into the natural place of conception, in manner of a pessarie, or mother suppositorie, bringeth foorth the birth, the flowers, and secundines. F

The roote brused or pounde very small, healeth al scuruie itche and manginesse, or foulenesse of the body, with spottes of diuers colours, especially layde to with vineger, as Dioscorides teacheth. G

The wilde Madder is not vsed in Medicine. H

Of Goosegrasse, or Cliuer. Chap.lxxiiij.

✤ *The Description.*

Cliuer or Goosegrasse hath many smal square branches, rough & sharpe, full of ioyntes, about whiche branches, at euery ioynt growe long narrowe leaues after the fashion of Starres, or lyke the leaues of Madder: but smaller and rougher, out of the same ioyntes grow litle branches, bearing white flowers, and afterwarde rounde rough seedes, most commonly two vppon a stemme. All the herbe, his branches, leaues, and sede, do cleaue and sticke fast to euery thing that it toucheth: it is so sharpe, that being drawen alongst the tongue, it wil make it to bleede.

✤ *The Place.*

This herbe groweth in all places in hedges and bushes.

✤ *The Tyme.*

It flowreth and beareth seede al the Sommer.

the Historie of Plantes. Aparine. 539

✻ *The Names.*

This herbe is called in Greke ἀπαρίνη, and of some φιλάνθρωπ⊙, καὶ ὀμφαλόκαρπος: in Latine, Aparine: in Frenche, Grateron: in high Douche, Klebkraut: in base Almaigne, Cleefcruyt: in Englishe, Goosegrasse, Cliuer, and Goosefhare.

✻ *The Nature.*

Clyuer is drie of complexion.

✻ *The Vertues.*

A　They drinke the iuyce of the leaues & sede of Goosegrasse, against the bitings, and stinginges of venemous beastes.

B　The same dropped into the eares, healeth the payne and ache of the same.

C　This herbe pounde, and layde vnto freshe woundes stoppeth the bleeding of the same, & pounde with Hogges grease, it dissolueth & consumeth the disease of ye necke, called the kinges euil, and al hard kernelles and wennes wheresoeuer they be, if it be laid therto, as Turner writeth.

Of Gallion. Chap. lxxv.

✻ *The Description.*

Gallion hath small, rounde, euen stemmes, with very small narrowe leaues, growing by spaces, at the ioyntes round about the stemme, starre fashion, and like Cliuer, but muche lesser, and gentler, very smothe, and without roughnesse. The flowers be yellow, and growe clustering about the toppes of the branches like to wilde Madder, the roote is tender, with hearie threddes or strings hanging at it.

¶ *The Place.*

This herbe groweth in vntoyled places, and hylly groundes, as vppon Roesselberch by Louaine.

✻ *The Tyme.*

It floureth in July, and August.

✻ *The Names.*

This Herbe is called in Greke γάλιοψ: and in Latine Gallium: of some Galation, & Galerium: in Spanish, Yerua Coaia leche: in French, Petit Muguet: in Douch, Walstroo: and as Matthiolus and Turner write, Unser Frauwen Wegstro, and of some Megerkraut: we may also name it Pety Muguet, Cheese runnning, or our Ladies bedstraw. ✻ *Th*

Gallion.

The fourth Booke of

❧ *The Nature.*

Gallion is hoate and drie of complexion.

❧ *The Vertues.*

The flowers of Gallion pounde, and layde vpon burninges, drawe foorth the inflammation and heate, and heale the sayde burninges.

The same layde vnto woundes, or put into the nose, stoppe bleeding.

The leaues of Gallion mingled with Oyle of rooses, and set in the Sonne, and afterwarde layde vpon wearied members, doo refreshe and comfort them. The rootes prouoke men to their naturall office in Matrimonie. The herbe may serue for Rennet to make Cheese: for as Matthiolus vpon Dioscorides writeth, the people of Tuscane or Hetruria doo vse it to turne their milke, bycause the Cheese that they vse to make of Peowes and Goates mylke, shoulde be the pleasanter and sweeter in taste.

Of Woodrow, or Woodrowel. Chap.lxxvi.

❧ *The Description.*   Asperula.

Woodrowe hath many square stalkes, full of ioyntes, at euery knot or ioynt, are seuen or eight long narrow leaues, set rounde about lyke a starre, almost like the leaues of Cliuer or Goosegrasse, but broader, and nothing rough. The flowers grow at the toppe of the stemmes or branches of a white color, and pleasant of smell (as all the herbe is.) The seede is round, and somwhat rough.

❧ *The Place.*

In this Countrie they plante it in all gardens, and it loueth darke shadowie places, and deliteth to be neare olde moyst walles.

❧ *The Tyme.*

Woodrowe flowreth in may, and then is the smell most delectable.

❦ *The Names.*

This herbe is called in Latine Asperula, Cordialis, Herba Stellaris, and Spergula odorata: in high Douch, Hertzfreydt, and Walmeyster: in base Almaigne, Walmeester: in Frenche, *Muguet*, by the whiche name it is best knowen in most places of Brabant. Some woulde haue it a kinde of Liuerwort, and therefore it is called of them in Latine Hepataria, Hepatica, Iecoraria, and in high Douche Leberkraut. The ignorant Apothecaries of this Countrie do call it Iua muscata, and do vse it in steede thereof, not without great errour.

❦ *The Nature.*

Woodrow taketh part of some heate, & dryness, not much vnlike to Gallion.

❧ *The Vertues.*

Woodrowe is counted a very good herbe to consolidate and glewe togither woundes, to be vsed in lyke maner, as those herbes we haue described in the ende of the first booke.

Some say, if it be put into the wine whiche men doo drinke, that it reioyseth the hart and comforteth the diseased liuer.

the Historie of Plantes. 541

Of Golden Croſworts, or Muguet. Chap.lxxvij.
Cruciata.

❧ *The Deſcription.*

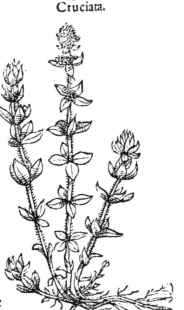

Croſwort is a pale greene herbe, drawing nere to a yellow Popingay colour, couered or set full of fine cotton or soft heares, hauing many square stalkes, ful of knottes or ioyntes. The leaues be litle, smal, and short, alwayes foure growing togither, standing one againſt another, in faſhion lyke to a Croſſe at euery ioynt: aboue the ſaid leaues growe vp from the ſayde ioyntes, many smal yellowiſh flowers, growing round about, & compaſſing the ſtem lyke Crownettes or garlands: and euery of the ſayd litle proper flowers, are parted againe into foure diuiſiós faſhioned like to a ſmal Croſſe. The rootes be nothing els, but a ſort of ſmal tender threds.

❧ *The Place.*

Croſwort groweth of his owne accorde, by trenches, and water courſes, and is founde vnder hedges in moyſt places.

❧ *The Tyme.*

Croſwort flowreth almoſt all the ſommer long, eſpecially from May vnto Auguſt.

❧ *The Names.*

This herbe is called of the Herboriſtes of theſe dayes in Latine Cruciata, that is to ſay, Croſwort: in Frenche, *Croyſée*: in high Douche, Golden Walmaiſter, that is, Golden Muguet: in baſe Almaigne, Cruſette.

❧ *The Nature.*

It is drie and aſtringent.

❧ *The Vertues.*

Cruciata hath a very good propertie to heale, ioyne, & close togither wounds, A agreeable for all manner of woundes both inwarde and outwarde, if it be ſo ſodde in wine and dronken.

They giue the wine of the decoction of this herbe, to folke that are burſten, B and lay the boyled herbe right againſt, or vpon the burſten place, as ſome, who haue made experience thereof, do affirme.

Of Buckes Beanes. Chap.lxxviij.

❧ *The Deſcription.*

Marriſhe Trefoyl hath brode, ſmothe, thicke leaues, alwayes three togither vpon one ſtemme, in faſhion, quantitie, thickneſſe, and proportion of leaues, lyke to the cómon beane. The ſtalke is ſmal, of a foote and a halfe, or two foote long, at þ top wherof grow white flowers, and afterwarde rounde huſkes or knoppes, conteyning a yellowiſhe browne ſeede. The roote is long, white, and full of ioyntes.

❧ *The Place.*

Marriſh Trefoyl groweth in lowe moyſt places, in pooles, and ſometyme on riuer ſydes.

❧ *The Tyme.*

It flowreth in May, and in June the ſeede is ripe.

Z z ❧ *The*

The Names.

This herbe is called of the writers nowe a dayes, Tritolium paluſtre: in Brabant, Bocrboonen, that is to say, Bockes Beanes: bycauſe it is like the leaues of the common Beane: it ſhoulde ſeeme to be ἰσόπυρον. Iſopyrum, whiche ſome doo alſo call Phaſiolon, bycauſe of the lykeneſſe it hath to Phaſiolos, as Dioſcorides writeth. Matthiolus confeſſeth that he neuer ſawe the right Iſopyron.

The Vertues.

The ſeede of Iſopyron is good againſt the cough, and other colde diſeaſes of the breaſt, to be taken with Meade or Hydromel: it is alſo good to be taken in like manner of ſuche as ſpet blood, and are lyuer ſicke.

Trifolium paluſtre.

Of Foxetayle. Chap.lxxix.

The Deſcription.

Foxetayle hath blades and helme almoſt lyke wheate, as Theophraſtus writeth, but ſmaller and better, like the blades & ſtems of Couche graſſe, at the top or end of the ſtemmes growe ſmall ſoft hearie eares or knoppes, very like to Foxetayle.

The Place.

Foxetayle groweth not in this Countrie: but in certayne places of Fraunce, in fieldes and alongſt the ſea coaſt.

The Tyme.

This herbe floureth in June and July.

The Names.

Theophraſt calleth this herbe in Greeke ἀλωπέκυρος, that is to ſay in Latine, Cauda vulpina: in Engliſhe, Foxetayle: in Frenche Queue de Renarde: in high Douche, Fuchs ſchuantz: in baſe Almaigne Voſſen ſteert.

Alopecuros.

The Nature and Vertues.

The Auncientes haue made no mention at all, of the nature, and vertues of this herbe.

Of

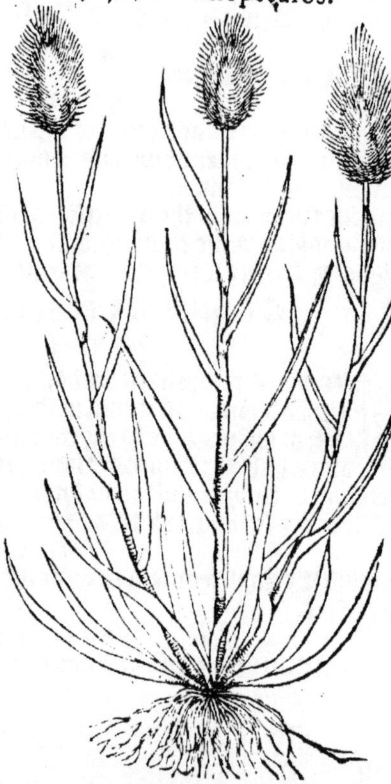

Of Tragacantha. Chap. lxxx.

✱ *The Description.* Tragacantha.

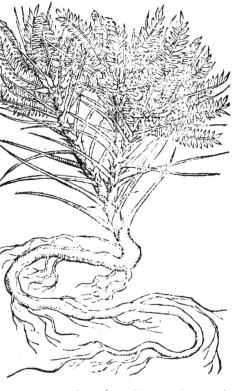

Tragacantha hath many branchie boughes and twigs, slender, and pliant, so spreade abrode vppon euery side, that one plante doth sometime occupie the roome or space of a foote, or a foote and a halfe in compasse. The leaues be as smal as the Lentil leaues, whitish, and somwhat mossie or heary, set in rewes, directly aunswering one leafe agaynst another, all alongst a small twigge or slender branche, neither greater nor lesse, but like the boughes and leaues of Lentilles. The flower is also lyke the blowing of þ Lentil, but much smaller, almost lyke the blossom or flower of Ciche peason, whitishe, and sometimes marked with purple lines or strakes. The seede is inclosed in smal huskes, almost like to the wild Lotus or Trefoyl. The whole plant on euery side is set ful of sharpe prickley thornes, harde white and strong. The roote stretcheth it selfe alongst, in length vnder the ground, like to the roote of the common Liquerise, yellowe within, and blacke without, tough and limmer, and harde to breake, the which roote being layde in some feruent hoate place, or in the Caniculer dayes laid in the Sonne, it getteth a white gumme, which is founde sticking fast vpon it.

✱ *The Place.*

Tragacantha groweth in Media, and Creta, as Plinie sayth: it is also found in other Countries, as in Prouince about Marselles, whereas I haue seene great store. ✱ *The Tyme.*

Tragacantha flowreth in April, the seede is ripe in June, & in the Caniculer dayes the gumme is founde cleauing to the roote.

✱ *The Names.*

This plant is called in Greeke ραγακάνθα: in Latine Tragacantha: and Hirci spina: vnknowen in Shoppes, euen amongst them where as it groweth.

The gumme also whiche commeth from it, is called in Greeke ραγακάνθα: in Latine Tragacanthæ lachryma: in Shoppes, Gummi Dragaganthi : in English, Gumme Dragagant. ✱ *The Nature.*

Tragacantha, as Galen writeth, is of nature like to gumme Arabique, that is to say, of a drie and clammie complexion.

✱ *The Vertues.*

Gumme Dragagant is good against the cough, the roughnesse of the throte & the hoarsenesse and roughnesse of the voyce, being licked in with honie. For the

Z z iij same

same purpose (that is to say for the roughnesse of the throte and sharpe Arterie or winde pipe) They make a certaine electuarie in shops, called Diatragaganthū.

They drinke it stieped in wine the quantitie of a dramme, against the paine of the kidneyes, and excoriation or knawing of the bladder, in putting thereto Hartes horne burnt and washed.

The sayd gumme is put into Collyres, and medicines that are made for the eyes, to take away the acrimonie and sharpnesse of the same: it doth also stoppe the pores and conduites of the skinne.

❊ *The Choise.*

You must chuse that whiche is cleare and shining, smal, firme, and close, well purified and cleene from al manner filth, and sweat.

Of Ficus Indica. Chap. lxxxi.

This strange kind of plante commeth foorth of one leafe set in the grounde, and sometimes it groweth high, and is named of Plinie Opuntia, nowe in these Dayes Ficus Indica.

That Euphorbium commeth foorth lykewise of one leafe, but yet it is separated from this kind, for the leaues of Euphorbium be long, rounde, and thick, fashioned like vnto Cucumbers, set on the sides with thornes. Of that Euphorbiū writeth Ioannes Leo in his historie of Aphrica, and is spoken of before in the second part of this booke in ỹ cxvi. Chap.

Ficus Indica.

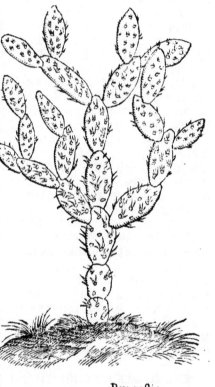

Of Bupreſtis. Chap. lxxxii.

This worme is called in Greeke Βούπρησις and in Latine Bupreſtis, in some places of the lowe Countrie he is called Veemol. And is called Bupreſtis, bycause it is hurtfull to cattel, as namely vnto Oxen and kyen. And is founde in certayne places of Holland, and lykewise somtimes in Brabant, and Flaunders: where the kyen sometimes are bitten of them.

This worme is of the kinde of Scarabeen or Horsewormes, the whiche are named Cantharides, or Spanishe Flyes, and hath winges lyke vnto these, and is of forme and bignesse suche as the figure doth shewe. And this figure haue we set here, bycause that some haue set foorth another worme, not lyke vnto the true Bupreſtis.

Bupreſtis.

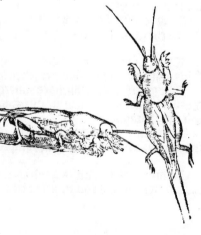

The end of the fourth Booke.

the Historie of Plantes. 545

¶ The fyfth part of the Historie

of Plantes, treating of the differences, fashions, names, vertues, and operations of herbes, rootes, and fruites, whiche are dayly vsed in meates:

Set foorth by Rembertus Dodonæus.

Of Orache.   Chap.i.

❧ *The Kindes.*

ORache as Dioscorides writeth, is of two sortes: the garden Orache, and the wilde Orache.

Atriplex satiua.
Garden Orache.

Atriplex syluestris.
Wylde Orache.

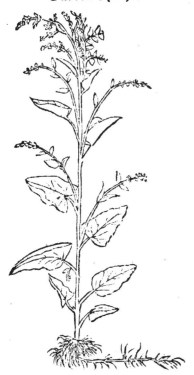

❧ *The Description.*

GArden Orache hath long straight stalkes, rounde next the roote, and square aboue with many branches. The leaues be (almost triangled) long and broade, of a feynt yellow, or white colour, as if they were ouerstrowen with meale or flower, especially those leaues that are yet yong and new sprong vp. The flowers growe at the top of the branches a number clustering togither, small and yellow, and afterwarde commeth the seede, which is broade, and couered with a litle skinne or rime. The roote is full of hearie stringes.

There is also another kinde of Garden Orache, whose leaues, stalkes, and

Z3 iij   flowers

flowers, be of a browne red colour, but in all thinges els lyke to the leaues, stalkes and flowers of the white Orache both in bignesse and proportion.

2 The wilde Orache hath also a long stalke moulded or crested, with leaues not muche vnlyke the leaues of the garden Orache, but somewhat lesser, and creuised or a little snipt rounde about. The flowers be yellowishe. The seede is harde, and groweth thicke clustering togither, lyke as the seede of the garden Orache. The roote is full of heares.

Of this wilde kinde, there is also founde another sort, the whiche groweth not very high, but remayneth lowe, and spreade abroade into many branches. It hath little long narrowe leaues nothing snipt or creuished about. The flowers, seede, and rootes are very muche lyke vnto the wilde kinde before described.

❧ The Place.

The garden Orache groweth amongst other pot herbes in gardens. The wilde Orache is founde alongst the feeldes and wayes.

❧ The Tyme.

Orache flowreth in June and July, and almost all the sommer.

❧ The Names.

It is called in Greeke ἀτράφαξις: in Latine Atriplex: of some Chrysolachanon, that is to say in Latine, Aureum olus: in Frenche, Arroches, or Bonnes Dames: in high Douche, Molten, and Milten: in base Almaigne, Melde: in English, Orache.

1 The garden Orache is called in Greeke ἀτράφαξις κηπωτή: in Latine Atriplex satiua, and Hortensis: in high Douche, Heymisch Molten, Zam Molten, and Garden Molten: in base Almaigne, Tam Melde.

2 The wilde is called ἀτράφαξις ἀγρία, and Atriplex syluestris: in high Douche, Wilde Molten, Ackermolten: in base Almaigne, Wilde Melde.

The lesser wilde kinde is called in high Douche, Kleyn Scheiszmilten: in base Almaigne, Cleyne Melde.

❧ The Nature.

Orache is colde in the first degree, and moyst in the seconde, especially garden Orache, the whiche is more colder and moyster, than the wilde Orache.

❧ The Vertues.

Orache eaten in pottage as other herbes, doth soften and loose the belly. A

The seede of Orache taken in Meade or Honied water, doth open and B comfort the stopped lyuer, and is good against the Jaundize, or Guelsought.

Greene Orache brused, is very good to be layde vppon inflammations and C hoate swellinges. that of the garden, at the beginning of the swelling or inflammation: and the wilde, at the ende or going away of the same.

With Saltpeter, honie & vineger, it is layd to Cholerique inflammations, D called wilde Fier (bycause it doth wast and consume the member it is in): and also to the gowte.

❧ The Danger.

The often vse of Orache engendreth many infirmities, ouerturneth the stomacke, and causeth diuers spottes, freckles, or pimples to arise in the face, and all the rest of the body. Also it is harde of digestion, as sayth Diocles, and Dionysius.

the Historie of Plantes. 547

Of Blites.   Chap.ij.

❧ *The Kindes.*

There be two sortes of Blites, the great and the small, and euery of them is diuided againe into two kindes, whereof the one is white, and the other redde, and both common in this Countrie.

❋ *The Description*

Blitum maius. **The great Blite.**
Blitum rubeum. **The red Blite.**
Blitum album. **The white Blite.**

A 1 THE great white Blite groweth two or three foote high, & hath grayish, or white rounde stalkes. The leaues be playne and smoth almost lyke the leaues of Orache, but not so soft, white, nor mealie. The flowers growe like Orache, and after them commeth the seede inclosed in little flat huskie skinnes.

B The great red Bleete is much lyke the other, sauing that his stalkes be very red, and the leaues of a browne greene color, changeable vpon redde, and so is the seede also.

A 2 The lesser Blite with the greene stalke, is full of branches, and groweth vp sodenly. The leaues be long and narrowe or smal, not much vnlike ẏ leaues of Beetes, sauing they be farre smaller. The flowers be browne turning towardes redde. The seede groweth clustering togither lyke Orache seede. The roote is full of hearie stringes.

B The smal red Blite hath stalkes red as blood, and so are his leaues and rootes, in so muche that with the iuyce of this herbe, one may write as faire a red, as with roset made of Brasill: otherwyse it is lyke the rest of the kindes of other Blites.

❡ *The Place.*

This herbe groweth wilde, and in some gardens amongst pot herbes, and where as it hath once taken roote, it commeth vp euery yeere, wherefore it is counted but a weede, or vnprofitable herbe.

❋ *The Tyme.*

It is founde most commonly in flower about midsomer.

❋ *The Names.*

This herbe is called in Greeke βλίτον: in Latine, Blitum: in Frenche, *Blette*, and *Pourée rouge*: in high and base Almaigne, Maier: in Englishe, Blite, and Blittes.

❡ *The Nature.*

This herbe is colde and moyst.

❋ *The Vertues.*

Blites eaten in pottage do soften the belly, but it hurteth the stomacke, and nourisheth not.

The fyfth Booke of

Of Goose foote. Chap.iij.

❊ *The Description*

Goose foote groweth a foote and a halfe high, or two foote in length, the stalke is straight and full of branches, the leaues be brode and deepely cut rounde about, almost like to a Ganders foote, wherefore it is so named. The flowers be small & reddish. The seede groweth clustering lyke the Orache seede. The roote is full of hearie threddes.

❊ *The Place.*

This herbe groweth wilde, and in vntopled places, alongst by the way sides, and is taken but as a weede or vnprofitable herbe. ❊ *The Tyme.*

You shal finde it flowring in June, and July. ❊ *The Names.*

This herbe is called of the writers in our tyme Pes Anserinus: in high Douche, Gensfusz: in Frenche, *Pied d'oyson*: in base Almaigne, Gansenvoet, and of some Schweinsztod, & Seutod, that is to say, Swines bane, bycause the Hogges eating of this herbe, are immediatly baned, or taken with the Murren, so that within short space they die. ❊ *The Nature.*

This herbe is cold almost in the third degree. ❊ *The Vertues.*

A This herbe in operation is much like Morel or Nightshade and may be vsed outwardly to all thinges wherevnto Nightshade is required.

Tragium Germanicum.

Of the ranke Goate, or stinking Motherworte. Chap. iiij.

❊ *The Description.*

This herbe also is somewhat lyke Orache, but in al thinges smaller. This is a little lowe tender herbe with many long branches trayling on the ground. The smal leaues are whitish, as though they were ouerstrowen with meale, lyke to y leaues of Orache, but muche smaller, neither muche greater then the leaues of Maierom gentil. The seede is smal and white, and groweth clustering togither like the seede of Orache. All the herbe stinketh like rotten corrupt fishe,

or lyke

the Historie of Plantes.　　　　549

or lyke stinking fishe broth, or lyke a ranke stinking Goate.

¶ *The Place.*

It groweth in this Countrie in sandie places by the way sides.

※ *The Tyme.*

You may finde it in flower and seede, about midsomer.

※ *The Names.*

This herbe hath no particuler Latine name, wherefore bycause of his stinking sauour, we do call it in Greeke τραγιον: in Latine Tragium, that is to say, Goates herbe. And bycause you shal reade in Dioscorides of two other herbes called Tragia, to make some difference betwixt them, we do name this Tragium Germanicum: in Frenche, *Blanche putain*: in base Alinaigne, Bocxcrupt: some call it Vuluaria, by whiche name it is knowen of the Herboristes of this Conntrie: Valerius Cordus calleth it Garolmos: I haue named it in Englishe, The ranke stinking Goate, or stinking Motherwort. And is taken of some to be that stinking herbe, that of Plautus is named Nautea.

※ *The Vertues.*

The smel of this herbe is good for women that are bexed with the rising vp of the mother: and for the same greefe, it is good to be layde vpon the nauell.

Of Beetes.　　Chap.v.

※ *The Kindes.*

There be two sortes of Beetes, the white and red. And of the red sorte are two kindes, the one hauing leaues and roote lyke to the white Beete, the other hath a great thicke roote, and is a stranger amongst vs.

Beta candida. **White Beete.**　　　　Beta nigra. **Redde Beete.**

℃ *The*

The fourth Booke of

✽ *The Description.*

1 THE white Beete hath great brode playne leaues, amongst the whiche riseth vp long crested or streked stalke. The flowers grow alongst by the stalkes one vpon another, like little Starres. The seede is rounde, harde, and rough. The roote is long & thicke, and white within.

2 The common redde Beete is muche lyke vnto the white, in leaues, stalkes, seede, and roote: sauing that his leaues and stalkes are not white, but of a swart browne red colour.

3 The strange red Beete is like to the common red Beete, in leaues, stalkes, seede, proportion, & color, sauing that his roote is muche thicker, and shorter, very well like to a Rape or Turnep, but very redde within, and sweeter in tast then any of the other two sortes.

✽ *The Place.*

They sowe the Beete in gardens amongst pot herbes. The strange redde Beete is to be founde planted in the gardens of Herboristes.

✽ *The Tyme.*

Beetes doo seede in August, a yeere after their first sowing.

Beta nigra Romana.
The strange red Beete.

✽ *The Names.*

Beetes are called in Greeke τεῦτλον, σεῦτλον: in Latine & in Shoppes, Beta: in Frenche, Bete, Iotte, Porée: in high Douch, Mangolt: in base Almaigne, Beete.

1 The white kind is called Sicula, and of some Sicelica, or Sicla, of the writers in our time, Beta candida. in Englishe, The white Beete: in French, Bete blanche: in high Douche, Weisser Sangolt: in base Almaigne, Witte Beete.

2 The common red Beete is called Beta nigra: in French, Bete rouge: in Douch, Roter Mangol, and Roode Beete.

3 The thirde is called Beta nigra Romana, that is to say, The Romayne or strange red Beete: in Frenche, Bete rouge Romaine, or Estrangere: in Douch, Roomsche roode Beete: of some, Rapa rubra: albeit this is no kynd of Rape or Naueau.

✽ *The Nature.*

Beetes are hoate, drie, and abstersiue, especially the white Beete, the which is of a more abstersiue and clensing nature.

✽ *The Vertues.*

The iuyce of the white Beete dronken, openeth the belly, and clenseth the stomacke, but it must not be vsed to often, for it hurteth the stomacke. A

The same with hony powred into a mans nose, purgeth the braine, and openeth the stoppinges of the nosethrilles, and swageth the headache. B

The same powred into the eares, taketh away the paynes in the same, and also the singing or humming noyse of the same. C

The rawe leaues of Beetes pounde and layde to, heale the white scurffe, so that the place be first rubbed well with Saltpeter. D

The

the Historie of Plantes.

¶ The same raw leaues pound are very good to be laid vpon spreading sores, & vpon the roome or noughty scales and scurffe, which causeth the heare to fal of.

¶ The leaues sodden, are layd to as an emplaister, vppon burninges and scaldinges, hoate inflammations, and wheales comming of choler and blood.

¶ The broth of Beetes scoureth away the scuruie scales, nittes, and lice of the head, being washed therwithall, and is good for mouled or kybed heeles, to be stued or soked in the same.

¶ The rootes of Beetes put as a suppositorie into the fundement, doth soften the belly.

¶ Beetes vsed in meates nourisheth but litle, but it is good for them that are splenitike: for being so vsed, it openeth ẏ stoppings of the liuer & melt or splene.

¶ The common red Beete boyled with Lentils, and taken before meate, stoppeth the belly.

¶ The roote of the Romaine or strange red Beete, is boyled and eaten with oyle and vineger before other meates, and sometimes with pepper, as they vse to eate the common Parsenep.

Of Colewurtes, and Cabbage Cole.   Chap.vi.

❧ The Kindes.

THere be diuers sortes of Colewurtes, not muche lyke one another, the which be al comprehended vnder two kindes, whereof one kinde is of the garden, and the other is wild. Agayne, these Colewurtes are diuided into other kindes. for of the garden Colewurtes, some be white, and some be red, and yet of them againe be diuers kindes.

Brassica Tritiana, siue Capitata.   Brassica Pompeiana, aut Cypria.

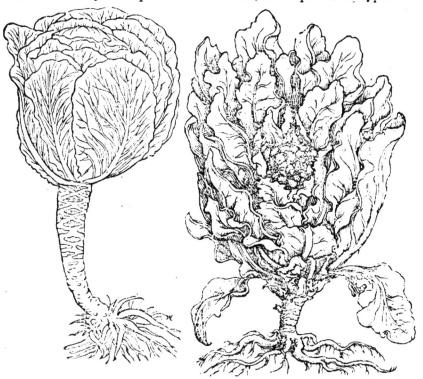

The fourth Booke of

❧ The Description.

1. THE first kinde of white Colewurtes, is the common white lofed or close Cabbage, þ which hath great large leaues, ful of grosse veynes, ribbes, or sinewes, whereof the first leaues before the closing of the Cabbage, are of a white greene colour, but the others folowing next vnder them, are as white as snow, the which do grow so closely layd, & folden harde one with, and vpon another, that they are lyke to a great globe, or round bowle. These Colewurtes (winter being once gone & past) do spreade abrode by vnfolding them selues, and doo bring foorth both flowers and seede, as the other Colewurtes doo.

2. The seconde kinde of white Colewurtes, is that whiche they call Sauoye Colewurtes. It is very much like to the white lofed Cabbage, & so it closeth, but nothing so firmely, neither is it so great nor so round as the aforesayd: but it abideth smaller, and in shape longer. This kind of Colewurtes cannot abide the colde, for most commonly it decayeth as soone as it beginneth to freese. Neuerthelesse the winter being caulme, as it was in the yeere of our Redeemer M.D.LX. after winter it bringeth foorth his stalke with fayre white flowers, and afterwarde his seede in small coddes lyke to the other Colewurtes.

3. The third kind of white Colewurtes is very strange, and is named Flowrie or Cypresse Colewurtes. It hath grayishe leaues at the beginning lyke to the white Colewurtes and afterwarde in the middle of the same leaues, in the steede of þ thicke Cabbaged, or lofed leaues, it putteth forth many smal white stemmes, grosse and gentle, with many short branches, growing for the most part al of one height, thicke set and fast throng togither. These little stemmes so growing togither, are named the flower of these Colewurtes.

4. The fourth kinde hath grayish or white greene leaues, as the other white Colewurtes haue, but they remayne still without closing or gathering to a rounde head or croppe: yet it beareth a great round knoppe like a Turnep, the which groweth right vnder the leaues, euen hard vpon the ground, & is white within lyke a Turnep, and is euen so drest and prepared to be eaten.

5. There is also a very strange kinde of Cole, whiche is also set amongst the white kindes of Colewurtes, and is now called swart, or blacke Colewurtes. It hath long high stemmes, and great, broade, swartgreene leaues, the which are vneuenly wrinkled, or crompled. The flowers be yellowe. The seede and coddes are very well lyke the other Colewurtes.

❧ The Description of the redde Colewurtes.

1. The first kinde of red Colewurtes, hath great, brode, and smoth, brownish, darke, red greene leaues, with reddish ribbes, or veynes going through them. The flowers be yellow, and the coddes or huskes be long and slender, the seede is small and round, browne without, and yellow within, muche like to Rape seede, but it is bigger.

2. The seconde kinde his leaues at the beginning are lyke to the leaues of the former, but afterwarde the middle leaues do gather them selues togither, and lie one vpon another like the white Cabbage or lofed Cole, the whiche be of a red or purple colour.

3. The thirde kinde of red Colewurtes his leaues be al to cut, and iagged, els it is like both in colour, flowers, and seede to the first.

4. The fourth kinde his leaues be ruft, crompled, and drawen togither or curled, the rest is lyke to the former red kindes.

5. The fifth kind of red Colewurts, is the least of them al, and almost like the wilde Cole, his stalkes and leaues are muche smaller, then the first, but in all thinges

the Historie of Plantes. 553

Brassica Cumana siue rubra.
Red Colewurtes.

Brassica Sabellica siue crispa.
Ruffed, or curled Colewurtes.

thinges els lyke. This sorte of Colewortes is not vsed in meates, but is sowen for the seede onely, from which they do drawe foorth an oyle, whiche is dayly and commonly solde for Rape oyle.

❧ *The Place.*

Al these kindes of Colewurtes, are planted in gardens of this Countrie. But the fifth kinde is sometimes sowen in the feeldes lyke Rapes.

❧ *The Time.*

The best Colewurtes, are they whiche be sowen in March, and planted againe in May: for they wil be redy to be eaten in winter, and if they abyde the winter, they wil flower in March and April, and the seede is ripe in May. But some kindes, especially the white Cabbage Cole, or losed Colewurtes, is also sowen in August, and planted againe in Nouember, & then it closeth or loseth in June, July, and August, and after that time it is good to be eaten.

❧ *The Names.*

Garden Colewurtes are called in Greeke κράμβαι ἥμεραι: in Latine Brassicæ satiuæ: in Shoppes, Coles: in high Douche, Kolen: in base Almaigne, Koolen.

1 The first kind of white Colewurtes, is called in Latine (of Plinie) Brassica Tritiana, of the writers in our dayes, Brassica sessilis capitata, and Imperialis: in Frenche, *Chous cabus*: in high Douche, Kappszkraut: in base Almaigne, Witte Sluptkoolen, & Kabuyskoolen: in Englishe, White Colewurtes, Losed Cabbage, and Great rounde Cabbage Cole.

2 The second kind is called of Plinie in Latine, Brassica Lacuturria: in french,

Aaa   *Chous*

The fyfth Booke of

Chous de Sauoye: in bafe Almaigne, Sauoy Koolen.

3 The thirde kinde is called in Latine, Brassica Pompeiana, of the writers in our time, Brassica Cypria: in Italian, Cauliflores: in Frenche, *Chous florys*: in bafe Almaigne, Bloemkoolen: in Englifh, Flowrie Cole, or Cypres Colewurtes.

4 The fourth kinde is nowe called Rapæ Caulis, that is to fay, Rape Cole: in Frenche, *Chou Naueau*: in bafe Almaigne, Raepkoolen.

5 The fifth kinde is called Caulis nigra: in Italian, Nigre Caules: that is to fay, Blacke Cole: in Frenche, *Chou noir*. in Douche, Swerte Koolen.

1 The firft kind of the red Cole is called of Cato in Greeke κράμβκ λεία: of Plinie in Latine, Brassica Cumana: in Frenche *Chous rouges & poly*: in high Douche, Breyter roterkolen: in bafe Almaigne, Groote roo koolen.

2 The fecond kinde is alfo called Brassica lacuturria: in French, *Chou cabu rouge*: in bafe Almaigne, Rooskens, and Roode Sluytkoolen.

3 The thirde kinde with the iagged leaues, is called in Greeke σελινοειδὲς: in Latine Brassica Apiana: in bafe Almaigne, Ghehackelde koolen: that is to fay, Cole with the iagged leaues.

4 The fourth kind of red Cole, is called Brassica Sabellica, and of fuch as write in thefe dayes, Brassica crispa: in Frenche, *Chous Crespues*: in high Douche, Kraufe r kol: in bafe Almaigne, Gherouckelde koolen: in Englifhe, Wrinckled or ruffed Cole.

5 The fifth and fmalleft, is called in high Douche, Kleinder kolen, that is to fay, the fmall and flender Cole: in Frenche *Petit Chou*: in bafe Almaigne, Slooren. This is the thirde kinde of Colewurtes defcribed by Cato, the whiche is properly called in Greeke κράμβκ, Crambe.

✸ The Nature.

Colewurtes are hoate and drie in the firft degree, and of a clenfing or fcouring facultie, efpecially the red kinde.

✸ The Vertues.

The iuyce of Colewurtes taken by it felfe, or with Saltpeter, fofteneth the belly, and caufeth one to go to the ftoole: the like propertie hath the firft water, wherin the Colewurtes haue ben boyled. A

The iuyce of Colewurtes dronken with wine, is good againft the bitinges of Serpentes. B

The fame layde to with the meale of Fenugrek, helpeth members troubled with the gowte. C

It doth clenfe and heale olde rotten fores. D

The fame put vp into the nofethrilles, purgeth the brayne and head. E

The fame mingled with vineger and put warme into the eares, is good againft deafeneffe, and againft the humming or ringing of the fame. F

The fame as a peffarie, put vp into the natural places of women, prouoketh the flowers. G

The fame boyled in a Syrupe with hony, & often licked in, is good againft hoarfeneffe and the cough. H

The decoction or broth of Colewurtes, efpecially of the firft kinde, and of the very worfte or meaneft forte of redde Colewurtes, haue all the aforefayde properties, the whiche taken eyther alone or with Sugar, doth both lightly and gently loofe and foften the belly, and prouoketh womens natural ficknesse. I

The fame broth is alfo good for all woundes: for if they be often wafhed therewith, it doth both mundifie and heale them. K

The

the Historie of Plantes. 555

¶ The young leaues eaten raw with vineger, or perboyled, do open the belly L
very gently, and cause to make water, and are very good also to be eaten of
suche as be splenitique.

¶ The same taken after meate or meale, in the same manner, do cure dron- M
kennesse, and the headache proceeding of the same.

¶ The same alone, or with parched Barley meale, are very good to be layde N
vnto blacke and blewe markes that come of stripes, and al other hoate inflammations or swellinges.

¶ The same leaues sod and layde to with hony, are good for consuming and O
filthy sores.

¶ The seede of Colewurtes taken in Meade or watered honie, doth kyll and P
expel al sortes of wormes.

¶ The stalkes burned to asshes, and mengled with old swines grease, is good Q
to be layde to the olde paynes or ache in the side.

※ *The Danger.*

Colewurtes eaten, engender grosse and melancholique blood, especially the
red kinde. The white are better to digest, and engender more agreeable and
better nourishment, especially when they haue ben twise boyled.

Of Wilde Colewurtes. Chap. vij.

※ *The Description.*   Brassica sylue stris.

WIlde Colewurtes in leaues and
flowers are much lyke to the small
Colewortes, or y‍e they cal Crambe,
sauing that his leaues and stalkes
be whiter and a litle hearie, & in taste much
bitterer.   ※ *The Place.*

¶ This Colewurt groweth in high rough
places by the sea side, as Dioscorides writeth. There is muche of it founde in many
places of Zealande vpon high bankes cast
vp by mans hand.   ※ *The Names.*

¶ This kinde of Cole is named in Greeke
κράμβη ἀγρία: of some Halmiridia: in Latine
Brassica sylue stris, and Brassica rustica: that is
to say, Wilde Colewurtes, or Countrie
Colewurtes: in base Almaigne, Zee Koolen, and wilde Zee Koolen: and of some
writers nowe a dayes, Caulis marinus, and
Brassica marina: albeit this is not that Brassica marina, whereof Dioscorides writeth,
whiche we haue described already in y‍e third
part of this worke, amongst those kinds of
plates called Windweeds, or bindweedes.

※ *The Nature.*

¶ This kind of Cole is very hoate and dry
of complexion, & stronger in working then
the great Colewurtes.   ※ *The Vertues.*

¶ The wild Cole in operation is lyke to the garden Colewurtes, but stronger A
and more abstersiue or scouring, and therefore nought to be vsed in meates.

¶ The leaues thereof newly gathered and stamped, do cure and heale greene B
woundes, and dissolue tumours and swellinges, being layde thereupon.

Of Spinache. Chap.viij.

Spinachea.

❧ *The Description.*

Spinache hath a long leafe, sharpe pointed, of a brownishe or greene colour, soft, gentle, ful of sap, and deeply cut with large slittes vpõ both sides about the largest parte or neather ende of the leafe. The stalke is round and holow within. Some of the plantes haue flowers clustering or thick set alongst the stalkes, and some bring foorth seede without flowers in thicke heapes or clusters full and plenteous, and for the most part prickley. ❧ *The Place.*

It is sowen in gardens amõgst pot herbes.

❧ *The Tyme.*

They vse to sow Spinache in March, and April, and it floureth and beareth seede within two monethes after the sowing. They also vse to sow it in September, & that continueth all the winter without bearing seede vntil the spring time. ❧ *The Names.*

This potherbe, or rather Salet herbe, is called of ÿ new writers Spanachea, Spinachea, Spinacheum olus, & of some Hispanicum olus: of Ruellius & certayne others Seutlomalache: of the Arabians, Hispanach: in Frẽch, *Espinars*: in high Douche, Spinet: in Neather Douche Spinagie: in Englishe, Spinache.

❧ *The Nature.*

Spinache is colde and moyst of complexion.

❧ *The Vertues.*

Spinache doth lose the belly, and the broth of the same is of lyke vertue. A

The same laid vnto hoate swellinges, taketh away the heate, and dissolueth B the swelling.

Of Dockes and Sorrel. Chap.ix.

❧ *The Kindes.*

Dioscorides setteth foorth foure kindes of Lapathum, bysides the fifth which groweth in ditches and standing waters, called Hippolapathum, the whiche shalbe described also in this Chapter.

❧ *The Description.*

1. The first kind of Lapathũ or Rumex hath long, narrow, hard, & sharpe pointed leaues, amõgst which come vp round holow browne stalkes with knees, ioyntes or knots, set and garnished with the like leaues. At the vpper part of the sayde stalkes grow many litle pale flowers one aboue another, and after them is found a blackish triangled seede, lapt in a thinne skinne. The roote is long, playne and yellow within.

2. The second kind called Patience, doth not differ much from the abouesaid, sauing that his leaues be greater, larger, softer, and not sharpe pointed. The stalkes be long and thicke, growing foure or fiue foote high. The flowers yellowish. The seede is red and triangled. The roote is long, smal and yellow.

The

the Historie of Plantes.

Oxylapathum **Sharpe poynted Patience.**   Lapathum satiuum. **Patience.**

3　The thirde kinde of Lapathum, is muche lyke to the first, yet for all that the leaues be shorter and larger most commonly layde alongst and spread vpon the ground, almost like the leaues of Plantayne, the stalke groweth not al so high.

Of this kinde is a red sort, the whiche hath faire red stalkes or purple, the leaues be browne and full of red veynes, out of the which (being bruised) commeth foorth a red iuyce or liquer, but els like to the other in stalkes, leaues, and seede.

4　The fourth kinde called Sorrel, hath long, narrow, sharpe poynted leaues, and broade next the stemme, very sharpe and eger in taste almost lyke vineger. The stalke is rounde and slender, vppon the whiche growe small flowers, of a browne red colour. The seede is browne, triangled and muche lyke the seede of poynted Patience. The roote is long and yellow.

Of this sort is found another kind called Romaine Sorrel, the which hath short leaues, in a manner round, somewhat cornered and whitish, almost lyke to Iuie leaues, but much smaller, and neither thicke nor harde. The stalkes be tender, vpon whiche groweth seede like the other.

There is yet another sort of Sorrel, whiche is smal and wild, and therfore called Sheepes Sorrell. The same in leaues, flowres, stalkes, and seede, is muche like to the great Sorrel, but altogither smaller. The leaues be very small, and the little stalkes are slender of a spanne long, the whiche sometimes both with his flowers and seede sheweth a blood red colour, and somtimes the leaues be red lykewise: sometimes also you shal finde them as white as snow.

5　The fifth kinde which groweth in waters and ditches, hath great leaues long and harde, muche like the leaues of poynted Patience, but muche larger.

The fyfth booke of

Oxalis. Sorrel.

Oxalis Romana. Tours Sorrel or Romayne Sorrel.

The stalkes be rounde growing, foure or fiue foote long or more, the sede is like to Patience. The roote is thicke and pale, of a faynt red colour within. ❧ *The Place.*

1 The sharpepoynted Docke or Patience, groweth in wette moyst medowes, & marshes.
2 The Docke called Patience, is planted in gardens.
3 The thirde kind groweth in dry places, and about wayes and pathes.
4 The red Patience is founde amongst potte herbes, growing in gardens.

Sorrel is commonly sowen in gardens, and is to be found also growing wylde in some medowes and shadowy places.

Sheepes Sorrel loueth dry soyles.

The fifth kinde groweth in ditches & standing waters, and is plentiful in this Countrie.
❧ *The Tyme.*

All these kindes of Lapathum, doo flower in June & July. ❧ *The Names.*

Al these herbes haue but one Greke name, that is λάπαθον, in Latine Rumex, and Lapathum: in Shoppes Lapatium.

Oxalis parua. Sheepes Sorrel.

the Historie of Plantes.

1. The first kinde is called in Greeke ὀξυλάπαθον: in Latine, Rumex acutus: in Shoppes, Lapatium acutum: in Frenche, *Parelle* in high Douch, Hegelwurtz, Grindtwurtz, Streiſſwurtz, Zitterwurtes: in baſe Almaigne, Patich, and Peerdick,

2. The second kinde is called λάπαθον ἥμερον: in Latine, Rumex ſatiuus, of ſome newe writers Rhabarbarum monachorum, of Galen alſo Hippolapathon: in Frenche, *Patience.* in baſe Almaigne, Patientie.

3. The third kind is called in Greeke λάπαθον ἄγριον: in Latine, Lapathū ſylueſtre, that is to ſay, Wilde Docke, or Patience: in baſe Almaigne, Wilde Patich.

The red kinde is called in Latine, Lapathum nigrum: and of ſome late writers, Sanguis Draconis: in Frenche *Sang de Dragon.* in Douche, Draken bloet: in Engliſhe, red Patience.

4. The fourth kinde is called in Greeke ὀξαλίς: in Latine, Oxalis: in Shoppes, Acetoſa: in Frenche, *Ozeille, vinette,* or *Salette:* in high Douche, Saur Impffer: in baſe Almaigne, Surckele: in Engliſh, Sorrell.

Romayne Sorrel is vndoubtedly a kinde of Oxalis: and it ſhoulde ſeeme to be that kind wherof the Auncients haue vſed and written moſt properly, called ὀξαλίς, Oxalis. The later writers do call it Oxalis Romana, and Acetoſa Romana: in Frenche, *Ozeille Romaine,* and *Ozeille de Tours:* in Douche, Roomſch Surckele.

The leaſt of theſe kindes is called Oxalis parua: in Shoppes Acetoſella: in Frenche, *Petit Ozeille,* and *Ozeille de brebis:* in high Douche, Klein Saurampffer: in Brabant, Schaeps Surckele, and Uelt Surckele: in Engliſhe, ſmall Sorrel, and Sheepes Sorrel.

5. The fifth kind, which groweth in ditches, is called in Greeke ἱπποlάπαθον: in Latine, Hippolapathum, or Lapathum magnum, or Rumex paluſtris: in Frēch, *Grande Parelle,* or *Parelle de marez* in high Douche, Waſſer Impffer: in baſe Almaigne, Groote Patick, or Water Patick: in Engliſhe, Great Sorrel, Water Sorrel, and Horſe Sorrel.

❧ *The Nature.*

Theſe herbes are of a reaſonable mixture betwixt colde and heate, but they be drie almoſt in the thirde degree, eſpecially the ſeede which is alſo aſtringent.

❧ *The Vertues.*

The leaues of all theſe herbes ſodden and eaten as meate, do looſe and ſoften A the belly gently, and the broth of them is of lyke vertue.

The greene leaues pounde with oyle of Roſes, and a little Saffron, do di- B geſt and diſſolue the impoſtumes and tumours of the head (called in Latine Meliceris) if it be layde thereunto.

The ſeede of Dockes and Sorrel dronken in water or wine, ſtoppeth the C laſke and blooddy flixe, and the wambling paynes of the ſtomacke.

The ſame is alſo good agaynſt the bitinges and ſtinginges of Scorpions, D ſo that if a man had firſt eaten of this ſeede, he ſhoulde feele no payne, albeit he were afterwarde ſtong of a Scorpion.

The rootes of this herbe boyled in wine & dronken, do heale the Iaundiſe, E prouoke vrine, and womens flowers, and do breake and driue foorth the ſtone and grauell.

The rootes of theſe herbes boyled in vineger, or bruſed rawe, doo heale all F ſcabbedneſſe and ſcuruie itche, and all outwarde manginesſe and deformitie of the ſkinne, being layde thereunto.

The decoction or broth of them, is alſo very good agaynſt all manginesſe, G wilde feſtering and conſuming ſcabbes, to make a ſtew or broth to waſhe in.

The wine of the decoction of them doth swage the tooth ache, to be kept in the mouth, and to washe the teeth therewith: it swageth also the payne of the eares, dropped therein.

The rootes also boyled and layd to the hard kernels, and swelling tumours behinde the eares do dissolue and consume them.

The same pounde with vineger doth heale and waste the hardnesseof the melt or splene, and pounde by them selues alone, and layde vpon the secrete places of women, doth stop the immoderate flure of the wombe, or flowres.

Some write that this roote hanged about the necke, doth helpe the kinges euill or swelling in the throte.

Of Lampsana. Chap.x.

❧ *The Description.*

Lampsana.

Lampsana is a wild worte or potte herbe, hauing large leaues of a whitishe or pale greene colour, deepely cutte vppon both sides like the leaues of Rape or Senuie, but a great deale smaller. The stalkes growe two foote high, & are diuided agayne into many small branches: at the toppe whereof growe many smal yellow flowers, almost lyke to the flowers of the least Hawkeweede. ❧ *The Place.*

Lampsana groweth most commonly in al places, by high way sides, and specially in the borders of gardens amongst wortes and potherbes.

❧ *The Tyme.*

It flowreth almost al the sommer.

❧ *The Names.*

This herbe is called in Grcke λαμψάνη: in Latine, Lampsana, & of some Napiũ.

❧ *The Nature.*

Lampsana is somewhat abstersiue or scouring. ❧ *The Vertues.*

Lampsana, as Galen writeth, taken in meate, engendreth euill iuice, and noughtie nourishment: yet Dioscorides sayth, that it nourisheth more, and is better for the stomacke, then the Docke or Patience.

Being layde to outwardly, it doth clense and mundifie the skinne. and therfore is good against the scuruie itche.

Of Algood. Chap.xi.

❧ *The Description.*

Algood, hath long large thicke leaues, almost like to the leaues of Sorrel, but shorter and broder, the stalke is grosse of a foote high, vpõ which groweth the seede clustering togither, almost like to Orache. The roote is great, long, thicke and yellow.

❧ *The Place.*

Algood groweth in vntoyled places, about wayes & pathes, & by hedges.

the Historie of Plantes. 561

❧ *The Tyme.*

You shall find it in flower in June and July.

❧ *The Names.*

This herbe is called in Latine Tota bona: & of some also Χρυσολάχανον, Chryſolachanó, that is to say in Latine, Aureũ olus, for his singuler vertue: in French, Toute bonne: in high Douch, Guter Herich, & Schmerbel: in base Almaigne, Goede Heinrich, Lammekens oore, and of some Algoede: in English, Good Henry, and Algood: of some it is taken for Mercurie.

❧ *The Nature.*

Algood is drie & abstersiue or scouring.

❧ *The Vertues.*

A   Algood taken as meate or broth, doth soften the belly, and prouoketh the stoole.

B   This herbe greene stamped, and layde to, healeth old sores, and greene woundes, and killeth and bringeth foorth wormes, that ingender in the same. Matthiolus. lib. 2. Dioscor. Chap. 162. Radicis succus illitus scabiem tollit, & Cutis maculas extergit, præsertim si cum aceto misceatur. Quidam eam quoque præferunt aduersus venenosorum animalium morsus.

Tota bona.

Of Endiue and Succory.   Chap.xij.

❧ *The Kyndes.*

ENdiue according to Dioscorides, and other Auncient writers of Physicke, is of two sortes, the one called Garden Endiue or Succorie: and the other wild Succorie. Wherof the garden Endiue or Succory is diuided againe into two sortes or kindes, one hauing brode white leaues, and the other narrowe iagged leaues. Likewise of the wilde kinde are two sortes, one kind hauing blew flowers, the other hath yellow flowers.

❧ *The Description.*

1   The white garden Succorie with the brode leaues, hath great, long, large, & soft, whitegreene leaues, not much vnlike the leaues of some sorte of Letuce. The stalke is rounde set with the like leaues, whiche growe vp sodenly, bearing most commonly blewe flowers, and sometimes also white. After the flowers foloweth the seede, whiche is white. The roote is white and long, the which withereth and starueth away, the seede being once ripe.

2   The second kind of garden Succorie hath long narrow leaues, sometimes creuished or slightly toothed about the edges. The stalke is round, the flowers blewe, lyke to the flowers of the aforesayde. The roote is white and long, full of sappe, and dieth not lightly, albeit it hath borne both his flowers and seede.

3   The thirde kinde called wilde Endiue, hath long leaues of a sad greene colour, and somewhat rough or hearie, the which be sometimes parted with reddish vaynes. The stalkes, flowers, & seede, are very much lyke to garden Succorie,

The fyfth Booke of

Intubum satiuum latifolium.
White Succory.

Intubum satiuum angustifolium.
Garden Succorie.

corie, and so is the roote, the which lasteth a long time, & doth not lightly perish.

4 The fourth kind, which is the wilde yellow Succorie, is also like to Succorie in stalkes and leaues, the stalkes be a cubite long or more, full of branches. The leaues be long, almost like the leaues of wilde Endiue, but larger. The flowers be yellow, fashioned like the flowers of *Dent de lyon*, but smaller. The roote is of a foote long, full of white sap or iuyce, which commeth foorth whan it is hurt. ✤ *The Place.*

1.2 The first and seconde kinde, are planted in the gardens of this Countrie.

3 The thirde groweth in drie, grassie, and vntoyled places, and somtimes also in moyst groundes.

4 The fourth kinde groweth in medowes, and moyst waterie places, about diches and waters. ✤ *The Tyme.*

These herbes flower at Midsomer, and sometimes sooner or rather, especially the white Endiue, the whiche being timely sowen in Marche, flowreth bytimes. Therefore the gardiners which would not haue it to flower, but are desirous to haue it great and large, do sowe it in July and August: for being so lately sowen, it flowreth not al that yeere, but waxeth large and great: a little before winter they plucke it vp from the ground, and bind togither the toppes, and burie it vnder sande, and so it waxeth all white, to be eaten in Salades with oyle and vineger. ✤ *The Names.*

These herbes be called in Greeke σερίδες: in Latine Intuba: of some πικρίδες, and Picridæ.

1 The first kinde is called Intubum satiuum latifolium: and of some Endiuia:

the Historie of Plantes.   563
Intubum ſyueſtre, Cichorium.   Hedypnois. Yellow Succorie.

in ſhoppes Scariola: in Frenche, *Scariole, Endiue*: in high Douch, Scariol: in baſe Almaigne, the common Countrie folke do call it Witte Endiue, the which are better acquainted with the right Endiue, then the ignorant Apothecaries, who in ſteede of Endiue, do vſe the wilde Letuce: in Engliſh, garden Succorie, or white Endiue with the brode leaues.

2   The ſecond is alſo a kind of garden Endiue, or Intubum ſatiuum, & is called Cichorium ſatiuum, & hortenſe: in ſhoppes Cicorea domeſtica: in Engliſh, garden Succorie: in Frenche, *Cichorée*: in high Douch, Zam Wegwarten: in baſe Almaigne, Tamme Cicorepe.

3   The thirde kinde is called in Greeke πικρὶς ἢ κιχώριον: in Latine, Cichorium, Intubum ſylueſtre, of ſome Ambubeia: in ſhoppes, Cicorea ſylueſtris: in french, *Endiue ſauuage*: in high Douche, Wilde Wegwarten : in baſe Almaigne, Wilde Cicorepe: in Engliſh, Wilde Endiue.

4   The fourth kind with the yellow flowers is called of Plinie Hedypnois: in high Douch, Geelwegwart: in french, *Cichorée iaulne*: in baſe Almaigne, Geel Cicorepe: in Engliſh, Yellow Succorie.

¶ *The Nature.*

Theſe herbes be colde and drie almoſt in third degree, eſpecially the wilde, which is more drie, and of a ſcouring or abſterſiue facultie.

❧ *The Vertues.*

Theſe herbes eaten, do comfort the weake and feeble ſtomacke, and do coole and refreſh the hoate ſtomacke, ſpecially the wild Endiue, which is moſt agreeable and meeteſt for the ſtomacke and inward partes.   The

The same boyled and eaten with vineger, stoppeth the laske or fluxe of the B belly proceeding of a hoate cause.

The iuyce or decoction of Succorie dronken is good for the heate of the li-C uer, against the Jaundise, and hoate feuers, and Tertians.

The greene leaues of Endiue and Succorie bruſed, are good againſt hoate D inflammations and impoſtumes, or gathering togither of euill humours of the ſtomacke, the trembling or ſhaking of the hart, the hoate gowte, and the great inflammation of the eyes, being layde outwardly to the places of the greefes.

The same layd to with parched Barley meale are good agaynſt choleriqueE inflammations, called Eryſipelas, and of ſome S. Antonies fier, or Phlegmon.

The iuyce of the leaues of Endiue and Succorie, layd to the forehead with F oyle of roſes and vineger, ſwageth headache.

The ſame with Ceruſe (that is, white leade) and vineger, is good for al tu-G mours, impoſtumes and inflammations whiche require cooling.

Of Sowthiſtel. Chap.xiij.

✣ The Kindes.

Sonchus is of two ſortes, the one more wilde, rough, and prickley, called Sowthiſtel, or milke Thiſtell, the other more ſoft and without prickles, which we may cal Hares Lettuce, or Connies milke Thiſtel.

Sonchus ſylueſtrior, aſpera.	Sonchus tenerior, non aſpera.
Rough milke Thiſtel.	**Tender or ſweete milke Thiſtel.**

❧ The Deſcription.

1 Sowthiſtell hath long brode leaues, very deepely cut in vpon both ſides, and armed with ſharpe prickles. The ſtalke is creſted, holowe within, ſpaced by ioyntes or knobbes, couered or ſet with the like leaues. At the toppe

toppe of the stalke growe double yellow flowers, lyke Dandelyon, but muche smaller: when they be past, there come vp white hoare knoppes or downie heades, which are caried away with the wind. The roote is long and yellow, full of hearie stringes.

2 The tender Milke thistel, is muche lyke to the aforesayd in leaues, stalkes, flowers and seede: but the leaues be somewhat broder, & not so deepely iagged or cut in vppon the borders, and they haue neither thornes nor sharpe prickles, but are al playne without any roughnesse.

❧ The Place.

These herbes doo growe of them selues both in gardens amongst other herbes, and also in the feeldes, and are taken but as weedes, and vnprofitable herbes.

❧ The Tyme.

Milke thistel and Sowthistel, do flower in June and July, and most commonly all the sommer.

❧ The Names.

These herbes be called in Greeke σόγχοι, in Latine, Sonchi: of the later writers Cicerbitæ, Lactucellæ, Lacterones: of Serapio and in shoppes, Taraxacon.

1 The first kinde is called Sonchus asperior, or sylueftrior: in high Douche, Genszdistel, Moszdistel: in Brabant, Gansendistel, & Welckweye: in Frenche, Laicteron, and Laceron in Englishe, Sowthistel, and rough Milke thistel.

2 The seconde kinde is called Sonchus non aspera, or Sonchus tenerior, of Apuleius Lactuca leporina: in Frenche, Palais de lieure in high Douche, Hasenkol: in base Almaigne, Hasen Lattouwe, Hasen struyck, Danwdistel, Canijnencrupt: in English, the tender or soft Milke thistel.

❧ The Nature.

These herbes be colde and drie of complexion, especially being greene and newe gathered: for being dry or long gathered, they are somewhat hoate, as Galen sayth.

❧ The Vertues.

The iuyce of eyther of these herbes dronken, swageth the gnawing paynes of the stomacke, prouoketh vrine, and breaketh the stone, and is of a soueraigne remedie against the strangurie and the Iaunders.

The same dronken, filleth the breastes of Nurses with good and holesome milke, and causeth the children whom they nourish, to be of a good colour. Of the same vertue is the brothe of the herbe dronken.

The iuyce of these herbes do coole and refreshe the heate of the fundement, and the priuie partes of the body, being layde thereto with cotton, and of the eares, being dropped in.

The greene leaues of Milke thistel, are good agaynst all hoate swellinges and impostumations, especially of the stomacke being brused & layd thervppon.

The roote with his leaues being pounde, and layde to as an emplayster, is good against the bytinges and stinginges of Scorpions.

Of Hawke weede. Chap.xliij.

❧ The Kindes.

Dioscorides setteth foorth two kinds of Hawke weede, the great and the smal: of the smaller are also three sortes.

❧ The Description.

1 THE great Hawkeweede putteth foorth a rough stalke somthing reddish, and holow within. The leaues be long, very muche iagged, and deepely cut vppon the sydes, eche cut standing wide, or a great waye one from another, and set with sharpe prickles, almost lyke the leaues

Hieracium maius.   Hieracium minus primum.
Great Hawkeweede.   Wilde Succorie.

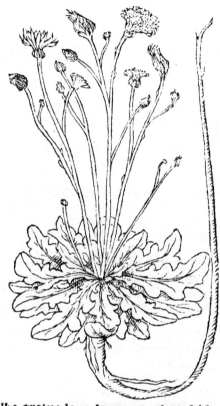

of milke Thistel, at the toppe of the stalke growe long knoppes, the whiche bringe foorth yellowe double flowers, lyke the flowers of milke Thistel, the whiche do change into rounde cotton or downie bawles, which are blowen away with the winde. The roote is not very long, but it hath threddy stringes hanging at it.

2   The first kinde of the lesse Hawkeweede hath long leaues, diuided and cut on the edges, almost lyke the leaues of Dandelyon, but not so bigge nor so deepely cut, and lying flat vpon the grounde, from amongst those leaues shooteth vp smoth naked brownish stalkes, bringing foorth double yellow flowers in the top, the whiche do turne into downe bawles or globes and do flee away with the winde. The roote is long and slender, smoothe, and white.

3   The seconde kinde of the lesser Hawkweede is lyke vnto the aforesayde in stalkes and flowers, the leaues do also lye spread vpon the ground, but they be smaller narrower and more deepely cut, then the leaues abouesayde. This Hawkweede hath no deepe downeright roote, but sheweth as though it were gnawen or bitten, lyke to the roote of Deuils bit, whereof we haue written in the first booke of this historie of Plantes, and it is full of stringes.

4   The third is the least of al three, his leaues be much lyke to the first Hawkweede, and so be his flowers, stalkes and rootes: but altogither lesse. The leaues be altogither smoothe and naked, and not so brownish as the leaues of the first Hawkweede.

☙ The

the Historie of Plantes.

❦ *The Place.*

These herbes grow in vntoyled places, as the borders of corne fieldes, in medowes, high wayes, and the brinkes of ditches.

❦ *The Tyme.*

These herbes doo flower from June to September.

❦ *The Names.*

This herbe is called in Greeke ἱεράκιον, of some σογχίτες: in Latine, Accipitrina: that is to say, Sperhawke herbe, or Hawkeweede, Apuleius calleth it Lactuca syluatica, picris, and Thridax agria.

1 The first kinde is called in Greeke ἱεράκιον τὸ μέγα: in Latine, Hieracium magnū: of some Sonchites, Lampuca, or Sitheleas: in Frenche, *Cichorée sauuage:* in high Douche, Grosz habichkraut, in base Almaigne, Groot hauickscruyt: That is to say, the great Hawkeweede.

2 The lesser kind is called in Greeke ἱεράκιον τὸ μίκρον: in Latine, Hieracium paruum: of some Intybum agreste, or Lactuca minor: in high Douche, Klein Habichkraut, that is to say, the lesser Hawkweede: in base Almaigne, Cleyn Hauickscruyt.

The seconde lesser kinde is also called of some Morsus Diaboli: in Douche Teuffels abbisz: that is to say in English, Diuels bit: and in Frenche, *Mors de Diable.* bycause his roote is eaten or bitten lyke the Scabiouse Diuels bit.

Hieracium minus alterum.

Yellow Deuils bit.

❦ *The Nature.*

These herbes be colde and drie.

❦ *The Vertues.*

These herbes in vertue and operation, are muche like to Sowe Thistel, or Sonchus, and being vsed after the like manner, be as good to al purposes.

They be also good for the eyesight, if the iuyce of them be dropped into the eyes, especially of that sort whiche is called Diuels bit.

Of Langdebeefe. Chap. xv.

❦ *The Description.*

THis herbe hath great broade leaues, greater and broader then the leaues of Borache, set ful of soft prickles, from whiche leaues commeth vp a tender weake brittle and triangled stalke set with leaues of the same sort, but smaller. At the toppe of the stalke growe many small leaues, thicke set and harde throng togither round about the stalke, from amongst whiche litle leaues commeth a rough round Thistely knoppe, bearing a purple flower, the whiche is caried away with the wind. The roote is thicke and crooked hauing many stringes.

❦ *The Place.*

This herbe groweth in the medowes of this Countrie, and in moyst places by water brookes or ditches.

✳ The Tyme.

This Thistel floweth in August.

✳ The Names.

This herbe is called in Greeke κρισιον ἢ κιρσιον: in Latine, Cirsium, of some Buglossum magnum, and Spina mollis: in Brabant, Groote Dauw distel, vnknowē in shoppes, some take Cirsion to be Lang-Debeese. T.lib.1.fol.143.

✳ The Nature.

It is colde and drie of vertue like Sonchus.

✳ The Vertues.

A Andreas the Herborist writeth that the roote of Cirsium tyed or bounde to the diseased place, swageth the ache of the veynes (called Varix) being to muche opened or enlarged and fylled with grosse blood.

Of Condrilla, Gumme Succorie. Chap. xvi.

✳ The Kindes.

There be two sortes of Condrilla, as Dioscorides writeth, the great and the small.

✳ The Description.

1 Condrilla is somewhat lyke to wylde Endiue: his leaues be long, grayish, and deepely cut vpon both sides, the stalke is small, of a foote long or somewhat more, in the litle stalkes of Condrilla, is founde a gumme lyke Masticke, of the bignesse of a beane, wherevpon growe round knoppes, which after their opening bringeth foorth faire flowers, whiche in collour and making are much like to the flowers of wild Endiue: but much smaller. The roote is long and white like to Succorie.

2 The other Condrilla hath long leaues deepely indented vppon both sides lyke to the leaues of the wilde Endiue, and for the most parte spreade abroade vpon the ground, amongst which leaues grow vp smal playne holow stalkes, carrying fayre yellowe double flowers the whiche past they turne into rounde blowballes, like to fine downe or cotton, and are carried away with the wind. The roote is long and slender yellowish and ful of milke, which commeth forth when it is cut or broken.

✳ The Place.

1 The great Condrilla is not common in this Countrie, but is to be founde in the gardens of Herboristes.

2 The lesser which is our Dandelion, groweth in al partes of this Countrie, in medowes and pastures.

✳ The Tyme.

The great Condrilla floweth in May, and in June. Dandelion floweth in April and August.

✳ The Names.

1 The first kinde of these herbes is called in Greeke κονδριλλη: in Latine Condrilla:

the Historie of Plantes. 569

Condrilla. Gumme Succorie.   Condrilla Dandelyon.

drilla: of Plinie Condrillon, and Condrillis: of some also Cichorion, and Scris: of the later writers Condrilla maior: in this Countrie Condrilla, and Gumme Succorie: in Douche, Condrilla.

2   The seconde kinde is called in Greeke κονδρίλλη ἑτέρα, in Latine, Condrilla altera, in shoppes, Dens leonis, and Rostrum porcinum: in Frenche, Pisse-en-lict: in high Douche, Kozlkraut, Pfaffenblat, Pfaffen rorin: in base Almaigne, Papencruyt, Hontstroosen, Canckerbloemen, and Schozstbloemen: in Englishe, Dandelyon.

✻ *The Nature.*

These herbes be colde and drie lyke Endiue and Succorie.

✻ *The Vertues.*

The iuyce of the great Condrilla taken by it selfe or with wine, stoppeth the laske, especially comming of the heate of the liuer.   A

The same bruised and eaten with his leaues & rootes, is very good agaynst the bitinges of venemous Serpentes.   B

The seede of Condrilla doth strengthen the stomacke, and causeth good digestion, as Dorotheus writeth.   C

Dandelyon in vertue and operation is much like Succorie, and it may be alwayes vsed in steede thereof.   D

It layeth downe the staring heares of the eyebrowes, and causeth newe heares to grow, if the iuyce be often layd to the place.   E

Of Groundswell.   Chap.xvij.

✻ *The Kindes.*

Although Dioscorides and other the Auncients haue set foorth but one sort of of Erigeron, yet for althat, the later learned writers do set out two kindes, the one great, and the other smal: vnto which we haue ioyned a third kind. Wherfore Erigeron is nowe to be counted of three sortes.

Erigeron primum, & secundum. Erigeron tertium.
The first & second kindes of Groundswel. The third kind of Groundswel.

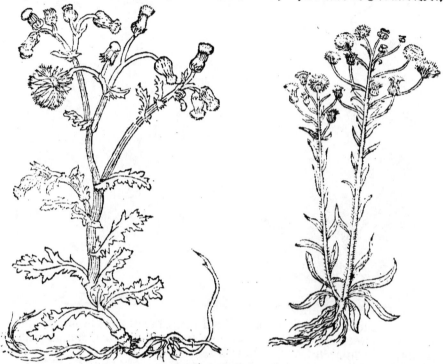

❧ *The Description.*

1. The great Groundswel, hath rough whitish leaues, deeply iagged and knawen vpon both sides, like to the leaues of white Mustard or senuie. The stalke is two foote high or more: at the top whereof growe smal knoppes, which do open into smal yellow flowers the which are sodenly gone, & changed into downie blowbawles like to the heades of Dantdelyon, and are blowen away with the winde. The roote is hearie, and the whole herbe is of a strange smell.

2. The lesser Groundswel hath greene leaues, whiche be also much torne, and deepely iagged vpon both sides like the leaues of the great groundswell, but a great deale smaller, greener, smother, and not so rough. The stalke is a spanne long, at the toppe whereof growe yellow flowers, whiche do also chaunge sodenly into hoare heades or blowbawles, and doo flye away with the winde. The roote is hearie, and hath no proper smell.

3. The third Groundswel hath a straight slender stemme, of a browne purple colour and set full of fine cotton or downie heares: the leaues be long and narrow. At the top of the stalkes grow smal knoppes, out of which come smal pale yellow flowers, the whiche incontinently after their opening do change, and become so sodenly gray or white, that he that taketh not the better heede, may thinke that they are so at the first opening of the knoppes: for euen the self same day,

day, and sometimes the very same houre of their opening, they become gray or hoare, and shortly after the knoppes do spreade abrode and open, and the gray heare with the seede, are blowen and carried away with the winde. The roote is small and very tender.

❧ *The Place.*

1 The great Groundswel groweth in sandy groundes, and alongst by wayes and pathes.
2 The lesser is often found amongst potherbes, and commonly in the feeldes.
3 The thirde groweth in darke shadowed wooddes, and dry Countries.

❧ *The Tyme.*

1 The great Groundswel floureth in June and July.
2 The lesser Groundswel floureth al the sommer, and somtimes also in winter, when it is milde and not to colde.
3 The thirde floureth at Midsomer.

❧ *The Names.*

This herbe is called in Greeke ἠριγέρων: in Latine, Senecio: of some Herbulū, or Erechtites: in Frenche, *Seneceon*, or *Senesson*: in high Douche, Grindtkraut: in English, Groundswel.

1 The first kinde is called Senecio maior, that is to say, Great Groundswell: in Brabant, groot Crupscrupt, and of some Silsom: in Frenche, *Grand Seneceon*.
2 The seconde is called in Latine Senecio minor, that is to say, the lesse Groundswel: in French, *Petit Senesson*: in Douch, Crupscrupt, or cleyn Crupscrupt, the whiche is well knowen.
3 The thirde sort is a right Erigeron, & Senecio, especially that which Theophraste describeth: for as it is abouesayde, his flowers waxe sodenly white hoare, from whēce it hath to name Erigeron. Conrade Gesner calleth it Ἠριγέρων, and placeth it with the kindes of Conyza.

❧ *The Nature.*

Erigeron, as Paulus writeth, hath somewhat a cooling nature, but yet digestiue.

❧ *The Vertues.*

The leaues and stalkes of Groundswell, boyled in water or sweete wine A and dronken, healeth the ache of the stomacke that riseth of choler.

The leaues and flowers alone, or stamped with a little wine, are good to be B layde to the burning heate or inflammation of the stones and fundement.

The same mingled with the fine powder of Frankencense, healeth all C woundes, especially of the sinewes, being layd thereto.

The downe of the flowers layde to with a litle Saffron & water, are good D for bleared and dropping eyes.

The same with a litle salt, doth wast & consume the kinges euil, or strumes E of the necke.

The small Groundswell is good to be eaten in Salades with oyle and vi- F neger, and is no euill or vnholsome foode.

Of Letuce. Chap.xviij.

❧ *The Kindes.*

OF Letuce are two sortes, the garden and wilde Letuce, and of the garden Letuce are sundrie sortes.

❧ *The Description.*

The first kind of garden Letuce, hath long brode leaues, euen playne and smothe, the whiche do neuer close, nor come togither: emongst which riseth a straight stalke full of white sappe lyke milke, of the height of two foote, the which diuideth it self at the top into sundry branches bearing yellow flowers,

The fyfth Booke of

Lactuca satiua. Garden Letuce. **Lactuca crispa. Curled or crispe Letuce.**

flowers, which do change into a graishe or white hoare bearde. The seede is white, long and smal. The roote is long & thicke like to a Carrot, but smaller.

2   The second kind of Letuce, hath crompled leaues, wrinckled and gathered or drawen togither almost like the Moquet or Chauden of a Calfe: otherwayes it is altogither like the aforesayd, in stalkes, flowers, seede, and rootes.

3   The third sort is the fairest and whitest kind: it hath great large leaues, the whiche do growe very thicke togither all from one roote, so that the first and nethermost leaues do spreade abrode vpon the ground, and the middelmost do growe and close togither one vppon another, losed and headed almost like to a Cabbage Cole: but the residue, as the stalkes, flowers, seede, and rootes, are like to the first. This kind is best beloued and most desired, and commonly vsed in meates.

4   Columella writeth of another kinde of Letuce, whose leaues be darke or browne, almost of a purple colour.

5   Yet there is another kinde whose leaues are reddish, plaine or smothe, very tender and sauerie: yet for al that both these kindes be vnknowen of the later writers.

❧ *The Place.*

They vse to sowe Letuce in gardens amongst potherbes in good fertile grounde, and they must be planted farre asunder one from the other, otherwise they will not spreade, nor growe to a rounde head or close Cabbage Letuce.

❧ *The Tyme.*

They sow Letuce early & late, al seasons of the yeere, but chiefely in March and April: and two or three monethes after the sowing, it bringeth forth both flower and seede, but then it is nothing worth to be eaten.

the Historie of Plantes. 573

❧ *The Names.*

The garden Letuce is called in Greke θρίδαξήμερος: in Latine, Lactuca satiua: in Shoppes, Lactuca: in high Douche, Lattich, or Lactuck: in base Almaigne, Lattowe: in English, Lettis, and Lettus.

1 The first kind of Letuce hath none other particuler name, but that general name Lactuca, Lettis.

2 The seconde kinde is called of Plinie, Lactuca crispa: in Englishe, Crispe or curled Lettis: in Frenche, *Laictue crespue* in high Douche, Krauser Lattich : in base Almaigne, Ghecronckelde Lattowe: in English, Crompled Lettis.

3 Plinie calleth the thirde kind Lactuca laconica, Lactuca sessilis : Columella calleth it, Lactuca betica : the later writers call it Lactuca capitata : in Frenche, *Laictue pommée*, or *Laictuca à pomme*: in Englishe, Losed, or Cabbage Lettis.

4 The fourth kinde is called of Columella, Lactuca Ceciliana.

5 The fifth kinde is called Lactuca Cypria, and of Plinie, Lactuca Græca.

❧ *The Nature.*

Garden Letuce is colde and moyst in the first or seconde degree.

❧ *The Vertues.*

The garden Letuce eaten in meate, engendreth better blood, and causeth A better digestion than the other wort or potherbe, especially beyng boyled and not eaten rawe.

It is good in meate agaynst the heate of the stomacke, and the wamblings B of the same, it slaketh thirste, and causeth good appetite, especially being eaten rawe in Salades.

The same taken in the same manner, causeth sound and sweete sleepe, it ma- C keth the belly good and soft, and engendreth abundance of milke : surely, it is very good for suche as cannot take their rest, and for Nurses, and for suche as giue sucke, whiche haue but small store of milke : but for that purpose it is better before it begynneth to shoote foorth his stalkes : for whan it putteth foorth his stalkes it wareth bitter and is not so good in meates as before.

The greene leaues of Letuce brused, are good to be layde vppon newe bur- D ninges and scaldinges before it riseth vp into wheales and blisters, and vppon all hoate swellinges and wilde fier, called Erisipiles.

Letuce seede being often vsed to be eaten a long space, drieth vp the natural E seede, and putteth away the desire to Lecherie.

And as Plinie writeth, it is good to be dronken in wyne agaynst the stin- F ginges of Scorpions.

Of wilde Letuce. Chap.xix.

❧ *The Description.*

THE wilde Letuce hath long leaues deepely cut vppon both edges, whitishe, and vnderneath the leafe the middle sinewe or ribbe is set full of sharpe prickles. The stalke is round and long, and groweth vp higher then the stalkes of the garden Letuce, it is rough and set with sharpe prickles, and leaues lyke the other but smaller : at the toppe of the stalke growe flowers lyke them of the garden Letuce. The seede is brownish, otherwise it is lyke the seede of the garden Letuce. The roote is small.

❧ *The Place.*

This herbe groweth in the borders of feeldes, alongst the wayes and such lyke vntopled places and sometimes in the gardens amongst potherbes: and where as it hath bene once sowen, it commeth agayne lightly without any more labour.

❧ *The*

The fyfth Booke of

Lactuca ſylueſtris.

※ *The Tyme.*

This Letuce flowreth in July and Auguſt.

※ *The Names.*

This herbe is called in Greeke θρίδαξ ἀγρία: in Latine Lactuca ſylueſtris: of Zoroaſtes, Pherumbrum: in ſhoppes Endiuia: albeit this is not the right Endiue: of ſome Seriola : in Frenche, *Laictue ſauuage* : in high Douche, Wilder Lattiche: in baſe Alemaigne, Wilde Lattouwe: in Engliſhe, Wilde Letuce, of Turner greene Endiue. And this is the herbe that the Iſraelites did eate with their Paſſeouer Lambe.

※ *The Nature.*

The wilde Letuce is partly colde and drie in the third degree, and partly ſharpe and abſterſiue or ſcouring, with ſome warmeneſſe.

※ *The Vertues.*

A The iuyce of the wilde Letuce dronken with Orimel, that is, honied vineger, ſcoureth by ſiege the waterie humours.

B It reconcileth ſleepe, and ſwageth al paynes: also it is good againſt the ſtinging of Scorpions, and the fielde Spider called Phalangium.

C It is alſo good with womans milke to be layde vnto burninges.

D The ſame dropped into the eyes, cleareth the ſight, and taketh away the clowdes & dimneſſe of the ſame.

E The ſeede of this Letuce alſo, abateth the force of Venus, and is of vertue like to the garden Letuce ſeede.

Of Purcelayne.. Chap.xx.

※ *The Kyndes.*

There be two kindes of Purcelayne, one of the garden, the other wilde: beſides theſe there is alſo a thirde kinde, the whiche groweth onely in ſalt groundes.

※ *The Deſcription.*

1 Garden Purcelayne hath groſſe ſtalkes, fat, round, and of a brownred colour, the which do grow vp to the length of a ſpan or more, vpõ the ſayd ſtalkes are ÿ thicke fat or fleſhie leaues, ſomthing long & brode, round before. The flowers grow betwixt the leaues and ſtalkes, and alſo at the higheſt of the ſtalkes, the which be very ſmal, & of a faynt yellowiſh colour. The ſame being paſt, there come little rounde cloſe huſkes, in whiche is founde ſmal blacke ſeede. The roote is tender and hearie.

2 The wild Purcelaine hath thicke fat round ſtalkes, like the garden Purcelayne, but tenderer, ſmaller, and redder, the which grow nothing at al vpright, but are ſpread abrode, and trayle vpon the ground. The leaues be ſmaller then the leaues of the other, but the flowers & ſede is like. Theſe two Purcelaynes are full of iuyce, and of a ſharpe or quicke taſte. They are vſed in the ſommer to be eaten in Salade, as they vſe Letuce.

3 The thirde kinde, the which groweth in ſalt ground, hath many ſmal, hard, and wooddy ſtalkes. The leaues be thicke, of a white greene or aſhe colour, very much like to the leaues of the other Purcelayne, but whiter and ſofter in hande.

the Historie of Plantes.   575

Portulaca hortensis.
Garden Purcelayne.

Portulaca sylueſtris.
Wilde Purcelayne.

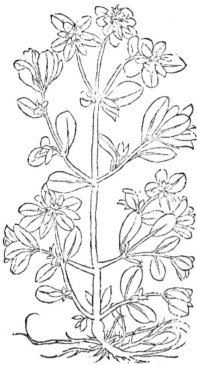

Portulaca marina.
Sea Purcelayne.

handeling, yet not ſo ſmoothe noꝛ ſhining. The flowers growe at the toppe of the ſtalkes, cluſtering togither lyke ye flowers of Oꝛache oꝛ Blite. The roote is long and of a wooddy ſubſtance, and liueth with his ſtalke, and certayne of his leaues all the winter.

¶ *The Place.*

1. The tame Purcelayne is ſowen in gardens.
2. The wild groweth of his owne accoꝛde in wayes and alies of gardens, & in ſome places it groweth vppon rockes, cleeues, and viniardes.
3. Sea Purcelayne groweth vpō bankes oꝛ walles caſt vp in places adioyning to the ſea: and great ſtoꝛe therof is founde in Zeeland. and byſides the Ile of Purbeck in Englande.

✚ *The Tyme.*

The garden & wilde Purcelayne, do flower from after the moneth of June, vntill September, and in this ſpace they yeelde their ſeede.

The

The fyfth Booke of

3   The sea Purcelayne flowreth in July.

❧ The Names.

Purcelayne is called in Greke ἀνδράχνη: in Latine and in shoppes Portulaca: in Frenche, *Pourpier*, or *Pourcelaine*: in high Douche, Burgel: in base Almaigne, Porceleyne: in English, Purcelayne.

1   The first kinde is called Portulaca sativa, or Hortensis: in Frenche, *Pourpier*, or *Pourcelaine domestique*, or *cultiuée*: in high Douche, Heymisch Burgel, or Burtzelkraut: in base Almaigne, Roomsche Porceleyne, or tamme Porceleyne: in English, garden and tame Purcelayne.

2   The seconde kinde is called of the newe writers, Portulaca sylvestris: in Frenche, *Pourpier sauuage*: in high Douche, Wildt Burtzel: in base Almaigne, Ghemeyne, or wilde Porceleyne: in English, Wild Purcelayne: but yet this is not that wild Purcelayne, which is described in some copies of Dioscorides, the which is of a hoate nature or complexion.

3   The thirde kinde of Purcelayne of the later writers, is called Portulaca marina: in Frenche, *Pourcelaine de mer*. In Douch, Zee Porceleyne. This seemeth to be that herbe which the Greekes call ἅλιμος, the Latinistes, Halimus, especially the seconde kinde described by Plinie.

❧ The Nature.

1.2   The garden and wilde Purcelayne are cold in the thirde degree, and moyst in the seconde.

3   Sea Purcelayne is playnely hoate and drie in the seconde degree.

❧ The Vertues.

They vse to eate the garden and wild Purcelayne in Salades and meates, A as they do Letuce, but it cooleth the blood, and maketh it waterie & nourisheth very litle: yet for all that, it is good for those that haue great heate in their stomackes and inwarde partes.

The same taken in lyke sort, stoppeth all defluxions and falling downe of B humours, and is good for the paynes of the bladder and kidneyes, & it healeth them, albeit they be exulcerated, fret or hurt.

Purcelayne comforteth the weake inflamed stomacke, & it taketh away the C imaginations, dreames, fansies, & the outragious desire to the lust of the body.

The iuyce of Purcelayne dronken hath the same vertue: also it is good a-D gainst burning feuers, & against the wormes that ingender in the body of man.

It is good for such as spit blood, it stoppeth the bloody flixe, the fluxe of the E Hemoroides, & al issues of blood. It hath the like vertue being boyled & eaten.

The iuyce of Purcelayne powred vpon the head with oyle & vineger roset, F swageth the head ache comming of heate, or of standing to long in the Sonne.

The same throwen vp into the mother or matrix, helpeth the burning in-G flammations, exulceratios, or gnawing frettings in the same, & powred in by a glister, it is good against the flixe of the guttes & exulceration of the bowelles.

The leaues of Purcelayne mingled with parched barley meale, and layde H to the inflammations of the eyes, easeth the same, and taketh away the hoate swelling: so it is likewise good against S. Antonies fier, called Erysipelas: against the heate and payne of the head, and against all hoate inflammations and tumours.

The same eaten rawe, are good against the teeth being set on edge, or asto- I nied, and it fasteneth them that be loose.

To conclude, Purcelayne cooleth all that is hoate, wherefore being layde K vpon woundes, eyther by it selfe or with the meale of parched barley, it preserueth woundes from inflammation.

The

the Hiſtorie of Plantes.

The ſeede of Purcelayne beyng taken, kylleth and driueth foorth wormes, and ſtoppeth the laſke.

The Sea Purcelayne is gathered in the ſommer, and is of ſome preſerued and kept in vineger for Salade, to be eaten at winter like Capers: for being ſo eaten, it doth heate and comfort the ſtomacke, cauſeth good appetite, or meate luſt, and prouoketh vrine.

If this Purcelayne be Halimus, the roote thereof is good againſt crampes and drawing awry of ſinewes, burſtinges and gnawinges in the belly, to be taken in Meade the waight of a dramme. It alſo cauſeth Nurſes to haue ſtore of milke.

Of Sampiere. Chap.xri.

Crithmum.   Crithmum ſpinoſum.

❧ *The Deſcription.*

1. Sampiere hath fat, thicke, long, ſmal leaues, almoſt lyke Purcelayne, the ſtalke is rounde of a foote, or a foote and a halfe long bearing round ſpokie tufts, which bring foorth litle white flowers, and a ſeede lyke Fenyll, but greater. The roote is thicke, and of a pleaſant ſauour. Searche the commentaries of Matthiolus in the ſeconde booke of Dioſcorides, there you ſhal finde three kindes more of Crithmum.

2. Of this is founde another kinde of Crithmus, whoſe leaues are lyke vnto the firſt, the crowne ſet about with harde pricking thornes, otherwiſe in all thinges like vnto the other.

3. Yet is there founde a thirde kinde of Crithmus, the whiche bringeth foorth many ſtalkes of one roote, ſet about with long ſmall leaues, the whiche are very thicke, vpon the top of the ſtalkes grow yellow flowers, almoſt lyke vnto the flowers of Chryſanthemũ, in the middes yellow, and round about ſet with yellow leaues. The roote is long. And this herbe is of taſte like vnto the firſt Crithmus, the whiche is very lyke to Creta marina.

578 The fyfth Booke of Crithmus Chryſanthemus.

❧ *The Place.*

This herbe groweth in ſalt ground by the ſea coaſt, and is found very plentifully in many places of Spayne, Fraunce, and England, alongſt the ſhoare or coaſt. The Herboriſtes of this Countrie doo plant it in their gardens.

❧ *The Tyme.*

Sampiere bloweth in this Countrie in Auguſt and September, but wher as it groweth of his owne kind, it flowreth more timely. ❧ *The Names.*

This herbe is called in Greeke κρίθμον, κ̃ κρίταμον: in Latine Crithmum, and Bati: in ſhoppes, Creta marina, by whiche name it is knowen in Brabant: in French, *Bacille, Crete marine,* and *Fenoil marin*: in Engliſhe, Sampier, and Creſtmarine.

❧ *The Nature.*

Creſtmarine is drie and ſcouring, and meetely warme. ❧ *The Vertues.*

A   The leaues, ſeede, or rootes, or al togither boyled in wine and dronken, prouoketh vrine and womens flowers: & helpeth muche againſt the Iaundiſe.

B   They keepe and preſerue the leaues & branches of Creſtmarin, or Sampier, in brine or pickle, to be eaten lyke Cappers: for being ſo eaten, they are good for ẏ ſtomacke and open the ſtoppinges of the liuer, the ſplene and the kidneyes.

Of Brookelime. Chap. xxij.

❧ *The Deſcription.*

BRookelime hath rounde fat ſtalkes, full of branches, & vppon the ſame fat thicke leaues: the which being bruiſed do yeelde a good ſauour. At the toppe of the ſtalkes and branches growe many fayre blewe flowers, not much vnlike the flowers of blewe Pimpernel. The roote is white & ful of hearie ſtringes. ❧ *The Place.*

This herbe groweth in ẏ borders & brinkes of ditches and pooles, and ſometimes alſo by running ſtreames, and brookes harde by the water, ſo that ſometimes it is ouerflowen and drenched in the ſame. ❧ *The Tyme.*

Brookelime flowreth in May, and Iune.

❧ *The Names.*

This herbe is called now in theſe dayes Anagallis aquatica, and Becabunga, and of ſome it is taken for that herbe that of Dioſcorides is named in Greeke κνπαία: in Latine, Cepæa: and it ſeemeth

Anagallis Aquatica.

seemeth to be a kinde of Soum, of the whiche is written by Cratenas: in high Douche, Wasserpungen, Bachpunghe, or Pungen: in base Almaigne, Waterpunghen: in English, Brookelyme.

✤ The Nature.

This herbe is hoate almost in the seconde degree.

✤ The Vertues.

Brookelime leaues dronken in wine do helpe the strangullion, & the inward scabbes of the bladder, especially if it be taken with the roote of Asparagus or Sperage.

They be also eaten with oyle and vineger, and are good for them that are troubled with the strangurie, and stone.

Of Earth Chesnut. Chap.xriij.

✤ The Description.

Bolbocastanon.

THE small Earth Chesnut hath euen crested stalkes, of a foote and a halfe long or more. The first leaues are lyke the leaues of common Parsely, but they be lesser, & smaller iagged & they that grow about the stemme, are not muche vnlyke the leaues of Dil, the flowers which are white, do growe in spokie tuftes lyke the toppes of Dyl. The seede is small of a flagrant smel, not much vnlyke the seede of Commin or Fenill, but a great deale smaller. The roote is rounde lyke a wherrow or wherle, or rather like a litle round appel, browne without and white within, in taste almost lyke to Carrottes.

❡ The Place.

This herbe groweth in many places of Hollande and Zeelande, in corne feeldes & alongst the wayes, there is good store of it in some places of Englande. The Herboristes of Brabant, do plant it in their gardens.

✤ The Time.

This herbe flowreth and deliuereth his seede in June.

✤ The Names.

This herbe is called in Zeelande, Cleyn Eerdtnoten, some Herboristes take it for Apios, others for Meum, and the thirde for Bulbina: but it hath no lykenesse vnto any of them three, it seemeth better in my iudgement to Βολβοκάςανον, Bolbocastanon, of Alexander Trallianus, the whiche the later Grecians do call ἀγριοκάςανον, Agriocastanon, whereunto it is very muche lyke: for the roote is lyke Bulbus, and in taste it is muche lyke to the Chesnut: in consyderation whereof, it may be well called Bolbocastanon, and Agriocastanon: in French, Noix-Chastaigne: in base Almaigne, Eerdtcastanien: in English, Earth Chesnut.

✤ The Nature.

Bolbocastanon is hoate almost in the seconde degree, and somewhat astringent, the seede is hoate and drie almost in the thirde degree.

✤ The Vertues.

In Sealande they eate this roote in meates, in whiche Countrie, it is not muche

muche differing in taste and vertue from Parsneppes and Carrottes: it prouoketh vrine, comforteth the stomacke, nourisheth indifferently, & is good for the bladder and kidneyes.

Bolbocastanon, as Alexander Trallianus writeth, is good to be eaten of them that spit blood.

The seede of the same causeth women to haue their natural sicknes, bringeth foorth the secondines, prouoketh vrine, and is very profitable for the reynes, the kidneyes, the bladder, and the spleene or milte being stopped.

Of Mallowes. Chap. xxiiij.

❧ *The Kindes.*

There be diuers sortes of Mallowes, whereof some be of the garden, and some be wilde, the whiche also be of diuers kindes. The garden Mallow, called the winter or beyondsea roose, is of diuers sortes, not only in leaues, stalkes, and growing, but in proportion, colour, & flowers: for some be single, some double, some white, some carnation, some of a cleare or light red, some of a darke redde, some gray, and speckled. The wilde Mallowes are also of two sortes, the great and the small.

Malua satiua.	Malua sylueſtris elatior.
Holyhocke or garden Mallow.	Wild Hocke or the greater wild Mallow.

❧ *The Description.*

1. THE great tame Mallow which beareth the beyondsea or winter rose, hath great round rough leaues, larger, whiter, and vneuener: then the leaues of the other Hockes or Mallowes. The stalke is rounde, and groweth

groweth fixe or seuen foote high or more: it beareth fayre great flowers of diuers coloures, in figure lyke to the common Mallowe or Hocke: but a great deale bigger, sometimes single, somtimes double. The flowers fallen the seede commeth vp lyke smal cheeses. The roote is great and long, and continueth a long time, putting foorth yerely newe leaues and stalkes.

2 The great wilde Mallow, hath leaues somewhat round, fat, and a litle cut or snipt rounde about the borders, but of a browner colour, smaller and euener then the leaues of the Hollyhocke. The stalke is rounde of two or three foote long, therupon grow the flowers in fashion like to the other, but much smaller, and parted into fiue leaues of a purple carnation colour, after whiche commeth the seede, whiche is rounde and flat, made lyke litle cheeses. The roote is long, and of a conuenient thicknesse.

Malua syluestris pumila.
The smal wild Mallow.

3 The smal wilde Mallow is very muche lyke to the great wilde Mallowe, sauing that his leaues be a litle rounder and smaller: the flowers be pale, & the stalkes grow not high, or vpright: but trayle alongest the grounde. The roote is lykewyse long and thicke. ❧ *The Place.*

The Hollyhocke or garden Mallowe, is sowen and planted in gardēs of this Countrie.

The wilde kindes growe in vntoyled 2.3. places, by path wayes, and pastures.
❧ *The Tyme.*

Hollyhocke flowreth in June, July, and and August. The wild beginneth to flower in June, & continueth flowring vntyl September, in the meane space it yeeldeth his seede. ❧ *The Names.*

Mallowes are called in Greeke, μαλάχη: in Latine, Malua: of Pythagoras, ἄνθεμα, Anthema, of Zoroastes, διάδημα, Diadema: of the Egyptians, Chocortis, of some Vrina muris: in Frenche, *Maulue:* in high Douche, Pappel: in base Almaigne, Maluwe: in Shoppes Malua: in Englishe, Hockes, and Mallowes.

1 The first kind of Mallowes, is called in Greeke, μαλάχη κηπευτή: in Latine, Malua satiua: of some Rosa vltramarina: that is to say, the Beyondesea Rose: in Frenche, *Maulue de iardin,* or *cultiuée* in hygh Douche, Garten Pappeln, Ernrosz, or Herbstrosz: in base Almaigne, Winterroosen: in English, Holyhockes, and great tame Mallow, or great Mallowes of the garden.

2 The wilde Mallow is called in Greeke, μαλάχι ἀγρία: in Latine, Malua syluestris: in high Almaigne, Gemeyn Pappeln: in base Almaigne, Maluwe, and Keeskenscruyt: wherof that sort which groweth vpright and highest, is called Malua elatior, that is the common Mallowe, or the tawle wilde Mallow, and the common Hockes.

°3 The second wild kind which is the least, is called Malua syluestris pumila, or Malua pumila, that is to say, the small wilde Hocke, or Dwarffe Mallowe: in Douche, Cleyn Maluwe.

The fyfth Booke of

❧ The Nature.

Mallowes are temperate in heate and moysture, of a digestiue and softening nature.

❧ The Vertues.

Mallowes taken in meate, nourish better then Letuce, and soften the belly: neuerthelesse they be hurtfull to the stomacke, for they loose and mollifie or relenт the same. A

The rawe leaues of Mallowes eaten with a litle salt, helpe the payne and exulceration of the kidneyes and bladder. B

For the same purpose and against the grauel and stone, Mallowes are good to be boyled in water or wine, and dronken. C

The decoction or broth of Mallowes with their rootes, are good agaynst al venome and poyson, to be taken incontinently after the poyson, so that it be vomited vp againe. D

It doth mollifie and supple the tumours and hardnes of the mother, if women bathe in the broth thereof. E

It is good against al going of, of the skin, excoriations, gnawings, roughnesse and fretting of the bladder, guttes, mother, and fundement, if it be put in with a glister. F

The seede of Mallowes dronken in wine, causeth abundance of milke, and is good for them that feele paine in the bladder, and are troubled with grauel. G

Mallowes are good to be layde to against the stinginges of waspes and Bees, and draw foorth thornes and splinters, if they be layde thereupon. H

The same raw or boyled, and pounde by them self, or with Swines grease, do supple, mollifie, rype, and dissolue all kindes of tumours, hoate and colde. I

The rootes of Mallowes rosted in the imbers or hoate asshes, and pounde very smal, are very good to be layd to as an implaister, against the exulceration and sorenesse of womens breastes. K

❧ The Choise.

The garden Mallow is wholsomer to be eaten, then the wilde Mallow: but in medicine, to soften hardnesse & dissolue swellinges or tumours, the wild kinde is better and of more vertue, then the garden Mallow.

Of Marrish Mallow, or white Mallow. Chap.xxv.

❧ The Description.

1 Marrish Mallow is muche like the other Mallowes, but a great deale whiter, and softer: his leaues be roundishe, white, softe, and almost frised or cottoned, whiche in proportion and quantitie, are almost like to the leaues of the common hocke or wilde Mallowe. The stalke is rounde and straight. The flowers are in figure like to the wilde Mallowe, after them commeth the seede, as in the other Mallowes. The roote is great and thicke, white within, and slymie.

2 The seconde kinde of white Mallow, whiche Theophrast describeth, hath roundish leaues, white and soft, and almost frised or Cottoned like the other white or Marrishe Mallowe, but farre greater, almost like in proportion and bignesse to the leaues of Gourde. The stalkes be long, thicke, and strong, vpon which betwixt the leaues and the stemme growe yellow flowers, & after them come crooked huskes (as though they were wrinckled) wherein is the seede.

❧ The Place.

1 Marshe Mallowe loueth fat and moyst grounde, adioyning to waters and ditches.

2 The second kind is a stranger in this Countrie: & therfore not to be founde but amongst certaine diligent herboristes.

the Historie of Plantes. 583

Althæa.
Marshe Mallowe, or
Mie Mallowe.

Ibiscus Theophasti. Abutilon Auicennæ.
Yellow Hibiscuus, or Abtilno.

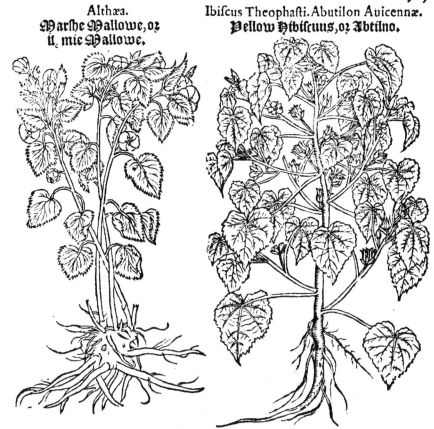

❧ The Time.

1. It flowreth togither with the other Mallowes.
2. The seconde sorte is sowen in Marche or Aprill, and deliuereth his flower and seede about the ende of Sommer.

❧ The Names.

1. These kindes of Mallowes are called in Greeke ἀλθαία: in Latine, Althæa, and Hibiscus: of Galen Anadendron, of some Aristalthæa: in shoppes Bismalua, and Maluauiscum: in French, Guymaulue: in high Douche, Ibisch, oder Eibisch: in base Almaigne, Witte Malue, or Witte Huemst: in English, Marrish Mallowe, and white Mallowe.
2. The seconde kind is called of Theophrastus also in Grecke ἀλθαία, καὶ μαλάχη ἀγρία: in Latine also Hibiscus, and to be knowen from the other Hibiscus Theophrasti: of Auicenne it is called Abutilon, by the whiche name it is knowen of the Herboristes.   ❧ The Nature.

Marshe Mallow is temperate in heate as the other Mallowes, but dryer euen in the first degree.   ❧ The Vertues.

The roote of Marsh Mallow boyled in wine and dronken, is good against A the paine and griefe of the grauel and stone, the bloody flire, the Sciatica, the trembling & shaking of any member, & for suche as are troubled with crampes and burstinges.

The same boyled in sweet new milke, healeth the cough, as Plinie wryteth. B

It

It is good also against the toothache: for it swageth the payne, being boyled in vineger and holden in the mouth.

The same boyled in wine or honyed water, and bruſed or pounde very smal doth cure and heale newe woundes, and it doth diſſolue and conſume all colde tumours and ſwellinges, as wennes and hard kernelles, also the impoſtumes that chaunce behinde the eares, and for the burning impoſtume of the pappes: it ſofteneth tumours, it ripeth, digeſteth, breaketh, and couereth with ſkinne, olde impoſtumes and blaſtinges or windie ſwellinges, it cureth the riftes and chappes of the fundament, and the trembling of the ſinewes, & ſinewie partes.

The same so prepared and pounde with Swines greaſe, Gooſe greaſe or Turpentine, doth mollifie and ſwage the impoſtumes and ſores of the mother, and openeth the ſtoppinges of the ſame, being put in as a peſſarie or mother ſuppoſitorie.

The leaues are good for all the greefes aforeſayde, being vſed in like manner, yet they be nothyng ſo vertuous as the roote.

The leaues of marſhe Mallow, beyng layde to with oyle, do heale the burninges and ſcaldinges with fire and water, and are good againſt the bytinges of men and Dogges, and againſt the ſtinginges of Bees and Waſpes.

The ſeede greene or dried, pounde and dronke, healeth the blooddy flyxe, and ſtoppeth the laſke, and all iſſue of blood.

The ſeede eyther greene or dry, layd to with vineger, taketh away freckles, or fowle ſpottes of the face both white and blacke, but ye muſt annoynt your ſelfe eyther in the hoate Sonne, or els in a hoate houſe or ſtewe.

The ſame boyled eyther in water, vineger, or wine, is good to be dronken of them whiche are ſtongue with Bees and Waſpes.

Of veruepne Mallow, or cut Mallow. Chap.xxvi.

❧ *The Deſcription.*   Alcea.

CUT Mallow, as witneſſeth Dioſcorides, is a kind of wild Mallow, whoſe leaues are more clouen, deeper cut, and diuided into ſundry partes, almoſt lyke þ leaues of Veruayne, but muche larger. The ſtalkes be round and ſtraight, two or three foote high. The flowers be of a cleare redde or incarnate colour, in figure like to the flowers of the other Mallowes, after the flowers commeth the ſeede alſo faſhioned lyke litle cheeſes. The roote is thicke and two foote long or more, white within.   **❧** *The Place.*

This herbe groweth in vntoyled places, in the borders of fieldes and hedges, and is not very common in this Countrie.

❧ *The Tyme.*

Cut Mallow flowreth at Midſomer, as the other wilde Mallowes or Hockes.

❧ *The Names.*

This herbe is called in Grecke, ἀλκέα: & in Latine, Alcea: vnknowen in ſhoppes: of ſome Herba Simeonis, & Herba Hungarica: in high Douche, Sigmarskraut, Sigmundſwurtz,

mundſwurtz, or Hochlenten: in Frenche, *Guymaulue ſauuage* in baſe Almaigne, Sigmaerts crupt: in Engliſh, Verueyn Mallow, or cut Mallowe, this is alſo a kinde of marſhe or ſlymie Mallow, Symons Mallow.

❁ *The Nature.*

Cut Mallow is temperate betwixt heate and colde, and hath ſomewhat a drying nature. ❁ *The Vertues.*

The roote of cut Mallowe, or Symons ſlymie Mallowe boyled in water or wine and dronken ſtoppeth the blooddy flyxe, and healeth, and glueth togither woundes and inwarde burſtinges.

Of Veniſſe Mallow. Chap.xxvij.

❁ *The Deſcription.* Alcea Veneta.

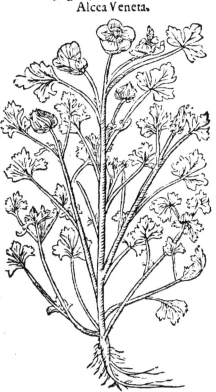

The Venitian Mallow, hath rounde tender ſtalkes, with handeſome branches, the leaues be of a darke greene, thicke or fat, clouen & iagged not much vnlyke the leaues of cut Mallow, or þ wild Guy Mallow, of a ſhining darke colour, not muche vnlyke the colour of the leaues of Acanthus. The flowers growe at the toppe of the ſtalkes, and are the fayreſt amongſt al the ſortes of Mallowes, almoſt lyke in making to the flowers of the other Mallowes, diuided alſo into fiue leaues, the extremitie & outſide of the leaues are white or pale, but the middle or inner part of the flower is of a browne red purple, with a yellowe Bodkin or Peſtil, lyke golde in the middle. Theſe flowers do not open at all vntyll three or foure houres after ſonne riſing, or an houre or two before noone, or there aboutes: and when they haue remayned open or ſpreade abrode the ſpace of an houre, or an houre & a halfe, they cloſe togither agayne, and fade or wither away, the whiche being paſt, there come in their

ſteede little huſkes or bladders, wherein are ſmal knoppes, or hearie pellettes, in whiche is a blacke ſeede. The roote is ſmal and tender, and periſheth yerely, ſo that it muſt be newe ſowen euery yeere.

❁ *The Place.*

This herbe is a ſtranger in this Countrie, and is not founde at all except in the gardens of ſome Herboriſtes, where as it is ſowen.

❁ *The Tyme.*

They ſowe it in Marche or Aprill, and it flowreth in June and July.

❁ *The Names.*

This herbe of the later writers, is taken for a kinde of Alcea, and is called Alcea Veneta, that is to ſay, The ſlymie or Mucculage Mallow of Vennis: of ſome Malua Theophraſti: in high Douch, Venediger Pappeln, or wetter Roſzlin: in baſe Almaigne, Veneetſche Maluwe. This is not Hypecoon, as Matthiolus tholus takes it, but it shoulde rather seeme to be Solanum Manicum, described in the xcij. Chapter of the thirde booke, whereunto it resembleth muche.

❧ The Nature.

The Mucculage Mallowe is hoate and moyst, lyke to the common Hocke or great wilde Mallow, we may well presume, that in operation and vertue it is lyke to the common Mallow, yet for al that we haue no certayne experience of the same. ❧ The Vertues.

Forasmuche as this Mallowe is hoate and moyst, we may well presume, that in operation and vertue, it is lyke to the common Mallowe, yet for al that we haue no certayne experience of the same.

Of Cucumbers. Chap. xxviij.

❧ The Kindes.

There be two sortes of Cucumbers, the garden and the wilde Cucumber. The garden Cucumber is vsed in meates. The wild kind is not good for that purpose, but serueth onely for medicine: we haue giuen you his description in the thirde booke of this historie the xl. Chapter.

❧ The Description.

Cucumis satiuus. Melopepon Galeni.
Cucumbers.

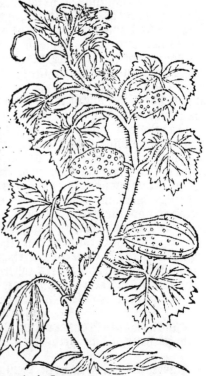

THE garden and eateable Cucumber, hath long rough branches, creeping alongest the grounde, vpon whiche growe rough roundishe leaues, and claspers or tendrelles. The flowers growe betwixt the leaues and the stalkes, of a faint yellowe colour, ẏ which being fallen away, the fruite foloweth after which is long, the outside thereof is sparckled, and set full of litle bowles or bosses, the coastes or sides be long, & greene at ẏ beginning, & afterward yellow, within the whiche groweth a broade or large white seede. The roote is of a competent length.

❧ The Place.

These Cucumbers are sowen in gardens, and loue places standing well in the Sonne.

❧ The Tyme.

The chiefest season, for the eating of Cucumbers, is in July and August, and they are ripe in September.

❧ The Names.

This kinde of Cucumber is called of the later writers in Greeke σίκυς ἥμερος: in Latine Cucumis satiuus, or Cucumer satiuus, of some Cucumis Anguinus, or Anguria: in shops, Cucumer, in French, Concombre: in high Douché, Cucumern, and Gurchen: in base Almaigne, Contomineren: and this seemeth to be the same, which Galen in libris de Alimentorum facultatibus, calleth μηλοπέπων, Melopepon.

❧ The Nature.

The Cucumber is colde and moyst in the seconde degree.

❧ The

the Historie of Plantes.   587

❧ *The Vertues.*

Cucumber taken in meates, is good for the stomacke and bowels that are troubled with heate: but it yeeldeth small nourishment & euil, insomuch that the immesurable vse therof, filleth the vaynes with colde noughtie humours, the whiche (bycause they may not be conuerted into good blood) doo at the length bryng foorth long and great agues and other diseases, as Galen writeth.   A

The seede dronken with milke or sweete wine looseth the belly gently, and is very good agaynst the exulceration, & rawnesse of the bladder, and inwarde stopping of the same.   B

The greene leaues stamped with wine and layde to, heale the bitinges of Dogges.   C

Of Melones and Pepones.   Chap.xxix.

❧ *The Kyndes.*

THE Pepon is a kinde of Cucumber, the whiche is nowe of diuers sortes, as the great, round, and flat: whereof the great is also of two sortes, that is white, and greene.

Pepones magni.   Pepones rotundi.
Great Melons or Pepons.   **Round Melons or Pepons.**

❧ *The Description.*

1   The great Pepon hath long, round, great, rough, and hollow branches, beset with short sharpe prickles. The leaues be great, broade, & rough, parted into foure or fiue deepe cuttes or iagges, much greater then the leaues of the Gourde: by the sayde leaues come foorth clasping tendrelles, whereby

The fyfth Booke of

whereby this Pepon groweth vp, and taketh holdfast by euery thyng. The flowers growe amongst the leaues, very great and hollowe within, iagged about the edges, and of a yellowe colour. The fruite is very bigge, thicke, and and long, one sort thereof is of a greenishe colour with many ribbes or costes, and the rinde is very harde: the other sorte is white, couered with a soft and tender rinde. The seede is inclosed in the fruite, and is white and broade, much larger then the seede of the Cucumber.

2   The seconde kind whose fruite is round, hath also prickly stalkes & leaues: the stalkes be smaller, and most commonly creepe alongst the grounde. The leaues be also smaller and not so deepe cut or rent. The flowers be yellow lyke the flowers of great Melon or Pepo. The fruite is rounde and somewhat flat, whereof one sorte is greene and the other white, wherin groweth the sede smaller than the sede of the other Pepone, and greater than the seede of the Cucumber.

The thirde kinde of Pepones is muche lyke to the seconde in creepyng branches, leaues, and flowers: but the stalkes be not so rough, the fruite is flat, brode, and round, couered with a soft and gentle rynde or coueryng, cronkeled & wrinckled about the borders or edgis, lyke to a buckler, wherin is the seede, lyke to the seede of the Cucumber but greater.

4   There is also a wilde kinde of Pepons, which are lyke ye tame Pepons, in stalkes and rough leaues: but the fruite is smaller, and altogither bitter lyke to Coloquintida, or the wilde Gourde, or wilde Cucumber, wherevnto this wilde kinde is agreeable in vertue and operation.

❧ *The Place.*

All these kindes of Melons, and Pepons, are sowen in gardens, and vsed in meates except the wilde kind.

❧ *The Tyme.*

The fruite is ripe in August, and sometimes sooner, if it be a hoate season, and a forwarde yere.

❧ *The Names.*

This fruit is called in Greke πίπονες: and in Latine, Pepones: of Galen also σικυοπέπονες, Sicyopepones, that is to say, Pepones Cucumerales: Cucumber Pepons.

The first kinde is called in English, Melons and Pepons: in Frenche, Pompons d'yuer, or Citroulen in high Douche, Pseben: in base Almaigne, Pepoenen: & of the newe writers in Latine, Magni Pepones, of some Cucumeres Turcici, & in Almaigne accordingly Turckischer Cucumeren, & Torcksche Cocommeren.

2   The seconde kinde of Pepons is called Pepo, or Cucumis marinus: of some

Zucco-

Pepones lati.

Brode Melons or Pepons.

Zuccomarin: in French, *Concombre marin, Pompons Turquins* in Douch, Zee Concommeren: in Englishe, Pompons, or Melons: we may also name them, Sea Cucumbers, or Turkie Pompons.

3 The thirde kinde whiche is the large Pompone, is for the same cause called Pepones lati, Broade Pepons: in Douche, Breede Pepoenen, and of some Torcksche Meloenen, that is to say, Turkie Melons.

✻ *The Nature.*

The garden Melons, or Pompons, are colde and moyst, but not so moyst as the Cucumbers.

✻ *The Vertues.*

The fruit of the garden Pepon is not often eaten raw, but wel boyled with good flesh or sweete milke, for being so prepared it is better and lesse hurtfull than the Cucumber, and is good for suche as haue a hoate stomacke.   A

The flesh or substance of Pepons finely stamped, doth swage and heale the inflammations of the eye, if it be layde vnto them, and being bound to the forehead, it stoppeth the falling downe of humours into the eyes.   B

The seede of Pepons powned with meale and their owne iuyce, doth beautifie the face, for it taketh away freckles and al spottes of the face, if the place be well rubbed with it in the Sonne.   C

The quantitie of a dramme of the dried roote taken with meade or honied water, maketh one to vomite.   D

The same layde to with honie, healeth the sores of the heate whiche be full of corruption and filthy matter.   E

Of Citrulle Cucumber. Chap.xxx.

✻ *The Description.*   Cucumis Citrulus.

The Citrul or Citrō Cucumber is also a kind of Cucumber hauing rounde rough stalkes, full of Capreoles or clasping tendrelles, whereby it taketh hold vpon hedges and stakes. The leaues be al iagged and rent, much lyke to the leaues of Coloquintida. The fruite is round and greene without, wherein groweth a flat blacke seede, lyke to a Melon or Pepon seede, but somwhat smaller.

✻ *The Place.*

This herbe is mainteyned in the gardens of some Herboristes.

✻ *The Tyme.*

The Citrull Cucumber is rype with Pompons or Melons, about the ende of Sommer.

✻ *The Names.*

1 This kind of Cucumber is called Cucumis Citrulus, of some Anguria: in shops Citrulum: and in Douch according to the same, Citrullen: in French *Concombre citrin*: in Englishe, Citrulles: and of some, Pome Citrulles.

2 The wilde kinde of this Cucumber, is the

The fyfth booke of

the right Coloquintida, described in the third booke of this historie of Plantes.

❧ *The Nature.*

The Citrull is of temperament, colde and moyst lyke the Pepon.

❧ *The Vertues.*

The Citrull Cucumber is muche lyke to the Melone in vertue and operation, whether it be taken in meate or medicine.

Of Melons. Chap.xxxi.

❧ *The Description.*   Cucumis Galeni, & Antiquorum.

The Melon trayleth alongst the grounde lyke the Cucumber, and hath tender branches with catching caprioles, and rounde rough leaues. The flowers be yellowe, lyke the flowers of the Cucumber. The fruite is long, and almost like to the Cucumber, but greater, and couered all ouer with soft heare, especially beyng yet young and tender, and yellowe within. The seede is muche inclosed in the inner parte of the fruite, and is muche lyke to the Cucumber seede.

❧ *The Place.*

Melons are sowen in gardens, and they require a fat & wel dounged ground, and also a drie grounde, standing well in the Sonne, for otherwise you scarse see them prosper in this Countrie.

❧ *The Tyme.*

The Melon is ripe in August & September.

❧ *The Names.*

Galen nameth this fruite in Greeke σίκυς, that is to say in Latine, Cucumis, & vndoubtedly it is the Cucumis of the Auncientes, wherof Cucumer Asininus, that is to say, the leaping Cucumber is the wilde kinde. Of the later writers at these dayes, it is called in Greeke μηλοπίπων: in Latine, Melopepo, of some Melo, and in some places of Italy, it is also called Citrulus, and Cucumis citrulus: in Frenche, *Melon* in high Douche, Melaunen: in base Almaigne, Meloenen: in Englishe, Melons, and muske Melons.

❧ *The Nature.*

The Melon in temperament is almost like to the Pepone, but not so moyst.

❧ *The Vertues.*

The Melon is in vertue like to the Pompon or Pepon, sauing that it doth not ingender so euill blood, neither doth it descende so quickly into the belly, wherefore it is by so much better then the Pepon.

Of Gourdes. Chap.xxxij.

❧ *The Kindes.*

The Gourde is of three sortes, that is to say, the great, the smal, & the long, which are muche lyke one another in leaues & branches, ouer and bysides the wilde kind which is described before in the third booke.   ❧ *The*

the Historie of Plantes. 591

Cucurbita cameraria maior.

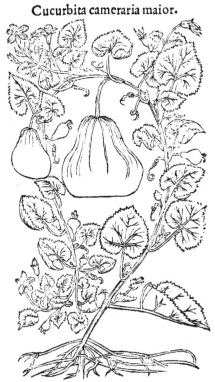

Cucurbita minor.

Cucurbita anguina.

❧ *The Description.*

1. THE Gourde hath long limmer stalkes, tender and full of branches and clasping tendrels or caprioles, whereby it taketh holde and climbeth vp, especially if it be set by perches, hedges, quick settes or trees, by the whiche it may take holde and wrap and winde it selfe for without such stayes & helpes the Gourde cannot climbe vp, but will lye alongst and growe harde by the grounde, and than it can not bring foorth his fruit. The leaues be rounde, whitishe, soft, and almost lyke veluet, drawing somewhat towardes the fashion of the great Clot Bur leaues, but smaller. The flowers be white, euery flower parted into fiue small leaues, after the flowers commeth the fruite, at the beginning greene, and ouerlayde or couered with a soft cotton or hearie downe, but after whan it turneth to ripenesse, it is of a yellowishe colour, and almost balde without heare or cotton. This first kinde is very great, rounde, thicke, and large. Within

Ddd ij this

this fruite is found a large long seede, with two peakes or corners at the ende of the same seede.

2 The seconde kinde is lyke to the first in stalkes, leaues, flowers, and seede, sauyng that the fruite is smaller, and lyke a rounde flagon or bottell with a long necke, which is the best fashion of Gourdes, for they be oftentimes vsed (especially of the Pilgrimes) in steede of flagons or bottelles, when they are made hollowe.

3 The thirde kinde is lyke to the aforesayde, sauyng that the fruit is neyther so short, nor so bigge as the fruite of the others, but most commonly is of three or foure foote long, and as bigge as ones legge or arme: the rest is lyke the others.

4 Bysides these three kinds of garden Gourdes (as some learned men write) there is found another sort whose fruite is very short and no bigger than ones finger, the residue, as the stalkes and leaues is lyke to the abouesayde.

5 Of this sorte is also a wilde kinde, whereof there is mention made in the Chapter of Coloquintida, in the thirde booke.

❊ *The Place.*

The three first kindes are planted in the gardens of this Countrie.

4 The fourth kinde groweth in some Countries in rough stony places.

❊ *The Tyme.*

The Gourde is ripe in this Countrie in August and September.

❊ *The Names.*

The Gourde is called in Greke κολόκυνθα καὶ κολόκυνθα ἰδοδιμϙ: in Latine and in the Shoppes, Cucurbita: in high Douche, Kurbs: in base Almaigne, Cauwoorde: in Frenche, Courge in Englishe, a Gourde, or Gourdes.

The three first kinds are called of Plinie Cucurbitæ cameraiæ, and of some also Perticales: bycause they growe vppon poles, rayles, and perches lyke vnto vines, whereof is sometimes made close herbours and vaultes or couerings.

1 The first kinde is nowe called of the later writers, Cucurbita magna, & maior: in Englishe, the great Gourde: in Frenche, Grande Courge: in high Douche, Grosz Kurbs: in base Almaigne, Groote Cauwoorden.

2 The seconde kinde is called Cucubita minor: in English, the lesser Gourde: in high Douch, Klein Kurbs: in base Almaigne, Cleyn Cauwoorden: in Frēch Petit Courge.

3 The third kind is called Cucurbita anguina and of some Cucurbita oblonga: in Frenche, Courge longue in high Douch, Lang Kurbs: in base Almaigne, Langhe Cauwoorden: in English, Long Gourdes.

4 The fourth kinde whiche is yet vnknowen in this Countrie, is called of Plinie in Greeke σομφός, Somphos: in Latine, Cucurbita barbarica, & marina.

❊ *The Nature.*

The Gourde is colde and moyst in the seconde degree.

❊ *The Vertues.*

The Gourde eaten rawe and vnprepared, is a very vnholsome foode, as A Galen sayth, for it cooleth, and chargeth, or lodeth the stomacke, and ouerturneth and hurteth the same by stirring vp the payne thereof.

But being boyled, backte, or otherwayes dressed, it is not so hurtfull, for it B doth coole and moysten the hoate and dry stomacke, slaketh thirste, and looseth the belly, neuerthelesse it nourisheth but litle.

The iuyce of the whole Gourde pressed out and boyled, and dronken with C a litle

the Historie of Plantes.

a litle hony and Saltpeter looseth or openeth the belly very gently.

The lyke vertue hath ye wine that hath stoode by the space of a whole night (abroade in the ayre) in a rawe holow Gourde, if it be dronken fasting.

The poulpe or inner substance of the Gourde pounde or brused doth slake and swage hoate swellinges and impostumes, the inflammations and rednes of the eyes, and especially the hoate payne of the gowte, being layd to the greeued places.

The iuyce of the Gourde with oyle of roses dropped into the eares, swageth the paynes of the same.

The same is very good to be layd to in the same sort, or by it selfe, vnto scaldings, burnings, and chafinges, and hoate Cholerique inflammations, called Erisipelas, or S. Antonies fier.

The croppes and tender branches, dronken with sweete wine and a little vineger, cureth the bloody flire.

The rinde or barke of the Gourde, burned into ashes, doth cure and make hoale the sores and blisters, that come of burning, and the old sores of the genitours, being strowed thereupon.

The seede of the Gourde is almost of the lyke vertue with the seede of the Cucumber.

Of Rapes and Turneps. Chap. xxxiij.

The Description.

Rapa.

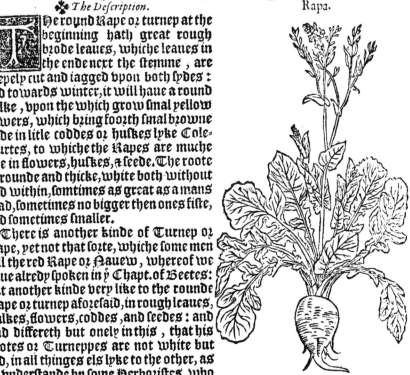

THe round Rape or turnep at the beginning hath great rough brode leaues, whiche leaues in the ende next the stemme, are deepely cut and iagged vpon both sydes: and towards winter, it will haue a round stalke, vpon the which grow smal yellow flowers, which bring foorth smal browne seede in litle coddes or huskes lyke Colewurtes, to whiche the Rapes are muche like in flowers, huskes, & seede. The roote is rounde and thicke, white both without and within, somtimes as great as a mans head, sometimes no bigger then ones fiste, and sometimes smaller.

There is another kinde of Turnep or Rape, yet not that sorte, whiche some men call the red Rape or Nauew, whereof we haue alredy spoken in ye Chapt. of Beetes: but another kinde very like to the rounde Rape or turnep aforesaid, in rough leaues, stalkes, flowers, coddes, and seedes: and and differeth but onely in this, that his rootes or Turneppes are not white but red, in all thinges els lyke to the other, as I vnderstande by some Herboristes, who haue declared vnto me, that the noble and famous Queene Douager of Hungarie and Bohem, doth cause them to be set and planted in her most ryche and pleasant gardens.

The Place.

The Turnep loueth an open place, it is sowen somwhere in vineyardes, as at Huygarden and the Countrie theraboutes, which do waxe very great: but they are most commonly sowen in feeldes, especially when the corne is ripe, but they become nothing so great. ❋ *The Tyme.*

They are sowen at the beginning of sommer, that they may waxe great: and in July and August after the cutting downe of corne: but the later sowing are neuer very great, & about April when sommer is at hand, they bring foorth stalkes, and flowers. The seede is ripe in May and June.

❋ *The Names.*

Rapes are called in Greeke γογγύλαι καὶ γογγυλίδες: in Latine, Rapæ: in French, *Naueaux:* in high Douch, Ruben: in base Almaigne, Rapen: in Englishe, Rapes and Turneps. ❡ *The Nature.*

Rapes are hoate and moyst of complexion.

❋ *The Vertues.*

The Turnep taken in meat nourisheth meetely wel, so that it be moderately taken, and wel digested, but if a man take so muche thereof as may not be well digested, it engendreth and stirreth vp much windynesse, & many superfluous humours in the body, especially when it is eaten rawe, for then it hurteth the stomacke, & causeth windinesse, blastings, and payne in the belly & small guttes. A

The same boyled in milke, swageth the payne of the gowt, being laid therto. B

Oyle of roses put into a Turnep made holow for the purpose, and then rosted vnder the hoate ashes or embers, healeth ye kibed heeles. The broth of Rapes is good for the same purpose, if the kibed heeles be washed and soked thereon, and so is the Nauew or Turnep it selfe, eyther baked or rosted, good to be layd vpon mouldy and kibed heeles. C

The croppes and young springes of Turneps, eaten, prouoke vrine, and are good for suche as are troubled with the stone. D

The seede of Turneps or Rapes, withstandeth all poyson, and therefore is put to the making of treacles, whiche are medicines ordayned agaynst all poyson, and for the swaging of paynes. E

The oyle of the same seede is of the same efficacie and working, and being taken rawe it expelleth the wormes that ingender in the body. F

The roote prepared and vsed as is beforesaid stirreth vp the pleasure of the body, the seede dronken is of the same vertue, the seede is also put into medicines, that are made for the beautifying of the face, and al the body, as Dioscorides, Galen, and other approued authours testifie. Rapes haue also a maruelous propertie to cleare the eye sight, as Auerrois the Philosopher (but enimie vnto Christ) writeth.

Of the long Rape, or Nauet gentle. Chap.xxxiiij.

❡ *The Kindes.*

The Nauew is of two sortes, tame and wilde.

❋ *The Description.*

1. Nuew gentle, or garden long Rape, hath great large leaues almost lyke the leaues of Turneps or round Nauewes, but muche smoother. The stalke is rounde of a cubite long, vpon the whiche growe flowers, huskes, and seede lyke to Turnep. The roote is very long, and thicke, in all thinges els like the Turnep or round Rape.

2. The wild Nauew is not much vnlyke the abouesayd, sauing that his leaues are more iagged from the neather part, euen vp to the top, and the roote is not so long, but shorter and rounder, almost lyke to a wilde peare.

The

the Historie of Plantes.

Napus hortenſis. **Garden Rape.** Napus ſylueſtris. **Wilde Rape.**

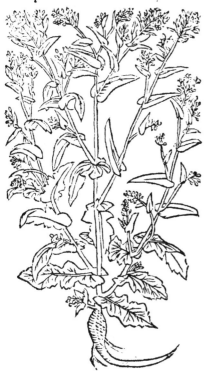

❦ *The Place.*

The Nauew gentle is much sowen in Fraunce, especially about Paris.
The wilde Nauew groweth in some Countries alongest by riuers and brookes, and such colde places.

❦ *The Tyme.*

The Nauew flowreth in the spring time, like the Turnep and Colewortes.

❦ *The Names.*

The Nauew is called in Greeke βουνιάδες: in Latine, Napi: in high Douche, Steckruben: in Brabant, Steckrapen, and Parijsche Rapen, that is to say, Long Rape, and Paris Nauewes.

1 Garden Nauew is called in Latine, Napus satinus: in high Douch, Truckē Steckruben: that is to say, the Drie Nauew: some do also cal it in English, Nauet, and Nauew gentle.

2 The wilde kinde is called Napus sylueſtris: in high Almaigne, Nas Steckruben, that is to say, the moyst or water Nauet.

❦ *The Nature.*

Nauewes are of complexion lyke to the Turneps, as Galen writeth.

❦ *The Vertues.*

The Nauew taken in meate, doth nourrish lesse then the Turnep, otherwise A in vertue and operation, it is much like to the rounde Rape or Turnep.

The seede thereof is very good against poyson, and therefore it is put into B treacles, and preseruatiues.

The fyfth booke of

Of Rampion or wilde Rapes. Chap.xxxv.

❧ *The Kindes.*

There be two sortes of Rampions or wilde Rapes, the great and the smal.

Rapum sylvestre paruum.	Rapum sylvestre aliud.
Litle Rampions.	**Wilde Rampions.**

❧ *The Description.*

1. THE smal common Rampion, his first leaues be roundishe, almost lyke the leaues of the March Violet, afterward it bringeth foorth a round harde stalke of two foote long, set about with long narrowe leaues, at the top of the stalkes growe pleasant flowers, very much lyke to the wild Bel flowers described in the seconde booke the xxiij. Chap. after the flowers come long cornered or square huskes, wherin the sede is inclosed which is very smal. The roote is long and white, sometimes as bigge as a mans litle finger, in tast almost like the Nauew gentle, the whiche in ye winter seaſon is vsed in salades.

2. The other Rampion, the whiche is not yet very well knowen his first leaues be brode, and they that grow vp afterward about ye stalke are narrowe: it hath one or two straight holow stems, in the top of the sayd stems groweth a great thicke bushie eare, ful of litle long smal flowers, which before their opening, are lyke litle crooked hornes, & being openly spread, are parted into foure litle narrowe leaues, of a blewe colour, purple, gray, or white. The flowers fallen, there appeare many rounde litle huskes, ioyning one to another, lyke to the huskes or cuppes of the other Rampion, but much smaller. The roote is great white & full of sap, in fashion & taste like the roote of the other Rampion.

3. The Marians Violet, and the Gauntelet, described in the second booke, are also of the kindes of Rampions.

the Historie of Plantes.   597

❧ *The Place.*

1 The little Rampion groweth in feeldes and pastures of this Countrie vnder hedges and bushes.
2 The other Rampion groweth most comonly in wooddes, in clay groundes, and other fat, moyst and darke places.

❧ *The Tyme.*

1 The litle Rampion flowreth in June and July.
2 The other flowreth in May.

❧ *The Names.*

1 Rampion is called in Greeke γογγύλη ἀγρία: in Latine, Rapa syluestris, that is to say, wilde Rapes.
2 The first kinde, is nowe called of the writers in these dayes, Rapontium, Rapunculum, and Rapunculum paruum: in French, *Raiponce*, and *Petite Raiponce*: in high Douche, klein Rapuntzeln: in base Almaigne, Cleyn or ghemeyne Raponcelen: in Englishe, Rampions and the litle Rampion.

The seconde is lykewise a kinde of Rampion, or wilde Rapes.

❧ *The Nature.*

Rampion is of nature somewhat like the Turnep.

❧ *The Vertues.*

The Rampion eaten with vineger and salt stirreth vp appetite or meate lust, A and prouoketh vrine, especially when it is but a litle boyled or parboyled.

Rampions mengled with the meale of Lupines or Iuray, doth clense and B beautifie the face and all other partes of the body, being layde therevnto.

The iuyce of the stalkes & leaues of Rampions, especially of the lesser kind, C dropped into the eyes with womens milke, cleareth the sight.

Of Radishe. Chap.rlrvi.

❧ *The Kindes.*

THere be two kindes of Radish, the tame, and the wilde, whereof the tame or garden Radish is of two sortes, the one with a round roote, like the Nauew or garden Rape, and is not very comon in Brabant. The other hath a very long white roote and is the common Radish of this Countrie. To this may be ioyned a thirde kinde of garden or tame Radishe, with the blacke roote whiche of late yeeres hath ben brought into Englande, and now beginneth also to waxe common.

❧ *The Description.*

1 The common Radishe hath great brode rough leaues, muche clouen or deepely cut in vpon both sides, not muche vnlyke the Turnep leaues. The stalkes be round, with many flowers of a purple or wan colour, euery flower parted into foure small leaues, the whiche being fallen, there come in their steede, long, rounde, sharpe poynted huskes, sometimes as bigge as ones little fingar, wherein is inclosed a rounde browne seede. The roote of the one kind of garden Radishe, is of a foote or foote and a halfe long, white both without and within, and of a sharpe taste. The roote of the other is short, and as bigge as a Nauew, and of a stronger and sharper taste then the longer roote. The third roote is blacke without and white within, in taste like to the others.

The wilde Radishe hath leaues like the common Radishe, but smaller and fuller of cuttes or iagges. The stalke is of a foote and a halfe long, or more, vpon which grow many yellow flowers, and afterward smal huskes, wherein the seede, which is very smal, is inclosed. The roote is as bigge as ones finger, in taste very lyke to a young Radishe, but stronger.

598   The fyfth Booke of

Radicula satiua. **Garden Radish.**   Radicula syluestris. **Wilde Radish.**

❊ *The Place.*
1. They sow Radish in gardens, and it requireth to be new sowen euery yere.
2. The wilde Radish groweth alongst by ditches sides, both by standing and running waters.

❊ *The Tyme.*
1. The garden Radish is sowen most commonly in June and July, and that will serue to be eaten at winter, and it floureth in Aprill and Maye: and that whiche is sowen in Marche floureth the selfe same yere in May or June, and is nothing worth for to eate.
2. The wilde floureth in June, and shortly after it yeeldeth his seede.

❧ *The Names.*
1. The first kind is called of the Athenienses, and other Auncientes in Greke ῥαφανίς καὶ ῥαφανίς ἥμερος in Latine Radicula, and Radicula satiua: of some Raphanus: and in Shoppes, Raphanus minor: in Frenche, *Raue & Raueforte:* in high Douche, Rettich: in base Almaigne, Radijs: in Englishe, Radish.
2. The seconde kinde is called in Greeke ῥαφανίς ἀγρία: in Latine, Radicula syluestris: of some Radicula palustris: in French, *Raue sauuage:* or *Raifort d'eaue:* in high Almaigne, Wilder Rettich: in base Almaigne, Wilde Radijs, and Water Radijs: in English, wilde Radish, and water Radish.

❊ *The Nature.*
1. Radishe is hoate in the thirde degree, and drie in the seconde.
2. The wilde Radish is stronger, and more biting than the garden Radishe.

❊ *The Vertues.*
Radish is now eaten with other meates, as they vsed in times past, Neuerthelesse

thelesse it is rather medicine then meate or nourishment, as witnesseth Galen: for it giueth very litle or no nourishment to the body, seing that it is sharpe and biting vpon the tongue.

The young stemmes and tender croppes or buddes of Radish, may be lykewyse eaten with oyle and vineger being first boyled, and they nourishe better then the rootes, although in deede they yeelde but litle nourishment.

Dioscorides sayth, that the roote of Radish is pleasant to the mouth, but euill for the stomacke: for it engendreth belching and windinesse, with a desire to vomit.

The same eaten before meate, lifteth vp the meate, and taken after meate or meale, it suppresseth the same, causing it to descende and digest.

It is good to be eaten before meale to cause vomit, especially the barke thereof, the whiche taken with Oximel (that is honied vineger) hath the greater strength to stirre vp vomiting, and purgeth tough and slymie fleme, and quickneth the wit and vnderstanding.

The decoction or broth of Radishe, dronken prouoketh vrine, breaketh the stone, and driueth it foorth.

The same rypeth tough fleme, and grosse humours, wherwithhall the brest and stomacke is charged, and causeth them to be spet out: it is also good against an olde cough, and the brest that is stuffed with grosse humours.

Radishe is good agaynst the Dropsie, and for them that be liuer sicke, and for them that haue any payne or stopping of the raynes, and eaten with vineger and mustarde, it is good against the Lethargie, whiche is a drowsie and forgetfull sicknesse.

It is also good for such as are sicke with eating Tadestooles or Mushrumes, or Henbane, or other venome, and for them that haue the cholique and griping paynes in their bellyes, as Plistonicus, and Praxagoras writeth.

It moueth womens flowers, and as Plinie writeth, causeth abundance of milke.

The roote stamped very smal with vineger, cureth the hardnesse of the melt or splene, being layde therevpon.

The same with hony stayeth fretting, festering and consuming sores, also it is good against scurffenesse, and scales of the head, and filleth vp agayne bare places with heare.

The same with the meale of Darnel or Juray, taketh away blewe spottes of brused places, and al blemishes and freckles of the face.

The seede thereof causeth one to vomit vehemently, and prouoketh vrine, and being dronken with hony and vineger, it kylleth & driueth foorth wormes of the body.

The same taken with vineger, wasteth the melt or splene, and slaketh the hardnesse therof.

The same sodden in honied vineger, is good to be often vsed hoate for a gargarisme against the Squinancie.

The wilde or water Radish hath the same vertue, and in working is like to the garden Radish, but altogither stronger, and is singuler to prouoke vrine.

Of Raifort or mountayne Radish. Chap.xxxvij.

❀ *The Description.*

Mountayne Radish or Rayfort hath great brode leaues, in fashion lyke to the great Docke, called Patience, but greater and rougher. The stalkes be tender, short, and small, at the top whereof are small white flowers, and after them very smal huskes, wherein is the seede. The roote

roote is long and thicke of a very sharpe taste, and biting vpon the tongue: & therefore it is pound or stamped very small to be eaten with meates, and specially fishe in steede of Mustarde.

The Place.

It is founde for the most part planted in gardens, and where as it hath ben once set, it remayneth a long season without perishing.

The Time.

The great Raifort springeth vp in April, and floweth in June.

The Names.

Raphanus magnus.

This herbe is called of the later writers, Raphanus magnus, & Raphanus montanus: in Frenche, Grand Raifort, & Raphanus: in high Douche, Meerretich, and Kern: in Brabant most commonly Raphanus, of some also Merradijs. Some of the learned sort of the later writers doo take it for ῥαφανῶ, Raphanus, of the Auncient Atheniens, the whiche as some write, is an enimie to the vine, but this is not ῥάφανος of Theophrast, or of the other Greekes their successours: Who take for Raphanus, Brassica Romanorum, whiche is our common Colewurtes. Some others iudge it to be Thlaspi, wherof Cratenas writeth, but their opinion is nothing like to the trueth.

The Nature.

The great Rayfort is hoate and drie almost in the thirde degree, especially the roote, in whiche is the cheefest vertue.

The Vertues.

The roote of the great Rayfort is in vertue muche like to Radishe, but it is hoater and stronger, but not so muche troubling the stomacke.

The same being very small grounde or stamped, may be serued to men in steede of Mustarde, or other sawce to eate fishe withall: for being so taken it warmeth the stomacke, and causeth good appetite, and digesteth fish very wel.

It hath bene also founde by experience, that the great Raifort doth hinder the growing of the vine, and being planted neare it, causeth the vine to starue and wither away, the whiche thing the later Greeke writers, & not the Atheniens, do ascribe to Colewurtes.

Of Carrottes. Chap. xxxviij.

The Kindes.

1 There be three sortes of Carrottes, yellowe and red, whereof two be tame and of the garden, the thirde is wilde growing of it selfe.

The Description.

The Yellow Carrot hath darke greene leaues, al cut and hackt, almost like the leaues of Cheruil, but a great deale browner, larger, stronger, and smaller cut. The stemmes be rounde, rough without, and hollowe within:

the Historie of Plantes. 601

Staphilinus luteus. Yellow Carrot.   Staphilinus niger. Red Carrot.

Staphilinus sylvestris. Wilde Carrot.

within: at the highest of the stems growe great shadowie tuftes, or spokie toppes, with white flowers, & after them rough seede, in proportion not muche vnlike Annys seede. The roote is thicke and long, yellowe both without and within, and is vsed to be eaten in meates.

2   The red Carrot is lyke to the aforesayde in the cuttes of his leaues, and in stalkes, flowers, and seede. The roote is lykewise long and thicke, but of a purple red colour both within and without.

3   The wilde is not much vnlyke the garden Carrot, in leaues, stalkes, & flowers. sauing the leaues be a little rougher, and not so much cut or iagged, & in the middle of the flowrie tuftes, amongst the white flowers groweth one or two little purple markes or speckes. The seede is rougher, and the roote smaller and harder then the other Carrottes.

❧ *The Place.*

⁂ 3. The manured or tame Carrot is sowen in gardens.

Eee   The

3 ❧ The wilde groweth in the borders of feeldes, by high wayes and pathes, and in rough vntoyled places.

❀ *The Tyme.*

Carrotes doo flower in June and July, and their seede is rype in August.

❦ *The Names.*

Carrottes are called in Grecke ςαφυλῖνοι: and in Latine Pastinacæ.

1 The first kinde is called ςαφυλῖνος ἥμερος: and Pastinaca satiua: of the later writers, Staphilinus Luteus: in high Douche, Zam Pastiney, Zam Pastinachen, and Geel Ruben: in French, *Pastinade iaulne*: in base Almaigne, Geel Peen, Pooten, and Geel wortelen: in Englishe, Yellowe Carrottes.

2 The second kinde is also Staphilinus satiuus, and is called Staphilinus niger: in Frenche, *Pastenade rouge*: in high Douch, Rot Pastiny: in base Almaigne, Caroten: in English, Red Carrottes.

And these two garden Carrottes are in sight lyke to δαῦκος, Daucus, described by Theophraste lib.ix. Chap.xv. and lyke to the herbe whiche Galen in his syxth booke of Symples nameth δαῦκος ςαφυλῖνος, that is to say, Daucus Pastinaca.

3 The wilde kinde is called in Greeke, ςαφυλῖνος ἄγριος: in Latine, Pastinaca syluestris: in Shoppes, Daucus, as we haue declared in the seconde booke, of some it is also named Pastinaca rustica, Carota, Babyron, and Sicha: in Frenche, *Des Panaz*, or *Pastenade sauuage*: in high Douche, Wild Pastnach, or wild Pastenep, and Uogelnest: in base Almaigne, Uogels nest, and Croonkens cruyt: in Englishe, Wilde Carrot.

❀ *The Nature.*

The roote of Carrottes is temperate in heate and drynesse. The seede therof, especially of the wilde kinde is hoate and drie in the second degree.

❀ *The Vertues.*

A Carrot rootes eaten in meates, nourishe indifferently well, and bycause it is somewhat aromaticall or of a spicelyke taste, it warmeth the inward partes, being eaten moderately: for when it is to muche and to often vsed, it engendreth euill blood.

B The rootes of Carrottes, especially of the wilde kinde, taken in what sorte soeuer it be, prouoke vrine, and the worke of veneri. And therefore Orpheus writeth, that this roote hath power to encrease loue.

C Carrot rootes made into powder, and dronken with Meade or honied water open the stoppinges of the liuer, the melt or splene, the kidneyes & raines, and are good against the Jaunders and grauel.

D The seede of wilde Carrot prouoketh womens flowers, and is very good agaynst the suffocation and stiflinges of the Matrix, being dronken in wine, or layde to outwardly in manner of a pessarie or mother suppositorie.

E It prouoketh vrine, and casteth foorth grauel, and is very good agaynst the strangurie, and Dropsie, and for suche as haue payne in the syde, the belly and raynes.

F It is good against all venome, and agaynst the bitinges and stinginges of venemous beastes.

G Some men write, that it maketh the women fruitfull that vse often to eate of the seede thereof.

H The greene leaues of Carrottes bruised with hony and layde to, doo clense and mundifie vncleane and fretting sores.

I The seede of the garden Carrot, is in vertue lyke to the wilde Carrot, but nothing so strong, but the roote of the garden Carrot, is more conuenient and better to be eaten.

Of Parseneppes. Chap.xxxix.

✤ *The Kindes.*

There be two sortes of Parseneppes, the garden and wilde Parsenep.

Pastinaca vulgaris.	Elaphoboscum.
Garden Parsenep.	Wilde Parsenep.

✤ *The Description.*

1. THE garden Parsenep hath great long leaues, made of diuers leaues set togither vpon one stemme, after the fashion or order of the leaues of the Walnut or Ashe tree, whereof eache single leafe is broade or somewhat large, and nickt or snipt round about the edges, the stalke groweth to the height of a man, channell straked and forrowed, hauing many ioyntes, lyke the stalke or stemme of Fenill: at the toppe growe spokie tuftes, bearing yellowe flowers, and flat seedes, almost lyke the seede of Dyll, but greater. The roote is great and long, of a pleasant taste, and good to be eaten.

2. The wylde Parsenep, in leaues flowers and seede is much lyke the garden Parsenep, sauing that his leaues be smaller, & his stalkes slenderer, the roote is also harder and smaller, and not so good to be eaten.

✤ *The Place.*

1. The manured and tame kinde is sowen in gardens.
2. The wilde groweth in this Countrie, about wayes and pathes.

✤ *The Tyme.*

Parseneppes doo flower in June and July: and the garden Parseneppes are best and most meete to be eaten, the winter before their flowring.

The fyfth Booke of

❧ *The Names.*

1   The first kind is called in the Shoppes of this Countrie, Pastinaca, and the neather Doucheman borowing of the Latine do cal it Pastinaken: in Englishe lykewise Parseney: in Frenche, *Grand Cheruy*: in high Douche, Moren, and Zam Moren, and according to the same the base Almaignes call it, Tamme Mooren. Some take it for σίσαρον, Sisarum, others take it for a kind of Staphilinus, and Pastinaca. And in deede it seemeth to be σταφυλίνος, that is, Pastinaca, whereof Galen writeth in his viii. booke of Simples.

2   The wilde kinde is called in some Shoppes, Branca leonina, or Baucia: in Frenche, *Cheruy sauuage*. in high Douch, Wild Moren: in base Almaigne, Wilde Moren: it is called in Greeke ἐλαφόβοσκον, of some, as witnesseth Dioscorides, ἐλάφικον, νέφριον, ὀφιγίνιον, ὀφιοκτόνον, λύμη: in Latine, Elaphoboscum, and Ceruiocellus: in Englishe, Wilde Parsenep.

❧ *The Nature.*

Parsenep is hoate and drie, especially the seede whiche is hoater and drier then the roote. ❧ *The Vertues.*

The roote of the garden Parsenep eaté in meates, as the Carrot, doth yeeld more and better nourishment then Carrot rootes, and is good for the lunges, the raynes, and the brest.

The same roote causeth one to make water well, and swageth the paynes of the sydes, and driueth away the windinesse of the belly, and is good for such as be bruysed, squat, or bursten.

The seede of the wilde Parsenep is good agaynst all poyson, and it healeth the bitinges and stinginges of all venemous beastes, being dronken in wine. And truely it is so excellent for this purpose, that it is left vs in writing, that when the Stagges or rather the wild Hartes haue eaten of this herbe, no venemous beasts may annoy or hurt them.

Of Skirwurtes.   Chap. xl.

❧ *The Description.*

THe Skirwurt hath rounde stalkes, the leaues be cut and snipt about lyke the teeth of a sawe, diuers set vppon one stemme not muche vnlyke þ leaues of garden Parsnep, but a great deale smaller & smoother. The flowers grow in round tuftes of spoky toppes, and are of a white colour, and after that cometh a seede somewhat broade, (*as I reade in my copie*) but the Skirworte that groweth *in my garden which agreeth in al things els with the description of this Skirwort, hath a little long crooked seede of a browne colour, the which being rubbed smelleth pleasantly, somewhat lyke the seede of* Gith, *or* Nigella Romana, *or lyke the sauour of Cypres wood*. The rootes are white of a finger length, diuers hanging togither, and as it were growing out of one roate, of a sweete taste, and pleasant in eating.

Sisarum.

❧ *The*

the Historie of Plantes.

¶ The Place.

These rootes are planted in gardens.

❉ The Tyme.

These rootes are digged out of the grounde to be eaten in March, and the least or smallest of them are at the same time planted agayne, the which be good and in season to serue agayne the yeere folowing to be eaten. But whan they be left in the grounde without remouing, they flower and are in seede in July, and August.

❉ The Names.

This roote is called in Greeke σισαρον: in Latine Siser, and Sisarum : & some men cal it Seruillum, Seruilla, or Cheruilla: in French, *Petit Cheruy*: in high Douch Gerlin, Gierlin, & of some Zam Rapuntzel: in base Almaigne, Suycker wortelkens, and Serillen: in English, Skyrwurt, and Skirwit rootes.

❉ The Nature.

Skirwurtes are hoate and drie in the seconde degree.

❉ The Vertues.

The roote of Skirrets boyled, is good for the stomacke, stirreth vp appetite, and prouoketh vrine. A

The iuyce of the roote dronken with Goates milke, stoppeth the laske. B

The same dronken with wine, driueth away windinesse, and gripinges of C the belly, and cureth the hicket or yexe.

Of Garden Parsely. Chap.cli.

❉ The Description. Apium hortense.

Garden Parsely hath greene leaues, jagged, & in diuers places deepe cut, and snypt rounde about lyke the teeth of a sawe. The stalkes be rounde, vppon the whiche growe crownes or small spokie toppes, with flowers of a pale yellowe colour, and after them a small seede somewhat rounde, and of a sharpe or biting tast, and good smell. The roote is white and long as the roote of Fenill, but a great deale smaller. *❉ The Place.*

Parsely is sowen in gardens amongst wurtes and potherbes, and loueth a fat and fruitfull grounde.

❉ The Time.

The common Parsely flowreth in June, & his seede is ripe in July, a yere after the first sowing of it.

❉ The Names.

The common Parsely is called in Greke σέλινον καὶ σέλινον κηπαῖον: in Latine, Apium, and Apium hortense: in shoppes, Petroselinum, and the Douchmen folowyng the same, calleth it Petersilgen, or Peterlin: in neather Douchland it is called Petercelie: in Frenche, *Persil*, or *Persil de iardin*: in Englishe, Parsely, and garden Parsely.

The Nature.

Garden Parsely is hoate in the seconde degree, and drie in the thirde, especially the seede whiche doth heate and drie more then the leaues or roote.

The Vertues.

Garden Parsely taken with meates is very wholesome and agreeable to the stomacke, it causeth good appetite, and digestion, and prouoketh vrine.

The broth or decoction of the roote of garden Parsely dronken, openeth the stopping of the liuer, the kidneyes, and all interior partes, it causeth to make water, it driueth foorth the stone and grauell, and is a remedie agaynst all poyson.

The seede of Parsely is good for all the aforesayde purposes, and is of greater vertue and efficacie then the roote: for it doth not only open al stoppinges, & resist poyson, but also it dispatcheth and driueth away all blastinges and windinesse, and therefore it is put into al preseruatiues and medicines, made to expell poyson.

It is also good against the cough, to be mixt with Electuaries & medicines made for that purpose.

The leaues or blades of Parsely pound with the crōbes of bread (or barley flower) is good to be layde to against the inflammations and rednesse of the eyes, and the swelling of the pappes, that commeth of clustered mylke.

Of Marish Parsely, March or Smallache. Chap.xlij.

The Description.
Elioselinon.

Smallache hath shynyng leaues, of a darke greene colour, muche diuided, and snipt rounde about with small cuttes or natches, muche greater and larger then the leaues of common garden Parsely. The stalkes be rounde and full of branches, vpon the which grow spoky tufts or litle shadowy toppes with white flowers, which afterwarde bring foorth a very small seede, lyke to garden Parsely seede, but smaller. The roote is small and set full of hearie threddes or stringes.

The Place.

Smalllache groweth in moyst places that stande lowe, and is sometimes planted in gardens.

The Tyme.

Smallache flowreth in June and yeldeth foorth his sede in July and August, a yere after the sowing thereof, euen lyke to garden Parsely.

The Names.

Smallache is called in Greke ἑλιοσέλινον: in Latine, Apium palustre, & Paludapium, that is to say, Marrish Parsely: of some ἱδροσέλινον ἄγριον, Hydroselinon agriō, that is, wilde water Parsely, and Apium rusticum: in shoppes, Apium. in Frenche, De L'ache: in high Douche, Epffich: in base

the Historie of Plantes.

base Almaigne, Iouffrouw merck, and of some after the Apothecaries Eppe: in Englishe, Marche, Smallache, and Marrishe Parsely.

※ *The Nature.*

Smallache is hoate and drie lyke garden Parsely.

※ *The Vertues.*

The seede and rootes of Smallage, in working are much like to the rootes A and seede of garden Parsely, as Dioscorides writeth.

The iuyce of Smallache doth mundifie and clense corrupt and festered sores, B especially of the mouth and throte, mingled with other stuffe, seruing to the same purpose.

Smallache, as Plinie writeth, is good against the poyson of Spiders. C

Of Mountayne Parsely. Chap. xliij.

Oroselinon.

※ *The Description.*

Amongst the kindes of Parsely, the Auncientes haue alwayes described a kinde whiche they name Mountayne Parsely. And albeit it be nowe growen out of knowledge, yet we haue thought it good to describe the same, to the intent that nothing should fayle of that, whiche apparteyneth to the kindes of Parsely, also we hope that this Parsely shalbe the sooner founde, bycause we do here expresse it by name. This Parsely, as writeth Dioscorides, hath smal tender stalkes of a span long, hauing litle branches, with smal spokie tops or crownets, lyke to Hemlocke, but much smaller, vpon the which groweth a litle seede somewhat long, like to the seede of Commin, smal, of a very good and aromatical sent, and sharpe vpon the tongue.

※ *The Place.*

This kinde of Parsely groweth in rough vntopled places, and vppon high stonie hylles, for the whiche consyderation it is called Mountayne Persely.

※ *The Names.*

This Parsely is called in Greeke, ὀρεοσέλινον: in Latine, Apium montanum, that is to say in Englishe, Hyll Parsely, or Mountayne Parsely: in Frenche, *Persil de montaigne:* in high Douch, Berch Epffich: in base Almaigne, berch Eppe.

※ *The Nature.*

This Persely is of complexion, or temperament lyke the other, but a great deale stronger, as witnesseth Galen.

※ *The Vertues.*

The seede and roote of hill, or mountayne Parsely dronken in wine, prouo- A keth vrine and womens flowers.

The seede with great proffite is put into preseruatiues and medicines pre- B pared to prouoke vrine.

Of stone Parsely. Chp. xliiij.

※ *The Description.*

This Parsely hath meetely large leaues, seuered into sundrie partes, or diuers smal leaues, the which vpō eache side are deepe cut and fynely hackt or snipt round about. The stalkes be small of two foote long, vpō whiche growe small spokie toppes with white flowers, and after them a seede somewhat browne, not muche vnlyke the seede of the garden Parsely, but better, and of an aromaticall sauour, & sharper taste. The roote is small with many hearie stringes hanging thereat. ※ *The Place.*

This kinde which is the right Parsely, groweth plentifully in Macedonia, in rough stony and vntopled places, and also in some places of Douchland, that be lykewise rough stony and vntopled. The Herboristes of this Countrie doo sow it in their gardens.

Eee iiij ¶ The

❧ The Time.

This Parsely floureth in July, and yeldeth his seede in August.

❧ The Names.

This strange (but yet the true Parsely) is called in Greeke πετροσέλινον, and bycause it groweth plentifully in Macedonia, πετροσέλινον μακεδονικὸν, Petroselinon Macedonicon: in Latine, Petrapium, Apium saxatile, and Petroselinum, that is to say in English, Stone Parsely, in high Douch, Stein Epffich, or Stein Peterlin: in base Almagne, Steen Eppe. It is also called of some ignorant Apothecaries Amomū: in Brabant they cal it, Uremde Peterselie, that is to say, Strange Parsly, the whiche without all doubt is the true Parsely, called by the name of the place, where as it groweth most plentifully, Parsely of Macedonie: the French men call it *Persil de Roches* and *Persil vray*.

❧ The Nature.

This Parsely is hoate and drie almost in the thirde degree.

❧ The Vertues.

A The seede of this Parsely moueth womens flowers, prouoketh vrine, breaketh and driueth foorth the stone and grauell togither with the vrine.

B It dispatcheth and dissolueth all windinesse and blastinges, and easeth the gripinges of the stomacke and bowels: it is also very excellent against all colde passions of the sides, the kidneyes, and bladder.

C It is also put with great profite in preparatiues, and medicines ordayned to prouoke vrine.

Of great Parsely or Alexander. Chap.xlv.

❧ The Description.

THE great Parsely hath large leaues, broade, and somewhat browne, not muche vnlyke the leaues of garden Parsely, but muche larger and blacker, almost lyke the leaues of Angelica. The stalke is rounde of three or foure foote high, at the toppe whereof it bringeth foorth round spokie tufts or circles with smal white flowers, and

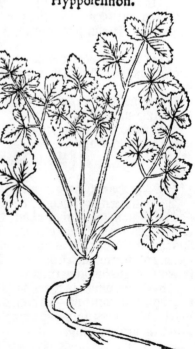

Petroselinum Macedonicum.

Hypposelinon.

and after them a blacke seede (somewhat long, and almost as bigge as the kernell of an Orenge) of a spicie sauour and bitterishe taste. The roote is white within, and blacke without, which being taken foorth of the ground, & broken in peeces putteth foorth a thicke liquer, or oylie gumme of a yellowishe colour, in taste very bitter and lyke to Myrrhe.

✻ The Place.

This Parsely groweth in some Countries in lowe shadowie places. The Herboristes of this Countrie do sowe it in their gardens.

✻ The Tyme.

This Parsely flowreth in July, and in August the seede is rype.

✻ The Names.

This Parsely is called in Greeke ὑπποσέλινον: in Latine, Equapium, and Olusatrum, of some σμύρνιον, Smyrnium: and ἀγριοσέλινον, that is to say, Apium syluestre: and of the later writers, Petroselinum Alexandrinum: in shoppes not without errour (Petroselinum Macedonicum) for it hath no similitude at all with the Parsely of Macedonie: in Frenche, Grand Persil, or Grand Ache, or Alexandre: in high Douche, Grosz Eppich, or Grosz Epffich: in base Almaigne, Groote Eppe: in English, Alexanders.

¶ The Nature.

This Parsely in temperament is hoate and drie, like the others.

✻ The Vertues.

The seede of the great Parsely dronken alone, or with honyed water, bringeth to women their desyred sicknesse, dissolueth windinesse, and grypinges of the belly, it warmeth the astonied members, or limmes taken with colde, and bruysing shiueringes or shakinges that come with extreame colde: and is good against the strangurie.

The roote of the great Parsely breaketh and driueth foorth the stone, causeth one to make water, and is good against the paines of the raines, and ache in the sides.

To conclude the seede of great Parsely is of lyke vertue to the seede of the garden Parsely, and in all thinges better and more conuenient then the common Parsely seede.

Of wilde Parsely. Chap.xlvi.

✻ The Description.

THE herbe which we (in folowing the auncient Theophrastus) do cal wilde Ache or Parsely, hath large leaues, al iagged, cut, and vittered, muche lyke the leaues of the wilde Carrot, but larger. The stalkes be rounde and holow of foure or fiue foote long, of a browne red colour next the grounde, at the top of them growe spokie rundels, or rounde tuffetes with white flowers, after them commeth a flat rough seede, not muche vnlyke the sede of Dyl, but greater. The roote is parted into two or three long rootes, the whiche doo growe very seldome downewardes, but most commonly are founde lying ouerthwarte and alongst, here and there, and are hoate and burning vpon the tongue. The whole herbe both stalkes & leaues, is full of white sappe, lyke to the Tithymales or Spurges, the whiche commeth foorth when it is broken or pluckt.

¶ The Place.

This herbe is founde in this Countrie in moyst places, about pondes, and alongst by diches, neuerthelesse it is not very common.

✻ The Tyme.

The wilde Parsely flowreth in June, and his seede is ripe in July.

The fyfth Booke of

Apium syluestre.

❧ The Names.

This herbe is called in Greeke, σίλινον ἄγριον, καὶ ὑδροσίλινον ἄγριον: in Latine, Apium syluestre, that is to say, wilde Parsely: in Frenche, Persil, or Ache sauuage: in high Douche, wilder Eppich, or Epffich: in base Almaigne, wilde Eppe. Of this herbe Theophrastus writeth, in his vij. booke the iiij. Chap. saying that þe wilde Parsely hath red stemmes. And Dioscorides in his third booke the lxvij. Chap. In some shops of this Countrie it is called Meum: & they vse the rootes of this Parsely in steede of Meum.

❧ The Nature.

The wilde Parsely and specially the roote thereof is hoate and drye in the thirde degree.

❧ The Vertues.

A   The roote of wilde Parsely holden in the mouth & chewed, appeaseth the rigour of the tooth ache, and draweth abundance of humours frō the braine.

Of water Parsly. Chap. lvij.

❧ The Kyndes.

There is founde in this Countrie two kyndes of this herbe, one great, the other smal, the which do differ but onely in figure, and that is long of the diuersitie of the places where as it groweth, for the one is changed into the other, whē as it is remoued frō one place to another. That is to say, that which groweth alwayes in the water, becommeth smal being planted vpon the lande or drie grounde: and on the contrarie, that whiche groweth vppon the drie land becommeth great, being planted in the water: so that to say the trueth, these two herbes are but all one, which doth not only happen to this herbe, but also to diuers others, that grow in the waters or moyst medowes.

❧ The Description.

1   THE great water Parsely, hath round, holow, smooth brittel stalkes, & long leaues made & fashioned of diuers little leaues standing directly one agaynst another, and spread abrode like winges, wherof each little leafe by it selfe is playne and smooth, and snipt about the edges lyke to a sawe. At the top of the stalkes growe litle spokie rundels with white flowers. The roote is ful of hearie threds, & it putteth foorth on the sides new springs, al the herbe is of a stronger & pleasanter sauour then any of the kindes of Parsely, & being bruised & rubbed betwixt the handes doth smell almost like Petrolium.

2   The lesser water Parsely, in sent is lyke to the abouesayde, his stalkes be lykewise holowe, but smaller. The leaues be not lyke to the greater, but drawing neare to the leaues of Cheruill, but yet more tenderer, and more mangled, pounsed or iagged. the smal flowers be white and do also growe in litle round tuftes, and shadowie or spokie circles growing thicke and neare throng togither. The roote is ful of threddy stringes, and doth lykewyse put foorth diuers

newe

the Historie of Plantes.

newe springes or branches, the whiche do stretche and spreade abroade vppon the grounde, and cleaue fast to the grounde taking roote here and there.

Lauer Crateuæ.
Great water Parsely.

Lauer minus.
Small water Parsely.
Iuncus adoratus.

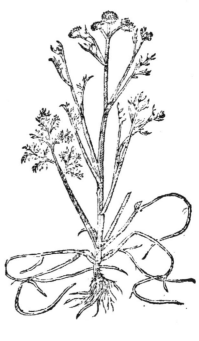

❀ *The Place.*

1   The greater water Parsely groweth in diches and pondes.
2   The lesser groweth in moyst medowes that stande lowe and waterie, not very farre from pooles, and standing waters, yet sometimes lykewise therein.

❀ *The Tyme.*

Water Parsely flowreth in June and July.

❀ *The Names.*

1   The first herbe shoulde seeme to be a kinde of that which is called in Greeke σιον: in Latine, Lauer, and Sium: in Frenche, *Berle*: in high Douche, Waller Epffich: in base Almaigne, Water Eppe, that is to say, Ache, or water Parsely. Turner and Cooper do call it, Sallade Parsely, Yellow water cresses, and Bell ragges.
2   The seconde is lykewise a kinde of Sium, as namely that whiche is called Iuncus odoratus. And yet it is not the vpright Iuncus, for this is but named for a likenesse vnto it, bycause that his stalkes be like rushes, and it hath a pleasant smell.

❀ *The Nature and Vertues.*

Without doubt this herbe is of complexion hoate and drie, and in vertue lyke to the other Sium.

Of Bastarde Parsley. Chap.xlviij.

The Description Caucalis.

Caucalis is a hearie herbe & somwhat rough, not much vnlike Carrot. The leaues be almost lyke the leaues of Coriander, but dismembred and parted into smaller iagges or frengis. At the toppe of the branches growe shadowy bushes or spoke rundels, with white flowers whose greatest blades or leaues are turned outwardes. The seede is long and rough like Carrot seede, but greater then Cominseede.

The Place.

This herbe is founde in this Countrie in the Meuze of Corne feeldes. *The Tyme.*

It flowreth in June, and within short space after the seede is ripe.

The Names.

This herbe is called in Greeke καυκαλίς: and also in Latine, Caucalis, of some Δαῦκ. ἄγρι.: that is to say, Daucus syluestris: vnknowen in shoppes: Cooper calleth it, Bastarde Parsley, and sayth it is an herbe lyke Fenill with a white flower and commeth of noughtie Parsly seede.

❧ *The Nature.*

Caucalis is hoate and drie.

The Vertues.

Caucalis prouoketh to make water like Daucus, wherunto Caucalis is much & muche like in vertues, as witnesseth Galen. Matthiolus attributeth many other excellent vertues to the herbe Caucalis, as you may see in his Commentaries vpon the seconde booke of Dioscorides.

Of Smyrnium. Chap.xlix.

The Description.

This herbe, as Dioscorides writeth, hath leaues lyke Parsley, and they bende downewarde, of a strong and pleasant Aromaticall smell with some sharpenes, and of a yellowish colour, greater and thicker then the leaues of Parsley: at the top of the stalkes grow smal spoky tuffets or rundels lyke Dyll, with yellowe flowers, and after them a small blacke seede, lyke the seede of Colewurtes, it is sharpe and bitter in taste like Myrrhe. The roote is of a good length, playne, and ful of iuyce, of a good smal and sharpe taste, blacke without and white within.

The Place.

Smyrnium, as saith Dioscorides, groweth in Cilicia vpon the mount Amanus,

the Hiſtorie of Plantes.

in ſtonie rough and drie grounde, but now ſome diligēt Herboriſtes do ſowe it in their gardens.

Smyrnion Dioſcorides.

❧ *The Names.*

This herbe is called in Greeke σμύρνιον: in Latine, Smyrnium: in Cilicia, Petroſelinon, and of ſome as Galen writeth, Hippoſelinon agreſte, that is wilde Alexander.

❧ *The Nature.*

Smyrnium is hoate and drie in the thirde degree.

❧ *The Vertues.*

A   The leaues and roote of Smyrnium doo appeaſe and mitigate the olde cough, and the hardneſſe in fetching breath, they ſtoppe the belly, and are very good agaynſt the bytinges and ſtingynges of venemous beaſtes, & agaynſt the payne to make water.

B   The leaues of Smyrnion layde to, doth diſſolue wennes and harde ſwellinges that be newe, it dryeth vp ſores, and exulcerations, and gleweth togither woundes.

C   The ſeede is good agaynſt the diſeaſes & ſtoppinges of the ſplene, the kidneyes, and the bladder, it moueth womens natural ſicknes, and driueth foorth the after birth or ſecondines.

D   To be dronken in wine it is good againſt the Sciatique, that is the diſeaſe of the hippes or hanche.

E   It ſtayeth the windineſſe and blaſtings of the ſtomacke, taken as is beforeſayde.

F   It prouoketh ſweat, and helpeth muche them that haue the Dropſie, and is good againſt the comming againe of ſuche feuers, as come by fittes.

Of Cheruill. Chap.I.

❧ *The Deſcription.*

Cheruill leaues are of a light greene colour, tender, brittel, much iagged and cut, ſomewhat hearie, and of good ſauour. The ſtalkes be rounde ſmal and holow, vpon the which grow rundels or ſpokie tuffetes with white flowers, and after them a long ſharpe browne ſeede. The roote is white and ſmall.

❧ *The Place.*

Cheruill is common in this Countrie, and is ſowen in al gardens amongſt wortes and potherbes.

❧ *The Tyme.*

The Cheruill that is ſowen in March or April flowreth bytimes, and deliuereth his ſeede in June and July, but that whiche is ſowen in Auguſt, abydeth the winter and flowreth not before April next folowing.

Fff   ❧ The

The Names.

This herbe is called of Columella, Chærophyllum, and Chærephyllum: of ye Apothecaries in our tune Cerefolium: in Frenche, Cerfueill: in high Douch, Korffelkraut, or Kerbelkraut: in base Almaigne, Keruel: in English, Cheruil, and Cheruel.

The Nature

This herbe is hoate and drie.

The Vertues.

A Cheruill eaten with other meates, is good for the stomacke, for it giueth a good taste to the meates, and stirreth vp meate lust.

B This herbe boyled in wine, is good for them that haue the stranguric, if the wine be dronken, and the herbe be layde as an implayster, vpon the place of the bladder.

C It is good for people that be dul, olde, and without courage, for it reioyceth and comforteth them, and increaseth theyr strength.

Cerefolium.

Gingidium.

Of Gingidium, in Spanish Visnaga. Chap. li.

The Description.

Gingidium, in leaues, flowers, knobby stalkes, and fashion, is lyke to the wilde Carrot, sauing that his leaues be tenderer, thicker set, and cut into smaller thrommes, or iagged frenges, and the stalkes be slenderer and playner, and the whole herbe is neyther rough nor hearie as the wylde Carrot is, but playne and smothe and of a bitter taste. The flowers be white and growe vppon spokie toppes or tuftes lyke the wilde Carrot: after them commeth the seede, the which being ripe, the stems with their spokie tuftes become stiffe, and waxe strong and harde, lyke small staues or little stickes, and the spokes or little stickes of the tuft of this herbe, the Italians and Spaniardes doo vse as toothpickes. For the whiche purpose it is maruelous good and excellent. The roote is white and bitter.

The Place.

This herbe groweth of his owne kind in Spayne, and as Dioscorides sayth, in Syria

Syria and Cilicia: it is not founde in this Countrie, but amongst certayne Herboristes. ❧ *The Tyme.*

This herbe flowreth in this Countrie in August, and deliuereth his seede in September. ❧ *The Names.*

This herbe is called in Grecke γιγγίδιοѵ: in Latine, Gingidium: in Syria, Lepidion: and of some also, as witnesseth Dioscorides, especially of the Romaynes, Bisacutum: therefore it is yet at this day called in Spayne, Visnaga: vnknowen in the Shoppes of Douchlande, Brabant, and this Countrie: it may be called Toothpicke Cheruill.

❧ *The Nature.*

Gingidium, as witnesseth Galen, is not so exceeding hoate, but it is drie in the seconde degree. ❧ *The Vertues.*

Gingidium eaten rawe or boyled with other meates, is very good for the stomacke, as Dioscorides sayth, bycause it is drie and comfortable, as Plinie writeth.

The same boyled in wine and dronken, is good for the bladder, prouoketh vrine, and is good against the grauell and the stone.

The harde stemmes of the great rundels or spokie tuftes are good to clense the teeth, bycause they be harde, and do easily take away such filth & baggage, as sticke fast in the teeth, without hurting the iawes or gummes: and bysides this they leaue a good sent or tast to the mouth.

Of Shepheardes Needel or wilde Cheruil. Chap.lij.

❧ *The Description.*   Scandix.

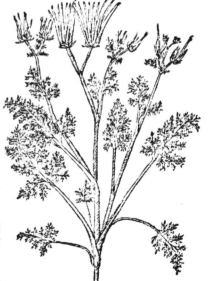

This herbe doth not muche differ in the quantitie of his stalkes, leaues and flowers from Cheruill, but it hath no pleasant smell. The stalkes be round and harde. The leaues be like the leaues of Cheruil, but greater and more finely cut, & of a browne grene colour. The flowers whiche be white grow vpon crownes or tuftes, after the whiche come vp long seedes, muche like to smal packe Needelles. The roote is white, and as long as ones finger.

❧ *The Place.*

Ye may finde it in this Countrie in fat and fertill feeldes.

❧ *The Time.*

Shepheardes Needell flowreth in May and June, and in shorte space after it yeeldeth his seede.

❧ *The Names.*

This herbe is called in Greke σκάνδιξ: in Latine, Scandix, herba scauaria, Acus pastoris, or Acula, bycause his seede is like to a needel, in Frenche, *Aguille de berger:* in Spanishe, Quixones: in base Almaigne, Naeldenkeruel: in Englishe, Shepheardes Needel, wilde Cheruel, and Needel Cheruill. ❧ *The Nature.*

Scandix is hoate and drie in the seconde degree.

❧ The Vertues.

Scandix eaten is good & wholesome for the stomacke and belly, & in times past hath bene a common herbe amongst the Greekes, but of smal estimation & value, & taken but onely for a wilde wurt or herbe. Aristophanes in times past by occasion of this herbe taunted Euripides, saying, that his mother was not a seller of wurtes or good potherbes, but onely of Scandix, as Plinie writeth.

The same boyled and dronken, openeth the stoppings of the liuer, kidneyes, and the bladder, and is good for all the inwarde partes, and bowels of man.

Of Myrrhis Cashes or Cares. Chap. liij.

❧ The Description.

Myrrhis.

Myrrhis in leaues and stalkes is somewhat lyke Hemlocke: it hath great large leaues, very much cut and iagged, & diuided into many partes, hauing sometime white speckles or spottes. The stalkes be rounde (somewhat crested) and two or three foote long: at the top of the stalkes growe rundels, or spokie tuftes with white flowers, and after them cometh a long seede. The roote is long & rounde, not much differing in taste and sauour from Carrot. The whole herbe, but especially the first leaues are beset with a soft downe or fine heare, and are in smell & sauour much lyke to Cheruil, and therfore it is called in base Almaigne wilde Keruel, that is to say, wilde Cheruel.

❧ The Place.

This herbe groweth of his owne kind in some medowes of Douchlande: in this Countrie the Herboristes doo sowe it in their gardens.

❧ The Tyme.

This herbe bloweth in May, and his seede is rype in Iune.

❧ The Names.

This herbe is called in Greeke μύῤῥις, and in Latine, Myrrhis, of some also μύῤῥα, Myrrha: and of the writers at these dayes, Cicutaria, bycause it doth somewhat resemble Hemlocke, whiche is named in Latine, Cicuta: in Frenche, Cicutaire, or Persil d'asne: in high Douche, Wilder Korssel: in base Almaigne, wilde Keruel: in Englishe, as Turner sayth, Cashes, or Cares, bycause Spinsters vse the stemmes both of this herbe and Hemlocke, for quilles and Cares, to winde yarne vpon, it may be called also wilde Cheruell, or mocke Cheruill.

❧ The Nature.

Myrrhis, especially the roote is hoate in the second degree, & of subtil partes.

❧ The Vertues.

The roote of Myrrhis dronken with wine prouoketh womens flowers, deliuereth the secondine & dead child, & purgeth & clenseth women after their deliuerance.

The same taken in lyke sort prouoketh vrine, & is good against the bitinges of feelde Spiders, and suche lyke venemous beastes.

The same boyled in the broth of flesshe, doth clense the breast from fleme and other corruption, and is very good for suche as are leane and vnlustie, or falling into consumption.

They

They say also that it is good to be dronken in wine, in the time of Pestilence, and that suche as haue dronken three or foure times of the same wine, shall not be infected with the plague.

Of Asparagus. Chap. liiij.

✤ *The Kindes.*

There be two sortes of Asparagus, the garden and wilde Asparagus.

Asparagus. Sperage.   Corruda. Wilde Sperage.

✤ *The Description.*

1. The Asparagus of the gardē at his first comming foorth of the ground, putteth foorth long shutes or tender stalkes, playne, rounde, without leaues, as bigge as ones finger, grosse, and thicke, hauing at the top a certayne bud or knop, the whiche afterwarde spreadeth abrode into many branches hanging lyke heares. The fruite groweth vpon the branches lyke round berries, first greene, and afterward of a yellowish red, euen of the colour of Coral, within that berrie is a blacke sede. The rootes be long and slender and interlaced or wouen one in another.

2. The wilde Asparagus in his first springes and fruite, is muche lyke to the garden Sparagus, the rest is altogither rough and pricking, for in steede of the long soft heares, wherewithal the garden Asparagus is couered, this hath nothing els but thornes, very smal, hard, short, & prickley, wherwithal the braches are furnished.

✤ *The Place.*

1. The manured or tame Asparagus groweth in Burgundie and some other Countries as in Almaigne, in stony places, where as is good earth, and fatte ground: in this Countrie it is planted in the gardens of Herborists.

2. The wilde kinde groweth in certayne places of Italy, and throughout all Languedoc.

618   The fyfth booke of

❀ *The Tyme.*

The bare ſtalkes or firſt tender ſpringes of Aſparagus ſhute vp in Aprill, at what time they be boyled & eaten in ſalade, with oyle, ſalt, & vineger. The fruit is ripe in Auguſt.   ❀ *The Names.*

1 Garden Aſparagus is called in greke ἀσπάραγος: in Latine, Aſparagus, & in ſhops Sparagꝰ: in high douch, Spargē: in baſe Almain Coraelcrupt: in engliſh ſperage.

2 The wild Aſparagus is called in Greke ἀσπάραγος πετραῖος, ἢ μυάκανθα: in Latine Aſparagus ſylueſtris, and Curruda: vnknowen in the ſhoppes of this Countrie.

¶ *The Nature.*

Aſparagus, eſpecially the rootes are temperate in heate and cold, taking part of a certaine dryneſſe.   ❀ *The Vertues.*

The firſt tender ſprings of Aſparagus parboyled & eaten with oyle & vineger, A prouoke vrine, and are good agaynſt the ſtrangurie, and they ſoften the belly.

The decoction or broth of Aſparagus, by it ſelfe (or with Ciche Peaſon) B dronken openeth the ſtoppinges of the liuer and kidneyes: and alſo it is good againſt the Iaundice, ſtopping of the water, ſtrangury, and the grauel & ſtone.

Some ſay, that if it be taken in the ſame maner, it eaſeth and conſumeth the C Sciatica and payne of members out of ioynt.

The roote boyled in wine is good for thē þ are bitten of any venemous beaſt. D

Of Senuie or Muſtarde.   Chap.lv.

❀ *The Kyndes.*

THere be two ſortes of Senuie, the tame & the wilde, wherof alſo the tame or garden Senuie is of two ſortes: the one with a great white ſeede, the other hauing a litle browne ſeede.

Sinapi hortenſe. Muſtarde ſeede.   Sinapi ſylueſtre. wilde Muſtarde ſeede.

❀ *The Deſcription.*

1 THe tame white Muſtarde hath great rough leaues, at the firſt not much vnlike þ leaues of Turnep, but after the firſt leanes there folowe other þ are ſmaller & more iagged, growing vpon the ſtalkes whiche be arie &

three

the Historie of Plantes.

three or foure foote long, & diuideth it self into many braches alongest ẏ which grow yellowish flowers, & after them long hearie huskes or coddes, wherin is the sede which is round & pale, greater then Rape sede, in taste sharpe & hoate.

2 The seconde kinde of tame Mustarde with the browne seede, whiche is the blacke Mustarde & common Senuy, is like to the aforesayd in leaues, stalkes and growing. The flowers be yellow. The sede is browne, smaller then Rape seede, and in taste also sharpe and hoate.

3 The wilde kind hath great large leaues, very much iagged and rough with stalkes like the other, but it groweth not so high. The flowers be of a pale yellow, fashioned like a crosse, after which commeth the seede which is reddish, enclosed in long round huskes. ❧ *The Place.*

1.2. Mustarde or Senuie is sowen in gardens and feeldes.

3 The wilde kind groweth of his owne nature, in stonie places, and waterie groundes, and alongst the high wayes. ❧ *The Tyme.*

The Mustarde and Charlock do flower in June and July, and during the same time, they yeelde their seede. ❧ *The Names.*

Mustarde is called in Greeke σίνηπι: in Latine, Sinapi, in shoppes Sinapis and Sinapium: in high Douche, Seuff: in base Almaigne, Mostaert: in English, Senuie and Mustarde.

1 The first kind is called σίνηπι κηπαῖον, Sinapi hortésè: & in ẏ shops of this Countrie Eruca: in Frenche, *Blanche Monstarde.* in high Douche, Weisser Seuff: in base Almaigne, Wit Mostaert: in English, White Senuie, & white Mustarde sede.

2 The seconde is also counted for a kind of Mustarde, and of the later writers is called Sinapi commune: in Frenche, *Seneue de iardin, ou Moustarde noire* in hygh Douch, Zamer Seuff: in base Almaigne, Ghemeyne Mostaert: in English, the common Senuie or Mustarde.

3 The wilde kinde is called of the later writers σίνηπι ἄγριον: Sinapi syluestre: in French, *Sanele:* in high Douch, Wilder Seuff: in base Almaigne, Wilden Mostaert. ❡ *The Nature.*

The Mustarde, especially the seede which men cal Senuie, is hoate and dry, almost in the fourth degree. ❧ *The Vertues.*

Senuie bruysed or ground with vineger is a wholesome sawce meete to be A eaten with harde & grosse meates, either flesh or fishe: for it helpeth their digestiō, and is good for the stomacke to warme the same, and prouoketh appetite.

It is good to be giuen in meates, to such as be short winded, & are stopped in B the breast: for it ripeth and causeth to cast foorth tough fleme, that troubleth or loadeth the stomacke and breast.

Mustarde seede chewed in the mouth draweth downe thinne fleme from C the head and brayne: appeaseth toothache: it hath the same vertue, if it be mingled with Meade, and holden in the mouth, and gargled.

They vse to make a good gargarisme with hony vineger & Mustarde seede, D against the tumours and swelling of the vuula and the Almondes about the throte, and roote of the tongue.

For the same intent, especially when suche tumours are become harde and E waxen old, they make a necessarie and profitable gargarisme with the iuyce of Mustarde seede & Meade, for it slaketh, wasteth, or consumeth such swellings and hardnesse of the Almondes and throte.

Senuie dronken with Hydromel or honyed water, is good agaynst the terrour and shaking of agues, prouoketh the flowers and vrine. F

The same sede snift vp into the nosethrilles, causeth one to sneese, helpeth the G that haue the falling sicknesse, and women that haue the strangling of the mother,

Fff iiij

The fyfth Booke of

ther, to waken them vp agayne.

H ❧ The same pound with figges, & layd to in manner of a playster, taketh away the homming noyse & ringing of the eares or head, & is good against deafnesse.

I ❧ The iuyce of the same dryed in the Sonne, and afterwarde delayed with hony cleareth the sight, and taketh away roughnesse of the eye browes.

K ❧ They make an emplayster with the same & figges, very good for to be layd vpon the heades of suche as are fallen into the Lethargie or Drowsie euill, and cannot waken them selues: it is likewise good against the Sciatica or payne of the hanche, the hardnesse of the splene or melte: and against the Dropsie, to be layde as an emplayster to the bellyes of suche as are greeued therewithall. To be short this emplayster is of great force agaynst all colde greefes and diseases, especially when they are waxen old, for it doth warme and bring heate agayne into the diseased partes, it digesteth colde humours and draweth them foorth.

L ❧ Senuie mingled with hony and newe grease, or with a Cerote made of waxe, cureth the noughtie scurffe or scales in the head whiche cause the heare to fall of, it scoureth the face from all freckles and spottes, and taketh away the blewe markes that come of brusing.

M ❧ If it be layde to with vineger, it is good for Lepres, wilde scabbes and running scurffe, and is good agaynst the bitinges of Serpentes.

N ❧ The parfume or sauor therof driueth away al venom, & venemous beastes.

Of Rapistrum, or Charlock. Chap.lvi.

❧ *The Description.* Rapistrum. Charlock.

Charlock hath great rough brode leaues, lyke the leaues of Turnep, the stalkes be rough & slender most commonly of a foote long, with many yellow flowers, coddes and seede like ye Turnep, but hoate or biting sharpe lyke to Mustarde seede. The roote is small and single. ❧ *The Place.*

Charlocke groweth in all places alongst the wayes, about old walles and ruynous places and oftentimes in the feeldes, especially there, where as Turneppes and Nauewes haue ben sowen, so that it shoulde seeme to be a corrupt & euill weede, or enimie to the Nauew.

❧ *The Time.*

Charlocke flowreth from Marche or April vntill midsomer, and the seede also rypeth from time to tyme in the meane space. ❧ *The Names.*

This herbe is called of the later writers Rapistrum, and of some also Synapi syluestre: in Frenche, Velar, or Tortelle: in high Douche, Hederich: in base Almaigne Hericke: in Englishe, Charlock.

❧ *The Nature.*

Charlock, and specially the seede is hoate and drie in the thirde degree, and of temperament lyke Senuie. ❧ *The Vertues.*

A ❧ This herbe of the later Physitions, is not vsed in medicine, but some with
this

this seede do make Mustarde, as with Senuie, the whiche they eate with meate in steede of Mustarde: whereby it is euident that the seede of this herbe doth not much differ from Senuie in vertue and operation, and that it may be taken in steede thereof, although it be not al thing so good, and therfore it was reckoned of Theophrast and Galen amongst those seedes, wherewithall men vsed commonly to prepare and dresse their meates.

Of Rockat. Chap.lvij.

❧ *The Kindes.*

Of this herbe be found two kindes, the one tame which is the common Rockat most vsed, the other is wilde.

Eruca. **Rockat.**   Erucasyluestris. **Wild Rockat.**

❧ *The Description.*

1. The tame Rockat hath leaues of a browne greene colour, very much and deepely iagged or rather torne vpon both sides, of a hoate biting taste, the stalkes be a foote lõg or somwhat more: vpon which grow many yellowe flowers, and after them little coddes, in whiche the seede is contayned. The roote is long with hearie stringes, and doth not lightly dye in winter, but putteth foorth newe stemmes euery yere.

2. The wylde kinde is muche lyke to the garden Rockat, sauing that it is altother smaller, especially the leaues and flowers, whiche be also yellower, and do bring foorth small coddes.

3. Bysides these two kindes, a man shall fynde in the gardens of this Countrie another kynde of Rockat, called Rockat gentle, or Romayne Rockat, in leaues and flowers much lyke to the wilde Mustarde, wherof we haue before spoken, sauing that his leaues be not so rough nor hearie, and are more conuenient to be beaten.

❧ *The*

¶ The Place.

1   The garden Rockat is planted in gardens, and is also found in this Countrie in certayne rude vntoyled and stonie places, and vpon olde broken walles.
2   The wild Rockat is found also in stony places about high wayes & pathes.

❀ The Tyme.

Rockat flowreth cheefely in Iune and Iuly.

¶ The Names.

Rockat is called in Greeke εὔζωμον: in Latine, Eruca: in Frenche, Roquette: in Douche, Roket: in base Almaigne, Rakette.

1.3.   The first and also the third kind is called Eruca satiua, & hortensis in French, Roquette domestique or cultiueé: in base Almaigne, Roomsche Rakette: in English, Garden or tame Rockat, and Rockat gentil.
2   The wilde is called Eruca sylueſtris, that is to say, wilde Rockat: in base Almaigne, wilde Rakette.

❀ The Nature.

Rockat is hoate and drie in the thirde degree.

❀ The Vertues.

Rockat is a good Salade herbe to be eaten with Letuce, Purcelayne, and other like colde herbes, for being so eaten it is good and wholesome for the stomacke, & causeth that such colde herbes do not hurt the stomacke: but if Rockat be eaten alone, it causeth headache, and heateth to much, therfore it must neuer be eaten alone, but alwayes with Letuce or Purcelayne.

The vse therof stirreth vp bodyly pleasure, especially of the seede, also it prouoketh vrine, and helpeth the digestion of the meates.

The seede thereof is good against the poyson of the Scorpion, & Shrowe and suche like venemous beastes.

The seede layd to with hony, taketh away freckles, lentils, & other faultes of the face, also it taketh away blacke and blewe spottes and scarres, layde to with the gawle of an Oxe.

Men say, that who so taketh the seede of Rockat before he be beaten or whipt, shalbe so hardened, that he shall easily endure the payne, according as Plinie writeth.

The roote boyled in water, draweth foorth shardes and splinters of broken bones being layde thereupon.

Of Tarragon or biting Dragon. Chap.lviij.

❀ The Description.

Tarragon hath long, narrow, darke, grene leaues, in taste very sharpe, and burning or biting the tongue almost like Rockat, not muche vnlyke the leaues of common Hysope, but muche longer, and somewhat larger. The stalkes be rounde of two foote hygh, parted into many branches, vpon whiche growe many small knoppes or litle buttons, the which at their opening shewe many small flowers, as yellowe as golde intermingled with blacke. They being past commeth the seede. The roote is long and small, very threddy creeping alongst the grounde hither and thither, & putteth foorth perely here and there newe stalkes and springes. Ruellius in his second booke Chap.xcvj. saith, that this herbe cometh of Lineseede put into a Radish roote, or within the scale of the sea Onyon, called Scylla in Latine, and so set into the grounde and planted, and therefore he saith, it hath part of both their natures, for it draweth partly towardes vineger, and partly towardes salt, as may be iudged by the taste.

the Historie of Plantes. 623

❊ *The Place.*

Tarragon is planted in gardens, but yet it is not very common.

❊ *The Tyme.*

Tarragon abideth greene, from the moneth of Marche, almost to winter, but it floureth in July.

❊ *The Names.*

This herbe hath not bene written of by any learned man before Ruellius tyme, neyther is it yet wel knowen, but in some places of Englande, France, and certayne Townes of this Countrie, as Anwarpe, Brucelles, Malines, &c. where as it was first brought out of France. And therfore it hath none other name, but that whiche was geuen first by the Frenchemen, who called it *Targon*, and *Dragon*: and according to the same it is called in Latine, Draco: and of some Dracunculus hortensis: that is the litle Dragon of the garden: it is also called in Englishe, Tarragon, whiche shoulde seeme to be borowed from the Frenche, neuerthelesse it was allowed a Denizon in England long before the time of Ruelius writing.

Draco.

❊ *The Nature.*

All this herbe is hoate and burning in the mouth and vpon the tongue, whereby it is certayne that it is hoate and dry in the thirde degree, and in temperature muche lyke to Rockat.

❊ *The Vertues.*

This herbe is also good to be eaten in Salade with Letuce, as Rockat, for it correcteth the coldenesse of Letuce and suche lyke colde herbes. Moreouer where this herbe is put into the Salade, there needeth not much vineger nor salt, for as Ruelius writeth, it is sharpe and salt ynough of it selfe.

Of Cresses. Chap.lix.

❊ *The Description.*

Garden Cresses haue small narrowe iagged leaues, of a sharpe burnyng taste: the stalkes be rounde of a foote long, and bring foorth many small white flowers, and after them little rounde flat huskes, within which the seede is contayned of a browne reddish colour.

❊ *The Place.*

Cresses are commonly sowen in all gardens of this Countrie.

❊ *The Tyme.*

Cresses that are timely sowen, bring foorth their seede bytime, but that whiche is later sowen, bringeth foorth flowers and seeede more lately.

❊ *The Names.*

This herbe is called in Greke καρδαμον: in Latine, Nasturtium of some later writers Cressio: in Frenche, *Cresson alnoys*, or *Nasitort*: in high Douche, Kresz, and Garten kresz: in base Almaigne, Kersse: in English, Cresses, Towne kars, or Towne Cresses.

The fyfth Booke of

Nasturtium.

❊ The Nature.

Cresses are hoate and dry almost in the fourth degree, especially the seede, and the herbe, when it is drie: for being but yet greene they do not heate nor dry so vehemently, but that they may be eaten with bread, as Galen saith.

❊ The Vertues.

A Cresses eaten in Salade with Letuce, is of vertue like to Rocket, a good amongst cold herbes, for eaten alone it ouerturneth the stomacke, and hurteth the same, bycause of his great heate and sharpenesse.

B The seede looseth the belly, and killeth, and driueth foorth wormes, it diminisheth the melte, prouoketh the flowers, and putteth foorth the secondine and the dead childe.

C It is good against Serpentes and venemous beastes, and the parfume of the same causeth them them auoyde.

D The same taken with the broth of a pullet or chicken, or any other lyke moyst meates, doth ripe and bring foorth tough fleme, wherewithall the breast is combred or charged.

E The same laide to with hony, cureth the hardnesse of the melte, scoureth away scuruinesse, and fowle spreading scabbes, dissolueth colde swellinges, and keepeth the heare from falling of.

F Being layd to with hony & vineger, it is good against the Sciatica, & payne in the hippes, and the head ache that is olde, and against all olde colde diseases.

G To conclude the seede of Cresses is in vertue very lyke Senuie, as Galen writeth.

Of Water Cresses. Chap.lx.

❡ The Kindes.

Water Cresses are of two sortes, great and small.

❊ The Description.

1 The great water Cresse hath rounde holowe stalkes of a foote and a halfe long, with long leaues made of diuers other litle roundish leaues standing togither vpon one stemme. The flowers be small and white, growing at the toppe of the branches alongst the stemmes, after whiche folow smal coddes or huskes, within which is the seede, which is small and yellowe. The roote is white and full of hearie laces or stringes.

2 The lesser water Cresse, at the first hath rounde leaues, then commeth the rounde stalke of a foote long, vpoyn the whiche growe long leaues iagged on both sides, almost like the leaues of Rocket. The flowers growe at the highest of the stalkes, of colour somewhat white, or of a light Carnation, after whiche come small huskes, wherein the seede lyeth.

❊ The Place.

1 The greater watercresse groweth in diches, standing waters, and fountaynes or springes.

2 The lesser watercresse groweth in moyst groundes and medowes that are

ouer-

the Historie of Plantes.

Sium Nasturtium aquaticum.
Great Watercresse.

Sisymbrium alterum cardamine.
Small watercresse.

ouerwhelmed and drenched with water in the winter season, also in standing waters and diches.

※ *The Tyme.*

1 The great watercresse floureth in July and August.
2 The lesser floureth in May, and almost vntill the ende of sommer.

※ *The Names.*

1 The first kinde is called in high Douche, Braun kerſʒ: in base Almaigne, waterkerſſe: in Shoppes also Nasturtium aquaticum and seemeth very wel to be that Sium of the which Cratenas maketh mention, in English, Water kars, and Water Cresse.

2 The seconde kinde is called in Greeke σισύμβριον ἕτερον, ἢ καρδαμίνη: in Latine, Sisymbrium alterum cardamine: of some also Sium: in Frenche, *Passerage sauuage*, or *Petit Cresson aquatique*: in high Douch, Gauchblum, wilder kreſʒ, and Wisen kreſʒ: in base Almaigne, Coeckoecxbloemen, and Cleyn Waterkerſſe: of the Herboristes, Flos cuculi, of some Nasturtium aquaticum: in Englishe, the lesser Watercresse, and Cocow flowers. This is no Iberis as some haue deemed it.

※ *The Nature.*

These two herbes are hoate and drie in the seconde degree.

※ *The Vertues.*

Water Cresse is good to be eaten in Salade, either by it selfe or with other herbes, for it causeth one to make water, it breaketh and bringeth foorth the grauel and stone, and is good for suche as haue the stranguric, and agaynſt all stoppinges of the kidneyes and bladder.

The lesser watercresse taketh away spottes and freckles from the face and al such blemishes, if it be laid therto in the euening & taken away in the morning.

The wilde Passerage boyled in lye, driueth away lyce, if the head or place where they be, are washed therwithall.

The kine feeding where as store of the wild Passerage or Cockow flowers growe, giue very good milke wherewithal is made excellent sweete butter.

Of winter Cresses. Chap.lxi.

❧ *The Description.* Pseudobunium. Barbaræa.

THIS herbe hath greene grosse leaues, broade, smooth, and somewhat round, not muche vnlyke the leaues of Smallage, or garden Rape, but greater and larger then Smallage leaues. The stalkes be rounde & full of branches aboue bringing forth many litle yellow flowers, and after them long rounde coddes, wherin is enclosed a litle seede. The roote is thicke and long.

❧ *The Place.*

This herbe groweth in the feeldes, & somtime also in gardens of potherbes, & places not toyled or husbanded.

❧ *The Tyme.*

This herbe is greene most commonly all the winter, but it flowreth & seedeth in May and June.

❧ *The Names.*

This herbe is called in Douche S. Barbarakraut: and according to the same in Latine, Sanctæ Barbarę herba: we haue named it Barbaræa the Frenchmen, *Herbe de S. Barbe:* in some places of Brabant they call it Steencrupt, bycause it is good against the stone and grauel: in Holland and other places Winterkersse, bycause they do vse to eate of it in the winter time in salades, in steede of Cresses, & therefore it is called Nasturtium, or Cardamum hybernum. This seemeth to be ψευδοβούνιον, Pseudobunium of Dioscorides: for surely this is not Sideritis latifolia, or Scopa regia, as some do take it: *Herbe Sainbarbe* ❧ *The Nature.*

This herbe is hoate and drie in the seconde degree.

❧ *The Vertues.*

Herbe S. Barbe is a good herbe for salade, and is vsed in the winter season for Salades like Cresses, for the whiche purpose it doth aswell as Cresses or Rockat.

It doth mundifie and clense corrupt woundes and vlcers, and consumeth dead flesh that groweth to fast, being either layde thereto, or the iuyce thereof dropped in.

Also it is certaynely proued by experience, that the seede of this herbe causeth one to make water, driueth forth grauel, and cureth the strangurie, which vertues be lykewise attributed to Pseudobunium.

the Historie of Plantes. 627

Of Thlaspi. Chap.lxij.
¶ The Kyndes.

There be foure kindes of wilde Cresse, or Thlaspi, the which are not muche vnlyke one another, nor vnlyke cresse in taste.

Thlaspi. The first kinde of Thlaspi.   Thlaspi alterum The seconde kinde of Thlaspi, or treacle Mustarde.

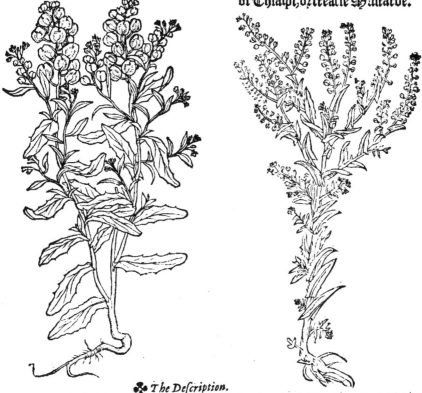

❀ *The Description.*

1. The first kinde of Thlaspi hath long narrowe leaues. The stemmes be hard and pliant or tough, of a foote and a halfe long, vpon which grow litle branches bringing foorth smal white flowers, and afterward flat huskes and round, with a certayne clouen brim, or edge all aboue at the vpmost part of eache huske, which chappe or clift, causeth the huske to resemble the hart of a man, within the sayde huskes is founde small seede the whiche is rounde, eger, and burning the mouth, and in the ende it tasteth and smacketh of garlike or onyons, and is of a brownish colour.

2. The seconde kinde hath long leaues and meetely large, longer and broader then the first, & iagged or cut about the edges. The stalkes be round of a foote long diuided into sundry smal branches, vpon which grow smal huskes, almost lyke the seede of Shepheardes pouche, within which huskes is likewise found a sharpe biting seede.

3. The thirde kinde of Thlaspi hath smaller stalkes and leaues then the aforesaid and hath more smal slender branches, vpon which grow flowers and seede lyke to the other, but altogither smaller.

4. The fourth kinde hath long, small, rough, white greene leaues, the stalkes be of a wooddy substance, round and tough or pliant, vpō the same grow smal

Ggg ij   white

white flowers, the whiche past, it bringeth foorth broade huskes or seede vessels, hauing a brownishe kinde of seede, very hoate in taste lyke to the seede of Cresses. ❦ *The Place.*

These herbes do grow in feeldes, and all alongst the same, in vntoyled places about wayes, & there is store growing togither, the one kinde in one place, and the other in another.

❦ *The Tyme.*

These herbes doo flower and are in seede at sommer from Maye to August.

❦ *The Names.*

This herbe is called in Greke θλάσπι, ἤ θλασπίδιον, καὶ σίναπι ἄγριον: in Latine, Thlaspi, Capsella, and Scandulaceum, of some also Myitis, Bytron, Dasmophon, Myopteron: in high Douche, Wilder Cretz: in Frenche, *Seneue sauuage*: in base Almaigne, Wilde Kersse: it may be also called in Englishe, Thlaspi.

1 The first kinde is the right Thlaspi of Dioscorides: and is called in base Almaigne, Uiselcrupt: and of some in Latine, Scordothlaspi: that is to say, Garliketh thlaspi.

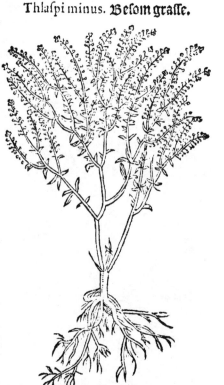

Thlaspi minus. Besom grasse.

2 The second kind is called of the later writers, Nasturtium rusticum, & Sinapi rusticum: in high Douche, Baurn seuff, or Baurn kretz, & the neather Douchmen in folowing the same call it, Boeren mostaert, or Boeren kersse, that is to say, Seneui, or Carles Cresse: or Churles Cresse: Turner calleth Thlaspi, treacle mustarde, Bowers mustarde, or Dishe mustarde: but I thinke it best next to Thlaspi, whiche is the Greeke name to call it Churles mustarde, both bycause of the strong and violent nature of this noughtie plant, as also in respect of the Boures, who began to be more mischieuous to the state of their Country, then this herbe is to mans nature.

3 The thirde kinde is called Thlaspi angustifolium, & Thlaspi minus: in high Douche, Bysemkraut: in base Almaigne Bessemcrupt: that is to say, Bessem weede, or y̆ herbe seruing for Bysoms. Turner calleth this Iberis Dioscoridis:

4 The fourth without all doubt is a kind of Thlaspi, but it hath no other particuler name. ❦ *The Nature.*

Thlaspi, especially the seede thereof, is hoate and drie almost in the fourth degree. ❦ *The Vertues.*

The seede of the first Thlaspi eaten, purgeth choler, both vpward & downewarde, it prouoketh womens flowers, and breaketh inwarde impostumes.

The same as a Clyster powred in at the fundement, helpeth the Sciatica. And it is good for the same purpose to be layde vppon the greeued place, lyke Mustarde seede. ❦ *The Danger.*

Seing the seede of Thlaspi is very hoate, and of a strong or vehement working, insomuche that being taken in to great a quantitie, it purgeth or scoureth euen vnto blood, and is very hurtful to women with child, therfore it may not rashly be giuen or minished inwardly.

Of

the Historie of Plantes.

Arabis siue Draba.

Of Candy Thlaspi. Chap.lxij.

❦ The Description.

This herbe groweth with narrow leaues, to the length of a foote, almost lyke to the leaues of Iberis. The flowers grow at the top of the plant in rounde tuftes lyke the flower of Elder, of a white or light Carnation colour: after them come flat huskes fashioned lyke the huskes of of the other Thlaspi, but muche smaller, within the whiche is contayned a seede of a sharpe biting taste, lyke the seede of the other Thlaspi.

❦ The Place.

This herbe is not found in this Countrie, but in the gardens of some diligent Herboristes.

❦ The Time.

It flowreth in Maye, and shortly after the seed is ripe.

❦ The Names.

This herbe is called in Greeke Ἀράβη: in Latine, Arabis & Draba: of Plinie as some men holde, Dryophonon: of þ Herboristes at these dayes, Thlaspi de Candie, vnknowen in shoppes.

❦ The Nature.

Candie Thlaspi is in complexion lyke to the other Thlaspies.

❦ The Vertues.

A They vse to eate the dryed sede of this herbe with meates in steede of Pepper, in the Countrie of Capadocia, as Dioscorides writeth.

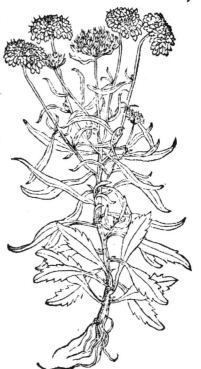

Of Erysimon Dioscorides. Chap.lxiij.

❦ The Description.

Erysimon hath lóg leaues deepely rent, & iagged vpō both sides, not muche vnlyke the leaues of of Rockat gentle or Romayne Rockat, or wilde Mustarde. The stalkes be smal, slender, and plyant, and wil twist and winde lyke Ozier withie, vppon the same stalkes or braunches grow many yellow flowers, & after thē come litle slender huskes, wherin also is a seede of a sharpe bitingt aste: the roote is long and thicke, with many smal strings or heariethreds.

❦ The Place.

This herbe groweth in all places of

Erysimon Dioscorides.liro.

Ggg iij   this

this Countrie alongſt the wayes, and in vntoyled ſtonie places.

❋ *The Time.*

Eryſimon flowreth very plentifully in this Countrie, in the moneth of June and July.

❋ *The Names.*

This herbe is called in Greeke ἐρύσιμον: in Latine, Irio: of ſome χαμαίπλιον: Chamæplion. This is the Eryſimum of Dioſcorides, and not of Theophraſtus, for the Eryſimum of Theophraſtus, is not all one with that of Dioſcorides, as we haue ſufficiently declared elſwhere. Cooper Engliſheth Irio: by the name of winter Creſſes.

❋ *The Nature.*

Eryſimon is hoate and drie lyke Creſſes.

❋ *The Vertues.*

The ſeede of Eryſimon taken with honie in manner of a Lohoc, and often licked, ripeth and cauſeth to ſpet out the tough and clammie fleme gathered within the breaſt & lunges: likewiſe it is good againſt the ſhortneſſe of breath, and the olde cough: it ſhalbe the more conuenient for the ſame purpoſe, if you ſteepe the ſeede firſt in faire water, and then drie it by the fire, or els lappe it in paſte and bake it. for els it wilbe to hoate.

The ſame ſeede ſo prepared and put into the medicines, is good agaynſt the Jaunders, and gripinges of the belly, againſt the Sciatica, and againſt all venome and poyſon.

The ſeede of Eryſimon mingled with hony and water auayleth much to be layd vnto hidden Cankers, harde ſwellinges, impoſtumes behinde the eares, the olde and harde impoſtumes of the breaſtes, and genitours: for it waſteth and conſumeth cold ſwellinges.

Of Iberis. Chap.lxv.

❋ *The Deſcription.*

Iberis. Sciatica Creſſe.

Iberis hath round ſtalkes of a cubite long, full of branches: the ſmall leaues be narrowe, yet a litle greater then the leaues of Creſſes. The flowers be ſmal & white, after which there folowe ſmall ſhelles or huſkes wherin the ſeede is, the roote is ſomwhat thicke & white, in taſte hoate & ſharpe.

❋ *The Place.*

Iberis groweth in Italy and other hoate Countries, about olde walles and other vntoyled places. The Herboriſtes of this Countrie do ſowe it in their gardens.

❋ *The Tyme.*

Iberis flowreth and is in ſeede at Midſomer.

❋ *The Names.*

This herbe is called in Greke ἴβηρις καὶ καρδαμαντικὴ, and of ſome λεπίδιον: in Latine, Iberis, Cardamantice, Lepidium, and of ſome Naſturtium ſylueſtre: in Engliſh, Iberis, and of Turner Sciatica Creſſe.

✤ *The*

the Historie of Plantes.

¶ The Nature.

Iberis is very hoate and drie, of nature lyke to Cresses.

⁂ The Vertues.

The Auncient Physitions, especially Damocrates, say that the roote of Iberis mengled with Swines grease, cureth the Sciatica gowt, if a man binde of this oyntmēt to his hanche huckle bone, or the aking place the space of foure houres, & the women two houres, but immediatly after the remouing of this oyntment, they must enter into a bathe. Reade Turner for the rest of this cure vnder the title Iberis.

Of Dittander Dittany, but rather Pepperwurt. Chap.lxvi.

⁂ The Description.

Piperitis.

Dittany whiche we may more rightly cal Pepperwurt, hath long brode leaues, not muche vnlyke the Baye tree leafe, but a great deale larger and longer, and a little natched or toothed about lyke a sawe. The stalkes and branches be round, vneasie or harde to be broken, and about two foote high: at the toppe whereof growe a number of small white flowers, and after them a small seede. The roote is long & single creeping vnder the earth, and putteth foorth yeerely in diuers places new springes and leaues.

⁂ The Place.

Dittany is sowen in some gardens of this Countrie, and where as it hath ben once set, it abideth or continueth well, so that afterwarde it cannot be easily destroyed.

⁂ The Tyme.

Dittany flowreth & is in Seede in June and July.

⁂ The Names.

This herbe is called of the later writers in these dayes, in Latine, Piperitis, of some also Syluestris Raphanus: in Frenche, Passerage: in high Alماigne, Pfefferkraut: in base Almaigne, Pepercruyt: This shoulde seeme to be λεπίδιον, Lepidium, of Paulus Aegineta, & of Plinie: yet for all that, this is not Lepidium of Dioscorides, neither yet Plinies Piperitis. although it be of some men, sometimes so called: it is fondly and vnlearnedly named in Englishe, Dittany. It were better in folowing the Douchemen to call it Pepperwurt.

¶ The Nature.

This herbe is hoate and drie in the thirde degree.

⁂ The Vertues.

Some in these dayes vse this herbe with meates, in steede of Pepper, bycause it hath the nature and taste of Pepper, wherof it tooke ye name Piperitis.

And bycause the roote of this herbe is very hoate and of complexion lyke to Mustarde or Rockat, it is therfore also very good agaynst the Sciatica, being applyed outwardly to the huckle bone or hanche, with some soft grease, as of the Goose or Capon.

The fyfth Booke of Hydropiper.

Of water Pepper. Chap.lxvij.

❧ *The Description.*

WAter Pepper, hath plaine, roũd, smooth, or naked stalkes & braunches, ful of ioyntes, ỹ leaues be long & narrow, not much vnlyke the leaues of withy, of a hoate burning taste, lyke Pepper, at the top of ỹ stalkes amongst the leaues growe the flowers vpon short stems, clustering or growyng thicke togither, almost lyke the flowers of Blite, smal and white, the whiche past there commeth a broade seede somewhat browne, which biteth the tongue, ỹ roote is hearie. ❧ *The Place.*

This herbe groweth in all this Countrie in pooles & Diches, standing waters and moyst places. ❧ *The Tyme.*

It floureth most commonly in July & August. ❧ *The Names.*

This herbe is called in Greke ὑδροπίπερι: in Latine Hydropiper, & Piper aquaticũ: in French, *Poyure aquatique*, or *Couraige*: in high Douch, Wasser Pfeffer, or Muckenkraut: in base Almaigne, Water Peper: in English, Water pepper, or Waterpepperwurt, and of some Curagie.

❧ *The Nature.*

Water Pepper is hoate and drie in the third degre. ❧ *The Vertues.*

A   The leaues & seede of water Pepper or Curaige, doth wast & consume colde swellinges and old hardnes, also it dissolueth & scattereth congeled or clotted blood ỹ commeth of stripes & bruses, being laid therto.

B   The dried leaues be made into powder, to be vsed with meate in steede of Pepper, as our Dyttanie, or Passerage is vsed.

Of Arselmart. Chap.lxviij.

❧ *The Description.*

THis herbe is lyke to water Pepper, in leaues, stalkes, & clustering flowers, but it is neither hoat nor sharpe, but most cōmonly without any manifest taste. The stalkes be round & haue many knobby ioyntes lyke knees. The leaues be long and narrowe lyke the leaues of water Pepper, but browner, with blackish spottes in the middle, which are not found in the leaues of water Pepper. The flowers be of a carnation or light red

Persicaria.

the Hiſtorie of Plantes. 633

red colour cluſtering togither in knops, after whiche commeth abrode browne ſeede. The roote is yellowe and hearie.

✤ The Place

This herbe groweth alſo in moyſt marriſhe places, and alongſt the water plaſſhettes, and is oftentimes founde growing neare to the water Pepper.

✤ The Tyme.

It flowreth in July and Auguſt, and ſhortly after it is in ſeede.

✤ The Names.

This herbe is called of the latter wryters in Latine, Perſicaria: in Frenche, Perſicaire, of ſome Curaige in high Douch, Perſichkraut, or Flochkraut: in baſe Almaigne, Perſickcruyt, and of ſome Wortcruyt: in Engliſhe, Arſſe-ſmart, or Ciderage.

✤ The Nature.

Arſeſmart is colde and dry of complexion.

✤ The Vertues.

The greene Arſeſmart pound, is good to be laid to greene or freſh woundes, for it doth coole and comfort them, and keepeth them both from inflammation and apoſtumation, and ſo doth the iuyce of the leaues dropped in.

Of Indian Pepper. Chap.lxix.

✤ The Kindes.

There be three ſortes of this Pepper, the one with huſkes of a meane length and greatneſſe, the others huſkes be long and narrow, and the third hath ſhort brode huſkes in al things els not much vnlyke one another, in figure and manner of growing.

Capſiacum.   Capſiacum oblongius.
Indian Pepper.   **Long Indian Pepper.**

¶ The

The Description.

Capsicum latum.
Large Pepper of Indie.

The Indian Pepper hath square stalkes somewhat browne of a foote high, vpon whiche growe brownish leaues, smooth & tender, almost lyke to the leaues of common Sorrel or Nightshade, but narrower & sharper poynted. Amongst the leaues growe flowers, vpõ short stemmes, with fiue or sixe smal leaues, of colour white, with a greene starre in the middle. After the flowers come smooth and playne huskes, whiche before they be rype are of a greene colour, and afterwarde red and purple. The huskes of the first kinde are of a finger length. The huskes of the second kind be lõger & narrower. They of the third kind are large, short and round. In the sayde huskes is founde the seede or graines, of a pale yellow color, brode, hoate, and of a biting taste lyke Pepper.

The Place.

This herbe groweth not of his owne kinde in this Countrie, but some Herboristes doo set and maintayne it in their gardens, with great care and diligence.

The Tyme.

The seede of this Pepper is ripe in this Countrie in Septēber & before winter.

The Names.

This strange herbe is called of Actuarius in Greeke καψικον: in Latine, Capsicum: of Auicen, Zingiber caninũ: of Plinie after the opinion of some men, Siliquastrum, and Piperitis: of such as write in these dayes, Piper Indianum, Piper Calecuthium, and Piper Hispanum: in high Douche, Indianischer Pfeffer, Calecutischer Pfeffer: in Frenche, Poyure d'Inde, or d'Espaigne: in base Alinaigne, Peper van Indien, and Bresilie Peper: in Englishe, Indian Pepper, or Calecute Pepper.

The Nature.

The Indian Pepper is hoate and drie in thirde degree.

The Vertues.

Indian Pepper is vsed in diuers places for the dressing of meates, for it hath the same vertue and taste as the vsual Pepper hath: furthermore it coloureth lyke Safron, and being taken in such sorte, it warmeth the stomacke, and helpeth greatly the digestion of meates.

The same doth also dissolue and consume the swelling about the throte called the kinges euyll, all kernelles, and al colde swellinges, and taketh away al spottes and Lentiles of the face, being layed therevnto with hony.

The Danger.

It is dangerous to be often vsed or in to great a quantitie: for this Pepper hath in it a certayne hidden euyll qualitie, whereby it killeth Dogges, if it be giuen them to eate.

the Historie of Plantes.   635

Of Pepper. Chap.lxx.

¶ *The Kindes.*

The old and ancient Physitions do describe and set foorth there kindes of Pepper, that is to say, the long, the white, and the blacke Pepper, ý which a man shal euen in these dayes find to be sold in the shops of the Apothecaries and Grossers.   ✱ *The Description.*

As touching the proportion & figure of the tree or plante that beareth Pepper, we haue nothing els to write, sauing that we haue found described of the Auncientes, and such as haue trauayled into India, and the Countries about Calecute: and bycause this is a strange kynde of of fruite, not growing amongst vs, we wyll write no more thereof, but as we haue gathered frō the writinges of the Ancientes, & others, which lately haue trauayled into those Countries, who notwithstanding be not yet all of one mind or opinion: for Plinie writeth that the tree which beareth Pepper is like to our Iuniper: Philostratus saith, the Pepper tree with his fruite, is lyke to Agnus castus Dioscorides with certayne others do write, ý Pepper groweth in India vpon a litle or smal tree. And that the long Pepper (the which is lyke to the knoppes or agglettes that hang in the Birche or Hasell trees before the comming foorth of the leaues) is as it were the first fruit which cōmeth foorth immediatly after the flowers, the which also in processe of time do waxe long, great and white bringing foorth many berries hanging togither, vpō one and the selfe same stem. The which berries being yet vnripe, are the white Pepper. and being ripe & blacke is our common blacke Pepper. Suche as trauell to the Indians, Calecute, & the Countries there aboutes do say, that Pepper groweth not vpon trees, but vpon a plante lyke Iuie or Bindweede, the which doth twist and wrap it selfe about trees and hedges, bringing foorth long weake stemmes, whereupon hang the Pepper cornes or berries, euen like the Ribes, or beyondsea Gooseberries, as ye may see in this Countrie: for Pepper is brought frō the Indians to Inwarpe preserued in comfiture with the stems, and foote stalkes hanging in it. The greene and vnripe berries, remayne white, and it is that we call white Pepper, but when they be through ripe they waxe blacke, & full of shriueled wrinckles, and that is our common blacke Pepper. The same authours or later trauaylers do affirme, that long Pepper is not the fruite of this plante, but that it groweth vppon other trees, lyke the thinges that you see hanging lyke Cattes tayles, or Agglettes, vpon the Nut trees and Birche trees in the winter, the which fruit they cal long Pepper, bycause in taste and working it is like Pepper.   ¶ *The Place.*

Pepper groweth in the Iles of the Indian seas, as Taprobane Sumatra, and certayne other Ilandes adioyning, from which Ilandes it is brought to Calecute, the which is the most famous and cheefest citie, as also the greatest marte towne of the Indians: and there it is solde not by waight, but by measures as they sel corne in this Countrie.   ✱ *The Names.*

Pepper is called in Greeke πίπερι: in Latine, Piper in high Douche, Pfeffer: in base Almaigne, Peper: and in English, Pepper.

1   Long Pepper is called in Greeke μακρόν πίπερι: in Latine, Piper longum: in Shoppes, Macropiper.

2   The white Pepper is called in Greeke λευκόν πίπερι in Latine, Piper album: in Shoppes, Leucopiper.

3   The blacke Pepper is called in Greeke, μελάν πίπερι: in Latine, Piper nigrum: in Shoppes, Melanopiper.   ✱ *The Nature.*

Pepper is hoate and drie in the thirde degree, especially the white and the
blacke,

blacke, for the long Pepper is not so drie, bycause it is partaker of a certayne moysture.

❧ *The Vertues.*

It is put into sauces to giue a good smacke & taste vnto meates, to prouoke appetite, and helpe digestion.

It prouoketh vrine, driueth forth windinesse, and paynes in the belly, to be taken with the tender leaues of Bay or Commin: it is also very good agaynst poyson, and the bitinges and stinginges of venemous beastes, and therefore it is put into treacles and preseruatiue medicines.

The same dronken before the coming of the fit of the Ague, or layde to & annointed outwardly with oyle, is good against y shakings & brusings of agues.

The same licked in with hony, is good agaynst the cough comming of a cold cause, and against all the colde infirmities of the breast and lunges.

The same chewed with Raysons, draweth downe from the head thinne fleme, and purgeth the brayne.

Layd to with hony it is good against the Squinancie, for it consumeth and wasteth the swellinges and tumours.

The same with Pitche dissolueth the kinges euill and kernels, and wennes or harde colde swellinges, and draweth foorth shardes and splinters.

Pepper, but especially long Pepper, is good to be mingled with eye medicines or Collyries made to cleare and strengthen the sight.

Of Garlike. Chap.lxxi.

❧ *The Kyndes.*

There be three sortes of Garlike, that is the common or garden Garlike, wilde Garlike, and Ramsons.

Allium satiuum. **Garden Garlike.**   Allium syluestre. **Crow Garlike.**   Allium vrsinum. **Ramsons.**

The

the Historie of Plantes. 637

Allium vrsinum. **Ramsons.**

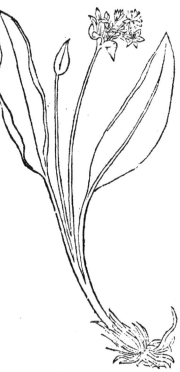

❀ *The Description.*

1 ¶ Arden Garlike hath leaues lyke grasse, or Leekes, amongst which (the yere after the sowing) come vp rounde holowe stems, whiche beare flowers and seede lyke to to the Onyon. The roote is rounde swelling out lyke the Onyon, heaped vp with many cloues or kernelles ioyned togither, vnder whiche hangeth a bearde or tassell of many small hearic stringes.

2 The wylde Garlyke hath no leaues, but in steede thereof it hath long, rounde, small, holowe, pyped blades, amongst whiche springeth vp a round hard stemme of two or three foote long, vppon whiche grow the flowers and seede. The roote is also round Bulbus fashion, without cloues or kernelles growing in it, yet sometimes it hath ioyned therevnto newe heades or or rootes, from which spring new plants.

Of this sorte is founde another kynde whiche is smaller, in all thinges els like the other, aswell in leaues, or blades, stemmes, and seede, as also in rootes, the whiche doo growe most commonly in medowes.

3 The thirde kinde of garlike (called Ramsons) hath most commonly two brode blades or large leaues, almost like the leaues of Liricumphancy, or May Lyllies: betwixt whiche commeth vp a stemme or twayne, bearing many smal white flowers. The roote is lyke to a young Garlyke head, of a very ranke sauour and taste.

❀ *The Place.*

1 Garden Garlike is planted in gardens.
2 The wilde Garlyke groweth by it selfe in feeldes, and hedges, and medowes, especially the smaller sort, for the bigger keepeth the feeldes & pastures most commonly.
3 Ramsons growe in moyst darke places.

❀ *The Tyme.*

2 The wilde Garlyke flowreth and is in seede in June and July.
3 Ramsons flowreth in Aprill and May.

❀ *The Names.*

Garlyke is called in Greeke σκόροδον: in Latine, Allium : in hygh Douche, Knobloch, or knoblouch: in base Almaigne, Loock.

1 The first kinde is called Allium satiuum : in Englishe, Garden Garlyke, and poore mens treacle: in Frenche, *Ail de iardin*: in Douche, Tam Loock, or Loock.

2 The seconde kind is called in Greeke ὁ φιοσκόροδον: in Latine, Allium anguinum & Allium syluestre: in French, *Ail sauuage* in high Douche, Wilder Knobloch, or feld Knobloch: in base Almaigne, Wilt Loock: in English, Crow Garlike, and wilde Garlike.

Hh h The

3 ¶ The thirde kinde is called of the later writers in Latine, Allium vrsinum: in Frenche, Ail d'ours: in high Douch, Waldt knoblauch: in base Almaigne, Das Loock: in English, Ramsons, Buckrammes, & Beares Garlike. This shoulde seeme to be that Garlyke, whiche Dioscorides calleth Scorodoprassum, or as some others thinke Ampeloprasum. ✥ *The Nature.*

Garlyke is hoate and drie almost in the fourth degree.

✥ *The Vertues.*

Garlyke eaten rawe, and fasting nourisheth not, but contrariewise it engen- A dreth euill blood, bycause of his exceeding heate: Neuerthelesse being boyled vntil it hath lost his sharpnesse, it engendreth not so euil blood, and although it nourisheth but litle, yet it nourisheth more then when it is eaten rawe.

It is good for suche people as are full of grosse, rawe, and tough humours, B for it wasteth and consumeth colde humours.

It dispatcheth windinesse, openeth al stoppinges, killeth and driueth foorth C brode wormes, and prouoketh vrine.

It is good against all venome & poyson, taken in meates or boyled in wine D and dronken, for of his owne nature it withstandeth al poyson: in so much that it driueth away all venemous beastes, from the place where it is. Therefore Galen prince of Physitions, called it poore mens Treacle.

It is layde with great profite to the bitinges of mad Dogges, and vpon the E bitinges & stinginges of venemous beastes, as Spiders, Scorpions, Vipers, and suche lyke: and for the same purpose it auayleth muche to drinke the decoction or broth of Garlyke sodde in wine.

It is also good to keepe such from danger of sicknesse, as are forced to drinke F of diuers sortes of corrupt waters.

The same eaten raw or boyled cleareth the voyce, cureth the old cough, and G is very good for them that haue the Dropsie: for it drieth the stomacke and consumeth the water: and doth not much alter nor distemper the body.

The decoction thereof made with Orygan and wine, being dronken, killeth H lyce and nittes.

It is very good against the tooth ache, for it slaketh the same, pounde with I vineger, & laid to the teeth: or boyled in water with a litle incense, & the mouth washed therewith, or put into the holownesse of the corrupt teeth. It is of the same vertue mixt with goose grease and powred into the eares.

The same brused betwixt the handes and layde to the temples, slaketh the K olde headache.

The same burned into ashes & mingled with hony, healeth the wild scabbe, L and scurffe of the head, and the falling of the heare, being layde thereupon.

Layde to in the same manner, it healeth blacke and blewe scarres, that re- M mayne after bruses and stripes.

It is also good against the fowle white scurffe, lepries, and running vlcers of N the head and all other manginesse, pounde with oyle and salte, and layde thereupon. Also it is good against the hoate inflammation called wilde fier, which is a spreading scabbe lyke a tetter.

With Swines grease it wasteth and dissolueth harde swellinges, and layd O to with Sulpher and Rosen, it draweth foorth the euill qualitie or noughtie humour from fistulas, as Plinie writeth.

It moueth womens natural sicknes, driueth foorth the secondine, if women P sit ouer the decoction thereof, or if it be cast vpon the quicke coles, and women receiue the fume of it through a fonnel or holow stole.

They cure the pipe or roupe of Pultrie and Chickens with Garlyke. Q

✥ *The*

the Historie of Plantes.

✱ *The Danger.*

Garlyke is hurtfull and nought for cholerique people, and suche as be of a hoate complexion, it hurteth the eyes and sight, the head and kidneyes. It is also nought for women with childe and suche as giue sucke to children.

Of Sauwce alone or Jacke by the hedge. Chap.lxij.

✱ *The Description.* Alliaria.

This herbe at his first springing vp, hath roundish leaues, almost lyke to Marche violettes, but much greater and larger, & of a paler color. Amongst those leaues cometh vp the stalke of two foote high, with longer and narrower leaues then the first were, and creuised or iagged about, not much vnlike the Nettle leaues, but greater. The whiche beyng brused betweene the fingers, haue the sauour and smell of Garlyke. About the highest of the stalke grow many small white flowers, and after them long coddes or huskes wherein is blacke seede. The roote is long & slender, and of wooddy substance.

❧ *The Place.*

This herbe delighteth to growe in lowe vntoyled places, as about the borders of medowes, and moyst pasture groundes, and somtimes in hedges, and vpon walles.

✱ *The Tyme.*

This herbe flowreth most commonly in May and June, and afterwarde cometh the seede. ✱ *The Names.*

This herbe is called of the later writers in the Latine tongue, Alliaria, of some also Scordotis: but this is not the true Scordotis, the whiche is also called Scordium, and is described in the first booke of this historie: Pandectarius calleth it Pes Asininus: it is named in French, Alliaire: in high Douch, Knoblochkraut, Leuchel, or Saßkraut: in base Almaigne, Loock sonder loock: in Englishe, Sauce alone, and Jacke by the hedge.

✱ *The Nature.*

This herbe is hoate and drie almost in (the thirde degree) fourth degree.

✱ *The Vertues.*

This herbe is not much vsed in medicine: but some do vse it with meates in steede of garlyke.

The ignorant Apothecaries doo vse this herbe for Scordium, not without errour, as it is manifest to all such as are learned in ye knowledge of Simples.

Of Onyons. Chap.lxiij.

❧ *The Kindes.*

There be diuers sortes of Onyons, some white, some red, some long, some rounde, some great, and some small: but al of one sauour and propertie, sauing that the one is a little stronger then the other. Yet they differ not in leaues, flowers, and seede.

Crommion, Cepa. Onyons.

The Description.

The Onyon hath leaues or blades almost like garlike, holow within. The stemmes be round, vpon whiche grow rounde bawles or heades, couered with little fine or tender white skinnes, out of which breake many white flowers lyke starres, whiche turne into smal pellettes or buttons, in whiche are contayned two or three blacke cornered seedes. The roote is rounde or long, made of many foldes, pylles, or coueringes, growing one vpon another, wherof the vpmost pilles or scales are thinnest. In the neather part of the roote is a bearde of hearie rootes, or stringes lyke a tassel.

The Place.

They are sowen in euery garden of this Countrie, but they loue a soft and gentle grounde.

The Tyme.

They are commonly sowen in February and March, and are ful growen in August, & are then pluckt out of the ground to be kept. And if they be plated againe in December, January, or February, then they wil blow in June, and bring foorth in July and August.

The Names.

The Onyon is called in Greeke κρόμμυον: in Latine, Cepa, and Cepe: in high Douche, Zwibel: in base Almaigne, Ceede Ayeuyn.

The Nature.

The Onyon is almost hoate in the fourth degree, and rather of grosse, then subtil partes.

The Vertues.

A. The Onyon engendreth windynesse, and causeth appetite, and it doth scatter, and make thinne grosse and clammie humours, without nourishing: especially to be eaten raw. But being boyled twise or thrise it is nothing so sharpe, and it nourisheth somewhat, but not muche.

B. Onyons eaten in meate, open the belly gently, and prouoke vrine plentifully.

C. They open the Hemorrhoides, so called in Greeke, layd to the fundement or siege with oyle or vineger, and so doth the iuyce or the whole Onyon mingled with rosted apples, and layde vpon the fundement with cotten.

D. Onyons sodden and layde to with Raysens and figges, do ripe and breake wennes and suche lyke colde swellinges.

E. The iuyce of them dropped into the eyes, cleareth the dimnesse of the sight, and at the beginning remoueth the spottes, cloudes, and hawes of the eyes.

F. The same iuyce dropped into the eares, is good agaynst deafenesse, and the humming noyse or ringing of the same, and is good to clense the eares from all filthinesse, and corrupt matter of the same.

G. The same powred or snift vp into the nosethrilles, causeth one to sneese, and purgeth the brayne.

H. Being put vnder in a pessarie, it bringeth out the flowers and secondine.

I. It is layde to ye bitings of dogges, with hony Rue & salt, with good successe.

K. It cureth the noughtie scabbe and itche, & the white spottes of all the body, and also the scurffe and scales of the head: and filleth agayne with heare the pylde places of the head, being layde thereto in the Sonne.

The

the Historie of Plantes.

¶ The same layde to with Capons greace, is good against the blisters of the feete, and against the chafing and gaulling of the shoe.

※ *The Danger.*

¶ The often vse of Onyons, causeth headache, and ouermuche sleepe, and is hurtfull to the eyes.

Of Leekes. Chap.lxxiiij.

※ *The Description.*

Porrum.

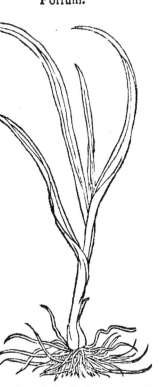

THE Leeke hath long brode blades, folden togither with a keele or crest in the backside, in taste and sauour not muche vnlyke the Onyon, betwixt which leaues in the second yere groweth a rounde stemme, whiche bringeth foorth a rounde head or bawle, with his flowers lyke the Onyon, and after the flowers it beareth seede, in fashion lyke to Onyon seede, but that it is of a grayishe colour. The roote is white and lesser then a meane Onyon, with a bearde or tassell of hearie stringes.

¶ *The Place.*

The Leeke is planted almost in euery garden of this Countrie, and is but seldome suffered to seede: but the blades are cut almost euery day harde by the grounde, to be daylye vsed in pottages, and other meates, and therefore it can vnethe or scarsely growe vp.

※ *The Tyme.*

The Leeke flowreth in Maye and June, a yeere after the sowing, if it hath not bene cut, for if it be continually cut, it beareth very seldome flowers or seede, and therefore some do write that the Leeke bringeth foorth neyther flowers nor seede, whiche is vntrue, for the Leeke whiche hath not bene cutte bryngeth foorth both flowers and seede.

¶ *The Names.*

The Leeke is called in Greeke πράσον: in Latine, Porrum: in Frenche, *Poureau*: in high Douche, Lauch: in base Almaigne, Paraye: in Englishe, a Leeke, or Leekes.

The vncut Leeke is called in Greeke πράσον κεφαλωτόν: in Latine, Porrum capitatum: that to say in Englishe, the headded or knopped Leeke.

The cut Leeke is called of Columella and of Palladius in Latine, Porrum sectiuum: in Englishe, Frenche Leeke, vnset Leeke, Mayden Leeke.

※ *The Nature.*

The Leeke is hoate and drie in the third degree, of Nature lyke the Onyon, but not so strong.

※ *The Vertues.*

Leekes engender grosse and euill blood, breede winde, and cause heauie dreames,

dreames, especially to be eaten raw: but boyled in water twise or thrise, it wilbe the better and more conuenient to be eaten.

It stirreth one to make water, it maketh the humours fine and thinne and softeneth the belly.

The iuyce of Leekes dronken with hony, is good agaynst the bitinges and stinginges of venemous beastes.

The iuyce of Leekes taken in an electuarie of Lohoc, doth mundifie & clense the breast, causeth one to spet out, and is good against hoarsenesse and the olde cough.

A bath of Leekes made with salt sea water, prouoketh womens flowers, openeth the stoppings of the Matrix, and doth mollifie and soften all hardnesse of the same, if they sit ouer the fume thereof.

The leaues, or as we say the the blades of Leekes will stanche bleeding, especially nosebleeding: the same vertue hath the iuyce mingled with vineger, and fine powder of Frankensence to be put into the nosethrilles.

The sede is good to be mingled and put into medicines, that serue to breake the stone.

It stoppeth and stancheth all superfluous bleeding to be taken with the like quantitie of Myrtill berries.

❧ *The Danger.*

Leekes engender euil humours, and windinesse: they cause heauy and terrible dreames, they darken the eye sight, and are very hurtful for them that haue any exulcerations or goyng of, of the skinne, of the bladder, or raynes.

Of Cyues, or Rushe Onions leekes. Chap.lxxv.

❧ *The Description.*

Cyues or Rushe Onyons, in the stede of leaues haue litle, smal, holowe, & slender piped blades, lyke to smal Rushes, growing thicke togither, in taste not much vnlyke the taste of Leekes. Amongst the Rushlyke leaues growe smal rounde stemmes, with smal bowles, or rounde knopped heades, like the bawle in the top of the seede Onyon, but much smaller, and ful of smal purple flowers. The rootes be lyke to small Oniōs, but a great deale smaller, growing close and thicke togither, ful of long hearie thredes or stringes, lyke the beard of the Onyons, or leekes,

❧ *The Place.*

It is set in gardens amongst potte herbes, or wurtes.

❧ *The Time.*

It flowreth in May and June a yere after ye sowing, new planting or setting.

❧ *The Names.*

This kinde of Leekes is called in English, Cyues, & of Turner in Latine, Cepa pallacana, & in greke Gethyū, which be En-

Schœnoprasum.

the Historie of Plantes. 643

Englisheth by al these names a Cyue, a Ciuet, a Chyue, or Sweth, and giueth to the same a very strange figure: but this kinde is called in French, *des Oignoncettes*, or *Porrettes*: in high Douche, Schnitlauch, Bryszlauch: in base Almaigne, Biesloock, that is to say, Rushe Garlike, bycause in steede of leaues it bringeth forth smal rushes like Crow Garlike. It hath neither Greke nor Latine name that I knowe. Therefore in folowing the Douche, we doo call it in Greeke Χοινοπράσον: and in Latine, Scœnopralum: whiche may be Englished, Rushe Leekes: and if any man had called it in Greeke Χοινοκρόμμυον, I without any presumption might haue called it Rushe Onyons. Some take it to be Porrum sectiuum: but it appeareth well by that whiche Columella and Palladius haue written, howe shamefully they erre, and by the same aucthoritie of Columella and Palladius we haue sufficiently proued in the former Chapter, that the cut Leeke, and the headed Leeke, whiche is our common Leeke are al one, and do come both of one seede, and do differ but only in this: that the one is suffered to growe and beare seede, and the other is oftentimes cut.

❧ *The Nature.*

Cyues are hoate and drie in the thirde degree, and of complexion or temperament lyke vnto Leekes.

❧ *The Vertues.*

Cyues are vsed in meates and pottages euen as Leekes, whiche they do resemble in operation and vertue.

Of Wilde Bulbus, or Wilde Onyon. Chap.lxxvi.

❧ *The Description.*   Bulbus syluestris.

This herbe hath long leaues or blades lyke Garlyke, but very seldome bringing foorth more then two blades, betwixt which springeth vp a rounde holow stemme of a spanne long at the top thereof growe many yellowe sterrelyke flowers, the whiche doo change into a three square or triangled huske or huskes: in which the seede is contayned. The roote is rounde as an Onyon.   ❧ *The Place.*

This Onyon groweth in diuers places of Almaigne, in sandy Countries in dales and vallyes about brookes and little streames, and sometimes also vnder hedges.

❧ *The Tyme.*

This kind of Bulbus flowreth in March, and is in seede in April, & in short space after it vanisheth away, so y in May folowing a man shal find neither stalkes neither leaues.

❧ *The Names.*

Howe this kinde of Bulbus hath bene called of the Auncientes or olde writers, is not certainly knowē, some think it to be Bulbina: some others would haue it βολβὸς ἐδώδιμος, that is to say, Bulbus esculentus, but as some learned men and I do thinke, this Bulbus is

neither

neyther the one nor the other. And therfore we call it Bulbus ſylueſtris: the high Douchemen do call it feldſwibel, Ackerzwibel: and there after it is called in baſe Almaigne, Velt Ayeuyn: in Frenche, *Oignon ſauuage*: that is to ſay, Wilde Onyon. Turner calleth it Bulbyne, wilde Lecke, and Corne Lecke, li. 1. to. 97. and in the firſt impreſſion. fol. 5. ❊ *The Nature.*

This wilde Onyon is hoate and drie in the ſeconde degree, the whiche is to be perceiued by his bitter taſte and rough aſtriction, or binding qualitie.

❊ *The Vertues.*

Suche as haue put this Bulbus in proofe, do affirme that it ſofteneth and driueth away harde ſwellinges being layde therevnto.

It is alſo (with great profite) applyed and layde vnto moyſt, corrupt, rotten, feſtered, fretting and conſuming ſores, being firſt roſted vnder imbers, and then pounde with hony and layde to.

Of the White ſelde Onyon. Chap. lxxvij.

Ornithogalum minus.
Bulbus Leucanthemus.

Ornithogalum maius.

❊ *The Deſcription.*

This kinde of Bulbus at the firſt ſpringing vp hath long ſmall narrow graſſie leaues or blades of a ſpan long: from amongſt which ſpringeth vp a rounde greene ſtemme, of a ſpan long or therabontes, bringing foorth foure or fiue ſmal flowers, greene without and white within, not much differing in proportion from the faſhion of the Lylie flower, eſpecially before they be fully ſpread abroade and opened, but they be much leſſer. The roote is rounde lyke an Onyon or Bulbe, white both within and without, and very ſlymie lyke Comfrey, when it is bruſed or broken in peeces: in taſte ſomewhat ſharpe. This agreeth not with Ornithogalum of Dioſcorides, for his

Ornithogalum is described to haue a certayne aglet, or a thing called Cachryos, growing vp in the middle of the flower: Neither is it lyke to be Matthiolus Ornithogalum: for that which he setteth betwixt Ornithogalum and Trasi, hath a roote blacke without and white within.

2 This Ornithogalum maius, is lyke the other, but much greater. The leaues of this be long and smal, but bigger then the first. The stalke groweth a foote & a halfe high, and is very euen. There grow vpō the top of the stalke faire pleasant flowers, of colour white, lyke vnto small Lylies, in the middle is a head lyke the seede that is named Cachrys. The roote is a Bulbus, the whiche lightly multiplyeth into many other.

❦ The Place.

This herbe groweth in sandy places that lye open to the ayre, and be manured or toyled, and is founde in many places of Brabant, especially about Malines or Mechelen almost in euery feelde.

❦ The Tyme.

The leaues of this Bulbus do spring vp first in March & Aprill, & the flowers in May, & about June they do so vanish, that they be not any longer to be seene or founde.

❦ The Names.

1 This herbe is called in Greeke ὀρνιθόγαλον: and in Latine, Ornithogalum: vnknowen in shoppes: in base Almaigne it is called, Wit velt Ipuepn, that is to say, the wilde white feelde Onyon: in some places of France, it is called Churles. It may lykewyse be very wel called, Bulbus Leucanthemus.

2 The other Bulbus, is lykewise an Ornithogalum, and is called of some nowe in these dayes, Lilium Alexandrinum, that is to say, Lylies Alexandria, bycause it is thought that it is first brought into knowledge in this Countrie from Alexandria.

❦ The Nature.

This Bulbus is temperate in heate and drynesse.

❦ The Vertues.

Dioscorides saith, that it may be eaten either rawe or rosted as ye liste. A

It is also very good to soulder and close vp fresh or greene wounds, being B layde vpon lyke Comfrey.

Of the Sea Vnyon called Squilla. Chap. lxxviij.

¶ The Kyndes.

AT this day there be found two kinds of Squilla, or Sea Union: the one bearing straight or narrow blades, the which is the right Squilla: the other hath brode blades, and is commonly vsed for Squilla.

❦ The Description.

1 The rounde bollens, or imbossed heades of the first & right Squilla, are very great and thicke, and whiter then the bollens or heades of the vsual & common Squilla. The blades be long and narrow, and of a white greene or grayish colour.

2 The common Squilla hath also great thicke heades or bollens, but they are most commonly redder, and the pilles or scales are thicker then the scales or coueringes of the other Squilla. The leaues be great and broade almost lyke to Lylie leaues. The flowers be smal and yellow growing at the highest and alongst the stalkes or stemmes, after them commeth the seede.

❦ The Place.

Squilla groweth not of his owne accorde in this Countrie, but is brought from Spayne hither to serue for medicine, wherof some is planted in gardens.

❦ The Names.

The first kind of this strange Vnyon, is called in Greeke σκίλλα: & in Latine, Scilla:

Scilla : in **Shoppes, Squilla**: in French, Scilla cōmunis. Pancratiū **Squilla**.
Stiboule, Squille, Oignon de mer. in high Douch **Meerzwibel**: in base Almaigne, **Zee Ayeuyn**: of Serapio, Cepe muris, that is to say, **Mowce Onyon** : in Englishe, **Squilla, and Sea Onyon.**

2   The second kind is taken of the greater number of Apothecaries for **Squilla**, albeit it is not the right kinde, but of that sort whiche the Grekes do cal παυκράτιον: the Latines Pancratiū, which is of nature lyke to **Squilla**, and therefore without any errour it may be used in steede of **Squilla**. And this kinde of the learned Peter Beion is counted to be Bulbus littoralis of Theophrastus, whereunto it is very muche lyke: for Dioscorides Pancratium, and Theophrastus Bulbus littoralis do seeme to be all one.

❧ *The Nature.*

Squilla is hoate in the second degree, and drie in the thirde degree, and of very subtile partes, also of a cutting and scouring nature.

❧ *The Vertues.*

Squilla (being first couered rounde about with dowe, or lapt in paste & baked in an ouen, or rosted vnder coles vntill it be soft or tender) then a spoonefull or two thereof taken, with the eight part of salt, causeth a man to go to the stoole, and putteth foorth plenty of tough and clammie humours.

The same rosted or prepared after the same manner, is good to be put into B medicines that prouoke vrine, and in suche medicines as are vsed agaynst the Dropsie, the Iaundise, belching or working vp of the stomacke, and gripinges or frettinges of the belly.

Taken with hony and oyle, it driueth foorth of the belly, both the long and C rounde wormes.

Prepared in manner aforesayde, it is put with great profite, into medicines D that are made against an old inueterate cough, and shortnesse of breath, which medicines do cause to spit out the tough and clammie flemes, that are gathered togither within the holownesse of the breast : for taken in the same manner it doth dissolue and loose grosse humours, and bringeth them foorth.

The same ordered with hony loseth the belly very gently : and the like ver- E tue hath the seede to be taken with figges or hony.

A scale or twaine of the roote of **Squilla** being yet greene and raw is good F to be layde vnder the tongue, to quenche the thirste of them that haue the Dropsie, as Plinie writeth.

Squilla sodden in vineger vntill it be tender, and pouned small, is good to G be layde as an emplayster, vpon the bitinges of Vipers and Adders, and suche other lyke venemous beastes.

The inner part of **Squilla** boyled in Oyle, or Turpentine is applyed with H great profite to the chappes or riftes of the feete, and also to kibed or moldye
                                                                                        heeles

the Historie of Plantes. 647

heeles, and hanginge wartes, especially when it is first rosted vnder the imbers.

In the same maner it healeth the running sores of the head, and the scurffie scales or bran of the head being layde therevnto.

The leaues of Squilla do dissolue and wast the kinges euyl and kernelles vnder and about the throte, beyng layd therevpon by the space of foure dayes.

Pythagoras saith, that if Squilla be hanged ouer the doore or chiefe entrie into the house, it keepeth the same from all mishap, witchcraft or sorcerie.

Bertius writeth that whan the flowers of Squilla be of a brownish colour and doo not soone fall, or vade away, that the yeere shalbe very fruitefull, and there shalbe great store of Corne.

Pancratium in vertue and working is muche lyke to Squilla, sauing that it is not so strong nor effectuall. And it may be vsed for want of the right Squilla in al things as witnesseth Galen, and is to be prepared in the lyke order as they prepare Squilla, as saith Dioscorides.

✱ *The Danger.*

Squilla is a very sharpe medicine, both subtil & wasting, hurtful and forsing the nature of man, when it is taken or vsed rawe: and therefore Galen saith, it ought not to be vsed or taken into the body without it be first sodde or rosted.

Of Affodyll. Chap.lxxix.

✱ *The Kyndes.*

There be three kindes of Affodill, that is to say, the male, and female, and a thirde sort with yellowe flowers.

Asphodelus mas.
The male Affodyll.

Asphodelus fœmina.
The female Affodyl.

Iii ij ❦ The

The fyfth Booke of

Asphodeli tertia species.
Yellow Affodyll.

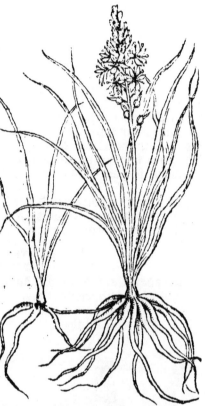

❦ *The Description.*

1. THE first kinde of Affodyll hath long narrow leaues, like Leeke blades, amongst which springeth vp a round stalke of a cubite, or cubite and a halfe long : vppon whiche from the middle vp to the toppe growe faire white flowers, or of a very pale carnation colour, which do begin to flower below, and do end their flowring aboue. The flowers past ther come smal huskes, round and writhed or turned about, and are found diuided and seuered into partes whē they waxe ripe: within the sayd huskes is a browne seede. The rootes do growe by great nūbers or companyes, a like to the rootes of the femal Pionie, eache one fashioned like to a lōg thicke kernell or somewhat longer, and within somwhat opē or spongie, in taste at the first somewhat astringent, and afterwarde bitter. Yet of no very strange taste, so that it is no maruell that men in tunes past dyd vse to of eate this roote as Hesiodus & certaine other do report.

2. The second kind of Affodyll hath narrow blades also lyke to the abouesayde, but smaller and shorter, amongst whiche springeth foorth a plaine straight stem of two foote high, from the middle of the top set with pale flowers, diuided into sixe partes, not much vnlike the flowers of the other Affodyl. They once past there appeare small triangled huskes, within the whiche lyeth the seede. The roote of this kinde is rounde as the head of an Onyon, almost lyke the roote of garden Bulbus, but somewhat bigger. To conclude, this Affodyll is not muche vnlyke the first kind but only in the roote, wherein is all the difference betwixt these two herbes: for they varie not much one from another in leaues, stalkes, flowers, and seedes, sauing that the leaues of this kind are shorter, the flowers stande further a sunder, and not so thicke set, or throng togither.

3. Bysides these two kindes there is found another Affodyl, whose leaues be longer & narrower then the leaues of the first kind, the stalkes be also round, & loden with pleasant yellow flowers, after whiche appeareth rounde huskes or knops lyke little heades, wherein the seede is contayned : it hath a number of rootes growing thicke togither like the first Affodyl, but euery roote is longer and smaller. The leaues of this Affodyl remayne greene al the winter, & do not vade and perish as the leaues of the other. And the rootes doo put foorth a certayne increase of newe springes and blades, wherby it incrocheth and winneth more grounde, and doth so multiply: that of one plant within a fewe yeres you shal get a number of others. ❦ *The Place.*

Affodyl is not founde growing of his owne kind in this Countrie, but in the gardens of Herborists, where as they do both sowe and playnt it.

❧ *The Time.*

1 The first kinde flowreth in May, and is in seede in June.
2 The seconde doth also flower and seede in June.

❧ *The Names.*

This herbe is called in Greke ἀσφόδελος: in Latine, Albucus, and Hastula regia: in shops Affodilus: in French, Hache royale, or Asphodel: of the common Herboristes of Brabant, Affodilen. The flower with his stemme is called in Greeke ἀνθέρικος, Anthericos: and in Latine, as Plinie sayth, Albucum: in English also Affodyl, and Daffodyll.

1 The first kinde is called Asphodelus mas, and Hastula regia mas, and is that same whiche Dioscorides describeth.
2 The seconde is called Alphodelus fœmina, and Hastula regia fœmina, and is that whiche Galen describeth, in lib. de alimentorum facultatibus.

⁋ *The Nature.*

1 Affodyl especially the roote of the first kind is hoate & dry in ye second degree.
2.3 The rootes of the other kinde, are hoate and dry almost in the thirde degree.

❧ *The Vertues.*

The roote of ye first kind boiled & dronke, prouoketh vrine, & womēs flowers. **A**

The waight of a dram therof taken with wine, healeth the payne in the side, **B** the cough, the shrinking of sinewes, crampes, and burstinges.

It is very good against the bitings of venemous beastes, to drinke the quan- **C** titie of three drammes therof with wine, and to lay vpon the wounde and hurted place the leaues, flowers and rootes beaten togither.

The seede & the flowers of the right Affodyl dronken in wine, are very good **D** against ye poyson of scorpiōs, & other venemous beastes, also they purge ye belly.

The roote boyled in the lyes of wine is good to be layd vpon corrupt festered **E** sores, and vpon olde vlcers, and the impostumes of the breastes and stones or genitours. It is also good against new swellings and impostemes that do but begin, being layde vpon in maner of an emplayster with parched barley meale.

The iuyce of the roote boyled with good olde wine, a litle Myrrhe and Saf- **F** fron, is a good medicine for the eyes, to cleare and sharpen the sight.

The same iuyce of it selfe, or mingled with Frankencense, hony, wine, and **G** Myrrhe, is good against the corrupt filth and mattering of the eares, when it is powred or dropped in.

The same prepared & ordered as is aforesaid, swageth the toothache powred **H** and dropped into the contrarie eare to the payne and greefe.

The ashes of the burned roote, and specialy of the seconde kind do cure and **I** heale scabbes and noughtie sores of the head, and doo restore agayne vnto the pilde head, the heare fallen away, being layde therebnto.

The oyle ye is sodden in the rootes being made holow, or the oyle in which the **K** rootes haue ben boyled, doth heale ye burnings with fire, mouldy or raw kibed heeles, & doth swage ye paine of the eares, & deafnesse, as Dioscorides writeth.

The rootes do cure the morphew or white spots in the flesh, if you rub them **L** first with a linnen cloth in the Sonne, & then annoynt the place with the iuyce of the roote, or lay the roote to the place.

Of the Vine. Chap. lxxx.

❧ *The Kyndes.*

There are diuers sortes of vines, but aboue all the rest there are two most notable: that is to say, the garden or husbanded vine, and the wilde vine, as writeth Dioscorides, and the Ancientes. The manured or husbanded vine is also of diuers sortes, both in fashion and colour, so that it is not easie to number or describe all the kindes: whereof it shalbe sufficient for vs to diuide the garden or husbanded vine into three kindes: whereof the first is very red, and yeeldeth a darke red liquer, the whiche is called of some Tinctura. The seconde is blewe, and yeeldeth a cleare white liquer, the which yet notwithstanding wareth redde, when it is suffered to setle in the vessel. The thirde vine is white and yeeldeth a white wine or liquer, the whiche continueth white. And all these sortes of the manured or garden vines are lyke one another in leaues, branches, wood, and timber.

✻ *The Description.* Vitis. **The manured vine.**

The vine hath many weake and slender branches, of a wooddy substance, ouercouered with a cloue barke, or chinking rinde (from which branches) groweth foorth new encrease of knottie shutes or springes, bringing foorth at euery knotte or ioynt, broade iagged leaues, diuided into fiue cuttes or partes, also it putteth foorth at the aforesayd ioyntes with the leaues certayne tendrelles, or clasping caprioles, & tying tagglets, wherewith al it taketh hold vpon trees, poles, and perches, and all thinges els that it may attayne vnto. The same new springes and branches, doo also bring foorth, for the most part, at the seconde, thirde, and fourth knotte or ioynt, first of all little bushie tuftes, with white blossoms or flowers, and after them pleasant clusters of many berries or grapes, thicke set and trussed togither, within whiche berries or grapes are founde small graynes or kernelles, whiche be the seede of the vine.

✻ *The Place.*

The vine delighteth to growe vpon mountaynes, that stande open to the South, in hoate Countries and Regions, as in Canarie, and the Ilandes adioyning in Barbaria, Spayne, Greece, Candie, Sicile, Italy, and diuers other hoate Regions. It groweth also in Fraunce, and Almaigne, by the riuer Rheyne, and in some places of Netherland, as Brabant, Haynau, and Liege: but that which groweth in these lower Countries do bring foorth very smal or thin wines, for none other cause but onely bycause ye Sonne is not so vehement, and the nightes be shorter. For (as Constantine Caesar writeth.) The Sonne must giue to the wine strength and vertue, & the night his sweetenesse, and the Moone shine his rypenesse. And therefore are the vines of Canarie, of Candie, and other the lyke hoate Countries, both sweete and strong: for the Sonne shineth vehemently in those Countries, and the nightes be longer then in this Countrie. And for this consyderation the wine of Rheyne, and of other the Septentrional or North Regions are weaker, and not so sweete & pleasant, bycause ye nights in those Countries be shorter, & the Sonne hath not so muche strength. And for the same cause also it groweth not in Norweigh, Swedlande, Denmarke, Westphale, Prusse, and other colde Countries: for the nightes be there in sommer short, and the power of the Sonne is but smal.

✻ *The*

✿ *The Time.*

The vine floweth in high and base Germanie or Almaigne, about the beginning of June, and the grapes be through ripe in September. A moneth after, that is to say in October, they presse foorth the wine, and put it into hogsheades, and vessels, fit for that purpose, and therefore they call the moneth of October in Douche, Wijnmaent.

✿ *The Names.*

The manured vine is called in Greeke ἄμπελος οἰνόφορος, καὶ ἄμπελος ἥμερος: in Latine, Vitis vinifera: in high Douche, Weinreb: in base Almaigne, Wijngaert: in Englishe, the garden or manured Vine or Grape.

✿ *The Nature.*

The leaues, branches, and tendrelles of the vine, are colde, drie, and astringent, and so be the greene berries or vnripe grapes: but the ripe grapes are hoate and moyst in the first degree, and the Raysen or dried grape is hoate and drie, as witnesseth Galen.

✿ *The Vertues.*

The iuyce of the greene leaues, branches, and tendrels of the vine dronken, A is good for them that vomit or spet blood, and is good against the bloddy flire, and for women with childe that are giuen to vomit. The same vertue haue the branches and clasping tendrelles to be taken alone by them selues: and so haue the kernelles, that are found within the fruit, to be boyled in water and dronken.

The same tagglettes or clasping tendrelles of the vine, pound with parched B barley meale, are good to be applyed to the headache comming of heate, and vpon the hoate vlcers of the stomacke.

The ashes of the drie boughes or cuttinges of the vine burnt, and layde to C with vineger, do cure the excrescence & swellings of the fundement, the which must first be scarrified or pared.

The same dissolued in oyle of roses and vineger, is good to be layde to the D bitinges of Serpentes, to dislocations or members out of ioynt, and to the inflammation, or heate of the splene or milte.

Greene grapes ingender windinesse in the belly and stomacke, and do loose E the belly.

The dryed Raysens are very good against the cough, and all diseases of the F lunges, the kidneyes and the bladder.

They be also very good (as Galen saith) against the stoppings and weake-G nesse of the liuer, for they both open the same, and strengthen it.

The broth of Raysen kernelles, is good agaynst the bloddy flire and the H laske, if it be altogither powred into the body at one glister.

It stoppeth also the superfluous course of womens flowers, if they bathe I them selues in the same brothe or decoction of the kernelles.

The same kernelles pounde very small and laide to with salt, doo consume K and waste harde swellinges, and swageth the blastinges and swellinges of womens breastes.

Of the wylde Vine or Grape.
Chap. lxxxi.

¶ *The Kindes.*

The wilde vine is of two sortes, as Dioscorides sayth, the one sorte hath flowers, & grapes which neuer come to ripenesse: and the other bringeth foorth small grapes or berries whiche come to ripenesse.

The Description.

1. The wilde vine is much like to the gardē vine, in branches, leaues, and clasping capreoles, wherof the first kind bringeth foorth first his flowers, and afterwarde his fruite lyke to the garden vine: but the fruite commeth not to ripenesse.

2. The second kind bringeth foorth smal clusters, ful of litle berries or grapes, the whiche do become ripe, and they drie them lyke Raysens. And of these are made the smal Raysens, which are commonly called Corantes, but more rightly Raysens of Corinthe.

The Names.

The wilde vine is called in Greeke ἄπελος ἀγρία: in Latine, Vitis syluestris, and Labrusca: in Englishe, the wilde grape or vine.

1. The decaying or fading fruite, of the first kinde of wilde vine, and also the flowers of the same, is called in Greeke οἰνάνθη: and in Latine, Oenanthe.

The iuyce whiche they presse out of the grapes of this vine, and of all other sortes of greene and vnripe grapes, aswel of the garden as of the wilde kind of vines, is called in Greke ὀμφάκιον: & in Latine, Omphacium: in shops, Agresta: in French, Verius, & of some Aigras: in base Almaigne, Verjus: in English, Verius.

2. The fruite of the seconde kinde is called in the Shoppes of this Countrie, Passulæ de Corintho: in Frenche, Raisins de Corinthe: in base Almaigne, Corinthen: in Englishe, Currantes, and small Raysens of Corynthe.

The Nature.

The leaues, branches, and clasping capreoles of the wilde vine, haue lyke power and vertue, as the leaues, branches, and clasping tendreiles of the manured or garden vine, & so hath the Verius of the same. The Raysens or Currantes are hoate and moyst of nature and complexion, not muche vnlyke the common frayle Raysens in operation.

The Vertues.

The leaues, branches, and tendrelles of the wilde vine, are of like vertue & operation, as the leaues, branches, & claspers of the garden vine, and do serue as wel to all purposes, as they of the garden vine. [A]

The flower with the vnripe and withering fruite of the first kinde of the wilde grape stoppeth the laske, and all other fluxe of blood. [B]

Being layde outwardly vpon the stomacke, they are good against the debilitie and weakenesse of the stomacke, and sower belchinges and lothsomnesse of the same, and they be also of the same effect to be eaten. [C]

It swageth headache, being layde vpon the same greene, or mingled with oyle of roses and vineger, and is muche profitable agaynst the spreading and fretting sores of the genitours or priuities. [D]

The Verius doth not much differ in operation and vertue, from the withered & vnripe grape, especially when it is dryed & made into powder: for being so prepared & occupyed, it is an excellent medicine agaynst the weakenesse and heate of the stomacke, for it doth both strengthen and refresh or coole the same, howsoeuer it be vsed, whether in meates or otherwyse. [E]

They make a syrupe with this Verius, sugar, or hony, the whiche is very good against thirste in hoate agues, and the wambling, vomiting, and turning vp of the stomacke, that commeth through heate of cholerique humours. [F]

It is also good for women with childe to stirre vp in them good appetite or meate lust and to take from them all inordinate lustes or vayne longing, and also to stop the wambling in their stomackes and parbreaking. [G]

Currantes or Raysens of Corinthe, do not much differ in vertue, from tapnet or frayle Raysens. [H]

The end of the fyfth part of the Historie of Plantes.

the Historie of Plantes.

¶ The sixth part of the Historie
of Plantes, contayning the description of Trees, Shrubbes, Bushes, and other Plantes of wooddy substance, with their fruites, Rosins, Gummes, and liquers: also of their kindes, fashions, Names, Natures, Uertues, and Operations.

By Rembertum Dodonæum.

Of the Rose. Chap.i.

¶ The Kyndes.

There be diuers kindes of Roses, whereof some are of the garden, sweete smelling, and are set, planted, and fauoured, the others are wilde, growing of their owne kinde without setting about hedges, and the borders of feeldes.

❉ The Description.

Rosa. The Rose.

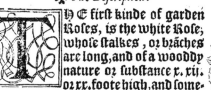

1. THE first kinde of garden Roses, is the white Rose, whose stalkes, or braches are long, and of a wooddy nature or substance x.xij. or xx.foote high, and sometimes longer, if they be staied vp or suckered. In many places set ful of sharpe hooked prickles, or thornes. The leaues be long, and made of fyue or seuen leaues, standing one against another, all vppon a stemme, whereof eache leafe by it selfe is rough, and snipt about the edges lyke to a sawe. The buddes doo growe emongst the leaues vppon short stemmes, closed in with fiue small leaues, whereof two are bearded vppon both sides, two haue no beardes, & the fifth is bearded but vppon one side. When these buddes do open and spreade, the sweete and pleasant Roses do muster and shewe foorth of colour white, with diuers yellowe heares or threddes in the middle. The flowers fallen there come vp rounde knoppes, and red when they be ripe, within which is a hard seede wrapped in heare or wooll. The roote of the Rose bushe is of a wooddy substance lyke the roote of other lowe trees and plantes.

2. The seconde kind of garden Roses be red, & are like to the white in leaues, shutes, and branches, but they neuer grow so high nor so great, neither are the branches so large. The flowers be of a pleasant sauour, of colour redde, and fashioned lyke the white Roses.

3. The third kind are they which some call Roses of Prouince, whose shutes

and ſpringes be lyke them of the red Roſe, ſauing that they growe vp higher, and yet for all that they grow not ſo high as the white Roſe, ſo that this Roſe ſhoulde ſeeme to be a middle ſort or meane kinde betwixt the red and the white Roſes, whiche thing the very colour of the flowers declare to be true, for they be neither redde nor white, but of a mixt colour betwixt red and white, almoſt carnation colour, in al thinges els lyke to the others.

4 The fourth kinde are the browne Roſes of Prouince, the whiche be almoſt lyke to the others in ſhutes ſprings and leaues. But their flowers be of a faire darke red colour, and of a very pleaſant ſauour or ſent, and theſe are beſt to be vſed in medicine.

5 The fyfth is a kinde of ſingle Roſes, whiche is ſmal and called Caſſia or Canel Roſe, or the Roſe ſmelling lyke Caſſia. The leaues wherof be ſmaller then the leaues of the other Roſes, the ſhutes and twigges be alſo ſmall and thicke ſet with thornie prickles, of a browne ruſſet colour, growing almoſt to ÿ height of the Prouince rooſes. The flowers be ſmal and ſingle, ſweet ſmelling, and of a pale red coloor, and ſometime Carnation.

6 The ſixth kinde of Roſes called Muſke Roſes, hath ſlender ſpringes and ſhutes, the leaues and flowers be ſmaller then the other Roſes, yet they grow vp almoſt as high as the Damaſke or Prouince Roſe. The flowers be ſmall and ſingle, and ſometimes double, of a white colour and pleaſant ſauour, in proportion not muche vnlyke the wilde Roſes, or Canel Roſes.

7 The wilde Roſe leaues be rough and prickley. The ſpringes, branches and ſhutes, are ful of ſharpe hookes or crooked prickles, like the white double Roſe of the gardẽ, but much leſſer, & the leaues be ſmaller, the flowers be alſo ſingle, white, & drawing towardes Carnation colour, & without ſauour. The which being fallen away, there riſe rounde knoppes or buttons, lyke as in the garden Roſe plant, within whiche redde knoppes and buttons, the ſeede is couched & laid, in a hearie downe or rough Cotton. Vpon this plant or buſhe is ſomtimes founde a ſpongious baule, rough heared, and of a greene colour turning towardes red, and is to be founde about the moneth of June.

8 Amongſt the kindes of wilde Roſes, there is founde a ſorte, whoſe ſhutes, twigges and branches, are couered all ouer with thicke ſmall thornie prickles. The flowers be ſmal ſingle & white, & of a very good ſauour. The whole plant is baſe and low, and the leaſt of al both of the garden and wilde kind of Roſes.

9 Byſides the Roſes aforeſayd, there is yet another kind of Roſe plant, which beareth yellowe Roſes, in al thinges els lyke to the wilde Roſe plante, as in ſhutes, twigges, and leaues.

10 The Eglentine or ſweete brier, may be alſo counted of the kindes of Roſes, for it is lyke to the wilde Roſe plante, in ſharpe and cruel ſhutes, ſpringes, and rough branches. The leaues alſo be not muche vnlyke, but greener and of a pleaſanter ſmel. The flowers be ſingle, ſmaller then the flowers of the wilde Roſe, moſt commonly white and ſometimes redde, after whiche there come alſo litle knappes or long red beries as in the other Roſes, in whiche the ſeede is couched.

❧ *The Place.*

The tame Roſes, & the Eglentine are planted in gardens. The wilde groweth in many places of Brabant and other Countries, alongſt by hedges and ditches, and other wilde places amongſt bryers and thornes. The other wilde kinde groweth in certayne places vppon rampers and bankes caſt vp by mans handes, and vpon the Sea coaſt of Flaunders.

❧ *The Time.*

The fiue firſt kindes of garden Roſes do flower in May and June, and ſo

Do

Do the wilde Roses & the Eglentine: but the Muske Roses do flower in May, and agayne in September, or there aboutes.

※ *The Names.*

The Rose is called in Greeke ῥόδον: in Latine, Rosa: in high Douche, Rose: in Neatherdouchelande, Roose: The leaues and flowers be called in Latine, Folia Rosarum, that is to say, Rose leaues.

The nayles, that is to say, the white endes of the leaues whereby they are fastened to the knappes (the whiche are cut of when they make Conserue or syrupe of Roses) is called in Latine, Vngues Rosarum, & in Greke, ὄνυχες τῶν ῥόδων.

The yellow heare whiche groweth in the middle of the Rose, is called in Greeke ἄνθος τῶν ῥόδων: in Latine, Flos Rosæ: in shops and of the Arabian Physitions Anthera, that is to say, the blowing of the Rose.

The bud of the Rose before the opening is called Calix.

The fiue litle leaues whiche stande rounde about the bud, or the beginning of Roses, are called in Latine, Cortices Rosarum, that is to say, the shelles or pilles of Roses: some do also cal them, the fiue brothers of the Roses, wherof, as is beforesayd, two haue beardes, and two haue none, and the fifth hath but halfe a one.

The rounde heades or little knoppes, vpon whiche the flowers do growe, and are fastened, and in whiche lyeth the seede, are called in Latine, Capita Rosarum: and in Greek κεφαλαὶ τῶν ῥόδων.

1 The first kinde of garden Roses is called in Italy, Rosa Damascena, in this Countrie, Rosa alba: in Frenche, *Rose blanche:* in high Douche, Weiss Rosen: in base Almaigne, Witte Roosen: in Englishe, White Roses. And this kinde seemeth to be that, which Plinie calleth in Latine, Campana Rosa.

2 The seconde kinde of Roses is called Rosa purpurea, and Rosa rubra: in Englishe, Red Roses, and of the common people, Double Roses: in Frenche, *Rose rouge* and *Roses francois:* in high Douche, Roter Rosen: in base Almaigne, Roode Roosen. And vnder this kinde are comprehended the Roses whiche Plinie calleth Trachinias, amongst whiche Rosæ Milesiæ are the deepest red.

3 The thirde kinde is called in Frenche, *Rosee de Prouinces:* in base Almaigne, Prouinsche Roosen: in high Douch, Liebfarbige Rosen: the which parauenture are they which Plinie calleth Alabandicas Rosas: we cal them in English, Roses of Prouince, and Damaske Roses.

4 The fourth kinde is also called in Frenche, *Rose de Prouins:* in base Almaigne, Prouinsch Roose, and Bruyn Prouinsche Roose: as a name of difference from the other, and these shoulde seeme to be Rosæ Milesiæ of Plinie.

5 The fifth kinde is called of the Herboristes of Brabant, Caneel Rooskens, that is to say, the Roses smelling lyke Canell or Cassia, and possible this is Rosa Prænestina of Plinie: some call it in Englishe, the Cyuet Rose, or Bastarde Muske Rose.

6 The sixth is named of Plinie in Latine, Rosa coroneola, of the writers at this daye Rosa sera, and Rosa autumnalis: in Frenche, *Rose Musquée,* and *Roses de Damas* in base Almaigne, Musket Rooskens: in Englishe also, Muske Roses, bycause of their pleasant sent.

7 The seuenth kinde is called in Greeke συνορόδον: in Latine, Rosa canina, and Rosa syluestris: in Frenche, *Rose sauuage:* in high Douche, Wilder Roosen, and Heckrosen: in base Almaigne, Wilde Rosen: in Englishe, the Bryer bushe, the wilde Rose, and Heptree. The spongious bawle or that rounde rough excrescence whiche is founde oftentimes growing both vppon the wilde Rose and Eglentine bushes, is called of som Apothecaries Bedegar: but wrongfully, for

Bedegar, is not that thistell which is commonly called Carlina. Examine Bedegar, lib. 4. fol. 361.

8 The eight is called of the neather Douchmen, Duyn Rooskens, of the place where as it is founde growing, and it shoulde seeme to be that which the Grecians call κυνόσβατον: in Latine, Canirubus, and Rubus canis, and of Plinie, Rosa spinosa.

9 The Ninth is called the yellow Rose: in French, *Roses iaulnes*.

10 The last is called of Plinie in Greke λυχνίς, Lychnis: in Latine, Rosa Græca: in Frenche, and base Almaigne, Eglantier: in Englishe, Eglantine.

✤ *The cause of the Name and historie thereof.*

The Rose is called in Greeke Rhodon, bycause it is of an excellent smel and pleasant sauour, as Plutarche writeth.

Ye shal also finde this writen of Roses, that at the first they were all white, and that they became red afterwarde with the blood of the Goddesse Venus, whiche was done in this sort.

Venus loued the younker Adonis better then the warrier Mars, (who loued Venus with all his force and might) but when Mars perceiued that Venus loued Adonis better then him, he slewe Adonis, thinking by this meanes, to cause Venus not onely to forgo, but also to forget her friende Adonis, and so to loue Mars only: of the whiche thing when Venus had warning howe and where it should be accomplished, she was suddenly moued & ran hastily to haue rescued Adonis, but taking no care of the way at a suddaine ere she was ware, she threw her selfe vpon a bed or thicket of white Roses, where as with sharpe and cruel thornes, her tender feete were so prickt and wounded, that the blood sprange out abundantly, wherwithal when the Roses were bedewed, & sprincled, they became al red, the which colour they do yet keepe (more or lesse) according to the quantitie of blood that fel vpon them) in remembrance of the cleare & pleasant Venus. Some others write that for very anger which she had conceiued against Mars, for the killing of her friende the faire Adonis, she gaue her tender body willingly to be spoyled and mangled: and in despite of Mars, she threwe her selfe into a bed or herbour of prickley Roses.

Some also say that Roses became red, with the casting downe of that heauenly drinke Nectar, whiche was shed by Cupide that wanton boy, who playing with the Goddes sitting at the table at a Banquet, with his winges ouerthrew the pot wherein the Nectar was. And therefore as Philostratus sayth, the Rose is the flower of Cupide, or Cupides flower.

✤ *The Nature.*

Rose leaues, that is to say of the flowers, be hoate of complexiō, & somwhat moyst, taking part of a binding qualitie. The flower that is to say, the litle yellowe heares that grow in the middle of the Rose, is manifestly drie and astringent: of the same nature are the buddes, knoppes, and fruite, with the rough rounde hearie bawle or excrescence that is founde growing vppon the wylde Rose.

✤ *The Vertues.*

A The iuyce of Roses, especially of them that are reddish, or the infusion or decoction of them is of the kinde of soft and gentle medicines, whiche loose and and ope the belly, and may be taken without danger. It purgeth downewarde cholerique humours, and openeth the stoppinges of the liuer, strengthning and clensing the same, also it is good agaynst hoate feuers, and agaynst the Jaunders.

B It is also good to be vsed against the shaking, beating, and trembling of the hart

hart, for it driueth foorth, and dispatcheth all corrupt and euyl humours, in and about the veynes of the hart.

It is lykewise good to be layd to the inflammation of the eyes, and al other hoate infirmities, and specially agaynst S. Antonies fier or wilde fire.

Roses pounde and beaten smal are good to be layde to the hoate inflammation or swelling of the breastes or pappes, & against the outragious heate of the Midriff & stomacke, also against S. Antonies fire, Erysipelas or Serpigo.

The wine wherein dryed Roses haue ben boyled, is good against the paine of the head, the eyes, the eares, the iawes or gummes, the bladder, the right gutte, and of the Mother or womens secretes, eyther powred in or annoynted with a fether.

The yellow growing in the middle of the Rose (which of some is called the seede & flower of the Rose) stayeth the superfluous course of womens flowers, and specially the white flowers, and all other issues of blood.

The fruite eaten stoppeth the laske, and al other issues of blood.

The wilde Rose powned with Beares grease (as Plinie sayth) is very excellent to annoynt the head against Alopecies, whiche some call the redde scall or falling away of the heare.

The rough spongeous bawle or excrescence that groweth in the wilde Rose bushe, is of great efficacie and vertue against the stone and strangurie: for it bringeth foorth the grauell and the stone, and prouoketh vrine.

Of Jasmine. Chap. ij.

✽ *The Description.*   Iasminum.

Iasmine groweth in maner of a hedge or quickeset, and must be led alongst and caried as the Rose or vine, it bringeth foorth many smal branches full of ioyntes or knottes, the shutes and twigges whereof are filled full of a spongie pith, lyke the pith of Elder. The leaues be of a darke greene colour, parted into fiue or seuē other litle leaues, (growing vppon a stem or foote stalke, like to the Ashe leafe) whereof eche little leafe by it selfe is smothe and somwhat long, nothing at all natched, or toothed about the edges. The flowers be white & long of a sweete and pleasant sauour, and do growe foure or fiue togither at the toppe of the branches.

❡ *The Place.*

Jasmyne groweth in some Countries of his own kind, as in Spaine and some places of England, in this Countrie it is planted in gardens.

✽ *The Tyme.*

Jasmyne flowreth in July and August, but the fruite in this Countrie commeth not to perfection.

❡ The

❦ The Names.

This plant is called of the Arabians Zambach & Iesemin, and accordingly it is called amongst the Herboristes of Englande, Fraunce, and Germanie Iasminum, and Ieseminum, and of some also Iosme, and Iosmenum. The later writers do call it also in Latine, Apiaria: bycause that Bees delight greatly to be about the flowers thereof: some call it also Leucanthemum.

❦ The Nature.

Serapio writeth, that Iasmin is hoate almost in the seconde degree, which a man may also very well perceiue by his bitter taste.

❦ The Vertues.

Iasmine cureth the fowle drie scurffe, and red spottes, it dissolueth cold swellinges, and wennes, or harde lompes, or gatheringes, when it is applyed and layde thereto.

The like vertue hath the oyle of Iasmine, the which put into ye nosethrilles or often smeld to, causeth nose bleeding, in them that are of hoate complexion, as Serapio and our Turner haue written.

Iasmine dryeth reumes or stilling downe of humours from the head, and the moystnesse of the brayne, and profiteth muche against the colde infirmities of the same.

Of Cistus.   Chap.iij.

¶ The Kindes.

There be two sortes of Cistus of Dioscorides, and the Auncientes.

The one is a kinde of plante whereof we do here geue you the figure.

The other plant is of wooddy substance, vppon whiche is founde that humor or fat liquor, whiche they call Ladanum.

1 The first kinde, whiche yeeldeth no Ladanum, is also of two sortes, that is to say, the male and female.

2 The male hath red flowers, and the female white, but in all thinges els one lyke the other.

❦ The Description.

1 THE first kinde of Cistus whiche beareth no Ladanum, hath rounde rough or hearishe stalkes, and stemmes with knobbed ioyntes, and full of branches. The leaues be roundishe and couered with a cotton or soft heare, not muche vnlyke the leaues of Sage, but shorter and rounder. The flowers grow at the top of the stalkes, of the fashion of a single Rose, whereof the male kinde is of colour red, and the femall white, at the last they change into knoppes or huskes in whiche the seede is conteyned.

Wheras Cistus groweth naturally of his owne kind, ther is foūd a certaine excrescence or outgrowing about ye roote of this plant, which is of colour somtimes yellow, sometimes white, and sometimes greene: out of the whiche is a certaine iuyce taken out by art, ye which they vse in shops, & is called Hypocistis.

2 The second kind of Cistus, which is also called Ledon, is a plant of a wooddy substance, growing like a litle tree or shrubbe, with soft leaues, in figure not muche vnlyke the others, but longer and browner.

Upon this plante is found a certayne fatnesse, wherof they make Ladanum the whiche about midsomer, and in the hoatest dayes, is found growing vpon the newe leaues of this Cistus, the whiche newe leaues (after that the seede with the old leaues are fallen of) do first bud foorth and spring in sommer. The sayde fat or grease is not onely taken from the beardes and feete of Goates, or Goate buckes whiche feede vpon the leaues and branches of this plante (as

Dioscorides

the Historie of Plantes.

Cistus non ladanifera. Cistus cum Hypocistide.

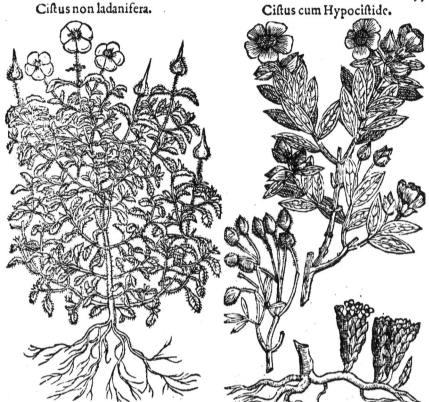

Dioscorides and the Auncientes do write) but also it is gathered & taken with thinges fit for that purpose, deuised by the industrie & diligence of man, as some of the learned writers of our time do report, especially yͤ learned Peter Belon the which hath much haunted and trauayled the Ilande of Crete or Candie.

✻ The Place.

The first kinde of Cistus, whose figure we set foorth here for your better vnderstanding, groweth in sundrie places of Italy, Sicile, Candie, Ciprus, Languedoc, & many other hoate Countries, in rough, stony & vntoyled places.

The seconde kinde is often found in Crete, Ciprus, and also in Languedoc.

✻ The Tyme.

1 The first kinde of Cistus flowreth in June, and sometimes sooner.

2 The seconde Cistus flowreth and bringeth foorth seede in the spring time, & immediatly after the leaues fal of. about sommer it recouereth newe leaues againe, vpon the whiche leaues about midsomer and in the hoatest dayes, is founde a certayne fatnesse, the which is diligently gathered and dried, to make that gumme whiche they call Ladanum.

✻ The Names.

1 The first kinde of these plantes is called in Greeke κίσ‑ καὶ κίσθ‑, of some κίσαρον καὶ κίθαρον: in Latine, Cistus, and Cistus non Ladanifera: of Scribonius Largus, Rosa syluatica.

That which groweth about the roote of Cistus, is called in Greeke ὑποκίσ‑: of some Erythanon and Cytinus, out of this they drawe foorth a sappe or liquor the which they call Hypocistis, and in shoppes Hypoquistidos.

2 ¶ The second kind of Cistus is called in Greke λῆδον ἢ λάδον: in Latine Ledum, Ladum, and of the later writers, Cistus Ladanifera.

The fat dewe or liquor, whiche is gathered from the leaues, is called in Greeke λάδανον: in Latine Ladanum: and in Shoppes Lapadanum.

❧ *The Nature.*

1 The flowers and leaues of Cistus are drie in the seconde degree, and somewhat astringent.

2 That whiche groweth about the rootes is of lyke temperature, but more astringent.

3 Ladanum is ful hoate in the first degree, and reacheth neare vnto the second, and is somewhat drie and astringent.

❧ *The Vertues.*

1 The flowers of Cistus boyled in wine and dronke, stoppeth the laske and all other issue of blood, and it dryeth vp all superfluous moysture, aswell of the stomacke as other partes of the belly.

The leaues of Cistus do cure & heale smal woundes, being laid therevpon. B

2 Hypocistis stoppeth all laskes and flures of the belly, & is of a stronger operation then the flowers or leaues of Cistus: wherfore it cureth the bloody flire and all other flures, especially the superfluous flowing of womens flowers. C

3 Ladanum dronken with olde wine, stoppeth the laske, and prouoketh vrine. D

It is very good agaynst the hardnesse of the matrix or mother, layde to in manner of a pessarie, and it draweth downe the secondes or after birth, when it is layde vpon quicke coles, and the fumigation or parfume therof be receiued vp into the body of women. E

The same applied to the head with Myrrhe and oyle of Myrrhe, cureth the scurffe, called Alopecia, and keepeth the heare from falling of, but wheras it is alredy fallen away, it will not cause the heare to growe agayne. F

Ladanum dropped into the eares with honyed water or oyle of Roses, healeth the payne of the same. G

If it be layde to with wine vpon the scarres or sores of woundes, it taketh them away. H

It is also very profitably mixt with al oyntmentes and playsters, that serue to heate, soften, and asswage paynes, and suche as be made to lay to the breast against the cough. I

Of the Bramble or Blackebery bushe. Chap. iiij.

¶ *The Kindes.*

The Bramble is of two sortes, as Ruelius writeth, the great and the smal.

❧ *The Description.*

1 THE great Bramble hath many long slender branches or shutes, full of sharpe prickley thornes, whereby it taketh holde, and teareth the garmentes of such as go neare about the. The leaues are not smoth but crompled or frompled, and deepely cut rounde about the edges, of colour white vnderneath, and browne aboue. The flowers be white, not much vnlyke the flowers of Strawberies: after commeth the fruit of a swart red colour at the first, but afterwarde it is blacke, and it consisteth of diuers beries clustering togither not muche vnlyke the Mulberie, but smaller, and ful of of a redde wynie sappe or iuyce.

2 The lesser Brambles are muche lyke to the greater, but this creepeth most commonly vppon the ground with his shutes and branches, and taketh roote easily in diuers places incroching grounde with the toppes of his branches.

The

the Historie of Plantes. 661

The branches or shutes of this Bramble be also set with prickley thornes, but the thornes or prickles be not so sharpe: the fruite is also like to a small Mulberie, but lesser then the fruite of the other. The rootes of both kinds do put foorth many slender shutes and branches, the whiche do creepe and trayle alongst the grounde.

Rubus. The Bramble.

✤ *The Place.*

Brambles do grow much in the feeldes and pastures of this Country, and in the wooddes and Copses, and such other couert places.

✤ *The Tyme.*

The Bramble bush floweth frō May to July, and the fruite is ripe in August.

✤ *The Names.*

1 The Bramble, especially the greater sort, is called in Greeke βάτος: in Latine, Rubus, and Sentis: in high Douche, Bremen: in base Almaigne, Breemē & Braemen: in Englishe, the Bramble or blacke berie bushe: in Frenche, *Rouce*.

2 The fruit of the same is called in Greke μόρον τῆς βάτου: in Latine, Morum rubi, & Vacinia: in shops, Mora bati, and of some ignorant people, Mora bassi: in Frenche, *Meure de Rouce*, or *Meurons*: in high Douch, Brombeer: in base Almaigne, Braebesien, and Haghebesien: in Englishe, Bramble beries, and blacke beries.

2 The lesser berie is called of Theophrastus in Greeke χαμαίβατος, Chamebatus, that is to say in Latine, Humirubus: and the fruite is called in Frenche, *Catherine*: in Englishe, a heare Bremble, or heath Bramble, a Cocolas panter, and of some a bryer. The fruite is called a Dewberie, or blackberie.

❡ *The Nature.*

The tender springes and newe leaues of the Bramble, are colde and drie almost in the thirde degree, and astringent or binding, and so is the vnripe fruite.

The ripe fruite is somewhat warme and astringent, but not so much as the vnrype fruite.

✤ *The Vertues.*

The newe springes of the Bramble do cure the euill sores and hoate vlcers A of the mouth and throte, also the swellinges of the gummes, Almondes of the throte, and the vuula, if they be holden in the mouth and often chewed vppon.

They do also fasten the teeth, when the mouth is washed with the iuyce or B decoction thereof. The vnripe fruite is good for the same purpose, to be vsed after the same manner.

The iuyce or decoction therof, is good to be dronken, to stoppe the laske, and C womens flowers and all other issue of blood.

The leaues be stamped, & with good effect are applyed to the region or place D of the stomacke against the trembling of the hart, the payne & loosenesse or ache of the stomacke.

They

They cure the Hemeroydes, and stay backe running, and consuming sores, being layde thereto.

The vnrype fruite stoppeth the belly, the bloody flixe, and all other issues of blood.

The iuyce of the same boyled with hony, is very good against all hoate vlcers, and swellinges of the mouth, the tongue, and throte.

The roote of the Bramble is good against the stone and prouoketh vrine.

Of Framboys, Raspis, or Hyndberie. Chap.v.

❧ *The Description.* Rubus Idæus.

THE Framboye is a kinde of bremble, whose leaues and branches are not muche vnlyke the other Bramble, but not so rough and prickley, nor set with so many sharpe prickles, and somtimes without prickles, especially the newe shutes and tender springes that be not aboue the age of a yeere. The fruite or berrie is redde, but otherwise it is lyke to the other. The roote is long creping in the ground, and putteth foorth euery yere new shutes or springes, the which the next yeere doo bring foorth their flowers and fruite.

❧ *The Place.*

The Framboye is founde in some places of Douchland in darke woods: and in this Countrie they plante it in gardens, and it loueth shadowye places, where as the Sonne shineth not often.

❧ *The Tyme.*

The Framboye flowreth in May and June, the fruite is ripe in July.

❧ *The Names.*

This Bramble is called in Greeke βάτⲟ ἰδαία: in Latine, Rubus Idæus, of the mountayne Ida, in Asia minor, or the lesser Asia, not farre from Troye, where as groweth abundance of this Bramble, and there it was first founde: it is called in Frenche, *Framboisier*: in Douch, Hinnebraemen: in Englissh, Framboys, Raspis, and Hindberrie. Joh. Agricola calleth it in Latine, Crispina.

The fruite of this Bramble is called in Greeke μόρον τῆς βάτου ἰδαίας: in Latine, Morum rubi Idæi: in Frenche, *Framboises*: in high Douch, Hymbeeren, and Horbeeren: in base Almaigne, Hinnebesien, ✻ Frambesien: in English, Raspis, and Framboys berries.

❧ *The Nature.*

The Framboye of complexion is somewhat lyke the blacke berrie, but it is not of so astringent nor drying qualitie.

❧ *The Vertues.*

The leaues, tender springes, fruit and roote of this Bramble, are not much vnlyke

vnlyke in vertue and working, to the leaues, shutes, fruite, and rootes of the other Bramble, as Dioscorides writeth.

¶ The flowers of Raspis are good to be bruysed with hony, and layde to the inflammations and hoate humours gathered togither in the eyes, and Erysipelas or wilde fire, for it quencheth such hoate burninges.

¶ They be also good to be dronken with water of them that haue weake stomackes.

Of Broome. Chap.vi.
¶ The Kyndes.

THE common Broome is of two sortes, the one high and tawle, the other lowe and small, vnder whiche groweth Broome Rape or Orabanche.

Genista. Broome.   Rapum genistæ.
                                          Broome Rape, or Orobanche.

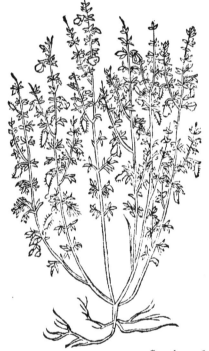 

❧ The Description.

1. THE great Broome putteth foorth first from his roote, harde, strong, and wooddish stemmes, with many smal, long, square, and limmer Branches or twigges lyke rushes, the whiche are easy to ploy and twist any way without breaking. Upon the same growe smal blackish leaues, amongst the which growe pleasant yellow flowers of a sweete smel, in fashion not vnlike the flowers of Pease. When the flowers be fallen there come flatte coddes, in whiche is found seede, that is hard, flat, and brownish. The roote is harde and wooddishe. This Broome groweth commonly to the length of a long or tawle man.

2. The smal Broome is much lyke to that aforesayd, in wooddish stalkes, smal branches, litle leaues, cods, & flowers, sauing that it is muche smaller, & groweth not to length, but abideth alwayes lowe, not exceeding the height of three foote.

3 Ye shall often finde at the roote of this smaller broome a plante which the Brabanders do call Bremrape, that is to say Broome Rape, the which is tackt and fastened at the roote with a long string or thredde, somtimes two or three foote of, or somewhat more from the principall or maister roote. It is almost like to a litle Turne or Peare, brode beneath, and narrow aboue, couered with litle scales or browne shales, and it groweth sometimes alone, and sometimes there are ioyning vnto it other smal Rapes. Frō the same groweth vp a holow brownishe stemme of a foote and a halfe long or more, whiche beareth a great company of lōg white browne flowers, clustering thicke togither round about the stemme, & are fashioned lyke to an open helmet in which there appeare fiue or sixe small threddes, the whiche ye shall perceiue to come foorth at the extremitie or vttermost part of the flower. The flowers past there commeth in their steede long rounde small huskes, in which is found a very small seede lyke vnto sande, of a whitishe colour, neuerthelesse it is both barren and vnprofitable.

4 There is also another plante, muche lyke to this Broome Rape or Naueau, whose stalkes are also brownishe, and they growe to the height of a foote, in flowers, forme, and colour lyke to Broome Rape: sauing that it hath not so great a roote or Naueau in the grounde, but for the most parte it hath a small roote or Naueau, and sometimes it hath no more but certayne hearie threddes or laces wrapped togither, especially that whiche groweth in drie and barren places.

❊ *The Place.*

1.2. The great and small Broome do grow in dry Countries and sandy places, alongst the wayes and sometimes in wooddes.

3 Broome Rape is also founde in dry barren and hungrie groundes, and in leane sandy places about the rootes of the lesser Broome, whiche neuer commeth to perfection, and beareth seldome or neuer any flowers.

4 The other kinde lyke to the aforesayde Naueau, is to be founde in certayne feeldes, amongst Otes, Pease, Beanes, Lentiles, and other grayne, where as there groweth no Broome at all, & also vpon drie banckes, and burned heapes alongst the sea coast.

❊ *The Tyme.*

1.2. Broome floureth in May and June. The coddes & seede are ripe in July.

3.4. Broome Rape is found in June and July. And so is the other plant that is lyke vnto it.

❊ *The Names.*

1.2. This plante is called of the later writers in Latine, Genesta, Genista, and Genestra: in Englishe, Broome: in French, *Genest*, and *Dugenet*, or *Geneste* in high Douche, Ginst, and Pfrimmen: in base Almaigne, Brem, and without doubt it is a kinde of Spartium.

3 That excrescence comming from the roote of Broome, is called of the Herboristes, and of some other in Douche, Brem Rape: in Latine, Rapum Genistæ, and Rapa Genestræ, that is to say, Broome Rape, and is without doubt a kinde of Orobanche and Limodorum.

4 The other kinde whiche is like to the aforesayde Naueau, is called of Dioscorides in Greeke ὀροβάχκη: in Latine, Orobanche, λειμόδωρον, Limodorū, of some other as of Photion also, ὀστρελίωρ, that is, Leguminū Leo. It hath no French nor Douche name that I knowe: Turner lib. 2. fol. 72. calleth it Orobanche, Choke fitche, Strāgle tare, Strangleweede, Orobstrangler, & Choke weede.

❊ *The Nature.*

Broome is full hoate in the seconde degree, and reacheth almost to the third degree, it is scouring and of subtil partes.

❊ *The Vertues.*

The leaues, branches, and croppes of Broome boyled in wine or water, are
good

good for them that haue the dropsie, and for all them that haue any stopping of the liuer, the splene or melt, the kidneyes or bladder: for partly it purgeth & driueth out of the belly, and partly it purgeth by vrine, all waterie, tough, and superfluous humours. The seede is of the same vertue to be taken the quantitie of a dramme, or a dramme and a halfe.

The same seede is very good to be mixt with all medicines whiche prouoke vrine and breake the stone, for by his subtill nature it helpeth the operation of other medicines, seruing to the same purpose.

Broome flowers mingled with swines grease, swageth the paynes of the gowte, being applyed thereto.

This Broome hath al the vertues of Spanish Broome, and it may be vsed against all such infirmities wherevnto Spanish Broome is required.

Broome Rape is counted of some Empiriques (or practisioners) in these Dayes, for an excellent medicine against the stone, & to prouoke vrine, to be first boyled in wine and giuen to drinke. for as they say, it openeth the stoppings of the kidneyes, prouoketh water, breaketh the stone, and driueth foorth grauell.

The freshe and greene iuyce of Broome Rape, doth cure and heale al newe woundes, and clenseth those that are corrupt & rotten: it may be lykewise vsed against other vlcers and corrupt sores, for it mundifieth and bringeth them to healing.

And for the better preseruation of the same iuyce, after it is pressed or taken out of the greene rootes, ye must set it in the Sonne vntil it waxe thicke, or ye must put to it a litle hony, & set it in the Sonne, for then it wilbe better, & more apt to mundifie & clense woundes and rotten vlcers: it may be also takē out of the rootes that be halfe dry, with oyle, & wil serue to al intentes, euē as ye iuyce.

The same oyle of Broome Rape doth scoure and driue away al spottes, lentiles, freckles, pimples, wheales, and pushes, as well from the face, as the rest of the body being often annoynted therewithall.

Dioscorides writeth, that Orabanche may be eaten, either rawe or boyled as the springes of Asparagus.

Of Spanish Broome. Chap. vij.

❀ *The Description.*

1 The Spanish Broome also, hath wooddish stemmes, from which grow foorth long slender plyant twigges, the whiche be bare & naked without leaues, or at least hauing very few small leaues, set here and there farre apart one from another. The flowers be yellow, not muche vnlyke the flowers of the common Broome, after which it hath coddes, wherin is the sede browne and flat, lyke the other Broome seede.

2 There may be wel placed with this Broome, a strange plant which beareth also long shutes or smal twigges, of a swarte colour & straight: and vpon them are smal browne greene leaues, alwaies three ioyned togither, lyke the leaues of Trefoyle, but smaller. The flowers be yellowe, rounde, and cut into fiue or sixe partes, in fashion not much vnlike the flowers of the common Buglosse, afterwarde they do bring foorth graines or berries, as bigge as a pease, & blacke when they be ripe, in which is found the seede, ye which is flat as a Lentil seede. The roote is long & smal, creping hither and thither vnder the earth, & putteth foorth new springes in sundry places. ❀ *The Place.*

This Broome groweth in drie places of Spayne, and Languedoc, and is not founde in this Countrie, but in the gardens of Herboristes.

❀ *The Time.*

This kinde of Broome flowreth in this Countrie in Iune, and somewhat after,

after, the seede is rype in August.

Genista Hispanica, siue Italica.
Spanish, and Italian Broome.

Genista peregrina Trifolia.
Trifolium fruticans.

❦ *The Names.*

This Broome is lykewise called in Latine, Genista: and sometime also Genistra, of the Herboristes of this Countrie, Genistra Hispanica: in base Almaigne, Spaensche Brem: in English, Spanish Broome: and it is not σπάρτον: in Latine, Spartum, whereof Dioscorides and Plinius do write.

The strange plante hath no name that I know: for albeit some would haue it to be Cytisus, this plant is nothing lyke thereto, and is lykewise named Trifolium fruticans.

❦ *The Nature.*

Spanish Broome is hoate and drie of complexion.

❦ *The Vertues.*

The flowers and seede of Spanishe Broome, are good to be dronken with mede or honyed water in the quantitie of a dram, to cause one to vomit strongly, euen as white Hellebor or Neesing powder, but yet without ieopardie.

The seede taken alone looseth the belly, & for the quantitie bringeth foorth great plentie of waterie and tough humours.

Out of the twigges or little braches steeped in water, is pressed forth a iuyce, the whiche taken in quantie of a Ciat or little glasse ful fasting, is good against the Squinansie, that is, a kind of swelling with heate and payne in the throte, putting the sicke body in danger of choking, also it is good against ye Sciatica.

Of

the Historie of Plantes. 667

Of baſe Broome, or Woodwaxen. Chap.viij.

❧ *The Deſcription.* Geniſta humilis.

His Broome is not muche vnlyke the common Broome, ſauing that it is not ſo high nor ſo ſtraight, but lyeth along almoſt vpon the grounde, with many ſmall branches, proceeding frō a wooddy ſtem, and ſet with litle long ſmall leaues, and at ỹ top with many faire yellow flowers not much vnlyke the flowers of the common Broome, but ſmaller: after them come narrow huſkes or coddes, wherein is a flatte ſeede. The roote is harde and of wooddiſh ſubſtance like to the others.

❧ *The Place.*

This kinde of Broome groweth in vntoyled places that ſtande lowe, and ſomtimes alſo in moyſt Clay groundes. It is founde about Anwarpe.

❧ *The Tyme.*

It floureth in July and Auguſt, and ſomtimes after, & ſhortly after the ſeede is rype.

❧ *The Names.*

This plante is doubtleſſe a kinde of Broome, and therefore it may be wel called in Latine, Geniſta humilis: in Italian Cerretta: that is, lowe and baſe Broome: in baſe Almaigne, Ackerbrem: the high Germaynes do make of it Flos tinctorius, that is to ſay, ỹ flower to ſtaine, or dye withal, & do terme it in their language, Ferbblumen, Geel Ferbblumen, and Heyden ſmucke, bycauſe the Dyers do vſe of it to dy their clothes yellow: in Engliſhe, Woodwaxen, and baſe Broome.

❧ *The Nature.*

This plante is of complexion hoate and drie.

❧ *The Vertues.*

Woodwaxen or baſe Broome in nature & operation is lyke to the common A Broome, but not ſo ſtrong.

Of Furze or Thorne Broome. Chap.ix.

❧ *The Deſcription.*

1 HE Furze or prickley Broome, hath many twigges or ſmal branches, of a wooddiſhe ſubſtance, the whiche in the beginning being yet but young and tender, are full of litle greene leaues, amongſt which grow ſmall thornes, the whiche be ſoft and tender, and not very prickley: but when as the twigges or branches, are aboue one yere old, then are they (for the moſt part) cleane without leaues, and then do their thornes waxe harde and ſharpe with cruel prickles. Amongſt the little ſmall leaues, are the flowers of a faynte or pale yellowe colour, and in ſhape and proportion like to Broome flowers, but muche ſmaller, after the whiche come ſmall coddes full of rounde reddiſhe ſeede. The roote is long and plyant.

2 The plant whiche the Brabanders do call Gaſpeldoren, ſhould ſeeme to be

Lll ij   a kinde

668 The ſyxth Booke of

a kinde of thornie Broome, the whiche is rough and very full of prickles, and bringeth foorth ſtraight ſpringes or ſhutes, of a wooddiſh ſubſtance, and without leaues, ſet thicke and ful of long ſharpe pinnes or prickles, very rough, boyſteous, harde and pricking, amongſt which growe ſmall yellowe flowers, and afterwarde coddes, like to the Broome flowers or coddes. The rootes be long growing ouerthwartly in the ground, and almoſt as plyant and limmer as the roote of Reſt harrow or Cammocke.

Geniſtilla. **Thorne Broome.**   Geniſta ſpinoſa. **Furze.**

❧ *The Place.*

1 Furze or thorne Broome groweth in vntoyled places, by the way ſides, and is founde in in many places of Brabant, and Englande.
2 The common or great Furze groweth alſo in the lyke places, and is founde in certayne places of Campany, Brabant, Italy, Fraunce, Buſcaye, and Englande.

❧ *The Time.*

Thorne Broome flowreth in May and June.
At the ſame time flowreth the common Furze.

❧ *The Names.*

1 The firſte plante is called of the later writers in Latine, Geniſtella, and Geneſtalla, that is to ſay, the ſmall Broome: in high Douche, Erdtfrymmen, of ſome, Klein Streichblumen, and Stechende Pfrymmen: in baſe Almaigne, Stekende Brem: in Engliſhe, Thornebroome.
2 And bycauſe the ſeconde kind in his flowers & cods is like Broome, it ſhould therefore

therfore seeme to be a prickley and wilde kind of Broome, wherfore it may be called in Latine, Genistaspinosa, and Genista syluestris: they call it in Frenche, *Du ionc marin* in base Almaigne, Gaspeldozen: in Englishe, the common Whyn, or great Furze. This is not Tragacantha, that is to say, Hirci spina, or Paliurus, as some do thinke: nor yet Nepa or Scorpius.

✻ *The Nature.*

Furze (but especially the leaues) are of nature drie and astringent.

✻ *The Vertues.*

The leaues of Furze boyled in wine or water, and dronken, do stop the excessiue course of womens flowers, and the laske also. [A]

The seede dronken in wine is good against the bitinges and stinginges of venemous beastes. [B]

Of Cammocke, Reste Harrow, or Pety Whyn. Chap. x.

✻ *The Description.* Anonis.

Cammocke or ground Furze hath many small, lythey, or weake branches, set full of swarte greene and roundish leaues, and sharpe, stiffe prickley thornes: amongst whiche are sweete smelling flowers lyke Pease flowers or blowinges, most commonly of a purple or carnatiō colour, somtimes all white, and sometimes yellowe lyke Broome flowers, but that it is very seldome seene or found: after the flowers come small coddes or huskes, ful of brode flat seede. The roote is long and very limmer, spreading his braches both large and long vnder the earth, and doth oftentimes let, hinder, & staye, both the plough and Oxen in toyling the ground, for they be so tough and limmer, that the share & colter of the plough cannot easily diuide, and cut them asunder. ✻ *The Place.*

Cammocke or ground Furze is found in some places of Brabant and England, about the borders of fertill feeldes, and good pastures. ✻ *The Tyme.*

It flowreth most commonly in June.

✻ *The Names.*

This herbe is called in Greke ἀνωνίς ἢ ὀνωνίς: & in Latine, Anonis, & Ononis: of the later writers Arrestabouis, Restabouis, & Remora aratri: of some also Acutella: of Cratenas Aegopyros: in frēch, *Arreste beuf* in high Douch, Hawhechel, Ochsenbrech, and Stalkraut: in base Almaigne, Prangwortel, & Stalcrupt: in Englishe Rest Harrow, Cammocke, Whyn, Pety Whyn, or ground Furze.

✻ *The Nature.*

The roote of Rest Harrow, is drie in the third degree, and somwhat hoate.

✻ *The Vertues.*

The barke of the roote taken with hony prouoketh vrine and breaketh the stone. [A]

The ſyxth Booke of

ſtone. The decoction or broth of the ſame ſodde in wine and dronken, hath the ſame vertue.

The ſame broth boyled in hony and vineger, is good to be dronken againſt B the falling euill, as Plinie writeth.

The ſame boyled in water and vineger and holden in the mouth, whyles it C is warme cureth the tooth ache.

The tender ſpringes and croppes before they bring foorth leaues, preſerued D and kept in brine or ſalt, are good to be eaten in ſalades, for they prouoke vrine, and bring foorth the ſtone and grauell being ſometimes vſed to be eaten.

Of Whortes and Whortelberies. Chap.xi.

❧ *The Kindes.*

There be two ſortes of Whortes, and Whortel beries, wherof the common ſort are blacke, and the other are red.   Vacinia rubra. **Red Whortes.**

Vacinia nigra.
Blacke Whortes.

❧ *The Deſcription.*

1. THE plant which bringeth foorth blacke Whortes, is baſe and lowe of a wooddiſh ſubſtance, bringing foorth many branches of the length of a foote or ſomwhat more: the leaues be round & of a darke greene colour, lyke to the leaues of Boxe or Myrtel, the which at the comming of winter do fall away as the leaues of other trees, and at the ſpring tyme there come forth agayne new leaues out of the ſame braches. The flowers be round and holowe, open before, and grow alongeſt the branches amongſt the leaues. The fruite is round, greene at the firſt, then red, and at the laſt when it is ripe, it is blacke and ful of liquer, of a good and pleaſant taſte. The roote is ſlender, long, and ſouple.

Of this ſorte there are founde ſome that beare white Berries when they be rype, howbeit they are but ſeldome ſeene.

the Historie of Plantes.

¶ The plant that bringeth foorth red wortes, in his growing and branches is like to that, which beareth the blacke berries or whortes, sauing that ye leaues be greater and harder, almost lyke the leaues of a great boxe bush, & they abide the winter without falling away or perishing. The flowers be of a Carnation colour, long, and round, and do growe in clusters at the toppe of the branches. The fruite is red, but els not muche vnlyke the other, in taste rough and astringent, or binding, and not altogither so full of liquer as the blacke whorte. The roote is of a wooddy substance and long.

Amongst these whortes or whortel berries we may recken those which the Germaynes or Almaignes dod call Veenbesien, that is to say, Marrishe or Fenberries, of whiche the stalkes be smal, short, limmer & tender creeping and almost layde flatte vpon the grounde, beset and deckt with smal narrow leaues, fashioned almost lyke to the leaues of ye commō Thime, but smaller, the beries grow vpon very smal stemmes at the ende or toppe of the litle branches, almost lyke the red whortes, but lōger and greater, of colour sometimes all red, and sometimes red speckled, in taste somewhat rough and astringent.

Vacinia palustria.
Marrish Whortes.

❧ The Place.

Whortes growe in certayne woods of Brabant and Englande. The blacke are very common and are founde in many places: but the red are dayntie, and founde but in fewe places.

Marrishe or Fen Whortes growe in many places of Holland, in low, moyst places. ❧ The Tyme.

Whortes do blowe in May, and their berries be ripe in June. Fen or Marrishe Whortes are ripe in July and August.

❧ The Names.

The two first fruites are called in some places of Fraunce, des Cusins, or des Morets. in high Douche, Heydelbeeren, Drumperbeeren, and Bruchbeeren, in Brabāt, Crakebesien, Postelbesien, & Hauerbesien. It may very well be called in Latine Vacinia, bycause they be little berries, in Latine, Baccæ: for as some learned men write, the word Vacinium, commeth of Baccinium, and was deriued of Bacca: and without doubt this name agreeth better with them, then the name of Myrtilli, the whiche some doo call them by: yet these berries be not the right Vacinia, whereof Virgil writeth saying, Alba ligustra cadunt, Vacinia nigra leguntur. Their true English name is Whortes, & of some Whortel beries.

The thirde kinde is called of the Hollanders accordyng to the place of their growing, Veenbesien, and Veencoren, that is to say, Marrishe beries, or Fenberies: and we bycause of the lykenesse betwixt them and the other Whortell beries, do cal them in Latine, Vacinia palustria, that is to say, Marrish Whorts, and Fenberies: for there is none other name knowen vnto vs, except it be Samolus of Plinie, or Oxyococron of Valerius Cordus.

❧ The Nature.

Whortes, but especially those that be blacke, do coole in the second degree, &

LLl iiij somewhat

ſomewhat they drie and are aſtringent. Of the lyke temperament are Marriſh whortes.

❈ *The Vertues.*

Whortes, and ſpecially thoſe that be blacke, eaten raw or ſtued with ſuger, are good for thoſe that haue hoate and burning feuers, and agaynſt the heate of the ſtomacke, the inflammation of the liuer, and interior partes.

They ſtoppe the belly, and put away the deſire or will to vomit. B

With the iuyce of them (eſpecially of the blacke kinde) is made a certayne C medicine called of the Apothecaries Rob, the whiche is good to be holden in the mouth againſt great drieth and thirſt in hoate agues, and is good for al the purpoſes whereunto the beries do ſerue.

Fen or Marriſhe Whortes doo alſo quenche thirſte, and are good againſt D hoate feuers or agues, and againſt all euil inflammation or heate of blood, and the inwarde partes, lyke to the other whortes whereunto they are much alike in vertue and operation.

To conclude the blacke and Marriſhe Whortes are muche lyke in nature, E vertue, and operation vnto Rybes, or the red, and beyond ſea gooſeberies, and may be taken and vſed in ſteede of them.

Of wilde Kuſhe, or Sumac. Chap.xij.

Rhus ſylueſtris Plinij.   Gratia Dei quibuſdam.
Plinies wilde Sumac.   **Hedge Hyſope.**

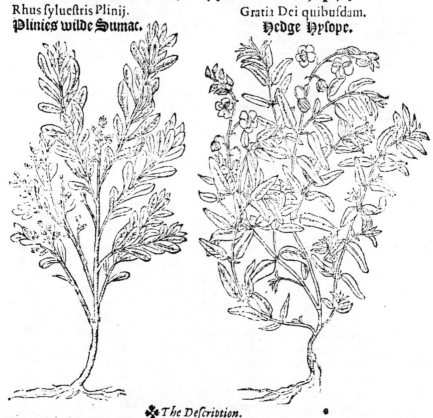

❈ *The Deſcription.*

This is a lowe ſhrub or wooddiſh plant, with many browne hard branches, vpon whiche grow leaues ſomewhat long, and not much vnlyke the leaues of the greater Boxe tree, but longer. Amongſt the leaued branches bꝛanches, come vp other litle bꝛanches, vpon whiche growe many ſpokie eares oꝛ tuftes, ful of many ſmall flowers, and after them ſtoꝛe of ſquare oꝛ coꝛnered ſeedes cluſtering togither. This ſeede is of a ſtrong ſauour and bitter taſte, and full of fat and Oylie ſap. The roote is hard as the roote of Whoꝛtes oꝛ Whoꝛtell plantes.

We may well ioyne to this, that wilde plant which Hierome Bocke calleth Hedge Hyſope, which bꝛingeth foorth from a wooddiſh roote, ſlender ſtalkes, ſpꝛeade abꝛode vpon the ground, couered with litle grayiſh leaues, ſomething rough, in faſhion lyke to garden Hyſope, but ſhoꝛter, at the top of whiche plant come foorth flowers faſhioned lyke to the flowers of wilde Tanſie, of colour ſomtimes a faint yellow, and ſomtimes white, after which come vp ſmal round knoppes oꝛ buttons, in whiche is founde a yellowe ſeede.

❧ The Place.

The firſt plant groweth in Bꝛabant, and in many places of the ſame Countrie about Kempen.

Hedge Hyſope is founde in certayne places of Germanie and Fraunce, in wilde vntoyled places and mountaynes.

❧ The Time.

This Rhus flowꝛeth in May and June, the ſeede is ripe in July & Auguſt. Hedge Hyſope flowꝛeth in June and July.

❧ The Names.

The firſt plant is called of the Bꝛabanders Gagel, & is of ſome Apothecaries called Myrtus, and the ſeede therof Myrtilli: notwithſtanding, it is not Myrtus. Wherefoꝛe it is called of ſome of the later wꝛiters, Pſeudomyrſine, and Myrtus Brabantica, and in ſome places of Almaigne they cal it Altſein, and Boꝛſt, ſome take it to be ἰλαγνο, Oleagnus, of Theophꝛaſtus, wherevnto it is not very muche lyke, but it ſeemeth to be that kinde of wilde Rhus, whiche Plinie ſpeaketh of in the xxiiij. Chapter of the xj. booke of his excellent woꝛke, called the Hiſtoꝛie of Nature.

Hedge Hyſope is called in high Douche, Heyden Pſop, Felde Pſop: in baſe Almaigne, Heyden Hyſope, bycauſe it groweth in Hedges, and wilde places. Some do call it in Latine, Gratia Dei, howbeit it is nothing lyke, Gratia Dei, oꝛ Gratiola, whiche is a kinde of the leſſe Centaurie, ſet foorth in the thirde part of this Hiſtoꝛie Chap. xlij. It ſeemeth to be Selago Plinij, Ualerius Coꝛdus calleth it Helianthemum. ❧ The Nature.

The wilde Rhus, oꝛ Sumac, eſpecially the ſeede is hoate and dꝛie almoſt in the thirde degree.

❧ The Vertues.

Wilde Rhus oꝛ Sumac is not vſed in medicine, but ſerueth to be layde in wardrobes and pꝛeſſes to keepe garmentes from mothes.

Of Kneeholme. Chap. xliij.

❧ The Deſcription.

Kneeholme is a lowe wooddiſhe plante, like the wilde Rhus oꝛ Sumacke, with rounde ſtalkes ful of bꝛanches, couered with a bꝛowniſh thicke barke oꝛ rinde, ſet full of blackiſhe leaues which are thicke and pꝛickley nothing differing frō the leaues of a myꝛtel tree, oꝛ the ſmaller Boxe, ſauing that eache leafe hath a ſharpe pꝛickle in the toppe. The fruite groweth in the middle vpon the leaues, the whiche is faire and red when it is rype, with a harde ſeede oꝛ kernell within. The roote is white and ſingle.

❧ The Place.

Kneeholme, groweth in Italy, Languedoc, and Bourgoyn, & in ſome places of

The sixth Booke of

of England, as in Essex, Kent, Barkeshire, and Hamshire, in many places it is planted in gardens.

❦ The Tyme.

This plant keepeth his leaues both winter and sommer, and in Italy and such lyke places where as it groweth of his owne accorde, it bringeth foorth his fruit in August, but in this Countrie it beareth no fruite.

❦ The Names.

This herbe is called in Greeke μυρσίνη ἀγρία, ὀξυμυρσίνη, μυρτάκανθα, ἡμυάκανθα: in Latine, Ruscum, Ruscus, & Myrtus sylvestris, and Scopa Regia, as Marcellus an Auncient writer sayth. In Shoppes it is called Ruscus: in English, Kneeholme, Kneehul, Butchers broome, and Petigree. also we may cal it þ wilde Myrtel: it is called in Frēch *Myrte sauuage*, of some *Buys poignant*, and *Housson*. in high Douche, Meuszdorn, and Keerbesien: in base Almaigne, Stekende palme, that is to say, Prickley Bore, bycause it is somewhat lyke Bore, the whiche they doo commonly call Palmboom: of some also MuysDorne.

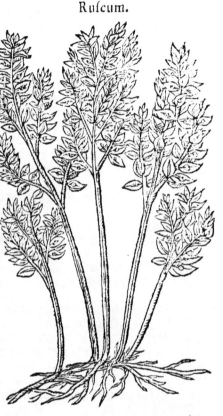

Ruscum.

❦ The Nature.

The rootes and leaues are hoate in the seconde degree and drie in the first.

❦ The Vertues.

The decoction of Kneeholme or Ruscus made in wine and dronken prouo keth vrine, breaketh the stone, and driueth foorth grauel: and is good for them that can not with ease make their water. [A]

It is good to be taken in the like maner against the Iaundise, the headache, and to prouoke womens flowers. [B]

The leaues and fruite be of the same working or facultie as the rootes be, but not so effectuall or strong, wherefore they be not much occupied or vsed. [C]

Of Horse tongue, Double tongue, and Laurus of Alexandria. Chap.xiiij.

❦ The Description.

1. Double tongue hath rounde stalkes lyke Salomons seale, of a foote and a halfe long, vpon which e grow vpon eache syde thicke brownish leaues, not muche vnlyke to Baye leaues, vppon the whiche there groweth in the midle of euery leafe another smal leafe fashioned like a tongue, and betwixt those smal and great leaues, there growe rounde redde beries as bigge as a pease or thereaboutes. The roote is tēder, white, long and of a good sauour.

2. There is founde another kinde of Double tongue, as some learned men write, the which also bringeth foorth his fruite vpon the leaues, and is lyke to the aforesayde, in stalkes, leaues, fruite, and rootes, sauing that there growe none other smal leaues by the fruit vpon the great leaues. The

the Historie of Plantes.

❧The learned Matthiolus setteth foorth a thirde kinde, the whiche is much lyke to the abouesayd in rootes and leaues: but the fruite thereof groweth not vpon the leaues as in the others, but euery berie groweth vppon a stemme by it selfe comming foorth betwixt the stemme and the leaues: the sayde beries be redde, and as bigge as Ciche Peasen.

Hippoglossum.
Horse tongue or double tongue.

Laurus Alexandrina.
Laurus of Alexandria.

❧ *The Place.*

Double tongue groweth in Hungarie and Austriche, and in some darke wooddes of Italy. The Herboristes of this Countrie doo plant it in their gardens.

❧ *The Tyme.*

It deliuereth his seede in September.

❧ *The Names.*

1 The first of these herbes is called in Greeke ἱππόγλωσσον, or ὑπόγλωσσον, or ἐπίγλωσσον, and as some write, ἐπιφυλλόκαρπον: in Latine also Hippoglossum, and Hypoglossum, of the later writers vuularia, Bonifacia, Lingua pagana, and Bislingua: in high Douche, Zapffkinkraut, Hauckblat, Auffenblat, Beerblat, & Zungenblat: and according to the same in base Almaigne, Keelcruyt, Tongebladt, and Tapkenscruyt, that is to say, Tongue herbe, or Tongue worte, also the Pagane or vplandishe tongue, Horse tongue, and double tongue, & tongue blade.

2 The seconde seemeth to be a kinde of Hippoglossum, and therefore some cal it Hippoglossum fœmina, and the first they call Hippoglossum mas.

The

The third is called in Greeke δάφνη ἀλεξάνδρεια, καὶ δάφνη ἰδαία: in Latine, Laurus Alexandrina, and Laurus Idæa, of some late writers Victoriola: in Frenche, Laurier Alexandrin in base Almaigne, Laurus van Alexandrien: in Englishe also, Laurus of Alexandria, or tongue Laurell.

❀ The Nature.

Tongueblade or double tongue his nature is to aſſwage payne, as Galen ſaith.

But the Laurel of Alexandria is hoate and drie of complexion.

❀ The Vertues.

The leaues and rootes of double tongue, are much commended againſt the ſwellinges of the throote, the vuula, and the kernelles vnder the tongue, and agaynſt the vlcers and ſores of the ſame, taken in a gargariſme.

Marcellus ſaith, that in Italy they vſe to hange this herbe about þ neckes of young children that are ſicke in the vuula: a garlande made therof & worne, or ſet next vpon þ bare head, is good for the headache, as Dioſcorides writeth. Baptiſta Sardus writeth, that this herbe is excellent for the diſeaſes of the mother, and that a ſpooneful of the powder of the leaues of double tongue cauſeth the ſtrangled matrix or mother to deſcende downe to his naturall place.

The roote of Alexandria Laurel boyled in wine and dronken, helpeth the ſtrangurie, prouoketh vrine & womens naturall ſicknes, eaſeth them that haue harde trauell, expelleth the ſecondine, and all other corruption of the matrix.

Of Tamariſk.   Chap.xv.

❡ The Kyndes.

Tamariſk is of two ſortes, as Dioſcorides ſaith, great and ſmall.

❀ The Deſcription.

Myrica humilis.

1 Tamariſk is a litle tree or plant, as long as a man, with many branches, of colour ſometimes pale greene, and ſometimes browniſh, vppon the whiche grow litle grayiſh leaues, almoſt like the leaues of Heath or Hather, or lyke to Sauine. The flowers be of a browne purple colour, and lyke wool or Cotton, the which at their falling of, are caried away with the winde.

2 The greater Tamariſk hath leaues lyke þ other, but it groweth much higher that is to ſay, to the length of other great trees, and beareth a fruit like to the leſſer Oke Apples or galles.

❀ The Place.

1 The ſmal or low Tamariſk groweth by ſlow ſtreames and ſtanding waters: and is founde in ſome places of Germanie, by the courſe or ſtreame of the riuer Rhene.

2 The greater Tamariſk groweth in Syria and Egypt, the whiche is yet vnknowen to them of our time.

❀ The Time.

The little Tamariſk flowreth in the ſpring

spring of the yeere, but especially in May.

The Names.

This plant is called in Greeke μυρίκη: in Latine, Myrica, and Tamarix: in the best Apothecaries Shops, Tamariscus, and according to the same in Englishe, Camarisk: in Frenche, Tamarix, of some Bruyere sauuage: in high Douche, Tamariscen holtz, of some Birtzenbertz: in base Almaigne, Tamarischboom.

The Nature.

The leaues and newe springes of Tamarisk, are somwhat warme and abstersiue, without any manifest drouth or drines. The fruite and the barke therof are drie and astringent, and of the nature of galles.

The Vertues.

Tamarisk is a medicine of excellent power and vertue agaynst the hardnes and stopping of the milt or Spleene, and for the same purpose it is so good and founde true by experience, that Swine whiche haue bene dayly fedde out of a trough or vessel made of the Tamarisk tree or timber, haue bene seene to haue no milt at al. And therefore it is good for them that are Splenitique to drinke out of a cup or dishe made of Tamarisk wood or timber.

The decoction of the leaues & young springes of Tamarisk boyled in wine with a little vineger and dronken, doth heale and vnstoppe the hardnesse and stoppings of the milt or splene. The same vertue hath the iuyce therof dronken in wine, as Plinie sayth.

Against the tooth ache, it is also very good to holde in the mouth the hoate decoction of the leaues and tender branches of Tamarisk boyled in wine.

The decoction of the leaues made in water, doth stay the superfluous course of womens flowers, if they sit or bath in the same whiles it is hoate.

The same decoction made with the young shutes and leaues killeth the lice and nittes, if the place whereas they be, be washed therewithal.

The fruite of the great Tamarisk is good against the spetting of blood, the superfluous course of womens flowers: against the laske and bitinges of venemous beastes.

They vse this fruite in steede of Galles in medicines, that are made for the disease of the mouth and eyes.

The barke of Tamarisk is of the same vertue as the fruite, and is good to stoppe laskes and all issue of blood.

Of Heath.   Chap.xvi.

The Kindes.

THere is in this Countrie two kindes of Heath, one whiche beareth his flowers alongst the stemmes, and is called lõg Heath. The other bearing his flowers in tutteys or tuftes at the toppes of the branches, the whiche is called smal Heath.

The Description.

Heath is a wooddish plant ful of branches, not much vnlyke the lesser Tamarisk, but much smaller, tenderer, and lower, it hath very small iagged leaues, not much vnlyke the leaues of garden Cypres (which is our Lauender Cotton) but browner and harder. The flowers be lyke smal knoppes or buttons parted in foure, of a fayre carnation colour, and sometimes (but very seldome) white, growing alongst the branches from the middle vpwarde euen to the top. The rootes be long and wooddishe, and of a darke red colour.

The second kind of Heath, is also a litle base plant, with many litle twigges, or small slender shutes comming from the roote, of a reddishe browne colour,

Erica. Heath.   Erica altera. Smal Heath.

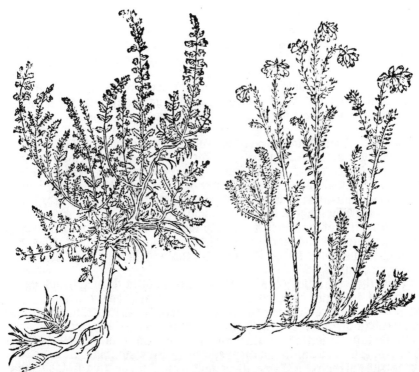

with very smal leaues, in fashion not vnlike the leaues of common Thyme, but muche smaller and tenderer, the flowers growe at the toppe of the strigges, or twigges, fiue or sixe in a company togither, hanging downewardes, of colour Carnation and red, of making long and rounde, hollowe within, and open at the ende lyke a little tonnell, smaller then a Cornell which is the fruite of a Cornell tree. The roote is tender, and creeping alongst, and putteth foorth in diuers places many newe twigges or strigges.

✻ *The Place*.

Heath groweth vpon mountaynes that be drie, hungrie and ἃ barren, and in playnes wooddes and wildernesse.

✻ *The Tyme*.

1 The first kinde of Heath floureth both at the beginning and the end of sommer vntyl September.

2 The seconde kinde floureth about midsommer.

✻ *The Names*.

1 Heath, Hather, and Lyng is called in high and base Almaigne, Heyden: and is thought of the later writers to be that plant which Dioscorides calleth in Greeke ἐρείκη: in Latine, Erice, and Erica.

2 The smaller kinde also without doubt is a Heath: and therfore it may truely be called in Latine, Erica altera: in Greeke ἐρείκη ἑτέρα.

✻ *The Nature*.

Both kindes of Heath haue a manifest and euident drynesse.

✻ *The Vertues*.

The iuyce of the leaues of Heath dropped into the eyes, doth heale ye paine of the same, taketh away the rednesse, and strengthneth the sight.

the Historie of Plantes. 679

If Heath be the true Erica of Dioscorides, the flowers and leaues thereof are good to be layed vpon the bitinges and stingings of Serpentes, and such lyke venemous beastes.

The learned Matthiolus in his Commentaries vppon Dioscorides lib.j. doubteth not of this plant but that it is Erica of Dioscorides, whereunto he hath set two other figures of strange Heath, sent vnto him by one Gabriel Falloppius a learned Physition. Moreouer he commendeth muche the decoction of our common Heath made with fayre water, to be dronke warme both morning and euening, in the quantitie of fiue vnces, three houres before meate, agaynst the stone in ye bladder, so that it be vsed by the space of thirtie dayes: but at the last the patient must enter into a bath made of the decoction of Heath, & whiles he is in the said bath, he must sit vpon some of the Heath that made the foresayde bath, the which bath must be oftentimes repeted and vsed. for by the vse of the sayd bath and diet or decoction he hath knowen many to be holpen, so that the stone hath come from them in very small peeces. Also Turner sayth, that for the diseases of the milte, it were better to vse the barkes of Heath (in steede of Tamarisk) then the barke of Quickbeme. Tur.li.1.fol.210.li.2.fol.59.

Of Cotton or Bombace. Chap.xvij.

❦ *The Description.* Xylon.

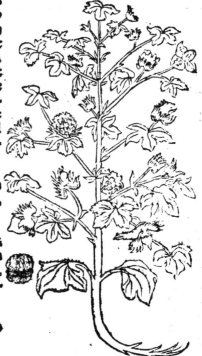

This plant is but a shrub or low tree that groweth not very high: the leaues be brode with deepe cuttes or slittes, smaller then vine leaues, but els somwhat lyke. The flowers be yellowe, and somewhat purple in the middes, iagged about the edges. The fruite is almost lyke to Fylbeardes, brode and flat, and full of fayre white cotton, or the downe that we call Bombace, in whiche the seede lyeth hydden.

❦ *The Place.*

Cotton tree groweth in Egypt and the Indias, and is planted in Candie, Maltha, and other suche Countries.

❦ *The Names.*

Cotton is called in Greeke ξύλον καὶ γοσσίπιον: and also in Latine, Xylum, and Gossipium: in shoppes, Cotum, Bombax, and Bombasum: in high Douche, Baumwol: in base Almaigne Boomwolle.

❦ *The Nature.*

The seede of Cotton, is hoate and moyst, as Serapio saith.

❦ *The Vertues.*

The seede of Cotton swageth the cough, and is good agaynst all colde diseases of the breast, augmenteth naturall strength, and encreaseth the seede of generation.

Of Capers. Chap.xlviij.

Capparis.

❊ *The Description.*

THE Caper is a prickley plant or bush almost lyke the Bramble, with many shutes or braunches spread abrode & stretched alongst the grounde, vppon whiche do grow hard sharpe and crooked prickles with blackishe rounde leaues, standing one against another, not muche vnlyke the leaues of Asarabacca, or folefoote, or the leaues of a Quince tree, as Dioscorides saith, but muche rounder. Amongst the leaues spring vp smal knops or buds, the whiche do open into faire starrelyke flowers, of a pleasant smell or sauour: afterwardes commeth the fruite whiche is long & round, smaller then an Olyue, & hath in it smal cornes or kernelles (lyke to them in the Pomgranate, as Turner saith.) The roote is long and wooddishe, couered with a white thicke barke or rinde, whereof they vse in Physicke.

❊ *The Place.*

Capers growe in rough vntoyled places, in stony sandy grounde, and in hedges: and it groweth plentifully in Spayne, Italy, Arabia, and other such hoate Countries: it groweth not in this Countrie, but the fruite and flowers are knowen vnto vs, bycause they be brought to vs from Spayne preserued in bryne or salt.

❊ *The Names.*

Capers are called in Greke κάππαρις: and in Latine, Capparis, of some also κυνόσβατος, Cynosbatos, that is to say in Latine, Rubus Canis, and Sentis Canis: in high Douche, Capperen: in base Almaigne, Cappers.

❊ *The Nature.*

The Capers that grow in Africa, Arabia, Lybia, & other hoate Countries, are very hoate euen almost in the third degree, causing wheales, pustulles, and vlcers in the mouth, consuming and eating the flesh euen to the bones, but they which growe in Italy and Spayne, be not so strong (and, as Simeon Sethy writeth, they be hoate and drie in the seconde degree) and therfore are fitter to be eaten, bycause they be moderately hoate, dry, & astringent, especial the barke of the roote which is most desired in Physicke: for the flowers & young leaues be not of so strong operation, and therefore doo serue better to be eaten with meates.

❊ *The Vertues.*

The barke of the roote of Capers is good against the hardnesse and stopping of the milt, to be taken with Oximel, or mingled with oyles & oyntments fit for that purpose, & applyed or layd to outwardly vpon the place of the milt.

Also they vse with great profite to giue of this roote in drinke, to suche as haue the Sciatica, the Palsie, and to them that are bruised or squat, or haue fallen from aboue,

It

It ſtirreth vp womens deſyred ſickneſſe, & doth ſo mightily prouoke vrine,
that it waxeth blooddy, if it be to muche vſed and in to great a quantitie.

It cleanſeth olde vlcers and rotten ſores that are harde to heale, and layde to with vineger, it taketh away fowle white ſpottes and morphew.

The fruite and leaues of Capers haue the lyke vertue as the rootes, but not ſo ſtrong, as Galen ſayth.

The ſeede of Capers boyled in vineger, and kept warme in the mouth, ſwageth toothe ache.

The iuyce of the leaues, flowers and young fruite of Capers, killeth the wormes of the eares when it is dropped in.

The Capers preſerued in ſalt or pickel, as they be brought into this Countrie, being waſhed, boyled, and eaten with vineger, are meate and medicine: for it ſtirreth vp appetite, openeth the ſtoppinges of the liuer and milt, conſumeth and waſteth the colde flemes that is gathered about the ſtomacke. Yet they nouriſhe very litle or nothing at all, as Galen ſaith.

Of Gooſeberies.   Chap. lix.

✤ *The Deſcription*   Vua Criſpa.

THE Gooſeberie buſhe is a wooddiſhe prickley plante growing to y height of two three, or foure foote, with many whitiſhe branches, ſet full of ſharpe prickles, and ſmothe leaues of a light greene colour, ſomewhat large and round, cut in, & ſnipt about almoſt like to vine leaues. Amongſt the leaues growe ſmal flowers, and after them rounde beries, the whiche are firſt greene, but when they waxe ripe, they are ſomewhat yellowe or reddiſhe and cleare through ſhining, of a pleaſant taſte ſomewhat ſweete. The roote is ſlender, harde, wooddiſhe, and full of hearie ſtringes.

✤ *The Place*.

The Gooſeberie is planted commonly almoſt in euery garden of this Countrie alongeſt the hedges & borders of the ſame.

✤ *The Tyme*.

The Gooſebery buſh ſpringeth by times, and waxeth greene in Marche, yea and ſometimes in Februarie, it floureth in April, and bringeth foorth his fruite in May the which is muche vſed in meates. The fruite is ripe at the ende of June.

✤ *The Names*.

The Gooſeberie is called of the later writers in Latine, Groſſularia. Geſnere thinketh it to be ἀκανθα κτανθος, Spina Ceanothos of Theophraſtus.

The fruite is called in Latine, Vua criſpa: of ſome Groſſula: of Matthiolus, Vua ſpina, whiche may be Engliſhed, Thorne grape: in Frenche, *des Groiſſelles*: in high Douche, Kreuszbeer, and Kruſelbeer: in baſe Almaigne Stekelbeſien, or Kroesbeſien, and of ſome alſo Knoeſelen.

682 The fyxth Booke of

❊ *The Nature.*

The fruite before it is ripe (for then it is most vsed) is colde and drie in the seconde degree, and binding, almost of the same nature that the vnripe grapes of the vine are.   ❊ *The Vertues.*

The vnripe Gooseberie stoppeth the belly, and all issue of blood, especially the iuyce of them pressed foorth and dried. A

The same greene Gooseberies or their iuyce, is very good to be layd vpon hoate inflammations, Erysipelas, and wilde fire. The leaues be likewise good for the same purpose, but not al thing so vertuous. B

The greene Gooseberie eaten with meates prouoketh appetite, & cooleth the behemēt heate of the stomacke and liuer, and doth swage and mitigate the inwarde heate of the same, and is good against agues. C

The young leaues eaten rawe, do prouoke vrine, and are good for suche as are troubled with the grauell and stone. D

Of redde Gooseberies.   Chap.xx.

¶ *The Kyndes.*

OF these beries there be two sortes in this Countrie: the one beareth a red fruite of a pleasant taste, the other beareth a blacke fruit of an vnpleasant taste.

❊ *The Description.*

Vua vrsi Galeni. Ribes, vulgò.

1 THE red beyondsea Gooseberie, hath woddishe pliant branches, couered with a brownish barke, and brode blackishe leaues, not muche vnlyke vine leaues, but smaller. The flowers growe amongst the leaues, vppon the young sprigges or sprayes clustering togither, and a great many hanging downeward by smal stringes or stemmes: whē those flowers be past there grow vpon euery syde of the said stringes many small greene berries at the first, ye which afterwarde waxe red, of a pleasant quicke and sharpe taste.

2 The blacke Gooseberies are lyke to the aforesayde, in branches, leaues, flowers, and fruite, sauing they be of a blacke colour and vnpleasant taste, and therefore not vsed.

❊ *The Place.*

1 Beyondsea Gooseberies are planted in diuers gardens, wherewithall they vse to make twisted hedges alongst by the allies and borders of gardens.

2 The blacke Goosederies growe of them selues in moyst vntoyled places, alongst by the ditches & water courses.

❊ *The Time.*

Beyondsea Gooseberies are most commonly rype in July.

❊ *The Names.*

This plant is called of the later writers in Latine, Grossularia rubra, Grossularia.

laria tranſmarina, Ribes, and Ribeſum: yet this is not right Ribes.

The fruite is also called of the later writers Groſſulę tranſmarinæ, and it ſhoulde ſeeme to be the fruite the whiche Galen lib. 7. de medicamentis ſecundùm loco, calleth ἄρκτου ϛαφυλαί, Vuæ vrſi: in ſhoppes they cal it Ribes: in french, Groiſelles d'outre mer: in high Douche, S. Johans treuble, or Treublin, and S. Johans beerlin: in baſe Almaigne, Beſiekens ouer zea, and Aelbeſiekens.

1 The firſt kinde is called Groſſulæ rubræ, Ribes rubrum: in Engliſhe, Redde Gooſeberies, Beyondſea Gooſeberies, Baſtard Corinthes, & common Ribes: in frenche, Groiſelles rouges: in baſe Almaigne, Roode Aelbeſien, and of this ſort onely they vſe in ſhoppes, and meates.

2 The ſecond kinde is called Ribes nigrum: in Engliſh, Blacke Gooſeberies, or blacke Ribes: in frenche, Groiſelles noires: in baſe Almaigne, Swerte Aelbeſien.

※ *The Nature.*

The red Gooſeberies are cold and drie in the ſecond degree, and aſtringent or binding.

※ *The Vertues.*

Red Gooſeberies do refreſhe and coole the hoate ſtomacke and liuer, and it is very good to be taken againſt al inflammation, & burning heate of the blood, and hoate agues.   A

The ſame holden in the mouth & chewed, is good againſt al inflammation, and hoate tumours in the mouth, and quencheth thirſt in hoate agues.   B

It ſtoppeth the laſke comming of a choleríque humour, and the bloodây flixe, eſpecially the Robbe or dried iuyce thereof.   C

The rob made with the iuyce of common Ribes and Sugar, is very good for all the diſeaſes aboueſayde, it ſtoppeth vomitinges, and the vpbreakinges of the ſtomacke, and is very good in hoate agues to be dronke with a litle cold water, or to be holden in the mouth againſt thirſt.   D

The blacke Gooſeberies are not vſed in Phyſicke.   E

Of Berberis. Chap.xxi.

※ *The Deſcription.*

THE Barberie plante, is a ſhrub or buſhe of ten or twelue foote high or more, bringing foorth many wooddiſh branches, ſet with ſharpe prickley thornes. The leaues be of a whitiſh greene & ſnipt round about, the edges like a ſaw ſet with fine prickles, of a ſharpe ſower taſt, & therfore is vſed in ſawces in ſteede of ſorrel. The flowers be ſmal, of a pale yellowiſhe colour, growing amongſt the leaues vpon ſhort cluſtering ſtems, after ỹ flowers there hang by the ſayd ſtemmes litle long round beries, red at the the firſt when they be ripe, but when they be dry, they are blackiſh, in taſte ſower & aſtringent, with a harde gray or blackiſhe kernel in the middle whiche is the ſeede. The roote is harde and long, diuided into many branches, very yellow within as al the reſt of the wood of this plant is, of taſte ſomewhat rough or ſowre binding.

※ *The Place.*

The Barberie buſhe is founde in Brabant about the borders of wooddes and hedges. It is also muche planted in gardens, eſpecially in the gardens of Herboriſtes.

※ *The Time.*

The Barberie buſhe putteth foorth newe leaues in April, as the moſt part of other trees doth: it flowreth in Maye, and the fruite is ripe in September.

Mm iiij   ※ *The*

The Names.

Crespinus Matthioli.

This plant is called in shops Berberis, especially the fruit thereof, ẏ which to them is best knowen: the learned Matthiolus calleth this plant in Latine, Crespinus: in English, Barberies, & the Barberie bushe or tree: in Frenche, *Espine vinette*: in high Douche, Paisselbeer, Saurich, Erbsel, Versich: in base Almaigne Sauseboom. This is a kinde of Amyrberis, that is to say, Oxyacantha, in Auicen and Serapio, the which do set out two kindes of Amyrberis: The one hauing a redde fruite, the whiche Dioscorides calleth Oxyacantha, & is described hereafter in the xxxj. Chap. the other with a long blackishe fruite, and is counted for the best Amyrberis, and is that whiche the later writers do call Berberis, it is also very lyke to be the Oxyacantha, described by Galen, lib. 2. de Alimentor. facultat. amongst those kindes of shrubbes or plantes whose young shutes and springes are good to be eaten.

The Nature.

The leaues and fruite of Barberies, are of complexion colde and drie in the second degree, & somewhat of subtil partes.

The Vertues.

With the greene leaues of the Barberie bush they make sawce to eate with meates as they do with Sorrel, the which doth refresh and prouoke appetite, and is good for hoate people and them that are vexed with burning agues. A

The fruite stoppeth the laske, and all superfluous fluxes of women, and al vnnaturall fluxe of blood. B

The roote thereof stieped in lye, maketh the heare yellow, if it be often washed therewithall. C

Of Acatia. Chap.xxij.

The Kyndes.

There be two sortes of Acatia, the one growing in Egypt. The other in the Countries of Pontus.

The Description.

1. The first kind of Acatia is a litle thornie tree or bushe with many branches, set full of sharpe prickles, amongst whiche do arise leaues parted into many other small leaues. The flowers are white. The seede is brode lyke Lupines, inclosed in long coddes, from out of whiche they drawe a iuyce or blacke liquor, the whiche is called Acatia. Matthiolus first figure of Acatia hath leaues like Asarabacca, and beareth timber of twelue cubites long, fit for buyldings, especially of shippes, some haue called it a thorne, bycause all the tree is set full of prickles.

2. The seconde kinde is also a thornie plant, set with long sharpe prickles, and the leaues be almost lyke to the leaues of common Rue. The fruite lykewise is inclosed in coddes, as the fruite of the first kinde.

the Historie of Plantes. 685

Acatia Aegyptia. Acatia altera.
Acatia of Egypt. **Acatia of Pontus.**

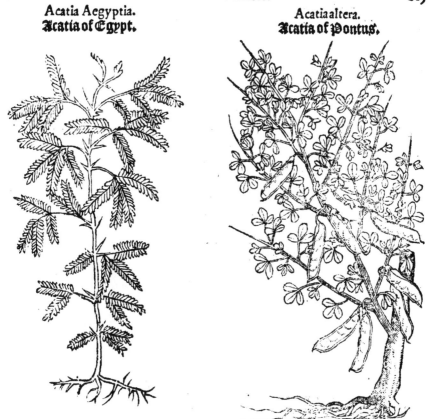

☙ *The Place.*

1 The first Acatia groweth in diuers places of Egypt, in the wildernesse or desertes.
2 The second groweth in Pontus and Cappadocia, as Dioscorides writeth.
☙ *The Names.*

1 This thornie tree or plante is called in Greeke ἀκακία: in Latine, Acatia, of Theophrastus ἄκανθ۔ ἡ ἄκανθα, in Latine, Spina. The gumme coming out of this tree is called in Shoppes Gummi Arabicum, & is wel knowen, howbeit the liquor or iuyce of Acatia, whiche is also called Acatia, is vnknowen: for in steede of Acatia, they vse in shoppes the iuyce of Sloos, or Snagges, whiche is the fruite of blacke thornes (called in base Almaigne, Sleen) and wrongly Acatia.

2 The other, whereof we haue giuen the figure as of the seconde Acatia, is taken of some learned men for ἀσπάλαθ۔, Aspalathus, and not for Acatia. Matthiolus setteth it foorth for the seconde kinde of Acatia, called Acatia Pontica, and Acatia altera. ❧ *The Nature.*

Acatia, especially the iuyce therfore (which the Ancientes vsed) is dry in the thirde degree, and colde in the first, as Galen saith.
❧ *The Vertues.*

The iuyce of Acatia stoppeth the laske, & the superfluous course of womens flowers: and bringeth backe agayne, staying and keeping in his natural place, the matrix or mother that is loosed and fallen downe, if the Acatia be dronken with red wine.
It

It is good to be layde to Serpigo, whiche is a disease of the skinne called B wilde fire, and vpon inflammations and hoate tumours: also it is good to be layde to the wheales or hoate blisters of the mouth.

It is also a very excellent medicine for the eyes, to heale the inflammation, C blastinges, and swelling out of the same, to be applyed therevnto.

Acatia maketh the heare blacke, if it be washed and often wet in the water D wherein it hath bene soked.

The leaues and tender croppes of Acatia do setle and strengthen members E out of ioynte, if they be bathed or soked in the hoate bath or stue made with the broth thereof.

Of the Myrtel tree. Chap. xriij.

❧ *The Kindes.*

There is nowe two sortes of Myrtell, the one called the great or common Myrtell, the other the fine or noble Myrtel.

❧ *The Description.*

Myrtus. Myrtel tree.

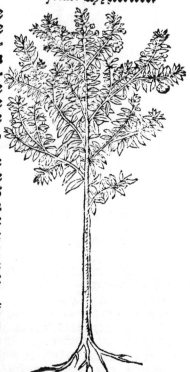

1 THE great Myrtell is a small tree growing in this Countrie to the height of a man, with many branches couered with blackish leaues, in fashion and quantitie almost lyke the leaues of Periuincle: amongst whiche leaues (in a hoate season) there is found in this Countrie, faire white and pleasant flowers, not much vnlike the flowers or blossoms of the Cherrie tree, but somewhat smaller.

2 The small or noble Myrtel is a litle lowe plante in proportion and making not muche vnlike ÿ other, but much smaller. The leaues be small & narrowe, smaller and straighter, or narrower than the leaues of Boxe, of colour not so blackishe as the leaues of the greater Myrtel. The flowers be also white, nothing differing from the others, sauing that they be somewhat smaller, and sometimes in leaues more doubble.

3 Also there is nowe founde a kinde of Myrtell whose leaues be greatest, which be almost as large as the leaues of Periuincle, called in Latine *Periuinca*, in all thinges els lyke to the others.

❧ *The Place.*

The Myrtell tree or bush, groweth plentifully in Spayne and Italy about Naples. It groweth not in this Counitre, but in the gardens of certaine Herboristes, the whiche do set it in paniers or baskettes, & with great heede and diligence they preserue it from the colde of winter: for it cannot indure the colde of this Countrie. The small Myrtell is more common in this Countrie, than the greater.

❧ *The Tyme.*

The Myrtell tree flowreth but seldome in this Countrie, except sometimes in a very hoate sommer: then it flowreth in June, without bearing either fruit or seede.

❧ *The*

the Historie of Plantes.   687

❧ The Names.

The Myrtell is called in Greeke μυρσίνη: in Latine, Myrtus: by the whiche name it is knowen in the shopppes of this Countrie.

The fruite of the Myrtel is called in shops Myrtilli.

❧ The cause of the Name.

The Myrtel is called in Greeke μυρσίνη, bycause of a young Mayden of Athenes named Myrsine: who in beautie excelled all the Maydens of that Citie, and in strength & actiuitie al the lustie laddes, or braue young men of Athenes, wherefore she was tenderly beloued of the Goddesse Pallas or Minerua. Who willed her to be alwayes present at tourney, and tilte, running, vauting, and other such playes of actiuitie or exercise: to the intent she should afterward as a iudge giue the garlande or Crowne of honour to suche as wan the price, and best deserued the same: but some of them who were vanquished, were so muche displeased with her iudgement, that they slue her. The whiche thing as soone as the Goddesse Minerue perceiued, she caused the sweete Myrtell to spring vp, and called it Myrsine, after the name of the Damosell Myrsine, to the honour and perpetuall memorie of her, whiche tree or plante she loueth asmuche as euer she loued the young Damosel Myrsine.

❧ The Nature.

Myrtell is drie in the thirde degree, and colde in the first.

❧ The Vertues.

Myrtel beries are good to be giuen them, which do spet, vomit or pisse blood, A for they stoppe all issue of blood, and the superfluous course of the menstruall flowers.

The same be also good against the laske, & the sores or vlcers of the blader. B

The dried iuyce of Myrtelles serueth well for all the aforesayde purposes, C and also for the weake and moyst stomacke, & agaynst the stingings of Scorpions and the felde spider.

The decoction of Myrtel beries maketh the heare blacke, & keepeth it from D falling, it cureth the euil sores of the head, and clenseth the same from roome, or scurtie scales, if the head be often washed therewithal.

It is good to washe outward vlcers and sores with the wine in which the E seede of Myrtel hath ben boyled. It is also profitable to be layd to the inflammations of the eyes, with a little fine flower: and agaynst the filthie matter or running of the eares, being dropped therein.

It keepeth from dronkennesse, if it be taken before hande.   F

The decoction of the seede and leaues of Myrtell, stoppeth the superfluous G course of the flowers, if you cause them to sit or bath in it.

It is good to washe suche members as haue bene burstten or out of ioynt: for H it doth strengthen and comfort them.

The greene leaues of Myrtell, are good to be layde vpon moyst sores, and I vpon all partes in whiche there is any great falling downe of humours.

The same with oyle of Roses, or any other of the same operation, is good a- K gainst consuming sores, and rotten vlcers, wilde fire, spreading tetters, & other such hoate scabbes or pustules.

The drie leaues of Myrtel layde to with conuenient oyntmentes or salues, L do heale the exulceration of the nayles, aswel of the handes as of the feete, and do take away the sweat of all the body.

Of the Bay tree.   Chap. xxiiij.

❧ The Kindes.

There are two sortes of Bay trees, the one with greene boughes & branches,

and

and harde thicke leaues, the other hath reddishe branches, especially when it is young, and softer leaues, and more gentle then the first.

✤ *The Description.*   Laurus. **Laurel or Bay tree.**

1. THE first kind of Bay groweth sometimes very high, with a harde or thicke stem, body, or tronke, the whiche parteth it selfe into many boughes & branches couered with a greene rinde or barke, and beareth leaues that be brode, log, hard, thicke & sweete smelling: amongst which there rise small white or yellowishe knoppes, the whiche doo open into flowers of an herbishe colour, and do change afterward into a long fruite, couered without with a thicke blacke browne pyll or barke, in which the kernell lyeth, of a whitish gray colour, fat & oylie, in taste sharpe & bitter.

2. The seconde kind of Bay is not much vnlyke the first, sauing that it groweth not so high, and it putteth foorth oftentimes newe shutes or branches from the roote, the whiche do often grow as high as the principall branches, so that this Bay doth seldome growe to the fashion or shape of a tree. The shutes & branches of this Bay are reddishe, and sometimes very red, and when they waxe olde, they are browne redde. The leaues be lyke to the others, sauing that they be more tender and soft, and as well smelling as the other.

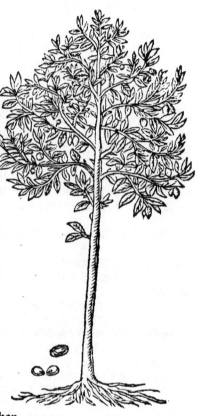

✤ *The Place.*

Bay groweth plentifully in Spayne and the lyke hoate Countries. in this lowe Countrie they plant it in gardens and defende it in the winter time from colde with great diligence, sauing Zealande, and by the Sea syde in saltishe groundes: for there it groweth well of his owne accorde, and dieth not in the winter season, as it doth in sweete groundes.

✤ *The Tyme.*

The Bay tree looseth not his leaues, but abydeth greene both winter and sommer. And about Marche or Aprill it putteth foorth new leaues & sprenges, it bringeth foorth no fruite in the lower Germanie, but in Englande it beareth plentie. ✤ *The Names.*

1. The Bay is called in Greeke δάφνη: in Latine, Laurus: in high Douche, Lorbeerbaum: in base Almaigne, Laurus boom: in Englishe, Bay or Laurel tree.

2. The fruit is called in Latine, Lauri baccę: in English, Bay beries: in French, Bayes, or *Graines de Laurier*: in high Douche, Lorbeeren: in base Almaigne, Bakeleers. ✤ *The cause of the Name.*

The Bay tree is called in Greeke δάφνη, by the name of a young Nymphe, called Daphne the daughter of Ladon, and the earth, whom the God Apollo loued, and was muche enamoured of her, so that he folowed her euery where

so

so long, that at the last he tooke hold of her and held her fast. But she not otherwise able to auoyde the importunate sewte of Apollo, sodaynely called for succour of her mother the earth, who presently opened, and swallowed in her daughter Daphne, and in steede of her brought foorth a fayre Bay tree. When Apollo sawe this change, he was much astonied, and named the tree Daphne, after the name of his beloued Daphne, and tooke a branch thereof, and twisted a garlande or cap, and set it on his head. Whereby from that time hitherto, the Bay hath still continued as a token of prophesie, and is dedicated to Apollo, that is to say, the Sonne. Therefore the Heathen say, that the Bay tree withstandeth all euill sprites and enchantmentes: so that in the house where as is but one branch of Bay, they affirme that neyther enchantements, lightninges, nor the falling euyll may hurt any body that is within. They say also, that the Bay or Laurell bringeth health. And for these causes (in times past) there was giuen a branche of Bay to the Romayne Senatours euery Newe yeres day. And for these causes also the Poetes were crowned with Garlandes of Bay, bycause that Poetrie, or the workes of Poetes, is a kind of prophesie or soothsaying, the whiche Apollo gouerneth and ruleth.

❧ *The Nature.*

The leaues and fruite of the Bay tree, are hoate and drie in the seconde degree, especially the fruite, the whiche is hoater then the leaues. The barke of the roote is hoate and dry in the thirde degree.

❧ *The Vertues.*

Bay beries taken with wine is good against the bytinges and stinging of Scorpions, and against all venome and poyson. A

The same pounde very small, and mingled with hony or some Syrope, and often licked, and kept in the mouth, is good for them that waxe drie, and are in consumption, and that haue the payne to fetche breath, and haue their breast charged with flegme. B

The decoction therof in wine, or the iuyce thereof dropped into the eares, cureth the singing or humming noyse of the same, and is good agaynst hardenesse of hearing and deafnesse. C

Bay beries are put into medicines that are made to refresh them that be tyred or weried, against crampes and drawing togither of sinewes, moyst and drie scuruinesse, being applyed with oyles or oyntmentes seruing to the same purpose. D

The oyle of Bay beries is of the same vertue: also it is good against bruses, and blacke and blewe markes, that chaunce after stripes or beatinges. E

The barke of the roote of Bay dronken in wine, prouoketh vrine, breaketh the stone, and driueth it foorth, and grauell also. F

The same taken in like maner, openeth the stoppinges of the liuer, the splene or milt, and to conclude, al other stoppinges of the inner partes: wherefore it is good agaynst the Jaundise, that is inueterate or rooted, the hardnesse of the splene or milt, the beginning of the Dropsie, and bringeth to women their desired sicknesse. G

Of Priuet. Chap. xxv.

❧ *The Description.*

PRiuet is a base plante, very seldome growing vpright, but is rather like to a bushe or hedge then a tree, with many slender twigges and branches, and leaues somewhat long, of a darke greene colour, lyke the leaues of Periwincle, but somewhat larger and longer. At the toppes of the branches

branches growe tuftes of white flowers, somwhat lyke the flowers of Eldren, after them come small beries, at the first greene, but afterwarde blacke.

❧ The Place.

Priuet groweth of his owne kinde in many places of Germanie and Englande, and is also planted in many gardens.

❧ The Time.

Priuet flowreth in May and June, and his fruit or beries are rype in September.

❧ The Names.

This plant is called in Greke (of Dioscorides κυπρος) who ioyneth φιλυρία next to Cypros: in Latine of Plinie, Ligustrum: yet this is not that Ligustrum, whereof Virgil and Columella haue written, whereof we haue treated before, li. 3. Cap. 52. in English, Priuet, or Primprint: in Frenche, Troesne: in high Douch, Beinholtzlin, Mundholtz, Reinweiden: in base Almaigne, Reynwilghen, Mondthout, and Keelcrupt.

❧ The Nature.

The leaues of Priuet are colde, dry, and astringent. The fruite hath a certayne warmenesse, but els in nature lyke to the leaues.

Phillyrea, Ligustrum.

❧ The Vertues.

The leaues of Priuet do cure the swellinges, apostumations, and vlcers of the mouth, and the sores, and pustules, or blisters of the throte, if the mouth be well washed, and the throte gargled with the decoction or iuyce thereof. A

The same leaues made into powder, are good to be strowed vppon hoate vlcers, and noughtie festering or consuming sores. And the fruite vsed in lyke maner, serueth to the same purposes. B

Whatsoeuer is burned or scalded with fire, may be healed with the brothe of Priuet leaues. The flowers layd to the forehead, swage the payne thereof. The oyle heateth and softeneth the sinewes, if it be mingled with things that are of a hoate nature, as Turner writeth, lib 2. fol. 32.

Of Agnus Castus. Chap.xxvi.

❧ The Description.

AGnus Castus groweth after the maner of a shrubby bush or tree, with many pliant twigges or branches, that wil bende and ploy without breaking. The leaues are most commonly parted into fiue or seuen partes, lyke to the leaues of Hempe, whereof eche part is long, and narrow, not much vnlike the Wythie leafe, but smaller. The flowers grow at the vpmost of the branches lyke to spikie eares clustering togither rounde about the branches, and are of colour sometimes purple, & sometimes of a light purple mixed with white. The fruite is rounde lyke Pepper cornes.

❧ The Place.

Agnus Castus (as Dioscorides sayth) groweth in rough vntoyled places alongst

alongſt by riuers, and watercourſes, in Italy and other hoate Countries, but here it is not to be founde, but in the gardens of ſome diligent Herboriſtes. ❧ *The Tyme.*

In this Countrie Agnus Caſtus flowreth in Auguſt.

❧ *The Names.*

This plante is called in Greeke ἄγνος, Agnos: of ſome λύγος ἡ ἄγνος: in Latine, Vitex, Salix marina, or Salix amerina: and of ſome Piperagreſte: in ſhoppes, Agnus caſtus: by the whiche name it is knowē of the Herboriſtes: in Engliſhe, Agnus Caſtus, Hempe tree or Chaſt tree : in Germanie it is called Schafmulle.

❧ *The Nature.*

Agnus Caſtus is hoate and drie in the thirde degree: & of nature very aſtringent. ❧ *The Vertues.*

A Agnus Caſtus is a ſinguler remedie and medicine for ſuch as woulde liue chaſte: for it withſtandeth al vncleanneſſe or the filthy deſire to lecherie, it conſumeth & drieth vp the ſeede of generation, in what ſorte ſoeuer it be taken, whether in powder, or in decoction, or the leaues alone layde on the bed to ſleepe vppon. And therefore it was named Caſtus, that is to ſay, Chaſte, cleane, and pure.

B The ſeede of Agnus Caſtus dronken, driueth away and diſſolueth all windineſſe and blaſtinges of the ſtomacke, entrayles, bowels, and mother: & from al other partes of the body, where as any windineſſe is gathered togither.

C The ſame openeth & cureth al hardnes & ſtoppings of the liuer & milt, and is good in the beginning of dropſies, dronken with wine in the quātity of a dram.

D It moueth womens natural ſickneſſe, to be taken by it ſelfe, or with Penny Ryal, or put vnder in manner of a peſſarie or mother ſuppoſitorie.

E They minge it profitably amongſt Oyles and oyntmentes that are made to heate, mollifie, and heale the harde or ſtiffe members, that are waxen dead, aſleepe, benummed, or weried : it cureth alſo the cliftes, or riftes of the fundement, and great gut, being layde to with water.

F Agnus Caſtus is good againſt al venemous beaſtes, it chaſeth and driueth away al Serpents, and other venemous beaſtes from the place where as it is ſtrowed or burned : it healeth all bitinges and ſtinginges of the ſame, if it be layde vpon the place grieued : the lyke vertue hath the ſeede thereof dronken.

G It helpeth the hardneſſe, ſtoppinges, apoſtumations, and vlcers of the matrix, if wemen be cauſed to ſit in the decoction, or broth thereof.

H The leaues therof with butter, do diſſolue and ſwage the ſwellinges of the genitours or coddes, being layde therevnto.

I Some write that if ſuch, as iourney or trauell, do carrie a branche or rod of Agnus Caſtus in their hand, it wil keepe them both frō chauffing & werineſſe.

The syxth Booke of
Of Coriers Sumach. Chap.xcvij.

❧ *The Description.* Rhus Coriaria.

Sumach groweth lyke a busshie shrub, about the height of a man, bringing foorth diuers branches, vpon which grow long soft heary or veluet leaues, with a red stem or sinewe in the middle, the whiche vpon euery syde hath sire or seuen litle leaues, standing one against another, toothed and snipt about the edges, lyke the leaues of Agrimonie, wherunto these leaues are muche lyke, the flowers growe amongst the leaues vpon long stemmes or footestalkes, clustering togither lyke the Cattes tayles, or blowinges of the Nut tree, of a white greene colour. The seede is flat and red, growing in rounde beries clustering togither lyke grapes.

❧ *The Place.*

It groweth abundantly in Spayne and other hoate Countries. It is not found in this Countrie, but amongst certayne diligent Herboristes.

❧ *The Tyme.*

Sumach flowreth in this Countrie in July. ❧ *The Names.*

This plant is called in Greke ῥοῦς, and of Hyppocrates, ῥοῶ: in Latine, Rhus, of some Rhos, of the Arabian Apothecaries and Physitions Sumach: in Brabant of the Coriers and Leather dressers, which for the most part do trimme and dresse Leather like Spanishe skinnes, Smack: in Englishe, Sumach, and Leather Sumach, or Coriers Sumack.

The seede of this Rhus is called in Greke ῥοῦς ὁ ἐπὶ τὰ ὄψα, and ἐρυθρός: in Latine Rhus obsoniorum: in Englishe, Meate Sumach, and Sauce Sumach.

The leaues are called in Greke ῥοῦς βυρσοδεψικὴ: in Latine, Rhus Coriaria, and with the same leaues they dresse and tanne skinnes in Spayne and Italy, as our Tanners do with the Barke of Oke.

❧ *The Nature.*

The leaues, iuyce, and beries of Sumach, are colde in the seconde degree, and drie in the thirde degree, and of a strong binding power.

❧ *The Vertues.*

The leaues of Sumach haue the same power as Acatia hath: wherefore they stop the laske and the disordered course of womens flowers, with al other issue of blood, to be first boyled in water or wine, and dronken. [A]

The water wherein the same leaues haue bene boyled, stoppeth the laske and blooddy flire, to be powred in as a glister, or to bathe in the same decoction: it drieth vp also the running water & filth of the eares, when it is dropped into the same, and it maketh the heare blacke, that is washed in the same decoction or broth. [B]

The seede of Sumach eaten in sauces with meate doth also stop all fluxes [C]

the Historie of Plantes.

of the belly, with the blooddy flixe and womens flowers, especially the white flowers.

The same layde vpon newe bruses, and squattes, that are blacke and blew, greene woundes and newe hurtes, defendeth the same from inflammation or deadly burning, appostumation or euil swelling, also from exulceration.

The same pounde with Oken coales, and layd to the Hemeroydes or flowing blood of the fundamēt healeth & drieth vp the same. The same vertue hath the decoction of the leaues or seede to wash or bathe the Hemeroydes therein.

Of Lycores. Chap. xxviij.

Glycyrrhiza. Radix dulcis.
Lycoryse.

Glycyrrhiza communis. Radix Scythica.
Common Lycorise.

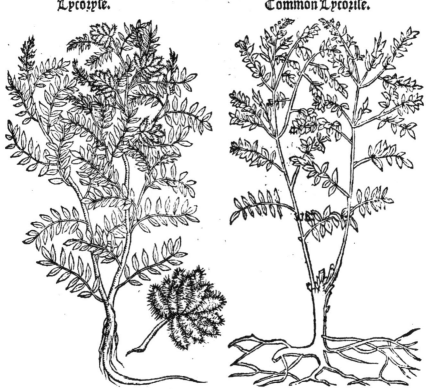

❧ *The Description.*

1. Lycoryse hath straight twigges and branches, of three or foure foote high, set with brownishe leaues, made of many smal leaues standing neare togither alongst the stemmes, one directly against another, lyke the leaues of þ Masticke tree, & Tragium or bastarde Dyctam, the flowers growe vpon short stemmes, betwixt the leaues and the branches, clustering togither lyke to small pellettes or balles, the which being past, there foloweth rounde rough prickley heades, made of diuers rough huskes clustered, or set thicke togither, in whiche is conteyned a flat seede. The roote is long and straight, yellow within and browne without, not much vnlyke the fashion of the roote of Gentian, but sweete in taste.

There is another kinde of Licoryse, whose stalkes and leaues be like to the aforesayde: but the flowers and coddes thereof growe not so thicke clustering
togither

togither in round heades or knoppes, but they grow togither lyke the flowers of Spike vpon small footestemmes, or lyke the flowers of Galega, or that kinde of wilde Fetche, whiche some iudge to be Onobrychis, or Medica Ruellij, in Frenche, Sainct Foin. The rootes of this Lycorise grow not straight, but trauersing ouerthwart with many branches, of a brownishe colour without, and yellowe within, in taste sweete, yea sweeter then the aforesayde.

The Place.

1 Lycoryse, as Dioscorides sayth, groweth in Pontus and Cappadocia.
2 The seconde sort is founde in certayne places of Italy and Germanie. In this Countrie they growe not of them selues, but planted in the gardens of some Herboristes: but the seconde sort is best knowen.

The Tyme.

Lycoryse flowreth in July, and in September the seede is rype.

The Names.

Lycoryse is called in Greeke γλυκυρρίζα: in Latine, Dulcis radix, and Dulci radix: in Shoppes, Liquiritia: in high Douche, Suszholtz, and Suszwurtsel: in base Almaigne, Suethout, Galissihout, and Calissihout: in Frenche, Riglice, Rigoliste, and Erculisse.

1 The first kinde of Lycorise or Glycyrrhiza, whereof Dioscorides writeth, may very well be called Glycyrrhiza vera, or Dioscorides Glycyrrhiza: that is Dioscorides Lycorise, and the right Licorise.
2 The second is Glycyrrhiza communis, or Glycyrrhyza Germanica, the which Lycoryse is common in the shops of this Countrie. This is that roote whiche Theophrastus calleth σκυθικὴ ῥίζα, and εὐτειγλυκεῖα: and of Plinie, Radix Scythica. Also this is the roote called in Greeke ἄλιμος, Alimos, without aspiration.

The Nature.

Lycorise is temperate in heate and moysture.

The Vertues.

The roote of Lycorise is good against the rough harshnesse of the throte and breast, it openeth and dischargeth the lunges that be stufte or loden, ripeth the cough, and bringeth foorth fleme being chewed and kept a certayne space in the mouth. The iuyce of the roote hath the same vertue to be taken for the same intent or purpose. A

For the same cause they vse to make a kinde of small cakes or bread in some Abbeys of Hollande against the cough, with the iuyce of Lycorise, mixt with Ginger and other spices, but the same serueth but against olde coughes & cold, and the like infirmities chauncing to the lunges and breast. B

The roote of Lycorise quencheth thirste, & doth coole and comfort the hoate and drie stomacke, & is good against the hoate diseases of the liuer, to be chewed in the mouth, or dronken in a decoction. C

The same is good against the vlcers of the kidneyes, and scabbes or sores of the bladder, it cureth the sharpenesse and smarting of vrine, and also the filthy corruption or mattering of the vrine, being boyled in water and often dronken. D

The same is good to be layde to with hony vppon the sores or vlcers of the outwarde partes: for it cureth the same, as Plinie writeth. E

To conclude, Lycorise and the iuyce therof is a very good and holsome medicine, fit to asswage payne, to soften, and make whole, very proper and agreable to the brest, the lunges, the raynes, the kidneyes, and bladder. F

the Historie of Plantes.

Of Rhamnus.   Chap.xxix.

¶ The Kyndes.

After the opinion of Dioscorides, there be three sortes of Rhamnus, one with long, fat and soft leaues: the other hath white leaues: and the third hath roundishe leaues and somewhat browne.

Rhamni secunda species.   Rhamni tertia species.
The seconde kinde of Rhamnus.   **The thirde kinde of Rhamnus.**

❧ The Description.

All the kindes of Rhamnus are plantes of a wooddishe substance, the whiche (as Dioscorides writeth) haue many straight twigges and branches, set with sharpe thornes and prickles, lyke the branches of white Thorne.

1. The first kinde of Rhamnus hath many long, narrowe, tender, flat leaues: amongst the whiche rise long, harde, and sharpe thornes.

2. The seconde kinde hath long narrow white leaues, in proportion not much vnlyke Olyue leaues, but muche smaller, amongst whiche there growe shorte thornes with stiffe prickles.

3. The thirde kinde hath leaues somewhat broade and almost rounde, of a brownishe colour drawing towardes red. The thornie prickles of this kinde, be neither so great, nor yet so strong, as the prickles of the first kinde. The flowers be yellowishe, the whiche past, there commeth vp the fruite whiche is large, and almost fashioned lyke to a wherrowe or buckler, in the whiche lyeth the seede.

Nn iiij   The

696　　　The ſyxth booke of

✼ *The Place.*

Rhamnus (as Dioſcorides writeth) groweth in hedges and buſhes.

1　The firſt kinde is not knowen in this Countrie, but in Languedoc there groweth plentie.

2　The ſeconde kinde groweth in ſome partes of Germanie vppon banckes or diches by the Sea ſide, eſpecially in Flaunders, where as in certayne places it groweth plentifully.

3　The thirde kinde is to be found in Brabant in the gardens of ſome Herboriſtes, and there is ſtore of it founde in the Countrie of Languedoc.

✼ *The Names.*

This kinde of buſhe is called in Greeke ῥάμνος: in Latine, Rhamnus: vnknowen in ſhoppes.

The thirde kinde of Rhamnus, is called in Italy, Chriſtes thorne.

✼ *The Nature.*

The leaues of Rhamnus are drie in the ſeconde degree, and colde almoſt in the firſt degree.

✼ *The Vertues.*

The leaues of Rhamnus do cure Eryſipelas, that is hoate, and choleryque inflammations, and conſuming ſores & fretting vlcers, when it is ſmal pounde and layde thereto.　A

The Phyſitions of Piemont haue found by experience, that the ſeede of the thirde kinde of Rhamnus, is very excellent againſt the grauell and the ſtone, to be taken in the decoction or otherwiſe.　B

Some hold, that the branches or bowes of Rhamnus ſtickte at mens dores and windowes, do driue away Sorcerie and Enchauntmentes that Witches and Sorcerers do vſe againſt men.

Of Bucke thorne, or Rheyn beries.　Chap.xxx.

✼ *The Deſcription.*

Rhamnus ſolutiuus recentiorum.

THIS plante groweth in manner of a ſhrub or ſmal tree, whereof the ſtemme is ofte̅times as bigge as ones thigh, the wood or timber whereof is yellow within, and the barke is of the colour of a Cheſnut almoſt like the barke of the Cherrie tree. The branches be ſet with ſharpe thornes both harde & prickley, and roundiſh leaues, ſomwhat like the leaues of a gribble, grabbe tree, or wilding, but ſmaller. The flowers are white, after whiche there come litle rounde berries, at the firſt greene, but afterwarde blacke.

✼ *The Place.*

This plante groweth in this Countrie, in feeldes, wooddes, and hedges.

✼ *The Tyme.*

It flowreth in Maye, and the fruite is ripe in September.

❧ The Names.

This thorne is called in Brabant, Rhijn besien dorrn: in French, Nerprun, or Bourg espine: in high Douche, Weghedorn, that is to say, Way Thorne: bycause it groweth alongst the high wayes and pathes: in Latine of Matthiolus, Spina infectoria, and of some others, Rhamnus solutiuus, the whiche name I doo subscribe vnto, bycause I knowe none other Latine name, albeit it is nothing lyke to Rhamnus of Dioscorides, or of Theophrastus, & therefore not the right Rhamnus. The Italians do cal it *Spino Merlo*, some cal it *Spino ceruino*, *Spin guerzo*, and of Valerius Cordus, Cerui spina: we may well call it in Englishe, Bucke Thorne.

The fruite of the same thorne is called in Brabant, Rhijnbesien, that is to say in Latine, Baccæ Rhenanæ: in English, Rheyn beries, bycause there is much of them founde alongst the riuer Rhene: in high Douche, Weghedornbeer, and Cruetzbeer.

❧ The Nature.

It is hoate and drie in the seconde degree.

❧ The Vertues.

The beries of Bucke thorne do purge downeward mightily, driuing foorth a tough flieme and cholerique humours, and that with great force, and violence, and excesse, so that they do very much trouble the body that receiueth the same, and oftentimes do cause vomit. Wherefore they be not meete to be ministred, but to young strong and lustie people of the Countrie, whiche do set more store of their money then their lyues. But for weake fine and tender people, these beries be very dangerous and hurtfull, bycause of their strong operation. And also bycause hitherto there is nothing founde, wherewithall to correct the violence thereof, or to make it lesse hurtfull.

Of the same beries before they be rype, soked, or delayed in Allom water, is they make a fayre yellowe colour, and when they be rype, they make a greene colour, the which is called in France, *Verd de vessie*: in high Douch, Safftgruu: in base Almaigne, Sapgruen: in English, Sappe greene.

Of the White Thorne, or Hawthorne tree. Chap.xxxi.

❧ The Description.

THE white Thorne most commonly groweth low and crooked, wrapped and tangled as a hedge, sometimes it groweth vpright after the manner and fashion of a tree: and then it wareth high as a Perrie, or wilde Peare tree, with a tronke or stemme of a conuenient bignesse, wrapped or couered in a barke of gray or ashe colour. The branches doo sometimes grow very long and vpright, especially when it groweth in hedges, and are set ful of long sharpe thornie prickles. The leaues be brode and deepe, cut in about the borders. The flowers be white & sweete smelling, in proportion lyke to the flowers of Cherrie trees, and Plomtrees: after the flowers commeth the fruite whiche is rounde and red. The roote is diuided into many wayes, and groweth deepe in the grounde.

❧ The Place.

White thorne groweth in hedges and the borders of feeldes, gardens, and woodes, and is very common in this Counttie.

❧ The Time.

It flowreth in May, and the fruite is rype in September.

❧ The Names.

This thorne is called in Greeke ὀξυάκανθα: in Latine, Spinaacuta, of some πυρίνα, Pyrina, and πυρυάυθκ, Pytianthe : it is Oxyacantha of Dioscorides, and the first kinde of Auicens Amyrberis : in Englishe, White Thorne, & Hawthorne : in French it is called *Aube espine* : in high Douche, Hagdorn : in base Almaigne, Haghedoren, and witte Haghedoren.

It seemeth also to be κυνοσβάτος, that is to say, Rubus canis, & Canina sentis, whereof Theophrastus, writeth lib.3. Cap.18.

¶ The Nature.

The fruite of White Thorne is drie and astringent.

❧ The Vertues.

A The fruit of this Thorne stoppeth the laske, and the flowers of women.

B And as some of the later writers affirme, it is good against the grauell and the stone.

Oxyacantha Dioscoridis.

Of Bore tree. Chap.xxxij.

¶ The Kindes.

There are two kindes of Bore, that is to say, the great & the smal, and both are meetely commō in this Countrie.

❧ The Description.

1 The great Bore, is a faire great tree, with a bigge body or stemme, that is harde, and meete for to make diuers and sundrie kindes of workes and instrumentes: for the timber therof is firme, hard, and thicke, very good to be wrought, and cut all manner wayes : and lasteth a long space without rotting or corruption. It hath many bowes and harde branches, as bigge as the armes and branches of some other trees, couered with many small darke greene leaues, the which do not fal away in the winter, but do remayne greene both winter and Sommer. The flowers growe amongst the leaues vpon the litle small branches, after whiche commeth the seede whiche is blacke, inclosed in round cuppes or huskes somewhat bigger then Coriander beries, of colour greene, with three feete or legges, like the fashion of a kitchin pot wherin meat is prepared and boyled, the whiche is very lyuely pictured in Matthiolus last edition.

2 The smaller Bore is a little bushe, not lightly exceeding the height of two foote, but spreadeth his branches abrode, the whiche most commonly do grow very thicke from the roote, and sometimes they growe out of a small tronke or stubbed stemme. The leaues of this kinde are of a clearer greene or lighter colour, and they be also rounder, and somewhat smaller than the leaues of the greater Bore, in all other partes lyke to the aforesayde.

¶ The Place.

Bore delighteth to growe vpon high colde mountaynes, as vpon the hilles
and

the Historie of Plantes.

and desertes of Switserland, and Sauoye and other lyke places, where as it groweth plentifully. In this Countrie they plante both kindes in some gardens.

※ *The Tyme.*

Boxe is planted at the beginning of Nouember, it flowreth in February & March, and in some Countries the seede is ripe in September.

※ *The Names.*

Boxe is called in Greke πύξος: in Latine, Buxus: in French, *Grand Buys*: in high Douch, Buxbaum: in base Almaigne, Buxboom, and of the common people Palmboom, that is to say, the Boxe tree, and Palme tree, bycause vpon Palme Sunday they carie it in their Churches, and sticke it rounde about in their houses.

The smal Boxe is called of some in Greke χαμαιπύξος: in Latine, Humi Buxus: that is to say, Ground Boxe, or Dwarffe Boxe: in Frenche, *Petit Buys*.

※ *The Nature.*

The leaues of Boxe are hoate, drie, and astringent, as the taste doth playnely declare.

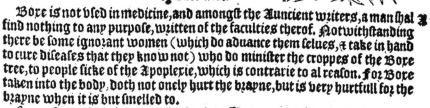

Boxe.

※ *The Vertues.*

Boxe is not vsed in medicine, and amongst the Auncient writers, a man shal find nothing to any purpose, written of the faculties therof. Notwithstanding there be some ignorant women (which do aduance them selues, & take in hand to cure diseases that they know not) who do minister the croppes of the Boxe tree, to people sicke of the Apoplexie, which is contrarie to al reason. For Boxe taken into the body, doth not onely hurt the brayne, but is very hurtfull for the brayne when it is but smelled to.

Some learned writers at this time do affirme, that the lye in which Boxen leaues haue bene stieped, maketh the heare yellow, if the head be often washed therewithall.

Of the prickley Boxe. Chap.xxxiij.

※ *The Description.*

PRickley Boxe is a tree not muche vnlyke to the other Boxe, with many great armes or branches of fiue or sixe foote long or more, the leaues be thicke and somewhat rounde, lyke Boxen leaues, and amongst them growe sharpe pricking Thornes, the flowers also growe amongst the leaues, and after them there commeth a blacke rounde seede, as bigge as a Pepper corne. The rootes are woddishe, and spreade muche abrode.

Of the smal branches and rootes of this tree, soked in water and boyled, or of the pressing foorth of the iuyce of the seede, they make Lycium, the whiche in times past was muche vsed of Physitions.

The syxth Booke of

Pyxacantha, Lycium.

❧ *The Place.*

This prickley Boxe groweth in Cappadocia and Lycia, and in some partes of Italie and Slauonia, it is yet vnknowen in this Countrie.

❧ *The Names.*

This thorne is called in Greke πυξάκανθα καὶ λύκιον: in Latine also, Pyxacantha and Lycium, of Theophrastus, ὀνόπυξος, that is to say, Buxus asinina: in Frenche, *Buys espineux*, or *Buys d'asne*: in base Almaigne, Burdozen, after the Greke: we may cal it in English, Boxe thorne, Asses Boxe tree, and prickley Boxe, also Lycium: Thorne Boxe.

❧ *The Nature.*

Lycium dried, is of subtil partes, and astringent, as Galen saith.

❧ *The Vertues.*

A   Lycium whiche is made of the branches, rootes, or seede of Boxe thorne, or prickley Boxe, helpeth them that haue the laske, and blooddy flire, as also those that spet blood, and haue the cough.

B   It stoppeth the inordinate course of the flowers, taken either inwardly, or applyed outwardly.

C   It is good against corrupt vlcers, and running scabbes, and sanious running eares, the inflammation of the gummes and kernelles, called the Almondes vnder the tongue, and against the choppes of the lippes, and fundament, to be layde therto.

D   It cleareth the sight, and cureth the scurffie festered sores of the eye liddes, and corners of the eyes.

Of Holme, Holly, or Huluer.   Chap.xxxiiij.

❧ *The Description.*

HOlme groweth sometimes after the maner of a hedge plant, amongst other thornes and bushes, and sometimes also it groweth vpright and straight, and becommeth a tall high and great tree, with a big stemme or body, and limbes and branches according to the same. The tymber of this tree is harde and heauie, and sinketh to the bottome of the water lyke Guaiacum, or Lignum sanctum, wherevnto our Holly in figure is not much vnlyke. The leaues of Holly are thicke and harde, of the quantitie of a Bay leafe, but full of sharpe poyntes or prickley corners. The whiche leaues remayne greene both winter and sommer, as the leaues of Boxe and Bay, and doo not lightly vade or wither. The beries or fruite of Holme is rounde, of the quantitie of a Pease, of colour red, and of an euill vnpleasant taste.

❧ *The Place.*

Holme groweth much in this Countrie in rough, stony, barraine & vntoyled places, alongst the wayes and in wooddes.

❧ The Tyme.

The same fruite or beries of Holme, are ripe in September; and hang fast vpon the tree a long tyme after without falling of.

❧ The Names.

Holme is called of some late writers in Greeke ὀξυμυρσίνη ἀγρία: in Latine, Ruscus syluestris: in high Douch, Walddistel oder Stehpalmen: in base Almaigne, Hulst: in Italian, Agrifolium, as Matthiolus writeth. And in sight it appeareth to be much lyke Plinies Aquifoliū, whiche is called of Theophrastus in Greeke κράταιγος, and κραταιγών, as witnesseth Plinie lib.17. Cap.7. Neither can it be Paliurus, as some do esteeme it: but it seemeth to be somwhat lyke ὀξυάκανθος. Oxycanthus of Theophrastus, the which is alwayes greene: in Englishe it is called Holme, Holly, and Huluer.

❧ The Nature.

The beries of Holme or Holly, are hoate.

❧ The Vertues.

Some boasting of their experience vpon Holme, do affirme that fiue beries therof taken inwardly, are good against the cholique, and prouoke to go to the stoole.

With the barkes of Holme they make Birdlyme: the order of making therof is very wel knowen, but if any be yet desirous to learne the same, let him seeke the thirde booke of Maister Turners herball Chap. lxxxj.

They vse the smal branches and leaues of Holme to clense and sweepe chimneyes, as they vse to do in Burgundie and other places, with Kneeholme or Butchers broome. Other then this we dare not affirme of Holme, bycause it serueth not in Physicke.

Agrifolium.

Of the Apple tree. Chap. xxxv.

❧ The Kyndes.

THere be diuers sortes of Apples, not onely differing in figure and proportion of making, but also in taste, quantitie, and colour, so that it is not possible, neither yet necessarie, to recite or number al the kindes, consydering that all Apple trees are almost lyke one another: and all sortes of Apples may be comprehended in a few kindes, for the playner declaration of their natures, faculties, or powers: as into sweet, sower, rough, astringent, waterish apples, and apples of a mixt temperature, as betwixt sweete and sower, &c.

❧ The Description.

THE Apple trees in continuance of time, do for the most part become high and great trees, with many armes & branches spread abroade. The leaues be greene and roundishe, more rounde then the Peare-tree leaues, and do fall of a litle before winter, and do spring and renew agayne in May. The flowers for the most part are white, and vpon some apple trees chaungeable, betwixt white and redde. The fruite is round and of

many

many fashions, in colour & taste as is abouesayde. In the middle of the apples are inclosed blacke kernels couered ouer with hard pilles or skinnes.

❧ *The Place.*

Apple trees are planted in gardens and Orchardes, and they delight in good fertil grounde.

❧ *The Tyme.*

Apple trees do most commonly blow, at the ende of Aprill and beginning of May. The fruit is ripe, of some in July, of some in August, and of the last sorte in September.

❧ *The Names.*

The Apple tree is called in Greeke μηλία: in Latine, Malus, & Pomus: in high Douche, Apffelbaum: in base Almaigne, Appelboom: in Frenche, *Pommier*. The fruite is called in Greeke, μλον: in Latine, Pomum, and Malum: in English, an Apple: in French, *Pomme* in high Douch, Apffel: in base Almaigne, Appel.

❧ *The Nature.*

Malus.

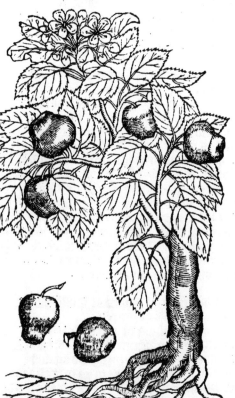

All sortes of Apples be colde and moyst, yet some more then the rest: those that be sower or sharpe, do dry more then the rest, especially if they be astringent or binding. Sweete Apples are not so colde, but rather of a meane temperature. The waterishe apples are moystiest, especially those that are neyther sower nor sweete but taking part of both tastes.

❧ *The Vertues.*

Apples do coole and comfort the hoate stomacke, especially those that be sowrish and astringent of taste, and they may be used in hoate agues, and other inflammations or heates of the stomacke, and against thirste: but otherwise they are hurtful to the stomacke, causing windinesse and blastinges in the belly.

Sower Apples boyled & eaten colde before meate, do lose the belly gently.

Apples eaten before meate do nourishe very litle, and do yeelde a moyst and noughty iuyce or nourishment: for they are soone corrupted in the stomacke, and turne to noughtie humours, especially the waterishe Apples.

The leaues of the Apple tree are good to be layde vpon the beginninges of phlegmons (that is hoate simple tumours or swellinges) and are good to be layde vpon woundes to keepe them from euyll heate and a postumation.

Of Orenges, Citrons, and Limons.
Chap. xxxvi.

❧ *The Kindes.*

There be at this present, three sortes of Apples or rather fruites, which of the
Auncientes

the Historie of Plantes.   703

Auncientes in times past were comprehended vnder the name of Citrium, wherof the first is called an Orenge, the seconde a Citron, the thirde a Limon.

❀ *The Description.*   Aurantia Mala. Medica Mala. Limonia Mala.
             Orenge.   Citrons.   Limons.

He trees y bring foorth Orenges, Citrons, and Limons, growe as high as other trees do, with many greene branches, in some places set with stiffe prickles, or sharpe thornes. The leaues be alwaies greene and thicke, not much vnlyke the Bay leaues. The fruite hath a very thicke pyl or rinde, within the rinde is a cleare through shining pulpe or moyst substance, full of iuyce & liquor, amongst the which is the seede or kernels.

1   The Orēge is round as an apple, with a thicke pyll, at the first greene without, but after when they be ripe, of a faire red or pleasant tawnie colour, or browne yellowe lyke Saffron, but the sayde pill is white within & spongious or somewhat open. The pulpe or inner pith is through shining cleare and ful of iuyce, the whiche in some is sower, & in others sweete. The seede or kernelles are most commonly as bigge as wheate cornes, & bitter in taste.

2   The Citron is long almost lyke a Cucumber, or somwhat longer and rugged, or wrinckled, the rinde or pil is thicke, yellow without, & white within. The inner part or substance is also cleare & through shining like y pulpe of the Orenge, wherin is also the seede or kernelles not much vnlyke Orenge kernelles.

3   The Limon in fashion is longer then the Orenge, but otherwayes not muche vnlike, sauing that the outsyde of the Limon pill is paler and smother, and the kernels smaller.

❧ *The Place.*

These fruites do now grow in Italy, Spayne, and some places of Fraunce. In this Countrie the Herboristes do set and plante the Orenge trees in their gardens, but they beare no fruite without they be wel kept and defended from colde, and yet for all that they beare very seldome.

❧ *The Names.*

The tree that beareth these fruites, is called in Greeke μηλία μηδική, in Latine, Malus medica, and Malus citria. And albeit the Citron and eche of the other are seuerall trees one from another, as it is playnely to be seene in Matthiolus Commentaries vpon Diosc. li. i. where also it is to be noted in the Citron tree, that his leafe is finely snipt about y edges or toothed lyke a saw, but the Limō and Orenge trees, whose leaues be euer greene lyke the Bay tree, are not indented, but smothe about the edges, so that at the first sight Citron, Orenge and Limon trees, do shew lyke Bay trees, but the pleasant sauour and smell of

Ooo ij   the

the leaues, be farre vnlyke the smell of the Bay leaues: these three trees, I say, be of the Auncientes, all contayned vnder the Citron tree.

The fruites also be all called of the Auncientes by one Greeke name μῆλα μηδικα: in Latine, Mala citria.

1 The first kinde is also called of the Auncientes in Greeke χρυσομήλον: in Latine, Aureū malum, & Malum Hespericum, of some also Nerantzium, of the later writers Anarantium, and Arantium: in Englishe, an Orenge: in French, Pomme d'Orenge: in high Douche, Pomerantsen: in base Almaigne, Arangie appelen: in Spanish, Naranzas, the whiche name seemeth to be taken from the worde, Narantzium, by the which the Apples were once called, as witnesseth Nicander.

2 The seconde kinde is called Cedromelon, and in this Countrie Citrones, & Mala citria: in Frenche, Citrons: in Englishe, Citrons: in high Douche, Citrinaten: in base Almaigne, Citroenen. This kind is called of the Italians, as Musa writeth, Limones.

3 The thirde kinde is called in the Shoppes of this Countrie Limones, and Malum Limonium: in Englishe, Limons, in Douche, Limoenen: in Frenche, Limons: Antonie Musa writeth, that the Italians doo call this fruite Citrium malum.

❊ The Nature.

The pill, especially the outwarde parte thereof is hoate and drie.

The pulpe with the iuyce is colde and drie in the thirde degree.

The seede is hoate and dry in the second degree, and the leaues be almost of the same nature.

❊ The Vertues.

A The iuyce of these fruites, and the inner substance wherein the iuyce is contayned, especially of the Orenges, is very good against contagiousnesse and corruption of the ayre, against the plague & other hoate feuers, and it doth not onely preserue and defende the people from suche dangerous sicknesse, but also it cureth the same.

B It comforteth the hart, & aboue al other the mouth of the stomacke: wherefore it is good against the weakenesse of the same, the trembling of the hart and pensiue heauinesse, wamblinges, vomitinges, and lothsomnesse, that happen in hoate agues and suche other diseases that trouble the stomacke.

C The same fruite with his iuyce quencheth thirst, and reuiueth the appetite.

D The syrupe that is made of the iuyce of this fruite, is almost of the same nature and operation that the iuyce is: but more fit and pleasant to be taken at the mouth.

E The pylles or barkes of these fruites condited or preserued with hony or sugar and eaten, do warme the stomacke and helpe digestion, wasting and driuing away all superfluities of the stomacke, and amending the stinking breath.

F The seede withstandeth all venome and poyson, and the bitinges and stinginges of all venemous beastes: it killeth and driueth foorth wormes, wherfore it is good to be giuen to children against the wormes.

Of Musa or Mose tree. Chap. xxxviij.

❊ The Description.

THE Mose tree leaues be so great and large, that one may easyly wrap a childe of twelue monethes old in them, so that as I thinke in seeking ouer the whole worlde a man shall not agayne finde a tree hauing so large

large a leafe. The fruite is lyke a Cucumber most sauerie & pleasant in taste aboue all other fraites of ẏ Countrie of Leuant.

¶ The Place.

This tree was found by a certayne fryer named Andro Theuet, in the Countrie of Syria, by the great towne Aleph, so called of the first letter of the Hebrue Alphabet, where as is great resort and traffique of marchants, aswell of Indians, Persians, & Venitians, as of diuers other strange nations.

¶ The Names.

This tree with his fruite is called of Auicen Chap. 495. Musa, & at this present in Syria Mose: And the Grekes and Christians of ẏ Countrie, as also ẏ Iewes, do say that this was the fruite whereof Adam dyd eate. This may be the tree which Plinie describeth lib. 12. Cap. 6. called Pala, whose fruite is called Ariene.

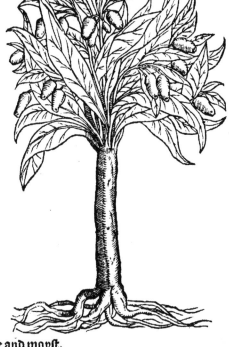

Musa.

¶ The Nature.

The fruite of Mose tree is hoate and moyst.

¶ The Vertues.

This fruite eaten nourisheth muche, and very quickly, as Auicen saith, but eaten in to great a quantitie, it stoppeth the liuer, and engendreth fleme and choler. [A]

It is also good for the breast, the stomacke, and the kidneyes, it mollifieth the roughnesse and sharpnesse of the throte, prouoketh vrine, and encreaseth naturall seede. [B]

Of the Pome Granate. Chap. xxxviij.

¶ The Kindes.

THere be two sortes of Pomegranates, the tame and the wilde : the fruite of the tame is three maner of wayes, ẏ one hauing a sowre iuyce or liquor, the other is sweete, and the thirde hath the taste of wine.

¶ The Description

THe tame Pomegranate is not very great, it hath many pliant bowes or branches, set with cruell thornes. The leaues be very greene and straight or narrowe, lyke vnto willow leaues, but shorter and thicker, with small litle red veynes going through them, & hanging vp a litle red foote stalke. The flowers be holow lyke a wine cup or goblet, cut about the brimmes after the fashiõ of a starre, of the colour Scarlet or Vermilion, after them commeth the fruite whiche is rounde, and within it is full of graynes of a Crimsin red colour, the whiche graynes haue corners or edges, lyke the stones called Granati, and within them lyeth small stones. The whiche graynes and beries

Malus punica. The Pomegranate.

(by the wonderful & maruelous worke of nature) are with certayne thinne and yellowish fyne belmes and skinnes, going betwixt, set and couched in very good order: from those graynes commeth the iuyce, the which is sower or sweete, or hauing the taste of wine. The shell or pyll of the Pomegranate is thinne and tender before it is dry, but being dried, it wareth harde, and of a woodish substance, yellow within, but without coloured lyke a Chesnut.

2 The wild Pomegranate tree is lykewise lyke vnto the aforesayde: but it bringeth foorth no fruite, and his flowers be very double, the whiche is the cause that it bringeth foorth no fruite.

❧ *The Place.*

The Pomegranates growe in hoate Countries, as Italy, Spayne, & diuers other places.

❋ *The Names.*

The Pomegranate is called in Greeke ῥοία καὶ ῥόα: in Latine, Malum punicum, & Malum Granatum: in Shoppes, Pomum Granatum: in Englishe, a Pomegranate: in high Douche, Granat apffel: in base Almaigne, Granate apple: in Frenche, *Pommes Granades*.

1 The flowers of the tame Pomegranate tree, is called in Greeke κύτινοι, and in Latine, Cytini.

2 The flowers of the wilde after Dioscorides, are called in Greeke βαλαύστιον: & accordingly in Latine, Balaustium: in French, *Des Balustres*. And these flowers are very double, and there foloweth no fruite after. The flowers that bring foorth fruite are single, and therefore they are named the tame.

The rinde or pill of the Pomegranate, is called in Greeke σίδιον: in Latine, Malicorium, and Sidium.

❧ *The Nature.*

Pomegranates be colde and somewhat astringent, but not al of a lyke sort. The sower are more drying and astringent. The sweete are not so much astringent, but more moyst then the others. Those that be in taste lyke wine, are indifferent.

❧ *The Vertues.*

The iuyce of the Pomegrate is very good for the stomacke comforting the same when it is weake and feeble, and cooling when it is to hoate or burning: it is good also against the weakenesse and wambling of the stomacke, lyke as the iuyce of Orenges and Citrons, and it is very good against al hoate agues, and the inflammation of the liuer and blood, especially the iuyce of the sower Pomegranates, and next to them such as be of winish taste: for the sweete

Pomegranates

Pomegranates (bycause they engender a litle heate and breede winde) are not very meete to be vsed in agues.

The blossomes both of the tame and wilde Pomegranate trees, as also the rinde or shell of the Pomegranate, made into powder and eaten, or boyled in red wine and dronken, are good against the blooddy flixe, and the inordinate course of the mother, not onely taken as is aforesayde, but also to sit or bath in the decoction of the same. B

The same barke or blossoms do stoppe the blood of greene woundes, if it be applied in what sort so euer it be. C

The same barke killeth wormes, and is a good remedie against the corruption in the stomacke and bowels. D

With the same barke or with the flowers of the Pomegranate, the moyst and weake gummes are healed, and it fasteneth loose teeth, if they be washed with the broth or decoction of the same. E

The barke (and as Turner saith the flowers) are good to be put into the playsters that are made against burstinges, that come by the falling downe of the guttes. F

The seede of Pomegranates dried in the Sonne, haue the lyke vertue as the flowers: it stoppeth the laske, & al issue of blood to be taken in the same maner. G

The same mingled with hony is good against the sores and vlcers of the mouth, the priuities and fundament. H

Some say, as Dioscorides writeth, that whosoeuer eateth three flowers of the tame Pomegranate, shalbe for one whole yere after preserued from dropping or bleared eyes. I

Of the Quince tree. Chap.xxxix.

Malus Cotonea.

¶ The Kyndes.

There be two sortes of Quinces: the one is rounde & called the Apple Quince: the other is greater, and fashioned lyke a Peare, and is called the Peare Quince.

❧ The Description.

THe Quince tree neuer groweth very high, but it bringeth foorth many braches as other trees do. The leaues be roundishe, greene vppon the vpper side, and white and soft vnder, the rest of the proportion, is lyke to the leaues of the common Apple tree. The flower changeth vpon purple mixed with white: after the flowers cōmeth the fruite of a pleasant smel, in proportion somtimes rounde as an Apple thruste togither, and sometimes long lyke a Peare, with certayne embowed or swellyng diuisions, somewhat resembling the fashion of a garlyke head, and when the hearie cotton or downe is rubbed of, they appeare as yellow as golde. In the middest of the fruite is the seede or kernelles lyke to other Apples.

Ooo iiij : ❧ The

✤ The Place.

Quince trees are planted in gardens, and they loue shadowy moyst places.

✤ The Tyme.

The Quince is ripe in September and October.

✤ The Names.

The Quince tree is called in Greeke μηλία κυδωνία: in Latine, Malus cotonea: in high Douche Quittenbaum, oder Kuttenbaum: in base Almaigne, Queappelboom: in Frenche, Coingnaciere.

The fruite is called in Greeke μῆλον κυδώνιον: in Latine, Malum Cotoneum: in Shoppes, Cytonium: in Frenche, Coing: in high Douch, Quitten opffel, and Kutten opffel: in base Almaigne, Queappel: in English, a Quince, & an Apple, or Peare Quince.

1 Some call the rounde fruite, Poma Citonia: in Englishe, Apples Quinces: in Frenche, Pomme de Coing, or Coing in base Almaigne, Queappelen.

2 The other fruite whiche hath the likenes of a Peare, Galen calleth στρυθία, Struthia: and it is called in Englishe, the Peare Quince: in Frenche, Pomme de Coing, Coignasse: in base Almaigne, Quepeeren, of some Pyra Cytonia.

✤ The Nature.

The Quince is colde in the first degree, and drie in the second, and astringent or binding.

✤ The Vertues.

The Quince stoppeth the laske or common flure of the belly, the Dysenterie, A & all flures of blood, and is good against the spitting of blood, especially when it is rawe: for when it is either boyled or rosted, it stoppeth not so muche, but it is than fitter to be eaten, and more pleasant to the taste.

The woman with childe that eateth of Quinces oftentimes, either in meate B or otherwayes, shal bring foorth wise children of good vnderstanding, as Simeon Sethy writeth.

The Codignac, or Marmelade made with honie (as it was wonte to be C made in times past) or with sugar, as they vse to make it nowe a dayes, is very good and profitable for the stomacke to strengthen the same, and to retaine and keepe the meates in the same, vntill they be perfectly digested.

Being taken before meate, it stoppeth the laske: and after meate it loseth the D belly, and closeth the mouth of the stomacke so fast, that no vapours can come foorth, nor ascende vp to the brayne: also it cureth the headache springing of suche vapours.

The decoction or broth of Quinces, hath the lyke vertue, and stoppeth the E belly and all flure of blood, with the violent running foorth of womens sickenesse.

With the same they vse to bathe the loose fundement, and falling downe of F the mother, to make them returne into their natural places.

They do very profitably mixe them with emplaysters, that be made to stop G the laske and vomiting. They be also layde vpon the inflammations, and hoate swellinges of the breastes and other partes.

The downe or heare Cotton that is founde vppon the Quinces, sodden in H wine, and layde therevnto healeth Carbuncles, as Plinie writeth.

The oyle of Quinces stayeth vomitinges, gripings in the belly or stomacke I with the casting vp of blood, if the stomacke be annoynted therewith.

The flowers of the Quince tree do stoppe the flure of the belly, the spetting K of blood, and the menstruall flowers. To conclude, it hath the same vertue as the Quinces them selues.

the Historie of Plantes. 709

Of the Peache and Abrecok trees. Chap.cl.

❧ *The Kindes.*

There be two kindes of Peaches, whereof the one kinde is late ripe, and most commonly white, and sometimes yellow, also there be some that are red. The other kindes are soner ripe, wherefore they be called Abrecot, or Aprecot.

Malus Persica.
The Peache tree.

Malus Armeniaca. The Aprecok tree.

❧ *The Description.*

1. THE Peache tree is more tender then other trees, and of long continuance, but doth perishe and die much sooner, then any other fruiteful trees. The leaues of Peache tree be long and lightly iagged about the edges, nothing differing from willowe leaues, sauing that they be somewhat shorter and bitterer. The flowers are of a reddishe skye colour, after whiche commeth the fruite whiche is rounde lyke an Apple, with a deepe and straight clift or forrow vpon one side, and couered ouer with a soft downe or hoare cotton, of colour sometimes white, sometimes greene, sometimes reddishe, and sometimes yellowe, and of a winishe taste, soft in feeling, and of a fleshy pulpe or substance, in the middest whereof is a rough harde stone, full of creastes and gutters, within whiche is a kernell lyke an Almonde.

2. The Abrecok in timber flowers and maner of growing is not much vnlyke the other Peache tree, sauing that his leaues be shorter & broder, and nothing like to the Peache leaues. The fruite is like to a Peache, but smaller, & sooner ripe.

The

✻ The Place.

They plante the Peache tree in gardens and vineyardes, and they loue a soft and gentle grounde standing wel in the Sonne.

✻ The Time.

The Peaches flower in Aprill, and the Abrecox are ripe in June, but the Peaches in September.

✻ The Names.

The Peache tree is called in Greeke περσικὴ μηλέα: in Latine, Malus Persica: in high Douche, Pfersichbaum: in base Almaigne, Perseboom: in French, *Vng Pescher*: in English, a Peache tree.

1 The fruite is called in Greeke μῆλον περσικόν: in Latine, Malum Persicum: in shops, Persicum: in French, *Pesches*: in high Douch, Pfersing: in base Almaigne, Persen: in Englishe, Peaches. That kinde whiche will not easily be separated from the stone, are called Duracina, in Frenche, *Des Presses*.

The Abrecok tree is called in Greeke μηλέα Ἀρμενιακή: in Latine, Malus Armeniaca: in Douche, Vroeghe Perseboom.

2 The fruite is called μῆλα Ἀρμενιακά: in Latine, Mala Armeniaca, Præcoqua, and Præcocia: in English, Abrecok, Aprecok, and Aprecox: in Frenche, *Abricoz*: in high Douch, Mollelin, and Molleten: in base Almaigne, Vroege Persekens, & Auant Perses: also of the high Douch men, S. Johans Pfersuch, which may be Englished, S. Johns Peaches, Hastie Peaches, and Midsomer Peaches.

The tree Persea with his fruite, is not to be reckoned amongst these kindes (as some thinke) for Persea is a great tree, like a Peare tree, alwayes greene and lode with fruit, as Theophrastus in his fourth booke the second Chapter writeth.

✻ The Nature.

The Peache is colde and moyste in the seconde degree. The leaues of the tree and the kernels of the fruite are hoate and drie, almost in the third degree, and of a scowring power by meanes of their bitternesse.

✻ The Vertues.

Peaches before they be ripe, do stoppe the laske, as Dioscorides saith.

But being ripe, they loose the belly, & engender noughtie humours: for they are soone corrupted in the stomacke, wherefore they ought not to be eaten after meates, but before, as Galen saith.

The leaues of the Peache tree, do open the stoppinges of the liuer, and doo gently loose the belly, and are good with other conuenient herbes, agaynst tertian feuers.

The same layde vpon the nauell, do kyl and driue out wormes, especially in young children.

The same dried and strawed vpon newe woundes, do cure and heale them.

The Peache kernel openeth all stoppinges of the liuer and lunges, and in vertue is much lyke to bitter Almondes.

It is good to recouer againe the speache of such as be taken with the Ipopletic, if it be stieped in the water of Penny Ryall.

Peache kernelles pownde or beaten very smal, and boyled in vineger vntil they dissolue or melte, and become lyke pappe, is good to be vsed against the Alopetiam: for it doth woonderfully restore the heare if the place be annoynted therewithall, as Matthiolus saith. There be other vertues attributed to the same kernelles, as ye may reade in Matthiolus and Myzalde.

Of the Almonde tree. Chap.xli.

¶ The Kindes.

There be two sortes of Almondes, that is to say, the sweete and bitter Almondes.

Amygdalus. Almondes.

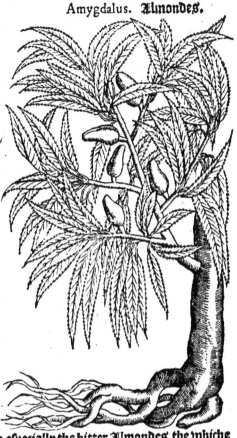

❧ The Description.

THE Almonde tree, in groth, and leaues, is lyke to the Peache tree, but it waxeth bigger, and stronger, & is of a longer continuance or lasting. The fruite is a harde nut like the Peache stone, but smooth without, and couered with an bitter huske or shale lyke the Walnut: within the inner shale is the Almonde, in taste bitter or sweete, as is abouesayde.

❧ The Tyme.

The Almonde tree flowreth bytimes, with the Peache tree. The fruite is ripe in June & July.

❧ The Names.

The Almonde tree is called in Greke ἀμυγδάλη: in Latine, Amygdalus: in high Douche, Mandelbaum, in base Almaigne, Amandelboom: in Frenche, *Amandier*.

The fruite is called in Greeke ἀμυγδαλόν καὶ ἀμυγδάλα: in Latine, Amygdala, and Amygdalum: in English, Almondes, or Almonde: in French, *Amand*: in high Douch, Mandel: in base Almaigne, Amandele.

❧ The Nature.

Almondes are somewhat hoate, especially the bitter Almondes, the whiche be not onely hoate, but also drie, and of clensing, and cutting power.

❧ The Vertues.

Almondes taken before meate, do stop the belly and nourishe but litle, especially being blanched or made cleane from their skinnes or huddes.

Bitter Almondes doo open the stopping of the lunges or lightes, the liuer, the melt, or splene, the kidneyes, & of al other inwarde partes: therefore they be good against the cough, the shortnes of wind, the inflammation & exulceration of lunges, to be mingled with Turpentine & licked in, as Dioscorides writeth.

Almondes are good for them that spet blood, to be taken in with the fine flower called Amylum.

The bitter Almondes taken with a litle sweete wine, as Muscadel or Bastarde, prouoke vrine, and do cure the hardnesse of the same, and painefulnes in making water, & are good for them that are troubled with the grauel & stone.

They vse to take fiue or sixe bitter Almondes fasting, to be preserued from dronkennesse al the same day.

They take away headache to be applied to the forehead with oyle of Roses, and vineger.

They are with great profite layde to with hony, vpon corrupt and noughty spreading sores, and the bitinges of mad Dogges.

They clense the skinne and face from al spottes, pimples and lentiles.

Of the Peare tree. Chap.xlij.

❧ *The Kindes.*

There be diuers sortes of Peares, aswell as there be kindes of Apples, whereof some be rathe ripe, some haue a later riping, and some be winter Peares, some perish quickly, some last a longer time and may be wel kept: some be sweete and full of sap or iuyce, some fat and grosse, and some harde and drie, &c. so that it is not possible to recite all the kindes of Peares: wherefore we do aduise the Readers to consyder the taste more then the proportion, or the time of the riping of Peares: for the taste doth best declare and giue notice of the qualitities and temperature of Peares.

❧ *The Description.*

Pirus. The Peare tree.

THE Peare tree is as great or greater then the Apple tree, and higher, with a great body or stemme, and manye great branches, the which for the most parte doo shute or mount vpright, & not one ouer another, as the branches of the Apple tree. The leaues be roundish, smoth, and very greene aboue: but vnderneath most commonly they be whitish. The fruite for the most part is long, brode beneath, and narrowe, and sharpe vpwarde towardes the stemme, very diuers or cōtrary, in colour, quantitie, proportion, and taste, as is aboue sayde. In the middle of the fruite there is a coare with kernels or peppins, lyke as in the middest of the Apples.

❧ *The Place.*

The Peare tree is planted in gardens and Orchardes: also it groweth sometimes in woods & wild vntoyled places, but they be none otherwyse esteemed, but as wildings or wild hedge Peares.

❧ *The Tyme.*

The Pearetree flowreth in Aprill or May, and the fruite is ripe in sommer and Autumne.

❧ *The Names.*

The Pearetree is called in Greeke ἄπιος: in Latine, Pirus: in Frenche, *Vng Poirier*: in high Douch, Byrbaum: in base Almaigne, Peerboom.

The fruite is called in Greke ἄπια in Latine, Pira, or as some do write Pyra: in French, *Poyres* in high Douch, Byren: in base Almaigne, Peeren, in English, Peares.

❧ *The Nature.*

All kindes of Peares are of a colde temperature, and the most part of them be

the Historie of Plantes.

be drie and binding, but not all alyke: for the wilde Peares, and others that be rough, binding, and chokely, do drie & stop a great deale more then the others. The sweete & grouse Peares, are moystier and very litle astringent or nothing at all. The middle sorte of Peares whiche are betwixt sweete and sower, are of complexion or temperature nearest to them vnto whom their taste draweth nearest.

The Peartree leaues are colde of complexion, drie and astringent, as Galen saith.

❧ The Vertues.

Peares taken before meate, do nourish but litle, yet they nourish more then Apples, especially those that be grouse and sweete.

The sower, rough, and chokely Peares, and others that are not waterie, to be eaten rawe or backte before meale, do stop the common laske or flowing of the belly, and do fortifie and strengthen the mouth of the stomacke.

They be also good to be laide to the beginnings of hoate tumours or phlegmons, and greene woundes.

The leaues are good for the same purpose, for they close togither and heale newe woundes.

Of the Medler tree.   Chap. xliij.

❧ The Kyndes.

DIoscorides setteth foorth two kindes of Medlars. The first kind growing vpon thornes. The second kinde is our common Medlars, the which also be of two sortes: for some be small and some great, but in fashio both lyke, and therefore some take them but for one kinde.

❧ The Description.

1 THE firste kinde is a thornie tree, with prickles and leaues, not muche vnlyke the hawthorne. The fruite of this plante is small and rounde, and, as Dioscorides saith, it hath three kernelles or stones in it: and they growe in clusters, fine or sixe, or more togither.

2 The common Medler is a tree in some places not altogither without prickles, growing almost lyke to the other trees. The leaues be somewhat long and narrowe, lesser then the leaues of the Apple tree, nothing at all dented or snipt about the edges. The flowers be white, and parted into fiue leaues. After the flowers groweth the fruite, whiche is of a browne russet colour, of a rounde proportion and somewhat broade or flat, of this kinde one is smal, the other great, yet they be alwayes lesser then Apples, with a great broade nauel or Crowne at the toppe, or ende, in the middle of the same fruite are fiue flatte stones, the whiche be the seede thereof.

❧ The Place.

1 The first kinde of Medler called Aronia, hath bene seene growing at Naples by the learned and famous Matthiolus: and is yet vnknowen to vs.

2 The common Medler is planted in gardens and Orchardes, & delighteth to growe in rough vntoyled places, about hedges and bushes.

❧ The Tyme.

Our common Medlers doo flower in Aprill and May, and are ripe at the ende of September.

❧ The Names.

1 The Medler is called in Greke μέσπιλος: in Latine, Mespilus: in high Douch, Nespelbaum: in base Almaigne, Mispelboom: in Frenche, *Nefflier*.

The syxth Booke of

Mespilus Aronia. **The Neapolitan Medler.** Mespilus altera. **The common Medlers.**

The fruite is called in Greeke μέσπιλον: in Latine, Mespilum: in Englishe, a Medle, or an open arsse: in French, Nessle: in Douch, Nespel: in neather Douchlande, Mispele.

1. The first kinde is called in Greeke μέσπιλ[ο] ἀρωνία, καὶ ῥίκοκκ[ο]: in Latine, Aronia, and Trigrania: at Naples Azærolo: we may call it also Azarola, the three grayne Medler, or the Neapolitan Medler.

2. The seconde kinde is called in Greeke ἐπιμηλίς, Epimelis, and of some σιτάνιον: Sitanium, or as some write Setanium.

The biggest of this late recited kinde is called in English, a great Medler: or the garden Medler: in French, Nessle cultiuée: in Brabant, Pote Mispelen.

✱ The Nature.

Medlers be colde, drie, and astringent. The leaues of the Medler tree, be of the same nature.

✱ The Vertues.

Medlers do stoppe the belly, especially being yet greene and harde, for after they haue bene a while kept, so that they become soft and tender, they doo not stoppe so muche: but then they are more conuenient to be eaten, yet they nourish but litle, or nothing at all.

The Medler stones made into powder and dronken, doo breake the stone and expulse grauel, as Antony Musa writeth.

Matthiolus & Mizalde, do intreate more largely of the vertues of this fruit.

Of the Mulberie tree. Chap.xliiij.

❧ The Description.

THE Mulberie tree is great and large, spreading his branches into breadth and length, his leaues be greene & large, snipt about the edges, after the maner of a sawe. The flower is smal with a fine hoare or soft cotton. The fruite consisteth of many beries growing togither like the fruite of ye Bramble, but it is larger and longer, of colour white at the beginning, after redde, and at the last blacke, of a winishe taste. The rootes be yellowishe, especially the barkes of them whiche be also bitter in taste.

❧ The Place.

The Mulberie tree reioyceth in the garden soyle, and other hoate and fat manured places.

❧ The Tyme.

The Mulberie tree bringeth foorth his newe leaues in May, a long time after other trees. And therefore it is called in the sayning of Poetes, the wisest of al other trees: for this tree only amongst al others bringeth foorth his leaues after ye colde frostes be past, so that by meanes therof it is not hurt or hindered, as other trees be.

Morus. Mulberies.

❧ The Names.

1 The Mulberie tree is called in Greeke μορία καὶ συκαμινία: in Latine, Morus: in some Shoppes, Morus Celsi: in high Douche, Maulbeerbaum: in base Almaigne, Moerbesieboom.

2 The fruite is called in Greke μόρον: in Latine, Morum: in Shoppes, Morum Celsi: in Englishe, a Mulberie, or Mulberies: in high Douche, Maulbeeren: in base Almaigne, Moerbesien: in Frenche, Meures.

❧ The Nature.

The vnripe Mulberies are cold and drie in the second degree, & astringent.

The ripe beries are of a temperate complexion.

The barke of the Mulberie especially of the roote, is hoate and drie in the seconde degree, and of a cutting, clensing, and abstersiue propertie.

❧ The Vertues.

The greene and vnripe Mulberies dried, do stoppe the belly, the blooddy flixe, and vomiting, to be dronken in redde wine. A

The rype beries do loose and moysten the belly, causing to go to the stoole, especially to be taken fasting, or before meate. B

The same taken after meate are soone corrupted in the stomacke, causing windinesse and blastinges in the same. C

Of the iuyce of ripe Mulberies is made a confection in manner of a syrupe, very good for the vlcers, and hoate swellinges of the tongue, the mouth, and the Almondes or kernelles in the throote. D

The leaues of the Mulberie tree layde to with oyle, healeth burninges. E

The barke of the roote of the Mulberie tree boyled & dronken, doth open the stoppings of the liuer, the milt, and it looseth the belly, and by the meanes therof, both long and flat wormes are expelled. **F**

The decoction of the leaues and rootes of the Mulberie tree, is good to holde in the mouth against the tooth ache. **G**

The roote being cut, nicked or scotched about the later ende of Haruest, putteth foorth a gumme or iuyce, whiche is exceeding good for the tooth ache, and it scattereth and driueth away swelling lumpes, and will purge the belly: but when you will haue this iuyce, you must first make a little furrowe about the roote you meane to scarrifie, and the next day after that you haue scarrified the roote, you shal finde the liquor clumpered or congeled togither in the furrowe. **H**

Of the Sycomore tree. Chap.xlv.

❧ *The Description.*   Sycomorus.

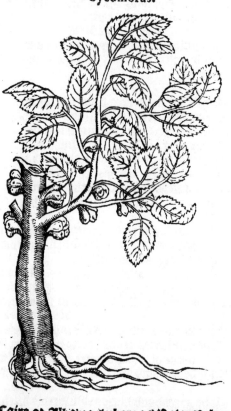

The Sycomore is a great tree lyke the Mulberie tree, with a great stem or tronke, & many great limmes & branches. The leaues be muche like to the leaues of ye Mulberie. The fruit is like to a wild figge, but it is without any smal sedes in it, and it groweth not vpō the young branches as the fruite of other trees groweth, but vppon the stocke or stem, & the greatest armes of the tree: also it neuer waxeth ripe vnlesse it be scraped with an iron toole.

Also there is a certayne gumme or liquor gathered frō out of the barkes of the young Sycomore trees, the whiche is gotten by pearsing the rinde or barkes of the young trees, before they haue borne any fruite.

❧ *The Place.*

The Sycomore tree, as Dioscorides writeth, groweth in Caria, and Rhodes, & in other places where as wheate groweth not. There is abundance of it planted in Egypt about the great Caire or Alkayre, where as Peter Belon hath seene it.

❧ *The Tyme.*

The trees be alwayes greene, and bring foorth fruite three or foure times a yere.

❧ *The Names.*

This tree is called in Greeke συκόμορⓈ, and of some συκάμυνⓈ: in Latine, Sycomorus: in Englishe, a Sycomore tree.

❧ *The Nature.*

The fruite of the Sycomore tree is somewhat temperate, the gumme therof hath power to make warme and to soften.

❧ The

the Historie of Plantes. 717

❧ *The Vertues.*

The Sycomore fruite is good to eate, but it yeeldeth small nourishment, it looseth the belly gently, and is not good for the stomacke.

The gumme is good for the hardnesse of the milt or Spleene, the payne of the stomacke, and bitinges of Serpentes, to be eyther taken inwardly, or layd to outwardly vpon the wounde.

It closeth woundes togither, and scattereth abrode olde gatheringes togither or collections.

Of the Figge tree. Chap.xlvi.

❧ *The Description.*

Ficus.

THE garden figge tree, whereof we shall nowe speake, hath many branches full of pith within, lyke the shutes or stalkes of Eldren, ouer couered with a smoth playne rinde or barke. The leaues be great and large, blackishe, and for the most part diuided in fiue. At the top of the branches groweth the fruite, the which is round and long, fashioned lyke Peares, sweete, and ful of smal kernelles or graines. Before the fruite be ripe, if it be hurt or scarrified, there commeth foorth a sappe or iuyce like milke, but being through ripe, the iuyce is lyke to hony.

⁋ *The Place.*

The figge trees are plentiful in Spayne and Italy, and are founde also sometimes in this Countrie, but very rare and seldom, they must be planted in warme places, that stand wel in the Sonne, and are defended from the North and Northeast windes.

❧ *The Time.*

The Figge trees in this Countrie are very long & late in waxing greene, for they begin to put foorth their leaues but at the end of May. Their fruite is rype about the ende of sommer.

❧ *The Names.*

1 The garden figge tree is called in Greeke συκῆ ἥμερος: in Latine, Ficus satiua: in high Douche, Feighenbaum, in base Almaigne, Vijghenboom: in Frenche, *Vng Figuier.* in Englishe, a Figge tree, or a garden Figge tree.

The fruite is called in Greeke σῦκον: in Latine, Ficus: by the whiche name it is knowen in Shoppes: in English, a Figge: in French, *Figue*: in high Douch, Ein Feigen; in base Almaigne, Een Vijghe: this fruite before it is ripe, is called in Greeke ὄλυνθος: in Latine, Grossus: and when it is drie, they call it in Latine Carica: in Greeke, ἰχάς, and not καρικὴ: for καρικὴ in Greeke Carice, is a kind of figge whiche groweth onely in Syria.

2 The wilde figge tree is called in Greeke, συκῆ ἀγρία καὶ ἐρινὸς: in Latine, Ficus syluestris, and Capriſicus.

The fruite of this figge tree, whiche neuer commeth to ripenesse, is named in Greeke as the vnripe fruite of the garden figge tree, ὄλυνθος: in Latine, Grossus: and of some also ἐρινεός, Erineus.

❧ The Nature.

The greene figges new gathered are a litle warme and somewhat moyst.

The drie figges are hoate almost in the second degree, and somthing moyst, and of subtill partes.

The mylkie iuyce of figges is hoate and drie almost in the third degree, and also sharpe and biting.

The leaues haue also some sharpenesse with an opening power, but not so strong, as the iuyce.

❧ The Vertues.

The newe gathered figges, nourishe more then the other fruites: but they ingender windinesse and blasting, and they lose the belly gently. A

They abate heate and quenche thirst, but taken into great a quantitie, they do hurt the stomacke making it weake and without meate lust. B

The drie figges do nourish better then the greene or newe figges, yet they ingender no very good blood, for those that feede muche vppon figges become lousie and full of vermine. C

Figges eaten before meat, do loose the belly, and are good for the kidneyes, for they driue foorth grauell with vrine. D

They prouoke sweate, and by the same meanes they sende foorth corrupt and stinking humours: wherefore they be very well giuen to young chyldren that are sicke of small Pockes, and wheales, or Meselo, for they bring them quickly foorth and without ieopardie. E

They be also good for the throte and lunges, they mitigate the cough, and are good for them that are short winded, they rype flegme causing the same to be easyly spet out, in what sort so euer they be taken, whither rawe, or rosted, or sodden with Hysope and dronken. F

The decoction of figges in water, is good to be dronken of those that haue fallen from high, and haue taken squattes or bruses, for they dispearse and scatter the congeled or clotted blood, and asswage or slake the payne. G

Figges pounde with Salt, Rue, and Nuttes, withstandeth all poyson and corruption of the ayre. And this was a secrete preseruatiue with Mithridates king of Pontus, vsed against all venome and poyson. H

The decoction of figges gargarised or holden in the mouth is good agaynst the sharpenesse and hoarsenesse of the throte: also agaynst all swellinges and impostumations of the mouth, the throte, the Almondes of the throte & iawes, and swelling of the tongue. I

Figges are good to be kept in the mouth, against the Ache and payne of the teeth and iawes. K

Being layde to with wheaten meale, they do soften and ripe impostumes, phlegmons (that is hoate and angrie swellinges) and tumours behinde the eares, especially if you put to it Lyneseede and Fenugreck, and if you put to it the rootes of Lyllies, it will ripe and breake impostumes and botches. L

Figges mixed with barley meale doo scatter and consume swellinges, as Galen writeth. M

The same sodden in wormewood wine with barley meale, are good to be layde to, as an Emplayster vpon the bellies of suche as haue the dropsie. N

Figges and Mustardseede pounde very small togither, doo helpe the ringing noyse and sounde of the eares, also they amende the hearing being layd to outwardly. O

In fewe wordes, the dried figges haue power to soften, consume, and make ₽ subtil, and may be very well vsed both outwardly and inwardly, whither to ripe or soften impostumes, or els to scatter and dissolue them.

The leaues of the figge tree do wast and consume away the kinges euil or Q swelling kernelles in the throte, and do mollifie and waste all other tumours, being small pounde and layde thereto.

The milkie iuyce of figges is good against all roughnesse of the skinne, Le-R pries, spreading sores, tetters, small pockes, meselles, pushes, freckles, lentiles, and other suche lyke spottes, and scuruinesse, both of the body and face, layde to with barley meale parched: also it taketh away wartes, if it be layde to with fatte or grease.

It cureth the tooth ache, if you dip a litle Cotton or Bombasie in the sayde S milke, and lay it to your tooth, or make a litle pellet thereof, and put it into the holownesse of the corrupt or aking tooth.

It openeth the vaynes of the Hemeroides, & looseth the belly being layd to T the fundement. The leaues haue the same vertue, if they be wrong in behinde at the fundement.

It is very good to be layde to with the meale of Fenugreke and vineger, V vpon the hoate gowte, especially the gowte of the feete.

It is good to powre of the same iuyce into the wound made with the sting X of the Scorpions, or the bitinges of madde Dogges.

The iuyce of figges turneth milke and causeth it to crudde, and agayne it ℞ scattereth, or dissolueth, or melteth the clustered crudde, or milke that is come to a crudde, as vineger doth.

The ashes of the Figge tree mixed with oyle of Roses and Waxe, cureth Z burninges, and the lye that is made with the ashes of the Figge healeth scuruinesse, and festred or fowle fretting sores, if they be washed therewithall.

Of Plum trees, Bullies, Slose, & Snagges. Chap. xlvij.

The Kindes.

First to speake generally of Plummes there be two kindes, whereof some apparteyne to the garden, and some are of a wilde kinde. The garden or tame sort of Plummes are of diuers kindes, some white some yellow, some blacke, some of the colour of a Chesnet, and some of a lyght or cleare redde: and some great, and some small, some sweete and dry, some fresh and sharpe, whereof eche kinde hath a particuler name. The wilde Plummes are the least of al, and are called Slose, Bullies, and Snagges.

The Description.

THE Plumme tree groweth vpright lyke other trees, especially if it be well guyded, and gouerned, and putteth foorth many branches, ouer couered with a smooth brownishe barke, from out of the which being scarrified or otherwise hurte, In sommer it putteth foorth gumme. The leaues are somewhat long, yet for all that almost rounde, and finely snipt or hackt about the edges lyke a sawe. The flowers be white lyke the blossoms of the Cherrie tree, and are also parted into fiue or sixe smal leaues. The fruite is most commonly somewhat long, whereof some are great, some small: of colour some be white, some yellowishe, some blacke, and some red. In the middle whereof is inclosed a litle long harde stone, hauing in it a nut or kernel of a bitter taste. The roote of this tree spreadeth muche abroade in the grounde, and putteth foorth in many places newe springes and scyons, the whiche will also growe vp to the height, if they be not cut of in season.

The ſyxth Booke of

Prunus.
The Plum tree.

Prunus ſylueſtris.
The Sloo or wilde Plum tree.

2 ⸿ The wilde Plumtree groweth not vp to the ſtature of a tree, but remaineth lowe by the grounde, lyke to a hedge buſhe, whereof it is a certayne kinde: it putteth vp many branches from one roote, ſet here and there with pricking thornes, & leaues like to the of the garden Plummes or Damſons, ſauing that they be ſmaller. The flowers be alſo white. The fruit is ſmal, a great deale leſſe then any other Plummes, in taſte ſower and binding, the roote alſo ſpreadeth farre abrode in the grounde or earth, very plyant, and of a woddiſhe ſubſtance.

❋ *The Place.*

1 ⸿ The Damſons and other of the garden kindes, are founde almoſt euery where in Orchardes.

2 ⸿ The wilde Plummes do grow in feeldes and wayes, and other vntoyled places and in hedges.

❋ *The Tyme.*

The Plumtrees do flower in Aprill or ſommer, eſpecially the wilde Plumtree, the whiche flowreth rather then the other.

The kindes of garden Plummes are ripe in Auguſt, the wilde moſt commonly in September.

❋ *The Names.*

1 ⸿ The Plumtree is called in Greeke κοκκιμηλἰα: in Latine, Prunus: in high Douche, Pflaumenbaum: in baſe Almaigne, Pruymboom: in Frenche, *Vng Prunier*.

The fruite is called in Greeke κοκκιμῆλον: in Latine, Prunum: in Engliſhe, a Plumme or Prune: in Frenche, *Vne Prune*: in high Douche, Ein Pflaume
oder

oder Praume: in base Almaigne, Een Pruyme.

The great sweete blewish Plummes, are called of Theocritus, βράβυλα, Brabyla: of the Latinistes, Pruna Damascena: in Englishe, Damaske Prunes: in Frenche, *Prunes de Damas*: in high Douche, Quetschen, and Blauw Spilling: in base Almaigne, Pruymen van Damasch.

The common browne blewe, and Crimsen Damsons, are called Hispanica.

The yellowishe Plummes are called in Latine, Cerea, and Cereola Pruna: in Englishe, the Wheaten, or white Plumme: in Frenche, *Prunes blanches*.

The great rounde redde Plummes, are called of some in Latine, Pruna asinina: in English, Horse Plummes: in French, *Prunes de Cheual*. in high Douch, Roßpflaumen: in base Almaigne, Rospruymen.

The least of all whiche be small and rounde, are called in Frenche, *Dauoines*: in high Douche, Herbstpflaumen, and in base Almaigne, Palloken.

The wilde Plumme tree, Blacke thorne, and Sloo tree, is called in Greeke κοκκιμηλέα ἀγρία, καὶ ἀγριοκοκκιμηλία: in Latine, Prunus syluestris: in high Douche, Schlehedorn: in base Almaigne, Slehedoren: in Frenche, *Prunier sauluage*.

The fruite is called in Greeke κοκκιμηλον ἄγριον, καὶ ἀγριοκοκκιμηλον: in Englishe, Slose, whereof that kinde whiche is founde growing vpon the blacke thorne, is called Catte Slose, and Snagges: in Frenche, *Prunelles*, or *Fourdeines*: in Latine, Prunum syluestre, Pruneolum, and Prunulum: in high Douche, Schlehen: in base Almaigne, Slehen.

The iuyce of Snagges or Catte Slose, is commonly called in Shoppes, Acatia, and is vsed in steede of Acacia.

❧ The Nature.

The garden Plummes do coole and moysten the stomacke and belly.

The Snagges and Catte Slose, are colde, drie, and astringent.

❧ The Vertues.

Plummes do nourishe but litle, and ingender noughtie blood: but they doo gently loose and open the belly, especially when they be freshe and newe gathered, after they be ripe.

Plummetree leaues are good against the swelling of the uuula, the throte, gummes, and kernelles vnder the iawes, for they stop the Reume, & flowing downe of humours, if a man will gargle with the decoction thereof made in wine, as Dioscorides writeth.

The gumme of the Plummetree dronken in wine, breaketh the stone and expelleth grauell, as some do write.

The wilde Plummes doo staye and binde the belly: and so doo the vnripe Plummes, and all others that be sower and astringent.

The iuyce of wilde Plummes or Snagges, doo stoppe the laske, womens flowers, and all issue of blood, and it may be very wel vsed in steede of Acatia.

Of Sebestens. Chap.xlviij.

❧ The Description.

Amongst the kindes of Plummes (those which are called in Shoppes Sebestens) may be accounted, the which do also growe vppon trees, whereof the body or stemme is couered with a white barke, and the branches with a greene pil or rind. The leaues be roundish. The fruit is not muche vnlyke the least kinde of Damsons or Plummes, but smaller, of a blewishe colour and sweete taste, white within, and of a viscus or clammie substance, in the middle whereof are small stones with kernelles in them, lyke to Plumme stones.

The Place.

This fruite groweth in Italy, and other hoate regions, from whence it is brought alredy dryed vnto vs.

The Names.

This fruite bycause of his clammishnesse and flyme, is called in Greke μύξα, Myxa, and μυξάρια, Myxaria: in Shoppes, Sebesten, & of some Sebastæ: of Matthiolus, Prunus Sebestena.

The Nature.

The complexion of Sebestens drawe towardes colde and moyst, and therfore they be of nature muche lyke to garden Plummes.

The Vertues.

Sebestens be good in hoate agues, especially when the belly is stopte or bounde. A

They be also very good against the cough, and flowing downe of hoate and salt Catarres and Reumes vpon the breast and lunges. B

They be also good against the inflammation of the bladder and kidneyes, and against the strangurie and hoate pisse, or the burninges of vrine. C

Of Iuiubes. Chap.xlix.

The Kindes.

There be two sortes of Iuiubes, as Columella writeth, that is to say, redde and white.

The Description.

Iuiubes is the fruit of a tree, as the Sebestens be, they be round and long, not muche vnlyke an Oliue, but smaller, of colour either white or red, in taste sweete, the whiche being long kept, waxe drie and full of wrinckles: eache Plumme or fruite hath a harde long stone in it, lyke in fashion to an Olyue stone, but muche lesser.

The Place.

Iuiubes do growe in hoate regions, as in Italy and other lyke places.

The Names.

This tree is called of Columella in Latine, Ziziphus: in English, the Iuiub tree: in Frenche, Iuiubier, and Guindoulier.

The fruite is called Zizipha: in shops, Iuiubæ: in English, Iuiubes: in Frēch, Iuiubes, and Guindoules: in high Douche, Brustbeerlin: some thinke that Galen called this fruite in Greeke, σηρικά, Serica.

The Nature.

Iuiubes are temperate in heate and moysture.

The Vertues.

Iuiubes eaten are harde of digestion, and nourishe very little, but taken in Electuaries, syrupes, and other medicines, they appease & mollifie the roughnesse of the throte, the breast and lunges, and are very good against the cough. A

In the selfe same maner they are very good for the reynes of the backe, the kidneyes, and the bladder, whether they be exulcerated or inflamed, or vexed with any sharpe and salt humour. B

Of Cherries. Chap.l.

The Kindes.

There be two sortes of Cherries, great and small: the small Cherries doo growe vpon high trees, and the greater Cherries vpon meane trees. And of euery kinde there be two sortes, some red, some blacke. Bysides these kindes

kindes there are Cherries that grow, three, foure, and fiue vpon a stemme, and also that hang in clusters lyke grapes, whereof the learned Matthiolus hath giuen vs the figures.

Cerasia. **Sweete Cherries.**
Cerasa Racemosa. **Cluster Cherries.**
Cerasa austera. **Sower Cherries.**

✤ *The Description.*

1 That tree which beareth the common small Cherries, or Mazars, is most commonly great, high, and thicke, lyke to other trees. The barke of this tree is playne & smooth of colour lyke the barke of the Chesnut tree, three or foure fold double, the which will suffer to be scaled, rinded, stripte, and pylde, like to the barke of the Birche tree. The leaues be great and somewhat long, hackt about the edges with teeth lyke a sawe. The flowers be white and parted into fiue or sixe small leaues. The fruite hath a sweete smak or taste, of colour sometimes redde, sometimes browne, in proportion like the greater Cherries, but smaller, yea sometimes very small. In the same there is founde a small stone, with a kernell therein closed.

2 The tree that beareth the great Cherries, is not very high, but most commonly of a meane stature, in al thinges els like þe other, both in leaues and flowers. The fruite is a litle long and rounde, and of a pleasant sweete taste, of colour somtimes of a faynt red, and almost halfe white and halfe red, sometimes browne, & wel neare all blacke, whereof the iuyce stayneth purple, or a faire Crimsen lyke to Brasill.

3 The sower Cherries are to be be accounted amongst the rest. This tree is most commonly weake and tender, neither high nor great : and therefore of no long continuance. The leaues be also smaller, but otherwayes lyke the leaues of the sweete Cherries, the flowers be white, the fruite is rounde and sower, sometimes red, and somtimes blacke, lyke the Mazar or Hurtel Cherries, this Cherrie hath also a stone in the middle of the fruite, but smaller and rounder then the Guyan or sweete Cherries.

Out of al these Cherrie trees, there issueth gumme like that of the Plumme trees, or Peache trees, especially when the rinde or barke is any wayes hurt or brused. ✤ *The Place.*

The tree that beareth the sweete Guyan Cherries, or the great Frenche Cherries is planted in gardens and Orchardes. But that whiche beareth the Mazars, or þe smal Cherries groweth in some places very plentifully in feeldes and wooddes.

Matthiolus writeth that about Trent a Citie in Italy, about the Citie of Prage in the Countrie of Bohem, & about Uienna in the Countrie of Austrige, there growe naturally wilde Cherries vpon lowe bushes or shrubbes, of little more then halfe a fa foote high, and their fruite is in all respectes aunswerable to the other small Cherries. ✤ *The*

The Time.

The great French Cherries, & the common Cherries do commonly flower in Aprill. The redde Cherries are rype in June, and continue vntyll July: but the blacke waxe ripe in July, and they may be kept fresh & whole vnto the ende of August.

The Names.

The Cherrie tree is called in Greeke κέρασ⊙: in Latine, Cerasus: in high Douch, Kirschenbaum: in base Almaigne, Kerseboō: in frēch, *Cerisier*: & *Guisnier*.

The fruite lykewise is called κέρασα, Cerasa: in English, Cherries, in French, *Guinnes* in high Douche, Kirschen: in base Almaigne, Kersen.

And for the better declaration both of the names and kindes of Cherries, I haue thought good to giue you to vnderstande, what I haue conceiued of this matter. I reade in Matthiolus, that the common people of Italy doo call the waterishe Cherries *Acquainola*. The famous learned man Robertus Stephanus in his Frenche Dictionarie, doth turne this French worde *Guisnes* into Latine, as foloweth (Aquitanica cerasa) whiche soundeth in English, Guyan Cherries, now whether the people of Italy do cal *Guyan Aquitan*, I referre that to them that be expert in that language. But the French word seemeth to haue his first originall of the Countrie Guyan, for they expound Cerasia, *Guisnes doulces*, Sweete Cherries.

Grosses guisnes, Duracina cerasa, Harde Cherries.
Guisnes noires, Cerasia Actiana, Small Cherries lyke Eldren beries.
Guisnes fort rondes, Cerasia Cæciliana, Rounde Cherries.
Guisnes fort rouges, Cerasia Aproniana, Grape or cluster Cherries: so that *Guisnes* is their proper worde for all sortes of Cherries, except sower Cherries, which they call *Griotes*: in Latine, Cerasia acida.

1　The first kind, especially that which beareth the smallest fruite, is the Cherrie tree described by Theophrastus.

2　The other is called of some χαμαικέρασ⊙, Chamæcerasus: yet it is not that Chamæcerasus whereof Asclepiades Myrleanus writeth, the fruite whereof maketh men dronken like wine. The Brabanders name this tree Spaensche Kerselaer, and the fruite Spaensche Kersen, that is to say, Spanish Cherries, or Cherrie tree: in Frenche, *Guinnier*, and *Guinnes*: in English, Frenche Cherries, and Spanish Cherries: they be also called in Frenche, *Cueurs*: and they that be halfe white *Bigarreans*.

3　The common sower Cherries is of the later writers taken to be a kinde of Cerasus, and therefore the fruite is lykewise called Cerasa, of some Merendæ, or Marenæ: Platina writeth of one Moretum ex Merendis, Cordus writeth of one compounde named Diamarenatum, and both these are made of Cherries.

The Nature.

All Cherries and Mazars, are colde and moyst of temperature, but aboue all the rest the sower Cherries do coole most, and specially those that be blacke, whiche are also astringent, especially beyng dryed.

The Vertues.

Cherries eaten first before other meates, do soften and loose the belly very gently, but they nourishe but little, and are hurtfull vnto moyst, vnhealthie stomackes: for they be soone putrified and corrupted within the same, especially the Mazers or small Cherries, the whiche do oftentimes ingender agues and other maladies.

The red sower Cherries, do lykewise loose the belly, and are more wholesome and conuenient for the stomacke: for they doo partly comfort, and partly slake or swage thirste.

The blacke sower Cherries doo strengthen the stomacke more then the rest, C
and being dried they stoppe the laske.

The Gumme of the Mazar or wilde Cherrie tree, of the Spanish Cherrie, D
and of our common sower Cherrie tree, is good to be dronken in wine of those
that are troubled with the grauell and the stone.

It is also good against the excoriation and roughnesse of the throte, lunges, E
and breast, and against the cough and hoarsenesse.

The water distilled of freshe and newe gathered Cherries, is good to be
powred into the mouthes of such as haue the falling euil, as oftē as the course
or fit troubleth them, for it is good against the rigour and violence of the same.

Of the Cornell tree. Chap.li.

❧ The Kindes.

THere be two sortes of the Cornell tree (as Theophrastus writeth) that is
to say, the tame and wilde.

Cornus mas.
The male or tame Cornell tree.

Cornus foemina.
Dogge berie, or Gatten tree.

❧ The Description.

1 THE tame Cornell tree, sometimes groweth vp handsomly, and
wareth meetely great lyke other meane trees: sometimes also it is
but low, and groweth lyke to a shrub or hedge bush: as diuers other
small trees doo. The wood or timber of this tree is very harde.

The flower is of a faynte yellowish colour. The fruite is very redde, and somewhat long almost lyke an Olyue, but smaller, with a long litle stone or kernell, thereinclosed like to the stone of an Olyue berie.

2 The wilde Cornell tree groweth not vp lyke a tree, but remayneth lowe as a hedge plant, the timber of this tree both of the young twigges and old branches is likewise very harde and plyant: the shutes and scorges, are full of knottes or ioyntes and within they be full of pith, lyke the shutes of Elder. The leaues are very lyke to them of the tame or male Cornell tree. The flowers be white and doo growe in tuftes, after them ryse small rounde beries, whiche are greene at the first, but afterwarde blacke when they be ripe.

The Place.

1 The tame Cornel tree is found growing wilde in many places of Almaigne like to other bushes: but in this Countrie it is not to be founde but in gardens and Orchardes.

2 The wylde Cornell tree is founde growing in hedges and alongest the feeldes.

The Tyme.

1 The tame Cornell tree flowreth bytime in Marche or sometyme rather: and afterwarde it bringeth foorth his leaues. The fruite is rype in August.

2 The wylde Cornell tree flowreth in Aprill and May: his beries be rype in September.

The Names.

1 The tame Cornell tree is called in Greeke κρανία: in Latine, Cornus: in Englishe, the Cornell tree, of some long Cherrie, or long Cherrie tree: in high Douche, Cornelbaum, Thierlinbaum, and Kucbeerbaum: in base Almaigne, Cornoelieboom.

2 The wilde Cornell tree, is called of Theophrastus in Greeke Θηλυκρανία: that is to say in Latine, Cornus fœmina: in Englishe, the female Cornel tree: Houndes tree, and Hounde berie, or Dogge berie tree, and the Pricke timber tree, bycause Butchers vse to make prickes of it: in high Douche, Hartriegel: it is called in Brabant of some Wilden Ulier, that is to say, Wylde Elder, bycause the pith of the young shutes is somewhat like Elder. Matthiolus calleth it Virga sanguinea.

The Nature.

The garden or tame Cornell tree or fruite is colde drie and astringent.

The Vertues.

The Cornell fruite (of the garden) taken in meate or otherwise, is good against the laske and bloody flixe, also they doo strengthen the weake and hoat stomacke. A

The leaues and tender croppes, will heale greene woundes, and stoppe the bleeding of the same, as Galen saith. B

The wilde Conell Berries are not vsed in medicine. C

Of the Sorbe tree.   Chap.lij.

The Kindes.

There be three sortes of Sorbus, wherof one kinde is rounde like Apples, the second is long after the fashion of Egges, and the thirde sorte is brode in the bottome, and not muche vnlyke the Peares.

Sorbus, Sorbe Apple tree.

❧ The Description.

THE Sorbe apple tree groweth high, with a straight body or stemme of a brownishe colour, and many branches, couered with long displaied leaues, which leaues are made of many slender leaues, standing ryght ouer one against another, all vppon one stemme, whereof eche of the litle leaues by them selues are lōg, and iagde about lyke to a sawe. The flowers be white, after them commeth the fruite, in figure sometimes rounde, sometimes long, and sointimes lyke to a Peare, and red vpon the syde next the Sonne.

❧ The Place.

The Sorbus tree delighteth in colde and moyst places, vppon mountaynes, but cheefely in stony places. It is founde in some places of Douchelande.

❧ The Tyme.

The Sorbus tree flowreth in March, and his fruite is ripe in September.

❧ The Names.

The tree whervpon this fruite groweth is called in Greeke ὄη καὶ ὄυη: in Latine, Sorbus: in Englishe, Sorbe Apple tree: and for the rest of the kindes of this tree, I referre you to the second part of Maister Turners herbal, fol. 143. This tree is called in high Douche, Spetwerbaum: and in base Almaigne, Sorbenboom.

The fruite is called in Greeke ὄοη καὶ ὄυοη: in Latine, Sorbum: in Englishe, Sorbe Apple: in Frenche, Corme, or Sorbe: in high Douche, Spiereling vnd Sporapfel: in base Almaigne, Sorben.

❧ The Nature.

The Sorbus fruite is colde, drie, and astringent, almost lyke to the Medlers.

❧ The Vertues.

The Sorbe Apples gathered before they be rype, & dryed in the Sonne or A otherwise, doo stoppe the laske, when they be eaten, or the decoction of them dronken.

To conclude, the Sorbe Apples or Seruice beries, are muche lyke to B Medlers, in vertue and operation, sauing that they be not althing so strong.

The barke of one kinde of Sorbus (whiche is our Quickbeme) is in some C places wrongfully vsurped in steede of the barke of Tamariske, for the diseases of the milte. Some also haue vsed to make dishes and drinking Cuppes of the tymber of Quickbeme to drinke out of as a remedie agaynst the Splene, but they are deceiued, for they shoulde make them of Tamariske timber.

Of the Arbute or Strawberie tree. Chap.liij.
Arbutus.

❧ *The Description.*

THE Arbute is a small tree not muche bygger then a Quince tree, the stemme or body whereof is couered with a reddish barke which is rough and scaly. The young branches are smooth and redde, set full of long broade and thicke leaues, hackt rounde about like a sawe. The flowers be white, small, & holow, and doo growe in clusters, after whiche comineth the fruite which is rounde, and of the fashion of a Strawberie, greene at the first, but afterwarde yellowishe, and at last red when it is ripe.

¶ *The Place.*

The Arbute tree groweth in many places of Italy and other Countries wild: but it is vnknowin this Countrie.

❧ *The Tyme.*

The Arbute tree flowreth in July and August: the fruit is ripe in September at the comming in of winter, after that it hath remained hanging vpon the tree by the space of a whole yere.

❧ *The Names.*

This tree is called in Greeke κόμαρος: in Latine, Arbutus, of some Vnedo, howbeit that name agreeth best with the fruite: in Frenche, *Arbousier*: in Englishe, the Arbute tree, and of some Strawberie tree.

The fruite is called in Greeke μεμαίκυλορ, or as some write, μεμάκυλορ: in Latine, Vnedo, and Memæcylon: in Frenche, *Arboses, or Arbousies.*

❊ *The Nature.*

The fruite of the Arbute tree is of a colde temperature.

❊ *The Danger.*

The fruite of the Arbute tree, hurteth the stomacke and causeth headache.

Of Lotus or Nettle tree. Chap.liiij.

❊ *The Description.*

LOtus is a great high tree, spreading abrode his branches, whiche be long and large. The leaues be also large and rough, cut round about the edges after the maner of a sawe. The fruite is rounde and bigger then Pepper, as Dioscorides writeth, hanging vpon long stemmes, at the first greene, then yellowe, and blacke when it is type and drie, and of a pleasant taste and sauour. ❊ *The Place.*

Lotus groweth plentifully in Africa, and is founde also in many places of Italy, and Languedoc.

❧ *The*

☙ The Tyme.

The fruite of Lotus is ripe in September, then it leeseth his leaues, and recouereth agayne newe togither with his flowers in the spring time.

☙ The Names.

This tree is called in Greeke λωτός: in Latine, Lotus, & Celtis: in some places of Italy, *Bagolaro*, & of some *Perlaro*: in Languedoc, *Micocoulier*, and the fruite *Micocoules*. Gesner saith that Celtis is called in French, *Algsiez*, or *Ledomier*. Peter Bellon calleth it also in French, *Fregolier*. Matthiolus saith that the Arabians call this tree Sadar, Sedar, or Alsadar: the Italians, *Loto Albero*: the Spaniardes, *Almez*: Turner calleth it in English, Lote tree, or Nettle tree, bycause it hath a leafe lyke a Nettle.

Cooper in his Dictionarie sayth, that the fruite of Celtis, or Lotos, is called in Latine, Faba Græca.

☙ The Nature.

The drie Lotus, is restrictiue, and of subtil partes.

☙ The Vertues.

The shauinges, or scrapinges of the shiuers, or wood of Lotus, boyled in water or wine stoppeth the laske, the blooddy flixe, and womens flowers or the flixe of the mother, to be eyther dronken, or taken in infuson.

The fruite doth also stop the belly, and is good to be eaten without hurt to the stomacke.

Of the Chesnut tree. Chap.lv.

☙ The Description.

The Chesnut tree, is a very great, high & thicke tree, not much vnlike the Walnut tree. The leaues be great & large, rough, and crompled, & snipt or iagged about like a saw, amongst the leaues at the top of y branches grow the Chesnuttes whiche are browne without, somewhat flat almost after the fashion of a hart, and playne and smooth pollished: they be also inclosed in shelles and very rough and prickley huskes lyke to a Hedgehogge or Urchin, the which huskes do open of their owne accorde when the Chesnuttes be ripe so that they fall out of their sayde huskes of their owne kinde.

☙ The Place.

The Chesnut delighteth in shadowie places and mountaynes whose situation is towardes the North. There is plentie growing about the riuer Rhene, in Swiserlande, and Daulphinie, also they growe plentifully in Kent, abrode in the feeldes and in many gardens of Englande.

Caſtanea. Cheſnut.

❧ The Time.

The Cheſnuttes be ripe about the end of September, and do laſt al the winter.

❧ The Names.

The Cheſnut tree is called in Greeke καςανα: in Latine, Caſtanea, and Nux Caſtanea: in high Douche, Keſtenbaum, & Caſtanibaum: in baſe Almaigne, Caſtanieboom: in Frenche, Caſtaignier.

The fruite is called in Greke Διὸς βάλανο, σαρδιανὰ βάλανο, λόπιμα, κασανίον κάςενον, καὶ κασάνιον: in Latine, Nux Caſtanea, Iouis glans, & Sardiana glans: in Engliſh, A Cheſnut: in Frenche, Caſtaigne: in high Douche, Keſten: in baſe Almaigne Caſtanie.

❧ The Nature.

The Cheſnuttes are drie and aſtringent, almoſt lyke the Akornes, or fruite of the Oke, & hoate in the firſt degree.

❧ The Vertues.

Amongſt all kindes of wilde fruites, the Cheſnut is beſt, and meeteſt for to be eaten, for they nouriſhe reaſonably wel, yet they be harde of digeſtion, and doo ſtoppe the belly.

They make an Electuarie with the meale of Cheſnuttes & hony, very good againſt the cough & ſpetting of blood.

The ſame made into powder & layd to as an emplaiſter with Barley meale and vineger, doo cure the vnnaturall blaſtinges, and ſwellinges of womens breaſtes.

The polliſhed red barke of the Cheſnut boyled and dronken, ſtoppeth the laſke, the bloodddy flire, and all other iſſue of blood.

Of the Walnut tree. Chap. lvi.

❧ The Deſcription.

THe Walnut tree is high and great, parted into many armes and branches, the whiche do ſpreade abroade in length and breadth: In the beginning of the ſpring time it bringeth foorth long tentes or yellowe ragged things compact of certayne ſcales, hanging vpon the tree, like ſmal Cattes tayles, almoſt like to that whiche hangeth vpon Wythie, but it is muche longer then the Chattons of Wythie, the whiche do vade and wither, and ſoone after they fall away. After theſe tentes or Catkens, the leaues begin to ſhowe, whiche be long and large, and of a good ſmell made of many leaues growing one againſt another alongſt a ribbe or ſinewe, whereof eache leafe is of lyke breadth and quantitie. The fruite groweth amongſt the leaues, two, three, or foure in a cluſter, couered with a greene huſke or ſhale, vnder whiche alſo there is another harde ſhale of a woddiſh ſubſtance, wherein is the braine, nut or kernell lapt in a ſoft and tender pill or ſkinne.

❧ The Place.

The Walnut tree loueth dry places & Mountaynes. They are planted in diuers places of this Countrie, and Almaigne, in Orchardes alongſt the feeldes.

the Historie of Plantes.

Nux. Walnuttes.

❧ The Tyme.

The ragged Catkens of the Nut tree, begin to spring out in Marche, or at the fardest in in April. The Nuttes be ripe about the ende of August.

❧ The Names.

The Nut tree is called in Greke καρύα: in Latine, Nux: in Frenche, Noyer: in high Douche, Nußbaum: in base Almaigne, Noteboom: in Englishe, the Walnut, and Walshe nut tree.

The ragged Catkens, whiche come foorth before the leaues, are called in Latine, Iuli nucum: in Douche, Catkens: in Englishe, Blossoms, Tentes, and Cattes taples.

The fruite is called in Greke κάρυον βασιλυκόν: in Latine, Nux regia, Nux iuglans, & Nux Persica: in shops, Nux: in Frenche, Noix: in high Douche, Welschnuß, and Baumnuß: in Brabant Okernoten: in Englishe, Walnuttes, Walshe Nuttes, and of some Frenche Nuttes.

❧ The Nature.

The Walnut being greene and newe gathered from the tree, is cold and moyst.

The Drie nuttes be hoate, and of a drying power, and subtill partes.

The greene huske or shale of the Walnut, dryeth muche and is of a binding power. The leaues be almost of the same temperature.

❧ The Vertues.

The newe greene Nuttes are much better to be eaten then the dry Nuttes, neuertheleffe they be harde of digestion, and do nourishe very litle. [A]

The dry Nuttes nourish lesse, and are yet of a harder digestion, they cause headache, and are hurtfull to the stomacke, and to them that are troubled with the cough, and the shortnesse of breath. [B]

A dried Nut or twayne taken fasting with a figge, and a litle Rue, withstandeth all poyson: also they are mingled with a litle Rue and a figge, to cure the vlcers of the pappes, and other colde impostumes. [C]

Dry Nuttes are good to be layd to the bitinges of mad Dogges with salt, hony, and Onyons. [D]

Olde Oylie Nuttes do heale the scurffe and scales, also they take away the blewe markes that come of stripes or bruses, being pounde very smal and layd thereupon. The same vertue hath the Oyle that is pressed out of them. [E]

They make a medicine with the greene barke or shale of the Walnut, the which is good against all tumours and vlcers, whiche do but begin to arise in the mouth, the throte, and Almondes, or kernelles vnder the tongue, to be gargeled. [F]

The decoction of the sayde greene huske (with hony) is good to gargell withall for the aforesayde purpose. And the leaues be almost of the same vertue. [G]

Ppp iiij   Of

Of the Nutmegge and Macis. Chap.lvij.

❧ *The Description.*

THE Nutmegge is the fruite of a certayne tree, which in growing and leaues is not much vnlike our common Peache tree. When this fruite is vpõ the tree, it is much lyke to a Walnut, sauing that it is somwhat bigger. First it hath in the outside a greene thicke huske or shale, lyke to the vtter shale of our Walnut, wherewithall it is couered all ouer, vnder the same there is founde certayne thinne skinnes, lyke to cawles or nettes, of a redde or yellowish colour, all iagged or pounsed of a very pleasant sauour (the whiche is the right Macis) and it lyeth fast couched vpon a harde wooddish shell, lyke to a Filberd shell: within that shel is inclosed the most Aromaticall and sweete smelling Nut, which is harde, thicke, and full of Oyle.

❧ *The Place.*

This Nut is founde principally in the Ile of Bandan, the whiche is in the Indian Sea: they grow there wilde in euery wood very pletifully, as Lewse the Romayne writeth. ❧ *The Names.*

These Nuttes be called in Greeke κάριον μυριστικὸν, and of some μοχοκάριον: in Latine, Nux myristica: in Shoppes, Nux moschata: in Englishe, a Nutmegge: in Frenche, *Noix muscade*, and *Noix musquette*: in high Douche, Moscaten: in base Almaigne, Note muscaten.

The litle thinne scale or pyll (whiche is found vnder the vtter shale, lying close vnto the harde wooddishe shel) is called in Greeke μάκις: in Latine, Macer, yet for all that this is not Macer of the Auncientes: it is called in English, and and in Shoppes Macis in French, *Macis*: in high Douch, Moscatenblumen: in base Almaigne, Foelie, and Moscaetbloemen.

❧ *The Nature.*

The Nutmegges be hoate and drie in the seconde degree: and of the same nature and complexion is Macis: moreouer they be somwhat astringent.

❧ *The Vertues.*

The Nutmegge doth heate and strengthen the stomacke which is cold and weake, especially the Orifice or mouth of ye stomacke, it maketh a sweet breath, it withstandeth vomiting, and taketh away the Hicket or Yeox, in what sorte soeuer it be taken. [A]

It is also good against the payne and windinesse of the belly, and against al the stoppinges of the lyuer and milt. [B]

The same pearched or dried at the fire stoppeth the laske, especially if it be taken with red wine. [C]

It is good for the mother, the kidneyes, the bladder, it remedieth the disease or greefe that letteth the due course of brine, and causeth that one cannot pisse, sauing by droppes, especially when the sayde disease springeth of a colde cause, it is good also for other hidden and secrete greeues both in men and women: it breaketh and driueth foorth grauell, especially being first soked and stieped in the Oyle of sweete Almondes. [D]

The Macis be almost in vertue lyke to the Nutmegges, and they doo not onely stoppe the laske, but also the blooddy flixe, and womens flowers. [E]

It is good also against the beating, trembling or shaking of the hart, and is muche better for al the cold greeues of the stomacke, then the Nutmegge it self. [F]

The oyle that is drawen out of Macis layde vpon the stomacke, cureth the infirmities of the same, taking away the desyre to vomit and the wambling of the stomacke, it causeth good appetite, and helpeth digestion. [G]

the Historie of Plantes.

Of the Hasel or Fylberde tree. Chap.lviij.

❧ The Kindes.

There be two sortes of Hasel, or wood Nut trees: the one kinde is set and planted in gardens, the other groweth wilde.

Corylus hortensis.
The Fylberde.

Corylus syluestris.
The Hasel Nut.

❧ The Description.

The Hasel and Filberde trees, are but small growing lyke to a hedge plante, and put foorth from the roote (whiche is muche displayed and spreade abroade) many straight roddes, shutes or springes, of whiche oftentimes some ware thicke and long and full of branches, and some ware long and slender, and are very fit to make roddes or poles to fish with, bycause they be firme and plyant, and wil not lightly breake. The leaues be broade and wrinckled somewhat hact or snipt round about, the which leaues spring foorth after the Catkins, agglettes, or blowinges, whiche hang vppon the Hasell tree be fallen of: betwixt the leaues commeth the fruit, growing thre or foure togither in a cluster, somewhat, but not altogither couered with a huske or pil. Their shales be harde and wooddishe, in whiche the rounde kernell or Nut is inclosed, and is ouercouered with a smooth tender huske or skinne, like to other Nuttes, the which is red in the Filberdes, and white or pale in Hasel Nuttes.

❧ The Place.

The Fylberdes are planted in gardens.

But the wilde groweth in wooddes and moyst places that be darke and shadowie.

The ſyxth Booke of

❧ *The Tyme.*

The Aglets or Catkens of Haſel, breake foorth in winter, and in the ſpring time they open into ſmal ragges or ſcales, ſhortly after the leaues appeare. The Nuttes be ripe in Auguſt.   ❧ *The Names.*

This tree or ſhrub is called in Greke καρύα ποντικὴ: in Latine, Nux auellana, & of Virgil, Corylus: in French, *Couldre,* & *Noiſetier:* in high Douch, Haſelſtrauch, & Haſelnutzbaum: in baſe Almaigne, Haſelaer: in Engliſh, Haſel or Filberd tree.

The Nut is called in Greke κάριον ποντικόν, και λεπτοκάριον: in Latine, Nux Pontica, Nux auellana, Nux præneſtina, & Heracleotica: in French the great & round kinde is called *Auelines,* and the ſmal and long kinde, is called *Noiſilles,* & *Noiſettes:* in Engliſh, the great and long kinde is called Filberdes, and the rounde kinde with the harde thicke ſhale, is called the Wood nut, or Haſelnut.

The red Filberdes are called in French, *Auelines rouges:* in high Douch, Rhurnutz, and Rotnutz: in baſe Almaigne, Roode Haſelnoten. They be the right Nuces Ponticæ deſcribed of the Auncientes.

❧ *The Nature.*

The Haſellnuttes and Fylberdes are in complexion not muche vnlyke the Walnuttes, but dryer although they be yet newe and greene: but when the be olde and drie, they be colder then Walnuttes.

❧ *The Vertues.*

Haſel Nuttes and Fylberdes nouriſhe very litle, and are harde of digeſtion, **A** they ingender windineſſe in the ſtomacke, and cauſe headache, if they be eaten in to great a quantitie.

The ſame dronken in Meade or watered honie, doo heale the olde cough: **B** and being roſted and taken with a litle pepper, they ripe the Cattar or Reume.

The ſame burned and layd to with hogges greaſe or Beares greaſe, doo heale the noughtie ſcurffe & ſcales of the head, & doo fil agayne with heare the balde or pilde places in the head.

They vſe of the ſhales or huſkes of Filberdes againſt ye Squinance euen as they vſe the huſkes of ye Walnuts.

Of Fiſtick Nuttes. Chap.lix.

❧ *The Deſcription.*

THE tree that bringeth foorth Fiſtick Nuts, hath long great leaues ſpread abrode, & made of fiue, ſeuē, or moe leaues, growing one againſt another all alongſt a reddiſh ribbe or ſinewe, whereof the laſt whiche is alone at the top of the leafe is the greateſt or largeſt, the fruite of this tree is muche lyke to ſmall Haſel Nuttes, & like the kernels of ye Pine Apple, in which lyeth ye kernel or nut.

❧ *The Place.*

This tree is a ſtranger in this Countrie, and is not founde but only in ye gardens of diligent Herboriſtes, but it commeth of plants in Syria, & other hoate Regions.   ❧ *The*

Arbor Piſtaciorum. **C**

D

✤ *The Names.*

These Nuttes are called in Greeke πιϛάκια: in Latine, Pistacia: in Shoppes, Fistici: in Brabant, Fisticen: in Frenche, Pistaces: in Englishe, Fistick Nuttes.

¶ *The Nature.*

Fistick Nuttes are of a meane or temperate heate, & somewhat astringent.

✤ *The Vertues.*

Fistickes are good against the stoppings of the liuer, and also to strengthen the same: they be also good for the stomacke: but to be takē as meate they nourish but litle. [A]

They vnstop the lunge pipes, & the breast, & are also good against the shortnesse of winde & payne to fetche breath, to be eaten either alone or with sugar. [B]

They be also vsed to be giuen with wine, as a preseruatiue or medicine against al ȳ bitings & stinginges of venemous beastes, as Dioscorides writeth. [C]

Of the Bladder Nut. Chap.lx.

✤ *The Description.*

Staphilodendron Plinij. Nux vesicaria.

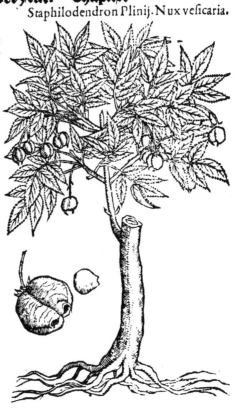

This kind of Nut is a wild fruite, whereof men make none accompt, growing vpon trees, which grow most commonly lyke shrubbes, or hedge bushes, as many other wild sortes of trees do. This tree his leaues are made of fiue blades or more, not muche vnlyke the Elder leafe, but smaller and grener. The flowers be white round and double, growing fiue or sixe togither, after them come the rounde holow bladders, diuided into two or thre partes, in whiche be founde most commonly two litle Nuttes, whereof the kernell is meetely sweete.

✤ *The Place.*

This plant is no where found, but growing wilde, there is plenty of it found wilde in Almaigne, and sometimes also in the hedges of this Countrie.

✤ *The Tyme.*

The small flowers doo blowe in May: and the Nuttes are ripe in September.

✤ *The Names.*

This wilde tree is called of Plinie in Greeke ϛαφυλοδένδρον: and in Latine, Staphilodendron: of them that write now Nux Vesicaria, and of some Pistacium Germanicum, although it is very litle lyke the Fistick Nuttes.

The fruite is called of the high Douchemen, Pimpernulz: of the base Almaignes, S. Intuenis Nootkens: ȳ is to say in English, S. Antonies Nuts.

✤ *The Nature and Vertues.*

As touching the naturall vertues and operations of this kinde of Nuttes, we can say nothing, bycause they serue to no purpose at al.

Colutea Theophrasti. Cytisus Latinorum.

Of Cytisus, or tree Trifoly. Chap.lxi.

❧ *The Description.*

CYtisus is a shrubbe or bush with leaues, not muche vnlyke Fenugreeke, or Sene, the flowers be faire and yellow, almost like to Broome flowers: ẏ which past there come holow huskes, puft vp & blowen lyke bladders, cleare and shining, the which do sound & rattell when they be shaken. In them is contayned the seede, whiche is flat, and swart, not much vnlyke Lentiles.

❧ *The Place.*

This plante is not founde growing in this Countrie, but in certayne gardens, & they plant it for Sene: but wrongfully.

❧ *The Tyme.*

It flowreth in May, & the sede is ripe in August. ❧ *The Names.*

This bushy shrubbe is named of Theophrastus in Greeke κολούτεα: of Theodor Gaza, Coloutea, or Colutea: in Englishe of some Cytisus bush, & tree Trifoly, but of the ignorant sort, it is falsly called Sene: in French, *Baguenaudier*, & *des Bagenaudes*: in high Douch, Welsch Linsen: in base Almaigne, Lombaertsche Linsen, and of the ignorant people vnproperly Seneboom. ❧ *The Nature.*

The leaues and seede of the Cytish bush are temperate of heate and moysture.

❧ *The Vertues.*

The fruit, that is to say the seede, & also ẏ leaues of Colutea, or Cytisus, as Theophrastus writeth, doth fat Sheepe very quickly, and causeth them yeelde abundance of milke.

Of the Date tree. Chap.lxij.

❧ *The Description.*

THE Date trees be great, with a straight thicke stemme or trucke, couered with a scaly barke. At the top thereof growe many long branches, with great plentie & store of long straight narrow leaues, or twigges lyke Reedes, so that the said branches seeme to be none other thing, but a bundel or sort of Reede leaues, growing thicke togither vpō one branch: amongst those branches groweth the

Palma.

the fruit clustering togither at the first, and lapped in a certayne long and brode forrell or couering lyke to a pillowe, the which afterwarde doth cleeue abrode and open it selfe, shewing foorth his fruite, standing alongst by certaine Sions or small springes, growing al out of a flatte and yellow branche like to the timber or wood of a Harpe: the same fruite is rounde and long, with a certayne long & very harde stone in the same. And it is to be noted that the male Palme tree bringeth foorth nothing els but the flower or blossom whiche vanisheth away, but the female beareth the fruite, which afterward commeth to ripenesse.

❊ *The Place.*

The Date tree groweth in Africa, Arabia, India, and Syria, Iudea, and other Countries of the East or Orient.

❊ *The Time.*

The Palme or Date tree is alwayes greene, & flowreth in the spring time: but the fruite in hoate Countries is ripe in Autumne.

❊ *The Names.*

The Date or Palme tree is called in Greeke φοῖνιξ: in Latine, Palma: in Almaigne, Dadelboom: and that is the right Palme.

The fruite is called in Greeke φοῖνιξ, Phœnix, and of Galien, Φοινικοβάλανος, Phœnicobalanos: in Latine, Palmula: in Shoppes, Dactylus: in Frenche, Dattes: in high Douch, Dactelen: in base Almaigne, Dadelen, and Daden: in English, Dates, and the fruite of the Palme tree.

¶ *The Nature.*

The branches and leaues of the Date tree are colde and astringent.

The fruite is hoate and drie almost in the seconde degree, & also astringent, especially when it is not yet throughly ripe.

❊ *The Vertues.*

A Dates be harde of digestion, they stoppe the liuer, and the milt: they engender windinesse in the belly, and headache, also they engender grosse blood, especially to be eaten greene and freshe, for when they be through ripe, they are not so hurtfull: and being well digested in a good stomacke, they nourishe indifferently.

B Drie Dates do stoppe the belly and stay vomiting, and wambling of the stomacke (especially of women with chylde) if they be layde as an emplayster to the belly or stomacke, or if they be mingled with other medicines and eaten.

C Also they do restore and strengthen the feeblenesse of the liuer and milte, to be mingled with medicines conuenient, either to be applyed outwardly, or to be ministred and taken inwardly.

D The leaues and branches of the Date tree, doo heale greene woundes, and soder or close vp vlcers, and doo refreshe and coole hoate inflammations: and therefore when as the Auncientes in olde time woulde make any emplayster for the purposes aforesayde, they dyd alwayes vse to stirre about their sayde playsters with some branche of the Palme tree, to the intent their sayde playsters and salues shoulde be of the more vertue and greater efficacie, as a man may see in the composition of the emplaister, named Diacalcitheos, in Galens first booke, De medicamentis secundum genera.

Of the Oliue tree. Chap.lxiij.

¶ *The Kindes.*

1 There be two sortes of Oliue trees, one called the garden or set Oliue tree, the other is the wilde Oliue tree.

❊ *The*

The Description.

Olea satiua. The garden Oliue tree.
Olea syluestris. The wilde Oliue tree.

1 THE garden Oliue tree groweth high & great, with many branches ful of long narrow leaues, not much vnlyke to Withy leaues, but narrower & smaller. The flowers be small and white, and growe in clusters. The fruite is somwhat long and rounde, almost of the making of a Damson, or Plumme, at the first greene without, but after they beginne to waxe ripe, they are blackish, in the middle whereof is a litle stone, which is hard and firme. Out of this fruite, that oyle is pressed, whiche we cal oyle Oliue.

2 The wilde Oliue tree is lyke to the garden or tame Oliue tree, sauing that the leaues therof be somwhat smaller, amongst which grow many prickley thornes. The beries or fruit also are smaller, & do seldom come to ripenes, insomuch as: that oyle which is pressed foorth of them abideth euer greene and vnripe.

The Place.

The Oliue tree delighteth to grow in dry vallies, and vpon smal hillockes or barrowes, & it groweth plentifully throughout Spayne and Italy, and other lyke regions.

The Tyme.

The Oliue tree floureth in Aprill, and about the beginning of May: but the Oliues are ripe in October.

The Names.

1 The Oliue tree is called in Greeke ἐλαία: in Latine, Olea: in high Douche, Oelbaum, and Oliuenbaum: in base Almaigne, Olijfboom.

2 The wilde Oliue tree is called in Greeke ἀγριελαία, of some κότινος, καὶ αἰθιοπικὴ ἐλαία: in Latine, Olcaster, Olea syluestris, and Olea Aethiopica.

The fruite also is called in Greeke ἐλαία: in Latine, Oliua: and according to the same it is called in Englishe, Frenche, and Douche, Oliue.

The Oliues condited in salt or brine, are called in Greeke κολυμβάδες, καὶ ἁλμάδες: in Latine, Colymbadæ.

The Nature.

The leaues & tender shutes of the Oliue tree, are cold, dry, & astringent. The grene vnripe oliues ar also cold & astringent, but being ripe thei be hoat & moist.

The Oyle that is made of vnripe Oliues, is colde and astringent: but that which is pressed out of the ripe Oliues, is hoate moyst and of subtil partes.

The Vertues.

The leaues of the Oliue tree laid to are good against Serpigo, or the disease A which is called wilde fire, bycause it creepeth hither and thither, fretting sores and consuming pore, and other suche hoate tumours or cholerique swellinges.

The same layd e to with hony, doo mundifie and clense vlcers, and doo also B
swage

the Historie of Plantes. 739

swage and slake all other swellinges and tumours.

They are good to be layde to againſt the vlcers, inflammations, and impoſtemes of the mouth, and gummes, eſpecially of childꝛen, if their mouthes be waſhed with the decoction thereof.

The iuyce of them ſtoppeth womens flowers, and all other fluxe of blood, with the laſke and bloody flixe, to be taken inwardly oꝛ applyed outwardly.

It is alſo good againſt the redneſſe, inflammation, and vlcers of the eyes to be put into Collyꝛes and medicines made foꝛ the ſame, and to clenſe the eares from filthy coꝛruption.

The greene and vnripe Olyues, do ſtrengthen the ſtomacke, and cauſe good appetite, eſpecially being condited in bꝛine, neuertheleſſe they be harde of digeſtion, and nouriſhe very litle.

The ripe Olyues doo ouerturne the ſtomacke, and cauſe wambling in the ſame, they alſo engender headache, and are hurtfull to the Eyes.

The Oyle of vnripe Oliues which is called Omphacinum, doth ſtay, & dꝛiue away the beginninges of tumours and inflammations, & doth coole the heate of burning vlcers, and exulcerations.

It is alſo good againſt the rotten ſoꝛes, and the exceſſiue & fylthy moyſture of the gummes, it faſteneth looſe teeth, to be laide vpon the gummes, with cotton oꝛ a little fine wooll.

The Ole of rype Oliues doth mollifie, it ſwageth payne, and diſſolueth tumours oꝛ ſwellinges, it is good againſt the ſtiffeneſſe of members & crampes, eſpecially when it is mixt oꝛ compounde with good herbes.

Oyle Oliue is very apt & pꝛofitable, to make al ſoꝛts of Oyles, whether they be of herbes oꝛ flowers: foꝛ it doth eaſyly, & redily dꝛaw vnto it the qualities and vertue of thoſe herbes & flowers, with the whiche it is ſet to be ſonned, oꝛ otherwiſe ſodde and pꝛepared.

Of the Carob tree. Chap.lxiiij.

The Deſcription.

This fruite groweth vppon great trees, whoſe bꝛanches are ſmall & couered with a round redde barke oꝛ rinde. The leaues be long and ſpꝛead abꝛoade lyke whinges, oꝛ after the maner of Aſhen leaues, and made of ſixe oꝛ ſeuen oꝛ eyght ſmall leaues, growing alongſt by a ribbe oꝛ ſtemme, and ſet one ouerright agaynſt another, whereof each blade oꝛ leafe is rounde, and of a ſadde oꝛ darke greene aboue, and of a light greene vnderneath. The fruite is certayne flat crooked cods oꝛ huſkes, ſomtimes of a foote & a half long, & as bꝛoade as ones thombe, ſweete, in which the ſeede is conteyned, the whiche is great, playne, and bꝛode and of a Cheſnut colour.

The Place.

Theſe huſkes oꝛ ſweete coddes, do grow in Spayne, Italie, & other hoate Regions oꝛ Countries. They growe not in this Countrie. Yet foꝛ all that they be ſometimes

Ceratonia Siliqua.

times founde in the gardens of some diligent Herboristes, but they be so small shrubbes, that they can neither bring foorth flowers nor fruite.

❧ The Names.

This tree is called in Greeke κερατωνια: in Latine also Ceratonia. The fruite is called in Greeke κερατιον: in Latine, Siliqua, and of some Siliqua dulcis: of the common Herboristes Carobe: in shoppes, Xylocaracta: in Frenche, Caronges, or Carobes: in high Douche, S. Johns brot: in base Almaigne, S. Jans broot: in English, a Carob tree, a Beane tree, the fruite also may be called Carobbes, and Carob beane coddes, or S. Johns bread.

❧ The Nature.

This fruite is somewhat hoate, drie, and astringent, especially when it is freshe and greene.

❧ The Vertues.

Fresh and greene Carobes eaten do loose the belly very gently: but they be hurtfull to the stomacke, harde of digestion, and nourishe but litle.

The same dried do stop the belly, prouoke vrine, and are not muche hurtfull to the stomacke, & are fitter to eate than the greene or fresh gathered Carobes.

Of Cassia Fistula. Chap.lxv.

❧ The Description.

Cassia Fistula.

THE tree whiche beareth Cassia Fistula, hath leaues not muche vnlyke Ashen leaues: they be great, lōg, & spreade abroade, made of many small leaues growing one against another, alongst by one stemme, whereof eache litle leafe is long and narrow. The fruite is long, round, blacke, hard, and with woodish huskes, or coddes most commonly two foote long, and as thicke as ones thombe or finger, parted in ÿ insyde, or seuered into diuers smal Celles or Chambers wherin the flat, and brownish sede is couched and layd togither with the pulpe or substance, which is blacke, soft, and sweete, & is called the flower, marrow or creame of Cassia: it is very expedient, and necessarie for Physicke or medicine.

❧ The Place.

Cassia groweth in Syria, Arabia, and suche lyke Regions.

❧ The Names.

Cassia is called of Actuarius, and of the later Greke Physitions κάσια μελαίνα in Greeke, that is to say, Cassia nigra in Latine: in shoppes and of the Arabian Physitions, Cassia Fistula.

❧ The Nature.

The blacke Pulpe or moyst substance of Cassia is hoate and moyst in the first degree.

❧ The

the Historie of Plantes. 741

※ *The Vertues.*

The inner pulpe of Cassia is a very sweete and pleasant medicine, the which A may be giuē without any danger to al weake people, as to women with child. It looseth the belly and purgeth cholerique humours cheefely. And sometime syme sieme gathered about the guttes, to be taken the waight of an ounce.

Cassia is very good for suche as be vexed with hoate agues, the Pleuresie, B Jaundise, or any other inflammation of the liuer, especially when it is mixed with waters, drinkes, or herbes that be of a cooling nature.

It is good for the raynes and kidneyes, it driueth foorth grauell, and the C stone, and is a preseruatiue against the stone, to be mingled with the decoction of liqueris or the rootes of Parsely, or Ciches, or a decoction made of all together, and dronken.

It is good to gargle with Cassia for to swage and mitigate the swellinges D of the throte, and to dissolue, ripe, and breake Apostemes and tumors.

Cassia layde to the member greeued with the gowte, swageth the payne, as E Auicen writeth.

Of Anagyris, Laburnum, and Arbor Iuda. Chap. lxvi.

Anagyris.   Laburnum.

※ *The Description.*

1. Anagyris is a litle lowe bush or shrub, with smal branches, vpon which growe small leaues, alwayes three togither, otherwayes almost lyke to the leaues of Agnus castus. The flowers be yellowe almost lyke to Broome flowers, whiche being past, there come vp long huskes or coddes,

coddes, in whiche is a flat fruite or seede that is harde & firme, almost lyke the kidney beanes, but somewhat smaller. The whole plant is of a strong ilfauoured stinking sauour, as it were the smell of Gladyn or Spurgewort.

2 There is also another litle bush or shrub founde lyke to Anagyris in leaues & growing. The flowers do grow very thicke togither hanging by a fine slender stemme, lyke to a spykie eare, but yellowe and somewhat resemblyng Broome flowers. The coddes or cases are rounder & smaller then the huskes of Anagyris, with a smaller fruite also. This plante is of no ranke smel, but his leaues be greater and larger then ý leaues of Anagyris.

3 Besydes the aforesayd there is founde another smal shrub or plant whiche bringeth foorth coddes or huskes also, the whiche being well ordered in ý growing vp, waxeth a tall tree. His branches are set with broade rounde leaues almost lyke to the leaues of Aristolochia clematitis, or Asarum, but stronger. The flowers be purple and redde, like to the flowers of garden or branche Peason, and the sayde flowers do not growe vpon the final branches, and betwixt the leaues lyke the blossoms, and flowers of other trees, but they growe about the lowest part of the great branches, the whiche afterwarde do change into long flatte coddes of colour somewhat blew or wanne, hauing a certayne flat seede within, which is harde and lyke to a Lentill.

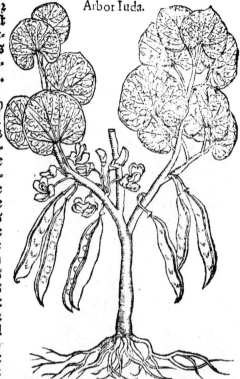

Arbor Iuda.

❊ *The Place.*

These plantes doo growe wilde in diuers places of Italy and Prouince, in wooddes and vpon the mountaynes. ❊ *The Time.*

Anagyris flowreth in Aprill and May: the other in May, and June: but Arbor Iuda in Marche. The fruite is ripe in September.

❊ *The Names.*

1 The first of these plantes is called in Greeke ἀναγυρις: in Latine, Anagyris: in Italian, *Eghelo* we may call it Beane trifoly, bycause the leaues grow three togither, & the seede is much like to a beane: the Frenche name may be *Bois puant.*

2 The second is thought to be Laburnum Plinij: This is not Anagyris, neither yet Lotus, as some do wrongfully iudge it.

3 The third is called of the Latine Herboristes, in Latine, Arbor Iuda, & Arbor Iudæ: this should seeme to be a kind of Laburnū, or as some men thinke κολυτέα, Colytea of Theophrastus, This is not that Cercis of Theophrastus, for Cercis is a kinde of Populer, the whiche Plinie calleth Populus Lybica.

❊ *The Nature.*

The leaues and seede of Anagyris are hoate and drie of complexion.

❊ *The*

✻ The Vertues.

They giue the waight of a dramme, of the leaues of Anagyris boyled in A wine, to moue womens flowers, and to driue foorth the secondine.

The young and tender leaues of this bushe, broken and layde to as an em-B playster, doth dissolue, and keepe downe colde swellinges.

The seede eaten causeth one to vomite sore and vehemently. C

Of Withy or Willow. Chap.lxvij.

❧ The Kindes.

There be two sortes of Withy very diuers. The one riseth vp very high & groweth to the bignesse and thicknesse of other trees: the other remaineth alwayes lowe, beareth Ozier roddes and twigges. The barke of the first sorte is sometimes reddish, sometimes white and sometimes yellowe.

✻ The Description.

Salix.

1. WITHY hath a great tronke, stocke, or stemme, out of whiche spring high branches or yeardes, which be long, straight, and full of boughes or twigges which be weake and plyant, and couered with a barke or rinde of a browne red colour, or white, or yellow, alongst the which branches and twigges grow the leaues which be long and narrow, greene aboue, and white or ashe colour vnderneath.

2. The seconde kind of Withy called the Franke Ozier hath no great stemme, but onely a great wride or head neare the ground, out of which spring many Sios, and slender twigges, or yeardes couered with a browne rinde or pyll: the whiche yeardes, twigges, or roddes, are very plyant, and easy to turne and twiste euery way. With this kinde of twigges or roddes they make Basketes, Chayres, Panniers, and suche lyke stuffe.

✻ The Place.

All kindes of Withy delight to growe in moyst places, along by diches and waters, but especially the Oziers.

✻ The Tyme.

Withy flowreth at the beginning of the spring time: his flower or blossom is lyke a fine throm or thicke set beluet heaped vp togither about a little stemme, the which when it openeth is soft in handling, and lyke downe or Cotton, and therefore the whole flower is called a Chatton, Kitekin or Catteken.

✻ The Names.

Withy is called in Greeke ἰτέα: in Latine, Salix: in Frenche, *Saulx*: in high Douche, Weydenbaum: in base Almaigne, Wilghe boom: Theophrastus doth surname it ἀλεσίκαρπος, Olesicarpos, that is to say in Latine, Frugiperda, bycause his Cattekins or blossoms do fall away before that his seede be scarse ripe.

That Withy or Willowe whiche groweth to a tree is called in Latine, Salix perticalis.

Of this sort, that whiche hath the reddish barke, is called Salix nigra, Salix purpurea, and Salix Gallica: in English, Red Withy, and the better sort therof is called Red sperte: in Frenche, L'ozier. in high Douche, Rotweiden: in base Almaigne, Roode wilghen, of some also, Salix viminalis: bycause the twigges be tough and plyant, and wilbe wrought and writhed more easily then any of the other kindes of Withy, insomuch that this kind of blacke or rather red Withy, is without doubt, of the selfe same kinde as the Franke Ozier is of: for if you plant it in lowe waterishe places, and cut it harde by the ground, it will turne to Ozier Withy.

The seconde sorte hath a white or gray barke, and is called Salix candida, and of some Salix Græca: in English, Dunne Withy, and Goore Withy.

The third kinde hath a yellow barke, and is called Salix vitellina: and after the minde of some, Salix amerina: these two kindes are called in high Douche, Weisz weiden: in base Almaigne, Witte wilghen, that is to say, White Withy: it is called about Parris, Du Bursauli: in English, Cane Withy.

The small lowe Withy is called in Latine, Salix pumila, and Salix viminalis: of Columella, Sabina salix, and Amerina salix: in Frenche, Franc Ozier: in high Douch, Klein weiden: in base Almaigne, Wijmen: in English, the smal Withy, the Osyar Withy, the Sperte or twigge Withy.

✤ *The Nature.*

The leaues, flowers, seede, and barkes of Withy, are colde and drie in the seconde degree, and astringent.

✤ *The Vertues.*

The leaues and barke of Withy, do stay the spitting of blood, the bomiting of blood, and all other fluxe of blood, with the inordinate course of womens flowers, to be boyled in wine and dronken.

The leaues and rindes of Withy boyled in wine, doo appease the payne of the sinewes, and do restore againe their strength, if they be nourished with the fomentation or natural heate thereof.

The greene leaues pounde very smal, and layde about the priuie members, do take away the desire to lecherie or Venus.

The ashes of the barke of willow mingled with vineger, causeth wartes to fall of, taketh away the harde skinne or brawne that is in the handes or feete whiche is gotten by labour, and the cornes in a mans toes or fingers, if it be layde therevpon.

Of the Oke tree. Chap.lxviij.

✤ *The Description.*

THe Oke is a great, brode, and thicke tree, most commonly spreading his great branches abrode, and also growing vp into height and length. The barke is gray and smooth whiles it is young, but thicke rough vneuen chapt and cracte when it is olde. The leaues be deepely cut and natched rounde about, vpon the which there is sometime founde growing in this Countrie little small Apples, called Oke Apples, lyke as in other Countries galles be found growing vpon the Oken leaues, whereof these little Apples be one kinde. The fruite of the Okes are certayne Mast or kernelles hanging foorth of rough huskes, whiche be rounde and hollowe lyke vnto cuppes or dishes. His roote spreadeth abrode very long and large.

Besydes these kindes of galles and Apples that are vpon the Oken leaues, there growe vppon the Oke diuers other thinges, as Theophrastus writeth, more at large, in his Historie of plantes the iij. booke, and viij. Chapter.

✤ *The Place.*

The Oke loueth sandy groundes, leane, and drie, as vpon playnes and heathes.

✤ *The Tyme.*

The Oke renueth his leaues in May. The Acornels or mast is ripe in August: the Oke apples do grow in sommer, and do begin to fall in September.

❡ *The Names.*

The Oke is called in Greeke δρῦς: in Latine, Quercus: in high Douche, Eichbaum: in base Almaigne, Eyckenboom.

The fruite is called in Greke βάλανος: in Latine, Glans: in English, an Akernel, or mast: in French, *Gland*: in high Douch, Eichel: in base Almaigne, Eeckel.

The round berie or apple which groweth vpon the leaues, is called in Greeke κηκίς: in Latine, Galla: in Frenche, *Noix galle*: in high Douch, Eichopffel, and Galopffel: in base Almaigne, Eycken apple, and Galnoten.

The shales or cuppes in whiche one part of the kernel is inclosed or couched, is called in Latine, Calices glandium: in shops, Cupulę glandiū. ✤ *The Nature.*

Quercus.

The leaues and barke of the Oke, as also the cuppes or shelles of the Acornes, are drie in the third degree, and astringent. The Acornes be almost of the same temperature, sauinge that they be warmer, and not so muche astringent.

The Gale is colde and drie in the thirde degree, and very astringent.

✤ *The Vertues.*

The leaues and barke of the Oke with the cuppes of the Acornes, do stop A and cure the spetting of blood, the pissing of blood, and all other fluxe of blood: the bloodely flixe and laske, being boyled in red wine and dronken.

The Oke leaues pounde very smal, do heale and close vp greene woundes, B and do stoppe the blood being layde thereupon.

The barke of the Oke made into powder, is good to be giuen to young chil- C dren, against the wormes and the inordinate laske.

The Cuppes of the Acornes with the barke of the tree, are good to be put D into medicines, oyntmentes, oyles, and emplaysters that serue to stay and kepe backe the fluxe of blood, or of other humours.

The Acornes are almost of the same vertue as the leaues and barke are, E but they stoppe not so muche, they prouoke vrine, and are good against all venome and poyson: and boyled in mylke they be excellent to be eaten against the bitinges and stinginges of venemous beastes.

The same pounde very smal, are very good to be laid to the beginninges of F phlegmons & inflammations: and pounde with salt, and Swines greale they cure, harde vlcers, and consuming sores.

The Gal is also very binding and stiptique. They be good against al fluxe of G
blood

blood, and laskes to be taken in what soeuer maner, whether they be ministred within the body, or mixt with oyles, oyntmentes and emplaisters to be layde outwardly.

They are also good against the excessiue moysture, & swelling of the iawes or gummines, and against the swellinges of the almondes or kernels of the throte, and also against the blistering sores of the mouth.

They staye the fluxe menstruall, and cause the mother that is fallen downe to returne agayne to his natural place, if women sit in the decoctiō of the same.

The same stieped or tempered in vineger or water, maketh the heare blacke: and doth eate and consume away superfluous and prowde fleshe beyng layde therevpon.

The same burned vpon coles & afterward quenched with wine or vineger, or as Turner saith, with brine made with vineger and salt, stoppeth all issue or fluxe of blood.

The Oke Apples or greater Galles, being broken in sonder, about the time of withering do forshewe the sequell of the yeere, as the expert husbandmen of Kent haue obserued by the liuing thinges that are founde within them: as if they finde an Ante, they iudge plentie of grayne: if a white worme lyke a Gentill, morreyne of beast: if a Spider, they presage pestilence, or some other lyke sicknesse to folowe amongst men. Whiche thing also the learned haue noted. For Matthiolus vpon Dioscorides saith, that before they be holed or pearsed they conteyne eyther a Flye, a Spider, or a Worme: if a Flye be founde, it is a prognostication of warre to folowe: if a creeping worme, the scarcitie of victual: if a running Spider the Pestilente sicknesse.

Of Missel or Misselto. Chap.lxix.

✿ *The Description.*   Viscum.

THIS plante hath many slender branches, the whiche are spread ouerthwart, and are wrapped or enterlaced one with another, couered with a barke of a light greene or Popingay colour. The leaues be thicke and of a darke or browne greene colour, greater and longer then the leaues of Boxe, but otherwise not much vnlike. The flowers be smal and yellow, the which being past there appeare small rounde and white beries, full of clammy moysture of which eche berie hath a blacke kernell, which is the seede.

✴ *The Place.*

Misselto groweth not vpō the ground, but vpon trees: and is oftentimes found growing vpō Apple trees, Peare trees, Wythies, and sometimes also vpon the Linden, Birche, and other trees: but the best and of greatest estimation, is that which groweth vpon the Oke.   ✴ *The Tyme.*

Misselto flowreth at the ende of May, and the fruite is ripe at the ende of September, the whiche remayneth all the winter.

☙ *The Names.*

This plante is called in Greeke ἰξός: in Latine, Viscum: in shoppes, Viscus quercinus:

the Historie of Plantes. 747

quercinus: in Englishe, Missell and Misselto: in Frenche, *Guy*: in high Douche, Mistel, and Eichen Mistel: in base Almaigne, Marentacken.

❋ *The Nature.*

The leaues and fruite of Misselto are hoate and drie, and of meetely subtill partes.

❋ *The Vertues.*

The leaues and fruite of Misselto, being laide to with Tarre, and Waxe, do soften, ripe, and consume away by the pores, harde swellinges and botches about the secrete partes, & other such rebellious impostemes & cold swellinges. A

The same leaues and fruite, with Frankensence, doo cure olde vlcers and sores, and great corrupt and euill impostemes. B

They also cure the felons or noughtie sores, which rise about the toppes of toes, and fingers endes to be layde to with Arsenik. C

The seede of Missell pounde with wine lyes, doth cure and waste the hard- D nesse of the Milt or splene to be applyed to the syde.

They say also that the wood of Misselto, that groweth vpon the Okes, and E not vpon any other tree, is very good against the falling euyll and Apoplexie, to be hange about the necke of the patient.

Of the Ashe tree. Chap.lxx.

❋ *The Kindes.*

After the mind of Theophrastus, there be two kindes of Ashe: the one called the Ashe tree, without any other addition. The other is called the wilde Ashe, or white Ashe.

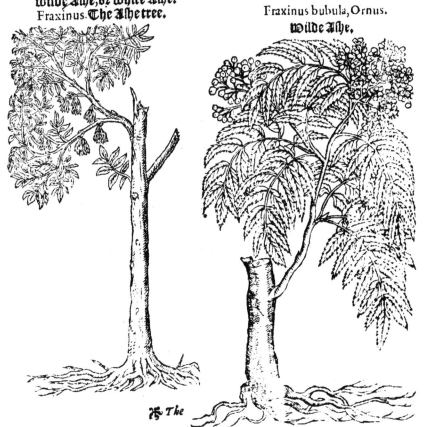

Fraxinus. The Ashe tree.   Fraxinus bubula, Ornus. Wilde Ashe.

❋ *The*

The syxth booke of

❧ The Description.

THe Asshe is a great high tree with many branches, whereof the young and newe sprong branches are ful of white pith, or a certayne soft substance, and they haue sundrie ioyntes, but when they waxe great and olde, those ioyntes are lost, and the pith is conuerted into timber. The barke of this tree (especially whereas it delighteth best to growe) is gray and smooth, but in other places it waxeth rough. The leaues be great long & large spread abrode after the fashion of winges made of many small leaues, growing one against another, all alongst one stemme or rib, whereof eache little leafe is long & somewhat snipt round about the edges. The fruite of the Asshe hangeth togither in clusters, and is nothing els but litle narrow huskes, wherein lieth the seede whiche is bitter.

2 The wilde Asshe also sometimes groweth to a great tree, but nothing lyke to the Asshe, for it is much smaller and slow in growing vp: whiche is the cause þ it is found so smal. The rinde or barke therof is browne, almost like to þ Aller rinde. The leaues be great & long, many growing alongst by one stem, rough and somwhat heary, much like to the leaues of Sorbe Apple tree. The flowers be white and growe in tuffets, the whiche do turne into rounde beries, greene at the first, but afterwarde red, and of an vnpleasant taste.

❧ The Place.

The Asshe delighteth in moyst places, as about the brinkes and borders of riuers, and running streames.

The wilde Asshe groweth vppon high mountaynes, and also in shadowy wooddes.

❧ The Tyme.

The Asshe seede is ripe at the ende of September.

The wilde Asshe flowreth in May, the fruite thereof is ripe in September.

❧ The Names.

1 The first tree is called in Greke μελία: in Latine, Fraxinus: in English, Asshe: in Frenche, *Fresne*: in high Douche, Eschernbaum, Eschernholtz, and Steyneschern: in base Almaigne, Esschen, and Esschenboom.

The huskes or fruite thereof are called in shoppes Lingua auis, and Lingua passerina: in English, Kytekayes.

2 The second kind is called of Theophrastus in Greke Βουμελία: Gaza calleth it in Latine: Fraxinus bubula: Plinie, and Columella calleth it Ornus, and Fraxinus syluestris: some of the later writers, calleth it Fraxinea arbor: and some call it Sorbus aucuparia: aswell bycause it hath leaues lyke vnto the Sorbe tree, as also bycause the Birders, and Fowlers doo vse the fruite thereof, as baite to take Birdes withal: in English, Quickebeame, feelde Asshe, wild Asshe, and white Asshe: in Frenche, *Fresne Champestre*, or *sauuage* : in high Douche, Malbaum, and grosser Malbaum, in base Almaigne, Haueresschen, and Qualster.

❧ The Nature.

The leaues and rinde of the Asshe, are of a temperate heate, & subtill partes or substance.

The seede is hoate and drie in the seconde degree.

The wilde Asshe leaues be also hoate and drie, and of subtill partes.

❧ The Vertues.

The leaues and barkes of the Asshe tree boyled in wine, and dronken, doo A open and comfort the liuer, & splene being stopped, and doo heale the discase of the sides: They haue the same vertue, to be boyled in oyle and layde to the side.

The leaues and barke with the tender croppes of the Asshe tree, are good to B be taken in the same maner against the dropsie, for they purge the water.

the Historie of Plantes.   749

For suche as are to grosse or fat, they vse to geue dayly three or foure asshen leaues to drinke in wine, to the intent to make them leane.

The iuyce of the leaues, barke, and tender croppes of the Asshe dronken in wine preserueth from al venome, especially against the bitinges and stingings of Serpentes and Vipers.

They say that the Asshe is of so great force against poyson, that in the circuitie or shadowe of the same there hath not bene knowen any maner of venemous beast to abyde.

The lye that is made with the asshes of the barkes of the Asshe tree, cureth the white scurffe, and suche other lyke roughnesse of the skinne.

The seede of the Asshe tree prouoketh vrine, increaseth naturall seede, and stirreth vp Venus, especially being take with a Nutmegge, as Isaac, Rhasis, Damascenus, and many other Arabian Phisitions do write.

The leaues of the wilde Asshe tree boyled in wine, are good against the payne of the syde, and the stopping of the liuer. And to be taken in the same maner, they slake the bellyes of suche as haue the dropsie.

Of the kindes of Popler and Aspe.   Chap.lxri.

❧ The Kindes.

THe Popler is of three sortes, as witnesseth Plinie: the one is called white, the other blacke, and the thirde is called Aspe: the which three kindes are very common in this Countrie.

Populus alba. **White Popler.**   Populus nigra. **Blacke Popler.**
Populus Lybica. **Aspe.**

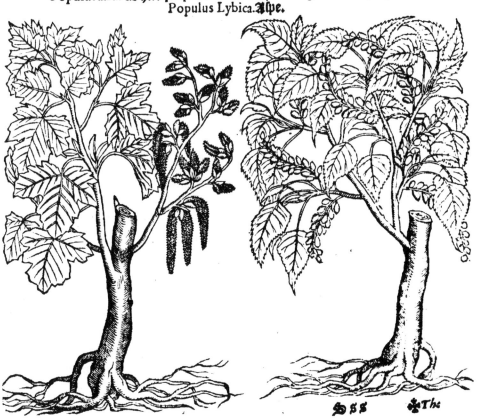

The Description.

1. The white Popler tree waxeth high, great & thick. The timber whereof is white, and not very harde to be wrought. The barke is smooth and whitishe, especially on the branches. The leaues be rounde with poynted corners, white, soft, and woolly vpon one side, and playne, smooth, & greene vpon the other side. Before it putteth foorth leaues, it beareth long woollishe tagglettes or Cattekens, of colour almost incarnate.

2. The blacke Popler also groweth high, great and thicke. The barke whereof is smooth, but browner, The leaues be somewhat long, and brode beneath towardes the stemme, and sharpe at the top, a litle snipt about the edges, but they be neither white, smooth, nor wollish. The Cattekens, or tagglets of these, doo turne into clusters with many round beries. The buddes which breake foorth before the leaues are of a sweete sauour, by reason of a certaine yellowish clammie oyle or grease which is contayned within them, of which is made the oyntment called Populeum.

3. The Ratling or trembling Aspe, is somewhat like to the blacke Popler: it waxeth as great as any of the other twayne. The ragges or Catkens of these are longer and browner, then the others, almost graye or Ashe colour browne. The leaues be somewhat roundishe, broade and shorte, browner and harder then the leaues of the blacke Popler, deeply indented round about the edges, the sayde leaues doo hang by a long, but a very small slender stemme, whiche is the cause of their continual shaking, and noysome clatter.

The Place.

These trees do growe in low moyst places, as in medowes, and neare vnto diches, standing waters, and riuers.

The Tyme.

The tagglettes or Catkens of the Popler do come foorth about the ende of Marche and Aprill, and then ye must gather the buddes to make Vnguentum Populeon.

The Names.

The white Popler, is called in Greeke λευκή: in Latine, Populus alba, and of some Farfarus: in Englishe, White Popler, or Pepler: in Frenche, *Aubeau*: in high Douche, Pappelbaum, Weiszalberbaum, & Weisz Popelweyden: in base Almaigne, Abeelboome, & of some ignorat people in Latine, Abies, & that very erroniously, for Abies is ȳ Pineapple tree, whereof we shal declare hereafter.

2. The seconde Popler is called in Greeke ἄιγειρος: in Latine, Populus nigra: in Englishe, Blacke Popler: in Frenche, *Peuplier*: in high Douche, Aspen, and Popelweiden: in base Almaigne, Populier, and Popelaere.

3. The thirde Popler is called of Plinie Populus Lybica: in French, *Tremble*: in base Almaigne, Rateleer: in English, Aspe.

The Nature.

The leaues and barke of Popler are temperate in heate and cold, neuerthelesse they be somewhat drie and abstersiue or clensing.

The buds of the blacke Popler, are hoate and drie in the first degree, and of subtill partes.

The Vertues.

The barke of the white Popler boyled in wine, prouoketh vrine, helpeth the stranguries, and them that haue the Sciatica, and payne in the hippe. **A**

The iuyce of the leaues swageth the payne of the eares, & healeth the vlcers of the same, to be dropped in. **B**

The leaues and young buddes of blacke Popler, doo swage the payne of the gowte in the handes and feete, being smal pounde and layde thereupon. **C**

The oyntment that is made of the buds, is good against al inflammations, and against all bruses, squattes, and falles, and against swellinges, to be layde thereupon. **D**

Of

Of the Elme. Chap.lxxij.

❧ The Kindes.

There be two sortes of Elme, as Theophrastus and Columella haue written: the one groweth in feeldes, and the other on mountaynes.

❧ The Description.

Vlmus.

1. THE first kinde of Elme, is a great high tree with many branches spread abroade at large. The timber therof is hard, brownishe, sinewie, & vneasie, to cleeue. The leaues be broade and wrinckled, somewhat snipt or cut about the edges, vpon the which there growe oftentimes certayne bladders or blisters, wherin is found a certayne slimie and clammie liquor, in whiche most commonly there be smal wormes: but when ye said liquor doth remayne, about the end of sommer you shal find it hardened by the force of the Sonne, euen lyke gumme. The seede of the Elme is broade, rounde, flat, smooth, & soft, not muche vnlyke Arache seede, but greater. The roote spreadeth far and broade, this kinde is very common in this Countrie.

2. The seconde kinde of Elme, is not muche vnlyke the aforesayd in leaues & timber, but it groweth much higher without spreading his branches so farre abroade, and it bringeth foorth seede very seldome. The leaues of this kind be more desyred and better lyked of cattell to feede vpon, then the leaues of the first kinde.

3. Bysides these two sortes of Elme, there are founde other trees drawing somewhat towardes the Elme, the which also doo waxe great and high, their timber is very tough & hard, and therefore it is much vsed to make wheeles & mylles, poullies, and such other instruments & engins for the carriage of great waightes and burthens. The leaues be likewise wrincled and somwhat snypt and toothed round about, much longer and narrower than the leaues of the other, of a faint greene colour vpon the contrarie syde, euen pollished, & shining, and of a good smel when they are drie: vppon these leaues there neuer growe any blisters or bladders, neither will the cattell so willingly eate of them, as they will doo of the Elme leaues. The seede of these is almost like the seede of the Elmes.

❧ The Place.

1. The first kinde groweth in lowe champion, and plaine feeldes, and delighteth the good fertill grounde, & is founde planted in diuers places of England and Brabant, by the high wayes, and feeldes.

2. The second kind loueth the hilles and mountaynes: yet you shal see some of it, in Westflaunder gardens which is compassed round with this kind of Elme planted in ranges, in very good order.

3. The third kind groweth plentifully in woods, as in the wood Soignie, and in other places alongst the feeldes.

752   The sixth Booke of
The Tyme.
The seede of the Elme groweth, and afterwarde falleth about the ende of Aprill, when the leaues beginne to spring.

The Names.
The Elme is called in Greke πτελέα: in Latine, Vlmus: in English, an Elme tree: in high Douche, Rustholtz, Rustbaum, Ulmenbaum, Lindbast, & Effenholtz: in base Almaigne, Olmboom: in Frenche, Ormee.

1   The first kinde is called of Theophrastus πτελέα, that is to say, Vlmus: of Plinie, Vlmus palustris: of Columella, Vlmus vernacula, and Vlmus nostras, that is to say, the Elme of Italie, and our common Elme.

2   The second is called of Theophrastus in Greeke ὀρειπτελέα: in Latine, Montiulmus: of Columella, Atinia, and Vlmus Gallica: in Picardie, & Artois, Ypreau:

The seede of the Elme is called in Latine, Samera.

The wormes that engender in the bladders or blisters of the Elme leaues, with the liquor that is conteined in the same, are called in Greke κνιπες, Cnipes: in Latine, Culices, and Muliones.

The liquor is called of the later writers, Gummi vlmi, that is to say, The gumme of the Elme.

3   The third tree is called in of Theophrastus in Greke κολυτέα, Colytea: this should seeme to be Vlmus syluestris, wherof Plinie maketh mentiō: in French, *Hestre*, it may be called also in Frenche, *Orme sauuage*: it is called in high Douche, Hanbuchen, and Bucheschern: in base Almaigne, Herseleer, and in some places Herenteer: I thinke this not to be the right Opulus: but the very tree, whiche we cal Witche, and Witche Hassel: in Frenche, *Opier*: & is the best kind of Elme to ioyne vines vnto, bycause his branches be faire and large of a goodly length but not so thicke. Reade more of Opier in the lxxx. Chapter of this booke.

The Nature.
The leaues and barke of Elme, are somewhat hoate, and astringent.

The liquor that is found in the bladders that grow vpon the leaues is dry, and of a clensing and scowring nature.

The Vertues.
Elme leaues do cure and heale greene woundes, being wel brused & layde A thervpon, the inner barke hath the lyke vertue, if it be bounde to the woundes as a swadling bande.

The broth of Elme leaues (or of the barke or roote, as Dioscorides saith) B is good to bath and soke the armes and legges that be broken and brused, for it speedyly healeth broken bones.

The leaues pounde with vineger & laid to, is good for the lepry & scurvines. C

The waight of an ounce of the vtter barke take with wine or water, putteth D foorth colde fleme and looseth the belly.

The liquor that is founde in the leaues, doth beautifie the skinne & the face, E and scoureth away all spottes, freckles, pimples, and spreading tetters, if it be layde thereto.

Also it healeth greene and fresh woundes, if it be powred in, as the writers F in these dayes haue founde by experience.

The leaues of Elme are good fodder for rother cattell, as Theophrastus G and Columella write.

Of the Linden tree.   Chap.lxxiij.
The Kindes.
Theophrastus describeth two sortes of Linden tree, that is the male and female. They are both to be found in this Countrie, but ye female is most common and better knowen.

                                                                    The

the Historie of Plantes. 753

✿ *The Description.*   Tilia foemina. **The female Linden tree.**

The common Linden tree, whiche is the female Tilia waxeth great and thicke, spreading foorth his branches long and large, and yeelding a great shadowe when the Sonne shineth, the barke is brownishe without, smooth, and playne, but next to the timber it is white moyst and tough, and wil easily be wrested, turned, and twisted euery way: wherfore it is the very stuffe wherof they make these cordes or halters of Barkes. The timber is whitish, playne, and without knottes, and very soft and gentle to handle: and therefore the coales that be made of this wood, are good to make gonpowder. The leaues be very greene and large, somewhat toothed or a little snipt rounde about the edges, otherwayes not muche vnlyke to Iuie leaues. The small flowers be whitishe and of a good sauour many hanging togither from out of the middle of a litle narrow white leafe. The fruite is none other but litle round beries or pellettes growing togither in litle clusters like to Iuie beries, in whiche is conteyned a small round seede, which is blackish, and falleth out, when the small pellettes or buttons do open and are rype.

2 The seconde kinde of Lynden tree, whiche is the male (called Tilia mas) groweth also great and thicke, and spreadeth abrode lyke the other Linden, the barke whereof is also tough and plyant and serueth to make cordes & halters: but it is rougher, thicker, and britteler, grayishe about the smal branches, but whiter then the barkes of the common Linden tree, yet not so white as the branches of Elme. The timber of this Linden is muche harder, more knottie, & yellower then the timber of the other, much like the timber or wood of Elme, the leaues be broade, not playne, nor euen, but rough and a little cut about the edges, very lyke to Elmen leaues, sauing that vpon them there neuer growe any smal bladders. This tree bringeth foorth fruite very seldome, and therefore some iudge it as barren, it bringeth foorth for his fruite, many things lyke to round flat huskes clustering togither, hauing a certayne clift or chinke at the end, much lyke in proportion and quantitie to the huskes of the right Thlaspie described in the fifth booke and lxij. Chapter, wherof eche hangeth alone vpon a stemme by it selfe.

✿ *The Place.*

The Linden tree loueth a good conuenient soyle, and it groweth lyghtly where as it is planted. One kinde of Linden groweth by Colchester in Essex, in the parke of one maister Bogges.

✿ *The Tyme.*

It flowreth in May, and the fruite is ripe at the ende of August.

✿ *The Names.*

This tree is called in Greeke φιλύρα: in Latine, Tilia: in Englishe, Linden: in

Sss iij   frenche,

The ſyxth Booke of

Frenche, Tillen, oꝛ Tillet in high Douche, Linden, and Lindenbaum: in baſe Almaigne, Linde and Lindeboom.

1 The firſt is called in Latine, Tilia fœmina, that is to ſay, the female Lynden.
2 The ſeconde is called, Tilia mas, that is the male Lynden: ſome call it in Frenche, Heſtre: and in baſe Almaigne, Ypelijne.

✿ *The Nature.*

The barke and leaues of Linden are of temperate heate, and ſomewhat dꝛying and aſtringent, almoſt in complexion lyke to the Elme.

✿ *The Vertues.*

The bꝛoth of the leaues of Lynden ſodde in water, cureth the noughtie A ulcers and bliſters of the mouthes of young childꝛen if they be waſhed therewithall.

The leaues pounde oꝛ bꝛuſed with water are good to be layde to the ſwelling of the feete. B

The barke of Lynden pounde with vineger, cureth the noughtie white C ſcurffe, and ſuche lyke euilfauoured ſpꝛeading ſcabbes, as Plinie wꝛiteth.

Of the Plane tree. Chap.lxxiij.

Platanus.
The Plane tree.

Aceris ſpecies, folio maiori.
A kind of Maple with the greater leafe.

✿ *The*

❧ The Description.

1 THe Plane is a strange tree, the whiche in time past hath bene of great estimation in Italie and Rome. In so much that ye may finde it written, howe they haue bedewed or watered it with wine. It groweth great and high, and spreadeth his branches and boughes very broade & wyde, the leaues be large, muche like in figure to the leaues of the vine, hanging by long reddish stemmes. The flowers be small and growe in little tuffetes. The fruite is rounde, rough, and somewhat woolly, of the quantitie of a Filberde.

2 There is founde in the Alpes in Almaigne, and some places of Brabant, a certayne tree, much lyke to the Plane tree. It hath brode leaues lyke the vine, hanging by long smal and red stemmes, but the flowers and fruite of this tree are nothing lyke the flowers and fruite of the Plane tree, but lyke the flowers and fruit of Maple (wherof this is a kind) which shalbe described in the lxxxj. Chapter of this booke.

❧ The Place.

1 The Plane tree groweth in many places of Greece: it is also to be founde planted in certayne places of Italie: it is vnknowen in this Countrie.

2 The tree whiche beareth leaues lyke the Plane, is founde vpon high mountaines in some places of Douchland and Brabant, and alongst the fseldes, but very seldome, and there is here and there a tree of it planted in Englande.

❧ The Tyme.

The Plane tree flowreth about the ende of March, & so doth the other also.

❧ The Names.

1 The Plane tree is called in Greeke πλάτανος: in Latine also, Platanus: in Frenche, Platane.

2 The tree that is lyke vnto it is called in English, the Planc tree, in Frenche, Plane: in high Douche, Ahorne, and Waldeschern. But it is not Platanus, but a kinde of Maple, and it shoulde seeme to be that kinde whith is called in Grcke ζύγια: in Latine, Carpinus. Yet the figure which Matthiolus hath giuen vs for Carpinus, is more like to a kind of Witch Hassel. Carpinus ab Accre differt, quod Accri candida atq; neruata materia, Carpino autem flaua crispaq, Theo. ca. 11.li.3.

❧ The Nature.

1 The Plane tree leaues are partakers of some colde and moysture.
2 The barke and fruit are more dryinge.

❧ The Vertues.

A The fruite of the Plane tree dronken with wine, helpeth them that are bitten of Serpentes.

B The same broken and mingled with grease, and layde to, healeth the burninges with fire.

C The Barke sodden in vineger, is good for to washe the teeth agaynst the tooth ache.

D The young and tender leaues wel pounde are good to be layde vpon swellinges and inflammations: and do stop the running & watering of the eyes.

Of the Aller. Chap.lxxv.

❧ The Description.

THE Aller is a high great tree, with many branches, the whiche wyll breake quickely, and will not lightly plop nor bende. The rinde of this tree is browne. The timber is meetely harde, and will last a long season vnder water, yea longer then any other kinde of timber: And therefore they make piles and postes for to lay fundations in fennes, & soft marrish grounds, also they are very good to make pipes, condites, and troughes for the leading along,

along, and carriage of water vnder grounde: but aboue ground water wil soone rot and consume it. This tii.ber wareth red, assoone as it is spoyled of his rinde, and lykewyse when it is old and dry. The leaues be somwhat clammie to handle, as though they were wet, with hony, of fashion roundish, and somewhat wrinckled, not muche vnlyke the leaues of the Hasell nuttes. The blowinges of Alder are long tagglets, almost like to the blowinges of Birche. The fruite is round lyke to small Oliue beris, and compacte or made of diuers scales, set close togither: the which being ripe and dry do open, so as the seede whiche is within them falleth out and is lost.

※ *The Place.*

The Aller delighteth to growe in low moyst woods, and waterish places.

※ *The Time.*

The Aller beginneth to bud, and to bring foorth newe leaues in Aprill as other trees do. The fruite is ripe in September.

※ *The Names.*

The Aller, or Alder is called in Greeke κλἡθρα: and in Latine, Alnus: in high Douche, Erlenbaum, and Ellernbaum: in base Almaigne, Ellenboom: in frenche, Aulne.

※ *The Nature.*

The barke and leaues of Alder, are cold, drie, and astringent.

※ *The Vertues.*

The barke or rinde of Alder, bycause of his astringent power, may be good against the impostumes, and swellinges of the throte, and kernelles, or Almondes vnder the tongue, euen as well as the shales or greene pilles of Walnuttes. But as yet it hath not bene vsed by any, sauing onely for the dyeing of certayne course cloth and cappes into a blacke colour, for the whiche purpose it is very fit.

The leaues be much vsed against hoate swellinges, vlcers, and al inwarde inflammations.

Of the Beeche tree. Chap.lxxvi.

※ *The Description.*

The Beeche is a great, high, thicke tree, whose leaues be soft, thinne, playne, smooth, and meetely large, almost like the leaues of Popler, but smaller. The blossoms therof are nought els, but smal yellowish Catkens, smaller then the Catkens of Birche, but otherwise like.
The

The fruite is triangled or three cornered Nuttes, in whiche are sweete kernels. These Nuttes be couered ouer with prickly huskes or shales, from out of whiche they fall downe when they be ripe.

⁋ *The Place.*

The Beeche loueth a playne open Countrie, and moysture.

✳ *The Tyme.*

The Beeche bloweth and breaketh foorth into newe leaues, at the ende of Aprill or Maye. The Nuttes be ripe in September euen with the Chesnuttes.

✳ *The Names.*

The Beeche tree is called in Greeke φηγὸς: in Latine, Fagus: in Frenche, *Fouteau*: in high Douche, Buchbaum, or Buche: in base Almaigne, Bueckenboë. The fruite is nowe called in Latine, Nuces Fagi: in frenche, *Faine*: in base Almaigne, Buecken nootkens: in English, Beeche maste.

✳ *The Nature.*

The leaues of Beech do coole. The kernell of the fruite is somewhat moyst and warme.

✳ *The Vertues.*

The leaues of Beeche are very profitably layde to the beginning of hoate A swellinges, blisters, and vlcers.

The water that is found in the holownesse of Beeches, doth cure the noughtie scurffe, and wilde tetters or scabbes of men, and horses, kyne, and sheepe, if they be washed therewithall. B

Men doo not yet gather these Nuttes for mans vse, yet they be sweete and good for to eate, and they doo almost serue to all those purposes, wherevnto the Nuttes of the Pine apple kernelles doo serue. C

Of Birche tree. Chap.lxxvij.

✳ *The Description.*

The Birche doth often grow to a great high tree, with many branches, which haue many smal roddes or twigges very limber and pliant, and most commonly hanging downewarde, and will abyde to be bowed easily any way that one list. The barke of the young twigges and branches is playne and smooth, and full of sappe, and of the colour of a Chesnut: but the barke of the body and greatest branches of the tree is harde without, white, rough, vneuen, and broken or clouen, vpon the branches that be of a meane sise or quantitie, the barke or rind is somwhat speckled: vnder the same barke, next ioyning to the wood or timber, there is founde another barke that is playne and smooth as paper, so that in times past it was vsed to write vppon, before

that

that Paper or Parchement were knowen or inuented. The leaues are meetely brode, and somwhat snipt about, smaller then Beechen leaues, but otherwise not muche vnlyke. The Birche tree hath tagglettes or Chattons for his blossome, lyke as the Hasell, but much smaller, in whiche the seede commeth.

Betula. Birche.

The Place.
Birche groweth in wooddes, and heathes, and drie commons, and also alongest the borders of Corne feeldes.

The Tyme.
Birche putteth foorth his new leaues in April: in September his small Catkens and seede is ripe.

The Names.
Birche is called in Greke σημύδα: in Latine, Betula: in Frenche, Bouleau. in high Douch, Birkébaum: in base Almaigne, Berckenboom.

The Nature and Vertues.
Birche is not vsed in medicine, wherfore his nature and vertues are not knowen: in old time they vsed the inner thin barke of Birche, in steede of Paper, & the young twigges and branches thereof to make roddes, and besoms, as they doo at this day.

Of blacke Aller. Chap.lxxviij.

The Description.

THE blacke Aller groweth not lyke a tree, neyther waxeth it very great, but it bringeth foorth many long straight roddes, whiche doo diuide them selues agayne in other small twigges couered with a thinne blacke rinde, vnder the whiche there is founde another yellowish rinde. The timber or wood of these twigges is whitish, with a browne red pith in the midle. The leaues be brode lyke the leaues of Aller, almost lyke to Cherrie tree leaues, but rounder and browner. The litle flowers be whitish after which come vp round beries, which are greene at the first, but afterwards red, and blacke when they are dried, of a strange vnpleasant taste.

The Place.
This kinde of wood groweth in lowe wooddes and moyst places.

The Tyme.
It flowreth in April, and the beries be ripe in August.

The Names.
This plant is called of the Brabanders, Sporckenhout, and of the chyldren of this Countrie, Pijlhout, that is to say, bolt timber, or arrow wood, bycause they make Arrowes with it, to shoote withal: in high Douch, Faulbaum, and Leuszbaum:

Leutzbaum: of some of y later writers, in Latine, Alnus nigra, that is to say in English, Blacke Aller.

❧ The Nature.

The inner barke of this wood is yellow, and of a drie complexion.

❧ The Vertues.

A The yellowe barke of Aller stieped in wine or bier, and dronken, causeth to vomit vehemently, and expelleth flemes, & corrupt humors contayned in the stomacke.

B The same boyled in vineger and holden in the mouth, swageth the tooth ache, & cureth the scurffe and fretting sores being layd thervpon.

C The leaues be good fodder, or feeding for kyne, and cause them to yeelde store of mylke.

Of Spindel tree, or Pricke timber. Chap.lxxix.

❧ The Description.

This plante groweth neyther high nor great lyke a tree, but remayneth small and lowe, putting foorth many braches. The stemmes of y olde branches are couered with a whitishe barke, and the younger branches are couered with a greene rinde or barke, hauing as it were foure straight lines running alongst the young shutes or branches, the whiche do make a quadrature, or a diuision of the said young branches into foure square partes or cliftes. The timber is harde of a whitishe yellow. The leaues be long, & somwhat large, soft, & tender. The small flowers be whitish, & hāging fiue or sixe togither, after them come small rounde huskes foure ioyning togither, ỹ which do opē when the fruit is ripe: In euery of the aforesayde huskes, is found a sede or kernel couered with a faire yellowe skinne, whiche being soked in water or any other liquor wil staine & die yellow.

❧ The Place.

Spindeltree groweth in this Countrie alongest the feeldes in hedges & woods.

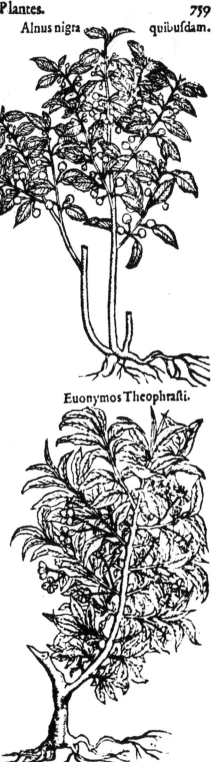

Alnus nigra quibusdam.

Euonymos Theophrasti.

❧ The

❧ *The Tyme.*

It floweth in April, and the fruite is ripe in September.

❧ *The Names.*

This plant seemeth to be that same, which Theophrastus calleth in Greke ἐυώνυμος, Euonymos: some call it in Latine, Fusaria, and Fusanum: in Englishe, Spindeltree, and Pricketimber: bycause the timber of this tree serueth very well to the making both of Prickes and Spindelles: in French, *Fusain, Couillon de Prestre,* and *Bois a fair Lardoires:* in high Douch, Spindelbaum, ᛫ Hanhoedlin: in Brabant, Papenhout. This is not ζυγία, Zygia, or Iugalis, or a kinde of Acer, as some do thinke.

❧ *The Vertues.*

Spindeltree, as Theophrastus writeth, is very hurtfull to all cattell, especially vnto Goates, for it killeth them, if they do not purge both vpwarde and downewarde.

Of Marris Elder, Ople, or Dwarffe Plane tree.   Chap.lxxx.

❧ *The Description.*   Sambucus palustris recentioribus.

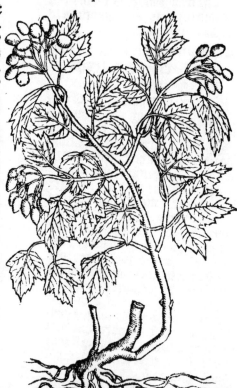

1 This plante is somewhat lyke Elder or Bourtree. The timber, but specially the young branches are ful of pith lyke Elder. The leaues be large, and fiue cornered, almost of the fashion of vine leaues, but smaller. The flowers be white, and grow in brode round shadowy tuffetes, whereof those in the middle are small, and they that stand al without about the border of the round spokie tuffettes, are great. The fruite is round beries, like the beries of Elder, but greater, and of a shining red colour.

2 There is yet another sort, which bringeth foorth flowers in round (but not flat) tuffets, in al things els lyke to the abouesayde.

❧ *The Place.*

This plant groweth by water courses, ᛫ in low waterish places.

❧ *The Tyme.*

It floweth in Maye, and the fruite is ripe in September.

❧ *The Names.*

This plant is called of the later writers in Latine, Sambucus palustris, and Sambucus aquatica, that is to saye, Marrishe Elder, or Water Elder, bycause of the flowers and fruite, also the timber is full of pith, lyke Elder. Cordus calleth it Lycostaphylos: some make it Chamæplatanus: it is called in Frenche, *Obiere,* or *Opiere.* in high Douche, Schwelder, and Bacholder: in Brabant, Swelken, ᛫ Swelkenhout. This is not Opulus, as some do thinke: it may be wel called in Englsih, Dwarffe Plans tree:

tree: I take this to be a shrub that is called in English, Whittentree, whereof are two kindes, one in all pointes agreeable with Sambucus palustris, the other altereth in leafe: for his leaues be like to Elme or Witche Hassel, and this kind is very tough and flexible. ❧ *The Vertues.*

Some will say, that the beries of Whittentree, taken into the body, will cause vomit and open the belly: but it hath not hitherto bene found true, of any learned and expert Doctours.

Of Frangula Matthioli.

Matthiolus hath ioyned to his Eldrens, a certayne plant, growing in Bohem, called Frangula, þ which I haue thought good also to place in this booke for the great proffite, which my Countrie men may haue by the knowledge of the same. ❧ *The Description.*

FRangula is a tree of a meane stature, the leaues are almost lyke to the Cornel, and Houndestree leaues, the barke is lyke to the barke or rind of Aller, speckled on the outside with white spottes: but the inner rinde is so yellow, that if it be chewed it will colour, and stayne yellow lyke saffron and Rubarbe. It putteth foorth white flowers, and small fruite or beries, of the bignesse of a pease so diuided in the midle, that it seemeth to be two beries by natures skill ioyned togither. At the first the fruite is greene, then red, and at last blacke, hauing within two small stones, almost lyke Lupines, but litle bigger then Lentilles, in whiche the kernels be. The substance of the timber is very brittle and frayle, whereof it tooke his name in Latine, Frangula. There is no small vertue in the barke or rind, both to loose and bind, for it looseth the belly and strengthneth the principall partes, euen lyke Rubarbe. It purgeth choller and fleme, and dispatcheth the water from suche as haue the dropsie. The sayde barke is boyled with common Eupatorie, Pontike wormwood, Agrimonie, Epithim, Hoppes, Cinamome, Fenill, Parsley, and both kindes of Endiue with their rootes, against the dropsie, and euill disposition of the body, & the Jaunders: it is giuen to them that be sicke of the aforesayde diseases, in the quantitie or waight of fiue ounces with singuler profite. But it shalbe very requisite, that first the superfluitie of humours, which lye in the stomacke, and the first vaines of the liuer be pourged. It looseth the belly without any danger, & doth very well purge and strengthen the liuer, so that such as haue bene greeued with the hardnesse of the milt and liuer, haue bene cured onely with this medicine: for it openeth all the stoppinges of the bowels and entrayles, and clenseth all the instrumental partes from grosse excrements. The vttermost barke is astringent: and the inner barke looseth. Both ought to be stript from the tree, in the very beginning of the spring time, and then to be dried in the shadowe for if it be occupied greene, it wil cause vomiting. The decoctiō that is made with it, ought to stande or rest two or three dayes before it be ministred, vntill the yellow colour be changed into blacke, els it may peraduenture cause vomiting. The same dronken before the vnloosing of the belly prouoketh appetite. And. Matthioli Comm. in lib. 4. Dioscoridi.

Of Maple. Chap. lxxxi.
❧ *The Description.*

MAple groweth somtimes lyke a tree, both high and thicke, with many great braunches: sometimes it groweth lowe, lyke a shrub. The barke is thicke and somewhat white. The timber is harde, and garnished with long streaming waues, or water vaynes. The leaues be brode, with fiue peakes or corners, lyke the leaues of Ople bushe, or Dwarffe Plane tree,

tree, but smaller and greener very lyke the leaues of Sanicle. The fruite is long, flat, and thinne, almost lyke to a feather of a small birde, or lyke the whing of a grashopper.

Aceris species, Folio minori.

❀ *The Place.*

Maple groweth in wooddes, where as it commeth to a great tree, and alongst by diches and running streames, where as it is but small.

❀ *The Time.*

It bloweth in Maye, and the seede is ripe in September.

❀ *The Names.*

This tree is called in Greeke σφίνδαμ۞: in Latine, Acer, and it shoulde be that kind which is called in Greeke πεδεινή, that is to say, Campestris, and of some, as Theophrastus writeth, γλεῖν۞, ἢ γλῖν۞, of Plinie, Gallica: in Englishe, Maple: in Frenche, Erable. in high Douch, Maszholder: in Brabant Booghout, and Aliethout.

❀ *The Vertues.*

A The rootes of Maple pounde in wine and dronken, are good against the paynes in the syde, as Serenus Samonicus hath written.

Of the Iuniper tree.   Chap.lxxxij.

❧ *The Kindes.*

There be two sortes of Iuniper, as Dioscorides writeth, whereof the one kinde groweth great and high. The other kind remayneth smal and low, and is well knowen in this Countrie.

❀ *The Description.*

1 The smal and common Iuniper sometimes groweth vp, and waxeth to the stature of some other trees, but most commonly it remayneth lowe, and groweth like a shrub or hedge plant. The branches of this Iuniper are couered with a thinne barke, the which will soone riue, or cleeue asunder: (especially in hoate Countries) after whiche chopping or chinking of the barke there commeth foorth a gumme or liquor, lyke Frankensence. The leaues be litle, small, and hard, growing alongst the stalkes and branches, & are alwayes greene without falling of in winter. The fruite is rounde litle beries, whiche be greene at the first and afterward blacke of a good sauour and sweet in taste, whiche at length waxe bitter.

2 The great Iuniper is a great high tree, and beareth beries as great as Filberdes, and sometimes as great as Walnuttes, as Dioscorides writeth.

❀ *The Place.*

Iuniper is found vpō high mountaynes, in shadowy woods, & low holow wayes, it loueth a cold stony ground.

❧ *The*

Iuniperus.

❋ The Tyme.

In the moneth of Maye there ariseth out of Iuniper a certayne yellow powder or dust, which is taken for the blowing or flower of Iuniper, after that you shal perceiue the smal beries to begin to grow vp, the whiche do waxe ripe in September, a yere after that they begin first to grow vp. Therefore ye shal finde vpon the Iuniper tree, beries both ripe and vnripe, great and small al togither.

❋ The Names.

Iuniper is called in Greke ἄρκυθος, and of some ἀκαταλίς: in Latine, Iuniperus: in Frenche, Ienéure, or Genéure: in high Douch, Weckholder, and Weckholterbaum: in Brabant, Geneuer.

The beries be called in Greke ἀρκυθίδες: in Latine, Baccæ Iuniperi: in shops, Grana Iuniperi: in Englishe, Iuniper beries: in Frenche, Graines de Genéure: in high Douch, Weckholterbeeren, and Kromerbeeren.

The gumme whiche sweateth out of this tree and his barke, is called in shops, Vernix: and in some places not without great and dangerous errour, Sandaraca: for the right Sandaraca is a gnawing, and venemous substance, whiche is founde in the mines of mettalles whereunto this gumme is nothing lyke.

❋ The Nature.

The Iuniper tree with all his partes, as leaues, barke, timber, fruite, and gumme is of complexion hoate and drie.

❋ The Vertues.

The fruite or beries of Iuniper, is good for the stomacke, lunges, liuer, and kidneyes: it cureth the olde cough, the gripinges and windinesse of the belly, and prouoketh vrine, to be boyled in wine or honied water and dronken. **A**

Also it is good for people that be bruised or squat by falling, to be taken in the aforesayde manner. **B**

The iuyce of the leaues doth withstand al benome, especially of Vipers and Serpentes: it is good to drinke the same, and to lay it outwardly vppon the woundes. The fruite is good for the same purpose, to be taken in what sort so euer ye list. **C**

Iuniper or the beries thereof burned, driueth away all venemous beastes, and all infection and corruption of the ayre: wherefore it is good to be burned in a plague time, in suche places where as the ayre is infected. **D**

The rind or barke of Iuniper burned, healeth the noughtie scurffe, and fretting scabbes, to be mingled with water and layde thereto. **E**

The gumme of Iuniper is good for them whose stomackes and bowelles are combred with colde flegmes: it expelleth all sortes of wormes, and stayeth the inordinate course of womens flowers. **F**

The parfume of Vernix, is good for the brayne, drieth vp the superfluous humors of the head, and stoppeth the falling downe of reume or humors from the same. **G**

The ſyxth Booke of

This gumme tempered with Oyle of Roſes, helpeth the riftes, cones, or chappinges of the handes and feete.

Of Cedar tree. Chap.lxxxiij.

❧ *The Kindes.*

There be two ſortes of Cedar, great & ſmall. The ſmal fruite alſo is of two ſortes: the one with ſharpe prickley leaues like Juniper: the other are not prickley at all.

❋ *The Deſcription.*

Oxycedrus.

1 The great Cedar wareth very ſtowte & tall, high, great & thicke, yea greater, & higher then the figge tree. The barke euen from the foote of the ſtem vnto the firſt branches is rough, and from thence foorth euen vp to the toppe, is very ſmoth &, layne, of a darke blew colour, o:.. of which there droppeth white Roſen of his owne kind, which is moyſt and odoriferant or ſweete ſmelling, the which by the heate of ſonne becommeth dry and harde. His limmes and branches be long, and ſtretched out into length & breadth and parted into many other ſmall branches, ſtanding directly or right one againſt another, lyke as in the Firre tree. The ſayde branches be clad and garniſhed with many ſmal litle leaues, thicke, ſhort, and ſweete ſmelling like þ leaues of Larix, or Larche tree. The fruit is like that of þ Firre tree, ſauing that it is greater, thicker, & harder, & the tree groweth ſtraight vpright like the Firre tree, as the painefull & diligent Peter Belon hath written. From the tronke or ſtemme of the Cedar tree, there commeth foorth a certayne cleare liquor, which the olde writers called Cedria.

2 The firſt kind of þ ſmaller Cedar, is much like to Juniper: but moſt cōmonly it is ſomwhat ſmaller. The ſtem is croked or writhed, & couered with a rough barke. The fruit is round beries, like Juniper beries, but ſomwhat greater, in colour at þ firſt greene, then yellow, & at laſt reddiſh, of an indifferent good taſt.

3 The ſecond kind of ſmal Cedar groweth not high, but remayneth alwayes ſmal and lowe lyke the other. The leaues of this kind are not prickley, but ſomwhat round & moſſie at the endes, almoſt lyke to the leaues of Tamariſke and Sauin. The fruite of this kinde alſo is rounde beries greene at the firſt, afterwarde yellow, and at laſt reddiſhe, in taſte bitter.

❋ *The Place.*

The great Cedar groweth in Africa and Syria, and as Vitruuius reporteth

teth also in Candie, vppon the high mountaynes, and places that be colde and moyst, whiche are commonly couered with snowe, as vppon the mountaynes Libanus, Amanus, and Taurus, as Belon wryteth.

2 The seconde groweth in Phoenicia, and certaine places of Italie, especially in Calabria, vpon the mount Garganus, and also in Languedoc.

3 The third groweth in Lycia, and is found in certayne places of Fraunce, as in Prouince and Languedoc.

❧ The Tyme.

1 The great Cedar tree bringeth foorth fruite of two yeres groth, and it is neuer without fruite, whiche is ripe at the beginning of winter.

2.3 The small Cedar trees be alwayes greene and loden with fruite, hauing at all times vpon them of fruite both ripe and vnripe lyke to Iuniper.

❧ The Names.

Cedar is called in Greeke κέδρος: in Latine Cedrus: in Englishe, Cedre, and Cedar: in Frenche, Cedre: in Douche, Cederboom.

1 The great Cedar is called of Plinie in Greeke κεδρελάτη: in Latine, Cedrus maior, and Cedrus Conifera: in Frenche, Grand Cedre.

The liquor that floweth out of this tree is called in Greeke κέδριον: in Latine also Cedria, and liquor Cedrinus: of Auicen, Serbin: of Serapio, Kitran: with this liquor in olde time they dyd vse to enbaulme the bodyes of dead men, the whiche at this time is taken out of the graues or Sepulchres, and solde in Shops in steede of Mumia, not without great and manifest errour. For that whiche the Arabians do cal Mumia, is called in Greeke Pissalphaltos.

2 The first smal Cedar is called in Greke ὀξύκεδρος, καὶ κέδρος Φοινίκη: in Latine, Acuta Cedrus, Cedrus Phœnicia, Oxycedrus, and Cedrula: in Frenche, Petit Cedre.

3 The seconde small Cedar is called κέδρος Λυκία: Cedrus Lycia: and in Prouince, as Peter Belon wryteth, Moruenic.

❧ The Nature.

1 The Cedar is hoate and drie in the third degree. The liquor Cedria, which runneth foorth of the great Cedar tree, is almost whoate in the fourth degree, and of subtil partes.

2.3 The fruite of the small Cedar, is also hoate and drie, but not so greatly.

❧ The Vertues.

Cedria that is the liquor of Cedar, swageth the tooth ache, being put into A the holownesse of the same.

Also it cleareth the sight, and taketh away the spots and scarres in the eyes, B being layde therevpon.

The same dropped into the eares with vineger, killeth the wormes of the C same, and with the wine of the decoction of Hysope, it cureth the noyse and ringing in the eares, and causeth the hearing to be good.

The Egyptians in times past, kept their dead bodyes with Cedria: for it D kepeth the same whole, and preserueth them from corruption: but it consumeth and corrupteth liuing flesh.

It kylleth Lyce and all suche vermine, wherefore whatsoeuer is annoyn- E ted with the same, Mothes, Wormes, and such other vermine, shal not hurt it at all.

The fruite of the same Cedar, is good to be eaten against the strangurie, it F prouoketh brine, and bringeth downe womens naturall sicknesse.

Of Sauine. Chap. lxxxiij.

❧ The Kindes.

There be two sortes of Sauine, one with leaues much like Tamariske, the other lyke to the leaues of Cypres.

❧ The Description.

Sabina. Sauin.

The Sauin tree that is knowen in this Countrie, groweth in maner of a small lowe shrubbe or tree, the stemme is sometimes as bigge as ones arme, the whiche diuideth it selfe into many branches, lightly spreading it selfe into length and breadth: those branches are diuided againe into other small branches: the which be yet againe parted into smal greene twigges or slender brushes, set full of small leaues almost like to Tamariske, but thicker and more prickley, remayning euer greene both winter and sommer, and of a rancke smell. The fruite is small blacke beries, not much vnlyke to Iuniper beries.

2 The other kinde of Sauine which is like to Cypres groweth to a competent height and quantitie, with a stemme greater then Cypres. It hath many branches spread abrode. The leaues be like Cypres. The fruit is round beries, greene at the first, and afterwarde blacke.

❧ The Place.

1 The first kinde of Sauin is founde planted in some gardens of this Countrie.
2 The second kind groweth in lesser Asia, & in the Greece, it is seldome found in this Countrie.

❧ The Tyme.

The fruite of the Sauin tree is ripe at the beginning of winter.

❧ The Names.

1 The first tree is called in Greeke Βραθυς: in Latine, Sabina: in Shoppes, Sauina, of some Sauimera: in English, Sauin: in Frenche, *Sauinier*: in high Douch, Seuenbaum: in base Almaigne, Saueleboom.
2 The seconde is also called of Dioscorides Βραθυς, and Sabina, and of Plinie, Cupressus cretica: it should seeme to be the tree which Theophrastus calleth in Greeke θυία, ἢ θύιον, Thuia, vel Thuium, and Plinie Bruthes, or Bruta, as Peter Belon, hath very wel left in writing.

❧ The Nature.

The leaues of Sauin which are most vsed in medicine, are hoate and drie in the thirde degree, and of subtill partes.

❧ The Vertues.

The leaues of Sauin boyled in wine and dronken, prouoke brine, and A driue it foorth so mightily that the blood doth folowe, it mooueth the flowers, driueth foorth the secondine and the dead birth, it hath the like vertue to be receiued vnder in a parfume,

The leaues pounde & layd to with hony, cureth vlcers, & stayeth spreading B and

the Historie of Plantes.

and eating sores: they do scoure and take away all spottes and speckles from the face or body of man.

They do also cause wartes to fal of, which grow about the yarde and other secrete places of man.

The wood or timber of Sauin is profitably mixt with hoate Oyles and oyntmentes, and it may be mixed in steede of Cinamome, taking double the waight, as witnesseth Galen and Plinie.

Of the Cypres tree. Chap.lxxv.

✽ *The Description.*

Cupressus.

THE Cypres tree hath a thicke, straight, long stemme, vppon whiche growe many slender branches, the whiche do not spread abroade, but grow vp in length towardes the toppe, so that y Cypres tree is not brode, but narrow, growing to a great height. The barke of the Cypres tree is browne, the timber yellowishe, harde, thicke, and close, and when it is drie, of a pleasant smell, especially being set neare the fire. The Cypres tree hath no particuler leaues, but the branches in steede of leaues bringe foorth short twigges greene and small, diuided againe into other smal twigges, the which be cut and snipt in many places, as if they were set about with many small leaues. The fruite is rounde almost as bigge, as a prune or plumme, the which being ripe doth open in diuers places, and hath in it a flat grayishe seede, the whiche is muche desyred of Emotes, Antes, or Pismiers.

✽ *The Place.*

The Cypres tree delighteth high mountaynes, and drie places. It will not lightly growe in lowe moyst places.

✽ *The Time.*

The leaues of Cypres be alwayes greene. The fruite is ripe in September euen at the beginning of winter.

✽ *The Names.*

This tree is called in Greeke κυπάρισσος: in Latine, Cupressus: in Shops, Cypressus: in Englishe, Cypres, and Cypres tree: in Frenche, *Cypres*: in high Douche, Cypressenbaum: in base Almaigne, Cypressenboom.

The fruite is called in Latine, Nuces Cupressi, Pilulæ Cupressi, and of some Galbuli: in Shoppes, Nuces Cupressi: in English, Cypres Nuttes: in French, *Noix de Cypres.*

✽ *The Nature.*

The fruite and leaues of Cypres are drie in the thirde degree, without any manifest heate, and astringent.

✽ *The Vertues.*

The fruite of Cypres taken into the body, stoppeth the laske & bloody flixe, and is good against the spetting of blood, and all other issue of blood. The decoction of the same made with water hath the same vertue.

The Oyle in whichthe fruite or leaues of Cypres haue boyled, doth strengthen the stomacke, stayeth vomiting, stoppeth the belly, and all other fluxes of the same, and cureth the excoriation or going of, of the skinne from the secrete partes or members.

Cypres Nuttes cure them that are bursten and haue their guttes fallen into their coddes, to be layde to outwardly. The leaues haue the same vertue

With the fruite of Cypres they cure and take away the corrupt flesh (called Polypus) growing in the Nose.

The same bruised with fat drie figges, doth cure the blastinges of the genitors: and if ye put leuen thereto, it dissolueth and wasteth botches and boyles being laide thereupon.

The leaues of Cypres, boyled in sweete wine, or Meade, doo helpe the stranguric, and issue of the bladder.

The same pounde very small, close vp greene or newe woundes, and stop the blood of them being layde therevnto.

They be also with great profits, layde with parched barly meale, to wilde fire, Carboncles, and other hoate vlcers, and fretting sores.

The leaues and fruite of Cypres, layde to with vineger, make the heare blacke.

Of the Yew tree. Chap. lxxxvi.

❦ *The Description.* Taxus.

THE Ewe is a great high tree, remayning alwayes greene, it hath a great stemme, couered with a graye barke, that is clouen and scabbed or scalye. The leaues be of a darke greene, long, & narrowe like a fether, set ful of smal leaues, growing al alongst a stemme, opposite or standing right ouer one against another, whereof eache leafe is narrow, and longer then the leaues of Rosemarie, otherwise not muche vnlyke. His fruite is faire rounde redde beries, somewhat bigger than whortes, but els not much vnlike.

❦ *The Place.*

Ewe groweth in Arcadia, Italy, Spayne, Fraunce, and Almaigne: also in the forrest of Ardein. In time past it was planted in gardēs for Tamariske. ❦ *The Time.*

The fruite of Ewe is ripe in September.

❦ *The Names.*

This tree is called in Greke σμίλαξ, of Theophrastus μίλ☉ and after Galen κάκτ☉, Cactos: in Latine, Taxus: of the ignorāt Apothecaries of this Countrie Tamariscus: in English, Ewe, or Yew: in Frenche, If: in high Douche, Ibenbaum: and accordingly in base Almaigne, Ibenboom. ❦ *The Nature.*

Ewe is altogither venemous, and against mans nature. ❦ *The Danger.*

Ewe is not profitable for mans body, for it is so hurtful and venemous, that suche as do but onely sleepe vnder the shadowe thereof become sicke, and sometimes they die, especially whē it bloweth. In Gascoigne it is most dangerous.

If any eate the fruite, it will cause the laske, the Birdes that eate the beries, do either dye, or cast their fethers.

The ignorant Apothecaries of this Countrie, do vse the barke of this tree, in steeds ſtede of the barke of Tamariſk, by this we may wel perceiue, what wickednes the ignorant Apothecaries do daily cómit by miniſtring of noughty hurtful medicines in ſteede of good to ye great perill & danger of the poore diſeaſed people.

Of the Pine tree. Chap.lrrrvij.

❦ *The Kindes.*

THere be two ſortes of Pine trees, as the noble Auncient Theophraſte writeth, that is to ſay, the garden, and wilde Pine trees, there be alſo diuers ſortes of the wilde Pine tree.   Pinus.

❦ *The Deſcription.*

THE Pine tree is high & great with many branches at the toppe, parted into other round braches, ſet round about with litle hard leaues & almoſt ſharpe pointed or prickly, very ſtraight or narrow, and of a greene whitiſh colour. The timber is red and heauy, & within about the harte, ful of ſappe and liquor. His fruite is great Boulleans or Bawles of a browne Cheſnut colour (and are called Pine Apples) in which grow ſmal nuttes, wherin is a ſweet white kernell, whiche is ſometimes vſed in medicine.

A   The firſt kind of the wild Pine trees, is high, great, and thicke, and yet not ſo high as the tame or garden Pine. The branches be ſpread abroade, with long ſharpe pointed leaues, the fruite is ſhort and not hard, opening eaſily, and falling quickly.

B   The ſecond kind groweth not ſo high, neither is the ſtemme growing ſtraight vp, but bringeth foorth many branches ſodainly frō the roote creeping by the ground, long, ſlender, & eaſily to be ployed or bente, inſomuche that thereof they make Circles and hoopes for wine hogſheades & tonnes, as the noble learned Matthiolus writeth. The fruite of this kinde is greater then the fruite of any of the other wilde Pine trees.

C   The third kind groweth ſtraight vpright, & wareth great & high, yet not ſo high as the other wild kindes. The branches of this do grow lyke the Pitche tree. The fruit is long & big, almoſt like the fruit of the Pitche tree, in the ſame is conteyned triangled ſmall nuttes, like to the nuttes of the Pineapple, but ſmaller & britler, with a kernell of good taſte, lyke the kernel of the tame Pine.

D   The fourth kind hath a long hard fruite, the which will not open eaſily nor fall lightly from the tree.

E   The fifth kinde hath ſmall rounde nuttes, not muche greater then Cypres nuttes, the whiche wil open and fal quickly.

From out of theſe trees commeth that liquor called Roſen, eſpecially from the wilde trees: and it runneth moſt cómonly out of the barke, or from the timber, but ſometimes alſo it is founde in the fruite or apples.

Frō theſe trees alſo cōmeth pitch both liquid & hard, ye which is drawē forth by burning of ye wood, as Theophraſtus teacheth. The which maner of drawing or melting of Pitch, is yet vſed in Candie, as Peter Belon writeth.   ❦ *The*

The Place.

1 The tame or garden Pine groweth in many places of Italy, Spayne, Grece, France, and England, in feeldes and gardens, whereas it hath ben planted.
2 The wilde Pines grow vpon mountaynes, & some of them vpon the highest mountaynes, whereas none other trees nor herbes doo growe, especially the first wilde kinde, the whiche is also founde in Douchlande, Liefland, Poland, and other colde regions.

The Tyme.

The fruite or Pine apples, are ripe in September.

The Names.

The Pine is called in Latine, Pinus: in Greeke not πίτυς, as diuers of our later writers do suppose, but πεύκη, as it is euident by Virgils verse in his vij. booke of Aeneidos.

Ipsa inter medias flagrantem feruida Pinum
Sustinet. Where as is to be vnderstanded by Flagrantem Pinum, teda pinea, as Seruius writeth.
  Ouidius in epistolis Heroidum.
Vt vidi, vt perij, nec notis ignibus ignibus arsi:
Ardet vt ad magnos pinea teda deos.
Item fastorum quarto.
Illic accendit geminas pro lampade Pinus.
 Hinc Cereris sacris nunc quoq; teda datur.
  Prudentius in hymno cerei pascalis.
Seu pinus piceam fert alimoniam.

By whiche verses one may knowe, that Teda commeth of the tree called in Latine Pinus, into the whiche, as Theophrastus writeth, πεύκη, Peuce is translated: so that by this one may knowe that Pinus, and Peuce, is but one tree.

1 The tame or garden kinde is called in Greeke, πεύκη ἥμερος: in Latine, Pinus satiua: in English, the garden Pine: in French, Pin: in high Douch, Hartzbaum, and Kinholtz: in base Almaigne, Pijnboom.

2 The wilde kinde is called in Greeke πεύκη ἀγρία: in Latine, Pinus syluestris, & Pinaster: in English, the wilde Pine: in French, Pin sauuage: in base Almaigne, Wilde Pijnboom: of the kindes of wilde Pine are those trees which be called in high Douche, Kijfferholtz, Forenholtz, or Fuerenholtz, Fichtenbaum, &c.

The first wilde kinde is called in French, Aleue, and Elue (as Peter Belon writeth) and it seemeth to be Pinus Tarentina, whereof Plinie writeth.

The seconde is called in Italian, Mughi: and it may be called in Greeke χαμαιπεύκη: in Latine, Humilis Pinus, or Pinus terrestris.

The thirde is called in some places, in the mountaynes betwixt Italy and Germanie, Cembri, & Cirmoli. This seemeth to be that (as Peter Belon writeth) whiche the Frenche men call Suiffe. This is not Sapinus, for Sapinus is the neather part of the stemme or tronke of the Firre tree, as we shall write hereafter.

The fourth is muche lyke to that, whiche Theophrastus calleth in Greeke πεύκη ἰδία: in Latine, Pinus Idea.

The fifth is called of Theophrastus πεύκη παραλία, that is to say, in Latine, Pinus marina: in Frenche, Pin marin.

The fruite of the Pine is called in Greke κῶνος: in Latine, Conus, and Nux Pinea: in Englishe, a Cone, or Pine Apple: in Frenche, Pomme de Pin: in high Douche, Zijrbel: in base Almaigne, Pijnappelen.

The Nuttes which are found in the Pine apples, are called in Greke στρόβιλοι, and of Hippocrates κογχάλοι: in Latine, Nuces pineæ: in Englishe, Pine apple kernels or Nuttes: in Frenche, Pignons.

the Historie of Plantes.

The hart or the midle of the timber which is full of liquor, and being kindled or burned lyke a tortche is called in Greeke δᾶς καὶ δάδιον: in Latine, Teda: And when the whole tree, or inner substance thereof, is become so fat and full of liquor, then Theophrastus saith, that it is changed into Teda: and then it dyeth bycause it is so full of fat or Oyle, euen lyke to a man or beast that is stuffed or rather stifled in grease and fat: and then are the said trees best, for the peelding or drawing foorth of the Pitche. Wherefore they be muche deceiued, that take Teda to be a kinde of tree by it selfe, and do not rather knowe it to be a kinde of corruption or maladie, insident to the Pine tree.

※ *The Nature.*

The barke of the Pine tree is drie and astringent, especially the scales of the Cones or apples, and the leaues be almost of the same complexion.

The kernell of the Nuttes is hoate and moyst, and somewhat astringent.

※ *The Vertues.*

The scales of the Pine apple with the barke of the Pine tree, do stoppe the laske, the bloody flire, and prouoke vrine, and the broth of the same dronken, hath the lyke propertie. A

The same is also good against al scorchinges and burninges with fire, to be pounde with the lytarge of syluer and frankensence: and if there be some Copperas mirt therewith, it will clense and heale consuming or fretting sores. B

The leaues of the Pine tree healeth greene woundes, & boyled in vineger, they swage the toothe ache. C

The kernels of the Nuttes which are founde in the Pine apples, are good for the lunges, they clense the breast, and cause the fleme to be spet out: also they nourish wel & ingender good blood, & for this cause they be good for suche as haue the cough, and begin to consume and drie away, in what sort soeuer they be taken. D

This fruite also doth vnstop the liuer and the milt, mitigateth the sharpnesse of vrine, and therfore is good for them that are troubled with the grauell & the stone.

The vertues of the Rosen and Pitche, shalbe declared hereafter.

Of the Pitch tree. Chap.lxxxviij.

※ *The Description.*

THe Pitche tree is also of an indifferent bignesse and talle stature, but not so great as the Pine tree, and is alwayes greene lyke the Pine and Firre trees, his timber or wood is nothing so red as the Firre tree. It is also fat and Roseny, peelding Rosen of diuers sortes. The branches be harde and parted into other spraies, most commonly crosse wise, vppon whiche growe small greene leaues, not round about the branches, but by euery syde, one right ouer against another lyke to litle feathers. The fruite is smaller then the fruit of the Pine tree.

Picea.

※ *The*

In burning of this wood there floweth out Pitche, euen lyke as out of the Pine tree, as witnesseth Dioscorides.

❧ The Place.

This tree groweth in many places of Grece, Italie, France, and Germany.

❧ The Tyme.

The fruite of this tree is also ripe in September.

❧ The Names.

This tree is called in Greeke πίτυς: in Latine, Picea: in high Douche, Rot thannen, and Rot dannebaum, that is to say, the red Firre tree: and accordingly in neather Douchlande, it is called Roode Denneboom.

That Pitys and Picea, are but one kinde of tree, Scribonius Largus doth sufficiently declare who in the CCj. Composition writeth after this manner. Resinæ pituinæ, id est, ex picea arbore.

❧ The Nature and Vertues.

The leaues, barke, fruite, kernelles or nuttes of this tree, are almost of the same nature, vertues, and operations, as the leaues, barkes, fruite, & kernels of the Pyne tree.

Of Rosen that commeth out of the Pine and Pitche trees. Chap. lxxxix.

❧ The Kindes.

1 The Rosen that runneth out of the Pine & Pitche trees is of three sortes, bysides the Pitche, which we will describe by it selfe in the next Chapter.

The one floweth out by force of the heate of the Sonne in the sommer time, from the wood or timber when it is broken or cut, but especially when it is cut.

2 The other is found both vpon and betwixt the barke of the Pine & Pitche trees, and most commonly whereas it is clouen or hurt.

3 The thirde kinde groweth betwixt the scales of the fruite.

❧ The Names.

All the kindes of Rosen are called in Greeke ῥητίνη: in Latine, Resina: in English, Rosen: in Frenche: *Resine*: in Douche, Herst.

1 The first kinde is called in Greeke ῥητίνη ὑγρά: in Latine, Resina liquida: in shops of this Countrie, Resina Pini: in base Almaigne, Rijnschen, or moruwen Herst, that is to say, liquid Rosen.

Of this kind is that Rosen called of the Ancients in Greeke ῥητίνη κολοφωνία: Resina Colophonia, whiche was so called, bycause in time past they brought it from Colophon (a Citie of Ionia in Greece, where was the temple of Apollo, called Clarius, and Homer the famous Poet was borne.) But nowe the ignorant Apothecaries, in y steede therof do vse a kind of drie Pitche to the great hurt of them that are greeued.

Of this sort is also the Rosen which the Brabanders do cal Spieghelherst, the which is molten with the Sonne in sommer, and remayneth drie, and may be made into powder: some call it Resina arida, that is to say, Dry Rosen, yet this is not Resina arida of the Ancientes.

2 The seconde Rosen is called in Greeke ῥητίνη ξηρά: in Latine, Resina arida: but that whiche sweateth out of the Pine tree, is called in Greeke ῥητίνη πυκίνη: Resina pinea: and that whiche commeth out of the Pitche tree, ῥητίνη πιτυίνη, Resina picea, and Resina piceę, of some Spagas, as witnesseth Plinie.

These two kindes of Rosen, and also the drie Rosen that sweateth out of the firre tree, are now a dayes without discretion, sold in shops for great incense, and

and is called of the ignorant Apothecaries Thus, of some Garipot, and they cal the right incense Olibanum, not knowing how that Olibanum, which is called in Greeke λίβανο., and Thus, be but one thing.

3 The thirde Rosen is called ῥητίνη ςροβιλίνη, Resina strobilina, this kinde vntill this time hath bene vnknowen in shoppes.

✤ *The Nature.*

All the kindes of Rosen are hoate and drie, of a clensing & scouring nature.

✤ *The Vertues.*

Rosen doth clense and heale newe woundes, therefore the same is for the A most part mengled withe all oyntmentes and emplaysters, that serue for newe woundes.

It softeneth hard swellinges, and is comfortable to brused partes or mem B bers, being applyed or layde to with oyles and oyntmentes agreable.

Of Pitche and Tarre. Chap.rc.

✤ *The Kindes.*

There be two sortes of Pitche: the one moyst, and is called liquid Pitche: the other is hard & dry, they do both run out of the Pine and Pitche trees, and of certayne other trees, as the Cedar, Turpentine, and Larche trees by burning of the wood and tymber of the same trees as apparteyneth, and as it hath bene before expressed.

✤ *The Names.*

Pitche is called in Greeke πίσσα: in Latine, Pix: in Frenche, Poix: in Douche, Peck.

1 The liquid Pitche is called in Greeke πίσσα υγρά: in Latine, Pix liquida: in Brabant, Teer: in Frenche, Poix de Bourgongne: in Englishe, Tarre.

2 The drie Pitche is called in Greke ξηρά πίσσα, παλιμπίσσα, καὶ πίσσα ξηρά: in Latine, Pix arida: in shoppes, Pix naualis: in English, ship Pitche, or stone Pitche: in Frenche, Poix seche: in base Almaigne, Steenpeck.

✤ *The Nature.*

Pitche is hoate and drie in the second degree, and of meetely subtile partes, but the stone Pitche is dryest, the liquid Pitche or Tarre is the hoater and of more subtil partes.

✤ *The Vertues.*

Liquid Pitche (as witnesseth Dioscorides and Galen) taken with honie, A doth clense the brest and is good to be licked in of those that haue the shortnesse of breath, whose brest is stuffed with corrupt matter.

It mollifieth and ripeth all harde swellinges, and is good to annoynt the B necke against the Squinansie or swelling of the throte. To be short, it is good to be put into softening playsters, anodines whiche take away payne & griefe, and maturatiues or riping medicines.

Layde to with Barley meale, it suppleth and softeneth the hardnesse of the C matrix and fundement.

Liquid Pitche mingled with Sulphur (or quicke Brimstone) repressseth fret-D ting vlcers and the noughtie scab, & foule scurffe, & if that salt be put thereto, it is good to be layd vpon the bytinges & stingings of Serpentes and Vipers.

It cureth the riftes and clouen chappes, that happen to the hands, feete, and E fundement, to be layde thereunto.

If it be powned very small with the fine powder of Frankensence it healeth F holowe vlcers or fistulas, filling them vp with flesh.

The stone or drie Pitche, hath the same vertue as the liquid Pitche, but not G so strong: but it is better, and apter to glew togither woundes, as Galen saith.

Of the Firre tree. Chap.rci.

The Description. Abies Firre.

THE Firre tree is great, high, & long, euer greene, growyng muche hygher then the Pine and Pitche trees. The stem is very euen or straight, plaine beneath, & without ioyntes, but with ioyntes and knoppes aboue, vpon whiche ioyntes grow the branches bearing leaues almost lyke Ewe, but smaller. The fruite is lyke to the Pine apple, but smaller and narrower, not hanging downe as ȳ Pine apple, but growing right vpward. With the timber of this tree they make Mastes for shippes, postes, and rayles for diuers other purposes.

B Fro out of the barke of ȳ young Firre tree is gathered a faire liquid Rosen, cleare & through shyning as the learned Matthiolus, and Peter Belon haue written, which is bitter and aromatical, in taste almost lyke to Citron pilles, or the barkes of Lemons condited.

C Also there is founde vpon this tree a Rosen or dry white gumme, lyke as there is founde vppon the Pine and Pitche trees, the whiche is solde for Thus, that is to say Francense, and so is esteemed of the common sorte.

The Place.

The Firre tree groweth vpon mountaynes: & is not only founde in Grece, Italy, Spayne, and Fraunce, but in Pruse, Pomeran, Liessande, and diuers other places of Germanie.

The Names.

This tree is called in Greeke ἐλάτη: in Latine, Abies: in Englishe, Firre: in Frenche, Sapin: in high Douche, Weſz Thannen, and Weiſz Dannenbaum: in base Almaigne, Witte Denneboom, and Mastboom.

The lower part of the stem of this tree whiche is without knots or ioyntes, is called in Latine, Sapinus, and the vpper part whiche is full of ioyntes and knottes, is called Fusterna, as witnesseth not onely Plinie, but also Uitruuius in his seconde booke of Architecture, or buyldinges.

B The liquid and cleare Rosen, running out of the barke of the young trees is called of the later writers Δάκρυον τῆς ἐλάτης: Lachryma abietis, Lachryma abiegna, and of some Abiegna resina liquida, and Abiegnum oleum: In Italian, Lagrimo: in Shoppes of this Countrie, Terebinthina veneta, and is solde for the right Turpentine: in English, Turpentine of Uenice: in French, Terebinthine de Venise: in base Almaigne, Veneetsche Terebenthijn, therebe some that thinke this Rosen to be ἐλαιώδης ῥητίνη, Oleaosa resina of Dioscorides.

the Historie of Plantes.

The drie white Rosen, is called ξητινη ελατινη, Resina abiegna, and is also solde in Shoppes for Thus, and Garipot, lyke the drie Rosen of the Pine tree.

⁋ The Nature.

The barke, as also the drie gumme or Rosen of this tree, are in nature and vertues, lyke to the barke and drie Rosen of the Pine tree, sauing that they be somewhat more aygre, and clensing.

The liquid or cleare Rosen, is hoate and drie in the seconde degree, and bycause of his aygre or sharpe qualitie, it hath a digestiue and clensing nature.

❋ The Vertues.

The cleare liquid Rosen of the Firre tree, taken about the waight of halfe an ounce, looseth the belly & driueth foorth hoate cholerique humours: it doth clense and mundifie the hurt kidneyes, prouoketh vrine, and driueth foorth the stone and grauel, and is good to be receiued oftentimes of such as are troubled with the gowte.

The same taken with Nutmegge and Sugar in quantitie of a nut, cureth the strangurie, and is very good against the excoriations, and going of, of the skinne, or fluxe of the secrete partes.

It is also excellent for all greene or fresh woundes, especially the woundes of the head, for it healeth and clenseth very muche.

Of the Larche or Larix tree. Chap.xcij.

❋ The Description.   Larix.

The Larix tree is great and thicke, spreading abroade his slender boughes or branches, whiche are very plyāt or limmer. The timber is reddish, thicke, waightie, and very hard, insomuch as fire cannot do it much harme, except it be burned in ye Furnis with other wood like chalke or white stone. The barke of this tree is smoother, then ye barke of the Firre tree. The leaues be greene, and small iagged, growing thicke togither in tuftes lyke tasselles, and do fall of, at the comming of winter. The fruit is like to Pine apples, sauing that it is muche smaller, and not muche greater then Cypres Nuttes.

From this tree commeth foorth a liquor, Rosen, or gumme, whiche is softe, moyst, whitishe, and darke, in substance lyke hony of Athens, as Vitruuius writeth.

There groweth in this kinde of trees a kinde of Mushrome or Tadstoole, that is to say, a fungeuse excrescence, called Agaricus, or Agarick, the whiche is a precious medicine and of great vertue. The best Agarick is that, which is whitest, very light, britle, and open or spongious. That which is otherwise, that is to say, blacke, thicke, close, clammie, and waightie, is not meete for medicine, but vnholesome and venemous.

❧ The Place.

This tree groweth in Lombardie, alongst by the riuer Padus, and in Silesia plentifully.

❧ The Tyme.

This tree hath newe leaues at the beginning of the spring time. The fruite is ripe in September.

❧ The Names.

This tree is called in Greeke λάριξ: in Latine, Larix: in some Shops, Larga: in high Douche, Lerchenbaum: in base Almaigne, Lorkenboom.

The Rosen of this tree is called in Greeke ῥητίνη λαρικίνη, ἡ λάριξ: in Latine, Resina laricea, and Resina larigna: in Shoppes, Terebinthina, not without errour, also in Douche it is called, Termenthijn, or Terbenthijn, that is to say, Terebinthin, or Turpentyn, & this is the common Turpentyn that we haue, whiche should rather be called Larche Rosen, or Larche Turpentyne.

The spongie excrescence whiche is founde in the Larche tree, is called in Greeke ἀγαρικόν: in Latine, Agaricum: in Shoppes, Agaricus: of some, Medicina familiæ: in Englishe, Agarik: in Frenche, *Agaric*.

❧ The Nature.

The Larche tree, his leaues, fruite, barke, and kernell, are of temperature almost lyke to the Pine and Firre trees, but not althing so vertuous, neyther yet so strong.

The Rosen of this tree is hoate and drie lyke the other Rosens, but it doth mundifie and clense better then the rest.

Agarick is hoate in the first degree, and drie in the seconde.

❧ The Vertues.

The Rosen of the Larche, or Larix trees, is as good as any of the other **A** Rosens, to be put into oyntmentes and implaisters, to glewe togither, clense and heale woundes.

To be licked in with hony it clenseth the breast, and looseth the belly, prouo- **B** keth vrine, and driueth out the stone and grauell, to be taken inwardly: to be short, it is of facultie very lyke to the right Turpentine, and may be vsed for the same, as Galen writeth, lib. de medicamentis secundum genera.

Agarick taken about the weight of a dramme, purgeth the belly from colde **C** slimie fleme, and other grosse and raw humours, whiche charge and stoppe the brayne, the sinewes, the lunges, the breast, the stomacke, the liuer, the splene, the kidneyes, the matrix, or any other the inwarde partes.

Agarick is good against the payne and swimming of the head, the falling **D** euill, and the impostumes of the brayne, to be taken with Syrupe Acetosus.

It is good against the shortnesse of breath called Asthma, ye hard continuall **E** cough or inueterate cough, it is good also for suche as haue taken falles, & are brused or squatte or hurt, or bursten inwardly, to be dronken in honyed wine when one hath no feuer, and with honyed water in a feuer.

It is also giuen with great profite, sodden in sweete wine to suche as haue **F** the Tysicke and consumption, and to them that spet and cast forth blood, when they loose their bellies, or go to the stoole.

It openeth the stoppinges of the liuer, and kidneyes, and preuayleth much **G** against the Iaundise, and suche as are euill coloured, for it putteth away the noughtie colour, and restoreth the faire naturall colour.

If it be taken with vineger, it openeth and cureth the stopping and harde- **H** nesse of the melt or splene.

The same taken drie without any liquor, dooth strengthen and comfort the **I** weake

weake and feeble stomacke, it cureth the wamblinges of the stomacke, and the sower belching out of the same, causing good digestion.

Agarick is a good medicine agaynst olde feuers, for to purge the body, and against wormes: it is also very profitably put into medicines, that are giuen against poyson or venome.

The Daunger.

Agarick is of slowe operation, and taken into great a quantitie, it feebleth the inwardes partes.

The Remedie.

Agarick is corrected, to be giuen either with Ginger, Sal geme, but chiefly with Oximell.

Of Turpentyne tree. Chap. xciij.

The Description.

Terebinthus.

THE Turpentine tree in some places is but shorte and base: and in some places it waxeth great and high, as Theophrastus writeth. It hath long leaues cósisting of many other leaues like to Baye leaues, growing one agaynst another alongst by one stem. The flowers be smal & reddish growing togither lyke grapes, afterwarde there come small rounde beries at the first greene, but afterward reddish, and when they be rype, they be blacke, clammie, or fat, and of a pleasant sauour. The rootes be long and growe deepe in the grounde, the timber is faire, blacke and thicke.

Out of this tree issueth the right Turpentine, the whiche is faire and cleare, thicker, then the liquid Rosen, whiche is gathered frō the barke, of the Firre tree.

The Place.

The Turpentine groweth plentifully in Syria, especially about Damascus, where as it waxeth very great: it is also found in Greece, and in some places of Italie and Languedocke.

The Time.

The Turpentine tree flowreth in the spring time, and is ripe about the end of sommer, euen with the grapes.

The Names.

This tree is called in Greeke τέρμινϑος· in Latine. Terebinthus: in Englishe, Turpentine tree: in Frenche, *Terebinthe*: in base Almaigne, Terebinthijn boom: and of the Arabian Physitions, Albotin.

The fruite is called of Auicen, Granum viride.

The gumme of Rosen is called in Greeke ῥητίνη τερμινϑίνη· in Latine, Resina Terebinthina: of Auicen Gluten albotin: vnknowen in ẏ shops of this countrie.

The Nature.

The leaues & barke of the Turpentine tree are hoate and drie in the seconde degree (especially being wel dried) they be also astringent.

The fruite is hoate and drie in the thirde degree.

The Rosen of this Turpentine, is hoate in the second degree, but not ouermuche drying: also it is clensing.

The Vertues.

The leaues and barke of the Turpentine tree, do stoppe the spetting foorth of blood, the bloody flire, and womens flowers. To conclude they be of power lyke to Acatia, and the leaues and barke of Lentiscus or the Masticke tree, if they be taken in lyke manner.

The fruite of this tree prouoketh vrine, stirreth vp fleshly lust, and is good against the bitinges of the feelde Spider, to be dronken in wine.

The Rosen of this tree whiche is the right Turpentine, looseth the belly, openeth the stoppinges of the liuer and melt or splene: it clenseth the kidneyes, prouoketh vrine, and driueth out grauel taken in the quantitie of a Walnut, as Iuicen writeth.

Turpentine in a lectuarie with hony, clenseth the breast and the lunges, ripeth flemes, and causeth the same to be spet out.

Turpentine is also good against the wilde scurffe, and euilfauoured manginesse & chappes or cliftes of the face: And it is much occupied in all emplaisters that serue to make smooth and soften.

Turpentine mingled with oyle and hony, is good to be dropped into the eares, against the matter running out of the same.

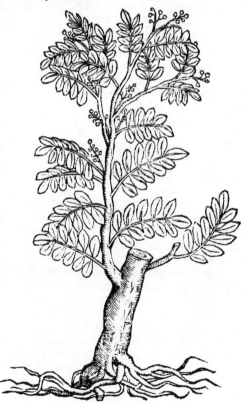

Lentiscus.

Of Lentiske, or Masticke tree. Chap.xciiij.

The Description.

THE Lentiske tree in some places is of a reasonable bignesse and stature, lyke to a tree of a meane sise: & in some places it putteth forth diuers springs or shutes from the roote lyke as the Hasel: the leaues which remaine alwayes greene, are lyke to the leaues of the Turpentine & Ashe tree, many growing togither alongst one stemme, but smaller then Turpentine leaues, of a darke greene colour and thicke. The barke is tough and plyant. The fruite is rounde, & groweth in clusters like the fruite of Turpentine.

Out of the Lentiske tree commeth foorth the noble and excellent gumme or Rosen called Masticke, the whiche is not liquid, neither growing togither as the other

other Rosens: but remayneth asunder in diuers smal graynes or partes, of the quantitie of wheate cornes. wherof the best Masticke is brought from the Ile Chio, it is faire, cleare, white, britle, and of a sweete sauour.

❧ The Place.

The Lentiske tree groweth abundantly in the Ile of Chios, whereas it is very wel husbanded and trimmed, for the Masticke whiche commeth from it: it is founde also in Italy, and certayne other Regions, but it yeeldeth very litle or no Masticke at all.

❧ The Names.

This tree is called in Greeke Χῖνος: in Latine, Lentiscus: in Englishe, the Lentiske, or Masticke tree: in Frenche, Lentisque, and Mastix: in Douche, Mastixboom.

The gumme or Rosen is called in Greeke ῥητίνη Χινίνη, καὶ μασίκη: in Latine, Resina Lentiscina, and Mastiche: in shoppes, Mastix.

❧ The Nature.

The leaues & barke of the Lentiske tree are of a meane or temperate heate, and are drie in the seconde degree, and somewhat astringent: and of the same temperature is the Masticke or gumme.

❧ The Vertues.

The leaues and barke of the Lentiske or Masticke tree stoppe the common laske, the bloody flire, the spetting of blood, the pissing of blood, the flure of the mother, and al other flure of blood: they be also good agaynst the fallyng downe of the mother and fundement.   A

The Masticke is also good against the spitting of blood, the common laske, the bloody flire, the inordinate flure of the Matrix, and all other issue of blood proceading from any parte of the body whatsoeuer.   B

Mastick is very goood for the stomacke, for it strengtheneth the same, and stayeth vomiting, swaging all the payne and greefe of the same, and reuiueth the appetite whiche was dulled.   C

The same holden in the mouth and chewed vppon, dooth drie and comfort the brayne, stayeth the falling downe of humours, and maketh a sweete breath.   D

They vse to rubbe the teeth with Masticke, to whiten the same, and to fasten them that be loose, and to comfort the iawes or gummes whiche be loose and weake.   E

The ende of the sixth and last parte of the Historie
of Plantes.

Bbb iiij   Index

Index Latinorum nominum, quibus Veteres & Viri docti in stirpium herbarumq; appellationibus, vtuntur.

Abiegnaresina liquida	774	Albucus	649	Anthropomorphos	438	Asphodelus	203.647.649
Abiegnum oleum	ibid.	Alcea	584	Anthycellon	13	Asphodelus foemina	647
Abies	750.774	Alcea Veneta	585	Anthyllion	ibid.	Asplenium	408
Abiga	28	Alchimilla	140	Anthyllis	13.500	Asplenium magnum	407
Abrotonum	2	Alcibiacum	10	Anthyllis altera	13	Asplenium sylueſtre	ibid.
Abrotonum foemina	ibid.	Alectorolophos	516	Anthyllis prior	116	Asplenum	408
Abrotonum mas	ibid.	Algæ	414	Anthyllon	13	After Atticus	36
Abrotonum siculum	ibid.	Alisma	334	Antirrhinum	180	Athanatos	158
Absynthium latifolium	5	Alliaria	639	Antimalum	438	Atractilis	532
Absynthium marinum	ibid.	Allium	637	Anydrou	448	Atractilis hirsutior	533
Absynthium Ponticum	ibid.	Allium anguinum	ibid.	Anydros	211	Astragalus	491
Absyathium Græca	ibid.	Allium satiuum	ibid.	Aononium	151	Atriplex	546
Absynthium seriphium	ibid.	Allium sylueſtre	ibid.	Anthericos	649	Atriplex hortenſis	ibid.
Absynthium ruſticum	ibid.	Allium vrsinum	638	Aparine	539	Atriplex satiua	ibid.
Absynthium santonicum	ibid.	Alnus	756	Aphace	485	Atriplex sylueſtris	ibid.
Acaaa	685.721	Alnus nigra	719	Aphedron	532	Auena	467
Acatia altera	ibid.	Aloë	353	Aptaria	658	Auena ſterilis	ibid.
Acatia pontica	ibid.	Aloë Gallica	332	Apiaſtrum	259.420	Auena herba	505
Acantha	527	Alopecuros	542	Apios	491.579	Aurelia	89
Acanthis	530	Alsine	52	Apium	605	Aureum malum	704
Acanthium	526	Alsine marina	ibid.	Apium hortenſe	ibid.	Aureum olus	546.561
Acanthus	527	Althæa	583	Apium montanum	607	Auricula leporis	63.506
Acanthus Germanica	528	Alyſſum	107.538	Apium paluſtre	420.606	Auricula muris	84.342
Acanthus sylueſtris	526	Almos	694	Apium ruſticum	420.666	**B**acca Iuniperi	763
Acatalis	763	Amaracus	19.234	Apium risus	ibid.	Baccæ renana	697
Acer	760.762	Amara dulcis	398	Apium saxatile	608	Bacca	671
Acer campeſtris	762	Amaranthus	168	Apium sylueſtre	420.610.617	Baccinum	ibid.
Accipitrina	567	Amaranthus luteus	84	Apolinum	369	Balauſtium	706
Acetabulum	38	Amaranthus purpureus	168	Apollinaris minor	447	Ballote	217
Acetabulum alterum	ibid.	Ambrosia	243	Apollinaris	450	Balaris	59
Achillea	18.144	Ambubeia	563	Aquifolia	701	Balſamita maior	250
Achillea sideritis	144	Ambuxum	386	Arabis	629	Balſamita minor	ibid.
Achimilla	140	Ammi	273	Arachus	485.484	Barba capri	48
Acydonium	153	Ammoniacum	308	Aracus	478.485	Barba hirci	167
Acinos	240	Ampeloprasum	638	Arbor Iuda	742	Barbula hirci	ibid.
Aconitum lycoctonum	429	Amygdala	711	Arbutus	728	Basilica	332
Aconitum pardalianches	426	Amygdalum	ibid.	Arcium	15	Basilicum	240
Acorum	198	Amygdalus	ibid.	Argemone	434	Batis	578
Acorus	514	Amyrberis	684.698	Ariene	705	Batrachij genera	419.421
Acron syluaticum	144	Amyrum	532	Aris	325	422. & inde.	
Aculeosa	521	Anagallis aquatica	578	Arisaris	ibid.	Batrachium Apulei	428
Acorna	531	Anagallis sylueſtris	180	Arisarum	ibid.	Bellis	170
Acula	615	Anagallis vtraq;	55	Ariſtaltea	383	Bellius	ibid.
Adianthana	409	Anagyris	742	Ariſtolochia genera	313.314	Bene olens	183
Ægilops	470	Anadendron	583	Aronia	714	Beta	550
Aglaophotis	338	Anarantium malum	704	Arthemiſia genera	16.18	Beta candida	ibid.
Æthiopicum seſeli	283	Anchuſa genera	9	Arum	323	Beta nigra	ibid.
Æthiopis	124	Anchuſa alcibiadium	ibid.	Arum maculatum	322	Beta nigra Romana	ibid.
Agnos	690	Anchuſa onochiles	ibid.	Arum paluſtre	ibid.	Betonica	26.291
Agaricum	776	Androſæmum	65.&66	Arundinis genera	514	Betonica Pauli	26
Ageratum	89.250	Anemone	423	Asarum	319	Betula	758
Agrosselinum	609	Anetum	270	Asclepias	317	Bipenula	138
Agriocinara	525	Anisum	276	Asyrum	65	Biſacutum	615
Agriocaſtanum	579	Ansnis	669	Asa	459	Biſtorta	23
Aiuga	28	Anonium	131	Aspalathum	346.685	Blattaria	122
Al num	90	Anthema	581	Asparagus	618	Blephara	182
Alabandica rosa	655	Anthemon	160	Asparagus sylueſtris	ibid.	Blitum	163.547
Albucum	649	Anthemon folioſum	ibid.	Aſſidion	532	Bolbocaſtanon	579
							Bonus

INDEX.

Bonus genius	298	Cantabrica	156	Cerasus	ibid.	Cissarum	850
Botrys	243	Capillaris	409	Ceratonia	740	Citrocation	361
Brabyla	721	Capillus Veneris	ibid.	Cerui spina	697	Citharon	659
Brassica genera	554 & inde	Capita rosarum	655	Cestrum	369	Citrago	259
Brassica marina	395.555	Capnium	24	Cherephyllum	614	Citrium malum	704
Brassica rustica	555	Capnos phragmites	23.316	Charophyllum	ibid.	Clauicula	388
Brassica syluestris	ibid.	Capnos Plinij	24	Chamebalanus	317.491	Clauus Veneris	181
Britannica	118	Capparis	680	Chamabatus	661	Clematis Ægyptia	33
Briza	459	Capraria	115	Chamecerasus	724	Clematis altera	386
Bromus	505	Caprificus	717	Chamæcissus	389.393	Clematis daphnoides	33
Bromus herba	ibid.	Capsella	628	Chamæcyparissus	29	Cleoma	425
Bruta	766	Capsici genera	633.634	Chamedaphne	33	Clethra	756
Bruthes	ibid.	Cardamantice	630	Chamadrys	25	Clynopodium	238
Bryonia alba	382	Cardamum hybernum	626	Chamairis	195	Cnecus	33
Bryonia nigra	ibid.	Carduus	522	Chamelea	369.370.371	Cnecus syluestris	132
Bryonia syluestris	384	Cardui syluestres	535	Chamelæa Germanica	ibid.	Cneoron	369
Bucinum	165	Carduus irinus	517	Chamelæa tricoccos	369.371	Cnicus	33
Buglossum	12	Carduus lacteus	ibid.	Chameleo vterq;	369.371	Cnidios coccos	517
Buglossum magnum	568	Carduus niger	ibid.	Chamameli genera	186	Cnipes	752
Bulbina	579	Carduus ramptarius	525	Chamamelum	173	Coccos gnidius	370
Bulbosa iris	198	Carduus satiuus	523	Chamamelū albū	183.186.517	Coscos	361.517
Bulbus	643	Carduus syluaticus	517	Chamamelum luteum	183.187	Caccus cnidius	ibid.
Bulbus agrestis	367	Carduus varinus	ibid.	Chameleo niger	517	Colchicum	367
Bulbus esculentus	643	Carduus Veneris	522	Chamamelum purpureū	ibid.	Colocasion	323
Bulbus littoralis	646	Careum	274	Chamapytis	28	Colocyntis	375
Bulbus syluestris	644	Carica	317.717	Chamamelum syluestre	186	Colubra	321
Bulbus vomitorius	211	Carpinus	755	Chamapelon	630	Columbaris	127
Bunium	287	Carum	274	Chelidonium vtranq;	31	Columbina vtraq;	ibid.
Buphthalmum	186.189.190	Caryotes	361	Chiliodynamis	334.345	Colophonium	396
Bupleurum	506	Casia	295	Chiliophyllon	144	Colus rustica	532
Buprestis	544	Casia nigra	740	Chironia	327	Colutea	736
Butomon	515	Cassytha	398	Chironia vitis	382	Colutea	ibid.
Buxus	699	Castanea	730	Chironion	332	Colymbada	738
Buxus asinina	700	Castor	216	Chrysanthemum	183.187.190	Colymbades	ibid.
Bytren	628	Catanance	508		420	Colytea	742.752
CAcalia	137	Caucalis	612	Chrysisceptrum	517	Coma	167
Cachrys	280	Caudamurina	96	Chrysitis	144	Condrilla	95.169
Cactos	768	Cauda muris	ibid.	Chrysemelon	704	Condrillis	ibid.
Calathiana viola	173	Cauda vulpina	542	Christophorina	382	Condrillon	ibid.
Calamagrostis	515	Caulias	304	Chrysolachanon	546.561	Coniugulum	393
Calami genera	514	Cedrelate	765	Cicer 479	Cicera 478	Coniugulum nigrum	ibid.
Calamintha	248	Cedria	ibid.	Cicercula	ibid.	Consiligo	189
Calamus odoratus	514	Cedromelon	704	Cicer columbinum	479	Consiligo Plinij	351
Calcifraga	116	Cedrula	765	Cicer satiuum	ibid.	Conuoluolus	394
Calices glandium	745	Cedrus	ibid.	Cicer syluestre	ibid.	Conus	770
Calendula	163	Cedrus conifera	ibid.	Cicer veneteum	ibid.	Conyza	35.575
Calicularis	450	Cedrus maior	765	Cichorium	563.569	Conyza mas	35
Calix rosarum	655	Cedrus Lycia	ibid.	Cichorium satiuum	563	Conyza fœmina	ibid.
Callion	443	Cedrus Phœnica	ibid.	Cicuta	451.616	Coriandrum	276
Callitrichum	409	Cedrinus liquor	765	Cinara	522.523	Cornu ceruinum	95
Caltha	163.189	Cedrus acuta	ibid.	Cinara acuta	524	Cornus	726
Calsbula	163	Celtis	729	Cinara rustica	522	Cornus fœmina	ibid.
Cammaron	426	Centauris	376	Cincinalis	127.409	Corona terra	389
Campana Rosa	655	Centaurium maius	327	Cinosbatos	680	Coronopodium	351
Campsanema	280	Centaurium minus	327.375	Circea	418	Coronopus Ruellij	95
Canchrys	ibid.	Centumcapita	519	Cirsium	368	Coronopi species	ibid.
Cania	129	Centuncularis	89.90	Cissanthemon	386	Corchorus	55
Canina sentis	698	Centunculus	ibid.	Cissophilon	ibid.	Corruda	618
Canina brassica	77	Cepa	640	Cistus	659	Corylus	734
Canirubus	656	Cepea	578	Cistus non ladanifers	ibid.	Cortices rosarum	655
Cannabis vtraq;	72	Cerasa	724	Cistus ladanifera	660	Costus niger	382

Co-

INDEX.

Cotyledon	38	Cynomorphos	216	Empetrum	116	Fagotriticum	468
Cotyledon aquatica	ibid.	Cynosorchis	222	Enoron	448	Fagus	757
Cotonea	768	Cyonos Phryce	283	Epabsynthion	399	Farclusinum	461
Crapula	527	Cynomazon	517	Eperingium	ibid.	Far	455.456
Crassula maior	39	Cynoxilon	ibid.	Epibaton	ibid.	Far venniculum album	455
Crataogonum	334.506	Cynozolon	ibid.	Epibryon	ibid.	Far venniculum rutilum	459
Crataonum	506	Cyperus	346	Epichamædrys	ibid.	Farrago	ibid.
Crataus	ibid.	Cytini	706	Epigenistron	ibid.	Faselus syluestris	473
Crambe	554	Cytinus	659	Epigetis	386	Faselus vterq;	ibid.
Crespinus	684	Cytisus	666	Epilinum	399	Fascolus	474
Crinita	409	Cytonium	763	Epimelis	714	Feria	127
Crista	516	D Actylitis	314	Epipactis	349	Fegopyrum	468
Crista gallinacea	127	Damasonium	334	Epithymbrum	399	Ferraria	127
Crithmum	378	Daphnoides	368	Epithymum	398	Ferulago	301.365
Crocodilion 522	Crocus 216	Dardana	15	Epitithymalos	399	Ferula syluestris	ibid.
Cruciata	334.541	Dasmophon	628	Equapium	609	Ferrum equinum	490
Cuculus Indus	370	Daucia	285	Equinalis	101	Festuca	471
Cucurbita	592	Daucum	ibid.	Equiseta	ibid.	Festuca altera	ibid.
Cucurbita anguina	ibid.	Daucum Creticum	ibid.	Equisetum maius	ibid.	Ficus	717
Cucurbita barbarica	ibid.	Daucus	287.602	Equisetum minus	ibid.	Ficus Indica	544
Cucurbita camerarie	ibid.	Daucus pastinaca	602	Equitium	ibid.	Ficus satina	717
Cucurbita magna & maior ib.		Daucus syluestris	612	Erechtites	571	Ficus syluestris	ibid.
Cucurbita marina	ibid.	Delphinium	165	Erica	678	Fidicula	410
Cucurbita minor	ibid.	Delphinium alterum	ibid.	Erica altera	ibid.	Filago	89
Cucurbita oblonga	ibid.	Demetria	127	Erice	ibid.	Filix fœmina	401
Cucurbita perticales	ibid.	Denticulus canis	203	Erigerum	571	Filix mas	ibid.
Cucurbita satiua	ibid.	Dens caninus	ibid.	Erineus	718	Filix querna	405
Cucurbita syluestris	375	Dictamnum	268	Erithranon	659	Filicula	403
Cucumis agrestis	373	Dictamnum Creticum	ibid.	Eriphion	261	Fistularia	516
Cucumis anguinus	373.586	Dictamnum nō Creticum ibid.		Eriphia Plinij	316	Flamma	149
Cucumis citrulus	589	Diadema	581	Erithales	114	Flammula	386.425
Cucumis erraticus	373	Diodela	144	Eruangina	664	Flos amellus	36
Cucumis Galeni	590	Dionysia	388	Eruca	622	Flos amoris	168
Cucumis satiuus	586	Diosciamus	450	Eruca hortensis	ibid.	Flos Aphricanus	176
Cucumer satiuus	ibid.	Dipsacum vtrunq;	522	Eruca satiua	ibid.	Flos Indianus	177
Cucumis syluestris	373	Dirclon 447	Dolichus 474	Eruca syluestris	ibid.	Flos rosa	655
Culices	752	Donax 514	Donacitis 525	Eruilia	476	Flos tinctorius	68
Cuminum Æthiopicum	273	Draba 629	Draco	623	Eruilia syluestris	ibid.	Fœniculum	269
Cuminum Alexādrinum ibid.		Dracunculus maior	321	Eruum	482	Fœniculum erraticum	ibid.
Cuminum latum	295	Dracunculus Matthioli	322	Erynge	519	Fœnum Græcum	490.492
Cuminum regium	273	Dracunculus minor	ibid.	Eryngium	ibid.	Folia rosarum	655
Cuminum rusticum	275	Dracunculus palustris	ibid.	Eryngium marinum	ibid.	Fontalis	106
Cuminum satiuum	ibid.	Dracunculus Plinij	22	Eryphia	316	Fontinalis	ibid.
Cuminum syluestre	ibid.	Dryophonum	629	Erysimum Diosc.	630	Fragra	85
Cunila	228.230.237	Dryopteris	405	Erysimum Theoph.	494	Fragaria	ibid.
Cunilagines	35	Dryopteris nigra	ibid.	Erysisceptrum 127.346.525		Fragula	ibid.
Cupressus	767	Dulcichimum	346	Erythanon	659	Fraxinea arbor	748
Cupressus Cretica	766	Dulcamera	398	Euonymus	760	Fraxinus	ibid.
Cyanus flos	161	Dulciradix	694	Eupatorium	57	Fraxinus bubula	ibid.
Cyanus maior	ibid.	Dulcisida	338	Eupatorium adalterinum	59	Fraxinus syluestris	ibid.
Cyclaminus	330	Dulcis radix	694	Eupatorium aquaticum	ibid.	Frugiperda	743
Cyclaminus altera	330.331.	Duracina Persica	710	Eupatorium Auicenna	ibid.	Fucus	414
384.386		E Bulus	380	Euphorbium	309.544	Fumaria	24
Cymbalium	38	Ebulum	ibid.	Euphrasia	40	Fusterna	774
Cyminalis	332	Echion	10.242	Euphrosine	ibid.	Fucus agrestis	532
Cynanthemis	186	Elaphoboscum	604	Exupera	127	G Alation	539
Cynara	523	Elaterium	373	F Aba	473	Galbanum	307
Cynobatae	186	Elatine	389	Faba lupina	450	Galbuli	767
Cynocrambe	77	Eleophyllon	78	Faba suilla	ibid.	Galax	49
Cynoglossa	ibid.	Eliochrysum	89.250	Fabulonia	ibid.	Galega	435.490
Cynoglossen Plinij	11	Elichryson	89	Fabulum	450	Galega altera	488

G 4

INDEX.

Galarion	839	Hemerocallis	203.204	Hypericum	64	Ladum	660
Galeopsis	44	Hepatorium	57	Hypocistis	659	Ladanum	ibid.
Galla 762 Gallica	745	Heraclea	131	Hypoglossum	675	Lagopus	562
Gallitricum	253	Heracleon	144	**I** Beris	625	Lamium	131
Gallium	539	Heranthemum	184.188	Ibiga	28	Lampada	159
Gariophyllata	134	Herba S. Barbara	626	Ibiscus	583	Lampsana	860
Garosmos	549	Herba casta	338	Idaus Dactylus	338	Lampuca	567
Geminalis	255	Herba coxendicum	38	Illecebra	115	Lanaria	119
Genista	664.666	Herba iniqua	89	Inguinalis	36	Lapathum	558
Gentiana	332.334	Herba iudaica	260	Intuba	562	Lapathum magnum	559
Geraniorum genera 47. & 48		Herba lutea	68	Intubum satiuum Latifoliū ibi.		Lapathum satiuum	ibid.
Gingidium	615	Herba pedicularis	372	Intybum agreste	563.567	Lapathum syluestre	ibid.
Gladiolus	197	Herba pulicaris	103	Intybum satiuum	ibid.	Larix 775 Laser	305
Gladiolus segetalis	ibid.	Herba Roberti	47	Intybum syluestre	ibid.	Laser Cyrenaicum	ibid.
Glans	745	Herba sanguinalis	127	Inula	336	Laserpitium	ibid.
Glandes terrestres	491	Herba Sardoa	420	Inula campana	ibid.	Laser Syriacum	ibid.
Glans sardiana	730	Herba Saracenica	314	Iouis faba	450	Latum cuminum	285
Glastum	67	Herba scanaria	615	Iouis flamma	724	Lathyris	362
Glaux	49.490.500	Herba scelerata	419	Iouis glans	730	Lathyrus	478
Glycyrrhiza	694	Herba stella	95	Iringus	519	Lauandula	265
Glycyrrhiza communis	ibid.	Herba vlicana	447	Iris	193	Lauandula foemina	ibid.
Glycyrrhiza Diosc.	ibid.	Herbulum	371	Iris cerulea	194	Lauandula mas	ibid.
Gnaphalium	90	Hermodactylus	367	Iris Germanica	193	Lauer	611
Glossypium	679	Herculis sanguis 216 327.347		Iris lutea	192	Lauer Cratena	ibid.
Gorgonion 290 Gramē	504	Herpacantha	527	Iris syluestris	193.196.199	Lauri bacca	688
Gramen arundinaceum	515	Hibiscus	583	Irio 630 Isophyllon	506	Laurus	ibid.
Gramen marinum	509	Hibiscus Thoprast.	ibid.	Isopyrum	542	Laurus Alexandrina	676
Gramen Parnasium	510	Hiera botane	127	Isatis vtraq.	67	Laurus idaa	ibid.
Granum cnidium	370	Hieracium	567	Isatis syluestris	11	Laurus rosea	430
Gratia dei	375	Hieracium magnum	ibid.	Ischias	530	Ledum	660
Gratiola	375.673	Hieracium paruum	ibid.	Iugalis	760	Leguminum leo	664
Grossus	717	Hieracopodium	159	Iunci	511	Lectipes 235 Lemnisis	327
Gruina	47.48	Hippoglossum	675	Iuncifolius	212	Lens	487
Gruinalis	ibid.	Hippolapatum	559	Iunci flos	511	Lens lacustris	106
Gummi Armeniacum	308	Hipposelinum	613	Iuncus acutus	ibid.	Lens palustris	ibid.
Gummi vlmi	752	Hipposelinon agreste	ibid.	Iuncus angulosus	346	Lenticula	487
Gymnocrithon	461	Hirci spina	543.669	Iuncus	511	Lentiscus	779
H alicacabon	447	Hirundinaria	31	Iuncus odoratus	ibid.	Lepidisma	615.630
Halicastrum	456	Hispanica pruna	721	Iuncus quadratus	346	Leptophyllos	361
Halmades	738	Holoconitis	346	Iuncus triangularis	ibid.	Lethe 432 Lethusa	ibid.
Halimus	576	Holoschœnus	511	Iuniperus	763	Leucanthemum	658
Harundinis genera	514	Holosteum 96 Hordeum	461	Iunonis rosa	200	Leucoion	151.216
Hastula regia	649	Hordeum cantherinum	ibid.	Iuls nucum	731	Leucoia lutea	151
Hebiscus	583	Hordeum galatinum	ibid.	Ixia	517	Lenisticum	295
Hedera folium	317	Hordeum nudum	ibid.	**L** Abrusca	384.652	Libanum	12
Hedera genera	388	Hordeum mundum	ibid.	Labrum Veneris	522	Libanotis	280
Hedera mollis	394	Horminum	253.255	Laburnum	742	Libanotides	264
Hedera terrestris	389.393	Horminum satiuum	255	Lachryma abiegna	774	Libanotis Theoph. ibid. & 281
Hederuticula	317.388	Horminum syluestre	ibid.	Lachryma abietis	ibid.	Libanotis	280
Hedypnois	563	Humibuxus	699	Lachryma tragacantha	543	Libanus	773
Hedysarum	489	Humirubus	661	Lachryma Iob	463	Lichen	411
Helenium	336	Hyacinthus	206	Lactaria	360	Licinia	127
Helianthemum	673	Hyacinthus autumnalis	ibid.	Lactaria foemina	361	Ligusticum	595
Heliotropium vtrunq.				Lactaria mascula	ibid.	Ligustrum	393.690
Helix	61 & 209	Hyacinthus Ouidij	202	Lactaria solsequia	ibid.	Ligustrum album	ibid.
Helleborastrum	388	Hydropiper	632	Lactaria latifolia	ibid.	Ligustrum nigrum	ibid.
Helleborine tenui folia	189	Hydroselimon agrion	606	Lactuca satiua genera fol. 573		Lilium	200
Helleborine	ibid.	Hyoscyamus	450	Lactuca leporina	565	Lilium marinum	204
Helleborus niger	224.349	Hyoscyamus albus	ibid.	Lactuca minor	567	Lilium rubrum	202
Helxine cissampelos	189 251	Hyoscyamus luteus	ibid.	Lactuca syluatica	ibid.	Lilium rufum	ibid.
Hemionitis	394	Hyoscyamus niger	406	Lactuca syluestris	574	Lilium syluestre	203
							Lilium

INDEX.

Lilium Theophrasti	178	Malum limonium	ibid.	Milax	393	Nigella Damascena	278
Limnesion	375	Malum punicum	706	Mollis hedera	394	Nigella domestica	ibid.
Limnites	327	Malum terræ	314.330	Milesia rosa	655	Nigella syluestris	ibid.
Limodorum	664.	Malus	702	Militaris	127.143.144	Norion	438
Limones	704	Malus Armeniaca	710	Militaris millefolia	13	Nuces cupressi	767
Limonium	535	Malus citria	703	Milium	463.465	Nuces fagi	757
Lingua	135	Malus cotonea	708	Milium Indicum	466	Nux	731
Lingua bouis	12	Malum Persicum	710	Milium saburrum	ibid.	Nux auellana	734
Lingua bubula	ibid.	Malus medica	703	Millefolium	106.144	Nux castanea	730
Lingua canis	11	Malus Persica	710	Mimmulus	516	Nux Heracleotica	734
Linguace	135	Marmoraria	527	Mithridatium	111	Nux iuglans	731
Lingua teruina	107	Mamiras	346	Moly	263.509	Nux myristica	732
Lingulace 135 Linum	70	Mandragoras	438	Moly Plinij	ibid.	Nux Persica	731
Linum triticeum	494	Mandragoras fœmina	ibid.	Montiulmus	752	Nuces pinea	770
Lithospermum	290	Mandragoras mas	ibid.	Moraria	522	Nux pinea	ibid.
Lobi 474 Lolium	409	Mandragoras Theoph.	446	Morum	715	Nux Pontica	734
Lonchitis aspera	407	Mandragoras Morion Dioscoridis		Morum rubi	661	Nux Prænestina	ibid.
Lotus syluestris minor	497	ridis	ibid.	Morum rubi idæi	662	Nux Regia	731
Lotus	729	Mania	450	Morus	715	Nymphæa vtraq;	181
Lotus satiua	495	Mariscus	511	Mose	705	Ocimastrum	242
Lotus syluestris	497	Maronium	327	Mula herba	406	Oculata	40
Lotus vrbana	495	Marrubiastrum	257	Muliones	752	Ocymum	240.468
Luparia	351.429	Marrubium	ibid.	Muralis	50	Ocymum maius	240
Lupinus	481	Marrubium nigrum	ibid.	Muralium	ibid.	Ocymum minus	ibid.
Lupinus satiuus	ibid.	Marrubium palustre	131.257	Musa	705	Ocymum gariophyllatum	ibid.
Lupinus syluestris	ibid.	Marum	234.235	Muscus	414	Ocymoides	517
Lupus salictarius	400	Masliche	779	Muscus marinus	ibid.	Oenanthe	652
Lupulus salictarius	ibid.	Mecon	361	Myagrum	494	Olea	738
Lustago	127	Medica	497.500	Myitis	628	Olea Æthiopica	ibid.
Luteum herba	67	Meleta	522	Myoctonon	426	Oleago	369
Lychnis	636	Melamphyllum	527	Myopteron	628	Oleagnus	673
Lychnis coronaria	158	Melampodium	351	Myrica humilis	677	Oleastellus	369
Lychnis satiua	ibid.	Melampyrum	164.469.494	Myrimorphos	144	Oleaster	738
Lychnis syluestris	159	Melanorhizon	351	Myriophyllon	ibid.	Olea syluestris	ibid.
Lychnitis	119.161.175	Melanthium	278	Myrrha	616	Oleoaresina	775
Lycium	700	Melanthium Damascenum ibi.		Myrrhis	ibid.	Olesicarpos	743
Lycoctonon	429	Melanthium satiuum	ibid.	Myrtaria	361	Olibanum	774
Lycoctonon Ponticum	ibid.	Melanthium syluestre	ibid.	Myrtus	673.687	Oliua	738
Lycopersium	177	Melica	466	Myrtus syluestris	674	Olusatrum	609
Lycopodion 414 Lycopsis	8	Melilotus	497	Myxa	722	Olyra	461
Lycopsis syluestris	9	Melilotus germanica	497.498	Myxaria	ibid.	Omphacium	652
Lysimachia	74	Melilotus Italica	ibid.	Napus	595	Onogra	74
Lysimachium	ibid.	Melilotus syluestris	ibid.	Napus satiuus	ibid.	Onitron	432
Lysimachium cæruleum	75	Melissophyllum	259	Napus syluestris	ibid.	Onobrychis	172.485
Lysimachium purpureum ibid.		Melittæna 259 Melo	590	Narce	332	Onocardion	522
Macer	732	Melocarpon	314	Narcissus	211.345	Ononis	669
Maiorana	234	Melopepo	590	Narcissus luteus	214	Ophris	224
Mala Armeniaca	710	Melopepo Galeni	586	Nardus rustica	134.340	Ophthalmica	40
Mala citria	703	Memacylon	728	Nardus syluestris	340	Opium	432
Malacocissos	394	Menta aquatica	244	Nasturtium	623	Opopanax	302
Mala insania	439	Menta Romana	245	Nasturtium hibernum	626	Opsago	447
Malicorium	706	Menta Saracenica	ibid.	Nasturtium rusticum	628	Opulus	760
Malua genera	581	Menta satiua	ibid.	Nasturtium syluestre	ibid.	Opuntia	544
Malua Theoph.	583.585	Mentastrum	245.248	Nastus	514	Orbicularis	330
Malum	702	Mercurialis	77	Nepa	116	Orobus	482
Malum anarantium	704	Mercurialis fœmina	ibid.	Nepita	248	Origana	234.235.237
Malum arantium	ibid.	Mercurialis mas	ibid.	Nerantzium	704	Origanum Heracleoticum	237
Malum aureum	ibid.	Mercurialis syluestris	ibid.	Nerium	430	Origanum onitis	234.237
Malum cotoneum	708	Mespilum	714	Neris	ibid.	Origanum Hispanicum	237
Malum granatum	706	Mespilus	713	Nession	327	Origanum syluestre	ibid.
Malum hespericum	704	Meum	337.579	Nigella	278	Orchios genera folio	221

Ornitho-

INDEX.

Ornithogalum	645	Pentadryon	448	Piper Indianum	634	Præsepium	532
Ornithopodium	487	Pæderota	527	Piper montanum	371	Prassium	257
Ornus	748	Pentaphylli genera	83	Piper album	635	Prætium	351
Orobanche	664	Peplion	363	Piper longum	ibid.	Proserpinaca	98
Orobieum	479	Peplis	ibid.	Leuconpiper	ibid.	Prosopon	432
Orontium	180	Peplus	ibid.	Leucopiper	ibid.	Prunum	720
Oriza Germanica	461	Pepones	588	Macronpiper	ibid.	Pruna asinina	721
Osmundi	483	Pepones cucumerales	ibid.	Melanopiper	ibid.	Pruna cerea	ibid.
Osyris	80	Pepones lati	ibid.	Melanpiper	ibid.	Pruna cereola	ibid.
Othonna	177	Pepones magni	ibid.	Piperitis	631.634	Pruna Damascena	ibid.
Oxalis parua	559	Perdicalis Percepier	50	Piper	635	Pruneolum	ibid.
Oxalis Romana	ibid.	Perdicium	ibid.	Pira	712	Prunulum	ibid.
Oxalis	ibid.	Periclymenum	391	Pirus	ibid.	Prunum sylueſtre	ibid.
Oxyacantha	684	Perpensa	319	Pissaphaltos	765	Prunus	720
Oxycedrus	765	Persion	448	Pistacia	735	Prunus sylueſtris	721
Oxsibænos	511	Personata	15	Pistacium Germanicum	ibid.	Pseudacanthus	528
Oxytriphillon	500	Personatia	ibid.	Pistolochia	314	Pseudoacorus	199
Oxys	503	Pes cornicis	95.351	Pisum	476	Pseudobunium	626
Oxytonon	432	Pes gallinaceus	24	Pithitis	432	Pseudocoronopus	95
Pæonia	338	Petasites	21	Pituitaria	372	Pseudodictamum	368.344
Pala	705	Petrapium	608	Pitys	772	Pseudohelleborus niger	351
Palalia	330	Petroselinum	608.613	Pityusa	361	Pseudohepatorium	59
Paludapium	606	Petroselinum Alexandr.	609	Pix	773	Pseudolinum	80
Palma	737	Petroselinum Macedonicū	287	Pix arida	ibid.	Pseudomoly	509
Palmula	ibid.		608	Pix liquida	ibid.	Pseudomyrsine	673
Pampinula	138	Peuce	770	Plantaginis genera	92	Pseudonarcissus	214
Paliurus	669.701	Peucedanum	298	Plantago aquatica	97	Pseudonardum	265
Panaces	293.295.334	Phalaris	465	Planta veris	316	Pseunophu	341
Panaces Asclepÿ	ibid.	Phaselus satiuus	473	Platanus	755	Psyllum	103
Panaces Centaurion	336	Phaselus sylueſtris	ibid.	Polemonia	345	Pulegium	231.232
Panaces Chironium	ibid.	Phasioli	474	Polemonium	ibid.	Pulegium montanum	231
Panaces Herculeum	293	Phasiolon	542	Policaria	35	Pulegium sylueſtre	248.268
Panax ibid. Pancarpon	517	Philyra	753	Polypodium	403	Pulmonalis	125
Pancratium	646	Pherusa	535	Polium	233	Pulicaria	35
Panicum	466	Philiterium	242	Polium montanum	ibid.	Pyonitis	119
Panicum peregrinum	ibid.	Phenix	504.737	Polyacanthus	521	Pyra	712
Panus procinus	330	Phœnicobalanos	737	Polyanthemum aquaticum	107	Pyra cytonia	708
Papauer	432	Pherubrum	574	421		Pyrethrum	843
Papauer album	ibid.	Phillyrea	609	Polyanthemum palustre	ibid.	Pyracantha	535
Papauer commune	436	Phu	304	Polygala	49.487.490.500	Pyrina	698
Papauer corniculatum	ibid.	Phu Græcum	ibid.	Polygonatum	103.530	Pytyanthe	ibid.
Papauer cornutum	432	Phu paruum	ibid.	Polygoni genera	99	Pyxacantha	700
Papauer erraticum	432.434	Phu sylueſtre	406	Polyrhizon	314.351		
Papauer fluidum	434	Phyllitis	78.530	Polytrichon	409	Quercus	745
Papauer nigrum	278.432	Phyllon	772	Polytrichon Apulei	414	Quinquefolium	83
Papauer palustre	181	Picea	347	Pomum	702		
Papauer rhœas	432.434	Pinatoxaris	767	Pomum cytonium	708	Radicula	598
Papauer satiuum	432	Pilula cupressi	138	Pomum Granatum	706	Radicula palustris	ibid.
Papauer spumeum	436	Pimpinella	770	Pomus	702	Radicula sylueſtris	ibid.
Papauer sylueſtre	ibid.	Pinaster	ibid.	Populus alba	750	Radicula satiua	ibid.
Papauer rubrum	ibid.	Pinus	511	Populus Libyca	ibid.	Radix caua	316
Papyrus		Pinus humilis	ibid.	Populus nigra	ibid.	Radix Rhodia	342
Paralios	361	Pinus Idea	ibid.	Porrum	415.641	Radix rosata	ibid.
Parietaria	50	Pinus satina	ibid.	Porrum capitotum	641.643	Radix Scytica	664
Parthenis	16	Pinus sylueſtris	ibid.	Porrum sectiuum	ibid.	Radix sylueſtris	ibid.
Parthenium	19	Pinus Tarentina	ibid.	Portulaca	576	Ranunculi genus	107
Pastinaca	602.604	Pinus terestris	ibid.	Portulaca marina	ibid.	Ranunculi genera	416.419
Pastinaca genera	602	Pinus marina	562	Portulaca sylueſtris	363.576	Ranunculi auricomus	420
Pedicularis	516	Pisrida	567	Præcocia	710	Rapa	594
Pelecinon	489	Picris	691	Præcoqua	ibid.	Rapa sylueſtris	597
Pelthronia	327	Piper agreste				Raponium	ibid.

Xxx Ra-

INDEX.

Rapontium paruum	ibid.	Ros	692	Rubia satiua 538	Scordothlaspi	628	Smilax lenis	393
Raphanus	598.600	Rubia sylueftris	ibid.	Scordoprassum	638	Smyrnium	609.613
Raphanus magnus	600	Rubus	661	Scorpioides	63	Solanum	443
Raphanus montanus	ibi.	Rub. canis	616.680.698	Scorpius	669	Solanum hortense	ibid.
Raphanus paruus	ibid.	Rubus Idæus	662	Scythica radix	694	Solanum lethale	446
Rapum sylueftre	174	Rumicis genera	519.& inde.	Secale	459	Solanum lignosum	398
Rapum porcinum	330.	Ruscum	674	Securidaca	489	Solanū manicū 446.448.585
Rapum terræ	ibid.	Ruscus sylueftris	701	Sedi genera	114	Solanum somniferum	447
Regium	240	Ruta	261	Selago Plinij	673	Solidago	133.145
Remenia	450	Ruta hortensis	ibid.	Semen	455	Solidago Sarracenica 133.141
Resina	772	Ruta muraria	409	Seminalis	98	Somphos	592
Resina abiegna	775	Ruta sylueftris	261.363	Sempervivi genera	114	Sonchi	565
Resina arida	772	S Abina	766	Sena	377	Sonchus aspera	ibid.
Resina colophonia	ibid.	Sabina salix	744	Senecio	571	Sonchus leuis	ibid.
Resina laricea	776	Sacopenium	306	Senecio maior	ibid.	Sonchus non aspera	ibid.
Resina larigna	ibid.	Sacra herba	127	Senecio minor	ibid.	Sonchus tenerior	ibid.
Resina lentiscina	779	Sagapeni succus	306	Sentis	661	Sonchus sylueftris	ibid.
Resina liquida	772	Sagapenium	ibid.	Sentis canis	680	Sorbum 727 Sorbus ibid.
Resina picea	ibid.	Sagapenum	ibid.	Septifolium	84.300	Sorghi	466
Resina pinea	ibid.	Salicaftrum	384	Serica	722	Spagas	773
Resina ftrobilina	775	Salicum genera	744	Seris	569	Sparganium	200
Resina terebinthina	777	Saltuaris	342	Seriphium	5	Spartum	666
Rha	329	Salix amerina	691.744	Serpentaria	10.22.321	Sphacelus	253
Rhabarbarum	ibid.	Salix equina	161	Serpyllum vulgare	231	Sphærocephalus	526
Rhacoma	ibid.	Salix marina	691	Sertula campana	498	Spina	685
Rhamnus	696	Saluia maior	252	Seseleos genera	283.284	Spina acuta	697
Ramnus solutiuus	697	Saluia minor	251	Setanium	453.714	Spina alba	325.530
Rha ponticum	329	Saluia nobilis	ibid.	Sisamoides paruum	95	Spina Arabica	ibid.
Rhecoma	ibid.	Saluia vsualis	ibid.	Sicula	550	Spina mollis	568
Rheum	ibid.	Saluia ibid.	Sambucus 378	Sicyonia	375	Spina peregrina	526
Rheum Indicum	ibid.	Sambucus humilis	380	Sicyopepones	388	Spinæ regia	525
Rheum ponticum	ibid.	Sambucus sylueftris	378	Sideritis	47.131	Spina Ceanothes	681
Rhizias	304	Samera	752	Sideritis altera	138	Spina hirci	543
Rhos	692	Samolus Plinij	671	Sideritis Heraclea	ibid.	Spinguerzo	697
Rhous	ibid.	Sampsycum	234	Sideritis latifolia	626	Spino merlo	ibid.
Rhus	ibid.	Sanguinalis	98	Sideritis prima	131.257	Spino ceruino	ibid.
Rhus coriariorum	ibid.	Sanicula	139	Sideritis tertia	47	Splenium	406
Rhus obsoniorum	ibid.	Sanguinaria	ibid.	Sidium	706	Spondyli	524
Rhus sylueftris Plinij	673	Santalum	461	Siligo	453.461	Spondylium	524.528
Ribes	683	Sapinus	770.774	Siliqua	492.740	Stachys	257
Ricinus	354	Sarcocolla	311	Siliqua dulcis	ibid.	Stichas	266
Robus	453	Sardiana glans	730	Siliqua	474	Stoechas	ibid.
Rosa	615	Satureia vulgaris	228	Siliquaftrum	634	Stellaria	95
Rosa canina	ibid.	Satyrium	225	Sinapi	619	Staphis agria	372
Rosa Iunonis	200	Satyrium basilicum	226	Sinapi commune	ibid.	Staphylinus luteus	602
Rosa mariana	158	Satyri. erythronium	225	Sinapi hortense	ibid.	Staphylinus	604
Rosa Græca	159.656	Satyrium regium	226	Sinapi Persicum	154	Staphylinus niger	602
Rosa Coronæa	655	Satyrium trifolium	225	Sinapi rufticum	628	Staphylinus satiuus	ibid.
Rosa autumnalis	ibid.	Saxifraga rubea	40	Sinapi sylueftre	619.620	Staphylinus sylueftris	ibid.
Rosa alba	ibid.	Scammonium	396	Sirica	722	Staphylodendron Plinij	735
Rosa Damascena	ibid.	Scandix	615	Sisarum	664.605	Stataria	298
Rosa Milesia	ibid.	Scandulaceum	628	Siser 605	Sison 287	Stellaria	521
Rosa praneftina	ibid.	Scœnophrasum	643	Sisgurichium	198	Sternutamentaria	343
Rosa purpurea	ibid.	Scœnuanthos	511	Sisymbrium	245	Stratiotes	143
Rosa rubra	ibid.	Sciara	522	Sisymbrium cardamine	625	Staechos	ibid.
Rosa sera	ibid.	Scilla	646	Sitanium	714	Stratoticon	144
Rosa sylueftris	ibid.	Scimbron	245	Sitheleas	567	Struthia mala	708
Rosa spinosa	656	Scolymus	335	Sium	611.625	Strution	335
Rosea arbor	430	Scopa regia	626.674	Sium Crateua	611	Succisa	110
Rosmarinum	280	Scordium	111	Smilax aspera	396	Succus Cyreniacus	304
Rosm. coronarium	264.265	Scordotis		Smilax hortensis	474	Succus Medicus	ibid.

Sucha-

INDEX.

Suchaha	530	Thymum Creticum	ibid.	Tuber terrae	330	Viola nigra	148
Supercilium terrae	409	Thymum durius	229	Tulipa, Tulpia, Tulpian.	213	Viola peregrina	153
Supercilium Veneris	144	Thymelaea	369	Tussilago	20	Viola tricolor	149
Sycomorus	716	Tilia	753	Typha	457.512	Viola purpurea	148
Sylua mater	391	Tilia foemina	754	Typha aquatica	513	Viperalis 261 Viperina	10
Symphitum magnum	145	Tilia mas	ibid.	Typha cerealis	457.513	Viscum 747 Vitalis	114
Symphoniaca	450	Tithymali omnes	362	Typha palustris	ibid.	Vitealis ibid. Vitex	691
Tamarix	677	Tithymalus	361.362	**V**asinia	661.671	Vitia	483
Tamus	384	Tomentitia	89.90	Vaccinium	148. 206	Vitis alba	382
Taxus	768	Topiaria	527	Vaccinia palustria	671	Vitis nigra	ibid.
Teda	771	Tordylium	284	Valeriana	340	Vitis syluestris	384.652
Telephium	118	Tormentilla	84	Valeriana genera	ibid.	Vitis vinifera	651
Teliphonon	426	Trachelium vtrunq3	172	Vatrasbion Apulei	419	Vlophonon 517 Vlmus	752
Terebinthus	777	Tragacantha	543.669	Veneris lauacrum	522	Vlmus campestris	ibid.
Terra capillus	409	Tragacantha lachryma	543	Veratrum adulterinum mi-		Vlmus Italica	ibid.
Terzola	59	Tragium	343.549	grum	351	Vlmus Gallica	ibid.
Thesium	316	Tragium Germanicum	ibid.	Veratrum album	347	Vlmus nostras	ibid.
Testiculus	222	Tragonatum	159	Veratrum nigrum	351	Vlmus syluestris	ibid.
Testiculus canis	ibid.	Tragorchis	222	Verbasci genera	119.122	Vmbilicus terrae	330
Testiculus hirci	ibid.	Tragoriganum	239	Verbascula	123	Vmbilicus Veneris	38
Testiculus leporis	ibid.	Tragopyron	468	Verbasculum minus	ibid.	Vmbilicus Veneris alter	ibid.
Testiculus odoratus	ibid.	Tragus	116	Verbena	127	Vnedo	725
Testiculus serapias	ibid.	Trasus	346	Verbenaca	ibid.	Vnesera	357
Testiculus vulpis	ibid.	Trachinia	655	Verbena recta	ibid.	Vngues rosarum	615
Teucrium	112	Tribulus aquaticus	536	Verbenaca supina	ibid.	Volucrum maius	391
Teuthrium	233	Tribulus	ibid.	Vernix 763 Veronica	26	Vrceolaris	50
Teuxinon	314	Tribulus terrestris	ibid.	Veruilago	517	Vrina muris	581
Thalictrum	117	Trichomanes	410.414	Vesicaria	445	Vrticae genera	129
Thalietrum	43.117	Trifolium	495.501	Vesicaria nigra	ibid.	Vrtica iners	131
Thapsia	365	Trifolium palustre	542	Vesicaria peregrina	ibid.	Vrtica labeo	44
Thlaspi	154.600.628	Trifolium bituminosum	501	Vesicula	ibid.	Vrtica mortua	131
Thlaspi Crateua	154	Trifolium foetidum	ibid.	Vetonica	156.291	Vstilago	471.517
Thlaspi alterum	628	Trifolium fruticans	666	Vetonica altilis	156	Vua taminea	384
Thlaspi angustifolium	ibid.	Trifolium odoratum	495.501	Vetonica coronaria	ibid.	Vua vrsi	683
Thlaspi minus	ibid.	Trifolium pratense	495	Vetonica coronaria minor	ib.	Vua spina	685
Thridacias	438	Trigrania	714	Vetonica syluestris	157	Vulgago	319
Thridax agria	567	Triorchis	222	Vicia	394	**X**anthium	15
Thryallis	161.175	Tripolium	364	Viola alba	152.153.154	Xylum	679
Thryon	448	Triticum	453	Viola alba Theoph.	216	Xyris	196
Thus	773.775	Triticum Romanum	457	Viola autumnalis	173	**Z**ea 280.455.456.457	
Thuia	766	Triticum Tiphinium	ibid.	Viola flammea	149	Zeopyron	458
Thuium	ibid.	Triticum boum	164	Viola hyemalis	153	Zea Monococcos	ibid.
Thylacitis	432	Triticum trimestre	453	Viola latifolia	ibid.	Zea simplex	459
Thymbra	230	Triticum vaccinum	164	Viola lutea	151	Zizipha	722
Thymum	229	Trixago	25.127	Viola Mariana	174	Ziziphus	ibid.
Thymum capita	ibid.	Trixago palustris	111	Viola matronalis	153	Zyga	760

Finis huius Indicis.

Index appellationum & nomenclaturarum omnium

Stirpium, hoc opere contentarum, quibus passim Officinæ Pharmacopolarum, Arabes, & nostri temporis Herbarij vtuntur.

Absynthium	5	Anthera 655	Anthos 264	Behen	ibid. Candelaria ibid.
Absynthiũ grçciæ	ibi.	Apium	606	Behen rubrum	ibid. Canicularis 450
Absynthium ponticũ	ibi.	Apium risus	420	Belle videre	175 Caniculata ibid.
Absynthiũ Romanũ	ibi.	Apolinum	369	Benedicta	134 Capillus Veneris 409
Abrotonum	2	Aprella	101	Berberis	684 Cappa monachi 429
Abutilon	583	Aquilegia	166	Besasa	263 Caprifolium 391
Acatia	721	Aquileia	ibid.	Beta	550 Cardiaca 131
Acetosa	559	Aquilina	ibid.	Betonica	291 Cardobenedictus 533
Acetosa Romana	ibid.	Arantium	704	Bibinella	286 Cardopatium 530
Acetosella	ibid.	Arbor vitis	735	Bifolium	224 Carduus asininus 535
Achimilla	140	Argentina	86	Bipennula	138,287 Carduus benedictus 533
Acorus	199	Aristologia	314	Bisacutum	615 Carduus fullonum 522
Acus moschata	47	Aristologia longa	ibid.	Bislingua	675 Crrduus Mariæ 525
Acus pastoris	ibid.& 615	Armoniacum	308	Bismalua	583 Carduus stellatus 521
Acutella	666	Arresta bouis	669	Bistorta vtraque	22 Carlina 530.531,656
Acylonium	158	Arocum	523	Blaptisecula	161 Carlina syluestris 531
Adiantum.	405.409	Arthanita	330	Bolbonac	154 Carobe 740
Affodillus	203,649	Arthemisia	16	Bombax	679 Carolina 530 Carota 602
Agaricus	776	Arthemisia vnicaulis	18	Bombasum	ibid. Carthamus 33
Agnus castus	691	Arthemisia Tragantes &		Bonifacia	675 Carthamus syluestris 532
Agresta	652	tragetes	ibid.	Borago	12 Carui 274
Agrifolium	701	Arthetica	123	Borda	14 Cassia fistula 740
Agrimonia	57	Articoca	523	Branca leonina	604 Cassilago 450
Agrimonia syluestris	86	Articocalus	ibid.	Branca	527,528 Castrangula 44
Albotin	777	Asa	304	Brunella	133 Cataputia maior 354
Alcea veneta	585	Asa odorata.	ibid.	Bruscus	674 Cataputia minor 362
Alchimilla	140	Asarum 319 Asia	419	Bryonia	382 Cauda equina 101
Alcocalum	523	Asperula	540	Bucheiden	226 Cauda murina 96
Alkakengi	445	Asprella	101	Buzeiden	ibid. Cauda muris ibid.
Alkali	116	Assa foetida	304	Bulbi species	209 Cauta 185,189
Alleluya	503	Asterancium	300	Buglossa	8 Caules marinæ 555
Alliaria	639	Astochodos	266	Buglossa domestica maior	Cauliflores 554
Allium vrsinum	637	Athanasia	18		ibid. Caulis ibid.
Aloës	353	Athanatos	158	Buglossa longifolia	ibid. Caulis nigra ibid.
Alectorolophos	253	Auellanada	346	Buglossa syluestris	9 Cembri 770
Alphesera	382	Axungia vitri	116	Bugula	133 Centauria minor 327
Alscebran	361	Azarolo	714	Bursa pastoris	80 Centrum galli 253
Altercangenum	450	B Abyron	602	C Achla	189 Centum capita 519
Altercum	ibid.	Bagolaro	729	Calcatrippa	165 Centumnodia 98
Alumen catinum	116	Balsamina	442	Calabrum	407 Centummorbia 78
Amaranthus luteus	89	Balsaminum	ibid.	Calamentum	248 Cepe 640
Amarella	19	Balsamita	250	Calamentũ montanũ	ibi. Cepe muris 646
Ambrosiana	253	Balsamita maior	ibid.	Calendula	163 Cerefolium 614
Ameos	272	Balsamita minor	ibid	Caltha	163.189 Cerui ocellus 604
Amyberis	684	Baptisecula	161	Calthula	163 Ceruicaria 172
Amomum	608	Barba Aron	323	Camomilla	183.186 Cestum morionis 513
Anogallus aquatica	578	Barba capri	41	Camomilla fatua	186 Ceterach 408
Anaphalis.	90	Barba Iouis	114	Camomilla foetida	ibid. Chamædryos 25,127
Anaxiton	ibid.	Babaræa	626	Camomilla inodora	ibid. Chamedrys 127
Anetum	270	Barbarica	156	Campana lazara	393 Chamedrys foemina 25
Angelica	297	Bardana maior	15	Campanula	173.394 Chamæpiteos 28
Angina lini	398	Baucia	604	Campanula cerulea	175 Chamæpitys 25
Anguria	586,589	Becabunga	578	Campanula syluestris	ibi. Charantia 442
Anisum	271	Bedegar	525,655	Cannabis	72 Chelidonia maior 31
Anserina	86	Beèn album	345	Candela regis	119 Chelidonia minor ibid.

Che-

INDEX.

Chelidonia palustre cordi	31	Cupulæ glandium 745	Flos amoris 168	Harmala 263
Cherua 354	Cuscuta 398 Cyamus 473	Flos Chrystalli 116	Halmiridia 355
Cheruilla 605	Cyclamen 330	Flos cuculi 625	Harmel 261.263.452
Chocortis 581	Cyminum 275	Flos Constantiopolitanus 157	Hedera 388
Citrullus 590	Cynoglossa 11	Flos Cyanus 161	Helleborus albus 347
Cicer 479	Cynoglossum ibid.	Flos equestris 165	Helleborus niger 351
Cicerbitæ 565	Cyperus 346	Flos Indianus 177	Hepataria 540
Cicorea domestica 563	Cypressus 767	Flos S. Iacobi 69	Hepatica 59.107.411
Cicorea sylustris ibid.	Cytonium 788	Flos regius 169	Hepatica alba 510
Cicutaria 616	Dactilus 737	Flos tinctorius 68.667	Hepatica aquatica 107
Cirmoli 770	Daucus 285.602	Fœnum Grecū 490.492	Hepatica palustris ibid.
Citocacium 369	Daucus Creticus 285	Fœniculus 269	Herba Serracenica 314
Citrones 704	Dens caballinus 450	Fœnicul' porcin' 280.298	Herba Benedicti 340
Citrullum 589	Dens leonis 569	Fraxinella 343	Herba S. Barbaræ 626
Cochlearia 118	Dèticulus canis Ruellij 96	Frumentum Asiaticū 464	Herba cancri 61
Coloquintida 375.590	Diagredium 396	Frumentum Turcicū ibi.	Herba clauellata 149
Colubrina 22	Diagridium ibid.	Fuga dæmonum 64	Herba fortis 141
Columbina recta 127	Diapensia 119	Fumus terræ 24	Herba Gerardi 300
Concordia 57	Dictamum 268	Funis arborum 393	Herba Hungarica 584
Condrilla 95.569	Digitalis 175	Galbanum 307	Herba giulia 250
Consolida 133	Digiti citrini 226	Galega 490	Herba S. Iacobi 69
Consolida maior 133.145	Diodela 144	Gallitricum 253	Herba impia 89
Consolida media 133	Draco 623	Gariophyllata 134	Herba Margarita 170
Consolida minor 170	Draculus hortensis ibi.	Garipot 773 Gelaso 90	Herba D. Mariæ 250
Consolida regalis 165	Dyptamum 268.344	Genesta 664	Herba paralysis 123
Consolida regia ibid.	Elacterium 373	Genestella 668	Herba Paris 425
Consolida Sarracenica 141	Elæophyllon 78	Genestra 664	Herba S. Petri 123
Corallina 414	Eghelo 742	Genestra Hispanica 666	Herba pinula 405
Cordialis 540	Endiuia 562.574	Genicularis 340	Herba Roberti 47
Coriandrum 276	Enula campana 356	Genista 664	Herba Simeonis 584
Corona regia 441	Eruca 619 Eschara 414	Genista humilis 667	Herba stellaris 95
Corona terræ 389	Esula 361 Euforbiū 309	Genista spinosa 669	Herba Trinitatis 59.149
Cornu ceruinum 95	Eufrasia 40	Genista sylustris ibid.	Herba tunica 157.345
Corota 602	Eupatorium 52	Genistella 668	Herba venti 423
Corrigiola 98	Eupatorium aquaticū ib.	Gentiana 332	Hermodactilus 367
Corsaluium 251	Eupatorium Mesue 250	Geranium supinum 47	Hepatorium Mesue 250
Cotula alba 186	Ezula 361	Githago 160	Hippia vtraque 52
Cotula fœtida ibid.	Ezula rotunda 363	Glaudes terrestres 491	Hires 90 Hirculus 265
Cotula non fœtida ibid.	Faba 473	Glandiola 13	Hirundinaria 317
Cotula lutea ibid.	Faba crassa 39	Gladiolus sylustris 197	Hispanach 556
Cotum 679	Fabaria ibid.	Gluten albotin 777	Hispanicum olus ibid.
Crassula maior 39	Faciens viduas 369	Grana Iuniperi 763	Hyacinthi genera 208
Crassula minor 114	Fagotriticum 468	Granatum pomum 706	Hypericum 64
Cressio 613	Farfara 20 Farfarus 750	Granum viride 777	Hypoquistidos 659
Creta marina 578	Febrifugia 327	Grassula 78	Hyssopus 227
Crista galli 516	Fel terræ ibid.	Gratia Dei 48.375.673	Iacea 149
Crista gallinacea 127.516	Ferde cauallo 490	Grossularia 681	Iacea nigra 109
Crocus hortensis 33	Feria 127 Ferraria 44.127	Grossularia rubra 681	Iacobea 69 Iaron 323
Crocus Sarracenicus ibid.	Ferraria minor 57	Grossularia trāsmarina ib.	Iasminum 658
Cruciata 314	Ficaria 31.44 Filago 89	Grossulæ 683	Iesemin ibid.
Cucullus monachi 429	Filicastrum 402	Grossulæ rubræ ibid.	Iecoraria 540
Cucumer 586	Filipendula 40	Grossulę trāsmarinæ ibid.	Ieseminum 658
Cucumer anguinus ibid.	Filius ante patrem 74	Gruinalis 47 Guadu 67	Iosmenum ibid.
Cucumer asininus 373.590	Filix aquatica 402	Gummi Arabicum 685	Iosme ibid.
Cucumer marinus 588	Fior de Cristallo 116	Gummi Armoniacum 308	Ireos 193 Iringus 319
Cucumer Turcicus ibid.	Fistici 735	Gummi benzui ibid.	Iris Florentina 193
Cucumus citrulli 589	Flammula 386.425	Gummi dragaganthi 543	Iris ibi. Iris Illyrica 195
Cucurbita 592	Flos Adonis 188.423	Gummi vlmi 752	Iua artetica 28
	Flos Amellus 3136		Xxx iiij Iua

INDEX.

Iua muscata	25,28.540	Marmorella	57	Nenuphar citrinum ibid.		cum	609
Iuiube	722	Marum	234.235	Nigella	278	Peucedanum	298
Kali	116	Materfilon	109	Nigellastrum	160	Philipendula	40
Karobe	740	Mater herbarum	16	Nola syluestris	175	Phyteuma	334
Keyri 151 Kitran	765	Mater sylua	391	Noli me tangere	78	Pigamum	43
Lacterones	565	Mater violarum	148	Nuces cypressi	767	Pilosella maior	87
Lactuca	573	Mastix	779	Nuces pineæ	770	Pilosella minor	ibid.
Lactucellæ	565	Matricaria	19	Nummularia	78	Pimpinella	138
Lagrimo	774	Matri salvia	253	Nux	731	Pinastellum	298
Lancea Christi	135	Medicina familię	776	Nux moschata	752	Pionia	338
Lanceola	92	Melampyrum	164.494	Nux vesicaria	735	Piper aquaticum	632
Lanceolata	ibid.	Melanopiper	635	Ocellus	156	Piper Calecuthium	634
Lapatium	558	Melica	466.497	Ocellus cerui	604	Piper Hispanum	ibid.
Lapatium acutum	559	Melega 466 Malegua ib.		Ocimū gariophillatū	240	Piperitis	631
Lapadanum	660	Melilotus	497.498	Oculus bouis	186.189	Pisareli	476
Lappa inuersa	15	Melissa	259	Oleander 430 Opiū	432	Pix naualis	773
Lappa maior	ibid.	Millefolium	144	Opoponacum	302	Plantago	92
Lappa minor	ibid.	Melo	590	Opulus 760 Organū	237	Planta leonis	140.358
Larga 776 Lassulata	250	Melospinus	441	Origanū Hispanicū	ibid.	Pneumonanthe	175
Lauandula vtraq;	265	Memirem	346	Osmunda	402.405	Podagra lini	398
Laureola	368	Memitha	436	Osmundi	483	Polipodium	405
Lenticula aquæ	106	Menta	244	Osteritium	300	Polytrichon	410
Lenticula	487	Menta aquatica	ibid.	Ostritium	ibid.	Poma amoris	440
Leontopodion	140	Menta Græca	250	Ostrutium	ibid.	Pomum amoris	ibid.
Leporis cuminum	502	Menta Romana	245	Palalia	330	Pomum aureum	ibid.
Leucanthemum	186	Menta Sarracenica	ibid.	Palma Christi	226.354	Pomum granatum	706
Leucopiper	635	Mercurialis	77	Pampinula	138	Pomum Hierosolymita-	
Leuisticum	195	Meum 610 Meu 337.610		Panis cuculi	503	num	442
Lilium	200	Mezereon	369	Panis porcinus	330	Pomum mirabile	ibid.
Lilium conuallium	178	Milium	463.465	Papauer album	432	Pomum spinosum	441
Lilium inter spinas	391	Milium soler	290	Papauer commune	ibid.	Porrum	641
Limones	704	Milius solis	ibid.	Papauer magnum	ibid.	Portulaca	576
Linaria 80 Linum	70	Millefolium	144	Papauer nigrum	278	Potentilla	86
Lingua auis	748	Millemorbia	44	Papauer rubrum	432.434	Prasium	239.257
Lingua bouis	208	Momordica	442	Paracoculi	441	Prasium fœtidum	257
Lingua canis	11	Mora bassi	662	Paritaria	50	Premula veris	123.170
Lingua ceruina	406	Mora bati	ibid.	Passulæ corintho	652	Premula veris minor ibid.	
Lingus pagana	675	Mora celsi	715	Pastinaca	604	Prunella 133 Psyllium 103	
Lingua passerina	748	Morella	443	Pastoria bursa	80	Pseudomelanthium	160
Lingua serpentis	135	Morsus diaboli	110.569	Pentadactylon	354	Pulegium	231.232
Liquiritia	694	Morsus gallinæ	52	Pera pastoris	80	Pulmonalis	125
Lolium rubrum	504	Morsus ranæ	106	Perfoliata	137	Pulmonaria	125.414
Longina 407 Lubia 474		Morus celsi	716	Perfoliatū ib.perforata 64		Pulsatilla	351.420
Lucciola 135 Luf 322		Mughi	:770	Perlaro 729 Perpensa 319		Pylocaracta	740
Lupha 323 Lunaria 78.136		Multiradix	327	Pentaphyllum genera	83	Pyra citonia	708
Lunaria	119	Mumia	765	Persicaria 633 Persicū 710		Pyrethrum	342
Lunaria maior	107.402	Myrtilli	671.673.687	Peruinca	33	Pyrola	135
Lunaria minor	136	Myrtus	673	Pes anserinus	548	Quercula minor	25
Lupinus 481 Lupul' 400		Myrtus Brabantica ibid.		Pes asininus	639	Quinquefolium	83
Luteum herba	67	Napellus	419	Pes columbæ	47	Quinquenerua	92
Macis	732	Napium	560	Pes cornicis	95.420	Rapa genestrę	664
Macropiper	634	Napus	429	Pes corui	420	Rapa rubra	550
Maiorana	234	Nardus rustica	134.319	Pes leonis	140	Rapecaulis	554
Mala insana	439	Naranzas	704	Pes leopardi	410	Raphanus minor	598
Malua	581	Nasturtium aquaticū	625	Pes leporis	502	Raphanus syluestris	631
Mala Theophrasti	583	Nasturtium	623	Pes lupi 414 Pes vituli 323		Rapiens vitam	369
Maluauiscum	ibid.	Negre caules	554	Petrasindula	286	Rapistrum	610
Mandragora	438	Nepita	248	Petroselinum	605 613	Rapontium	597
Marinella	340	Nenuphar	181	Petroselinum Macedoni-		Rapum genistæ	664

Ra

INDEX.

Rapunculum	597	Sanguis draconis	519	Sorgho	ibid.	Trixago palustris	111
Rapunculum maius	ibid.	Sanguisorba	138	Spanachea	556	Turbith Mesue	365
Rapunculum paruū	ibid.	Sanicula 139	Satolina 69	Sparta parilla	396	Turbith Serap.	ibid.
Raued	329	Saponaria	159.334	Spartula foetida	196	Typha aquatica	515
Rauediseni	ibid.	Sarcocolla	311	Spelta 455 Sparagus	618	Valeriana	340
Raued Turcicum	ibid.	Sarratula	25	Spergula	56,540	Valeriana domestica	
Rauet	ibid.	Saturegia	228	Spergula odorata	540	ibidem	
Rauetseeni	ibid.	Satyriones 222. & inde.		Spica Celtica	414	Valeriana hortensis	ibid.
Regina prati	41	Sauimera	766	Spicata 106	Spica 265	Verbena	127
Remora aratri	669	Sauina ib.	Saxifraga 288	Spicanardi	ibid.	Vermicularis	114
Resina arida	772	Saxifraga alba	ibid.	Spinachea	556	Veronica foemina	26
Resina pini	ibid.	Sauifraga aurea	ibid.	Spinacheum olus	ibid.	Vernix	765
Resta bouis	669	Saxifragæ albæ semen	286	Spina mollis	568	Victoriola	676
Rhabarbarum monacho-		Saxifraga lutea	286.498	Squilla	646	Vinca peruinca	33
rum	519	Saxifraga rubea	40	Squinantum	511	Viola	148
Rhapontica	327	Saxifraga maior	286	Staphisagria	372	Viola palustris	106
Rheu	329	Saxifraga minor	ibid.	Stataria	298	Violaria	148
Rheubarbarum	ibid.	Scabiofæ genera	109	Stramonia	441	Viperina	10
Ribes	683	Scammonea	396	Stellaria	95,140	Virga aurea	141
Ribes nigrum	ibid.	Scariola 563	Scarlea 253	Sticados Arabicum	266	Virga pastoris	522
Ribes rubrum	ibid.	Scatum coeli	38	Sticados citrinum	89	Viscus quercinus	747
Ribesum	ibid.	Scatum cellus	ibid.	Sticas Arabica	266	Visnaga	615
Robertiana	47	Sceuola 101	Scirpus 197	Sticas citrina	89	Vitalba	386
Rosa Græca	159	Scolopendria	406	Stoecas citrina	ibid.	Vitealis	394
Rosa mariana	153	Scordium	111	Stoecados citrinum	ibid.	Viticella	582
Rosa vltramarina	581	Scrophularia	44	Struchion	157	Vlmaria	41
Rosmarinus	264	Scorodonia	253	Succisa 110 Suchaha	550	Vmbilicus Veneris	38
Ros solis	414	Scrophularia maior ibid.		Sumach	691	Vngula caballina	20
Rostrum ciconiæ		& 31		Superba	156	Vnifolium	178
Rostrum porcinum	569	Scrophularia minor	31	Symphytum syluestre	125	Volubilis	395
Rorella	414	Sebastæ	722	Tamariscus	677.708	Volubilis maior	ibid.
Ruberta	47	Sebesten	ibid.	Tanacetum	18	Volubilis minor	ibid.
Rubia tinctorum	538	Segala 459	Sena 377	Tanacetum maius	ibid.	Volubilis acuta	396
Ruta capraria	490	Serapinū 306 Serbin	765	Tanacetum minus	ibid.	Volubilis media	394
Ruta	261	Serpentaria	10.78.321	Tanacetum syluestre	219	Volubilis pungens	396
Ruscus	674	Serpentaria maior	321	Tapsus barbatus	119	Vrinalis	80
Saggina	466	Serpillum	251	Taraxacon	565	Vsnea	414
Salicaria	116	Septifolium	300	Terebinthina	777	Vstilago	471.517
Salicornia		Serratula 25 Seruilla	605	Terebina Veneta	774	Vua crispa	681
Sal alkali	ibid.	Seruillum	ibid.	Testiculus hirci	242	Vua lupina	425.443
Salsirora	414	Seutlomalache	556	Testiculus leporis	ibid.	Vua versa	425
Saluia agrestis	253	Sicla 550	Sicelica ibid.	Testiculus vulpis	ibid.	Vua vulpis	443
Saluia maior	251	Sicha	601	Tetrahil	260	Vulgago	39
Saluia minor	ibid.	Sigillum beatæ Mariæ	384	Tetrahit	ibid.	Vuluaria	549
Saluia nobilis	ibid.	Sigillum Salomonis	103	Thus	773.774	Vuularia	172.675
Saluia Romana	250	Siler montanum	293.295	Thymus	229.219	X	
Saluia vsualis	251	Sinapis	619	Tinearia	89	Xylocaracta	740
Salusandria	278	Sinapium	ibid.	Tota bona	561	Y	
Sambucus	378	Sinapi Persicum	154	Tomentitia	89.90	Ysopus	227
Sambucus aquatica	760	Solanum letale	446	Tormentilla	84	Yreos	193
Sambucus ceruinus	378	Solatrum	443	Tragopyrum	468	Z	
Sambucus humilis	380	Solatrum mortale	446	Trali	346	Zahasaran	216
Sambucus palustris	760	Solbastrella	138	Trasci	ibid.	Zambach	658
Sambucus syluestris	378	Soldanella	395.414	Trifolium acetosum	503	Zarfa parilla	396
Sanamunda	134	Solidago	133	Trifolium bituminosū	501	Zinziber caninum	634
Sancti Iacobi flos	69	Solidago Sarracenica	141	Trifolium humile	502	Zizania	469
Sandaraca	763	Sophia 117	Sorgi 466	Trifolium odoratum	495	Zuccomarin	588

FINIS.

The Englishe Table conteyning the names and syr-names of all the Herbes, Trees, and Plantes, of this present Booke, or Herball.

A

Acatia	685	Lowe Basill	142	Great clote Burre	ibid.	Wilde Clarie	255
Ache	611	Basill royall/&c.	140	Diche Burre	ibid.	Smelling Clauer	105
Aconit	416	Crispe Baulme	245	Lowse Burre	ibid.	Clauer gentle	ibid.
Adders grasse	222	Baulme	259	Butter Burre	11	Rough Clauer	502
Adders tongue	135	Beanes	473	Butterflowers	4.12	Garde or sallet Clauer	496
Adder-wurt	23	Bockes Beanes	542	Byrthwort	314	Treacle Clauer	501
Aethiopis	114	Beane tree	740			Clarye or Cleare-eye	253
Alsodyl	649	Garden Beane	473	**L**		Clematis altera	385
Agrimonie	57	Great Beanes	473	LOsed Cabbage	553	Cliuer	539
Noble Agrimonie	59	Ridney Beanes	474	Great round Cabbage cole	ibid.	Coccow flowers	625
Wilde Agrimonie	86	Romaine Beanes	ibid.	Calfes foote	323	Cockle	160
Agnus castus	691	Wilde & blacke Beanes	473	Calfes snowte	180	Cochowes meate	503
Aichweede	300	Our Ladies Bedstraw	539	Calamynte	248	Cocolas panter	661
Algood	561	Beeche maste	757	Rough Calaminte	ibid.	White Colewurtes	553
Alkakengie	445	Beeche tree	ibid.	Calathian violets	173	Wilde Colewurtes	555
Alkanet	9	Beetes	550	Caltrop	521	Lyppes Colewurtes	554
Blacke Alle	759	Behen or Been album	345	Caltha	190	Countrie Colewurtes	555
Aller tree or Alder	756	Belflowers	172	Cameline	494	Prickled or ruffed Cole	554
Almondes or Almonde	711	Belroin or Benroin	304	Camels strawe	511	Coliander	276
Ambros	253	Blew belles	174	Cammocke	669	Coloquintida	375
Amelcorne	456	Bay beries	688	Camomill	183	Coltes foote	20
Ammoniacum	308	Blacke beries	661	Stincking Camomill	186	Columbine	166
Amaus Apples	439	Bramble beries	ibid.	Common Camomill	ibid.	Comfrey & Comferie	145
Marche Betil or Pestil	513			Purple Camomill	184.188	Compn or Compn	175
Ameos or Ammi	172	Betony or Betayne	291	Yelow & white Camomill	183	Cone or Pine apple	770
Anemone	413	Paules Betony	26	Wilde Campion	159	Consounde	
Angelica	297	Water Betony	44	Spanish Canes	514	Great Conyza	165
Anthyllis	13	White Beete	550	Larges brode Compn	183	Bastard Copinthes	683
Anyse	271	Bindeweede	393	Sugar Cane	514	Cornerose	
Apple tree	702	Rough Bindweede	396	Canterbury belles	172	Cornflowre wylde and great	434
Apples of loue	439.440	Birche tree	758	Capers	680		161
Apples of Perowe	441	Byrdes foote or fowle foote	487	Capsifoyle	391	Cornell tree	726
Female balsam Apple	441			Carob tree	740	Coriander	276
Prickle Apples	441	Blacke beric bushe	661	Wilde Caroline	531	Coronations or Cornations	
Thornie Apples	ibid.	Bladder nut	735	Wilde Carthamus	532		156
Male balsam Apple	442	Blewblaw	161	Carrottes and of all his kinde	602	Golden Cotula	187
Blacke Archangell	157	Blewoottel	ibid.			White Cotula without vorie	186	Cotton	679
Arbute tree	728	Blighted	471	Wilde Carrot	285.528	Couch & Couch grasse	504
Arbor Iuda	741	Blites and Blittes	547	Cataphilago	90	Cowslippes	113
Argentine	526	Bloodwort	380	Caruway	274	Cowslip	125
Aristologia/&c.	314	May Blossoms	178	Caseweede	81	Yellow Craw	411
Arisaron 314 Iron	312	Blood strange	96	Cassia fistula	740	Creame or flower of Cristall	
Iesse-smart	633	Bockwheate	468	Casidonie	266		116
Artechock	524	Bolbanack or strange violets	153	Casshes or Capes	616	Cranes bil 48 Cresses	613
Wilde Artechock	535			Catanance	508	Sciatica Cresse	630
Asarabacca	319	Bombace	679	Cattes tayle	513	Water Cresse	625
Asclepias	317	Small Bombase	90	Cedar and Cedre tree	765	Winter Cresses	626
Ashe tree/&c.	748	Borage	11	Celandyne		Yellow water Cresses	611
Aspe	750	Boxe thorne	700	Centorie great & smal	317	Crest-marine	578
Asses boxe tree	700	Boxe tree	699	Ceterach	408	Crompled lettis	577
Assa foetida	304	Boyes Mercury or Philon 78 Brake	401	Chafeweed	90	Croswort	541
Auens	134			Thistel Chameleon	517	Yellow Crow belles	174
Autumne Belflowers	173	Bramble	661	Charlock	610	Crowe sope	159
Arewort	489	Brookelyme	579	Cheele running	539	Crowfoote	419
Arsich	429	Broome	664	Cheries and of al his kinde and fruite	714	Water or marrishe Crowfoote	
Sea Iygreene	353	Byer bushe	655				410
		White & blacke Briony	387	Winter Cheries	445	Heath Crowfoote	419

B

BAchelers Buttons	412	Buckhorne	95	Long Cherie tree	716	White Crowfoote 107.410
Bay or Laurel tree	688	Buckrammes	636	Cheruil and Cheruel	614	Crowtoes	206
Ballock grasse	222	Bugle	133	Wilde Cheruel	615	Wylde Cucumber	373
Fooles Ballore	222	Bugloffe	8	Toothpicke Cheruil	ibid.	Cucumbers	586
Hares Ballore	212	Bugloffe the lesser	10	Chesnut tree	730	Leaping Cucumber	373
Balsampnte	250	Vipers Bugloffe	ibid.	Middle Chickeweede	53	Cudweed 90 Curagie	632
Yellow woolfs Bane &c.	429	White Bulbus violet	216	Choke fitche	664	Currantes	651
Banewort	415	Bulbyne	644	Chokeweed	664	Cuscuta/&c.	390
The Barberie bushe or tree		Bulfoote	20	Chyne or Sweth	643	Lyues Liuet	641
	684	Bupleuros	506	Licheling	478	Cytisus bush	736
Barberies	ibid.	Bupcestis	544	Wilde Liches	479	Lyppes tree	767
Barleys of al his kind	461	Burned	471	Lidrage	633	Lyppes nuttes	ibid.
Bards Mercury or Philon	78	Burnet	138	Cinquefoyle or Sinkefoyle 83 Cistus	659	Feelde Lyppes	28
Base Broome	667	Butchers broome	674				
		Great Burre	15	Citros 704 Citrulles	589	Garden Lyppes	19

Dasio.

The Table.

D
Daffodill	649
White Daffodill	10
Dandelyon	569
Darnell	469. 504
Danewort	320
Date or Palme tree	737
Daurus of Candy	285
Daysies	170
Flower Deluce	193
The smallest flowre Deluce	195
White flower Deluce	53
Yellow flower Deluce	199
Deuils bit	110
Dewberie or blackberie	661
Dictam or Dictamnum of Candie	168
Dill	270
Bastarde or false Dictā	344
Dittany	631
Diuels bit 567 Docks	558
Doder	398
Dogge berie tree	726
Dogges Call	77
Dogges Camomill	186
Dogges Leekes	209
Dogges Tooth	203
Doue foote	47
Double tongue	675
Dragon biting	622
Water or Marshe Dragon	322
Dragons and Dragon wurt, &c.	322
Dzauck	471
Dubble leafe	214
Duches meate	107
Dunche downe	513
Dwale	446
Dwarffe Plame tree	760

E
Erthnuttes	491
Earth Chestnut	579
Eglantine	656
Water Elder	760
Elder or Bourtree	378
Elecampane	336
Marshe Elder	760
Wilde white Elleboz	349
Elme tree	751
Greene Endiue	574
Wilde Endiue	563
White Endiue with the brode leaues	563
Erysimon	630
Esula and Ezula	361
Euphorbium	309
Ewe or Yew tree	768
Eyebright	40

F
Fit or floure of glasse	116
Fenberies	671
Felworte	332
Fenell	269
Wilde, & great Fenell	269
Dogge Fenell	186
Fenell Giant	269
Fenegreek	492
Ferne male & female	401
Stone Ferne	408
Oke Ferne	403, 405
Perle Ferne	405
Ferula	19
Feuerfew	544
Ficus Indica	717
Figge tree	408
Finger Ferne	31
Figworts	

Fistich Nuttes	375
Water Flagges	199
Wilde Flagges	ibid.
Flax or Lyn	70
Corne flag	197
Tode and wilde Flax	80
Fleabane	104
Fleawurte	ibid.
Flebane	35
Bloody Flewurte	89
Flixwort	117
Floramor	168
Our Ladyes flower	209
Floure Gentill	168
Floure Constantinople	157
Flourie dole	554
Folefoote	20, 319
Forget me not	28
Fore gloue	175
Foxetayle	542
Framboys	662
Francke	56
Fumetorie	24
Hedge Fumetorie	ibid.
Great Furze	669
Ground Furze	ibid.
Fiuelcaued grasse	83

G
Galangal	346
Galbanum	307
Gallowgrasse	72
Garden woad	67
Garden flagges	193
Garlyke	637
Crow & wilde Garlike	ibid.
Garden Garlyke	637
Beares Garlike	638
Rushe Garlike	643
Garlikethlaspi	628
Gentian	332
Ballinet Geranium	48
Dwarffe Plame tree	760
Smal or dwarf Gentiā	334
Crofoote Geranium	481
Germander	25
Water Germander	111
Wilde Germander	112
Water Gilloffer	106
Yellow Gillofers	151
Casteel Gilloffer	152
Stocke Gillofers	ibid.
Rogues Gillofers	153
Cloaue Gillofers	156
Garden Gillofers	ibid
Feathered Gillofers	156
Cloue Gillofers	ibid.
Marshe Gillofers	157
Cockow Gillofers	ibid.
Mocke Gilloffer	335
Gingidium	615
Stinking Gladin	196
Corne Gladin	197
Right Gladin	ibid.
Ranke stinking Goate	549
Goates bearde	41, 167
Goates Cullions	111
Goldcuppes	422
Golden Appels	440
Golde floure	89, 190
Golden Floure of Perrowe	192
Golden flower	410
Goldknoppes	411
Goose foote	548
Golden Rodde	141
Goldnappe	411
Goosenest	224
Goose-grasse	539
Gooseshare	ibid.
Good Henry	561

Goe to bed at noone	167
Gourde	592
Long Gourdes	ibid
Gooseberies	683
Blacke Gooseberies	683
Beyondsea Gooseberies	683
Redde Gooseberies	ibid.
Grace of God	48
Sea Grape	116
Wilde Grape or Vine	651
Gratia Dei	375
Grasse comforting the eyes	506
Threeleaued Grasse	406
The grasse of Parnassus	510
S. Ihons Grasse	64
Square S. Johns grass	
Gremill	65, 290
Gromel	ibid.
Ground Pyne	28
Groundswel	570
Gumme Dragagant	543

H
Hires foote	502
Harmall	263
Hartes ease	149, 151
Hartwurt	314
Hashewurt	171
Hasel or Filberd tree	734
Haselwort	319
Hauer	505
Hawkeweede	567
Hawthorne	698
Heare brēble	661
Hearons byll	47
Deathmouse eare	87
Heath	678
Heath bramble	661
Blacke Hellebor	351
White Hellebor	347
Hempe	71
Hepe tree or chast tree	691
Yellow and white Henbane	450
Hemlocke	451
Hennes foote	14
Herbe Aloe	353
Herbe Bennet	134
Carpenters Herbe	133
S. Christophers Herbe	402
Herbe grace	261
Herbe Jue	28, 95
Judaical Herbe	260
Herbe Paris	425
Herbe Robert	47
Herbe twopence	78
Vipers Herbe	10
Hermodactil	367
Hepeтree	655
Hindberie	662
Hirse	463
Huluer	519
Hurt Sicle	161
Husucr	701
Hyacinthe	206
Autumne Hyacinthe	209
Bush or tuft Hyacinthe	ib.
Hygtaper	110
Hysope	277
Pepper Hyssope	330
Hockes	581
Smal wilde Hocke	581
Hooke heale	135
Holewurte	316
Holly	701
Holyhockes	581
Common Hockes	ibid.
Sea Holly	519
Holme	701

Sea Holme	519
Holowcroote	316
Smal honesties	156
Honysuckle	391
Hoppe	400
Horehounde and of all his kinde	257
Horestrange	298
Marshe or water Horehounde	257
Horseflowre	164
Horseheele	336
Horse houe	20
Horse tayle & sauegrasse	102
Horse tongue	675
Hound-s tree	726
Hound berie tree	ibid.
Housleeke	114

J
Icke by the hedge	639
Jasmine	658
S. James worte	70
S. Johns worte	64
Ioeries	630
Wall Ierne	403, 404
Josephs flowr	167
Narrowe bladed Ireos	194
Dwarffe Ireos	195
Wilde Ireos	196
Bulbus Ireos	198
Iris	193
Yellow wilde Iris	199
Italian ferche	490
Iuiub tree	712
Iungfraw hare	414
Iuniper beries	763
Iuniper tree	ibid.
Juraye	469
Iuybindweede	394
Iuye black and small	388
Gronude Iuye	389

K
Water kars	625
Prickled Ball	116
Kneeholme	674
Knapweede	109
Knecuyl	674
Knights milfoyle	143
Kynde	252
Knights water Hengreene	143
Knights worte	ibid.

L
Ladies mantell	140
Langdebeek	568
Lampsana	560
Larckes claw	165
Larckes spurre	ibid.
Laser	303
Laserpitium	ibid.
Lauender cotton	19
Lauender gentle	166
French Lauender	ibid.
Lauers, or Leuers	199
Lauriel, or Lowrye	695
Laurus of Alexandria	676
Frenche Leeke / vnset Leeke	641
Leeke or Leekes	641
The bradded or knopped Leeke	641
Hayden Leeke	ibid.
Rushe Leekes	643
Wild & Corne Leeke	644
Lentiles	487
Leopardes foote	420
Lettis	573
Water Lentils	107
Losed, or Cabbage Lettis	

The Table.

His 573 Lettus ibid.	Mew or Meon 337	Rushe Onyons 643	Water Pepper 632
Wilde Lettuce 574	Mezereon 369,370	Wilde Onyon 644	Water Pepperwurt ibid.
Liblong 39	Middell Consounde 133	White feelde Onyon 645	Indian Pepper 634
White Lillie 200	Middle Comfrey ibid.	Sea Onyon 646	Calecute Pepper ibid.
Wilde Lillie 203	Milfoyle 18,144	Opopanax 302	Periploca 318
Lillie nor. Bulbus 204	Milhewurte 49	Drache 546 Orchis 222	Perwincle 33
Lillie of Alexandria 205	Mill 463 Millet ibid.	Bastard Orchis 224	Biting or bushe Perwincle 325 S. Peters wortes
White water Lillie 182	Mistell and Mistelto 747	Orenge 704	
Limons 704	Mistwast 406 Moly 509	Origanie 232 Origan 237	Petigree 674
Linden tree 753	Bastarde Moly ibid.	Wilde Origan ibid.	Pety cotton 90
Lillynarcissus 213	Momordica 442	Goates Origan 239	Pilcorne 467
Lingwort 347	Monywort 79	Mobstrangler 664	Great Pilosella 87
Liuelong 39	Moonewort 136	Dipyne 39 Orchanet 9	Pimpernell 55
Liuewort 411	Morell 443	Olnside the Waterman 402	Pimpinell 138
Stone Liuerwort ibid.	Petie Morel ibid.	Osmunde Baldepate 402	Pinkeneedell 47
Lote tree 729	Moschata 47	Small Osmunde ibid.	Pine tree and of all his kind 770 Cockowpitel 323
Louein idlenes 249	Mosse of the Sea 414	Pylde Osmunde ibid.	
Lousepowder 372	Motherworde 89	Otes 467 Pilde Otes ib.	Pitche 771 Plane tree 755
Louage 295	Motherwort 131	Pour Otes 471	Plantayne 92
Lungwort 414	Stinking Motherwort 549	Wilde Otes 505	Sea Plantayne ibid.
Lunarie 136 Lupines 482	Mothe Mulleyn 122	Ote grasse 505	Coronop Plantayne 95
Lungwurt 115	Mouse eare 54,87	Orecye 189 Orelips 123	Water Plantayne 97
Lycorose 694	Mouse tayle 96	Oretongue 8,12	Plumme tree 720
Lyllie Conuall 178	Mugworte 16	Oxytriphyllon 501	Prickley bore 700
May Lyllies ibid.	Pety Muguet 539		Prickmadam 114
Lyrconfancy ibid.	Golden Muguet 541	P	Priche timber tree 726
Lysimachia 74	Mulleyne 120	Pidelion 140	Priest pintell 222,323
Blewe or azured Lysimachus 75	Mulberie tree 715	Pagane or vplandishe tongue 675	Primeroses 123
	Musa or Mose tree 705	Palma Christi 216,354	Primerose picrelesse 121
Three leafe Liuerwurt 59	Mustarde 619	Palme tree 737 699	Primprint 690
	White Mustarde ibid.	Panar 293 Pances 149	Priuet ibid. Prunel 133
M	Myagrum 494	Panick 466	Pylewort 31
Reede Mace 513	Milt waste 408	Petie Panick 465	Pynchens 156
Madder 538	Myntes &c. 244	Parsely 605	Poleruhe 512
Male knot grasse 99	Horse Mynte and of all his kynde 245	Garden Parsely ibid.	Polemonium 345
Mallowes &c. 581		Hyll Parsely 607	Poley/&c. 233
Dwarfe Mallowe ibid.	Corne Mynte 248	Mountayne Parsely ibid.	Polypody 403
Common cleyn and tawle wilde Mallow ibid.	Wilde Myrtel 674	Marche Smallache and Warrishe Parsely 607	Goldylockes Polytrichon 414
Marish Mallowe 583	Myrtell tree 687		
White Mallowe ibid.		Stone Parsely 608	Pomegranate 706
Cut Mallowe 585	**N**	Great Parsely or Alexander 610 Wild Parsely 610	Turkie Pompons 589
Symons Mallowe ibid.	Narcissus 211		Pondeweede 105
Veruayn Mallow ibid.	Rush Narcissus 212		Poores mens treacle 637
The slymie or Mucculage Mallow of Vennis 525	Bastarde Narcissus 214	Sallade Parsely 611	Popler or Pepler 750
	Yellow Narcissus ibid.	Water Parsely ibid.	Poppie 432
Male Mandrage 438	Narcissus violet 216	Bastard Parsely 612	Blacke & wilde Poppie ib.
Female Mandrage ibid.	Nauet 595	Parsnep 604	Redde Poppie 434
White Mandrake 438	The moyst or water Nauet 595 Nauew geele ib.	Wilde Parsenep ibid.	Horned Poppie 436
Blacke Mandrake ibid.		Medow Parsenep 528	Pudding grasse 232
Maple 761 March 453	Paris Nawes ibid.	Cowe Parsenep ibid.	Pultall mountayne 232
Maricrom / and of his kynde 234.236	Neckwee	Parietary 49 Partizan 75	Puliol Royall 232
	Needel Cheruill 615	Passe flower 122.410	Purcelayne 576
Bastarde Marierom 237	Shepheardes Neel ibid.	Pastel 67	Wilde Purcelayne ibid.
Coast Marie 250	White and Yellow Nenuphar 182	Redde Patience 559	Garden and tame Purcelayne 576
Marsche Marigolde 31		Peache tree 710	
Marygoldes & Ruddes 163	Neppe & Cat Mynte 248	Pearleplante 290	Purple 122
Wilde Marygolde 190	Nelewurte 347	Peare tree 712	Purple veluet flower 168
Marril, & beries 671	Smal Nettel 179	Common Peason 476	Red purple lillie 202
Mastic 235	Great common Nettel ibid.	Garden Peason ibid.	
Iuperatoria Mayster wort 298 Mastersion 109	Romayne or Greeke Nettel 120 Dead Nettel 131	Wilde Peason ibid.	**Q**
		Branche Peason ibid.	Qince tree 708
Mathers 186	Blind Nettel and Archangel ib.	Great Peason ibid.	Quickbeame 748
Foolish Mathes ibid.		Brode or flat Pease 478	Our ladies quishio 509
Redde Mathes 122	Nettle tree 729	Ciche Peason 479	
Mawdelein 250	Fielde Nigella 160	Sheepes riche Pease ibid.	**R**
Mayden Mercury 78	Garden Nigella 278	Pellitorie of the wall 49	Radish 598
Maydenheare 410	Wilde Nigella ibid.	Pellitorie of Spayne 300	Wild or water Radish ib.
Medewerte 41	Wooddy Nightshade 398	Bastard Pellitorie or Bertram 341	Bell Ragges 611
Medewurte ibid.	Nightshade 443.445.446		Ragwort 111
Medick & yellow fitche 485	Noose bleede 144	Wilde Pellitorie 343	Raifort or mountayne Radish 599 Rampions 597
Medow Shruegrasse 99	Nutmegge & macis 732	Pellamountayne 231	
Medler tree 713	Water Nuttes 536	Penny Royall 231	Ramsons 638 Rapes 594
Germaines Melilot 497		Wilde Penny Royall 248	Countrie Rapes 174
The common and best knowen Melilot 498	**O**	Sheepe kyllyng Pennye grasse 38	Rape Crowfoote 595
	Oculi Christi 255		Long Rape 595
Melons 588.590	One tree 745	Great Pennywurt ibid.	Wilde Rapes 497
Muske Melons 590	Oke of Hierusalem and Oke of Paradise 343	Mountaine Penywort ib.	Smal Rasens of Cornih the 652 Raspis 651
Turkie Melons 589		Peonie male & female 338	
Mercury 77	Oleader 430 Oliue tree 737	Pepons 588 Pepper 635	Raspis and Framboys beries ib. Red Rattel 516
	One blade 178	Pepperwurt 631	
	One leafe ibid.		Raye 45 Red-Ray 504
	One berie 425 Onyon 640		Pot

The Table.

Pole Reede 514
Cane Reede ibid.
Indian Reede ibid.
Sugar Reede ibid.
Reede grasse 515
Rest harrow 669
Reubarbe or Rhabarba 372
Bastard Reubarbe 43
Rhamnus 696
Blacke Ribes 683
Common Ribes ibid.
Right Scolopendria 408
Rice 461 Rheyn beries 697
Garden or tame and gentill
Rockat 612 Rockat ibid.
Water Rose 181
Rose tree 430 Rose 655
Cyuet Rose ibid.
Wilde Rose ibid.
White Roses ibid.
Muske Roses ibid.
Damaske Roses ibid.
Rose of Prouince ibid.
Rose Campion 158
Wilde Rose Campion 159
Rosemary 164
Libanotis Rosmarie 279
Rose Baye tree 430
Rosen that cometh out of
the Pine and Pitche of
trees 771 Rosewurt 341
Rue of the garden 261
Wilde Rue 263
Goates Rue 490
Rue of the wall 409
Rushes 511
Bul Rush or panier rush ib.
Rush candle ibid.
Mat or frayle Rushe ibid.
Wilde Rushe 671 Rye 459

S

Saffron 217
Bastard Saffron 33
Mede & wild Saffron 367
Wilde bastaerd Saffron 531
Sagapenium 306
Sage & of his como sage 251
Sage of Ierusalem 115
Wood and wilde Sage 253
Haligot 536
Salomons seale 103
Salt wort 116 Sapier 578
Hanamunda 134
Sanicle or Sanikell 139
Great Sanicle 140
Sarapias stones 221
Sarcocoll 311
Sarrasines Comfery 141
Sarrasines consound ibid.
Satyrion 225
Bastard Satyrion 225,226
Red & Syrian Satyrion 225
Three leaued Satyrion 225
Satyrion royal or noble 226
Sawce Sumach 692
Sauin tree 766
Commo garde Sauorie 112
Somer Sauorie 109
Winter Sauorie 230
Sauce alone 639
Saxifrage 186
Golden and white Saxi-
frage 188 Scabious 109
Seabwort 336
Scaleferne 408
Scamonie 396 Scorpio 111
Scorpion wurt 63
Scorpion ibid.
Scorpioides ibid.

Sea cawle 594
Our Ladies Seale or Sig-
net 384 Sebestens 712
Helfe heale 133 Senuie 619
White Senuie ibid.
Sene 576 Sengreene 114
Setfoyle 24
Serpentes tongue 135
Seseli 281
Seseli of Candie 184
Setwal or Sydwal 340
Sharewurt 36
Sheapherds purse 81
Singleleafe 178
Wilde Skirwit rootes 605
Skirwurt ibid.
Sleeping Nightshade 447
Sloo tree 711
Garden Smilax 474
Smyrnium 613
Cat Sloses & Snagges 711
Snakeweede 23
Snapdragon 180
Soldanella 395
Sowbread 330
Sonnedeaw 414
Sophia 117 Soopewort 335
Sorbe apple tree 727
Sorrel 559
Great Sorrell ibid.
Sheepes Sorrell ibid.
Small Sorrell ibid.
Water Sorrell ibid.
Horse Sorrel ibid.
Souldiers yerrow 143
Southrenwood 1
Female Southrenwood ib.
Great Southrenwood ib.
Smal Southrenwood ib.
Sowfenill 298
Sowthistel 565
Spanish broome 666
Spanish or canary sede 465
Spearwurt 340
Single Spelt 458
Spelt or Seia 455
Sperage 474,618
Sperhawke herbe 567
Sperwort 425 Spier 514
Spike and Lauender 165
Spinache 556
Broade or large Splene-
wort 406
Wilde or rough Splene-
wort 407 Spoonewort 118
Spourgewort 196
Spourge and of all his
kinde 361 Spurrie 56
Squilla 646 Squinat 511
Stachis 257
Standelwort 112
Eunuches Staedergrasse 114
Standergrasse 222,225
Stannewort 80
Starre of Hierusalem 167
Golden Stechados 29
Starrewort 36
Stichwurt 525
Stone-breake 188
Great stone croppe 114
Stonehore 115
Storkes byll 47
Strangleweede 664
Strangle tare ibid.
Strawberie tree 718
Strawberie or Strawbery
plant 85 Lousestrife 1
Stubwort 503
Garden Succorie 563

Gumme Succorie 569
Yellow Succorie 563
Sulpherwurt 298
Sumac 673
Sumach 692
Losiers Sumach ibid.
Leather Sumach ibid.
Meate Sumach ibid.
Indian Sunne 192
Swallowurt 13,317
Swinescresses 95
Sycomore tree 716

T

Tansie 12
Wilde Tansie 86
Tamarisk 677 Tares 426
Tarragon 623
Fullers Teasell 522
Tetterwurt 31
Hundred headed Thistel 519
Starre Thistel 522
Carde Thistell 522
Our Ladies Thistel 525
Globe Thistel 526
Cotton Thistel ibid.
Ote Thistel ibid.
White cotton Thistel ibid.
Wilde white Thistel ibid.
Siluer Thistel ibid.
Carline Thistel 530
Blessed Thistel 533
Wilde Thistels 535
Lowe Thistell ibid.
Rough milke Thistel 565
The tender or soft milke
Thistel ib. Thlaspi 618
Candy Thlaspi 619
Bucke Thorne 697
Thorne broome 668
White Thorne 698
Thorne grape 681
Thorne boxe 700
Blacke Thorne 711
Thorow ware 137
Thorow leafe ibid.
Thotewurt 171
Spurge time 363
Dogges Tongue 11
Houndes Tongue ibid.
Sheepes Tongue 9
Stone hartes Togue 406
Tongueherbe 675
Tonguewort ibid.
Tongue blade ibid.
Tongue Laurel 676
Torches 520
Water Torche 521
Cromett 24 Cornsol 61
Towne Cresses 613
Towne kars ibid.
Base Trefoyl 501
Sea Tryfoly 49
Common Trefoyle 495
Medowe Trefoyle ibid.
Sweet Trefoyle 496
Wilde yellow Trefoyl 497
Horned Trefoyle or cla-
uer 500
Spanish Trefoyl ibid.
Stinking Trefoyle 501
Pitche Trefoyle ibid.
The right Trefoyle ibid.
Tree Tryfoly 746
Tulpia or Tulipa 123
Iesues Turbith capsia 365
Scrapions Turbith 364
Turbith corne 464
Turkie Gilloflers 176

Turneps 594
Tursan or parke leaues 66
Twayblade 224
Tyme &c. 229
Running Tyme 231
Wild Time ib. Tymbra 230

V

Great wilde Valeris 340
Wall Barley 504
Wall flowers 151
Walnut / and walshe nut
tree 731 Walwort 380
The lesser watercresse 625
Waterferne 402
Water spike 106
Wartwurt 361,363
Way Benuet 504
Bitter Uetche or Ers 482
Base or flat Ueruayne 117
Wilde Uetche 484
Weede winde 394
Dyers weede 68
Weede ibid. 186
Windweede 394
Wetche or wetches 483
Wheate and of all his kind
453,455,457.
Low wheate 164
Dre wheate ibid.
Typhe wheate 457
Bearded wheate 456
Spelt wheate 458
Indian wheate 464
Whiteroote 103
Whitewurt 19
Whittree 761 Whorts 671
Whortel beries ibid.
Whyn 669 Pety whyn ib.
The common whyn ib.
Wilde yellow lotus 497
Sweete Williams 156
Wilde Williams 157
Willow herbe 74
Withywinde 393
Withy or Willow 743
Woodbine 391
Woodsowe / or Woodso-
wel 540 Woodsorel 541
Woodwaxen 667
Woolfes clawe 414
Worme grasse 114
Wormwood 5
Sea wormwood ibid.
Lauender wormwood 6
Narrowe leaued worme-
wood ibid. Wulleyn 170
Wountwurt 44
S. Peterswurt ibid.
Wrdowaple 369
Uaencrupt 351
Uenus bath or Bason 522
Neruaene or Uaruayn 117
Uiolets 148
Marche Uiolet ibid.
Earnesee Uiolets 151
Damaske Uiolets 153
Marians Uiolets 174
Theophrastus white Uio-
let 716 Wilde Uine 324
Uinceroricke 127 Uioine 186
The garden or manured
Uine or grape 651
Smal wild Uetchlings 485
Ansuerie Camomil 186
Blanke Ursine 517
Douche branch Ursine 518

Y

Yerrow 144

A Table vvherein is conteyned the Nature, Vertue, and Dangers, of al the Herbes, Trees, and Plantes, of the vvhich are spoken in this present booke, or Herball.

A

TO drawe away the Afterbirth / vide Secondine.
To drawe downe the Afterbirth / vide dead Childe.
For the Ague / 3.b/18.c/59.d/115.g/133.b/148.b/157.a/ 170.a/187.d/302.a/329.a.
For hoate Agues or Feuers / 104.a/135.b/146.a/149.h/ 182.d/319.d/396.a/564.c/576.b.656.a/672.a.d/683.a/ 684.a/702.a/704.a/706.a/712.a/741.b.
For the tertian Ague / 11.c/59.a/61.c/64.b/83.a/93.a/99. c/118.l/133.d/411.a/501.d.
For long colde Agues or Feuers / 287.d/301.b.
Agaynst olde Agues / 399.b/777.h.
To engender or cause Agues or Feuers / 397.c/724.a.
To quench the thirste of hoate Agues / 683.b.
To driue away shakinges & shiueringes of Agues / 301.a/ 305.b/306.c/315.c/391.c/619.f/636.c.
For S. Antonies fyre or wilde fyre / 19.c/10.a/38.a/50.a/ 99.g/104.c/115.c/117.c/144.c/201.g/217.d/276.f/319.f/ 355.c/411.c/433.d/444.a/446.a/488.i/535.b/546.d/ 564.c/573.d/576.b/593.g/638.n/656.c/663.b/682.b/ 686.b/737.d/768.h/778.e.
To restore and cause good Appetite / 301.c/481.g/503.a/ 573.b/597.a/600.b/605.a/605.a/614.a/619.a/ 636.a/640.a/681.b/682.a/684.a/704.c/732.a/739.e.
For the Apoplexi / 178.a/187.d/306.a/375.b/382.c.
To bring their speache agayne to them that are taken with the Apoplexie / 187.f/310.d/710.g/747.e
For the falling downe of the Arsegutte / 37.c/151.b.
Against euil infected Ayres / 161.b/187.c/297.a/299.a/ 335.b/704.a/718.h/763.d.

B

Earde doe growe speedily / 1.c.
To keepe Bees togither / and cause other Bees to come in company / 160.c.
To kyll Bees and waspes / 179.d.
To lose or purge the Belly very gently / 331.c/352.b/ 400.c/437.a/473.a/471.a/474.b/482.b/487.a/493.1/ 554.f/559.a/587.b/593.c/640.a/646.e/649.d/651.c/ 702.b/710.c/717.a/718.a.d/721.a/714.a.b/739.a.775.a
To open the Belly mightily, and purge grosse fleames / 278.a/321.b/432.a/550.a/697.a.
To ope and lose the Belly / 34.b/43.a/78.a/149.h/170.b/ 238.b/239.a/291.g/292.a/308.a/353.a/355.a/361.a/377.a/ 378.a/383.d/394.a/546.a/547.a/551.c/554.a/573.c/581. a/591.a/612.a/656.a/702.d/7.o.b/715.b/716.f/741.a/ 721.d/726.b/778.c.
To cause blastinges and payne in the Belly / 594.a.
To stop the fluxe of the Belly / 83.b/84.b/93.b/99.d/101.a 115.a/199.a/206.a/223.d/236.c/239.a/172.b/305.h/332.b/ 347.f/426.a/409.c/433.c/467.c/473.b/485.a/491.a.b 501.a/504.a/576.c/583.b/584.b.585.a/593.h/613.a/660. c/661.b/681.a/692.c/798.a/711.a/714.a/715.a/721.d.c/ 729.b/730.a/740.b/767.b/779.a.
For the windinesse & blastinges of the Belly / 6.d/71.a/ 301.c.363.b/636.b.737.a.
For the griping paynes of the Belly / 35.b/71.f/117.b/ 230.b/231.a/246.b.111/250.a/252.h.c/171.b/173.a/175.b/ 281.a/183.a/185.b/187.c/196.b/198.a/301.a.b/317.a 319.a/338.a/380.a/403.a/466.a/504.a/510.b/533.b/ 577.h/599.h/600.c/608.b/639.b/646.b/707.1/732.b/ 763.a.
To kyll and spoyle wylde and tame Beastes / as Ryen Swine, wolues and Dogges / &c.410.d/411.a/415.a/ 417.a/430.a/431.a.
To driue away all venemous Beastes / 1.c/6.f.
Agaynst greeuous Beatinges / vide Falles.

To preserue Bier from sowring / 14.c
The inwarde scuruinesse or hurt of the Bladder / 301.a.
The exulceration or rawnesse of the Bladder / 587.b/ 712.c/768.b.
Hurtfull to the Bladder / 158.1.
The inward scabbes of the Bladder / 579.a/687.a/694.c
For the paine or stoppinges of the Bladder / 15.a/18.d/41. a/80.b/93.d/101.b/141.b/182.b.c/184.c/166.a/187.a 299.b/337.a/344.a/399.a/444.b/446.b/514.d/544.a/ 582.b/613.b/616.b/625.a/651.f/657.c 732.
To coole hoate Blood / 576.a/672.b/683.a.
To stoppe al issues of Blood / 42.a/93.b/99.a/115.a/132.a 210.a/176.c/319.d/4.o.c/505.b/517.a/530.f/540.b/576.e 584.h/642.b/652.b/657.f/660.a.c/661.c/661.f/657.b/ 681.b/687.a/691.a/707.g/708.a/721.c/730.d/744.a/ 745.a/746.i.767.a/779.a.b.
To stanche the Blood of greene woundes / 48.c/68.a/ 75.c/83.b/132.a/138.c/144.d/145.c.
To stoppe the Blood of al woundes / 252.d/354.d/411.a/ 415.d/492.b/504.b/510.c/539.c/540.b/707.c/726.b/ 745.b/768.g.
To purge Blood from all corrupt humours / 400.a.c.
For the inflammation of Blood / 411.a/672.b/706.a.
To engender grosse Blood and humours / 482.g.
To engender grosse & melancholike Blood / 555.f/641.a.
To engender euyll Blood / 638.a/718.c/721.b/737.a.
Against the Bloody flixe / 23.a/33.a/57.c/75.a/81.a/83.a/ 84.b/80.a/90.a/93.b/99.a/101.a/104.b/107.b/111.c/115. a/117.a/120.a/138.a/145.a/121.a/199.a/206.a/250.a/232.c 329.a/345.a/466.a/487.b/559.c/651.a.b/683.c/691.b.c 700.a/707.b.715.a/716.a/729.a/730.d/732.a/739.d/ 746.g/771.a/776.f/778.a/779.a.b.
Agaynst the spettyng of Blood or corrupt matter / 16.c/ 33.a/81.a/84.a/99.a/111.f/138.a/139.b/145.a/130.d.246.c/ 252.d/258.a.b/191.f/327.a/354.c/409.c/411.c/454.f/455.a 525.a/576.c/580.b/651.a/577.f/687.a/700.a/708.a/711.c 730.b/744.a/745.a/767.a/776.f/778.a/779.a.b.
To stop the pissing of Blood / 57.b/81.a/84.b/93.b/96.a/ 99.a/138.a/111.a/115.c/537.f/687.a/744.a/745.a/779.a
Cause to pisse Blood / 132.b.
Bodyly lust / vide Fleshly desyre.
To strengthen the Body / 111.a/225.a.
Obstructions and stoppinges of the Body / 15.a.
To dissolue clottie or congealed blood in the Body / 28.f 86.b/111.h/133.a/134.a/229.a/131.c/248.c/263.b/262.c/ 273.b/304.c/333.c/306.c/381.c/384.b/398.b/538.f/ 632.a/718.g.
Agaynst windinesse or ventositie of the Body / 278.a/ 691.b/701.a.
Hurtfull to the Body / 348.1/397.c.
Botches / vide impostemes.
To strengthen the Bowelles / 86.d/248.c/510.a/761.a.
Inflammation and heat exulcerations of the Bowelles or entrayles / 104.a/134.b/495.a/506.g.
To dissolue windinesse and blastinges of the Bowels / 691.b/763.a.
Good to purge & mundifie the Braynes / 23.g/131.b/ 235.f/138.f/187.f/301.b/306.a/308.a/311.a/331.b/368.b/ 373.b/382.a/328.d/550.b/554.c/636.b/640.g.
To cōfort the Braynes / 241.a/164.b/165.b/266.d/179.f 371.d/618.c/779.d.
To dry the Braynes / 779.d.
To warme & dry the Braynes / 250.b/341.a/460.a/658.c.
Slymie fleame from the Braynes / 343.a/348.c.
Troubleth the Braynes / 154.a.
Impostemes of the Brayne / 775.b.
To cleanse the the Breast / 71.g/110.a/111.d/124.a/120.a/ 134.b/146.c/194.d/229.a/232.b/258.a/272.f/241.a/303.c/

403.f.

A Table of the Nature, Vertue, and Dangers.

304.f/306.c/308.b/335.a/336.b/381.a/382.a/492.b/533.c/
599.g/604.a/616.a/630.a/642.d/694.f/705.b/735.
b/771.d/776.b/778.d.
Obstruction or stopping of the Breast/2.a/34.a/121.f/
127.a/143.a/266.b/275.b/308.b/311.b/365.a/492.b/619.b
694.b/721.a/773.a.
The exulcerations or swellinges of womens Breastes
or pappes/154.i/455.c/583.c/582.b/651.k/657.d.
To dry vp womens Breastes/319.f.
For impostumes of the Breast/10.b/110.a/317.b/443.k/
630.c/649.c/651.b/707.g/721.a.
For olde diseases of the Breast/181.b/410.i.
Agaynst great paynes vpon the Breast/450.c.
To cure vnnatural swellings of womens Breasts/730.a.
To keepe maydens Breastes small/452.a.
To heale the hardnesse of womens Breastes/18.d/246.g.
Hurtfull for them that are short vpon the Breast/731.b.
For the shortnesse of Breath/2.a/20.b/25.d/129.a/152.a/
194.b/217.a/229.a/239.c/241.a/248.c/261.b/278.a/283.a/
298.a/303.c/306.c/307.c/308.b/313.b/319.b/322.b/327.a/
365.a/373.a/381.d/401.a/533.b/613.a/619.b/630.a/646.
b.711.c/718.f/735.b/775.a/776.a.
Agaynst payne in fetching of Breath/34.a/689.a/733.b.
Cause to haue a good sweete Breath/304.a/731.a/779.a.
To amende the stinking Breath/704.e.
For suche as are Broken/528.a.
Agaynst scalding or Burning with fyre or water/9.a/
16.g/64.d/65.b/121.b/135.b/201.g/211.b/388.b/g/513.a/
528.b/540.a/551.f/573.d/584.g/593.a/649.b/690.c/715.c
719.a/755.b/777.b.
For them that are Bursten or bruised inwardly/319.c/
403.a/538.c.
For all Burstinges/87.a/101.b/m.b/135.a/146.e/224.d/
336.a/315.d/327.a/443.c/514.c/530.b/542.b/604.b/649.
d/768.c.
For Burstinges of young chyldren/104.d/107.d/137.b/
139.c/299.f.

C

To heale Canckers/44.b/93.f/130.g/322.f/479.c/
630.f.
To prouoke Carnall Copulation, and hinder the entisements agaynst it/3.f. vide fleshly desyre.
Carnall Copulation/ vide fleshly desyre.
To dry vp Catharres or Reume/74.b.
For falling downe Catarres or humours/93.a/433.b/
450.c/530.c/721.b/722.b.
For pestilent Carbuncles/sores/or botches/111.c/148.f/
707.b/768.b.
Hurtfull to all Cattell/760.a.
For suche as are sicke with eating of Champions/ or
Todestooles/6.c/261.c/509.f.
For to delyuer the dead Childe/28.f/35.a/163.b/184.a/
229.a/231.a/236.b/158.b/d/h/261.f/265.a/268.a/283.b/
285.c/287.b/294.b/199.b/303.f/304.b/307.c/308.b
314.b/327.b/331.c/342.b/352.c/373.c/382.c/396.d/438.b.
482.a/613.c/616.a/614.c/660.f/766.a.
To engender male Children/78.b/113.c/60.b.
To engender female Children/78.b/113.c.
For Chyldren troubled with the crampe, or drawing of
any member/525.c.
To dye or colour Clothes yellow/697.b
To dye or colour Clothes greene/68.a/697.b.
To dye or colour Clothes blacke/755.a.
For the Colerike humours/6.b/14.b/24.c/104.a/377.
a/733.a/775.a.
For Colerike inflammations/19.c/50.a/117.b/246.b.
For the Colicke/34.a/134.c/184.b/187.b/246.d/285.b/
287.c/355.a/365.a/375.b/403.a/443.b/510.b/701.a.
To take away the good Colour/ and bring palenesse/
273.c/155.f.
To take away euyll Colour/479.a/481.b.
For suche as fall into Consumption/221.a/689.b/771.
d/776.f.

For the Cough/15.a/34.a/93.a/148.b/152.a/239.a/252.e/
262.h/275.b/291.a/299.c/305.f/30.a/330.b/346.b/
409.a/433.b/440.c/455.b/534.b/581.b/60.b/649.b/651.
f/660.f/711.b/712.f/722.e/725.e.
Agaynst the olde Cough/17.c/20.e/71.g/79.a/110.a/c/
246.b/158.a/261.b/270.c/272.e/283.b/283.b/c/
c/311.a/322.a/381.a/592.b/513.a/b/c/a/542.b/646.b/c/694.
b/730.b/771.b/777.e.
For the Cornes which be on the toes and feete/101.c/
305.f/744.b.
For the Crampe, or drawing togither of sinewes/132.k
236.a/287.b/292.b/299.f/301.b/305.l/306.b/307.a/310.a/
319.a/321.a/329.a/443.b/510.b/528.a/530.b/577.n/581.a/
649.b/689.b.

D

Agaynst the Deafenesse/390.a/583.a/554.f/620.b/
649.k/689.c/718.b/764.c.
To bring and cause Deafenesse/514.c/140.f.
For the Disenteria, or dangerous flixe also called/189.h
To make good Digestion of meate/174.a/c/a/287.c/
291.b/296.c/301.b/336.b/507.c/606.b/619.a/604.b/
634.a/636.a/104.c/712.g.
Agaynst the bytinges of mad Dogges/15.c/91.f/130.g/
246.b/158.c/261.b/270.f/317.b/301.c/305.h/127.c/407.b
454.a/452.c/584.g/587.c/638.c/711.g/719.b/731.b.
To keepe a man from Dreaming a starting/172.i/574.e.
Good for melancolike Dreames/71.b.
To cause heauy Dreames/44.b/641.a/642.f.
To keepe a man from Dronkennesse that day/6.c/117.a
687.f/711.c.
To cure Dronkenesse/555.m.
To cause Dronkenesse/164.a/111.a.
Against Droppisse/vide Stranguria.
To helpe the Dropsie/6.b/15.a/34.a/111.b/78.a/147.a/
141.b/194.a/234.a/261.i/p/216.b/111.c/221.b/311.c/
346.a/347.a/352.g/355.c/361.a/c/d/408.b/173.a/179.c
380.b/384.a/395.a/448.a/514.b/517.b/601.c/613.b/626.b.
638.g/646.b/668.a/681.g/711.n/741.b/769.a.
To slake the belly of such as haue the Dropsie/240.b.
To make wyues or maydens Dugges harde/140.b.
For them that are heauy and Dull/177.c/614.c.

E

For payne into the Eares/50.b/73.c/71.c/87.c/93.b/
99.c/104.c/217.b/236.g/246.i/261.b/255.b/312.c/352.
c/374.i/400.b/433.f/444.c/410.a/440.b/539.b/550.
c/551.b/591.f/b/57.c/66.c/738.c/718.b/750.b.
Impostumes behinde the Eares/1.a/g/444.c/484.f/
560.i/514.b/610.c.
Impostumes in the Eares/70.a/93.b/172.i/307.b/281.g
444.g/464.c/640.g.
Good for runing Eares/104.c/328.g/687.c/691.b/700.
a/739.c/778.c.
For wormes in the Eares/124.c/270.i/611.g/765.c.
To clense stopping into the Eares/104.c/641.f.
For singing or humming of the Eares/100.a/400.b/
550.c/554.b/620.b/641.f.e/p/712.b/765.c.
Inflammations or rednesse of the Eyes/f.36.b/52.a/
56.b/81.b/b/105.c/148.f.161.c/176.c/178.c/193.b/
337.c/356.c/444.b/450.a/513.g/564.b/576.b/580.b/
581.c/606.c/656.c/671.a/686.c/707.c/715.c.
Dimnesse of the Eyes/6.g/84.c/111.c/141.c/211.a/263.b/
281.b/187.b/306.i/311.b/328.b/354.g/411.b/457.b.
For bloodshot or blacke spottes Eyes/6.g/31.a/161.b/
271.c/281.b/301.f/p.
For the paynfull bleared Eyes/6.g/229.b/450.a/571.b
557.c/707.b.
To dryue away hawe or pearle from the Eyes/15.c/161.
b/306.q/311.g/424.b/406.b/490.c/640.c/715.b.
Agaynst fistulas and vlcers in the corners of the Eyes/
54.a/114.g/374.g/471.b/421.b/497.b/607.c/b/c.
To preserue Eyes from flowyng downe of humours/
117.c/241.c/444.b/45.a/281.b/b.b.
To take away roughnesse of the Eye browes/610.i.
Hurtfull for Eyes and sight/130.a/64.a/641.b/739.g.
For the payne of the Eyes/657.c/678.a.

To

A Table of the Nature,

To stop the running and watering of the Eyes/755.d.
To sharpe and quicken the Eye sight/24.a/.d./32.a/40.a/71.d/88.d/154.a/158.a/161.k/b/163.b/281.d/296.g/303.c/305.g/306.i/308.g/331.g/348.f/510.d/557.b/574.d/597.c/610.i/636.g/740.c/649.f/700.d/765.b.
For Enchantmentes or witching/102.c/121.l/509.b/695.t.
For Epilepsie/vide Fallyng sicknesse.

F

For the rednesse of the Face/44.c/84.c.
To take away spottes and lentiles, and clense the Face/skinne/or the body/70.c/86.d/103.b/116.b/181.c/194.i/211.c/261.d/279.b/287.d/296.g/308.g/310.c/329.c/331.i/333.c/355.d/384.b/391.d/454.d/467.b/428.d/561.b/584.i/589.c/599.n/610.b/612.d/626.b/634.b/665.b/711.h/752.c/767.c.
To beautifie the Face and Skinne/589.c/594.g/597.b/751.c.
To cause diuers spottes/freckles/pimpels/to aryse in the Face/546.c.
For the Falling euyll/670.b/715.f/747.e/776.d.
For such as are fallen aloft/and are bruysed or beaten/253.a/301.d/302.a/333.c/398.b/538.c/680.b/718.g/750.d/763.b/776.i.
Agaynst grieuous falles/319.c/403.a.
For suche as are faynt and fallen in a sounde/241.e.
For the Feuer/vide Agues.
Quartayne feuer/12.c/vide Ague/61.c/64.b/83.e/93.c/305.d/403.a/408.a/501.d/522.d/533.c.
Cornes on the Feete and handes/101.e/305.f.
Chappes or riftes of the feete/646.h/778.e.
To kyll Fishe/362.e.
To the Fistulas/83.g/93.f/144.d/223.e/325.a/328.m/348.f/361.d/471.a/773.f.
For them that are Flegmatique/311.a.
To dryue away Fleas/6.i/36.g/104.f/133.n.
To take away olde nature Fleshly desyre/ or Carnall copulation of Uenus/182.c/111.b/576.b/691.a/744.c.
To prouoke Fleshly desyre or Uenus/197.c/117.a/221.a/225.a/153.a/251.a/271.e/271.e/524.b/d/544.g/602.b/617.b/749.g/778.b.
To take away superfluous proude Flesh/746.k.
To dryue away Flyes/6.i/175.d/179.d/348.g.
For to prouoke and bring downe the natural Flowers of women/1.a/28.b/35.a/64.a/101.a/111.a/117.e/130.k/151.d/152.b/163.b/183.d/184.a/194.c/201.b/229.a/230.a/231.a/234.a/235.d/236.a/243.b/248.b/253.a/258.b/261.a/263.a/265.a/268.a/278.a/281.a/283.b/284.a/285.a/287.b/292.b/294.a/296.b/c/299.b/303.g/305.i/306.f/307.c/314.a/319.d/327.b/331.b/336.a/338.a/341.a/344.a/346.a/342.b/351.c/373.c/381.f/396.d/481.c/501.a/510.c/533.b/538.g/554.g.i/555.e/578.a/580.c/599.b/601.b/607.a/609.a/613.c/619.c/614.b/618.a/630.b/640.b/641.e/649.a/674.b/676.d/681.c/691.d/729.a/765.i/766.a.
To stoppe the inordinate or ouermuch flowing of womens flowers or termes/13.b/37.a/75.b/84.b/85.a/86.a/93.b/99.d/c/101.c/115.b/138.a/145.b/181.b/241.c/276.b/275.d/333.d/347.f/415.a/c/432.i/477.b/c/487.b/491.c/505.a/511.d/517.a/530.c/580.b/651.i/657.c/660.c/661.c/669.a/677.d/684.b/685.a/687.a.g/692.a/c/698.g/700.b/708.f/731.c/739.d/744.a/746.i/763.g/778.a.
Good for the white floud or Flowers of women/86.a/181.b/495.b/657.f/691.c.
To purge Melancholy Fleumes/19.a/377.a/403.a/437.b.
To rype Fleumes/219.f/217.f/316.b/381.a/409.a/414.a/499.f/599.g/631.a/691.a/718.f/771.d/778.d.
To purge colde Fleumes of the stomacke/331.681.b/751.d./769.a/763.f.
To engender Fleumes and choler/705.a.
Cause a man to fall into Frentic/176.g.
Agaynst Frensie/396.a.
Dead Fruite/vide dead Childe.

To take away all outgrowynges in the Fundament/306.c.354.c.
To settell the Fundament fallen out of his place/331.f/707.f/779.a.
For the swelling of the Fundament/37.d/271.c/354.d.
To heale chappes, riftes, and fistulas of the Fundament/512.a/691.c/700.c/775.c.

G

To open the stoppinges of the Galle/399.a.
For the blastinges and swellinges of the Genitors/262.n/593.i/768.c.
Agaynst spreadyng and fretting sores of the Genitors, or priuie members/652.d.
To dryue away Gnattes/36.g/71.d.
For the swelling of the Goute/374.f.
To asswage the payne of the Goute/450.d/454.b/488.f/554.c/593.c/594.f/665.c/741.750.c.
For the Goute in the hand & feete/50.c/61.c/63.g/115.c/174.a/219.c/232.b/291.g/301.f/303.f/304.d/308.a/325.a/331.m/347.a/365.a/379.b/415.b/433.g/454.b/528.b/546.b/564.b/719.b.
To bring foorth and dryue out Grauel and Stone/14.h/19.b/50.c/96.a/184.c/217.c/231.a/304.b/433.b/447.b/581.c/602.c/c/602.a/615.b/618.b/616.c/657.i/661.c/670.a/674.a/687.c/694.b/698.711.d/714.b/718.d/711.c/725.c/731.d/741.c/771.c/775.c/776.b.
To strengthen the Gummes/331.i/657.e/707.e.
For swellinges of the Gummes/657.c/661.a/700.c/721.b/739.c/779.c.
For the fylthie moysture of the Gummes/739.i/746.h.
For them that haue their Guttes fallen into their coddes/768.c.
Corruption or scraping of the Guttes/375.a/576.g.

H

For the heauynesse of the Harte/9.d/138.d/540.b.
To dryue away all venome from the Harte/84.a.
Stitches or griping tormentes about the Harte/238.b.
Hurtfull to the Harte/371.c/389.l.
To comfort the Harte/157.a/178.a/141.a/252.a/260.a.
For the trembling or shaking of the Harte/164.d/265.b/564.d/556.b.661.d/732.f.
For the scuruie heate or itche of the Handes/53.d.
To helpe riftes or chappynges of the Handes and feete/764.h/773.c.
To cure the falling of, or the Heare/331.i/354.i/624.c/632.l/557.b/687.d.
To restore the Heare fallen from the head/2.d/409.b/99.m/649.i/710.b/734.c.
To make yellow Heare/111.h/684.c/699.b.
To make blacke Heare/380.c/686.d/687.d/692.b/746.k/768.i.
To restore Heare beyng burned or scalded/101.f.
Cause Heare to fall/361.b/405.a/220.l.
Good for the Headache/104.c/115.d/117.d/148.c/181.d/194.h/130.c/140.m/252.a/261.m/266.b/271.h/279.c/291.g/299.c/306.a/308.a/310.a/319.f/341.d/354.b/373.a/377.c/410.d/433.c/444.b/499.d/533.a/5.o.b/56.c/576.f/h/614.f/651.b/652.b/b/2.b/676.b/708.b/711.f.
For turning or giddinesse and swymmyng in the Head/19.a/133.h/324.g/165.b/342.a/375.b/381.c/393.n/531.a/776.d.
For the naughtie scurffe of the Head/181.f/101.c/410.c/511.c/ Scurffe.
To purge naughtie fleame or humours of the Head/31.f/56.b/554.c/163.g.
Impostumes and tumours of the Head/291.c/559.b.
Good for the dryenesse of the Head/1.8.c.
Cause the Head to be dull and heauie/4.1.b.
Cause Headache/154.a/253.a/34.c.f./421.g.493.b.k/

Vertue, and Dangers.

511.a/639.a/641.a/718.a/731.b/734.a/737.a/739.g/
Olde payne or greeues of the Head/347.a/389.f/397.c/
462.b/638.k.
To dry humours of the Head/658.c/763.g.
To heale woundes of the Head/775.c.
For kybed Heeles/38.a/306.y/331.m/482.d/551.g/594.c/
646.h/649.k.
To heale the inwarde and outwarde Hemerrhoides/
11.D./31.c/44.D/115.h/131.D/118.c/301.D/304.c/316.c/354.c
443.g/576.b/640.c/551.c/661.b/693.c/719.c.
For the Hernies/vide Burstynges/and Ruptures/
87.a.
For the Hicket/108.a/146.a/m/246.b/271.D/315.c/329.a/
408.b/605.c/731.a.
For Hydropsie/134.a/238.c/241.a.
To wast waterishe Humours/373.c/308.a/384.c/395.a/
574.g/665.a/666.b/761.0.
Dissolue and waste al colde Humours/3.g/263.a/266.a/
307.f/620.h/608.b.

I
Iaunders/84.c/93.d/117.h/187.h/189.b/190.a/b/134.a/
238.a/241.a/264.a/265.b/281.b/305.b/310.a/329.D/354.
c/355.c/398.a/399.b/408.b/446.b/449.b/538.a/564.c/
555.a/601.c/630.b/616.b/655.a/689.g/761.a/778.g.
For the euyll colour remaynyng after the Iaundise/
373.b/776.g.
For the Iaundise/or yellow soght/6.b/28.a/32.c/34.c/
35.D/40.c/113.h/D/80.a/133.b/141.b/248.c/317.g/331.c/352.
m/390.a/546.b/559.c/674.b/741.b.
For corruptions or swellynges of the Iawes/vide
Gummes/537.7.a.l/711.b/779.c.
To breake inwarde Impostemes/628.a.
For all Impostemes about pryuie members/or Geni
tors or Vulua/36.a/201.D/271.b/275.c/315.c/347.c/354.
D/380.c/473.c/473.c/498.a/564.b/565.c/571.b/ 630.c/
649.c/691.c/707.c/711.b/747.c/755.D.
For al Impostemes/68.b/70.b/93.b/187.c/374.g/415.h/
420.f/42.c/564.g/750.c.d/775.b.
To soften hoate Impostemes about the fundament/
148.f/154.h/158.g/171.c/281.c/428.i/498.a/775.b.
To rype and breake harde Impostemes/or swelling ul
cers/111.c/383.k/461.a/469.c/483.c/493.c/718.l/719.p/
745.f/775.b.
For all inwarde Inflammations/249.b/253.b.
To cure hoate Inflammations or impostemes/11.b/20.a
38.c/99.g/115.c/128.f/131.c/132.b/144.c/146.c/148.f/176.c
438.D/444.a/446.a/495.1/514.b/ 533.i/ 545.c/ 551.f/
555.n/564.b/593.b/682.b/623 b/694.b/696.c/737.c.
Beginning of impostemes or Inflammations/739.h/
745.f/756.a/757.a.
For ache or payne in the Ioyntes/wits haue ben before
broken/16.c/104.c/111.f/131.a.
For the partes beyng out of Ioynt/331.m/514.a.
To helpe and cure the Itche of scuruinesse/217.c/355.c/
372.b/377.c/383.l/59.b/560.b.
To the Itche of pryuie members/225.g.

K
Kyll the body/276.g.
To heale broken or hollowe Hyppes/513.b.
The Kinges euyll or harde swelling about the throte/
252.0./276.c/287.a/304.c/507.b/537.b/549.c/560.l/571.c/
634.b/636.f/647.k/719.p.
Ulcerations and hurtes of the Kidneyes/139.b/144.a/
284.b/521.b/597.D/775.a.
Hurtfull to the Kidneyes/218.i/639.a.
To mundifie and clense the Kidneyes/289.a/304.g.
Stopping and paine of the Kidneyes/27.c/51.a/80.b/
93.D/101.b/114.c/170.c/291.c/299.b/319.a/ 317.a/ 331.b/
399.a/400.a/446.b/479.c/4 9.f/ 520.c/ 538.a/ 544.a/
576.b/578.a/580.a/601.a/604.a/606.b/608.b/613.b/616.b/

618.b/625.a/631.f/695.a/705.b/711.b/712.D/722.c/b/732.
D/741.c/763.a/776.g/778.c.
To dryue away Knattes/or gnattes/36.g/75.b.

L
Lameness/310.b.
To stoppe the Laske/23.a/33.a/57.c/64.a/81.a/84.a
85.a/86.a/96.b/101.a/104.b/107.b/ 117.a/130.a/ 144.a/
168.a/182.a/206.a/213.D/230.c/ 241.c/ 261.a/ 271.a/b/
276.b/285.c/329.a/406.a/415.c/ 433.c/ 466.a/ 487.c/
492.a/492.a/511.a/559.c/564.b/569.a/577.l/584.b/ 605.
b/651.a/652.b/657.f/ 660.b/663.c/699.a/677.f/ 684.c/
685.a/691.a/698.a/700.a/707.g/708.a/D.g/ 710.c/
713.c/721.c/723.c/726.b/727.c/729.a/731.c/ c/ 745.a/
c/746.g/767.a/771.a.
To stoppe Laske commyng of cholerique humours/
683.c.
For them that are leane/and unlusty/616.c.
To make them leane that are grosse and fat/749.c.
To heale legges or armes that be broken/751.b.
For sores that runne in the Legges/107.a.
The Lethargie/or the sleepyng and forgetful sicknesse/
262.q/199.b/310.b/610.k.
Cause the Lethargie/433.b/488.l.
Good for Lasar and leper/248.c/348.b/361.D/383.l/386.
b/469.b/620.m/638.n/719.r/751.c.
Cause Leper/482.l.
Chappes of the Lyppes/706.c.
For Lice and nittes/89.b/389.k/551.g/626.c/638.h.
To dryue away Lice from the head/apparell/and body/
371.b/c/677.c/765.c.
Inflammation of the Liuer/7.p/59.a/b/111.a/143.a/149.
h/170.a/421.b/444.b/672.a/681.c/694.c/706.a.
To strengthen the Liuer/57.a/83.b/158.a/510.c/540.b/
546.b/656.a/735.a/761.a.
Hurtful to the Liuer/375.
Stoppe the Liuer/705.c/737.a.
For stopping of the Liuer/18.b/31.c/34.c/55.a/57.a/71.
b/80.a/84.c/93.D/133.b/134.c/141.b/151.c/ 158.a/ 165.a
270.c/272.b/291.c/317.a/319.a/331.c/338.b/349.a/377.c
398.a/400.a/408.a/409.b/ 411.a/437.a/446.b/ 479.c/
481.b/538.a/546.b/551.f/564.c/ 578.c/601.c/ 606.b/615.
b/618.b/651.b/656.a/c/b/b/606.b/ 681.b/ 617.g/ 691.c/
710.c/711.b/716.f/731.b/735.a/747.a/ 749.h/ 763.a/
771.c/776.g/778.c.
The payne of the Loynes/10.b.
To encrease Loue/601.b.
Inflammation of the Lungue/114.a/142.b/415.b/711.b.
To clense the Lunges/149.b/194.b/231.b/338.D/ 243.a/
311.a/481.a/694.a.
Roughnesse of the Lunges/718.f/722.a/715.c.
For the dryness and harmes of the Lunges/17.c/79.a/
83.D/84.c/110.a/111.c/114.b/139.D/146.b/ 161.h/ 166.b/
604.a/651.f/694.c/710.c/711.b/718.f/722.a/735.a/771.b.
778.b.
Bodyly lust/vide Fleshly desyre.
To take away al inordinate Lustes or vayne longinges
of women with childe/651.b.

M
Agaynst Madnesse/191.g/347.a/351.a.
Make Madnesse/482.l.
Poyson hurtfull to Man/and kylleth the body/420.D/
421.a/425.a/430.a/431.k/433.b/432.h/ 447.a/ 448.c/
451.a/452.c/768.a.
Blacke Markes commyng out of stripes or beatyng/
217.c/235.c/138.i/149.g/262.a/170.g/113.D/304.c/354.c/
365.c/383.b/398.b/555.n/ 559.n/ 610.l/621.b/ 631.
638.m/689.c/731.c.
Take away Markes with hoate irons/411.c.
Take away Markes of the small pockes and Meselles/
331.l.
Mundifie the Matrix/304.b/314.b/338.a/676.D.
To close up the Matrix/252.b/779.h
JPP 4   windinesse

A Table of the Nature,

Windinesse in the Matrix/174.a/691.b.
Blastinges and windinesse of the Matrix or Mother/ 48.a/ 3 .m/301.g/363.b.
Good for the payne of the Matrix or mother/14.b/19.d 33.b/36.d/71.f./11.d/127.c/291.b.329.a/381.g/443.c/450. a/657.c/712.d.
Settell the Matrice in his naturall place that is risen out/18.b/204.a/676.c/685.d/703.f/746.i/779.a
Suffocation of the strangling of the Mother or Matrix/11.a/137.f/170.d/171.f/192.d/199.b/306.5/306.g/ 307.c/318.c/381.n.601.d/619.g.
For stopping or hardnesse of the Mother or Matrix/ 19.c/201.b/258.d/315.b/327.g/331.i/345.d/346.c/380.d/ 388.f/491.b/c/501.a/509.a/558.a/c/641.e/660.c/691.g.
Against Melancholie/12.a/19.a/84.c/148.e/229.c/241.a/ 260.a.

Members that are waxen dead/691.e.
Mollifie harde and stiffe Members/691.e/739.k.
Shrinking of any Member/327.a/691.f.
To warme all cold partes of Members/261.f.
Swollen Members/92.g/vide Ioynt.
Dislocation or displacing Members out of ioynt/51.e/ 194.f/101.c/110.b/238.i/96.b/512.b/651.b/686.c.
To mortifie and take away a Member/451.e.
To strengthen and comfort the Memorie/40.b/178.b/ 264.b/266.d/533.a.
To cause a man to be glad and Mery/12.a/128.k/246.k.
Against Meselo/17.b/217.c/718.l.
To cause plenty of Milke in womens breastes/10.c/49. a/111.b/269.d/171.a/171.a/278.a/ 479.a/ 563.b/ 573.c/ 577.a/582.g/599.k.
To cause wen to yelde store of Milke/56.a.
To dry Milke in womens breastes/71.a/261.g.
For clottered or cluttered Milke in womens breastes/ 481.h.

Open the Milt and splene/148.a.
Wast the swelling or inflammation of the Milt/ 298.a/ 444.b/538.c/651.b.
For the payne and stopping of the Milt or splene/194.f 261.f/101.c/407.a/479.c/481.b/551.f/578.b/580.c/665.a/ 677.b/680.d/681.h/689.g.
Diminish the Milt/624.b.
Hardnesse of the Milte or splene/15.c/59.a/ 68.d/ 80.a/ 111.a/112.a/130.h/152.f/181.b/156.a/232.l/234.c/234.d/216. a/303.d/306.b/307.g/3.8.d/314.b/363. b/371.c/ 381.b/ 389.c/399.b/402.b/403.a/4:6.b/407.a/407.a/408.a/409.b/493. c/538.a/560.h/599.f/601.c/h/624.c/677.a/631.c/ 716.f/ 717.b/72.b/747.a/748.c/761.b/771.c/776.b/778.c.
Against hoate and harde impostemes of the Mother/ 151.k/ 3.7.b/ 443.c/ 498.a/ 576.g/ 582.c/ 584.c/ 691.f/ 773.c.
For the rising vp of the Mother/54.a.
To keepe cloth and garment from Mothes/6.i/89.c/ 19 .d/143.d/673.a/765.c.
Against the old vlcers and greevances of the Mouth/ 11.b/c/d/43.c/83.a/84.c/85.b/86.c/91.b/110.b/117.a/ b/ 133.c/h/139.b/151.c/172.a/184.c/225.c/127.c/216.c/379.c/ 341.c/354.h/503.b/537.d/601.b/601.a/685.b/690.a/707. b/715.b/ ~18.i/~21.f/739.c/745.h.
Amende stinking of the Mouth/48.e/85.b/116.a/141.c/ 371.d/503.b/505.b.
Against vlcers of young childrens Mouthes/ 754.a.
For the Murren of Hogges or Swines/315.d.

N

TO dry vp Nature and sede of generation/71.a/181.c 2 2.g.3.1.b/573.c/4.c.
To encrease the seede of generation/ or Nature/705.b 749.e.
For the going out of the Nauell/104.d/137.c/299.f.
For the Nauell of young children/49.b.
Agnaples growing about the roote of the Naples/383.k
For corrupt euyll Naples of handes & feete/71.c/70.c/

83.c/194.h/258.f/420.a/687.k.
For the harde impostemes of the Necke/ or kings euil/ 70.a/171.a/258.c/461.b/507.b.
Cause the Nesing/241.d.
Cause the Nose bleede/170.i/658.b.
Take away stenche or smell of the Nose/305.a.
Stanche Nose bleeding/33.d/75.c/99.c/101.c/130.i/262.l 271.c/411.a/540.b/641.f.
To open the conductes of the Nose/56.b.
To heale the superfluous flesh growing in the Nostrilles called Polypus/404.c/768.d.

O

Good fodder to fat Oxen/482.f/752.g/759.c.

P

AGaynst members taken with the Paulsie/166.b/ b.
Against the Paulsie/ 299. f/ 301.b/ 306.b/ 310.a/b/ 392.d/680. b.
Bring or cause the Paulsie/433.k.
To take away Parbrake, or stay vomiting/23.c/84. f/ 91.a/212.b/270.b/291.b/482.g.510.a/672.b/683.b/707.g 732.a/f/737.b/767.b/779.c.
Good for Parbrake and wambling of the stomacke of women with childe/65.
To swage all Paynes/443.a/450.d/e.
Inwarde Paynes/433.c.443.a.
To dissolue Pestilent Carboncles/300.a/ 303.h/ 305.k/ 481.f/533.i/
To lap vpon Phlegmons/148.f.
For suche as haue the Phtisick, or consumption/291.a.
Cause to Disse well/187.a/711.d.
Against the hoate Pisse/ 14.a/35.b/ 273.a/283.b/408.b/ 712.c.
For them that can not Pisse, but by droppes/18.d/80.b 217.f/235.a/542.h/674.b/732.a.
To preserue from the Plague, or infection of the Pestilence/207.c/617.d/704.a/763.d.
Agaynst the Plague and Pestilent Feuers/21.a/ 84.a/ 110.a/157.a/154.b/178.b/161.b/287.c/297.a/300.a/335.b. 341.b/197.d/530.c/553.f/704.a.
Against Pleurelie/114.a/129.a/148.b/238.d/305.i/390.a/
For great or Frenche Pockes/21.b/ 24.b/ 44.b/ 93.f/ 311.a/411.c/737.a.
Against weaknes or debilitie coming from the French Pockes/310.g.
For small Pockes/17.b/217.c/718.c/c.
For the Podagra/vide Goute in the feete.
Against Poyson of Serpents and Vipers/ 9.b/ 83.d/ 84.a/291.d/294.a.
Against all Poyson, vide Venome/ 115.g/ 178.d/ 230.c/ 234.b/235.g/216.b/168.a/285.c/287.d/291.b/297.a/304.a 314.a/319.b.331.b/333.a/335.b/341.a/349.a/396.a/ 416.a. 491.a/501.c/518. c/533.g/582.b/594.c/595.b/604.c/606.b 630.c/636.b/638.b/641.d/629.a/704.c/718.b/751.c/745.c
For al vlcers/inflammations of the Pulme, or lunges 27.a.

Purge clammy fleame and thicke humours/365.a/375. a/741.a.
Purge by vrine/391.a/399.a/515.a.
Purge women after their deliuerance/161.f.
Purge rawe and grosse fleame/311.a/331.a.
Purge hoate cholerique humours/160.b/139.a/ 296.a/ 310.a/317.c/329.b/361.a/363.a/373.a/378.a/386.a/396.a/ 399.a/656.b/607.a/761.a.
Purge hoate melancholy humours/232.c/235.b/352.i/377 a/378.c.
Purge choler both vpwarde and downewarde/628.a.
Purge by siege downewarde/ 34.a/61. a/ 78.c/ 148.a/ 194.a/201.b/214.a/117.b/219.b/296.a/302.b/306.a/310.a/ 311.a/351.a/i/367.a/386.c/560.b/646.a.

Rage

Vertue, and Dangers.

R

Rage or madnesse caused by the biting of a mad dogge/108.b.

Cokrill Rattes and Myce/348.g.

Make to raue/and mad/448.e/451.a

Against rauing/or frensie/230.e/310.d.

For rawe and without skinne places/493.h.

For the payne of Raynes/vide Kidneyes/10.b/14.a/284.b/289.a/437.a/476.b/514.b/580.c/694.b/711.b/741.c.

Refresh a man/684.a.

Against subtil Reumes and catharres/433.b/530.f.

For all Ruptures/87.a/vide Burstinges.

S

Against running and spreading Scabbes and sores/50.a/347.d/381.h/411.c/420.b/455.c/469.b/479.b/518.c/551.c/559.g/607.b/614.c/647.h/662.c/687.h/700.c/711.g/719.c/754.c/763.c/766.c/771.b/775.d.

Against Scabbes or Scuruinesse/27.b/110.b/238.h/161.b/312.c/352.h/374.b/386.b/396.c/400.a.

Against Schalding with fyre/vide Burning.

Against Sciatica/1.a/15.d/28.a/64.c/65.a/83.b/114.a/129.a/232.h/261.h/299.f/303.h/304.b/308.b/319.c/336.c/341.a/352.a/395.b/307.469.c/481.b/530.a/613.b/618.c/620.b/614.c/618.b/630.b/631.b/666.c/680.b/750.a.

Against stinging of Scorpions/19.a/61.b/63.a/148.g/ng.a/164.a/165.a/194.c/235.b/238.a/261.b/270.c/305.d/...b/346.b/417.455.c/531.a/559.b/565.c/573.c/622.c/629.a.

Against white noughtie Scuruinesse/110.b/201.c.

Noughtie white Scuruinesse of the head/305.b/331.m/361.d/365.c/459.b/479.b/497.c/540.b/620.l/638.l/640.f/647.i/649.i/660.f/687.b/734.c/749.c/754.c/759.b.

Noughtie Scurffe/or Tetters of Kyen/Sheepe and Horses/757.b.

Against the foule Scurffe/tetter/gaule/and scabbes/410.b/454.m/518.c/535.b/620.m/687.a/719.c/731.c/752.c/757.d.

Against the drie Scurffe and manginesse/281.h/305.b/310.c/329.c/361.d/374.h/377.c/383.L/454.m/518.c/530.c/551.c/559.c/599.m/638.h/647.i/458.a/763.c/778.c.

Seede of generation/vide Nature.

To driue away Serpentes/75.d/299.g/307.c/380.f/614.c/691.f.

Against the biting or Serpentes/10.a/15.b/118.c/29.a/111.a/184.d/201.g/236.a/142.a/258.b/261.a/281.a/291.b/294.c/296.a/301.c/380.c/406.a/523.c/554.b/569.b/620.m/651.b/679.b/691.c/749.b/755.b/783.c.

To driue away the Secundine/or afterbirth/163.b/229.a/232.a/236.b/253.a/258.b/b/262.b/265.a/268.a/285.a/287.b/299.b/303.g/304.b/314.b/319.b/381.f/538.g/580.c/611.b/615.a/614.b/638.p/640.h/660.c/641.a/766.a.

For the falling Sickeuesse/14.b/35.c/37.d/43.e/148.b/206.b/210.i/212.c/287.b/283.c/291.g/305.m/306.b/307.c/308.a/315.c/335.a/338.c/342.a/347.a/375.b/377.c/381.b/501.a/520.f.

Against the falling Sickenesse of young children/172.g

For the excoriation or goyng of the Skinne of the secrete partes/767.a/775.b.

To make blisters and holes in the Skinne/415.e/420.a

To take away harde Skinne of handes or feete gotten by labour/744.d

Roughnesse of the Skinne/348.b/351.h/361.b/396.b/506.b/719.c/749.f.

Make a man ouermuche Sleepe/641.a.

To prouoke a quiet Sleepe/148.e/182.h/271.i/433.a/b/c/b/b/h/i/435.a/438.c/450.f/573.c.

For them that are very Sleepie/310.b.

To restore the Smelling being lost/179.c.

To prouoke Sniffing/407.c/619.b/640.g.

Olde Sores/9.a/t.b/16.g/24.b/71.b/101.b/115.b/258.b/354.b/561.b/687.c.

For filthy fretting rotten Sores/11.b/27.b/32.b/44.b/e/50.b/56.c/68.c/83.c/g/86.c/197.a/115.e/128.f/223.e/294.c/315.c/321.f/333.f/361.b/437.c/469.b/554.b/555.a/599.m/607.h/649.c/690.b/737.a/745.f.

To dry up Sores and apostumations/613.b.

Splinters vide Thornes.

Cause to Spit blood/348.f.

Hardnesse of the Splene/vide Milte.

For dulnesse or heauinesse of Spirite/148.e/614.c/vide Dull.

To cure the Squinance/305.f/316.a/373.d/599.q/636.c/666.c/vide Swelling in the Throte.

Against Stone/vide Grauell/14.h/19.b/48.a/50.c/64.b/86.d/96.a/187.h/188.a/217.f/329.c/330.a/341.h/169.b/285.a/408.a/446.a/507.b/510.d/521.h/536.a/559.c/581.c/662.h/665.c/682.c/696.b/698.b/715.b.

To breake and driue foorth the Stone/38.b/41.a/141.b/194.b/232.a/252.a/277.a/289.b/340.a/241.i/305.b/344.a/409.b/449.f/510.b/520.c/559.c/565.a/594.b/599.f/606.a/608.a/609.b/615.a/617.c/641.g/665.b/669.a/b/674.a/679.c/687.c/689.f/714.b/721.c/741.c/775.a.

Hurtful to the Stomacke/3.i/348.i/375.b/379.c/395.b/546.c/591.a/594.a/599.c/611.a/624.a/715.c/718.b/714.a/728.a/731.b/740.a.

To strengthen and comfort the Stomacke/134.b/217.a/246.a/176.a/183.a/187.f/191.h/301.f/314.b/354.a/b/510.a/524.b/569.c/576.c/577.m/580.a/612.a/701.a/704.a/706.a/708.b/732.a/c/735.a/739.c/767.b/776.i/779.c.

Good for payne of the Stomacke/6.a/m.c/232.b/331.b/246.b/329.a/341.a/354.a/b/455.c/499.c/565.a/571.a/608.b/620.b/763.a/779.c.

For the boyling and wambling of the Stomacke/6.b/270.b/329.a/355.a/559.c/573.b/651.f/661.b/704.b/706.a.i/737.b/779.h.

For the inflammation of the Stomacke/7.p/36.a/84.c/181.c/444.b/565.a/567.b/573.b/576.a/657.b/671.a/682.c/701.a/706.a/716.a.

To warme the Stomacke/196.b/354.b/503.a/600.a/619.a/704.c/731.a/f.

To refreshe the hoate Stomacke/38.a/174.a/383.m/563.a/587.a/589.a/694.c/701.a/706.a.

Against cold windinesse & blastinges of the Stomacke/41.b/117.b/174.a/175.a/295.a/298.a/336.c/337.a/608.b/613.c/636.c/691.b/763.f.

To strengthen the weakenes & ouercasting of the Stomacke/488.b/563.a/646.b/652.c/c/663.c/704.b/723.c/716.a/732.a/f/739.f.

Ouerturne the Stomacke/739.g.

Engender windinesse in the Stomacke/488.l/715.c/734.a/739.f.

Comfort the mouth of the Stomacke/608.b/706.a/718.b/718.b/732.a.

To purge the Stomacke from fleame/354.b.

Clense the Stomacke/550.a.

Against Stinging of Bees and waspes/246.o/161.b/581.h/184.g.

Against Stranguric/or droppisse/14.a/15.a/35.b/41.a/99.h/111.a/283.b/184.o/285.a/187.a/289.a/319.a/337.a/344.a/345.a/408.b/501.a/514.b/518.b/521.a/565.a/602.c/609.a/614.b/618.a/622.a/626.c/657.i/676.b/711.c/750.a/765.f/768.f/775.b.

Prouoke and cause Sweating/18.g/24.a/151.b/248.b/281.f/199.n/301.b/342.b/533.b/613.f/718.c.

Against al hardnesse and Swellinges/44.a/70.a/394.c/b/499.c/651.h/658.a/739.b/745.c/755.b/775.b.

Against all hoate Swellinges/104.c/107.c/141.b/379.b/411.h/433.b/438.b/498.a/525.c/537.c/555.h/582.i/638.b/738.a/775.b.

To dissolue and breake al colde and harde Swellinges/189.a/196.c/197.b/217.c/235.c/247.c/255.b/279.b/281.c/303.c/304.c/306.h/307.c/308.c/374.b/454.c/l/460.b/461.a/473.c/493.c/496.a/613.b/637.a/630.c/719.p/775.b.

Against cold Swellinges/3.g/83.c/112.g/129.g/139.b/285.b/396.b/522.a/582.i/584.b/624.a/687.b/713.c/731.c/743.b/747.a.

A Table of the Nature,

For all Swellings about the siege or fundamentall/82.c/ vide Fundament.
To take away Sweating/687.l.
For hardnesse and shrinking of Sinewes/1.a/72.d/ 187.c/193.f/201.i/239.b/222.f/306.b/307.a/311.a/319.a/ 329.f/424.b/443.d/454.c/517.a/649.b.
Good for drawing, shakinges, and ache of Sinewes, 14.b/104.f/205.l/328/554.a/325.b/528.a/571.c/744.b.
Payne or swellinges of Sinewes/18.b/212.b/310.a/327.h
To ioyne Sinewes togither that are cut/201.c/571.c.
Cause to draw and shrinke the Sinewes into the body, 348.i/351.p/488.b.
To appease the payne of the Syde/111.f/167.a/229.c/ 252.c/261.h/301.a.3 7.g/313.c/341.a/365.a/501.a/530. a/ 555.q/ 604.a/ 605.b/ 649.b/ 747.d/ 743.a/ 749.h.

To dry the moyst Stomacke/438.d/638.g/687.c.

T

TO beautifie and clense the Teeth/311.c/615.c.
Make fast loose Teeth/12.d/26.d/117.b/264.c/315.e 576.i/661.b/707.c/738.i/779.c.
Womens Termes/vide Flowers.
Spreading Tetters/17.b
Swelling in the Throte, or Squinancie/148.d/227.c/ 373.d/437.d/676.a/718.i/756.a/775.b.
Strumes or swellinges of the Throte / 4 0.c/ 607.b/ 661.a/663.g/690.a/ 713.d/ 718.i/ 7.i.b/ 731.f/ 747.b/ 748.b.
Against roughnesse of the Throte/454.c/o/455.b/718.f i/712.a.
Against the roughnesse and hoarsenesse of the Throte, 17.b,194.f/228.c/694.a.
Sores and inflammations of the Throte/12.b/70.b/93. b/ 10.d/ 141.c/ 136.c/ 171.h/ 316.a/341.b/ 354.b/607.a/ 619.d.
For the Tooth ache/32.d/31.c/55.c/82.a/86.d/93.i/120.c/ 127.b/217.c/187.f/ 299.h/ 301.i/ 305.h/307.i/ 341.c/ 341.d/348.f/ 361.c/372.c/374.f 389.i/443.c/448.c/450.c/ 518.f/525.d/530.d/560.h/584.c/612.c/638.f/649.h/ 670.c/677.c/681.f/716.g/h/ 718.h/729.c/ 755.c/ 759.b/ 765.h/771.c.
To draw foorth Thornes, or splinters that sticke into the flesh/56.c/70.b/194.b/156.c/197.a/211.c/155.b/168.f 196.l/107.f 315.D/144.b/83.a/ 454.h/ 460.b/469.c/514. a/522.h/622.f/676.f.
Take away the asperities & roughnesse of the Tongue/ 246.i/543.a/725.c.
To cure kernels vnder the Tongue/354.k/700.c.
Almondes or vlcers about the roote of the Tongue, 619.b/662.g/676.a/715.d/718.i/731.f/750.a.
Slake the Thirste/ 573.b/ 591.b/672.c/ 683.d/689.b/694. c.702.a/704.c/718.b/724.b.
For them that are sicke of eating of Todestooles/599.i vide Champions.
Tumors/vide Swellinges and Impostumations.

V

Against Venemous shot of dartes and arrowes/105.p/ 307.b/315.b.
To driue away Venemous beastes/ 78.d/ 230.f/ 234.b/ 248.a/ 258.c/ 179.d/ 307.d/ 380.f/ 620.n/ 638.d/691.f/ 743.b.
Against Venome dronken or eaten/.2.b/6.f/117.f/307.b/ 520.c/599.i/601.b/662.d/654.a/720.d/763.c.
Against al Venome of wilde beastes/o.b/17.d/33.c/36.g 99.b/117.c/120.a/168.a/185.c/119.b.5.6.b/601.f.
Against byting of vipers, snakes, & venemous beastes/ 13.b/16.h/22.c/30.c. 55.a/57.a/85.h/ 111.b/ 112.b/ 117.a/ 184.d/194.c/205.a/230.c/234.b/235.a/241.a/248.a/252.d/ 258.d/260.a/268.a/279.c/281.a/285.c/297.d/301.c/305.p/

306.f/314.a/311.b/337.a/341.a/400.c/438.c/454.c/501.b/ 507.c/520.g/525.c/533.i/539.a/601.f. 604.c/ 611.a/ 616.b/ 611.d/622.c.e.b/631.c/ 642.c/ 646.g/ 649.c/ 669.b/ 677.f/704.f/735.c/745.c.

Driue away Ventosities and windinesse/ 41.a/187.c/ 196.a.
Engender windes and Ventosities/473.a/474.a/480.f 594.a/649.a/641.a/542.i/651.c/7.8.a.
To keepe cloth and garments from Vermine/6.i/243.b
To clense and mundifie old rotten Vlcers/123.c/28./b/ 299.l/311.b/315.c/322.a/352.D/3 1.d/454.g/448.g/ 615.b/ 665.f/ 687. c/k. 738.b/ 739.i/ 750.b/ 766.b/ 768.h/ 773.f.
For newe Vlcers/44.b/183.h/756.a/757.a.
To lose and cure corrupt filthy Vlcers / vide Sores, 28.c/ 44.b/c/48. b/d/ 50.c/64.b/68.d/ 70.b/118.d/117.e 184.g/194.b/201.g/h/235.c/258.f/181.c/294.c/125.a/28.l 331.n/335.c/388.a/444.a/448.d/454.n/589.c/638.n/549. c/681.b/690.b/691.a/700.a/737.b/773.d.
Hoate Vlcers in priuie places or partes/48.d/53.b/66.c 69.f/133.c/146.b/661.a/ vide Impostumes.
For the desyre of Vomite/vide Parbrake.
Desyre of Vomite vpon the sea/139.a.
To cause Vomite, and cast out easyly slymie flegmes, and cholerique humours/115.g/119.c/211.a/212.a/ 223.a 231.a/322.i/319.c/335.a/361.a/372.b/590.o/ 697.a/ 759.a.
Vomit with great force/347.a/666.a/743.c/759.a.
To cleare the Voyce/638.g.
To take away roughnesse of the Voyce.543.a.
Sharpnesse of water or Vrine/694. d.
Filthy corruption and matter of Vrine.694.b.
To stoppe the inuoluntarie running of Vrine/ 405.a.
To prouoke Vrine or water/15.a/38.b/41.a/48.a/ 64. k. 99.h/129.b/152.b/184.a/c/187.b/191.b/196.a/206.a/217.f 229.a/234.a/236.a/241.a/243.b/252.a/161.a/262.c/265. a/ 269.b/171.a/181.a/284.a/ 185.a/ 289.a/ 291.a/ 296.b/ 299.b/305.q/319.a/336.a/337.a/341.a/344.a/382. a/109.b 437.a/442.a/479.a/499. f/501.a/509.a/510.b/521.a/525.b 528.a/533.b/565.a/577.m/ 578.a/ 594.b/ 597.a/ 599.f/ 604.b/605.a/605.a/607.a/608.a/ 615.b/ 616.b/ 618.a/ 619.f/622.b/625.a/626.c/638.c/ 640.b/ 641.b/ 646.b/ 649.a/660.b/662.h/ 665.b/c/ 669.a/b/ 674.a/ 676.b/ 689.c/711.b/740.b/749.g/750.a/ 703.a/ 765.f/766.a/ 771.a/775.a/b/778.b/c.
Difficultie or stopping of Vrine/2.a/28.b/50.c.

W

VVArtes growing about the Iarde and secret places 747.d.
Wartes take away/61.d/ g/229.g/235.b/304.c/361.b/420.a 719.c/744.d.
The paynefull making of Water/285.a/445.b/613.a.
Against drinking of corrupt stinking noughtie water/ 232.f/638.f.
For Weales comming of choler and blood/551.f/ 665.f/ 719.c.
To refresh Weary members/540.c/689.d.
That trauelers shall not be weary/21.b/691.f.
For the Wilde fyre/11.a/19.c/137.c/ vide S. Antonies fyre.
Biting of wilde beastes/482.b.
For the shortnesse of Winde/vide Breath/2.a/18.a/20.b/ 217.b/243.a/389.c/391.a.
To dissolue blasting and windinesse of the belly/ 285. b/ 520.b/606.c/602.b/609.a.
For Witching/vide Enchantements.
For Women with childe giuen to vomit/651.a.
Cause women to haue easie deliuerance of childe/199.b 441.c/676.D.
To purge and clense women after their deliuerance of childe/616.a.
Dangerous for women with childe/331.p/383.p/389.l/ 391.c/401.c/618.
Kyll or destroy wormes/710.b/745.c/763.f/777.k.

Es